FUNDAMENTALS OF MATHEMATICS

VOLUME I

Foundations of Mathematics
The Real Number System and Algebra

Fundamentals of Mathematics

Volume I

Foundations of Mathematics
The Real Number System and Algebra

Volume II

Geometry

Volume III

Analysis

FUNDAMENTALS OF MATHEMATICS

VOLUME I

Foundations of Mathematics
The Real Number System and Algebra

Edited by
H. Behnke
F. Bachmann
K. Fladt
W. Süss

with the assistance of
H. Gerike
F. Hohenberg
G. Pickert
H. Rau

Translated by
S. H. Gould

The MIT Press Cambridge, Massachusetts, and London, England

Originally published by Vandenhoeck & Ruprecht, Göttingen, Germany, under the title *Grundzüge der Mathematik*. The publication was sponsored by the German section of the International Commission for Mathematical Instruction. The translation of this volume is based upon the second German edition of 1962.

First MIT Press paperback edition, 1983

Printed and bound in the United States of America.

ISBN 0-262-02048-3 (hardcover)
0-262-52093-1 (paperback)
Library of Congress catalog card number: 68–14446

Contents

Translator's Foreword ix

From the Preface (to the 1958 Edition),
Heinrich Behnke and Kuno Fladt x

PART A 1
FOUNDATIONS OF MATHEMATICS
H. Hermes and W. Markwald

1. Conceptions of the Nature of Mathematics 3
2. Logical Analysis of Propositions 9
3. The Concept of a Consequence 20
4. Axiomatization 26
5. The Concept of an Algorithm 32
6. Proofs 41
7. Theory of Sets 50
8. Theory of Relations 61
9. Boolean Algebra 66
10. Axiomatization of the Natural Numbers 71
11. Antinomies 80

Bibliography 86

PART B 89
ARITHMETIC AND ALGEBRA
Introduction, W. Gröbner 91
CHAPTER I
Construction of the System of Real Numbers, G. Pickert and L. Görke 93

1. The Natural Numbers 93

2. The Integers 105
3. The Rational Numbers 121
4. The Real Numbers 129
Appendix: Ordinal Numbers, D. Kurepa and A. Aymanns 153

CHAPTER 2
Groups, W. Gaschütz and H. Noack 166

1. Axioms and Examples 167
2. Immediate Consequences of the Axioms for a Group 178
3. Methods of Investigating the Structure of Groups 182
4. Isomorphisms 188
5. Cyclic Groups 191
6. Normal Subgroups and Factor Groups 194
7. The Commutator Group 197
8. Direct Products 198
9. Abelian Groups 199
10. The Homomorphism Theorem 212
11. The Isomorphism Theorem 214
12. Composition Series, Jordan-Hölder Theorem 215
13. Normalizer, Centralizer, Center 217
14. p-Groups 219
15. Permutation Groups 220
16. Some Remarks on More General Infinite Groups 230

CHAPTER 3
Linear Algebra, H. Gericke and H. Wäsche 233

1. The Concept of a Vector Space 235
2. Linear Transformations of Vector Spaces 246
3. Products of Vectors 266

CHAPTER 4
Polynomials, G. Pickert and W. Rückert 291

1. Entire Rational Functions 291
2. Polynomials 296
3. The Use of Indeterminates as a Method of Proof 312

CHAPTER 5
Rings and Ideals, W. Gröbner and P. Lesky 316

1. Rings, Integral Domains, Fields 316

2. Divisibility in Integral Domains 327
3. Ideals in Commutative Rings, Principal Ideal Rings, Residue Class Rings 338
4. Divisibility in Polynomial Rings Elimination 346

CHAPTER 6
Theory of Numbers, H.-H. Ostmann and H. Liermann 355

1. Introduction 355
2. Divisibility Theory 355
3. Continued Fractions 372
4. Congruences 380
5. Some Number-Theoretic Functions; The Möbius Inversion Formula 388
6. The Chinese Remainder Theorem; Direct Decomposition of $\mathfrak{C}/(m)$ 391
7. Diophantine Equations; Algebraic Congruences 395
8. Algebraic Numbers 401
9. Additive Number Theory 405

CHAPTER 7
Algebraic Extensions of a Field, O. Haupt and P. Sengenhorst 409

1. The Splitting Field of a Polynomial 410
2. Finite Extensions 418
3. Normal Extensions 420
4. Separable Extensions 422
5. Roots of Unity 425
6. Isomorphic Mappings of Separable Finite Extensions 431
7. Normal Fields and the Automorphism Group (Galois Group) 433
8. Finite Fields 438
9. Irreducibility of the Cyclotomic Polynomial and Structure of the Galois Group of the Cyclotomic Field over the Field of Rational Numbers 448
10. Solvability by Radicals. Equations of the Third and Fourth Degree 452

CHAPTER 8
Complex Numbers and Quaternions, G. Pickert and H.-G. Steiner 456

1. The Complex Numbers 456

2. Algebraic Closedness of the Field of Complex Numbers 462
3. Quaternions 467

CHAPTER 9
Lattices, H. Gericke and H. Martens 483

1. Properties of the Power Set 485
2. Examples 490
3. Lattices of Finite Length 495
4. Distributive Lattices 497
5. Modular Lattices 501
6. Projective Geometry 505

CHAPTER 10
Some Basic Concepts for a Theory of Structure, H. Gericke and H. Martens 508

1. Configurations 509
2. Structure 515

CHAPTER 11
Zorn's Lemma and the High Chain Principle, H. Wolff and H. Noack 522

1. Ordered Sets 522
2. Zorn's Lemma 524
3. Examples of the Application of Zorn's Lemma 525
4. Proof of Zorn's Lemma from the Axiom of Choice 529
5. Questions Concerning the Foundations of Mathematics 534

Bibliography 536

Index 537

Translator's Foreword

The pleasant task of translating this unique work has now extended over several years, in the course of which I have received invaluable assistance from many sources. Fortunately I had the opportunity, in personal conversation or in correspondence, of discussing the entire translation with the original authors, many of whom suggested improvements, supplied exercises, or made changes and additions in the German text, wherever they seemed desirable to bring the discussion up to date, for example, on the continuum hypothesis, Zorn's lemma, or groups of odd order. To all these authors I express my gratitude.

For technical and clerical help I am especially indebted to Linda Shepard, of the Law School at the University of Utah, for her expert typing and discriminating knowledge of English; to Diane Houle, supervisor of the Varitype Section of the American Mathematical Society, for her unrivaled skill and experience in the typing of mathematical translations; to Linda Rinaldi and Ingeborg Menz, secretaries, respectively, of the Translations Department of the Society and the firm Vandenhoeck and Ruprecht, for keeping straight a long and complicated correspondence; to the staff of The MIT Press for their customary technical expertness; and to my wife, Katherine Gould, for help too varied and too substantial to be readily described.

S. H. Gould
Institute of Mathematics
Academia Sinica
Taipei, Taiwan
Republic of China

September 1973

From the Preface

Volume One was begun as the first contribution, by the German section of the International Commission for Mathematical Instruction, to the topic of the scientific foundations of instruction in mathematics, which was one of the topics chosen by the Commission, at a meeting in Paris in October 1954, in preparation for the International Congress of Mathematicians in Edinburgh in 1958. Originally we kept chiefly in mind the needs and interests of the instructor in mathematics, but as our cooperative effort continued from year to year, it became clear that the material in our book was equally important for mathematicians in science, government, and industry. For the best realization of our general purposes, each chapter has been written by two authors, one of them a university professor, the other an instructor with long experience in teaching. In addition to these specifically named authors, of whom there will eventually be more than one hundred, from Germany, Yugoslavia, the Netherlands, Austria, and Switzerland, important contributions have been made to each chapter, in joint semiannual sessions, by the other members of our large group of coworkers.

H. Behnke
K. Fladt

PART A

FOUNDATIONS OF MATHEMATICS

1. Conceptions of the Nature of Mathematics

1.1. *Mathematics and Its Foundations*

In this section, which is an introduction to the work as a whole, we shall be discussing the foundations of mathematics. In other words, we are not *doing* mathematics here; we are talking about mathematics. We are engaged in a scientific activity that has received the appropriate name of *metamathematics*.

Metamathematics forms a bridge between mathematics and philosophy. Some of its investigations can be carried out by mathematical methods, and to this extent the subject shares the exactness of mathematics, the most precise of all sciences. But other parts of metamathematics, among them the most fundamental, are not of a mathematical nature, so that we cannot expect them to have the absolute clarity of mathematics. As in all other branches of philosophy, the answers to many questions are to some extent a matter of subjective attitude and even of faith, and in any given period the attitude predominantly adopted is determined in part by the general spirit of the age. Fundamental philosophical concepts, such as idealism, realism, and nominalism, which for centuries have contended with one another with varying success, are reflected in the different views about the nature of mathematics. Apparently there is no hope of progress in an attempt to refute any one of these views scientifically; rather we try to characterize them as precisely and clearly as possible and in this way keep them apart.

Studies about the foundations of mathematics have experienced a tremendous upsurge during the past hundred years, especially since the turn of the century. The chief impetus for these investigations was provided by the discovery of contradictions in the theory of sets, a mathematical discipline created during the nineteenth century in connection with eventually successful attempts to clear up the nature of the real numbers. Since many of these paradoxes had already become apparent in antiquity, it is natural to ask why we are now able to deal with them successfully, whereas the ancients found them completely intractable. The answer is that the paradoxes necessarily remained intractable as long as they were expressed in one of the natural languages, such as English. On the shaky ground of such an imprecise language it is impossible to deal with questions of great subtlety, and our present-day successes are entirely due to a new instrument, the thoroughgoing formalization of mathematics. With this new tool it has at last become possible to construct metamathematical theories (for example, that of "classical" logic) which are just as exact as the theories of ordinary mathematics. These new metamathematical theories are regarded by many mathematicians as the essential hallmark of present-day mathematics.

In the following pages we shall describe some of the various conceptions of the nature of mathematics, but it must be remembered that they are only ex post facto idealizations of the nature of mathematics. All idealizations are extreme in one direction or another, so that scarcely any mathematician will agree with every detail of any of the positions that we shall describe. Mathematics as it exists today is in fact the creation of scientists whose inspiration has come from the most varied sources. It is to this variety that mathematics owes its immense vitality.

1.2. *The Genetic Conception of Mathematics*

We first describe a conception of mathematics in which the central role is played by the human being and his capabilities, so that mathematics may almost be said to be a branch of psychology. For example, let us consider the subject of geometry. It is certainly true that the earliest knowledge of geometry, say among the Babylonians, depended on the empirical results of practical surveyors; it is easy to imagine, for instance, how the Pythagorean theorem could arise from individual observations. Yet at this stage the theorem can hardly be called *mathematical*, since the characteristic difference between a natural science and the purely abstract science of mathematics is considered to be that the statements of a natural science can be tested (directly or indirectly) by observation, whereas for mathematical statements such a test is regarded (for widely varying reasons), as meaningless; mathematics is an a priori science in the sense of Kant. Consequently, geometry was in its origins a natural science, and was not "raised" to the position of an abstract, and therefore mathematical, science until the time of the Greeks. It was they who under the influence of Plato distinguished between axioms and the theorems derived from them. In their view the axioms were self-evident (cf. §1.3), and the theorems were derived by the process of logical deduction. It is probable that the Greek mathematicians took the same attitude toward logic as is taken today by most "naive" mathematicians: in principle, the ability to reason logically is inborn but can be improved with practice.

Arithmetic and many other branches of mathematics may well have begun like geometry as a collection of empirical facts, which was gradually raised to the status of a mathematical science.

But mathematical sciences can arise in another way, which may be called *intramathematical*, to distinguish it from the natural sciences. One of the strongest impulses here is the inborn urge, experienced by most mathematicians and particularly well-developed among the Greeks, toward the sort of beauty that manifests itself in simplicity and symmetry. The mathematician feels compelled, while continuing to observe the demands of logic, *to do away with exceptions*. The desire to make the operations of subtraction and division universally applicable led to the rational numbers. Exceptions in the operation of passing to the limit no longer arose in the field of real numbers. The exceptional case of parallel lines was removed by the introduction of "infinitely distant" points, and in recent times the many exceptional cases arising from the existence of nondifferentiable functions have been avoided by the introduction of *distributions* (cf. Vol. III, chap. 3, §3), which had already turned up among the physicists, in the form of the Dirac δ-function.

Most of these new mathematical entities, created to avoid the necessity for exceptional cases, were in the first place introduced more or less uncritically to meet the demands of each given case. But subsequently there arose a desire to establish the actual existence of such entities. A powerful tool here is the process of *abstraction*, which may be described as follows. Let there be given a set of entities which agree in many of their properties but differ in others. By an act that is in essence arbitrary, we shall declare that some of these properties, depending on the context in which we make the decision, are *essential* while all others are not essential. The act of "abstraction" from the nonessential properties consists of *identifying* (i.e., regarding as identical) those entities that differ only in nonessential properties. A set of such entities thus becomes a single unit and in this way a new entity is created (cf. §8.5). This act of creation, familar to every present-day mathematician, may be regarded as a general human capability. Here we shall only remark that in modern mathematics the process of abstraction, in conjunction with the search for simplicity, has led to the general structures that are to be found, for example, in the theory of groups (cf. §4.3, and Vol. IB, chap. 2).

1.3. *The Extent to Which Mathematical Propositions Are Self-Evident*

As mentioned before, the Greeks divided valid mathematical propositions into axioms and theorems derived therefrom. The axioms were considered self-evident, immediately obvious to everyone, "neither in need of proof nor admitting proof." The theorems, on the other hand, were not immediately obvious in themselves but became evident by being derived from the axioms through a series of arguments, each of which was obviously valid. But today, as a result of the discovery of non-Euclidean geometries, hardly any mathematician holds to the obviousness of Euclidean geometry. The axioms of group theory, field theory, lattice theory, and so forth are no longer considered obvious. At most, the theorems of arithmetic, logic, and perhaps the theory of sets may appear evident (either directly or indirectly) to certain mathematicians. For example, the *intuitionists*, following L. E. J. Brouwer, require that every mathematical construction shall be so immediately apparent to the human mind, and the result so clear, that no further proof is necessary. In §4.7 we shall discuss the attempts that have been made to show that mathematics is free of inconsistencies. Clearly such a proof of consistency will be more widely accepted if it can be based on concepts intuitively apparent to everyone.

To clarify these remarks, let us give an example of a statement that will be considered self-evident by many readers. Let there be given two distinct symbols, neither of which can be divided into meaningful parts. Then it will be considered self-evident that the two "words" obtained by writing these symbols, first in the one order and then in the other, are distinct from each other.

1.4. *The Meaning of Mathematical Propositions*

In general, mathematicians are convinced that their propositions are meaningful, the extreme position in this respect being that of the so-called *formalists*, who consider mathematics to be a mere game with symbols, the rules of which, in the last analysis, are chosen arbitrarily (*conventionalism*). Formalism was introduced by Hilbert as a methodological principle whereby the concept of a proof of consistency could be clearly stated. The formalistic point of view can also be applied to physics if with H. Hertz[1] we define the task of theoretical physics as follows: "Within our own minds we create images or symbols of the external objects, and we construct them in such a way that the logically necessary consequences of the images are again the images of the physically necessary consequences of the objects." In other words, we construct a process parallel to the process of nature. But the essential feature here

[1] *Die Prinzipien der Mechanik*, Ambrosius Barth, Leipzig (1894), Introduction.

is not that this process involves "logical thought" but rather that it runs parallel to the process of nature. Thus we could equally well have chosen a purely formalistic process, which of course would have to be suitably constructed.

Although, as was stated before, the majority of mathematicians hold to the belief that mathematical propositions are not meaningless, they hold widely different opinions about their meaning. It is impossible to go into details here about these varied opinions, and we shall content ourselves with discussing a fundamental dividing line among them, having to do with the concept of infinity. If we adopt the *concept of actual* (*completed*) *infinity*, we may speak of the totality of all natural numbers just as readily, for example, as of the totality of natural numbers between 10 and 100. But those who hold to the *concept of potential infinity* emphasize that the infinite totality of all natural numbers as a set is not immediately available to us, and that we can only approach it step by step, by means of successive constructions, such as are indicated by

$$|, ||, |||, \ldots.$$

This is the so-called *constructive* point of view; compare the concept of an algorithm described in §5.

If we examine these concepts further, certain other differences come to light, one of which we will now illustrate by an example. For any given natural number, we can determine in a finite number of steps whether the number is perfect or not.[2] The proposition:

(1.1) *either there exists an odd perfect number between 10 and 100, or else there exists no odd perfect number between 10 and 100*

is acceptable from either the actual or the potential point of view. But matters are quite different for the proposition:

(1.2) *either there exists an odd perfect number, or else there exists no odd perfect number.*

From the *actual* point of view, there is no essential difference between these two propositions. In each case the argument runs as follows: either there exists an odd perfect number between 10 and 100 (or in the set of all natural numbers), in which case (1.1) and (1.2) are correct, or else there is no such number, and in this case also (1.1) and (1.2) are correct.

But in case (1.2) an adherent of the *constructivist* school will argue as follows: the assertion that an odd perfect number exists is meaningful only if such a number has been found (constructed). On the other hand, the assertion that no odd perfect number exists is meaningful only after

[2] A natural number is called *perfect* if it is equal to half the sum of its divisors; for example, 6 is perfect. It is not known whether an odd perfect number exists.

we have shown that the assumption of the existence of such a number leads to a contradiction (i.e., that we can construct a contradiction on the basis of this assumption). But in the present state of our knowledge we cannot make either of these assertions and thus we have no reason to conclude that case (1.2) is true.

Propositions like (1.1) and (1.2) are special cases of the so-called *law of the excluded middle* (tertium non datur). The actual point of view, in contrast to the potential, accepts this law in every case.

The constructive mathematician is an *inventor*; by means of his constructions he creates new entities. On the other hand, the classical mathematician, who regards the infinite as given, is a *discoverer*. The only entities he can *find* are those that already exist.

It is customary nowadays to give the name *classical* to the actual point of view, although the potential attitude can also be traced back to antiquity.

1.5. *Remarks on the Following Sections*

These and other differences in the various conceptions of mathematics have given rise to a great diversity of opinion about the foundations and nature of mathematics, particularly with regard to where the boundary should be drawn between mathematics and logic. Within the space at our disposal it is impossible to discuss all these questions from every point of view. In the following sections we give preference to the classical position, with an occasional reference to the constructivist point of view, when the difference between them is important. Our reasons for giving preference to the classical position are as follows: (1) the greater part of established present-day mathematics is based more or less on the classical conception, whereas many parts of constructive mathematics are still in the process of being built up; (2) the constructive mathematics appears to be far more complicated than the classical. For example, it is not possible to speak simply of the real numbers. These numbers fall into various "levels," and for each level there exist still more complicated numbers.

In the present chapter we have no intention of giving an encyclopedic survey. We have given priority to such questions as are naturally related to college instruction. In some cases the treatment is more detailed because the authors believe that the subject is suitable for discussion by undergraduates in a mathematics club.

The material has been arranged as follows: mathematical proof depends on the fact that propositions have a certain structure (§2); from the classical point of view the basic concept of logic and mathematics is that of a consequence (§3), which plays a fundamental role in the axiomatic method (§4); in practice, the mathematician obtains consequences by carrying out proofs (§6), a process which has been analyzed in a profound way in the theory of calculi (§5). The next three sections deal with the theory of sets (§7), Boolean algebra (§8), and the theory of relations (§9). A system of axioms of great importance for the mathematician is the Peano system for the natural numbers (§10). Finally, we give an analysis of some of the best-known antinomies (§11).

Bibliography

The bibliography at the end of the present chapter contains several textbooks of mathematical logic dealing with the various problems discussed in the following sections. Let us mention here, once and for all: Beth [1], Curry [1], Kneebone [1], Novikov [1], Rosser [1], Wang [1], and the article on "Logic" by Church [2] in the Encyclopaedia Britannica. On intuitionism see Heyting [1] and Lorenzen [1], and on the history of logic see Kneale [1].

2. Logical Analysis of Propositions

2.1. *The Language of Mathematics*

The results of mathematics, like those of any other science, must be communicable. The communication may take place in either spoken or written form, but for mathematics the difference between them is of no great importance. In studying the foundations of mathematics it is customary to use written symbols.

Communication is ordinarily carried on in one of the natural languages, such as English. But a natural language decays and renews itself like an organism, so that we are engaged in a rather risky business if we wish to entrust "eternal, unchanging truths" of mathematics to such a changing language. Everyone knows how easily misunderstandings arise in the ordinary spoken language. So to attain clarity in his science, the mathematician must try to eliminate the ambiguities of such a language, although the attempt involves a laborious process of evolution and cannot be completely successful. One method of producing greater clarity lies in *formalization*. In the ordinary mathematical literature this process is only partly carried out, as can be seen by a glance at any mathematical text, but in studies of the foundations of mathematics, ordinary speech has been completely replaced by formalized languages. To some extent these artificial languages have been abstracted from ordinary language by a process of analyzing the statements of the latter and retaining only what is logically important. Let us now undertake this process of *logical analysis*. The reader will note a certain resemblance to grammatical analysis, but many of the distinctions made in grammar have no significance in logic. As a result, technical terms common to logic and grammar do not necessarily have the same meaning. Finally, let us emphasize once and for all that the process of logical analysis is not uniquely determined and could just as well be undertaken in a manner different from the one adopted here.

2.2. *Propositions*

Many combinations of letters are called *propositions*. For example:

(2.1) *Every even number is the sum of two odd numbers.*

(2.2) *Every odd number is the sum of two even numbers.*

(2.3) *Every positive even number, with the exception of the number two, is the sum of two prime numbers.*

In classical logic, which goes back to Aristotle, propositions are divided into *true propositions* and *false propositions*. The *principle of two-valuedness* states that every proposition is either true or false, although it is not required that we should always be able to decide which is the case. For example, it remains unknown at the present time whether the *Goldbach conjecture* (2.3) is true or false, but in classical logic it is assumed that statement (2.3) is in fact either true or false.

Thus the classical logic recognizes two truth values, *true* and *false* (often represented by T and F). Today attention is also paid to *many-valued logics*, and attempts are being made to apply them in quantum mechanics.

The classical point of view has often been criticized (cf. §1.4). But even if we adopt a different attitude, we still *accept* certain propositions, for example (2.1) and *reject* others, for example (2.2); and in general there will be propositions which, at least up to now, have been neither accepted nor rejected, for example proposition (2.3).

It must be emphasized that in the terminology adopted here, which is customary in modern researches in the foundations of mathematics, a proposition is simply a set of written symbols, so that it becomes essential to distinguish between the proposition itself and the state of affairs which it describes. Since this distinction will be of importance in the following sections, let us point out that one of the most profound thinkers in modern logic, G. Frege (1848–1925), distinguishes between the *sense* (Sinn) and the *denotation* (Bedeutung) of a proposition. By the *denotation* of a proposition, Frege means its truth value. Thus the propositions "$1 + 1 = 2$" and "$2 + 2 = 4$" have the same denotation, namely *true*.[3] But these propositions have different *senses*. Similarly, the desig-

[3] One must distinguish between a proposition and a *name* for the proposition, and when we speak of an object, we must have a name for it. Thus we shall make frequent use of the following convention: we obtain a name for a proposition (or more generally for a set of written symbols) if we enclose the proposition (the set of written symbols) in quotation marks. In the present section we shall strictly observe this convention, but later it will be convenient, as frequently in mathematics, to let a set of written symbols stand as a name for itself (*autonomous notation*).

nations (not propositions) "$2 \cdot 2$" and "2^2" have the same denotation, namely the number four, but they too have different senses.

2.3. *Propositional Forms*

In mathematics, we frequently encounter, in addition to the propositions, sets of symbols of the following sort:

(2.4) $x + 3 = y$, (2.5) $f(2, 3) = 5$,

(2.6) $f(x, y) = z$, (2.7) $P2$.

We are not dealing here with propositions, since it is obviously meaningless to ask, for example, whether (2.4) is true or false. The characteristic feature of these new formations is that they contain *variables*, namely "x," "y," "f," "P." Variables are letters that do not refer to any definite entity but rather to a definite range of entities, whose names can be substituted for these variables; the range of the variables must be determined in each case. Thus in (2.4) and (2.6) the "x," "y," and "z" are number variables; for the "x," "y," "z" we may substitute the names of numbers, e.g., "3" or "π." In examples (2.5) and (2.6) the "f" is a function variable, for which we may, for example, substitute "+" and in this way[4] convert (2.5) into the proposition "$2 + 3 = 5$." In particular, the range to which the variable refers may consist of sets of linguistic expressions, when we may allow the entities themselves (and not their names) to be substituted for the variables. A case of this sort occurs in (2.7), where "P" is a *predicate variable*, referring to predicates. An example of a predicate is the set of written symbols "is a prime number." When this predicate is substituted for "P," the expression (2.7) becomes the proposition "2 is a prime number." Written symbols like "2" are called subjects (cf. 2.5), so that "x," "y," "z" are *subject variables.*

In order to indicate that a variable "x" has the real numbers for its range, mathematicians often say that x is an *indeterminate real number*, but phrases of this sort are misleading and should be avoided.

After replacement of the variables by objects in their specified ranges, expressions (2.4) through (2.6) become propositions. Consequently, such sets of symbols are called *propositional forms* (formulas).[5] If we agree, as is often done, to extend the meaning of a propositional form to include the propositions themselves, then the latter are propositional forms *without* free variables.

When a proposition is analyzed logically step by step, we usually encounter intermediate forms that are no longer propositions but are still propositional forms. For example, consider the *Fermat conjecture*:

[4] Strictly speaking, of course, this proposition should read "$+ (2, 3) = 5$," but we will permit ourselves to make such changes tacitly.

[5] See the footnote in §4.1.

(2.8) *There do not exist natural numbers* x, y, z, n, *for which* $x \cdot y \cdot z \neq 0$ *and* $2 < n$ *and* $x^n + y^n = z^n$,

where it is natural to regard the propositional form

(2.9) $$x \cdot y \cdot z \neq 0 \quad and \quad 2 < n \quad and \quad x^n + y^n = z^n,$$

as a logically important part of (2.8).

Let us therefore examine propositions and propositional forms simultaneously. In the analysis of propositional forms we find, in addition to the variables, two types of elements. *First* there are such frequently repeated words (or groups of words) as "not," "and," "or," "for all," which in a certain sense are the logical framework of a proposition. The most important of these are the propositional constants (§2.4) and the quantifiers (§2.6). *Secondly* there are the words (or groups of words) that are characteristic of the mathematical theory under examination at the moment and do not occur, in general, in other theories. Examples are "2," "π," "is a prime number," "lies on," "+." The most important types here are subjects, predicates, and function signs (§2.5).

In the following sections we shall examine these elements more closely. They should be compared with the operator of set formation in §7.7, the notation for functions in §8.4, and the description operator in §2.7.

2.4. *The Propositional Constants*

These serve the purpose of combining propositional forms in order to construct new propositional forms. A simple example is "and."

The two propositional forms

(2.10) 2 *divides* x (2.11) 3 *divides* x

are combined by "and" into the *one* propositional form

(2.12) 2 *divides* x *and* 3 *divides* x.

The propositional form (2.12) is called the *conjunction* of (2.10) and (2.11). The conjunction of two *propositions* is again a *proposition*, which is true (accepted) if and only if both the components united by the "and" are true (accepted). This fact is expressed by the

Truth table (logical matrix) for conjunction

	T	*F*
T	*T*	*F*
F	*F*	*F*

For example, the conjunction of a false proposition (the "*F*" in the left column of the above table) with a true proposition (the "*T*" of the top row) is a false proposition (the "*F*" at the intersection of the given row and column).

Another propositional constant is "not," as in

(2.13) 8 *is not a perfect square.*

In a logical systematization of the language it is customary to put the "not" at the beginning and to write:

(2.13′) *Not* 8 *is a perfect square.*

The proposition (2.13′) is called the *negation* of "8 is a perfect square." In the nonclassical schools of logic, negation is either completely banned or, if admitted, it is variously interpreted by the various schools. *One* possibility consists of accepting the negation of a proposition *a* if from *a* we can derive a contradiction (i.e., a proposition that is always rejected). If negation is admitted at all, it is always subject to the condition that no proposition is accepted together with its negative. In the classical two-valued logic it follows that "not" reverses the truth value. Thus we have:

Truth table (logical matrix) for negation

T	*F*
F	*T*

Another important propositional constant is "*or.*" The word "or," which in everyday English has several different meanings, is almost always used in mathematics in the nonexclusive sense of the Latin "*vel,*" for example:

(2.14) *Every natural number greater than two is a prime number or has a prime factor.*

The combination of two propositions by the nonexclusive "or" is called an *alternative* (or also a *disjunction*, although it would be more correct to reserve the word "disjunction" for the combination of propositions expressed by "either–or"). An alternative is true (accepted) if and only if at least one of its components is true (accepted):

Truth table (logical matrix) for the alternative (disjunction)

	T	*F*
T	*T*	*T*
F	*T*	*F*

The "either–or" is used like the Latin "*aut,*" as indicated in the following table:

Truth table (logical matrix) for the strict disjunction

	T	*F*
T	*F*	*T*
F	*T*	*F*

Among the other constants of the propositional calculus we shall mention here only *implication* (and its consequence *equivalence*), which in the English language is represented by the words "*if—then*." For the "if—then" of ordinary spoken language, the logicians have distinguished, in the course of the centuries, several essentially different meanings. We shall restrict ourselves here to describing the one which appears most often in classical logic and mathematics and can be traced back to the Stoics (Philon, ca. 300 B.C.). If a reader feels that he cannot reconcile the "if—then" of the following truth table with his everyday spoken language, he is referred to §3.

Let us now take up the task of constructing a truth table for "if—then." [The four entries will be determined as soon as we have fixed on the truth value of the following four propositions:

(2.15) *If* $1 + 1 = 2$, *then* $1 + 1 = 2$.

(2.16) *If* $1 + 1 = 2$, *then* $1 + 1 = 3$.

(2.17) *If* $-2 = 2$, *then* $(-2)^2 = 2^2$.

(2.18) *If* $1 + 1 = 3$, *then* $1 + 1 = 3$.]

We regard (2.15) and (2.18) as true, and (2.16) as false. As for (2.17), we can argue as follows: The proposition

(2.17′) *For arbitrary real numbers* x, y *it is true that, if* $x = y$, *then* $x^2 = y^2$

is true. A statement that holds for arbitrary real numbers x, y, holds in particular for $x = -2$ and $y = 2$. Thus we recognize (2.17) as a true proposition. Consequently we have the

Truth table (logical matrix) for implication

	T	F
T	T	F
F	T	T

We establish the *convention* that in discussing the classical logic we shall use "if—then" in the above sense. It should be noted that there is no inherent connection between the two parts of an implication defined in this way. For example, the following proposition is true: "*if* $7 + 4 = 11$, *then a triangle with three equal angles has three equal sides*."

An *equivalence* ("*if and only if*") may be defined as a conjunction of reciprocal implications (see below). Thus we have the

Truth table (logical matrix) for equivalence

	T	F
T	T	F
F	F	T

The propositional constants "not," "and," "or," "if—then," "if and only if" occur so frequently in mathematics that it is worthwhile to introduce *symbols* for them. Usage in present-day logic is not yet uniform. In the following table the symbol given first is the one used in this article.

List of Propositional Symbols

Connective	Everyday English	Symbol
Negation	not	$\neg$ (suggests "—"), over-lining
Conjunction	and	$\wedge$ (dual to "$\vee$"), &, ., immediate juxtaposition
Alternative	or	$\vee$ (suggests "vel")
Implication	if—then	$\rightarrow$, $\supset$
Equivalence	if and only if	$\leftrightarrow$ (combination of "$\rightarrow$" and "$\leftarrow$"), $\equiv$

As already mentioned, we may consider an equivalence as the conjunction of two reciprocal implications. But then we may also say that the equivalence is *defined* by this conjunction. If we introduce the *propositional variables* "p," "q," it is easy to calculate from the tables that we may put "$p \leftrightarrow q$" in place of "$(p \rightarrow q) \wedge (q \rightarrow p)$," as may be seen by calculating the four cases $p, q = T, T; T, F; F, T; F, F$. To state a definition we use the sign "$\Leftrightarrow$," thus, in the present case:

(2.19) $$p \leftrightarrow q \Leftrightarrow (p \rightarrow q) \wedge (q \rightarrow p).$$

Similarly, we can justify the following definitions

(2.20) $$p \rightarrow q \Leftrightarrow \neg\, p \vee q,$$

(2.21) $$p \vee q \Leftrightarrow \neg\, (\neg\, p \wedge \neg\, q),$$

(2.22) $$p \wedge q \Leftrightarrow \neg\, (\neg\, p \vee \neg\, q).$$

2.5. *Subjects, Predicates, and Function Signs*

If we examine the following propositions and propositional forms:

(2.23) 4 *is a prime number*, (2.24) x *lies between* 2 *and* 9,

(2.25) $3 < x$, (2.26) $2 + 4 = 8$,

we see that in addition to the variable "x" they contain the following elements:

the *subjects* "2," "3," "4," "8," "9,"

the *predicates* "*is a prime number*," "*lies between—and*," "$<$," "$=$,"

the *function sign* "$+$."

In the proposition "6 exceeds 3," it is true that from the grammatical point of view "6" is the subject and "3" is the object, but in logic both the "6" and the "3" are subjects.

The above predicates are successively 1, 3, 2, 2-place predicates, and the function sign "+" is a two-place predicate.

Higher-place predicates also occur in mathematics: e.g., the four-place predicate in the propositional form "the point-pair A, B separates the point-pair C, D." A k-place predicate becomes a proposition through the adjunction of k subjects, and in agreement with this manner of speaking we shall sometimes say that the propositions are *0-place predicates.*

In principle, function signs can be dispensed with entirely, being replaced by predicates. For example, the "+" is superfluous if we introduce the three-place predicate "is the sum of ... and." For then in (2.26) we may write: "8 is the sum of 2 and 4." Since function signs can be eliminated in this way, it is a common practice in purely logical investigations to confine oneself to predicates, and in §3 we will take advantage of this simplification. But mathematicians would be unwilling to give up the functional notation, which is a very suggestive one.

The importance of subjects and predicates will be discussed below in §3.3.

2.6. *Operators in the Calculus of Predicates; Bound Variables*

The proposition

(2.27) *All positive numbers are squares*

contains the operator (or *quantifier*) "all" of predicate logic, which we may analyze in the following way (although there are other possibilities): we are dealing here with the one-place predicates "is a positive number" and "is a square," which we may make more prominent by reformulating the proposition:

(2.27′) *For all entities: If an entity is a positive number, then this entity is a square.*

Here the repeated word "entity" obviously has the task of indicating the places to which the operator "all" shall refer. The same task may be performed in a clear and simple way if we insert one and the same sign in each of these places; for example, the letter "z." Thus we get the standard form:

(2.27″) *For all z: If z is a positive number, then z is a square.*

The letter "z" serves only to mark the place; instead we could use any other letter, e.g., "y." It must be noted that "z" is not a variable of the kind considered in §2.5, since (2.27″) is a genuine proposition, as distinct from a propositional form. If "z" is replaced in (2.27″) by the name of a number, e.g., "2," we do not obtain a proposition, but rather the linguistic gibberish: "for all 2: If 2 is a ... ."

It is customary to call the letter "z," as used in (2.27″), a *bound variable*, whereas the variables considered earlier are *free variables*. In the propositional form

(2.28) *If z is a positive number, then z is a square*

the letter "z" is a free variable, and (2.27″) is obtained from (2.28) by binding the "z" with the quantifier "all." In this way, a free variable becomes bound.

Bound variables refer, in the same way as free variables, to a given range; in the present case, for example, to the set of real numbers.

The quantifier "all" is called the *universal quantifier*, and (2.27″) is the *universal quantification* of (2.28). A synonym for "all" is, e.g., "every," and a phrase like "for no z" means "for all z not."

A second operator in the calculus of predicates is the *existential quantifier* "there exists" or "there exist" or "for some," as in the following example.

(2.29) *There exist prime numbers.*

(2.29′) *There exists a y, such that y is a prime number.*

(2.29″) *For some x: x is a prime number.*

(2.30) *x is a prime number.*

The propositions (2.29′) and (2.29″) are variants of (2.29). The existential quantifier can also be used to bind variables. In this connection (2.29) is called an existential quantification of (2.30).

If the range of the variable is finite (for example, the natural numbers from 1 to 9), then the universal and existential quantifiers are, respectively, equivalent to a multiple conjunction and a multiple alternative. Thus if "P" stands for "is a prime number," then "*Every number is a prime number*" is equivalent to "$P1 \wedge P2 \wedge \cdots \wedge P9$" and "*There exists a prime number*" is equivalent to "$P1 \vee P2 \vee \cdots \vee P9$." Consequently we speak of a generalized conjunction or alternative and introduce the symbols "$\bigwedge$" for the universal quantifier and "$\bigvee$" for the existential. Then the familiar Cauchy definition of the continuity of a function f in an interval I takes the following easily understood form:[6]

(2.31) $\bigwedge_x \bigwedge_\epsilon (x \in I \rightarrow \bigvee_\delta \bigwedge_y (y \in I \wedge |x - y| < \delta \rightarrow |f(x) - f(y)| < \epsilon))$.

[6] Here the variables x, y refer to real numbers, and the variables ϵ, δ to positive real numbers. Without the latter convention, the statement of (2.31) would be somewhat more complicated.

For every x and every ϵ there exists, if x is an element of the interval I, a δ such that for every element y of the interval I whose distance from x is less than δ the difference between the functional values $f(x)$ and $f(y)$ is less than ϵ.

It is essential to note that if, as in the present case, several quantifiers appear in the same proposition (or propositional form), then the bound variables used (here x, y, ϵ, δ) must be *distinct* from one another.

In classical logic (but only there!) either of the above quantifiers can be defined in terms of the other. For if H is an arbitrary propositional form, we can write:

$$\bigwedge_x \mathsf{H} \Leftrightarrow \neg \bigvee_x \neg \mathsf{H}, \tag{2.32}$$

$$\bigvee_x \mathsf{H} \Leftrightarrow \neg \bigwedge_x \neg \mathsf{H}. \tag{2.33}$$

These definitions indicate a certain "duality" between $\bigwedge$ and $\bigvee$, which corresponds to a duality between $\wedge$ and $\vee$ (cf. also §9.2).

The following notations are to be found in the literature:

Universal quantifier: $\bigwedge_x \mathsf{H}$, $(x)\mathsf{H}$, $\forall x\mathsf{H}$, $\prod_x \mathsf{H}$,

Existential quantifier: $\bigvee_x \mathsf{H}$, $(\exists x)\mathsf{H}$, $(Ex)\mathsf{H}$, $\exists x\mathsf{H}$, $\sum_x \mathsf{H}$.

2.7. *Identity and Description*

The notation $x = y$ ($x \equiv y$) means that x and y are the same entity. With this sign for identity we can formulate the statement that the property denoted by a given predicate is possessed by *exactly one* entity. If we let "$\mathfrak{P}$" stand for the predicate "is an even prime number," then the fact that there exists exactly one even prime number can be represented by the proposition:

$$\bigvee_x \mathfrak{P}x \wedge \bigwedge_x \bigwedge_y (\mathfrak{P}x \wedge \mathfrak{P}y \rightarrow x = y).$$

If the property indicated by a predicate holds for exactly one entity, we may speak of *the* entity which has this property. Here we need the *description operator*, represented in ordinary English by some such words as "that—which" and usually denoted in logic by the symbol (ιx). Thus $(\iota x)\, \mathfrak{P}x$ is a name for the number 2, in which x occurs as a bound variable (cf. §2.6). If we are given an arbitrary predicate $\mathfrak{Q}$, e.g., "is divisible by 2," then $\mathfrak{Q}(\iota x)\, \mathfrak{P}x$ means that the property indicated by $\mathfrak{Q}$ is possessed by that unique entity for which $\mathfrak{P}x$ holds. The expression $\mathfrak{Q}(\iota x)\, \mathfrak{P}x$ is often used by Russell as an abbreviation for the proposition: There exists exactly one entity which possesses the property indicated by $\mathfrak{P}$, and all entities which have this property also have the property indicated by $\mathfrak{Q}$; or, expressed in symbols:

$$\bigvee_x \mathfrak{P}x \wedge \bigwedge_x \bigwedge_y (\mathfrak{P}x \wedge \mathfrak{P}y \rightarrow x = y) \wedge \bigwedge_x (\mathfrak{P}x \rightarrow \mathfrak{Q}x).$$

This proposition is still meaningful (though false), if the property indicated by $\mathfrak{P}$ does not hold for exactly one entity.

Exercises for §2

1. Set up the truth table for "neither-nor." Represent this connective in terms of

   (a) $\neg$ and $\wedge$,

   (b) $\neg$ and $\vee$.

2. Calculate the truth tables for the propositional forms:

   (a) $(p \wedge q) \rightarrow \neg p$

   (b) $(p \rightarrow q) \rightarrow (\neg p \rightarrow \neg q)$

   (c) $(p \vee \neg q) \wedge \neg (q \rightarrow p)$

   (d) $[(p \rightarrow q) \wedge (q \rightarrow r)] \rightarrow (p \rightarrow r)$

3. Express (2.14) in formal language, with the following definitions:

   $Nx \Leftrightarrow x$ is a natural number,

   $Px \Leftrightarrow x$ is a prime number,

   $Gxy \Leftrightarrow x$ is greater than y,

   $Rxy \Leftrightarrow x$ divides y.

4. Translate into English:

$$\bigvee_x (Nx \wedge \bigwedge_y (Gxy \wedge Ny \rightarrow (Ryx \vee Py))).$$

5. What does

$$\bigvee x\,(31 < x \wedge \neg R2x \wedge \neg R3x \wedge x < 37)$$

   mean?

6. Formulate the axioms of a system of axioms for geometry in the symbolism of the predicate logic.

Bibliography

On the technical use of symbols see Carnap [2]. Information on the use of symbols can also be found in many textbooks of mathematical logic (see the bibliographies for §1 and §6). Especially interesting to mathematicians are Tarski [1] and Frege [2].

3. The Concept of a Consequence

3.1. *Semantics*

In this section we discuss a concept which must be considered as basic in the classical treatment of mathematics and particularly of axiomatization. We wish to investigate the connection that exists between, for example, the Euclidean axioms and the theorem of Pythagoras, a connection which is usually expressed in the form: the theorem of Pythagoras is a consequence of the Euclidean axioms. In the present section we think of this connection as being *static*: if the Euclidean axioms are given, then the Pythagorean theorem is in some sense given at the same time. But we may also think of the situation as a *dynamic* one: given the Euclidean axioms, how can we proceed, step by step, to derive the theorem from them. We will return to this question in §6.

The connection between the theorem and the axioms established by saying that the theorem is a *consequence* of the axioms, can be described as follows: the language in which we formulate our mathematical theorems stands in a certain relation to the actual "world," which is to some extent described by the language. In other words, the actual world provides an *interpretation* of the language. The science that deals with such questions is today called *semantics*. Some of the concepts of semantics can be traced as far back as Aristotle and were important in the work of Bolzano, which remained to a great extent unrecognized in his time. The *modern* science of semantics is due to A. Tarski.

In contrast to semantics, investigations of a language that have nothing to do with any interpretation of it are called *syntax*.

3.2. *Definitions*[7]

In the construction of a mathematical theory we not only formulate and prove theorems but also make definitions. A *definition* is an abbreviation. For example, "x is a prime number" stands for "x is a natural number which is different from 1 and has no factor other than 1 and itself." Although the importance of definitions is largely a practical one, it must not be underestimated. If it were not for such abbreviations, the majority of mathematical theorems would be so cumbersome as to be completely unintelligible.

In our study of the concept of a consequence, we must take the definitions into account. It would be simplest, of course, to eliminate them entirely by replacing them with the expressions for which they stand. In the Pythagorean theorem, for example, the expression "is a right-angled triangle" would be replaced by some expression involving only the fundamental concepts of geometry. If, for convenience, we allow the

[7] For the so-called recursive definitions see §7.4.

definitions to stand as they are, it would be more precise to say: the theorem of Pythagoras is a consequence of the Euclidean axioms *and* the definitions that are used in the formulation of the theorem.

3.3. *The Ontological Assumptions of Semantics*

Let us examine more carefully the ideas underlying this attempt to define a consequence more precisely, since in the semantic construction of mathematics it is assumed that such ideas are "understood." If we ask for the meaning of the linguistic expressions we have called *subjects* and *predicates*, we see that a subject is a name for an *individual*, and a predicate is a name for an *attribute* (a *property*). Subjects in ordinary speech, such as "Lincoln" or "New York," name individuals that have a "real existence." Many mathematicians hold the view that individuals such as those named by the subjects "2" and "π" have an "ideal existence," being of different kinds according to the branch of mathematics under consideration. In real analysis, for example, they are the real numbers; and in the theory of functions of a complex variable they are the complex numbers. The individuals investigated in any given context are regarded as forming a *domain of individuals*: for example, the domain of natural numbers. The domain of individuals may have finitely or infinitely many elements but is assumed to have at least one element.

It is also assumed that together with any given domain of individuals the totality of all relevant properties is also given. In this connection a property is relevant if for each individual in the domain the answer to the question whether or not the individual possesses the property is in the nature of things well defined, even though we may not be able to decide whether it is "yes" or "no." This is the ontological basis of the Aristotelian principle of two-valuedness (cf. §2.2). In addition to the one-place properties, such as the one described by the predicate "is a prime number," we also consider many-place properties, e.g., the two-place property (or *relation*) denoted by "$<$." For an n-place property (or relation), it is assumed that for every ordered n-tuple of individuals from the domain under consideration it is determined in the nature of things whether the individuals in the given order stand in the given relation or not.

3.4. *Mathematical Axioms as Propositional Forms*

The concept of a mathematical consequence has been developed chiefly in connection with geometry, above all in researches on the independence of the parallel postulate. We shall therefore take geometry as the starting point for our discussion. The modern attitude toward the axioms of geometry was described in a drastic way by Hilbert when he said: "We must always be able to replace the words 'point,' 'line,' and 'plane' by 'table,' 'chair,' and 'beer-mug.' "

Of course, Hilbert does not mean that the theorems of geometry will remain *true* if we make the suggested change, but only that *for mathematics*, which has the problem of determining *consequences* of the axioms of geometry, it is of no importance whether we speak of points, etc., or of tables, etc. In other words: if a geometrical proposition is a consequence of the Euclidean axioms, then the proposition that arises from it through Hilbert's suggested change in terminology is a consequence of the corresponding axioms arising from the change. In the epigrammatic phrase of Bertrand Russell, "a mathematician does not need to know what he is talking about, or whether what he says is true."

Since in geometry (as in any purely mathematical science; cf. §1.2) we have no interest in the meaning of the predicate "is a point," we may replace it (and correspondingly the other geometric predicates) by a predicate *variable*, thereby concentrating our attention on what is mathematically essential and doing away with everything else. If we write "P" for "is a point," "G" for "is a line," and "L" for "lies on," the first Euclidean axiom (in Hilbert's formulation)

(3.1) *Given any two points A, B, there exists a line a which corresponds to each of the two points A, B.*

Given two points A, B, there is not more than one line which corresponds to each of the given points A, B,

becomes, in the logical symbols introduced in our earlier sections:

$$(3.2)\qquad \bigwedge_x \bigwedge_y ((Px \wedge Py \wedge x \neq y) \rightarrow \bigvee_g (Gg \wedge Lxg \wedge Lyg))$$

$$\bigwedge_x \bigwedge_y \bigwedge_g \bigwedge_h (Px \wedge Py \wedge x \neq y \wedge Gg \wedge Gh \wedge Lxg$$

$$\wedge\, Lxh \wedge Lyg \wedge Lyh \rightarrow g = h).$$

Thus we see that for the pure mathematician it is more precise to regard the geometric axioms as *propositional forms* [like (3.2)] than as propositions [like (3.1)]. The so-called *fundamental concepts* of a given mathematical theory, i.e., the subjects and predicates appearing in its axioms, are in this sense simply linguistic paraphrases for subject variables and predicate variables. When the axioms are regarded as propositional forms, they cannot be said to be either true or false. They become true or false only after the variables occurring in them (i.e., the fundamental concepts of the given mathematical theory) have been given an *interpretation*; that is, only when to each (free) subject variable we have assigned an individual of the underlying domain of individuals and to each predicate variable a property (with the same number of places) of the elements of the domain. When propositional variables occur, they are to be interpreted by means of propositions. Then it becomes meaningful to say that a given propositional form is *true* or *false* in this interpretation. The fact that a propositional form H is true in the interpretation $\mathfrak{I}$ is expressed by saying: $\mathfrak{I}$ satisfies H, $\mathfrak{I}$ is a model of H, $\mathfrak{I}$ verifies H, or H is true in $\mathfrak{I}$. As an example let us choose the domain of natural numbers and consider the propositional forms

$$(3.3)\quad Px \qquad\qquad (3.4)\quad \neg Px$$

$$(3.5)\quad Px \wedge Qx \qquad\qquad (3.6)\quad Px \vee Qx$$

$$(3.7)\quad Px \wedge \neg Px \qquad\qquad (3.8)\quad Px \vee \neg Px$$

$$(3.9)\quad \bigvee_x (Px \wedge Qxy) \qquad\qquad (3.10)\quad \bigwedge_x Px \leftrightarrow \bigwedge_y Py.$$

The form (3.3) is true in the interpretation which to the variable x assigns the number 4, and to the variable P the property of being even; in other words, 4 has this property. The form (3.3) is not true if P is interpreted as before while x is interpreted as 5. The form (3.4) is the opposite of (3.3). The form (3.6) is true, and (3.5) is not true, if x is interpreted as 4, while P is interpreted as the property of being prime, and Q as that of being a perfect square. The form (3.8) is true for any interpretation, and (3.7) for none. Consequently, (3.8) is said to be *valid* or a *tautology*, and (3.7) is *contradictory* or a *contradiction*. In (3.9) only the P, Q, y require interpretation and in (3.10) only the P, since the other variables are *bound* (cf. §2.6). The form (3.9) is true if y is interpreted as 10, P as the property of being prime, and Q as the relation of "smaller than," since there exists at least one number which is both prime and smaller than 10. The form (3.10) is a tautology, expressing the fact that a bound variable may be renamed at will. See also the examples in §3.8.

3.5. *The Artificial Language of the Predicate Logic*

The propositional forms (3.2), (3.3), ... (3.10) contain, apart from brackets, only logical symbols and subject and predicate variables. These propositional forms are called *expressions in the predicate logic*. Here it is important that only the subject variables, and not the predicate variables, can be bound by quantifiers.[8] The language of this predicate logic is an artificial language capable of expressing a great part of mathematics. As soon as we have chosen a domain of individuals, we can interpret the subject variables and the predicate variables and can then give an exact definition of what it means to say that a proposition is true in this interpretation. It is most convenient to construct a definition inductively by

[8] If we also allow the predicate variables to be bound, we are in the so-called "logic of the second order," or "extended predicate logic," cf. §10.2.

proceeding from simpler to more complicated expressions. Through lack of space we must content ourselves with this remark and with the above examples.

3.6. *The Concept of a Consequence*

Now let $\mathfrak{A}$ be the set of axioms and H a theorem in a mathematical theory, e.g., in Euclidean geometry. We then say that H is a consequence of $\mathfrak{A}$. If we now take H and the elements of $\mathfrak{A}$ to be propositional forms and interpret the fundamental concepts in such a way that all the axioms are true, it is reasonable to expect that in the given interpretation H will also be true. Thus we have a necessary condition which the concept of a consequence must satisfy. In order to give the widest possible meaning to the concept, we agree to regard this necessary condition as being also sufficient. In this way we arrive at the following

Definition of a Consequence: *The propositional form* H *follows from the set* $\mathfrak{A}$ *of propositional forms* (H *is a consequence of* $\mathfrak{A}$) *if every model common to all the propositional forms of* $\mathfrak{A}$ *is also a model of* H.

Examples: Py, and also Qy follow from $Py \wedge Qy$; and $\bigvee_x Px$ follows from $\bigwedge_x Px$ (here it must be noted that by §3.3 a domain of individuals contains at least one element). Also, $\bigwedge_y Py$ follows from $\bigwedge_x Px$ and conversely. Every propositional form follows from a contradictory propositional form. A tautology follows from any propositional form.

3.7. *Consequence and Tautology*

If the number of axioms is finite, we can reduce the concept of a consequence to that of a tautology. For this purpose we first form the conjunction Θ of all the axioms in $\mathfrak{A}$. Then we have the important theorem:

H *follows from* $\mathfrak{A}$ *if and only if* $\Theta \rightarrow \mathsf{H}$ *is a tautology.*

This theorem expresses the relation between "follows from" and "if—then." The theorem is proved as follows:

(*a*) We first assume that H follows from $\mathfrak{A}$. Then we must show that $\Theta \rightarrow \mathsf{H}$ is a tautology, i.e., that $\Theta \rightarrow \mathsf{H}$ is true for every interpretation over an arbitrary domain of individuals. To this end we make an arbitrary interpretation $\mathfrak{D}$ of the given domain. In case Θ is false in $\mathfrak{D}$, then $\Theta \rightarrow \mathsf{H}$ is certainly true in $\mathfrak{D}$ (cf. the logical matrix in §2.4); and in case Θ is true in $\mathfrak{D}$, then H must also be true in $\mathfrak{D}$, in view of the hypothesis that H is a consequence of $\mathfrak{D}$; thus in this case also $\Theta \rightarrow \mathsf{H}$ is true in $\mathfrak{D}$.

(*b*) We now assume that $\Theta \rightarrow \mathsf{H}$ is a tautology and must show that H follows from $\mathfrak{A}$. But if this were not the case, then there would be an interpretation $\mathfrak{D}$ for which all the propositional forms in $\mathfrak{A}$ (and therefore Θ) would be true but H would be false. Then $\mathfrak{D}$ falsifies $\Theta \rightarrow \mathsf{H}$, in contradiction to the hypothesis that $\Theta \rightarrow \mathsf{H}$ is a tautology.

3.8. *Examples of Tautologies*

$$\neg \neg p \leftrightarrow p, \tag{3.11}$$

$$\neg (p \wedge p) \leftrightarrow (\neg p \vee \neg q), \tag{3.12}$$

$$\neg (p \vee q) \leftrightarrow (\neg p \wedge \neg q), \tag{3.13}$$

$$\neg (p \rightarrow q) \leftrightarrow (p \wedge \neg q), \tag{3.14}$$

$$\neg (p \leftrightarrow q) \leftrightarrow (p \leftrightarrow \neg q), \tag{3.15}$$

$$\neg \bigwedge_x \mathsf{H} \leftrightarrow \bigvee_x \neg \mathsf{H}, \tag{3.16}$$

$$\neg \bigvee_x \mathsf{H} \leftrightarrow \bigwedge_x \neg \mathsf{H}. \tag{3.17}$$

These tautologies form the basis for the *technique of negation.* We obtain a simple application of the theorem in 3.7 if we weaken (3.14) to

$$\neg (p \rightarrow q) \rightarrow (p \wedge \neg q). \tag{3.14'}$$

Then $p \wedge \neg q$ follows from $\neg (p \rightarrow q)$.

Exercises for §3

1. Which of the following propositional forms are tautologies and which are contradictions?

   (a) $\bigvee_x Px \vee \bigwedge_x \neg Px$,

   (b) $\bigwedge_x Px \vee \bigvee_x Px$,

   (c) $\neg \bigwedge_x Px \wedge \neg \bigvee_x \neg Px$,

   (d) $\bigvee_x \neg Px \rightarrow \bigvee_y (Py \rightarrow Qy)$,

   (e) $\bigwedge_x \bigwedge_y (Rxy \wedge Ryx \rightarrow x = y)$.

2. $H_1(=) \bigwedge_x \bigwedge_y (Rxy \wedge Ryz \rightarrow Rxz)$

   $H_2 (=) \bigwedge_x \bigwedge_y (Rxy \vee Ryx)$

   Over the domain of individuals $\{1, 2, ..., 10\}$ give interpretations which will falsify, or verify,

   (a) H_1

   (b) II_2

   (c) $H_1 \wedge H_2$

3. $H_1 \,(=) \bigvee_y \bigwedge_x Rxy$

$H_2 \,(=) \bigwedge_x \bigwedge_y Rxy$

$H_3 \,(=) \bigwedge_x \bigwedge_y Rxy$

$H_4 \,(=) \bigvee_x \bigvee_y Rxy$

$H_5 \,(=) \bigwedge_y \bigwedge_x Rxy$

$H_6 \,(=) \bigvee_y \bigvee_x Rxy$

Which expressions follow from which? In the cases in which H_i does not follow from H_k give a counterexample, i.e., a model of H_i which is not a model of H_k.

Bibliography

On the foundations of geometry see Borsuk-Szmielew [1], and on modern semantics see Carnap [1], Linsky [1] and Tarski [2].

4. Axiomatization

4.1. *The Origin of Systems of Axioms*

It is today customary to construct a mathematical science axiomatically, that is, by first choosing a set of propositions[9] as the axioms and then drawing consequences from them. The subjects and predicates that occur in the axioms are called the *fundamental concepts* of the system of axioms. From the modern point of view these axioms are considered to be variables, as explained in §3.4. In general, the number of axioms is finite, although infinite systems of axioms are sometimes admitted if their structure is immediately clear. For example, we might take for axioms all the propositions of a certain form. In this case we sometimes speak of an *axiom schema* (for examples, see §10.3 and §11.2).

If the fundamental concepts occurring in the axioms are taken to be variables, so that the axioms themselves become propositional forms, we can no longer regard them as "self-evident." Moreover, if two systems of axioms are *equivalent* (that is, if each of them is a consequence of the other), then in principle they are on an equal footing, even though one of them may be preferred on more or less subjective grounds, e.g., because of its greater logical clarity.

Theoretically, we could use any propositions at all to form our set of axioms, but it turns out that in modern mathematics relatively few systems

[9] By the arguments in §3.4 we should really say "propositional forms" instead of "propositions," but here we wish to conform to the ordinary mathematical usage, in which axioms and theorems are called "propositions."

are in actual use. So it is natural to ask about the motives for choosing these particular systems. We shall confine ourselves to a discussion of this question from the following point of view: It is an established fact that in many cases the theory had a prior existence (at least to a great extent) and the axioms for it were chosen later. But in many cases the axioms are primary; and the theory is to a certain extent secondary, since it has been created and defined by the axioms themselves. We shall distinguish the two cases by speaking of an *heteronomous* and an *autonomous system of axioms*, but it must be emphasized that these concepts are idealizations; in fact, it is often very hard to decide how a given system of axioms actually arose.

4.2. *Heteronomous Systems of Axioms (Subsequent Choice of Axioms)*

In general, we are dealing here with the following problem: we are given a set $\mathfrak{B}$, usually large, of preassigned propositions and we must find a system of axioms (as simple and clear as possible) from which all these propositions follow.[10]

A characteristic example is provided by any theory in physics, or in any other science based on observation. Here the preassigned set $\mathfrak{B}$ consists of a large number of empirical facts, perhaps accompanied by certain hypotheses, and it is our task to find a system of axioms $\mathfrak{A}$ that will provide an economical description of the whole relevant body of knowledge $\mathfrak{B}$. Assuming that we have found such a system of axioms $\mathfrak{A}$, we obtain a *mathematical science* if we ask what are the consequences that follow from $\mathfrak{A}$; but if we then proceed to ask whether these consequences (so far as they can be tested) are in agreement with observation, we are in the domain of *theoretical physics*. Here again the distinction is clear between a *natural science* (cf. §1.2) and mathematics as a purely *abstract science*. When a mathematical system of axioms $\mathfrak{A}$ has arisen in this way, we shall say that the theory determined by $\mathfrak{A}$ has a physical (or, more generally, an empirical) origin. It seems reasonable to believe that Euclidean geometry is such a science. Basic geometrical concepts, like point and line, originated from the need to describe physical data, and consequently the first geometrical propositions were of a physical nature. An example is the theorem of Pythagoras, already well known to the Babylonians in 1700 B.C. This physical origin of geometry becomes particularly clear when we reflect that "experiments" are often made in school to convince the students that the sum of the angles in a triangle is 180°. The axiomatization of geometry was begun by the Greeks, who from the time of Thales (about 590 B.C.) showed that certain geometrical propositions could be made to depend upon others. In relinquishing all

[10] The system of axioms must, of course, be consistent. See §4.7.

recourse to experience, they became the creators of mathematics in the strict sense of the word. The name of Euclid (about 300 B.C.) marks the completion (for the time being) of the axiomatization of geometry.

It is also reasonable to suppose that arithmetic and, to take a modern example, the theory of sets have an empirical origin. The axiomatization of these sciences will be discussed in §10 and §7.

4.3. *Autonomous Systems of Axioms (Systems of Axioms as Sources of New Theories)*

The mathematical theories discussed above were already in existence, at least in a certain sense, long before the corresponding systems of axioms, as becomes quite clear when we recall that in the schoolroom these sciences are often presented without reference to any system of axioms at all; for example, Euclidean geometry in the secondary school and the infinitesimal calculus or naive set theory in the university are often taught in this way. But the situation is completely different in modern mathematical sciences like group theory, ring theory, or lattice theory. These theories cannot be separated from their axioms, since it is only through the axioms that they have come into existence at all. A typical example is the *theory of groups*. In the development of mathematics it has often happened that widely diverse subjects have been seen to depend on lines of argument that are surprisingly similar to one another; e.g., the period of the decimal expansion of a given rational number compared with the number of times a dodecahedron must be rotated in order to bring it back to its original position. It would clearly be more economical not to repeat such arguments at every new occasion but to present them once and for all in such a form that they are immediately applicable to every special case. But an even more important advantage is the fact that by proceeding in this way we concentrate on the essential features of the situation, thereby gaining a deeper insight into the connections among its various parts. In group theory such a program has been carried out. It is possible to formulate a small number of axioms with only *one* fundamental concept, namely *group multiplication*, such that the theory is defined, or so to speak created, by the axioms themselves. The consequences of these axioms are called the *theorems of group theory*. Then by interpreting the group multiplication in various ways, each of which must satisfy the axioms, we at once obtain the original theorems in the various branches of mathematics that led us originally to create the theory of groups. The whole of modern mathematics is characterized (cf. III14) by the attempt to give an increasingly central role to such systems of axioms as those of group theory.

Several of the more modern studies of geometry consist of examining the consequences of a part of the Euclidean axioms, e.g., the axioms of connection or the axioms of order. Systems of axioms of this sort can also be called autonomous. Similarly, the autonomous systems for algebra are to be considered as arising from the heteronomous system for arithmetic.

4.4. *Independence of a System of Axioms*

A system of axioms is said to be *independent* if no axiom is a consequence of the others. In general, independence is desirable but not altogether necessary; often it is an advantage that can be obtained only at the cost of great complication.

The independence of a given system of axioms is most simply demonstrated by finding for each axiom H an interpretation in which H is false but all the other axioms are true. As a simple example let us consider the three axioms that define an *equivalence relation* R:

(4.1) $\bigwedge_x Rxx$ (reflexivity)

(4.2) $\bigwedge_x \bigwedge_y (Rxy \rightarrow Ryx)$ (symmetry)

(4.3) $\bigwedge_x \bigwedge_y \bigwedge_z ((Rxy \wedge Ryz) \rightarrow Rxz)$ (transitivity)

In each case let us choose as domain of individuals the set of three natural numbers $\{1, 2, 3\}$.

Independence of (1): we interpret R as the empty relation (i.e., the relation that holds for no pair).
Independence of (2): we interpret R as the $\leqslant$ relation.
Independence of (3): we interpret R as the relation that holds between two elements x, y of the domain of individuals if and only if $|x - y| \leqslant 1$.

The fact that the parallel axiom is independent of the other Euclidean axioms can also be proved by this method (see II2, §2).

4.5. *Completeness of a System of Axioms*

Let there be given an axiom system $\mathfrak{A}$. A proposition that contains only subjects and predicates already occurring in $\mathfrak{A}$ will be called a *relevant proposition* and $\mathfrak{A}$ is said to be *complete*[11] if for every relevant proposition H, either H follows from $\mathfrak{A}$ or $\neg$ H follows from $\mathfrak{A}$. This is of course, different from saying that H $\vee \neg$ H follows from $\mathfrak{A}$; the latter proposition is always true, since H $\vee \neg$ H is a tautology (tertium non datur).

[11] Other definitions of completeness can also be found in the literature; cf. §6.2.

Autonomous systems of axioms are in general incomplete as a result of their inherent nature (cf. §4.6). E.g., from the system of axioms for group theory it is impossible, as can be easily shown by examples, to deduce either

$$(4.3) \quad \bigwedge_x x = x^{-1} \qquad \text{or} \qquad (4.4) \quad \neg \bigwedge_x x = x^{-1}$$

On the other hand it is natural to expect, in general, that heteronomous systems of axioms will be complete in view of their physical origin. For suppose we have a relevant proposition H such that neither H nor $\neg$ H follows from the axioms. Then the physicist will at once attempt to obtain *experimental* evidence of the correctness or falsity of this proposition and, if successful, will add either H or $\neg$ H to the set of axioms. Thus, physicists are always striving to complete their systems of axioms, so that it is natural to expect completeness in a well developed theory.

Examples of a complete system of axioms are the system for Euclidean geometry or the Peano system for the natural numbers (cf. §10.2).

4.6. *Monomorphy of a System of Axioms*

The concept of isomorphy, familiar to every mathematician from group theory (see, e.g., IB2, §4.2), can be generalized (we omit the definition here; cf. IB10, §1.3) in such a way that one may speak of *isomorphic interpretations* of a system of axioms. To take an example from geometry: The "natural" interpretation of the Euclidean system of axioms, in which the points are "idealized actual points" and the lines are "idealized actual lines," etc. is isomorphic to the interpretation provided by analytical geometry, in which the points are triples of numbers, the lines are the coefficients of the Hesse normal form, etc.

If a given system of axioms is valid in one interpretation, it is also valid in any isomorphic interpretation. For example, if a given structure is a group, then every isomorphic structure is also a group.

Consequently, it is impossible to characterize a given model completely by means of a system of axioms. The most that can be attained in this direction is to characterize the model "up to isomorphism." A system of axioms is said to be monomorphic (categorical) if any two models are isomorphic.

Autonomous systems of axioms are intended to have a wide range of application and therefore, in general, they are not monomorphic; in fact, there exist nonisomorphic groups, nonisomorphic rings, etc. On the other hand, the heteronomous systems of Euclidean geometry and arithmetic (cf. §10) are monomorphic.

Every monomorphic system of axioms $\mathfrak{A}$ *is complete*: Let H be a relevant proposition. Then we must show that H or $\neg$ H follows from $\mathfrak{A}$. We proceed indirectly by assuming that neither H nor $\neg$ H follows from $\mathfrak{A}$. By the definition of a consequence given in §3, there exists an interpretation $\mathfrak{D}_1$ in which, since H does not follow from $\mathfrak{A}$, all the axioms of $\mathfrak{A}$ are true but H is false. In the same way, there exists an interpretation $\mathfrak{D}_2$ in which, since $\neg$ H does not follow from $\mathfrak{A}$, all the axioms of $\mathfrak{A}$ are true but $\neg$ H is false, and therefore (by the tertium non datur) H is true. On account of the assumed monomorphy of $\mathfrak{A}$, the two interpretations $\mathfrak{D}_1$ and $\mathfrak{D}_2$ are isomorphic, and since H is true in $\mathfrak{D}_2$, it follows that H must also be true in $\mathfrak{D}_1$. But this contradiction refutes the assumption.

4.7. *Consistency of a System of Axioms*

Here we discuss the concept of *semantic consistency*, to be distinguished from *syntactic consistency* (see §5.7), which is another extremely important concept in modern studies in the foundations of mathematics. A system of axioms is said to be (*semantically*) *consistent* if it has at least one model.

In view of the physical origin of many heteronomous systems of axioms, it is natural to regard them as being consistent. But it must always be kept in mind that the consistency of a system of axioms is not, in general, an established fact but only a belief based on confidence in our intuitions. Particularly problematical is the consistency of a set of axioms that can only be interpreted in a domain with infinitely many individuals.

The question of the consistency of a given system of axioms can often be reduced to the same question for another system, in which case we speak of a *proof of relative consistency*. Thus, by means of analytical geometry we can show that the system of axioms for Euclidean geometry is consistent if the system for real analysis is consistent. The most interesting proof of relative consistency is due to Gödel, who proved that a system of axioms for set theory which includes the axiom of choice and the continuum hypothesis (see §7) is consistent relative to the same system without these axioms.

In fact it is well known that belief in the existence of a suitable interpretation can be quite mistaken, for example, in naive set theory (see §7 and §11).

A system of axioms $\mathfrak{A}$ *is inconsistent (self-contradictory) if and only if the proposition* H $\wedge \neg$ H *follows from* $\mathfrak{A}$ *for every relevant proposition* H. For if $\mathfrak{A}$ is inconsistent, then $\mathfrak{A}$ has no model. Thus, every model of $\mathfrak{A}$ is also a model of any arbitrary relevant proposition H, and in particular of H $\wedge \neg$ H; that is, H $\wedge \neg$ H follows from $\mathfrak{A}$. On the other hand, if H $\wedge \neg$ H follows from $\mathfrak{A}$, every model of $\mathfrak{A}$ must also be a model of H $\wedge \neg$ H; but H $\wedge \neg$ H is unrealizable, and therefore $\mathfrak{A}$ has no model.

Exercises for §4

1. The order and the successor relation for the natural numbers can be described by the following axioms (Peano-Hilbert-Bernays):

$$
\begin{aligned}
&{}_{x}\,(-x < x)\\
&{}_{xyz}\,((x < y \wedge y < z) \rightarrow x < z)\\
&{}_{x}\,x < x'\\
&{}_{x}\,-x' = 0\\
&{}_{xy}\,(x' = y' \rightarrow x = y)
\end{aligned}
$$

Show by means of suitable models that this system of axioms is independent.

Bibliography

See the textbooks listed in the other sections.

5. The Concept of an Algorithm

5.1. *Examples of Algorithms*

Mathematicians are interested not only in theoretical insight and profound theorems but also in general methods for solving problems, methods whereby certain classes of problems can be handled in such a systematic way that the actual process of solution becomes, so to speak, automatic. Every newly discovered method represents an advance in mathematics, although the problems that are solvable by this method thereby become trivial and cease to form an interesting part of creative mathematics.

A general method of this sort is often called a *calculus*, the name being derived from the small stones or calculi formerly used in computation. Another word with the same meaning is *algorithm*, derived from the name of the Arabic mathematician al-Khuwārizmi (about A.D. 800).

Let us give some examples of algorithms: (*a*) the usual methods of addition, subtraction, multiplication, and division of integers in the decimal notation; (*b*) the Euclidean algorithm for the highest common factor of two integers; (*c*) the well-known procedures for solving linear and quadratic equations with integral coefficients; (*d*) the method of extracting a square root by computing successive decimal places; (*e*) integration of rational functions by means of partial fractions.

The essential feature of an algorithm is that it requires no inspiration or inventiveness but only the ability to recognize sets of symbols and to combine them and break them up according to rules prescribed in advance;

in other words, to carry out elementary procedures that can in principle be entrusted to a machine.

An algorithm proceeds step by step. Some algorithms, when applied to a concrete problem, break off after a finite number of steps, as in the above examples (*a*), (*b*), (*c*), (*e*). Others do not come to an end but can be carried out as far as we like, as in the extraction of a square root, example (*d*). In the above examples every step is, in general, uniquely determined. But in other algorithms it may happen that each step depends upon a free choice among several (finitely many) possibilities. For example, consider the algorithm (*f*) which, when applied to two prescribed integers a, b (in decimal notation), leaves open at each step a free choice between two possibilities: when two numbers (including a and b) are already found, we may take (1) their sum or (2) their difference. This algorithm enables us to find all the numbers in the module (a, b) generated[12] by the two numbers a and b.

A set of numbers (i.e., a row of symbols) which, as in this example, can be determined by an algorithm, is said to be *recursively enumerable*. Of course as long as the word "algorithm" is being used in an intuitive way, the meaning of "recursively enumerable" also remains intuitive; precise definitions are given in §5.5.

Algorithm (*f*) has two initial "formulas," a and b, to be thought of as given in their decimal notation, since an algorithm is restricted by its very definition to operating with rows of symbols (or equivalent objects). The above possibilities (1) and (2) for proceeding from one step to the next are called the *rules of the algorithm*.

The initial formulas of an algorithm are sometimes called *axioms* and its rules are *rules of inference*. A finite sequence of formulas, in which each formula is an axiom or arises from the preceding formulas by application of one of the rules, is called a *derivation* or a *proof*. These terms are borrowed from logic but are used here in a much more general sense.

5.2. *Examples of "Arithmetical" Algorithms*

An algorithm for the enumeration of finite sets of strokes (or, as we may say, "of natural numbers") can be described by one axiom

$$|, \tag{5.1}$$

and one rule

$$\frac{e}{e\,|}. \tag{5.2}$$

[12] For the concept of a module, see IB1, §2.3.

(Here a/b is to be read: from a we may proceed to b.) This rule contains a *proper variable* e, to be interpreted as follows: any expression already derived may be substituted for e, and then a stroke may be added to the right of it. For example, in the algorithm defined by (5.1) and (5.2), the following expressions are derivable: | (as an axiom), ||, |||, ||||.

For the expressions derived in a given algorithm, we may use variables, say n, m, p, q, for the rows of symbols in the algorithm just described, and then these variables can be used to describe further algorithms. For example, we can define an algorithm for the addition of natural numbers (sequences of strokes). As an axiom we take

$$n + | = n\,|, \tag{5.3}$$

which is more precisely an *axiom schema*. Then n can be replaced by any one of the rows of symbols, e.g., ||, that are derivable in the algorithm defined by (5.1) and (5.2). As a specialization of (5.3), we obtain the axiom:

$$|| + | = |||. \tag{5.3$'$}$$

As the only rule in the new algorithm we take

$$\frac{n + m = p}{n + m\,| = p\,|}\,. \tag{5.4}$$

By setting || for n, | for m, and ||| for p, we obtain from (5.3′) the formula

$$|| + || = ||||. \tag{5.4$'$}$$

In order to construct an algorithm for multiplication, we adjoin the further axiom (axiom schema):

$$n \times | = n, \tag{5.5}$$

and the rule (now with two "premisses"):

$$\frac{\begin{matrix} n \times m = p, \\ p + n = q \end{matrix}}{n \times m\,| = q}\,. \tag{5.6}$$

As a special case of (5.5) we obtain

$$|| \times | = ||. \tag{5.5$'$}$$

If we apply the rule (5.6) to (5.4′) and (5.5′), we have

$$|| \times || = ||||. \tag{5.6$'$}$$

5.3. *Recursively Enumerable and Decidable Sets*

Although it has been possible to set up algorithms for the solution of many mathematical problems, others have continued to resist every attack of this kind, a prominent example being the "word problem" in group theory (cf. IB2, §16.1). As a result, mathematicians finally began to suspect that certain problems cannot be solved by any algorithm whatever. It is obvious that a theorem of this sort will become meaningful, and we can proceed to demonstrate it, only when we have given an exact definition of the concept of an algorithm.

More precisely, we need only know what we mean by saying that a given set of rows of symbols is recursively enumerable, i.e., can be found by means of an algorithm. Here we must realize that more is expected from such a definition than, for example, from the definition of continuity of a function. In the latter case we are quite satisfied with the simple, well-known definition of Cauchy, since it is to a great extent in agreement with our intuition, although everyone knows, from certain striking examples, that this agreement is by no means complete. But for a recursively enumerable set, where we are dealing with the question of what can or cannot be accomplished in an actual computation, the definition must agree to the greatest possible extent with our basic intuitive notion of what is meant by effective calculation of the answer to a given problem. The assertion that a given set is not recursively enumerable, i.e., that it is impossible to construct an algorithm for finding the elements of the set, is of interest only to the extent to which our formal definition of an algorithm is in agreement with our intuitive notion of a process of computation.

Several different definitions have been suggested for enumerability (the first one of them by Church in 1936), but in spite of the fact that they originated in very different settings, they are all equivalent to one another. Consequently, many logicians and mathematicians are convinced that these definitions correspond completely to our intuitive notion of computability. They are to be considered from the classical point of view, since they make use of the nonconstructive phrase "there exists." If they have been criticized, it is usually by mathematicians who do not share the classical point of view and therefore assert that the definitions include more than our original intuitive notions. However, a proof of non-enumerability based on too broad a definition retains its validity when the definition is restricted.

After a preparatory section, we shall give in §5.5 a definition of algorithm (or alternatively of recursive enumerability) which is based on the concept of a recursive function. We could set up an alternative definition by generalizing the procedure in §5.2; and a third method stems from the fact that, in principle, every algorithm can be entrusted to a machine

(Turing). There are further possibilities but we omit them here for lack of space.

A property $\mathfrak{E}$ of formulas is said to be *decidable* if the set of formulas that have the property $\mathfrak{E}$ and also the set of formulas that do not have it are recursively enumerable. The decidability of several-place properties (relations) is defined correspondingly. In the case of a decidable property we can decide, by any of the three methods of recursive enumeration mentioned above, whether a given formula ζ has the property or not.

5.4. *Gödelization*

The formulas that can be written in a given finite or countably infinite alphabet can be characterized in various ways by natural numbers (or by the sequences of strokes that correspond to them). We now describe one such method, taking as an example the formulas (words) that can be written with the twenty-six letters of the Latin alphabet. We first enumerate the letters; e.g., 1. *a*, 2. *b*, ..., 26. *z*. Now consider a given n-letter word (i.e., a formula) in the alphabet, and let the numerals assigned to the successive letters of this word be $\nu_1, \ldots, \nu_n$. Also let $p_1 = 2, p_2 = 3, p_3 = 5, \ldots$ be the sequence of prime numbers. Then the given word can be characterized by the number (Gödel index)

$$p_1^{\nu_1} \cdot p_2^{\nu_2} \cdots p_n^{\nu_n}. \tag{5.7}$$

For example, the word "cab" will receive the number $600 = 2^3 \cdot 3^1 \cdot 5^2$. Distinct words correspond to distinct numbers but not every number corresponds to a word. If the number of a word is known, the word itself can be recovered.

A transition of this sort from the words to the corresponding numbers is called *arithmetization* or *Gödelization*. In all questions concerning algorithms, it makes no difference whether we discuss the original formulas or their Gödel numbers.

A recursively enumerable set of words is transformed in this way into a recursively enumerable set of natural numbers and vice versa. It therefore makes no difference, in principle, whether the desired exact definition of recursive enumerability is expressed in terms of words or of natural numbers. Since the natural numbers have a somewhat simpler structure and are more familiar to mathematicians, we will now proceed to define the concept of recursive enumerability for a set of natural numbers.

5.5. *Computable Functions and Recursively Enumerable Sets*

Instead of giving a direct definition of a recursively enumerable set, we shall first define the concept of a computable function, to which the concept of recursive enumerability can be reduced.

We consider functions, with one or more arguments ranging over the entire set of natural numbers, whose values are also natural numbers. Such a function is said to be *computable* (in the intuitive sense) if, for arbitrarily preassigned arguments, there exists a procedure for calculating the value of the function in a finite number of steps. Examples of computable functions are the sum of two numbers, and their product. The following example defines a function f about which we do not know at the present time whether it is computable or not:

$$(5.8)\quad f(n) = \begin{cases} 0, \text{ in case there exist natural numbers } x, y, z \text{ such that} \\ x \cdot y \cdot z \neq 0 \quad \text{and} \quad x^n + y^n = z^n, \\ 1 \text{ otherwise.} \end{cases}$$

At present we know only a few of the values of this function, e.g.,

$$f(1) = f(2) = 0, \qquad f(3) = f(4) = \cdots = f(100) = 1.$$

If the Fermat conjecture is true, then $f(n) = 1$ for $n \geqslant 3$.

The following argument shows that the computable functions are exceptional. There cannot exist a greater number of computable functions than there are methods for computing them. Every method of computation must be capable of being described. A description consists of a finite number of symbols. It follows that there are only countably many possible descriptions, and therefore only countably many computable functions. On the other hand, the total of number of functions is uncountable, as may be proved by the same diagonal procedure as the uncountability of the continuum (see §7).

The concept of a recursively enumerable set can be reduced to that of a computable function. For we have the theorem:

A non-empty set of natural numbers is recursively enumerable if and only if it is the range of values of a computable function.[13]

To prove this theorem we argue as follows: A set which is the range of values of a computable function f can be recursively enumerated by calculating the successive values $f(0)$, $f(1)$, $f(2)$, ..., as may be done in each case in a finite number of steps. We thus obtain an algorithm that produces all the elements of the set (in general, of course, they will not be obtained in order of magnitude, but that is not necessary).

On the other hand, let there be given a non-empty, recursively enumerable set M, so that M contains at least one element n_0. Now the successive steps of an algorithm for the recursive enumeration of M can be arranged (if necessary by the adjunction of certain rules) in a unique

[13] It is customary to say that the empty set is also recursively enumerable.

sequence, with a zeroth, first, second step, etc. Every step produces an element of M, or at least an intermediate stage toward the production of such an element. We now define a function f as follows:

$$f(n) = \begin{cases} n_0\text{, in case the } n\text{th step provides only an} \\ \text{intermediate stage,} \\ k\text{, in case the } n\text{th step provides an element of } M \\ \text{and this element is } k. \end{cases}$$

From the definition of f it is clear that f is computable and that the range of values of f coincides with the set M.

Thus it is only necessary to give a precise definition of the concept of a computable function. This precise definition is provided by the *recursive functions* as defined in the next section.

5.6. *Recursive Functions*

In the domain of natural numbers the sum function is determined by two equations (cf. §5.2):

$$x + 0 = x, \tag{5.9}$$

$$x + y' = (x + y)', \tag{5.10}$$

where the successor of y is denoted by y'. These equations enable us to calculate the sum $u + v$ of any pair of natural numbers u, v in a purely formal way. For this purpose we require only two rules: (a) for the variables occurring in (5.9) and (5.10) we may substitute numerals $(0, 0'(=1), 0''(=2), \ldots)$, and (b) if for these numerals we have already derived the result $z_1 + z_2 = z_3$, then on the right-hand side of any subsequently derived equation we may replace $z_1 + z_2$ by z_3. Corresponding rules hold for the product function, except that in this case the set of two equations (5.9) and (5.10) must be augmented by two further equations

$$x \cdot 0 = 0, \tag{5.11}$$

$$x \cdot y' = x \cdot y + x. \tag{5.12}$$

Thus the sum plays the role of an auxiliary function for the product.

The concept of a recursive function, as defined by Herbrand and Gödel, is based on a generalization of the above procedure. An n-place function ϕ is said to be *recursive* if there exists a finite system of equations $\sum$ containing a function symbol f corresponding to ϕ and also in general, containing function symbols g, h, ... for auxiliary functions, such that for every choice of $n + 1$ numbers $k_1, \ldots, k_n, k$ we have the following result:[14] if $z_1, \ldots, z_n, z$ are the numerals corresponding to the numbers $k_1, \ldots k_n, k$, then the equation $f(z_1, \ldots, z_n) = z$ can be derived from $\sum$ if and only if $\phi(k_1, \ldots, k_n) = k$. In the process of derivation we may make use of two rules corresponding to the ones given above: (*a*) in every equation of $\sum$ we may substitute numerals for the variables; (*b*) if for any given numerals $Z_1, \ldots, Z_n, Z$ and function symbol F we have already derived an equation $F(Z_1, \ldots, Z_n) = Z$, then on the right-hand side of any subsequently derived equation we may replace $F(Z_1, \ldots, Z_n)$ by Z.

The precise concept of a recursive function is to be regarded as corresponding to the intuitive concept of a computable function. In particular, the functions x^y, $x!$, $| x - y |$ are recursive.

5.7. *Consistency of an Algorithm and Consistency of Mathematics*

The formulas that can be derived by an algorithm consist of rows of single symbols (not necessarily letters in the ordinary sense of the word) from a given *alphabet*. In general, it will not be possible to derive all the various formulas that could be constructed from this alphabet. There will be at least one formula Λ whose derivability is "undesirable." Such a formula might, for example, be $Px \wedge \neg Px$ (cf. §4.7), or $x \not\equiv x$, or $| \equiv ||$.[15] An algorithm K is called *consistent* with respect to a formula Λ of this sort if Λ is not derivable. We are speaking here of the *syntactical consistency* already mentioned in §4. A *consistency proof* for K consists in a demonstration that Λ is not derivable. A consistency proof in the constructive sense must employ only self-evident assertions and must avoid all ideas that are problematical from the semantic point of view, e.g., the idea of the actual-infinite, since such ideas are not accepted by all mathematicians. On the other hand, it is considered acceptable to make use of inductive proofs concerning the structure of an algorithm. Let us give a simple example: the alphabet of the algorithm K consists of the two letters $\circ$ and $|$. There is a single axiom

$$(5.13) \qquad \circ.$$

As a rule of inference (with the proper variable e) we take

$$(5.14) \qquad \frac{e}{e\,|}.$$

[14] Let us note the difference between numbers and numerals. It is customary to regard numbers as some sort of ideal entities that are represented in writing by symbols called numerals. In order to make the discussion more systematic, it is better here not to use the ordinary Arabic numerals for the numbers but, as was mentioned above, to represent the number 4, for example, by the numeral $0''''$. Numbers cannot be written down, but numerals can.

[15] $x \equiv y$ means that x and y are the same formulas.

In this algorithm the formula | is not derivable: that is, K is consistent with respect to |. The proof is inductive: we cannot derive | from (5.13), since | is different from ○. Also, we cannot derive | from (5.14) since every formula that can be derived from (5.14) must consist of more than a single letter.

For many of the important algorithms in mathematics, it has been possible to derive their consistency by "acceptable" methods of this sort, sometimes called "finitary." Moreover, the researches of Hilbert, Gentzen, Ackermann, Schütte, Lorenzen, and others have proved the consistency of the so-called *ramified analysis* closely connected with constructive mathematics (cf. §1, Nr. 4 and 5). On the other hand, no one has yet succeeded in proving the consistency of classical analysis.

Even though algorithms are of great importance for mathematics, it is still the opinion of many researchers that the whole of mathematics itself cannot be regarded as an algorithm (cf. "Incompleteness of Arithmetic," §10.5). In this case it makes no sense to speak of the syntactical consistency of mathematics as a whole.

For the *constructivist school* of mathematics, as represented, for example, by Curry and Lorenzen (§1.4, 5), all mathematical theorems are evident in the above sense. For the adherents of this school the whole of mathematics is a priori as reliable as a consistent algorithm.

Exercises for §5

1. From the functional equations

(5.9)

(5.10)

(5.11)

(5.12)

and the rules given in the text prove that

$$0''' \cdot 0'' = 0'''''' \qquad (3 \cdot 2 = 6).$$

2. Give recursion equations for the function x^y. From them prove that

$$(0'')^{0''} = 0'''' \qquad (2^2 = 4).$$

3. Introduce the functions

$x!$

$\mathfrak{P}(x)$ (predecessor of X; 0 if $x = 0$)

$a \dot{-} b$ (difference; 0 if $a < b$)

by recursion equations. Assume $a + b$, $a \cdot b$ and functions already defined.

4. Show that the calculus determined by the equations (5.9), (5.10), together with the rules (a) and (b) given for them in the text, is syntactically consistent.

5. If a set of natural numbers arranged in order of increasing magnitude is recursively enumerable, then it is also decidable.

Bibliography

For recursive functions and the concepts related to them see Davis [1], Hermes [1], and Kleene [1].

6. Proofs

6.1. *Rules of Inference and Proofs*

Let there be given a system of axioms, say the axioms of Euclidean geometry. The theorem of Pythagoras is a consequence of these axioms, but that fact is not immediately obvious; it becomes so only step by step. Each step consists of the application of a rule of inference. A *rule of inference* is an instruction concerning a possible transition from certain preceding formulas (the premisses) to a subsequent formula (the conclusion). A simple example with two premisses is the *modus ponens* (the *rule of separation*)

$$\begin{array}{c} \mathsf{H} \rightarrow \Theta \\ \mathsf{H} \\ \hline \Theta \end{array} \tag{6.1}$$

This rule enables us to make the transition from the two premisses $\mathsf{H} \rightarrow \Theta$ and H to the conclusion Θ. An *inference* is a transition in accordance with a rule of inference. A *proof* (*derivation*, *deduction*) is a finite sequence of expressions each of which (unless it is an axiom) can be derived from the preceding expressions by means of the rules of inference.

6.2. *A Complete System of Inference*

Although it is clear that there exist an infinite number of different rules of inference, in actual practice the mathematician makes use of only a very few of them, which recur again and again in many different arrangements. So we naturally ask whether it is possible to find a *finite* system of rules of inference by means of which we can deduce *all* the consequences of an arbitrary system of axioms. Such a system may be called a *complete system of rules of inference*, and it is one of the basic discoveries of modern logic that, within certain limitations, complete systems of rules of inference actually exist. The limitations in each case depend on how much the given system of logic is able to express. For example, a complete system can be found if we confine ourselves to axioms and to consequences expressible in the language of predicate logic, which is sufficient for many parts of mathematics. But the situation is different if we admit quantification of predicate variables. See the "Incompleteness of the Extended Predicate Logic" (§10).

The fact that within the framework of predicate logic every consequence can be derived by a finite system of rules of inference is described by saying that the predicate calculus determined by these rules is *complete*. The existence of such a calculus was foreseen by Leibniz in his demand for an *ars inveniendi*; to a certain extent it was experimentally verified by Whitehead and Russell in their monumental work *Principia Mathematica* (1910–1913) (based on the preliminary work of various logicians; in particular, Boole's *Algebra of Logic*, 1847), and finally, in 1930, it was proved by Gödel in his famous Gödel *completeness theorem*.

In the terminology of the foregoing section the Gödel completeness theorem asserts the existence of an algorithm for recursively enumerating all consequences of an arbitrary system of axioms that can be stated in the language of predicate logic.

6.3. *The Complete System of Rules of Inference of Gentzen (1934) and Quine (1950)*

Several different complete systems of rules of inference are known today but here we must restrict ourselves to the one which, since it is closely related to the ordinary reasoning of mathematicians, is called the "*system of natural inference*." The advantage of close relationship with ordinary mathematical practice is gained at the expense of unnecessary loss of symmetry and formal elegance, so that in purely logical investigations it is customary to use other systems.

For a greater clarity let us make some preliminary remarks. A characteristic feature of mathematical reasoning is the use of *assumptions*. Among the assumptions introduced during the course of a proof in any given mathematical theory we must include the axioms of the theory, or at any rate those axioms that are referred to in the proof. But in addition to the axioms, a mathematician will often introduce further (unproved) assumptions, on the basis of which the proof then proceeds. Of course, all assumptions that are made in this way must later be eliminated.

A special case of the introduction of assumptions occurs in an *indirect proof*. Here we arbitrarily assume the negative of the theorem to be proved.[16] Then in the course of the proof we try to reduce this assumption

[16] In case we wish to prove $\neg$H, we arbitrarily assume the proposition H (cf. the last example in §6.6).

ad absurdum, that is, we try to deduce from it two mutually contradictory results. It should be emphasized that, at least from the point of view of the classical logic under discussion here, an indirect proof is just as good as any other (although the situation is different for other schools of logic; see §6.7).

Another characteristic feature of mathematical reasoning is the introduction of variables for entities whose existence is already known. Consider, for example, two nonparallel lines g and h in a plane. We know that g and h have at least one point in common (in particular if the two lines coincide), and then the mathematician will say something like, "let a be a point common to the two lines." But the variable a here has no independent significance; it is meaningful only with respect to the proposition asserting its existence, a fact that must be kept in mind during the course of the proof. Variables of this sort also occur in the system of Gentzen and Quine, where they are called *flagged variables*. In order to avoid the danger of misunderstanding and consequent mistakes, it is not permissible to introduce the same variable for different entities during the course of a proof; this restriction is called the *restriction against flagging the same variable twice*. In general, a flagged variable will "depend" on other variables that have already appeared in the proof (in our example, a depends on g and h), in which case we stipulate that no variable may depend (even indirectly) on a second variable which in turn depends on the first; this restriction is called the *restriction against circularity*.

6.4. *List of the Rules of Gentzen and Quine*

For an explanation of these rules see §6.5, and the example of §6.6. Most of the rules have to do with the *introduction* or the *elimination* of a logical constant.

Two further rules without premisses (cf. §6.5):

a. the rule for *introduction of assumptions*,

b. the rule of *tertium non datur*.

6.5. *Explanation of the Rules*

By a *proof* we shall mean, here and in the rest of this section, a finite sequence of expressions that follow one another according to these two rules. Here it must be emphasized that this precise definition of a proof is altogether necessary in studies of the foundations of mathematics, in contrast to the situation in ordinary unformalized mathematics, where it is not customary to state the rules of inference being used. The lines in a given proof can now be numbered. Each line consists of finitely many *assumptions* (perhaps none) and an *assertion*. As a typical example we take the rule for $\wedge$-induction. This rule allows us to proceed from a line numbered i and a line numbered k ($i \gtreqless k$) to a line numbered l (with $l > i, l > k$), whose assertion is the conjunction of the assertions of the ith and kth lines, and whose assumptions consist of the "juxtaposition" of the assumptions of the ith and kth lines; i.e., an expression is an

The Rules of Gentzen and Quine for the Predicate Calculus

Logical Constant	*Introduction*	*Elimination*
$\wedge$	$\dfrac{\begin{array}{c}H\\ \Theta\end{array}}{H \wedge \Theta}$	$\dfrac{H \wedge \Theta}{H} \quad \dfrac{H \wedge \Theta}{\Theta}$
$\vee$	$\dfrac{H}{H \vee \Theta} \quad \dfrac{\Theta}{H \vee \Theta}$	$\dfrac{\begin{array}{c}H \vee \Theta\\ H \rightarrow Z\\ \Theta \rightarrow Z\end{array}}{Z}$
$\leftrightarrow$	$\dfrac{\begin{array}{c}H \rightarrow \Theta\\ \Theta \rightarrow H\end{array}}{H \leftrightarrow \Theta}$	$\dfrac{H \leftrightarrow \Theta}{H \rightarrow \Theta} \quad \dfrac{H \leftrightarrow \Theta}{\Theta \rightarrow H}$
$\neg$	$\dfrac{\begin{array}{c}H \rightarrow \Theta\\ H \rightarrow \neg\, \Theta\end{array}}{\neg\, H}$	$\dfrac{\begin{array}{c}H\\ \neg\, H\end{array}}{\Theta}$
$\rightarrow$	$\dfrac{\Theta}{H \rightarrow \Theta}$	$\dfrac{\begin{array}{c}H \rightarrow \Theta\\ H\end{array}}{\Theta}$
$\bigvee$	$\dfrac{\Theta\ ^{17}}{\bigvee_x H}$	$\dfrac{\bigvee_x H\ ^{17}}{\Theta}$
$\bigwedge$	$\dfrac{\Theta\ ^{18}}{\bigwedge_x H}$	$\dfrac{\bigwedge_x H\ ^{18}}{\Theta}$

[17] Assumption: H becomes Θ by free renaming of x to a variable y, and conversely Θ becomes H by the reverse renaming of y to x. *The variable y must be flagged* with respect to the free variables occurring in $\bigwedge_x$ H and $\bigvee_x$ H.

[18] Assumption: H becomes Θ by *free renaming* of the variable x to a variable y. (An exact definition of free renaming cannot be given here. We shall merely give a typical example: $H = (\bigwedge_u Pxu \wedge Qxy)$ becomes $\Theta = (\bigwedge_u Pyu \wedge Qyy)$ by free renaming of x to y.) Here y may also coincide with x.

assumption of the lth line if it is an assumption of the ith or of the kth line (the order in which the assumptions are written is of no importance, and an assumption which occurs several times may be written only once); schematically:

Line Number	*Assumptions*	*Assertion*
...	...	...
i	$A_1, \ldots, A_r$	H
...	...	...
k	$B_1, \ldots, B_s$	Θ
...	...	...
l	$A_1, \ldots, A_r, B_1, \ldots, B_s$	$\mathsf{H} \wedge \Theta$

When use is made of the rules of V-elimination (elimination of the existential quantifier) or of V-introduction (introduction of the universal quantifier), it is mandatory to *flag a variable* with a statement of the variables on which it depends. For example, if $u_1, \ldots, u_n$ are the free variables in $\mathsf{V}_x H$, then in making use of the rule of V-elimination we must write the new line l as follows:

Line	*Flagged Variables*	*Assumptions*	*Assertion*
l	$y(u_1, \ldots, u_n)$	$A_1, \ldots, A_r$	Θ.

The procedure for the rule of Λ-introduction is analogous.[19]

The rule for →-introduction may also be called *assumption-elimination*; for if H is an arbitrary assumption of the initial line (see the list of rules), then H will no longer occur as an assumption in the final line of the proof. In contrast to the rules described up to now, which allow us to pass from one, two, or three lines of the proof to a new line, the two rules of assumption-introduction and tertium non datur allow us to write down a line in the proof without making use of any preceding line. The rule of *assumption-introduction* consists simply of writing down an arbitrary proposition both as assumption and as assertion:

Line Number	*Assumptions*	*Assertion*
l	H	H

[19] In this rule the necessity for flagging is perhaps not immediately obvious; let us motivate it by the remark that the rule for Λ-introduction is *dual* to the rule for V-elimination.

The rule of *tertium non datur* allows us to write down any particular case of tertium non datur without assumptions:

Line Number	*Assumptions*	*Assertion*
l	—	$\mathsf{H} \vee \neg \mathsf{H}$

The last line of a *finished proof* must not contain any flagged variable as a free variable, since such a variable has no independent significance. Also, after constructing such a proof, we must verify that we have observed the restrictions against flagging a variable twice and against circularity.[20]

It can be proved that the assertion of the last line of a finished proof is a consequence (in the sense of §3.6) *of the assumptions of the last line of the proof. Conversely, if Θ is a consequence of $\mathsf{H}_1, \ldots, \mathsf{H}_n$, then there always exists a finished proof with a last line whose assumptions are $\mathsf{H}_1, \ldots, \mathsf{H}_n$ and whose assertion is Θ.*

In the present sense of the word, a proof is analogous to a schematic procedure for making a computation. Thus the process of proof has all the advantages and disadvantages of other schematic procedures that have been developed in mathematics. The *advantage* lies in the fact that in a mechanical procedure of this sort it is no longer necessary to do any thinking, or at least not as much as before, although this advantage can only be gained at the cost of considerable training in the art of carrying out the procedure. The *disadvantage* of a schematic procedure is that the rules which are simplest from the formal point of view are not always the ones that are most immediately obvious to the human mind.

On the other hand, if we wish to explain why exactly these formal rules were chosen, and no others, our explanation must be based on arguments whose meaning is intuitively clear. For lack of space we cannot give a detailed explanation here and will merely make a few remarks: the rules for $\vee$-introduction express the fact that if we have proved an assertion H under certain assumptions, then under the same assumptions we may make the weaker assertion $\mathsf{H} \vee \Theta$ or $\Theta \vee \mathsf{H}$. This rule is used in arithmetic, for example, in making approximations where we proceed from an already proved assertion of the form $x < 1$ to the weaker assertion $x \leqslant 1$ (i.e., $x < 1 \vee x = 1$). The rule for $\neg$-elimination expresses the following fact: if the assertion H follows from certain assumptions, and the assertion $\neg \mathsf{H}$ from certain other assumptions, then the two sets of assumptions taken together form an inconsistent system from which an arbitrary proposition Θ follows trivially. The rule for $\rightarrow$-introduction means only that if a proposition Θ follows from certain assumptions, including in particular the assumption H, then the proposition *if* H *then* Θ follows from the same set of assumptions excluding H.

We now give two examples of proofs. The reader is advised to direct his attention less to the actual meaning of the steps in the proof than to the question

[20] The restriction against flagging a variable twice prevents us from proceeding from $\bigvee_x \mathsf{H}$ through H to $\bigwedge_x \mathsf{H}$, since x would have to be flagged twice; and even if we introduce a new variable y, we cannot pass from $\bigvee_x \mathsf{H}$ to $\bigwedge_x \mathsf{H}$ without double flagging.

whether the above formal rules have been correctly applied. Of course, this will cost him some effort, comparable to the effort required when one undertakes for the first time to solve a quadratic equation by some formal procedure.

6.6. *Two Examples of Proofs*[21]

We begin with a proof that H follows from $\neg\neg$ H. This fact, which is valid only in classical logic, makes use of the tertium non datur. In the right-hand column we indicate the rule and the preceding lines that justify the step taken in each line.

Line Number	*Assumptions*	*Assertion*	*Rule Used*
1	$\neg\neg$ H	$\neg\neg$ H	introduction of assumption
2	$\neg$ H	$\neg$ H	introduction of assumption
3	$\neg\neg$ H, $\neg$ H	H	$\neg$-elimination (2, 1)
4	$\neg\neg$ H	$\neg$ H $\rightarrow$ H	elimination of assumption (3)
5	H	H	introduction of assumption
6		H $\rightarrow$ H	elimination of assumption (5)
7		H $\vee$ $\neg$ H	tertium non datur
8	$\neg\neg$ H	H	$\bigvee$-elimination (7, 6, 4)

Since we have used only rules from the propositional calculus, there has been no need to flag variables.

In the same way we can prove the four rules of contraposition, by which we mean the following steps: (1) from H $\rightarrow \Theta$ to $\neg\,\Theta \rightarrow \neg$ H, (2) from H $\rightarrow \neg\,\Theta$ to $\Theta \rightarrow \neg$ H, (3) from $\neg$ H $\rightarrow \Theta$ to $\neg\,\Theta \rightarrow$ H, (4) from $\neg$ H $\rightarrow \neg\,\Theta$ to $\Theta \rightarrow$ H.[22]

As a second example (see page 48) we wish to give part of an indirect proof and choose for this purpose the proof of the irrationality of $\sqrt{2}$. We use the variables $p, q, r, s, t, u, x, y, z$ for positive integers and take advantage of the fact that a rational number can be represented as the quotient of two natural numbers which have no factor in common and thus, in particular, are not both even. Then our problem is to prove the proposition

$$\neg \bigvee_p \bigvee_q (2q^2 = p^2 \wedge \neg (2 \mid p \wedge 2 \mid q)). \tag{6.2}$$

Since $2 \mid p$ is only an abbreviation for $\bigvee_s 2s = p$, we may rewrite (6.2) in the form

$$\neg \bigvee_p \bigvee_q (2q^2 = p^2 \wedge \neg (\bigvee_s 2s = p \wedge \bigvee_s 2s = q)), \tag{6.3}$$

[21] A further example is given in §11.2.

[22] The last two rules are not valid in the logic of intuitionism, which also rejects the step from $\neg\neg$ H to H.

Line Number	*Flagged Variables*	*Assumptions*	*Assertion*	*Rule Used*
1		H_0	$\bigvee_p \bigvee_q (2q^2 = p^2 \wedge \neg (\bigvee_s 2s = p \wedge \bigvee_s 2s = q))$	introduction of assumption
2	p	H_0	$\bigvee_q (2q^2 = p^2 \wedge \neg (\bigvee_s 2s = p \wedge \bigvee_s 2s = q))$	$\vee$-elimination (1)
3	$q(p)$	H_0	$2q^2 = p^2 \wedge \neg (\bigvee_s 2s = p \wedge \bigvee_s 2s = q)$	$\vee$-elimination (2)
4		H_0	$2q^2 = p^2$	$\wedge$-elimination (3)
5		H_0	$\bigvee_t 2t = p^2$[23]	$\vee$-introduction (4)
6		$A_1, \ldots, A_n$	$\bigwedge_u (\bigvee_t 2t = u^2 \rightarrow \bigvee_s 2s = u)$	(arithmetic)
7		$A_1, \ldots, A_n$	$\bigvee_t 2t = p^2 \rightarrow \bigvee_s 2s = p$	$\wedge$-elimination (6)
8		$\mathsf{H}_0, A_1, \ldots, A_n$	$\bigvee_s 2s = p$	$\rightarrow$-elimination (7, 5)
9	$s(p)$	$\mathsf{H}_0, A_1, \ldots, A_n$	$2s = p$	$\vee$-elimination (8)
10		$\mathsf{H}_0, A_1, \ldots, A_n$	$2q^2 = p^2 \wedge 2s = p$	$\wedge$-introduction (4, 9)
11		$A_1, \ldots, A_n$	$\bigwedge_x \bigwedge_y \bigwedge_z (2x = y^2 \wedge 2z = y \rightarrow 2z = x)$	(arithmetic)
12		$A_1, \ldots, A_n$	$\bigwedge_y \bigwedge_z (2q^2 = y^2 \wedge 2z = y \rightarrow 2z^2 = q^2)$[24]	$\wedge$-elimination (11)
13		$A_1, \ldots, A_n$	$\bigwedge_z (2q^2 = p^2 \wedge 2z = p \rightarrow 2z^2 = q^2)$	$\wedge$-elimination (12)
14		$A_1, \ldots, A_n$	$2q^2 = p^2 \wedge 2s = p \rightarrow 2s^2 = q^2$	$\wedge$-elimination (13)
15		$\mathsf{H}_0, A_1, \ldots, A_n$	$2s^2 = q^2$	$\rightarrow$-elimination (14, 10)
16		$\mathsf{H}_0, A_1, \ldots, A_n$	$\bigvee_t 2t = q^2$	$\vee$-introduction (15)
17		$A_1, \ldots, A_n$	$\bigvee_t 2t = q^2 \rightarrow \bigvee_s 2s = q$	$\wedge$-elimination (6)
18		$\mathsf{H}_0, A_1, \ldots, A_n$	$\bigvee_s 2s = q$	$\rightarrow$-elimination (17, 16)
19		$\mathsf{H}_0, A_1, \ldots, A_n$	$\bigvee_s 2s = p \wedge \bigvee_s 2s = q$	$\wedge$-introduction (8, 18)
20		H_0	$\neg (\bigvee_s 2s = p \wedge \bigvee_s 2s = q)$	$\wedge$-elimination (3)
21		$\mathsf{H}_0, A_1, \ldots, A_n$	$\neg \mathsf{H}_0$	$\neg$-introduction (19, 20)
22		$A_1, \ldots, A_n$	$\mathsf{H}_0 \rightarrow \neg \mathsf{H}_0$	elimination of assumption (21)
23			$\mathsf{H}_0 \rightarrow \mathsf{H}_0$	elimination of assumption (1)
24		$A_1, \ldots, A_n$	$\neg \mathsf{H}_0$	$\neg$-introduction (23, 22)

[23] Strictly interpreted, the rule of existence-introduction in §6.4 allows us to go from $\bigvee_t 2t = p^2$ to $2q^2 = p^2$ by introducing a suitable variable for the variable t. But that is not exactly what we are doing here, since we must replace t by q^2, and q^2 is not a variable. The difficulty lies in the fact that for simplicity in the above example, and for consistency with the nomenclature of ordinary mathematics, we have used the functional notation, which, in principle, we could have avoided, as we have seen in §2.5.

[24] Here we have replaced x by q^2. Cf. the preceding note.

which we shall now abbreviate to $\neg$ H_0. Here the axioms of arithmetic are indicated simply by $A_1, ..., A_n$. From $A_1, ..., A_n$ it follows that an arbitrary positive integer u is even if its square is even. We have made use of this fact in the second line and, strictly speaking, we should give a complete proof of it. The same remark applies to line 11, which expresses an elementary result from arithmetic.

It is easy to verify that we have now constructed a finished proof in which we have respected the restrictions against flagging the variable twice and against circularity.

From this example it is clear that proofs in the precise sense in which we are now using the word are generally much longer than the "proofs" of ordinary mathematics. This fact should cause no surprise, since we are employing only a few rules of inference of a very elementary character.

6.7. *Recursive Enumerability and Decidability in the Predicate Logic*

The calculus discussed above has provided us with a procedure (an *ars inveniendi*) for recursively enumerating the theorems of any theory that is axiomatized in the language of the predicate logic. The verification of the correctness of any proof can be carried out, at least in principle, by a machine, since we are dealing here only with simple formal relationships among rows of symbols. Thus, it is a *decidable* question whether or not a given sequence of expressions is a proof.

But it must be emphasized that such a calculus does not enable us, for an arbitrary finite system of axioms $\mathfrak{A}$ and an arbitrarily given expression H, to decide whether or not H follows from $\mathfrak{A}$. To decide such a question would require an *ars iudicandi* in the sense of Leibniz, and since 1936 it is known (Church) that for the predicate logic such a decision procedure cannot exist.

6.8. *Nonclassical Systems of Rules*

As was pointed out in §6.5, the rules given in §6.4 for the predicate logic can be established semantically. But the nonclassical conceptions of logic can lead to corresponding systems of rules that are not necessarily equivalent to the system described here. For example, the rule of *tertium non datur* is not valid for a potential interpretation of infinity (cf. §1.4).

Exercise for §6

Let the axioms for a group be given in the following form:

M (Multiplication) $\bigwedge_x \bigwedge_y \bigvee_z xy = z$

A (Associative law) $\bigwedge_x \bigwedge_y \bigwedge_z x(yz) = (xy)z$

U (Unity) $\bigwedge_x xe = z$

J (Inverse) $\bigwedge_x \bigwedge_y xy = 0$

E_1 (Equality) $T = T$

E_2 (Equality) $H(T_1) \wedge T_1 = T_2 \rightarrow H(T_2)$

Here T, T_1, T_2 denote terms, e.g., ab, $(ab)c$ and so forth, and $H(T_1)$ is an arbitrary term-equation containing the term T_1. Also, E_1 and E_2 are axiom-schemes (4.1). The axiom E_2 can be represented more conveniently, and equivalently, by the additional rule of inference

$$E_2 \qquad \frac{H(T_1),\ T_1 - T_2}{H(T_2)}$$

From these axioms construct a proof for the propositional form

$$\bigwedge_a \bigvee_b ba = e$$

(existence of a left inverse).

Bibliography

The following textbooks of logic discuss the methods of proof for the predicate logic, but on the basis of rules of inference quite different from those described in the present section: Church [1], Hilbert-Ackermann [1], Kalish-Montague [1], Kleene [1], Quine [1], Quine [2], Rosenbloom [1].

7. Theory of Sets

7.1. *Introductory Remarks*

Many definitions and theorems contain such expressions as *set*, *totality*, *class*, *domain*, and so forth. For example, in the definition of a real number by means of a Dedekind cut (see IB1, §4.3) the *totality* of the rational numbers is divided into two non-empty *classes*, a first or lower and a second or upper class. An ordered *set* (cf. §7.2) is said to be well-ordered if every non-empty *subset* contains a smallest element. Again, we may visualize a real function as the *set* of points of a curve and may speak of its *domain* and *range* (§8.3). Finally, we have already spoken of a *domain* of individuals in our definition of the concept of a mathematical consequence (§3.6).

The concept of a *set*, which is thus seen to be of fundamental importance, was for a long time regarded as being so intuitively clear as to need no further discussion. Cantor (1845–1918) was the first to subject it to systematic study. His definition of a set (not a definition in the strict mathematical sense of the word but only a useful hint in the right direction) runs as follows: *A "set" is any assemblage, regarded as one entity M, of definite and separate objects m of our perception or thought.*

The Cantor theory of sets developed rapidly and soon exercised a great influence on many branches of mathematics, the theory of point sets, real functions, topology, and so forth.

But with the discovery of contradictions—the so-called *antinomies of the theory of sets* (cf. §7.2 and §11)—the foundations of the theory, and therewith of the whole of classical mathematics, were placed in jeopardy. The discussion of this problem, which is still continuing, has contributed in an essential way to the development of modern research on the foundations of mathematics. The various schools of thought have made several suggestions for the construction of a theory of sets; let us mention a few of the most important.

The *naive* or *intuitive theory of sets* simply attempts to avoid the introduction of contradictory concepts. Frege and Russell tried (logicism) to reduce the theory of sets to logic. Zermelo, von Neumann, and others have introduced systems of axioms for the theory of sets from which it is possible to deduce many of the theorems of the naive theory. The consistency of these systems of axioms remains an open question (cf. §4.7). Still other authors insist that a set must be explicitly definable by a linguistic expression (a propositional form with a free variable), which must then satisfy certain additional conditions, depending on the school of thought to which the author belongs.

In Sections 7, 8, and 9 we deal chiefly with the naive theory of sets; as for the axiomatic theory, we confine ourselves to a brief description of *one* of the various systems in use. The three sections are closely related to one another in subject matter and are separated here only for convenience.

7.2. *Naive Theory of Sets*

The Cantor definition of a set makes it natural for us to gather into one set all the entities that have a given property; for example: (1) the set of chairs in the room (these are objects of our perception), or (2) the set of even numbers (objects of our thought). To denote variables for sets and their elements we use the Latin letters a, b, c, ... M, N, ..., and so forth. To express the fact that *y is an element of x* we write $y \in x$, and for $\neg\, y \in x$ we also write $y \notin x$. It is possible for one set to be an element of another set. Sets that contain the same elements are regarded as being equal, i.e.,

$$(7.1) \qquad \bigwedge_{x} (x \in a \leftrightarrow x \in b) \rightarrow a = b.$$

This requirement is called the *principle of extensionality*. Thus a set is determined by the elements "contained" in it, by its *content* or *extension*.

The property of being a prime number between eight and ten defines a set that contains no element. By the principle of extensionality there can be only *one* such set, which is called the *empty set*, and is here denoted by 0, although some authors use the special symbol $\emptyset$.

Let us now define the simplest set-theoretic concepts: A set a is called a *subset* of b (*a is contained in b*, $a \subseteq b$) if $\bigwedge_x (x \in a \rightarrow x \in b)$. If $a \neq b$, then a is a *proper subset* of b or is *properly contained* in $b (a \subset b)$. The set c is called the *union* of a and $b (c = a \cup b)$ if $\bigwedge_x (x \in c \leftrightarrow x \in a \vee x \in b)$. The set c is the *intersection* of a and $b (c = a \cap b)$ if $\bigwedge_x (x \in c \leftrightarrow x \in a \wedge x \in b)$. Two sets a, b are *disjoint* if they have no element in common, i.e., $a \cap b = 0$. The *complement* $\bar{x}$[25] of a set x is the set of all elements which are not elements of x. But here we must be careful, since the complement of the empty set is then the "universal set," which easily leads to contradictions (cf. §7.5). These contradictions can be avoided if we consider only subsets of a certain fixed set M. Then $\bar{x}$ is the set of y with $y \in M \wedge y \notin x$.

It is convenient to illustrate these concepts with sets of points in the plane:

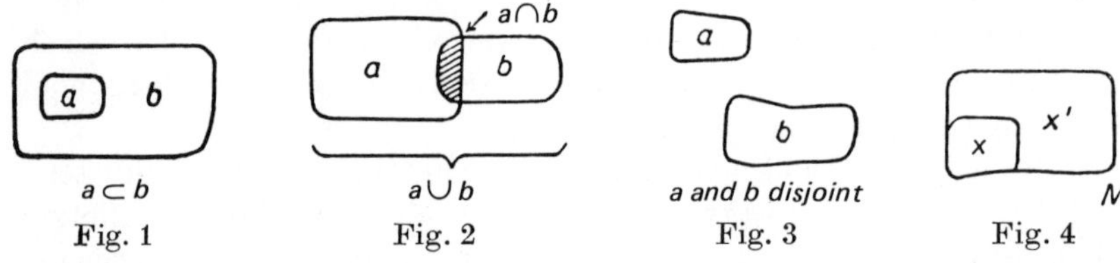

Fig. 1 Fig. 2 Fig. 3 Fig. 4

By the *power set* $\mathfrak{P}a$ of a set a we mean the set of all subsets of a: $\bigwedge_x (x \in \mathfrak{P}a \leftrightarrow x \subseteq a)$. The set that contains x as its single element is written $\{x\}$,[26] and correspondingly $\{x, y\}$ is the set containing exactly the two elements x and y, and so forth. For example, $\{0\}$ contains exactly one element, namely the empty set, whereas 0 contains no element at all. In a set-theoretic treatment of functions (cf. §8) an important role is played by the *ordered pairs* $\langle x, y\rangle$,[27] defined by

$$(7.2) \qquad \langle x, y\rangle = \{\{x\}, \{x, y\}\}.$$

From $\langle x, y\rangle = \langle u, v\rangle$ follows $x = u \wedge y = v$. Thus the order of the components in an ordered pair is significant.[28]

[25] The complement of x is often denoted by "x'."

[26] $\{x\}$ and x differ from each other, since in general x does not have x as its only element. Nevertheless, in cases where no confusion can arise, it is customary to write x for $\{x\}$.

[27] Ordered pairs are also denoted by (x, y).

[28] For sequences of symbols the construction of ordered pairs may be carried out simply by means of juxtaposition and a suitable symbol for separation.

It is easy to prove the following rules, which lead us to speak of an *algebra of sets* (cf. §9) or of a *field of sets*.

Laws for $\cap$ *and* $\cup$:

(1) *The commutative laws:*

$$a \cap b = b \cap a, \qquad a \cup b = b \cup a.$$

(2) *The associative laws:*

$$a \cap (b \cap c) = (a \cap b) \cap c, \qquad a \cup (b \cup c) = (a \cup b) \cup c.$$

(3) *The absorption laws:*

$$a \cap (a \cup b) = a, \qquad a \cup (a \cap b) = a.$$

(4) *The distributive laws:*

$$a \cap (b \cup c) = (a \cap b) \cup (a \cap c),$$
$$a \cup (b \cap c) = (a \cup b) \cap (a \cup c).$$

Laws for $\subseteq$:

(1) *The reflexive law:*

$$a \subseteq a.$$

(2) *The identitive law:*

$$a \subseteq b \wedge b \subseteq a \rightarrow a = b.$$

(3) *The transitive law:*

$$a \subseteq b \wedge b \subseteq c \rightarrow a \subseteq c.$$

Thus, the relation $\subseteq$ is a partial ordering (in the sense of §8.3).

Laws for $\subseteq$, $\cap$, $\cup$:

(1) $a \subseteq b \leftrightarrow a \cap b = a, \qquad a \subseteq b \leftrightarrow b \cup a = b,$

(2) $a \subseteq b \cap c \leftrightarrow a \subseteq b \wedge a \subseteq c, \qquad a \cup b \subseteq c \leftrightarrow a \subseteq c \wedge b \subseteq c.$

Laws for complementation (a, b are subsets of m):

(1) $a = b \leftrightarrow \bar{a} = \bar{b}$, (2) $\overline{(\bar{a})} = a$,

(3) $a \subseteq \bar{b} \leftrightarrow \bar{b} \subseteq \bar{a}$, (4) $\bar{0} = m, \quad \bar{m} = 0$,

(5) $\overline{(a \cap b)} = \bar{a} \cup \bar{b}, \quad \overline{(a \cup b)} = \bar{a} \cap \bar{b}$,

(6) $a \subseteq b \leftrightarrow a \cap \bar{b} = 0, \quad a \subseteq b \leftrightarrow \bar{a} \cup b = m.$

Laws for 0 *and* m (a is a subset of m):

(1) $a \cup 0 = a, \qquad a \cap m = a,$

(2) $a \cap 0 = 0, \qquad a \cup m = m.$

Up to now we have introduced the concept of union for two sets only, but it is often necessary to consider the union of arbitrarily many sets. Let M be a set of sets. Then by $\bigcup_{x \in M} x$ we denote the set of elements y belonging to at least one x in M. Correspondingly, as a generalization of the intersection of two sets, we write $\bigcap_{x \in M} x$ for the set of those elements of y which belong to every x in M.

7.3. *Cardinal Numbers in the Naive Theory of Sets*

One of the most important concepts introduced by Cantor is that of the *power* or *cardinality* of a set. It represents an extension to infinite sets of the number of objects in a finite set. Two sets x, y are said to be *equivalent* ($x \sim y$) if a one-to-one correspondence can be set up between the elements of x and those of y. For example, the set $\{1, 2, 3\}$ and $\{0, \{0\}, \{\{0\}\}\}$ are equivalent; moreover, the set of natural numbers and the set of squares are equivalent, as is shown by the following correspondence between them:

$$\begin{array}{cccccc} 0 & 1 & 2 & 3 & 4 & \dots \\ \updownarrow & \updownarrow & \updownarrow & \updownarrow & \updownarrow & \\ 0 & 1 & 4 & 9 & 16 & \dots . \end{array}$$

This example shows that an "infinite" set a can be equivalent to a proper part of itself, a property which is usually taken as the definition of infinity *(Dedekind definition of infinity)*. The *cardinal number* $\tilde{x}$[29] of a set x is then regarded as representing "that which is common" to all sets that are equivalent to x. Thus, we might say that the cardinal number of x is simply the set of all sets that are equivalent to x, although such a definition is problematical on account of its relationship to the universal set. On the other hand, among all the sets that are equivalent to x we could choose one definite set as a representative of x and then say that this set is the cardinal number of x. But the problematical feature of such a definition is that we do not know how to decide which set should be chosen as the representative. In any case we have

$$x \sim y \Leftrightarrow \tilde{x} = \tilde{y}. \tag{7.3}$$

The cardinal number of a finite set can simply be identified with the number of elements in the set.

For all sets, finite or infinite, we have the *Bernstein equivalence theorem:*

If $x \subseteq y$ and $y \subseteq z$ and $x \sim z$, then $y \sim z$.

An ordering $\leqslant$ for the cardinal numbers (cf. §8.3) can be defined by setting $\tilde{x} \leqslant \tilde{z} \Leftrightarrow \bigvee_y (y \subseteq z \wedge x \sim y)$, $\tilde{x} < \tilde{z} \Leftrightarrow \tilde{x} \leqslant \tilde{z} \wedge \tilde{x} \neq \tilde{z}$ (cf. §7.4).

[29] Cardinal numbers are also often denoted by "$\bar{\bar{x}}$."

The cardinal number of the set of natural numbers is denoted by $\aleph_0$ (pronounced aleph-zero). If $\tilde{x} < \aleph_0$, the cardinal number $\tilde{x}$ is said to be *finite*, but if $\tilde{x} \geqslant \aleph_0$, then $\tilde{x}$ is *transfinite*. If $\tilde{x} = \aleph_0$, then $\tilde{x}$ is *countable*, and if $\tilde{x} \leqslant \aleph_0$, then $\tilde{x}$ is *at most countable*.[30] A transfinite cardinal number that is not countable is said to be *uncountable*. A set is called *countable*, *at most countable*, or *uncountable* if its cardinal number has the corresponding property. Finite cardinal numbers correspond to finite sets, and transfinite cardinal numbers to infinite sets.

The set of rational numbers is countable. The truth of this assertion is evident from the following *schema* in which every "positive" rational number occurs at least once *(first Cantor diagonal procedure):*

$$
\begin{array}{ccccccccccc}
0 & \rightarrow & 1 & & 2 & \rightarrow & 3 & & 4 & \rightarrow & 5 \; \cdots \\
& & \downarrow & & & & & & & & \\
& & \frac{1}{2} & \nearrow & \frac{2}{2} & \swarrow & \frac{3}{2} & \nearrow & \frac{4}{2} & \swarrow & \frac{5}{2} \; \cdots \\
& & & & & & & & & & \\
& & \frac{1}{3} & \swarrow & \frac{2}{3} & \nearrow & \frac{3}{3} & \swarrow & \frac{4}{3} & & \frac{5}{3} \; \cdots \\
& & \downarrow & & & & & & & & \\
& & \frac{1}{4} & \nearrow & \frac{2}{4} & \swarrow & \frac{3}{4} & & \frac{4}{4} & & \frac{5}{4} \; \cdots \\
& & & \swarrow & & & & & & & \\
& & \cdot & \cdot & \cdot & \cdot & \cdot & \cdot & \cdot & \cdot & \cdots \; .
\end{array}
$$

The existence of uncountable sets was first proved by Cantor by his *second diagonal procedure: the set of real numbers* α with $0 < \alpha < 1$ is uncountable. Proof: let us assume that we have set up a one-to-one correspondence between these numbers and the positive integers:

$$
\begin{aligned}
\alpha_1 &= 0.\, a_{11}a_{12}a_{13} \cdots \\
\alpha_2 &= 0.\, a_{21}a_{22}a_{23} \cdots \\
\alpha_3 &= 0.\, a_{31}a_{32}a_{33} \cdots \\
&\cdots\cdots\cdots\cdots .
\end{aligned}
$$

Here the real numbers have been written as infinite decimals, so that $0 \leqslant a_{ik} \leqslant 9$. Now let us form the number $\alpha' = 0, a_1'a_2'a_3' \ldots$, where $a_i' = 1$ if $a_{ii} \neq 1$, and $a_i' = 2$ if $a_{ii} = 1$. Then α' differs from every number listed above in at least one decimal place, since it differs from α_n in the nth place, and thus α' is not included in the list. Since $0 < \alpha' < 1$, our assumption is wrong and the theorem is proved.

The correspondence set up in Figure 5 shows that the set of all real numbers, often called the *continuum*, has the same power as the set of real numbers in the interval just considered.

[30] The terms *countable* and *countably infinite* are often used in the sense of our "at most countable" and "countable," respectively.

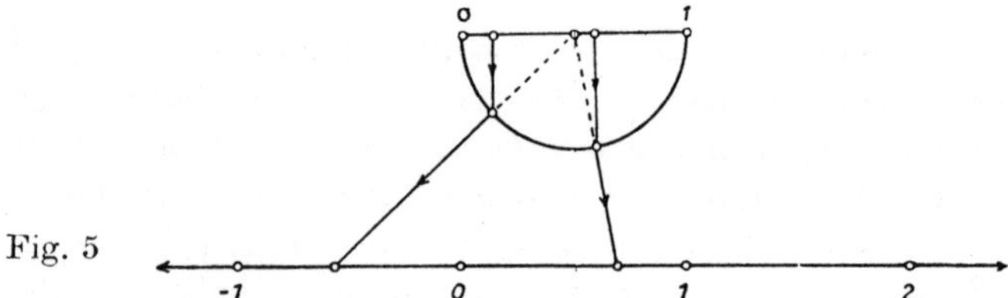

Fig. 5

7.4. *Ordinal Numbers in the Naive Theory of Sets*

A set x for which an order (cf. §8) has been defined is called an *ordered set*. Two ordered sets that can be put into one-to-one correspondence with each other with preservation of the order (so that they are isomorphic in the sense of §8.4) are said to be *similar*. By the *order type* $|x|$ we mean "that which is common" to all sets similar to the given ordered set x (cf. the remarks on the concept of a cardinal number in §7.3). If the ordering of x is a well-ordering in the sense of §8.3, then $|x|$ is an *ordinal number*. For the ordinal numbers we can define an ordering $>$, which turns out to be a well-ordering, by setting $|x| < |y| \Leftrightarrow \bigvee_z (z \subseteq y \wedge |x| = |z|) \wedge |x| \neq |y|$. For every ordinal number β the set of ordinal numbers with $\alpha < \beta$ in the ordering $<$ is itself a representative of β. The *well-ordering theorem*, which can be proved on the basis of the axiom of choice (cf. §7.6), states that *every set can be well-ordered*. Only by means of this theorem can we prove that the relation $\leqslant$ defined for the cardinal numbers in §7.3 is an ordering and in fact a well-ordering.

A non-empty set S of ordinal numbers is called a *number class* if (1) any two members of the set are equivalent (§7.3), and (2) every ordinal that is equivalent to S belongs to S. Thus, every cardinal number determines a number class. To every finite cardinal number corresponds exactly one ordinal number, so that the corresponding class has only one element. But the number classes corresponding to transfinite cardinal numbers have infinitely many elements.

The natural numbers can be identified with the finite ordinal numbers, or also with the finite cardinal numbers. Then the empty set 0 corresponds to the number 0, the class of sets with a single element to the number 1, and so forth. The cardinal number of the set $\{0, ..., n-1\}$ is n. In this way we can construct a theory of natural numbers on the basis of the theory of sets; and in particular, we obtain a model for the Peano axioms (cf. §10).

If to a representative a of a given ordinal number we adjoin another element x, which thus becomes the "last" element in the sense of the ordering, the set $b = a \cup \{x\}$ thus created represents an ordinal number $|b|$, which is called the *successor* of $|a|$ and is denoted by $|a|'$. Thus there is no ordinal number between $|a|$ and $|a|'$. Ordinal numbers (except 0) which, unlike $|b|$, have no immediate predecessor are called *limit numbers*.

Every non-empty set of ordinal numbers contains a smallest element (since the ordinal numbers are well-ordered). Thus we may state the principle of *transfinite induction* [a generalization of induction for the natural numbers (cf. §10.2)]: if w is a well-ordering for a set a, then a property H holds for all $x \in a$ if it satisfies the following conditions:

(1) H holds for the w-smallest element of a.

(2) If H holds for all x that are w-smaller than $y (y \in a)$, then H also holds for y.

The ordinal number of the set of natural numbers, well-ordered in the usual way, is denoted by ω, which is thus the smallest transfinite ordinal number. For a general discussion of the transfinite ordinal numbers, cf. IB1, Appendix.

Functions whose domain is a transfinite set of ordinal numbers are often defined inductively by means of three conditions; for example, as follows (α, β are arbitrary ordinal numbers, λ is an arbitrary limit number and $\lim_{\beta \in \lambda} f(\alpha, \beta)$ is the smallest ordinal number γ with $f(\alpha, \beta) \in \gamma$ for all $\beta \in \lambda$):

$$(1) \quad f(\alpha, 0) = \alpha,$$
$$(2) \quad f(\alpha, \beta') = f(\alpha, \beta)',$$
$$(3) \quad f(\alpha, \lambda) = \lim_{\beta \in \lambda} f(\alpha, \beta).$$

This is not an explicit definition, since in (2) and (3) the symbol "f" to be defined occurs on the right-hand side, but by transfinite induction we can show that there exists exactly one function f with the properties (1), (2), (3), and then we can write $\alpha + \beta$ for $f(\alpha, \beta)$. A schema of the form (1), (2), (3) is called a *transfinite inductive definition*. If condition (3) is omitted, the result is a recursive definition, for functions whose arguments are natural numbers. For the justification of such a recursive definition we need only the usual *complete induction* (cf. §10.2).

7.5. *Antinomies in the Naive Theory of Sets*

It is easy to show that the power set of any set x has a greater cardinal number than x itself: $\tilde{x} < \widetilde{\mathfrak{P}x}$. For finite sets x we have $\widetilde{\mathfrak{P}x} = 2^{\tilde{x}}$, which leads us to write $2^{\tilde{x}}$ for $\widetilde{\mathfrak{P}x}$ in the case of infinite sets as well. The power of the continuum is $2^{\aleph_0}$. If we form the set A of all sets (the so-called *universal set*), we first of all have $\tilde{A} < \widetilde{\mathfrak{P}A}$. On the other hand $\mathfrak{P}A$ is certainly equivalent to a subset of A, in view of the definition of A; thus $\widetilde{\mathfrak{P}A} \leqslant \tilde{A}$, in contradiction to the fact that $\leqslant$ is an ordering. This is the *antinomy of the universal set*.

Another example of a contradictory concept is the *set Ω of all ordinal numbers (antinomy of Burali-Forti)*. Like every set of ordinal numbers, this set is well-ordered by $<$, and thus it has an ordinal number $|\Omega|$. By the definition of a successor we have $|\Omega| < |\Omega|'$, but by the definition of Ω we also have $|\Omega|' \leqslant |\Omega|$, in contradiction to the fact that $<$ is a well-ordering.

These examples show that caution must be exercised in the formation of sets. (Cf. also the Russell antinomy in §11.)

7.6. *Axiomatic Theory of Sets*

The antinomies of the naive-set theory mostly arise from the fact that *arbitrary* properties, described by propositional forms $\mathsf{H}(x)$, are admitted for the definition of sets. Thus the trouble arises from assuming that for every propositional form $\mathsf{H}(x)$ there exists a set a described by the axiom schema $\bigwedge_x (x \in a \leftrightarrow \mathsf{H}(x))$. In the axiomatization of von Neumann, Bernays, and others, to which we now turn, this axiom scheme (axiom of comprehension) is suitably restricted.

The system deals with objects $x, y, z, \ldots$, called *classes*, between which a two-place relation $\in$ can exist. Thus $x \in y$ is read: *class x is an element of class y*. There is no formal distinction between classes and elements. Certain classes are called *sets:* namely those which are elements of at least one class

$$Mx \Leftrightarrow \bigvee_u x \in u. \tag{7.4}$$

Our first task is to define *equality* of classes. It is clear that two classes may be regarded as identical if (1) they contain the same elements and if (2) whenever either one of them is an element of a class, the other is an element of the same class.

$$a = b \Leftrightarrow \bigwedge_x (x \in a \leftrightarrow x \in b) \wedge \bigwedge_x (a \in x \leftrightarrow b \in x). \tag{7.5}$$

For our first axiom we may take the principle of extensionality (7.1) from the naive theory of sets:

$$\bigwedge_x (x \in a \leftrightarrow x \in b) \rightarrow a = b. \tag{7.6}$$

Thus a class is completely determined by its elements.

Now let $\mathsf{H}(x)$ be a relevant propositional form (see §4.5); for example, $x = x$ or $x \in y \vee x \in z$. The restricted *axiom of comprehension* is

$$\bigwedge_x (\mathsf{H}(x) \rightarrow Mx) \rightarrow \bigvee_u \bigwedge_x (x \in u \leftrightarrow \mathsf{H}(x)), \tag{7.7}$$

where $\mathsf{H}(x)$ does not contain u as a free variable.

Thus a property $\mathsf{H}(x)$ can be used as the definition of a class only if it refers exclusively to sets, that is to classes which can be an element of some class [cf. also (7.9)]. Then the class defined by $\mathsf{H}(x)$ is uniquely determined by (7.6) and can be given a name appropriate to its definition.

Let us now try to prove, for example, the Russell antinomy (see §11) by setting $x \notin x$ for $\mathsf{H}(x)$, so that from (7.7) we obtain

$$\bigwedge_x (x \notin x \rightarrow Mx) \rightarrow \bigvee_u \bigwedge_x (x \in u \leftrightarrow x \notin x).$$

In particular, for $x = u$

$$\bigwedge_x (x \notin x \rightarrow Mx) \rightarrow \bigvee_u (u \in u \leftrightarrow u \notin u). \tag{7.8}$$

The right-hand side is obviously false, and therefore, by *logical rules* the left-hand side is also false. Thus $\neg \bigwedge_x (x \notin x \rightarrow Mx)$, and consequently

$$\bigvee_x (x \notin x \wedge \neg Mx). \tag{7.9}$$

Instead of a contradiction we have obtained the (acceptable) proposition that there exists a class x (with $x \notin x$) which is not a set.

Let us now examine certain properties to see whether they are suitable for the definition of a class.

(1) Mx for $\mathsf{H}(x)$. The premiss for (7.7) then reads $\bigwedge_x (Mx \rightarrow Mx)$ and thus is satisfied. Consequently, there exists a class A which includes *all sets* and which we therefore call the *universal class:*

$$x \in A \Leftrightarrow Mx. \tag{7.10}$$

(2) $x = x$ for $\mathsf{H}(x)$. Because of $\bigwedge_x x = x$, the proposition $\bigwedge_x (x = x \rightarrow Mx)$ would then lead to $\bigwedge_x Mx$, which contradicts (7.9). Thus there is no class that includes all *classes* as its elements, and in this way we have avoided the antinomy of the universal set.

(3) $x \neq x$ for $\mathsf{H}(x)$. This expression is always false, so that we always have $\mathsf{H}(x) \rightarrow Mx$. Thus $x \neq x$ defines a class which obviously contains no element: it is the empty class 0.

(4) $x \in y \vee x \in z$ for $\mathsf{H}(x)$. Here Mx follows from $x \in y$ and also from $x \in z$, so that the premiss of (7.7) is satisfied. Thus $\mathsf{H}(x)$ defines a class that depends only on y and z, namely their *union* $y \cup z$.

Other classes can now be defined as in the naive theory of sets; for example, the *intersection* $a \cap b$ of two sets a and b, the class containing one element $\{a\}$, and the class of pairs $\{a, b\}$ and $\langle a, b \rangle$. The theorems in the algebra of classes can then be proved in the same way as in the naive theory of sets and, to a great extent, the theory of cardinal and ordinal numbers can be developed analogously. For this purpose we must introduce step by step the following axioms, which for the most part require that certain classes shall be sets.

The axiom for the empty set: $M0$.

The axiom for sets with one element: $Mx \to M\{x\}$.

The first axiom for unions: $Mx \wedge My \to M(x \cup y)$.

The axiom of infinity: MNz (*where* Nz is the class of natural numbers).

The second axiom for unions: $Mx \to M \bigcup x$ ($\bigcup x$ is the union of all the elements of x).

The replacement axiom: If the domain of a function (8.3) is a set, then its range is also a set. This axiom enables us to prove that for every set a there exists a *power class* $\mathfrak{P}a$.

The power set axiom: $Mx \to M\mathfrak{P}x$.

The axiom of choice: If a is a class of non-empty sets x, there exists a function (§8.3) f such that $f(x) \in x$ for all $x \in a$. (Thus from every set $x \in a$ the function f "chooses" an element $f(x)$.) Here also the axiom of choice is an essential instrument in the proof of the well-ordering theorem.

The continuum hypothesis: Between the cardinal number of an infinite set x and the cardinal number of its power set $\mathfrak{P}x$ there is no other cardinal number. The particular case $\tilde{x} = \aleph_0$ is the *special continuum hypothesis.* From the special hypothesis it follows that every uncountable subset of the set of real numbers has the power of the continuum.

7.7. *Independence of the Axiom of Choice and the Continuum Hypothesis*

In §4.7 we have mentioned the Gödel proof of relative consistency. Gödel's result can be formulated as follows: let $\mathfrak{A}$ be the system of axioms for set theory as stated just above (§7.6), but without the axiom of choice A and the continuum hypothesis K. Let it be assumed that $\mathfrak{A}$ is consistent (although it is still unknown today whether this assumption is true). Then $\neg A$ cannot be deduced from $\mathfrak{A}$ nor K from $\mathfrak{A} \cup \{A\}$.

In 1963 Cohen proved further that (if $\mathfrak{A}$ is consistent) it is also impossible to deduce A from $\mathfrak{A}$ or K from $\mathfrak{A} \cup \{A\}$. Thus we have shown (see also §4.4) that A is independent of $\mathfrak{A}$ and K is independent of $\mathfrak{A} \cup \{A\}$.

7.8. *Symbols for Sets*

If we are given a propositional form $\cdots x \cdots$, it is convenient to have a symbol for the set of those x which possess the property corresponding to this propositional form. Several notations are customary in the literature:

$$\hat{x}(\cdots x \cdots), \qquad \{x; \cdots x \cdots\}, \qquad \{x \mid \cdots x \cdots\},$$

all of which are read: the set of x with the property $\cdots x \cdots$ Let us note that the set in question could be denoted just as well by $\hat{y}(\cdots y \cdots)$; in other words, we are dealing here with a bound variable (cf. §2.6).

Exercises for §7

1. Let (n) be the set of rational integers divisible by n. Illustrate the sets (3), (6), (9), (15) by point sets in the plane (as in figures 1–4) in such a way that the proper inclusions are correctly represented. What is the number-theoretic significance of the various intersections and unions?

2. Show by dual representations that the set of real numbers x with $0 < x < 1$ has the same power as the set of points $\langle x, y\rangle$ of the square $(0 < x < 1;\ 0 < y < 1)$.

3. Let x' be defined by $xu\{x\}$ (cf. 7.4); also let

$$(*)\ n \in N \Leftrightarrow \bigwedge_x [(0 \in x\ {}_r(r \in x \rightarrow r' \in x)) \rightarrow n \in x].$$

   Assume the principle of extensionality, the restricted axiom of comprehension, the axiom of the empty set, the axiom for sets with one element, and the first axiom for unions.

   Prove:

   (a) The right-hand side of (*) defines a class N.

   (b) $0 \in N$

   (c) $0' \in N$

   (d) $\bigwedge_x (x \in N \rightarrow x' \in N)$

   (e) $\bigwedge_x (x' \in N \rightarrow x \in N)$

   (*f*) $\bigwedge_x (x' \in N \rightarrow 0 \in x')$

Bibliography

Relatively elementary is Kleene [1]. Let us also mention Bernays [1], Fraenkel [1], Fraenkel and Bar-Hillel [1], Halmos [1], Sierpiński [1] and Suppes [1].

8. Theory of Relations

8.1. *The Concept of a Relation*

We may consider *relations* as properties of ordered pairs (§7.2). For example, $3 < 4$ (3 stands in the $<$-relation to 4) states that the property "smaller than" holds for the ordered pair $\langle 3, 4\rangle$. Or: *the point A lies on the line g* states that the pair $\langle A, g\rangle$ has the property described by the predicate *lies on.*

Analogously, we may regard an *n-place relation* as a property of ordered n-tuples. For example, the expression $x + y = z$ defines a three-place relation for the natural numbers. Except when otherwise noted, we shall always take the word *relation* to mean a two-place relation.

Relations have the same fundamental importance in mathematics as sets. Many of the basic concepts of mathematics are to be defined by relations (e.g., function, congruence, order) or at least can be understood in terms of relations (e.g., group, lattice, factor group, cf. §8.5).

For simplicity we here take the naive point of view (cf. §7.1), so that relations may simply be regarded as *sets of ordered pairs*. Thus instead of saying: x is in the relation r to y (abbreviated xry), we can equally well say: the ordered pair $\langle x, y\rangle$ is an element of the set r:

$$xry \Leftrightarrow \langle x, y\rangle \in r. \tag{8.1}$$

The elements of the pairs are assumed to belong to a fixed *ground set* M in which the relations are defined, and $r, s, t, f, g, h, \ldots$ are variables for them. For example, if M is the set Nz of natural numbers, then $\langle m, n\rangle$ belongs to the relation $\leqslant$ if and only if $m \leqslant n$. Thus $\leqslant$ consists of the pairs $\langle 0, 0\rangle, \langle 0, 1\rangle, \ldots, \langle 1, 1\rangle, \langle 1, 2\rangle, \ldots$ and so forth. By the *first domain* $\theta_1(r)$ of a relation r we mean the set defined by $\bigvee_y xry$, and by the *second domain* $\theta_2(r)$ we mean the set defined by $\bigvee_y yrx$. For example, $\theta_1(<) = \{0, 1, 2, \ldots\}$, $\theta_2(<) = \{1, 2, 3, \ldots\}$. The set $\theta_1(r) \cup \theta_2(r)$ is called the *domain* of the relation r.

An important relation is the *identity* I, defined by $xIy \Leftrightarrow x = y$. For the class of natural numbers it consists of the pairs $\langle 0, 0\rangle, \langle 1, 1\rangle$, and so forth. The *empty* or *void* relation, which contains no pair at all, will be denoted here by $\dot{0}$. It is identical with the empty set 0 (§7.2). The *universal relation*, which contains *every* pair with elements from M, will be denoted by $\dot{1}$. It is to be distinguished from the "universal set." Obviously we have $\bigwedge_x \bigwedge_y \neg\, x\dot{0}y$, $\bigwedge_x \bigwedge_y x\dot{1}y$.

8.2. *Combination of Relations (Algebra of Relations)*

Since the relations are defined as sets, it is clear what we mean by the *intersection* $r \cap s$, the *union* $r \cup s$, and the *complement* $\bar{r}$:

$$\begin{gathered} x(r \cap s)\, y \Leftrightarrow xry \wedge xsy, \qquad x(r \cup s)\, y \Leftrightarrow xry \vee xsy, \\ x\bar{r}\,y \Leftrightarrow \neg\, xr\,y. \end{gathered} \tag{8.2}$$

Similarly, the *inclusion* $r \subseteq s$ is defined by $\bigwedge_x \bigwedge_y (xry \rightarrow xsy)$.

In addition to these purely set-theoretic constructions, there are two other important ways of combining relations: the *converse relation* $\breve{r}$ and the *relative product* rs. The *converse relation* is defined by:

$$x\breve{r}y \Leftrightarrow yrx. \tag{8.3}$$

Thus the *converse* $\breve{r}$ of r arises from r through "reversal" of all the pairs.

The *relative product* is defined by:

(8.4) $$x(rs)\,y \Leftrightarrow \bigvee_z (xrz \wedge zsy).$$

Thus the *relative product* rs of r and s arises, roughly speaking, from "juxtaposition" of r and s. As may be shown by simple examples, this operation is not commutative. For rr it is customary to write r^2. Thus, if M is the class of natural numbers, we have: $I \subset \leqslant$, $< \cap I = \dot{0}$, $rI = Ir = r$ for every r, $\breve{<} = >$, $I^2 = I$. A set of relations which is closed (cf. IB10, §2.2) with respect to all these operations is called a *field of relations*. For computation with relations we have the same rules as for the algebra of sets (§7.2), and also certain other rules, which are easily proved directly from the definitions; for example,

(8.5) $$(r \cap s)^{\smile} = \breve{r} \cap \breve{s}, \qquad (r \cup s)^{\smile} = \breve{r} \cup \breve{s}, \qquad \breve{\breve{r}} = r, \qquad \overline{(\breve{r})} = (\bar{r})^{\smile},$$
$$r(s \cup t) = (rs) \cup (rt), \qquad r(s \cap t) \subseteq (rs) \cap (rt).$$

8.3. *Special Properties of Relations*

A relation r is *symmetric* if $\bigwedge_x \bigwedge_y (xry \rightarrow yrx)$, a requirement which by (8.3) may also be written in the shorter form $r \subseteq \breve{r}$. Definitions like this last one, which make no reference to the elements of the ground set, are often more concise. In what follows we shall give the definitions, wherever possible, in both forms, leaving to the reader the task of proving that they are equivalent. In the examples, M is the class of natural numbers, unless otherwise noted.

If xrx for all x, the relation r is *reflexive* ($I \subseteq r$). Example: $x \geqslant y$.

A *transitive* relation is defined by $\bigwedge_x \bigwedge_y \bigwedge_z ((xry \wedge yrz) \rightarrow xrz)$. (Alternatively written $r^2 \subseteq r$.) Example: $x < y$.

A relation is *identitive* if $\bigwedge_x \bigwedge_y ((xry \wedge yrx) \rightarrow x = y)$. (In the shorter form, $r \cap \breve{r} \subseteq I$.) Example: x is a factor of y.

A relation is *connex* if $\bigwedge_x \bigwedge_y (xry \vee yrx)$. (In the shorter form, $r \cup \breve{r} = \dot{1}$.) Example: $x \leqslant y$.

Relations which are transitive, identitive, and connex are called *orderings in the sense of* $\leqslant$ (example: $x \leqslant y$). For *orderings in the sense of* $<$ the requirements of identitivity and connexity are replaced by $\bigwedge_x \neg\, xrx$ and $\bigwedge_x \bigwedge_y (x \neq y \rightarrow xry \vee yrx)$ (example: $x < y$).

If we discard connexity altogether, we obtain the so-called *partial orderings* (*in the sense of* $\leqslant$ *or in the sense of* $<$), which are sometimes called *semi-orderings*. Examples are: inclusion, and strict inclusion (cf. §7.2), for the set of all subsets of a given set. In more recent literature, partial orderings in the above sense are sometimes called orderings, and orderings are called total or complete orderings.

If an ordering $<$ contains no infinite "decreasing" sequence $\cdots < x_3 < x_2 < x_1$ ($x_{i+1} \neq x_i$), it is called a *well-ordering* (of M), where a distinction is to be made between well-orderings in the sense of $<$ and well-orderings in the sense of $\leqslant$. Thus an ordering is a well-ordering if and only if every non-empty subset M_1 of its field has a *minimal element* in the sense of the ordering, i.e., an element for which there is no smaller element in M_1. By discarding the requirement of connexity, we obtain the *partial well-orderings*.

A set M is said to be *directed* with respect to a relation r if r is transitive and if for every $x, y \in M$ there exists a $z \in M$ such that xrz and yrz.

8.4. *Functions*

An important class of relations consists of the *functions*, defined by the requirement of uniqueness $\bigwedge_x \bigwedge_y \bigwedge_z ((xry \wedge xrz) \rightarrow y = z)$. (In the shorter form, $\breve{r}r \subseteq I$.) For functions it is customary to write $f(x) = y$ in place of xfy. The function f is a *mapping* of the first domain $\theta_1(f)$ *onto* the second domain $\theta_2(f)$; if $\theta_2(f)$ is contained in a set A, we say that f is a mapping *into* A. If $(f)x = y$, we say that y is the *image* of x (under f) and that x is the *pre-image* of y. If $\breve{f}$ is also a function (that is, $f\breve{f} \subseteq I$), then f is a one-to-one (*invertible*) mapping of $\theta_1(f)$ onto $\theta_2(f)$, and $\breve{f}$ is called the *inverse function* of f. Functions whose domain is the set of natural numbers are also called *sequences*. On the basis of the definition (7.1) for equality of sets, two functions are equal (or *identical*) if they have the same domain and if for every element in that domain the two functions have the same values.

As an example, let us formulate the Dedekind definition of an infinite set (§7.3) in the language of the theory of relations:

$$\text{Infinite} \quad a \Leftrightarrow \bigvee_f (\breve{f}f \subseteq I \wedge f\breve{f} \subseteq I \wedge \theta_1(f) = a \wedge \theta_2(f) \subset a).$$

In words: There exists a one-to-one mapping of a onto a proper subset of a.

Two relations r, s are said to be *isomorphic* if there exists a one-to-one mapping f of their fields onto each other such that $\bigwedge_x \bigwedge_y (xry \leftrightarrow f(x)\, sf(y))$.

In mathematical literature a function f is often written in the form $f(x)$, but this notation is essentially incorrect, since it appears to mean that the variable x is free. If we wish to use the variable x as part of the notation for a function, we must indicate that this variable is bound. Acceptable notations are $\lambda x f(x)$ or $x \rightarrow f(x)^1$).[31]

[31] The second of these is more common in recent literature, but it is to be noted that the arrow here has nothing to do with the symbol for implication in §2.4.

8.5. *Equivalence and Congruence Relations*

Relations which are symmetric, reflexive, and transitive are called *equivalence relations* (e.g., the identity I), cf. §4.4. They play an important role in mathematics, especially in algebra.

If we assume reflexivity, we may replace the requirement of transitivity and symmetry by that of *comparativity:* $x \sim z \wedge y \sim z \rightarrow x \sim y$.

Let $\sim$ be an equivalence relation in M, and let $\tilde{z}$ denote the set defined by $x \in \tilde{z} \Leftrightarrow x \sim z$. This set is called the *equivalence class* generated by z or corresponding to z. We have

(8.6) $$z \in \tilde{z},$$

(8.7) $$(x \in \tilde{z} \wedge y \in \tilde{z}) \rightarrow x \sim y,$$

(8.8) $$(u \in \tilde{x} \wedge u \in \tilde{y}) \rightarrow \tilde{x} = \tilde{y}.$$

All the elements of an equivalence class are thus equivalent to one another, i.e., they are related by $\sim$. Two equivalence classes are either identical or without common element, so that every equivalence relation generates a partition of its field M into disjoint classes. Conversely, every such partition of M into classes generates an equivalence relation in M; for if M is the union of disjoint subclasses, we define: $x \sim y \Leftrightarrow$ (x and y lie in the same subclass).

An equivalence relation defined in a ground set M gives rise to a *process of abstraction* (cf. §1.2), which means that elements of the same equivalence class are regarded as indistinguishable; in other words, we abstract from their distinguishing features. Conversely, every process of abstraction in M gives rise to an equivalence relation in the field M.

If for a ground set M there are given finitely many k-place functions $f_1 \ldots, f_n$ with values in M, then $\langle M, f_1, \ldots, f_n\rangle$ is called an *abstract algebra* (cf. IB10, §1.2). For example, let there be given a two-place function f, whose value for the arguments x, y we shall write in the form $x \cdot y$. Then it is clear that we shall usually be interested in those abstractions that preserve the operation $x \cdot y$; that is, if we denote the new equality by $\sim$, we must be able to define $\tilde{x} \cdot \tilde{y}$ as $\widetilde{x \cdot y}$. This will be possible if

(8.9) $$\bigwedge_{x_1} \bigwedge_{x_2} \bigwedge_{y_1} \bigwedge_{y_2} ((x_1 \sim x_2 \wedge y_1 \sim y_2) \rightarrow x_1 \cdot y_1 \sim x_2 \cdot y_2).$$

In this case the equivalence relation $\sim$ is called a *congruence relation* (with respect to the operation $x \cdot y$). The situation can also be described in the following way: a congruence relation is an equivalence relation that is *consistent* with the operations of the abstract algebra. For example, in the ring of rational integers, $x \equiv y \pmod 6$ is a congruence with respect to addition and multiplication (cf. IB6, §4.1).

If the algebra has a unit element e such that $\wedge_x (x \cdot e = x)$, then the set of x with $x \sim e$ forms a subalgebra N, since from $x \sim e$, $y \sim e$ it follows that $x \cdot y \sim e \cdot e = e$. Let $x \cdot N$ be the set of products of x with arbitrary elements of N. For every x we have $x \cdot N \subseteq \tilde{x}$. The congruence classes (complete classes of mutually congruent elements) form an algebra of the same "type." If $\wedge_x x \cdot N = \tilde{x}$, then N is a "normal factor." In this way many algebraic concepts and theorems (e.g., the theorem of Jordan-Hölder; see IB2, §12.1) can be interpreted as concepts and theorems in the theory of relations.

Exercises for §8

1. Prove (cf. §§8.3 and 8.4) that

r is reflexive	$\leftrightarrow I \subseteq r,$
r is transitive	$\leftrightarrow r^2 \subseteq r,$
r is identitive	$\leftrightarrow r \cap \breve{r} \subseteq I,$
r is connex	$\leftrightarrow r \cup \breve{r} = I,$
r is a function	$\leftrightarrow \breve{r}r \subseteq I,$
r is a one-to-one mapping	$\leftrightarrow \breve{r}r \cup r\breve{r} \subseteq I.$

2. State the axiom of choice and the well-ordering theorem in the symbolic language developed in §§7 and 8.

Bibliography

For the concepts and applications of the theory of relations see Carnap [2].

9. Boolean Algebra

9.1. *Preliminary Remarks*

In the present section we are interested in certain phenomena that first came to light in the study of the propositional calculus (§2); the fact that they are essentially algebraic in nature was first recognized by G. Boole (1847).

Let us consider the one-place predicates P, Q, ... for a fixed domain of individuals M (cf. §3). These predicates can be put in one-to-one correspondence with the subsets p, q, ... of M by assigning x to p if and only if P holds for x; that is,

$$x \in p \leftrightarrow Px. \tag{9.1}$$

The conjunction of two predicates obviously corresponds to the intersection of two sets; similarly, the alternative corresponds to their union:

(9.2) $$x \in p \cap q \leftrightarrow Px \wedge Qx, \qquad x \in p \cup q \leftrightarrow Px \vee Qx.$$

The distributive, associative, and other laws for $\cap$ and $\cup$ correspond to the same laws for $\wedge$ and $\vee$. Negation corresponds to complementation ($x \in \bar{p} \leftrightarrow \neg Px$), where (cf. §2.2):

(9.3) $$\begin{aligned} p \cup \bar{p} &= M, & Px \vee \neg Px &\leftrightarrow W, \\ p \cap \bar{p} &= 0, & Px \wedge \neg Px &\leftrightarrow F. \end{aligned}$$

Logical implication corresponds to set-theoretic inclusion:

(9.4) $$p \subseteq q \leftrightarrow \bigwedge_x (Px \rightarrow Qx).$$

We see that the domain of predicates for M has the same "structure" as the domain of subsets of M; the two domains are isomorphic. For the general study of such domains it is therefore natural to introduce an abstract algebra by means of axioms. The system of axioms will be *autonomous* in the sense of §4.

9.2. *Boolean Lattices*

A set M of elements $a, b, \ldots$ with operations $\cap, \cup, -$ is called a *Boolean lattice* if the following axioms are satisfied:

B0. $a \cap b$, $a \cup b$, $\bar{a}$ are defined for all elements of M and are themselves elements of M.

B11. $a \cap b = b \cap a$
B12. $a \cup b = b \cup a$ (Commutative laws)

B21. $a \cap (b \cap c) = (a \cap b) \cap c$
B22. $a \cup (b \cup c) = (a \cup b) \cup c$ (Associative laws)

B31. $a \cap (a \cup b) = a$
B32. $a \cup (a \cap b) = a$ (Absorption laws)

B41. $a \cap (b \cup c) = (a \cap b) \cup (a \cap c)$
B42. $a \cup (b \cap c) = (a \cup b) \cap (a \cup c)$ (Distributive laws)

There exist elements 0 and 1 in M such that for every a in M

B51. $a \cap \bar{a} = 0$
B52. $a \cup \bar{a} = 1$ (Complementation laws)

In the set-theoretic interpretation, $a \cap b$ and $a \cup b$ are read as *intersection of a and b* and *union of a and b*, respectively, and in the logical interpretation, as *a and b* and *a or b*. This system of axioms is denoted by B.

Looking through the list of axioms in B, we see that for every axiom there exists a *dual* axiom, formed by interchanging $\cap$ with $\cup$ and 0 with 1. Thus for every theorem there is also a dual theorem, whose statement and proof arise from the given theorem by these interchanges (*principle of duality* for Boolean algebra). A corresponding principle of duality holds for the predicate logic, if we interchange T and F. For example, the theorem $\bigwedge_x Px \vee \bigvee_x \neg Px \leftrightarrow T$ is dual to $\bigvee_x Px \wedge \bigwedge_x \neg Px \leftrightarrow F$.

Let us state a few easily proved theorems for Boolean lattices:

$$\begin{aligned} &a \cap a = a, \quad a \cup a = a, \\ &a \cap 0 = 0, \quad a \cup 1 = 1, \quad a \cup 0 = a, \quad a \cap 1 = a, \\ &a \cup b = b \rightarrow a \cap b = a, \quad a \cap b = b \rightarrow a \cup b = a, \\ &\bar{0} = 1, \quad \bar{1} = 0. \end{aligned} \tag{9.5}$$

A domain with operations $\cap$ and $\cup$, for which only the axioms B0 (without complementation), B1, B2, and B3 are required, is called a *lattice*. Boolean lattices are distributive and complemented (cf. IB9, §1).

9.3. *Inclusion in Boolean Lattices*

Inclusion can be defined by

$$a \subseteq b \Leftrightarrow a = a \cap b, \tag{9.6}$$

which corresponds to the set-relation, or equivalently by (§7.2) $b \subseteq a \Leftrightarrow a = a \cup b$.

Let $a \subset b$ signify that $a \subseteq b$ and $a \neq b$. It is easy to show that the relation $\subseteq$ is reflexive, transitive, and identitive, and is thus a partial ordering in the sense of $\leqslant$ (cf. §8.3). Also,

$$a \cap b \subseteq a, \qquad a \subseteq a \cup b, \qquad a \subseteq 1, \qquad 0 \subseteq a, \tag{9.7}$$

$$(a \subseteq b \wedge a \subseteq c) \rightarrow a \subseteq b \cap c, \qquad (b \subseteq a \wedge c \subseteq a) \rightarrow b \cup c \subseteq a. \tag{9.8}$$

From (9.8) it follows that $b \cap c$ and $b \cup c$ may serve as *greatest lower bound* and *least upper bound* of b and c with respect to $\subseteq$. Every element that is contained in b and c is also contained in the greatest lower bound $b\ c$, and every element that contains b and c also contains the least upper bound $b \cup c$ of b and c. The greatest lower bound of all the elements is 0, and their least upper bound is 1. Thus every lattice is partially ordered, with a least upper bound and a greatest lower bound for arbitrary a and b. Conversely, the above properties of inclusion may be used to construct a lattice from a partial ordering with least upper bound and greatest lower bound. For example, we may define $x = a \cap b$ by

$$x = a \cap b \Leftrightarrow \bigwedge ((z \subseteq a \wedge z \subseteq b) \leftrightarrow z \subseteq x). \tag{9.9}$$

If we note that for $a \cap b = c \cup d$ we may also write $\bigvee_x (x = a \cap b \wedge x = c \cup d)$, we see that the axioms of $\mathfrak{B}$ may be at once translated into axioms for $\subseteq$. For $a = 0$ we write $\bigwedge_x (a \subseteq x)$; for $a = 1$ we write $\bigwedge_x (x \subseteq a)$; and for $a = \bar{b}$ we write $\bigwedge_x (x \subseteq a \cup b) \wedge \bigwedge_x (a \cap b \subseteq x)$.

9.4. *Boolean Rings*

A third possibility for the description of Boolean algebra lies in the theory of rings. We define

$$(9.10) \qquad a \cdot b \Leftrightarrow a \cap b, \qquad a + b \Leftrightarrow (a \cap b') \cup (a' \cap b).$$

Then we can easily show

$$(9.11) \qquad \begin{aligned} &a \cdot b = b \cdot a, \qquad (a \cdot b) \cdot c = a \cdot (b \cdot c), \qquad a + b = b + a, \\ &a + (b + c) = (a + b) + c, \qquad a \cdot (b + c) = a \cdot b + a \cdot c, \\ &a \cdot 1 = a, \qquad a + 0 = a. \end{aligned}$$

$$(9.12) \qquad a \cdot a = a, \qquad a + a = 0.$$

These are the axioms for a commutative idempotent ring with unity element.[32] Such a ring is called a *Boolean ring*. Conversely, from a Boolean ring we can form a Boolean lattice by setting

$$(9.13) \qquad a \cap b \Leftrightarrow a \cdot b, \qquad a \cup b \Leftrightarrow a + b + a \cdot b, \qquad \bar{a} \Leftrightarrow 1 + a.$$

9.5. *Finite Boolean Lattices*

The subsets of a finite set M form a finite Boolean lattice with respect to the set-theoretic operations. Here the empty set represents the element 0 and the whole set M represents the element 1. If M has n elements, then the lattice has 2^n elements (cf. §7.2). Thus every finite Boolean lattice has 2^n elements ($n = 0, 1, 2, \ldots$), since we can show that every finite Boolean lattice is isomorphic to a lattice of subsets. The proof of this theorem rests on the fact that every element of a finite Boolean lattice is the union of atoms in the lattice, where an element a is called an *atom* if $a \neq 0$ and if from $x \subset a$ it follows that $x = 0$. The atoms of a lattice of subsets are the sets with one element $\{x\}$ (see §7.2). The finite Boolean lattices can be very clearly illustrated by diagrams in which the elements are represented by points in a plane in such a way that if $a \subset b$ and $\neg \bigvee_c (a \subset c \wedge c \subset b)$, then a lies below b and is joined to b by a line segment. Thus if the number of elements is 2^0, 2^1, 2^2, 2^3, we obtain the following figures:

[32] For the concept of rings see IB5, §1.5 ff.; a ring is *idempotent* if $a \cdot a = a$ for each of its element.

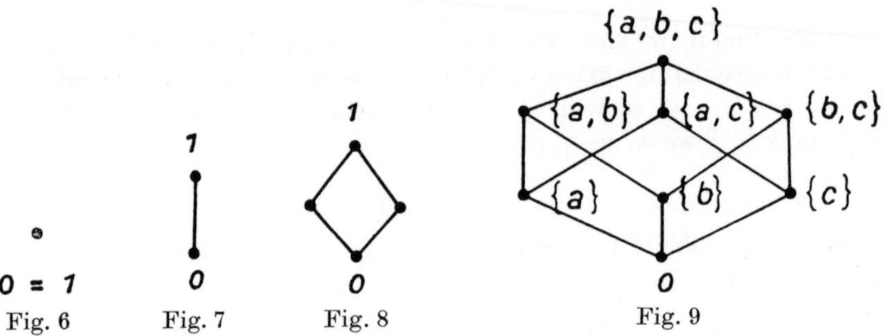

Fig. 6 Fig. 7 Fig. 8 Fig. 9

The diagram for the lattice with 2^3 elements shows the subsets of the 3-element set $M = \{a, b, c\}$.

Exercises for §9

1. Prove
   (a) (9.5) from the system of axioms $\mathfrak{B}$,
   (b) (9.7) and (9.8) from $\mathfrak{B}$ and (9.6),
   (c) (9.11) and (9.12) from $\mathfrak{B}$ and (9.10).

2. Consider propositional forms constructed from countably many (cf. 7.3) propositional variables p, q, ... (cf. 2.4) by the connectives $\neg$, $\wedge$, $\vee$, $\rightarrow$, $\leftrightarrow$ (cf. 2.4) of the propositional calculus. Define

$$H \sim \Theta \Leftrightarrow H \leftrightarrow \Theta \text{ is a tautology (3.4).}$$

   Now prove

   (a) $\sim$ is an equivalence relation

   (b) $\sim$ is consistent with the functions K, A, N defined on the set of propositional forms as follows:

$$K(H, \Theta) = H \wedge \Theta$$
$$A(H, \Theta) = H \vee \Theta$$
$$N(H) = \neg H.$$

   (c) The equivalence classes form a Boolean algebra under the following definitions:

   (1) $\tilde{H} \cap \tilde{\Theta} = \widetilde{H \wedge \Theta}$

   (2) $\tilde{H} \cup \tilde{\Theta} = \widetilde{H \vee \Theta}$

   (3) $\bar{\tilde{H}} = \widetilde{\neg H}$

   (4) $0 = \widetilde{p \wedge \neg p}$

   (5) $1 = \widetilde{p \vee \neg p}$

By b) the definitions (1)–(3) are independent of the representatives of the equivalence classes. Show that in (4) and (5) the definitions are independent of the choice of the propositional variable p.

(d) If the number of propositional variables is finite, then the Boolean algebra is also finite. If n is the number of variables, then the number of elements in the Boolean algebra is 2^{2^n}.

Bibliography

For Boolean algebra, see Goodstein [1].

10. Axiomatization of the Natural Numbers

10.1. *Preliminary Remarks*

The theory of natural numbers occupies an especially important place in studies in the foundations of mathematics. In the first place, the arithmetic of natural numbers offers a simple and important example of a theory with an infinite domain of individuals, in which the problems connected with the concept of *infinity* can be studied. Secondly, it has turned out that many other interesting metamathematical questions can be reduced to arithmetic (cf. the arithmetization in §5.4). Finally, the results of Gödel on arithmetical algorithms have had a lasting influence on the whole program of metamathematics. Let us discuss these remarks in greater detail.

The "leap to infinity" involved in recognizing the domain of the natural numbers is already adequate for all the ontological needs of the predicate logic (cf. §3); this is the meaning of the fundamental theorem of Löwenheim and Skolem, which essentially states that in order to investigate the concept of a consequence there is no need to use any domain of individuals other than the natural numbers.

Since in a system of axioms $\mathfrak{S}$ the means of expression (variables, logical symbols, and so forth), are obviously *countable*, it is clear that the obtainable expressions are also countable. Thus the expressions can be "numbered" constructively (see §5.4). For every expression the resulting index is computable and, conversely, for every number we can decide whether or not it is the index of an expression; if it is, then the expression can be recovered. As a result, certain metamathematical properties like *...is an expression*, *...is the conjunction of... and...* *...is true* are transformed into number-theoretical properties. Thus all questions of decidability can be translated into the corresponding questions for arithmetic. Moreover, if the system $\mathfrak{S}$ includes an arithmetical system of axioms, many of the metamathematical propositions about $\mathfrak{S}$ can be formulated in $\mathfrak{S}$ itself, and in this way it is possible to obtain extremely general theorems about mathematical systems of axioms (cf. §10.5).

For a long time the concept of the (infinite!) *totality of natural numbers* was held to be intuitively clear, and indeed quite self-evident [cf. the similar situation for the concept of a set (§7.1)]. It was Frege (1884) who first pointed out the necessity for an exact definition of a *natural number*. In his attempt to reduce arithmetic to logic he defined the number 1, for example, as the totality of all one-place predicates that hold for exactly one individual. This definition is closely related to the set-theoretical introduction of the natural numbers and leads to the same kind of difficulties as the naive theory of sets (§7.3). Thus we naturally seek, as in that theory, to characterize the natural numbers by a system of axioms. The best-known system of axioms for the natural numbers is due to Dedekind (1888) but is named after Peano (1889). In §10.3 we shall discuss a somewhat modified system, formulated in the language of predicate logic. The question of axiomatizing the whole of arithmetic (§10.4) then leads us to the well-known Incompleteness Theorem of Gödel (§10.5). The present section closes with some remarks on the operational construction of arithmetic recently proposed by Lorenzen.

10.2. *The Peano Axioms*

The Peano axioms (with unimportant changes):

(a) 0 *is a natural number.*[33]

(b) *If n is a natural number, then so is n'.*

(c) *If $m' = n'$, then $m = n$.*

(d) *There is no number n for which $n' = 0$.*

(e) *Axiom of complete induction:*

If a property P of the natural numbers satisfies the following two conditions, then P holds for every natural number:

(1) *P holds for* 0.

(2) *For every natural number n, if P holds for n, then P holds for n'.*

These axioms can be stated in a formal language consisting, as before, of formulas or rows of symbols, but now, in view of the fact that the axiom (e) speaks of an *arbitrary property*, we must make use of a generalized *predicate* variable; that is, a predicate variable bound by the universal quantifier. Expressions with quantified predicate variables are regarded as belonging to *logic of the second order*, or to the *extended predicate logic*. Expressions in which only subject variables are quantified are said to

[33] The sequence of natural numbers is often taken to begin with 1.

belong to *logic of the first order*, or to *elementary predicate logic*. For the extended predicate logic, as well as for the elementary (§3), it is possible to give a semantic definition of the concept of a consequence.

Except for axiom (e) we will continue to confine our arithmetical expressions to the elementary logic. *In particular, in questions of completeness and decidability we shall consider only relevant expressions of the first order*. The fundamental concepts of our system of axioms are: (1) an individual variable for zero; as such we take the traditional symbol 0; (2) a predicate variable for the relation of successor; we make use of the functional notation and denote the successor of x by x' (cf. §2.5); (3) a predicate variable for identity; as such we use the traditional symbol $=$. There is no need to mention the axioms (a) and (b), since we do not admit any individuals other than the natural numbers. In a supplementary axiom we express the conditions that must be satisfied by the identity.

The Peano system of axioms $\mathfrak{P}$ *in the extended predicate logic.*[34]

$$x' = y' \rightarrow x = y, \tag{P1}$$

$$\neg\, x' = 0, \tag{P2}$$

$$\bigwedge_P (P0 \wedge \bigwedge_y (Py \rightarrow Py') \rightarrow \bigwedge_x Px), \tag{Ind}$$

$$x = y \leftrightarrow \bigwedge_P (Px \rightarrow Py). \tag{G}$$

The semantic consistency (cf. §4.7) of $\mathfrak{P}$ is obvious for anyone who feels convinced of the "existence" of the natural numbers. But for the extended predicate logic we have not yet defined a concept of deducibility, so that for the time being the question of syntactical consistency (cf. §5.7) does not arise.

The system $\mathfrak{P}$ *is monomorphic* (§4.6) and thus, as desired, it characterizes the natural numbers. Let us outline the proof.

Let M and $\overline{M}$ be arbitrary models (cf. §3) of $\mathfrak{P}$. Then M contains a domain of individuals J, a function f (for x') defined on J and a fixed element n (representing 0) in J. We denote the corresponding objects for $\overline{M}$ by $\bar{J}$, $\bar{f}$, $\bar{n}$. We must now show that M and $\overline{M}$ are isomorphic (§8.4); that is, we must demonstrate the existence of a mapping Φ of J onto $\bar{J}$ with the properties of an isomorphism.

$$\Phi(n) = \bar{n}, \tag{10.1}$$

$$\Phi(f(x)) = \bar{f}(\Phi(x)). \tag{10.2}$$

[34] For clarity, we have emphasized here that P is generalized, i.e., bound by the universal quantifier. Of course, x and y are also to be considered as generalized.

First we define inductively a *relation* Φ by

(10.3) $$n\Phi\bar{n},$$

(10.4) $$\bigwedge_x \bigwedge_y ((x \in J \wedge y \in \bar{J}) \rightarrow (x\Phi y \rightarrow f(x)\, \Phi \bar{f}(y)))^{\cdot},$$

(10.5) Let $x\Phi y$ hold only as required by (10.3) or (10.4).

We now prove step by step [with tacit use of the axiom of equality (G)].

(1) The first domain of Φ is J (proof by the axiom of induction for the model M).

(2) The second domain of Φ is $\bar{J}$ (proof by the induction axiom for the model $\overline{M}$).

(3) There is no x in J with $n = f(x)$ [proof by the axiom (P2) for M].

(4) There is no x in J with $f(x)\, \Phi\bar{n}$ [proof by (3) and (10.3, 4, 5)].

(5) If $x\Phi\bar{n}$ and $y\Phi\bar{n}$, then $x = y$ [proof by (4) and (1)].

(6) If $x\Phi z$ and $y\Phi z$, then $x = y$; that is, $\breve{\Phi}$ is a function (8.3) [proof by induction for $\overline{M}$, (5) and (P1)].

(7) Φ is a function [proof analogous to (6)].

Thus we have shown that Φ is a one-to-one mapping of J onto $\bar{J}$, from which the properties of an isomorphism follow immediately by (10.3) and (10.4).

We must note, however, that this proof can be attacked on the ground that it is based in an essential way on semantic ideas that are closely associated with the naive theory of sets. For in fact the "totality of all properties" referred to in (G) and (Ind) is uncountable. From the monomorphy of $\mathfrak{P}$ it follows that $\mathfrak{P}$ is complete (cf. §4.5).

10.3. *The Peano Axiom with Restricted Axiom of Induction*

We now turn to an axiom system $\mathfrak{P}_1$, which completely avoids the extended predicate logic. In order to exclude quantification of predicate variables, we must first make some change in the axiom of equality (G). Let us replace it by the two axioms

(G1) $$x = x,$$

(G2) $$x = y \rightarrow (\mathsf{H}(x) \rightarrow \mathsf{H}(y)).$$

Since for $\mathsf{H}(x)$ we may write any expression of the elementary predicate logic, it follows that, strictly speaking, (G2) is not an axiom but an *axiom schema* (§4) which in an obvious way represents countably many axioms.

The axioms (P1) and (P2) remain unchanged, but for (Ind) we must also introduce an axiom schema:

(Ind$_1$) $\mathsf{H}(0) \wedge \bigwedge_y (\mathsf{H}(y) \rightarrow \mathsf{H}(y')) \rightarrow \mathsf{H}(x)$ (*induction schema*).

The system (G1, G2, P1, P2, Ind$_1$) will be denoted by $\mathfrak{P}_1$. Like $\mathfrak{P}$, the system $\mathfrak{P}_1$ is of course semantically consistent. On the other hand, monomorphy is lost in the transition from $\mathfrak{P}$ to $\mathfrak{P}_1$. For we see that the proof of monomorphy for P cannot simply be repeated for $\mathfrak{P}_1$, since the properties to which (Ind) was applied in that proof are not necessarily capable of formulation (and in fact cannot be formulated) in the elementary predicate logic (cf. §10.2). In §10.5 we shall see that P_1 actually admits nonisomorphic models. It can be shown that the set of deductions from $\mathfrak{P}_1$ or from $\mathfrak{P}$ is decidable.[35] These systems are therefore complete and their theorems can be obtained by algorithms.

10.4. *Systems $\mathfrak{Z}$ and $\mathfrak{Z}_1$ for Arithmetic*

For the construction of arithmetic it is clear that the successor function alone is not enough. We also need addition and multiplication. These functions, as we know, can be defined recursively (§5.6), and the equations defining them can be adjoined to the axioms. Let us first state the axioms for addition:

(10.6) $$x + 0 = x,$$

(10.7) $$x + n' = (x + n)'.$$

From $\mathfrak{P}$ and $\mathfrak{P}_1$ we thus obtain axiom systems $\mathfrak{D}$ and $\mathfrak{D}_1$, respectively, to which the properties of monomorphy and nonmonomorphy, of completeness and decidability, are transferred. But these advantages are offset by a certain poverty in our means of expression. To be sure, we can still express such number-theoretical concepts as $x < y$ or 3 *is a factor of* x:

(10.8) $$x < y \Leftrightarrow \bigvee_z (z \neq 0 \wedge x + z = y),$$

(10.9) $$3 \mid x \Leftrightarrow \bigvee_z (z + z + z = x).$$

But it can be shown that other important concepts like $x \mid y$ or x *is a prime number* cannot be defined, so that many interesting number-theoretical problems cannot be formulated and thus cannot be decided within the framework of these theories.

[35] It must be noted that in these formal systems multiplication does not occur and cannot be (explicitly) defined.

In order to enrich our means of expression we adjoin the recursive definition of multiplication to the axioms of $\mathfrak{D}$ and $\mathfrak{D}_1$:

(10.10) $$x \cdot 0 = 0,$$

(10.11) $$x \cdot (n') = (x \cdot n) + x.$$

The resulting systems will be denoted by $\mathfrak{Z}$ and $\mathfrak{Z}_1$. In these systems we can define, for example, the following arithmetical concepts:

(10.12) $$x \mid y \Leftrightarrow \bigvee_z (y = x \cdot z),$$

(10.13) $$\text{Prime } x \Leftrightarrow x \neq 0 \wedge x \neq 0' \wedge \bigwedge_z (z \mid x \rightarrow (z = 0' \vee z = x)).$$

Gödel has shown, although we have no space for his proof here, that all decidable properties and relations (§5.4), e.g., $z = xy$, are now definable: The system $\mathfrak{Z}$ (or $\mathfrak{Z}_1$) includes the complete recursive theory of numbers.

The (syntactical) consistency of $\mathfrak{Z}_1$ was proved by Gentzen in 1936.

In comparison with the preceding systems, the investigation of $\mathfrak{Z}$ and $\mathfrak{Z}_1$ gives rise to considerably greater difficulties. Consider, for example, the existence of such unsolved number-theoretical problems as the Goldbach conjecture:

(10.14) $$\bigwedge_z (2 \mid z \wedge z \neq 2 \rightarrow \bigvee_{x,y} (\text{Prime } x \wedge \text{Prime } y \wedge z = x + y)).$$

Such problems make it plausible, as is in fact the case, that in these systems the set of consequences is not decidable. The truth of this statement results from the following theorem of Gödel, which is one of the most important discoveries in the whole theory of the foundations of mathematics.

10.5. *The Gödel Incompleteness Theorem:* $\mathfrak{Z}_1$ *Is Incomplete* (*Even Essentially Incomplete*; cf. End of the Present Subsection)

Although it will be impossible to include many of the details, we wish to give an outline here of the proof of this theorem, partly on account of its great importance, but also in order that the reader may see how an argument which in a natural language leads to a contradiction (namely to the Antinomy of the Liar described in §11.3) can in a formal language be put to good use, namely, to prove the incompleteness of $\mathfrak{Z}_1$.

An important instrument in the proof is the arithmetization described in §5.4, where we have shown that a procedure can be set up whereby the formulas of the language are characterized by their so-called Gödel numbers. Since it is decidable whether or not a given formula is a relevant expression,[36] it is also decidable whether a given natural number is the

[36] A *relevant expression* here is the same as a *relevant* proposition in §4.5.

Gödel number of some relevant expression. Since we have shown in §10.4 that all decidable properties can be defined in $\mathfrak{Z}_1$, there exists a relevant expression $A(x)$ which in the natural interpretation (i.e., the interpretation in which 0 corresponds to zero, and so forth) holds for a natural number if and only if this number is the Gödel number of an expression in $\mathfrak{Z}_1$.

Finite sequences of relevant expressions can be represented by numbers in the same way as the expressions themselves, so that, in particular, proofs can be expressed by numbers, since they are merely special sequences of expressions. Since it is decidable whether a given rule of inference has been correctly used, we can now find a relevant expression $C(p, q)$ which in the natural interpretation is true for p and q if and only if p is the number of a relevant expression H and q is the number of a proof of H in $\mathfrak{Z}_1$.

We now proceed to construct a relevant expression E, containing no free variables, which in the natural interpretation states that E (in other words, the expression itself) is unprovable (cf. the Paradox of the Liar in §11.3). If we assume that E is provable, we then have the following situation: Every model of $\mathfrak{Z}_1$, and consequently also the natural interpretation, satisfies E and therefore states, in contradiction to our assumption, that E is unprovable. On the other hand, if we assume that $\neg E$ is provable, the natural model will satisfy $\neg E$, and therefore falsify E; that is, E is provable, a result which, taken together with the provability of $\neg E$, contradicts the consistency of $\mathfrak{Z}_1$. Thus neither E nor $\neg E$ is provable.

This syntactical result, when reformulated in semantic language, states that neither E nor $\neg E$ is a consequence of $\mathfrak{Z}_1$. In other words, $\mathfrak{Z}_1$ is incomplete, as asserted.

The expression E, which asserts its own unprovability, is constructed as follows: If n is the Gödel number of an expression with exactly one free variable x, let us denote this expression by $A_n(x)$ and call n an A number. We construct the propositional form

(10.15) *x is an A number and y is the Gödel number of a proof of $A_x(x)$.*

By means of the arithmetization, this propositional form can be represented by an expression $B(x, y)$ in $\mathfrak{Z}_1$ with the two free variables x and y. Now let p be the Gödel number of the expression $\bigwedge_y \neg B(x, y)$. We form the expression $A_p(p)$ obtained by replacing x with p in $A_p(x)$. By (10.15) this expression states: *for every y, the number y is not the Gödel number* of a proof of $A_p(p)$. Thus $A_p(p)$ is a proposition E of the desired kind.

This theorem can obviously be extended to all axiomatic theories that have constructive definitions for their expressions and rules of inference, and that include a sufficiently large part of arithmetic.

The incompleteness theorem has some remarkable consequences:

(1) There exist arithmetical propositions (e.g., E) that are true for the natural numbers but are not provable in $\mathfrak{Z}_1$. It is conceivable, for example, that the Fermat conjecture or the proposition (10.14) is true but cannot be deduced by means of the familiar rules of inference in $\mathfrak{Z}_1$.

(2) From the incompleteness of $\mathfrak{Z}_1$ it follows by §4.6 that $\mathfrak{Z}_1$ *is not monomorphic*. For example, the proposition E is true for the model of the natural numbers but certainly untrue for some other model of $\mathfrak{Z}_1$, since E is not a consequence of $\mathfrak{Z}_1$.

(3) If we introduce into $\mathfrak{Z}$ certain natural rules of inference (it is to be noted that the language in which $\mathfrak{Z}$ is formulated goes beyond the means of expression available in the predicate logic), we can prove, just as for $\mathfrak{Z}_1$, that there exists in $\mathfrak{Z}$ a proposition E such that neither E nor $\neg E$ is deducible. Then we could proceed, again just as for $\mathfrak{Z}_1$ (see above), to prove that $\mathfrak{Z}$ is incomplete, provided we were allowed, as is the case in $\mathfrak{Z}_1$, to replace the concept of provability by the concept of a consequence. But we know that $\mathfrak{Z}$ is complete, as may be proved in exactly the same way as for $\mathfrak{P}$ in §10.2. Thus we have the important result that in $\mathfrak{Z}$, and more generally in the logics of higher order as contrasted with the predicate logic, the concept of a consequence cannot be reduced to an algorithm.

One might think that the incompleteness of $\mathfrak{Z}_1$ could be removed by the introduction of further axioms that would leave the system consistent. But so long as we are dealing with finitely many axioms (or more generally with a decidable schema of axioms), the concept of provability remains decidable, so that the above argument can be applied to the enlarged system of axioms. Thus we are dealing here with an *essential, nonremovable incompleteness*.

These results for $\mathfrak{Z}_1$ and $\mathfrak{Z}_2$ can also be obtained in the following way. We can show that in any sufficiently expressive arithmetical language there always exists, for any given recursively enumerable set (§5.3) M of arithmetical theorems [i.e., arithmetical propositions that are valid in the natural interpretation (§10.5)], an arithmetical proposition E which, together with its negation, does not belong to M. Thus we have:

(*a*) Since the set of deductions in $\mathfrak{Z}_1$ is recursively enumerable (§6.2), the system $\mathfrak{Z}_1$ is incomplete;

(*b*) The system $\mathfrak{Z}$, like $\mathfrak{P}$, is monomorphic and therefore complete (§10.2). Thus the set of deductions in $\mathfrak{Z}$ is not recursively enumerable, and therefore certainly not decidable.

For a system of axioms $\mathfrak{S}$ that includes arithmetic we can also construct, by means of our arithmetization, a proposition W expressing the syntactical consistency of $\mathfrak{S}$. Then the Gödel theorem leads to the result that W

is not deducible in $\mathfrak{S}$, provided $\mathfrak{S}$ is consistent. Consequently, in order to prove the consistency of $\mathfrak{S}$ we must make use of methods that lie outside $\mathfrak{S}$.

10.6. *The Operational Construction of Arithmetic*

In this construction the theorems of arithmetic and of other branches of mathematics are regarded, without reference to any possible semantic interpretation, as statements concerning the application of certain rules of *operation with finite systems*, which may consist of numerals or of concrete objects of any kind. If we study these systems (which are made up of finitely many "atoms" or indivisible systems), we can distinguish them according to their "length," and in this way we necessarily arrive at the conception of a number. By "abstraction" from systems of the same length we obtain the fundamental numbers, which can be uniquely represented by systems such as |, ||, |||, ... (Lorenzen). Propositions, rules of inference, sets, and so forth are again merely systems or "terms" (possibly with certain rules of transition from one system to another). The fundamental rules of operation are given in the form of *algorithms*, on the basis of which further systems and rules can be "deduced." However, this "deducibility" must be of an obviously "constructive" nature; in his "protologic," Lorenzen gives a number of principles of deduction that can be considered constructive.

The operative construction of arithmetic can only be briefly indicated here (see also §5.2). The system for generating the numerals is defined by an algorithm with one axiom and one rule, involving the proper variable e (cf. §5.2):

(10.16) $$|,$$

(10.17) $$\frac{e}{e\,|}\,.$$

Equality is defined by the following algorithm (k, l are variables for numerals):

(10.18) $$| = |,$$

(10.19) $$\frac{k = l}{k\,| = l\,|}\,.$$

By various principles of deduction we now realize that:

(10.20) $$k\,| = l\,| \rightarrow k = l,$$

(10.21) $$k\,| \neq k,$$

(10.22) $k = l \wedge A(k) \rightarrow A(l)$ (so-called principle of equality).

(10.23) $A(|) \wedge \bigwedge_k (A(k) \rightarrow A(k\,|)) \rightarrow A(l)$ (so-called principle of induction).

The significance of (10.20) is that the rule

$$\frac{k\,|=l\,|}{k=l} \tag{10.24}$$

is superfluous, i.e., in the algorithm for equality nothing can be deduced with this rule that cannot be deduced without it, as follows from the so-called principle of *inversion:* since $k\,| = l\,|$ can be obtained only from $k = l$, it follows that $k = l$ must also be deducible.

The atom $\wedge$ is introduced by a rule which is identical with the rule for $\wedge$-introduction in §6.4, but the rules in §6.4 for the elimination of $\wedge$ are not required here, since the principle of inversion shows that they are superfluous.

The systems (10.20)–(10.23) correspond to the Peano axioms; but in the present case they are not "postulated" but follow from certain "protological" theorems applied to the arithmetical algorithm.

One advantage of this construction of mathematics lies in the fact that by its very nature it leads only to propositions that can be seen intuitively to be true and therefore cannot involve contradictions.

Exercises for §10

1. On the basis of the axioms ($P1$), ($P2$), (Ind) and (G) prove the following theorem

$$\bigwedge_P (P0 \wedge P0' \wedge \bigwedge_x (Px \rightarrow Px'') \rightarrow \bigwedge_x Px).$$

2. To ($P1$), ($P2$), (Ind), (G), 10.6 and 10.7 adjoin the axioms

$$0^2 = 0$$
$$(x')^2 = x^2 + x + x + 0'.$$

Then show that in the resulting system it is possible to define the relation that holds for x, y and z if and only if $x \cdot y = z$.

Bibliography

Elementary problems in the foundations of arithmetic are discussed in Tarski [1]. For the theory of the systems Z and Z_1 see Russell [1]. On the concept of arithmetic itself see Frege [1].

11. Antinomies

11.1. *Classification of the Antinomies*

A proposition (or a propositional form) together with its negation form a *contradiction.* By an *antinomy* or *paradox* we mean an argument that leads to a contradiction.

It is natural to ask what could be the nature of such an argument. This question is most easily answered if we are dealing with an algorithm, since an antinomy then consists in the deduction of a proposition and of its negation. Since only formal processes are involved here, we speak of a *syntactic antinomy* (for the concepts of syntax and semantics cf. §3.1).

But it can also be the case that an argument which leads to a contradiction is not truly formal but depends on the meaning of the propositions (or of parts of them) that are used in the argument. In this case we speak of a *semantic antinomy*.

Since algorithms in the strict sense of the word are very recent inventions, it is not surprising that syntactic antinomies have been known for a relatively short time. On the other hand, many semantic antinomies were already discussed in antiquity.

If we can deduce a proposition and its negation, then by the rule of $\neg$-elimination (see §6.4) we can deduce *any* proposition. But if we can deduce *everything*, there is no interest in constructing arguments. As a result, we reject any algorithm that leads to a syntactic antinomy. As for semantic arguments leading to a semantic contradiction, we must make up our minds to revise at least one detail of the intuitive truths "inserted" into the argument, but it is often very difficult to accomplish this change in a convincing way.

Syntactic antinomies can also lead, at least indirectly, to a revision of our intuitive ideas. In general, an algorithm is not set up arbitrarily but is based on certain of our intuitive conceptions, which it presents in a concentrated form. Thus, if we find an antinomy in such an algorithm, we must realize either that the conceptions are not adequately represented in the algorithm, or else that they must be rejected, at least to some extent.

We confine ourselves here to a detailed description of two antinomies: the Russell Antinomy, as a characteristic example of a syntactic antinomy, and the Antinomy of the Liar, as a characteristic example of a semantic antinomy.

11.2. *The Russell Antinomy*

We are dealing here with a system of axioms in the language of predicate logic, so that the deductions can be obtained by means of an algorithm. The intuitive conceptions at the basis of this system of axioms are of a set-theoretical nature (cf. §7). Let us describe them briefly: there exists a property defined by the predicate "x is an element of the set y." We represent this predicate by the symbol Exy (that is, we use the symbol Exy of predicate logic to mean $x \in y$). Sets are represented by propositional forms with one variable; for example, the set of even numbers is represented by the propositional form

(11.1) $$2 \mid x,$$

and the set of prime numbers by the propositional form

$$(11.2) \qquad x > 1 \wedge \bigwedge_y (y \mid x \rightarrow y = 1 \vee y = x).$$

Now if we assume, as seems natural, that every propositional form H with a variable x corresponds to a set y containing exactly those objects which satisfy H, we are led to require as part of our system of axioms that

$$(11.3) \qquad \bigvee_y \bigwedge_x (Exy \leftrightarrow \mathsf{H}).$$

It is to be noted that this requirement is not a single axiom but an axiom schema (cf. §4.1), since (11.3) is a prerequisite for every propositional form H containing x (but not y) as a free variable.

The Russell Antinomy now consists of showing that this schema of axioms, within the framework of predicate calculus, leads to a contradiction.

The contradiction is obtained by taking for H the propositional form $\neg\, Exx$. Then the set y, whose existence is required by (11.3) (and whose uniqueness, unimportant here, follows from the principle of extensionality), is *the set consisting of every set that does not contain itself as an element*. But this set y gives rise to a contradiction if we ask whether or not y is a member of itself. For if y is an element of itself, then y, precisely because it is an element of itself, cannot, by definition, be an element of itself. On the other hand, if y is not an element of itself, then, again by the definition of y, it must be an element of itself. Let us deduce the contradiction by a formal argument. In addition to the rules in §6, our set of axioms now includes all the special cases of (11.3) (see the following table).

Line Number	*Flagged Variable*	*Assumptions*	*Assertion*	*Rule Used*
1			$\bigvee_y \bigwedge_x (Exy \leftrightarrow \neg\, Exx)$	axiom
2	y		$\bigwedge_x (Exy \leftrightarrow \neg\, Exx)$	$\bigvee$-elimination (1)
3			$Eyy \leftrightarrow \neg\, Eyy$	$\bigwedge$-elimination (2)
4			$Eyy \rightarrow \neg\, Eyy$	$\leftrightarrow$-elimination (3)
5			$\neg\, Eyy \rightarrow Eyy$	$\leftrightarrow$-elimination (3)
6		Eyy	Eyy	introduction of assumption
7			$Eyy \rightarrow Eyy$	elimination of assumption (6)
8		$\neg\, Eyy$	$\neg\, Eyy$	introduction of assumption
9			$\neg\, Eyy \rightarrow \neg\, Eyy$	elimination of assumption (8)
10			$Eyy \vee \neg\, Eyy$	excluded middle
11			Eyy	$\vee$-elimination (5, 7, 10)
12			$\neg\, Eyy$	$\vee$-elimination (4, 9, 10)
13			Ezz	$\neg$-elimination (11, 12)
14			$\neg\, Ezz$	$\neg$-elimination (11, 12)

Lines 1–13 provide a finished proof for Ezz, and lines 1–14 for $\neg\, Ezz$, so that the contradiction is proved.[37]

This antinomy indicates that we must in some way revise the set-theoretical conceptions underlying the axioms (11.3). As a result, it is no longer assumed today that *every* propositional form defines a set (cf. §7.6).

11.3. *The Antinomy of the Liar*

This antinomy, already well known in antiquity, makes use of the concept of truth (cf. also §3) and is thus a semantic antinomy. We begin with the stipulation already stated in precise form by Aristotle, that a proposition is true if and only if it describes an actual state of affairs. As a concrete example, let us consider the proposition "it is snowing." Then we can say:

(11.4) "it is snowing" is true if and only if it is snowing.

But this proposition, consisting of the whole of line (11.4), remains true if we replace the proposition "it is snowing" by any other proposition. Thus we are led to recognize the validity of all propositions of the following form:

(11.5) ... is true if and only if - - -

where in place of "- - -" we may put an arbitrary proposition, provided that at the same time we put a name of this proposition in place of "...". In order to obtain the Antinomy of the Liar we consider the particular proposition:

(11.6) The proposition that follows "(11.6)" is not true.

In other words, the proposition asserts its own falsity. We now insert this proposition in (11.5) in place of "- - -" and at the same time we insert a name for this proposition in place of "...". For such a name we choose: "the proposition that follows '(11.6)'." Then as a special case of (11.5) we obtain:

(11.7) The proposition that follows "(11.6)" is true if and only if the proposition that follows "(11.6)" is not true.

But from (11.7) it is easy to obtain a contradiction (cf. the Russell Antinomy starting from line 3 of the proof).

This contradiction cannot be avoided as long as we agree to the following conditions: we accept the Aristotelian criterion of truth (11.5), we admit

[37] We could not stop with line 11 or 12, since they still contain a free occurrence of the flagged variable y.

that what is contained in the line (11.6) is a proposition and that "the proposition that follows '(11.6)'" is a name for this proposition, and finally we accept the elementary logical deductions that lead from (11.7) to an actual contradiction.

If now, faced with this contradiction, we ask at what stage we should change our point of view, it would be natural to look first at the Aristotelian criterion of truth (11.5). Yet it must be admitted that proposition (11.5) seems almost self-evident and that we would never have felt any doubt about it if the antinomy had not been brought to our attention. Moreover, we must take note of the fact that in a certain respect we have already made use of this criterion in §3.4, where we discussed the validity, in a certain interpretation, of an elementary propositional form $Px_1, \ldots, x_n$. For we can express the Aristotelian criterion, as applied to that special case, in the form:

(11.8) If we replace x by 3 and P by the property of being a prime number, then Px is true if and only if 3 has the property of being a prime number.

The similarity with (11.5) is unmistakable.

But this comparison indicates how we can attack the Antinomy of the Liar. In (11.8) the problem at issue is to define what is meant by saying that a given propositional form is true in a given interpretation. Now the propositional form Px belongs to the language of predicate logic but the desired definition will be given, *not* in the language of predicate logic, but in some other language, namely whatever language we use for talking about predicate logic. Our choice for such a language is everyday English, cautiously used in a somewhat refined form. The predicate "is true" introduced in (11.8) belongs to this everyday language but refers not to propositions of everyday language, but to propositional forms in the language of predicate logic (in conjunction with the given interpretations).

Thus the difference between (11.5) and (11.8) is essentially as follows: in (11.8) we are dealing not only with a given language (the language of predicate logic) but also with a *metalanguage* (everyday English), in which we speak about the first language. The predicate "is true" in (11.8) is a predicate in the metalanguage. But it refers not to propositions in the metalanguage, but to expressions in the first language. In (11.5), on the other hand, there is only *one* language, namely everyday English. The predicate "is true" occurring there belongs to this everyday language and also refers to propositions in the same language.

Now it is easy to see that in (11.8) no antinomy is to be feared (or at any rate we cannot so easily construct one as in the Antinomy of the Liar). For the Antinomy of the Liar is based on a proposition that states its own falsity. But such a situation is not possible (or at any rate not immediately possible) if we distinguish between language and metalanguage. For in that case we cannot form a proposition that states its own falsity. Such a proposition, call it α, must belong to the metalanguage, since it contains the word "true" (or "false"); but the word "true" in the metalanguage refers to propositions of the initial language and therefore cannot refer to α.

In summary, we may say: we can escape from the Antinomy of the Liar by distinguishing between language and metalanguage and by speaking about the truth of the propositions in a given language—not in that language itself but in a metalanguage. Such distinctions between a formal language and a metalanguage, or a meta-metalanguage and so forth, are common in modern logical investigations. Since the natural languages of the world are "universal" and fail to make this distinction, in the sense that they use the word "true" for arbitrary propositions expressible in them, many investigators consider these natural languages to be inevitably self-contradictory.

As a final remark, let us point out that the other semantic antinomies can be avoided when we make the distinction between language and metalanguage. Consider, for example, the antinomy of the smallest natural number that cannot be described in English in fewer than a hundred words. The antinomy arises from the fact that, precisely in the definition just given, this number has nevertheless been described in fewer than a hundred words. But the above definition refers to all possible descriptions and thus, since it speaks of these descriptions, it must belong to a language that is a metalanguage with respect to the language to which the descriptions belong. Consequently, we obtain *in the metalanguage* a description for the number which is shorter than any possible description *in the initial language*. But this result is not a contradiction.

Exercises for §11

1. An adjective A is said to be *autologic* if A has the property described by A, and otherwise A is *heterologic*. Examples of autologic adjectives are: "seventeenlettered," "English," "pentasyllabic." Consider the word "heterologic." Is it heterologic or autologic? Explain and resolve the antinomy (Grelling's antinomy).
2. If the definition of an object or element m depends on a set M and if m is then assigned to M as an element, the definition of m is said to be *impredicative*.
   (a) Show that the antinomies mentioned in the text make use of impredicative definitions.
   (b) Show that the definition of the least upper bound of a set M of real numbers, as given in real analysis, is impredicative.

Bibliography

On the antinomies in the theory of sets see Beth [1] and Linsky [1].

Bibliography

Bernays, R.: [1] Axiomatic Set Theory. With a historical introduction by Abraham A. Fraenkel. North-Holland Publishing Company, Amsterdam, 1958, VIII + 226 pp.

Beth, E. W.: [1] The Foundations of Mathematics. North-Holland Publishing Company, Amsterdam, 1965, XXVIII + 741 pp.

Bocheński, J. M., [1] A Précis of Mathematical Logic. Translated from the French and German Editions by Otto Bird. D. Reidel, Dordrecht, Holland; Gordon and Breach, New York, VII + 100 pp.

Borsuk, K. and W. Szmielew: [1] Foundations of Geometry. North-Holland Publishing Company, Amsterdam, 1960, XIV + 444 pp.

Carnap, R.: [1] Introduction to Semantics. Harvard University Press, Cambridge, Mass., 1946, 2. Druck, XII + 263 pp.

Carnap, R.: [2] Introduction to Symbolic Logic and its Applications. Dover Publications, Inc., New York, 1958, XIV + 241 pp.

Church, A.: [1] Introduction to Mathematical Logic. Princeton University Press, Princeton, N.J., 1956, IX + 376 pp.

Church, A.: [2] Logic. Article in Encyclopaedia Britannica, Vol. 14. Encyclopaedia Britannica, Ltd., London, Chicago, 1963, pp. 295–305.

Cohen, P. J.: [1] The Independence of the Continuum Hypothesis. Proceedings of the National Academy of Sciences, Vol. 50, pp. 1143–1148 (1963) and Vol. 51, pp. 105–110 (1964).

Cohen, P. J.: [2] Set Theory and the Continuum Hypothesis. W. A. Benjamin, Inc., New York, 1966, VI + 154 pp.

Curry, H. B.: [1] Outlines of a Formalist Philosophy of Mathematics. North-Holland Publishing Company, Amsterdam, 1951, VII + 75 pp.

Curry, H. B.: [2] Foundations of Mathematical Logic. McGraw-Hill Book Company, Inc., New York, 1963, XII + 408 pp.

Davis, M.: [1] Computability and Unsolvability. McGraw-Hill Book Company, Inc., New York, Toronto, London, 1958, XXV + 210 pp.

Fraenkel, A. A.: [1] Abstract Set Theory. North-Holland Publishing Company, Amsterdam, 1953, XII + 479 pp.

Fraenkel, A. A. and Y. Bar-Hillel: [1] Foundations of Set Theory. North-Holland Publishing Company, Amsterdam, 1958, X + 415 pp.

Frege, G.: [1] The Foundations of Arithmetic. Transl. by J. L. Austin. Basil Blackwell, Oxford, 1950, xii + xii^e + XI + XIe + 119 + 119^e pp.

Frege, G.: [2] Translations from the Philosophical Writings of Gottlob Frege. Edited by P. Geach and M. Black. Basil Blackwell, Oxford, 1952, X + 244 pp.

Goodstein, R. L.: [1] Boolean Algebra. The Macmillan Company, New York, 1963, VI + 140 pp.

Halmos, P. R.: [1] Naive Set Theory. D. Van Nostrand Company, Inc., Princeton N.J., 1960, VII + 104 pp.

Hermes, H.: [1] Enumerability, Decidability, Computability. Springer-Verlag, Berlin, Heidelberg, New York, 1965, IX + 245 pp.

Heyting, A.: [1] Intuitionism. North-Holland Publishing Company, Amsterdam, 1956, VIII + 132 pp.

Hilbert, D. and W. Ackermann: [1] Principles of Mathematical Logic. Chelsea Publishing Co., New York, 1950, XII + 172 pp.

Kalish, D. and R. Montague: [1] Logic. Techniques of Formal Reasoning. Harcourt, Brace & World, Inc., New York, 1964, X + 350 pp.

Keene, G. B.: [1] Abstract Sets and Finite Ordinals. Pergamon Press, Oxford, 1961, X + 106 pp.

Kleene, S. C.: [1] Introduction to Metamathematics. D. Van Nostrand Company, Inc., Princeton, N.J., 1952, X + 550 pp.

Kleene, S. C.: [2] Mathematical Logic. John Wiley & Sons, Inc., New York, 1967, XIII + 398 pp.

Kneale, W. and M. Kneale: [1] The Development of Logic. Clarendon Press, Oxford, 1962, VIII + 761 pp.

Kneebone, G. T.: [1] Mathematical Logic and the Foundations of Mathematics. D. Van Nostrand Company Limited, Princeton, N. J., 1963, XIV + 435 pp.

Kreisel, G.: [1] Mathematical Logic. Lectures on Modern Mathematics, Vol. III (edited by T. L. Saaty), pp. 95–195. John Wiley & Sons, Inc., New York, 1965.

Linsky, L. (editor): [1] Semantics and the Philosophy of Language. The University of Illinois Press, Urbana, 1952, IX + 289 pp.

Lorenzen, P.: [1] Formal Logic. D. Reidel Publishing Company, Dordrecht, 1965, VIII + 123 pp.

Nagel, E., and J. R. Newman: [1] Gödel's Proof. New York University Press, New York, 1958, IX + 118 pp.

Novikov, P. S.: [1] Elements of Mathematical Logic. Oliver & Boyd, Edinburgh, 1964, XI + 296 pp.

Quine, W. V.: [1] Elementary Logic. Ginn and Company, Boston, 1941, VI + 170 pp.

Quine, W. V.: [2] Mathematical Logic. Harper & Row, New York, 1962, XII + 346 pp.

Robinson, A.: [1] Introduction to Model Theory and to the Metamathematics of Algebra. North-Holland Publishing Company, Amsterdam, 1963, IX + 284 pp.

Rosenbloom, P. C.: [1] The Elements of Mathematical Logic. Dover Publications, Inc., New York, 1950, IV + 214 pp.

Rosser, J. B.: [1] Logic for Mathematicians. McGraw-Hill Book Company, Inc., New York, 1953, XIV + 530 pp.

Russell, B.: [1] Introduction to Mathematical Philosophy. The Macmillan Co., New York, 1924, VIII + 208 pp.

Sierpiński, W.: [1] Cardinal and Ordinal Numbers. Państwowe Wydawnictwo Naukowe, Warsaw, 1958, 487 pp.

Suppes, P.: [1] Introduction to Logic. D. Van Nostrand Company, Inc., Princeton, N.J., 1957, XVIII + 312 pp.

Suppes, P.: [2] Axiomatic Set Theory. D. Van Nostrand Company, Inc., Princeton, N.J., 1960, XII + 265 pp.

Tarski, A.: [1] Introduction to Logic and to the Methodology of Deductive Sciences. Oxford University Press, New York, 1941, XVIII + 239 pp.

Tarski, A.: [2] Logic, Semantics, Metamathematics. Clarendon Press, Oxford, 1956, XIV + 471 pp.

Wang, H.: [1] A Survey of Mathematical Logic. North-Holland Publishing Company, Amsterdam, 1964, X + 651 pp.

Wilder, R. L.: [1] Introduction to the Foundations of Mathematics. Second edition. John Wiley & Sons, Inc., New York, 1965, XII + 327 pp.

PART B

ARITHMETIC AND ALGEBRA

Introduction

The development of modern algebra since the beginning of the present century has been a process of continually increasing abstraction, so that the subject was often called *abstract algebra.* It was realized that important simplifications could be gained, both in concept and in method, if for the various fields of arithmetic, the theory of numbers, algebraic equations, functions of a complex variable, and so forth we establish as clearly as possible what is common to these subjects and then present it in a form that is valid for all of them. For it often happens that theorems that have been discovered and proved in widely different fields of mathematics are found to be identical from the logical point of view, so that the proof can be carried out quite independently of the various interpretations in one field or another. In fact, the proof is generally much simpler and clearer when these particular interpretations are set aside; moreover, we can spare ourselves the trouble of proving the same theorem over and over again, since the general "abstract" proof is valid for all the "concrete" cases.

Since mathematics is in itself a very abstract science, the reader may feel surprised that certain branches of it are described as "abstract." Let us examine the situation.

The concept of a natural number is already the result of a complicated process of abstraction by no means easy to retrace (cf. IA, §10.1, and IB1, §1.1), and we are scarcely conscious of it in everyday calculations. But the immense intellectual effort involved in first making this abstraction has been richly rewarded by our being able to apply the simple rules of arithmetic to problems dealing with any kind of objects—stones, trees, lengths, weights, and so forth.

The same remark can be made about geometry. The concept of a triangle, which underlies the theorems of geometry, is already extremely abstract; it means only that we are dealing with a figure consisting of three points and the lines that join them. But then the theorems we deduce, e.g., that the sum of the angles of a triangle is equal to two right angles or that the sum of two sides is greater[1] than the third side, are valid for all possible triangles, regardless of their size, shape, or origin. The task of setting up the abstract geometric concept of a triangle demands a massive intellectual effort, but this effort is far more than offset by the simplicity and generality of the resulting theorems, which can now be applied to all possible triangles.

Now it is reasonable to expect a similar advantage from what we may call a second stage of abstraction, namely, from the fact that certain concepts and methods in the various branches of present-day mathematics can be identified with one another if we make an abstraction from their interpretations in various special fields.

The objects of study in *abstract algebra* are sets of an extremely general nature; their elements may be numbers, polynomials, functions, vectors, transformations, or any conceivable entities, whose meaning in any particular branch of mathematics is quite irrelevant. These sets have an *algebraic structure* consisting in certain relations or laws of combination among the elements within the set, in the existence of certain distinguished subsets, and so forth (see also IB10). Examples are *groups* (IB2) together with their *subgroups*, *modules* (IB1, §2.3), and *lattices* (IB9); Chapter IB5 will deal with general (commutative) *rings* and *integral domains*, whose structure is characterized by the presence of certain distinguished subrings, namely, the *ideals*. Special rings also occur in other chapters; for example, the *ring* or *integral domain of rational integers* (IB1, IB6), the *field of rational numbers* (IB1, IB6), *rings of algebraic numbers* and the *field of algebraic numbers* (IB6, IB7), *rings of polynomials* (IB4), *rings of matrices* (IB3), *rings of groups* (IB2), *rings of endomorphisms* (IB1, 2.4), and so forth.

[1] Or at most equal, if we allow the three vertices to lie on one line.

CHAPTER 1

Construction of the System of Real Numbers

1. The Natural Numbers

1.1. *The Peano System of Axioms*

The simplest approach to the natural numbers (in the present section they are simply called numbers) is provided by the common practice of counting objects by making marks on paper, so that the number of objects is represented by a row of marks, for example ||||. This procedure suggests that we define the natural numbers as the diagrams obtained by writing vertical strokes one after the other. The number | is also written in the form 1 and is called "one." The number formed by writing a vertical stroke to the right of the number a is called the *successor* of a; in the present §1 (but only here) we write this number[1] in the form a'. Equality of two numbers is defined as follows: beginning from the right-hand end (many other procedures would also be possible), we attempt to make a one-to-one correspondence between the two sets of strokes. If such a correspondence can be set up (as in the diagram) we say that the numbers are equal; otherwise they are unequal.

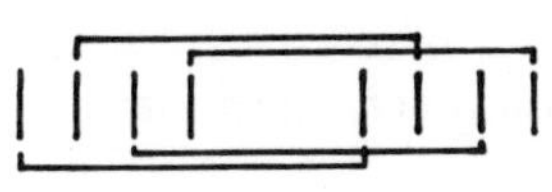

We see at once that the logical requirements for a definition of equality (see §2.2) are satisfied here, that $a' \neq 1$ for every natural number a, and finally that $a' = b'$ is equivalent to $a = b$. Every number can be formed from the number 1 by repeated construction of a successor. Consequently, any property that belongs to the number 1 and is *hereditary*, i.e., is bequeathed by each number to its successor, belongs to every

[1] The symbol a | would be quite adequate but we do not adopt it here, partly for typographical reasons and partly because we want to keep our notation independent of any particular method of introducing the natural numbers.

number. Let us summarize this information in the following system of axioms:

I. *1 is a number.*

II. *To every number a there corresponds a unique number a′, called its successor.*

III. *If* $a' = b'$, *then* $a = b$.

IV. $a' \neq 1$ *for every number a.*

V. *Let* $A(x)$ *be a proposition containing*[2] *the variable x. If* $A(1)$ *holds and if* $A(n')$ *follows from* $A(n)$ *for every number n, then* $A(x)$ *holds for every number x.*

From this system of axioms (which is usually named after Peano; cf. IA, §10) we shall see that by logical reasoning we can derive any theorem about the natural numbers without further reference to the way in which they were introduced. Thus a reader who for any reason is dissatisfied with our definition of natural numbers may adopt any other definition that leads again to I–V, and then he can follow our further developments. Our reason for setting up a system of axioms is not that there is anything inexact about the procedure[3] using vertical strokes; the system of axioms simply sets us free from this particular procedure. For example, we could define cardinal numbers as classes of equivalent sets (see IA, §7.3) whereupon[4] we would quickly arrive at I–IV; then axiom V serves to distinguish the natural numbers among all the cardinal numbers: a cardinal number is a natural number if and only if it possesses every hereditary property that belongs to the number 1.

Axiom V is called the *axiom of induction*, or the *principle of complete (or mathematical) induction* (on n) or also the *argument from* n *to* $n + 1$. The "complete" induction of mathematics is thus in sharp contrast with the "incomplete" induction of the experimental sciences, where a general law is derived from (finitely many) individual cases. This unfortunate choice of name must not be allowed to obscure the fact that in complete induction we are dealing with a deductive principle and not with the verification of a proposition $A(x)$ for a finite number of x values; for in fact, in applying the principle, we are required to show that for an arbitrary

[2] Thus $A(n)$ is the proposition that is formed when x is replaced by n. Strictly speaking, $A(x)$ is a propositional form (see IA, §2.3).

[3] Any apparent inexactness is due to the brevity of these introductory remarks. A complete description of the operational method of introducing numbers can be found in P. Lorenzen [1]. See also IA, §10.6.

[4] The number 1 is now the class of those sets that contain only one element; and if a is the class of sets that are equivalent to a given set M, its successor a' is the class of sets equivalent to M', where M' is formed from M by adjoining an element not yet in M.

n the proposition $A(n')$ *always* follows from $A(n)$, and such a proof can only depend on some general procedure, not on any special knowledge for a given number n. We refer to $A(1)$ as the *initial case*, to the argument from $A(n)$ to $A(n')$ as the *induction step*, and to $A(n)$ as the *induction hypothesis*. If N is the set of natural numbers, axiom V can also be expressed as a proposition about an arbitrary set M:

If $1 \in M$ *and if* $n' \in M$ *follows from* $n \in M$ *for every natural number* n, *then* $N \subseteq M$;[5] for we may write any proposition $A(x)$ in the form $x \in M$, where M is the set of those elements x that have the property $A(x)$.

It should also be mentioned that the choice of axioms is to a great extent arbitrary; it is only necessary that they imply exactly the same consequences as can be deduced from I–V; that is, they must imply the axioms I–V and be implied by them. Instead of "one" and "successor" we may introduce other fundamental concepts, e.g., the ordering defined later in §1.4 (the relation of "smaller than").[5a]

In the lower grades at school the natural numbers occur in the form of cardinal numbers; in other words, the number 3 is introduced by abstraction from sets of three objects (persons, marks, points or the like). The essential identity of (finite) cardinal and ordinal numbers is brought out by arranging objects in rows. Addition arises as the mathematical expression for putting sets of objects together (forming their union) or by extending the rows of objects (this process is recognizable in the recursive definition of addition in §1.3). The other rules for calculating with natural numbers are based on addition. Further work with natural numbers depends on the familiar rules of calculation (commutative and associative laws of addition and multiplication, distributive law, monotone laws), which in the following pages are derived from axioms I–V but in early school years are learned by experience without any explicit formulation. Thus, in early instruction these rules play the role of axioms; much later, in the more advanced grades, they are supplemented by the principle of complete induction.

1.2. *Recursive Definitions*

In order that the sum of the numbers of elements in two disjoint finite sets (for these concepts see §1.5) may be equal to the number of elements in the union of the two sets, the following equations must obviously be satisfied:

$$a + 1 = a', \tag{1}$$

$$a + b' = (a + b)'. \tag{2}$$

[5] The notation $M \subset M'$ means that $x \in M'$ follows from $x \in M$ but $M \neq M'$. In this case M is called a *proper subset* of M', but in the case $M \subseteq M'$ (that is, if equality is also possible) M is simply called a *subset* of M' (cf. IA, §7.2).

[5a] See, e.g., Feigl and Rohrbach [1].

So we have the task of introducing for every $a \in N$ a function f which is defined in N and has the properties

$$(3) \qquad f(1) = a', \qquad f(x') = f(x)' \qquad \text{for all} \quad x \in N;$$

for then we can simply define $a + b$ as $f(b)$. The fact that for every number a there exists exactly one function f with the properties (3) is a special case of the following general theorem:

Let c be a number and let F be a function of two arguments defined in N and with values in N. Then there exists exactly one function f defined in N such that

$$(4) \qquad f(1) = c, \qquad f(x') = F(x, f(x)) \qquad \textit{for all} \quad x \in N.$$

It is clear that (3) is obtained from (4) by setting $c = a'$ and $F(x, y) = y'$. The definition of a function f by the conditions (4), which is possible in view of the general theorem, is called a *recursive definition*, since the determination of $f(x')$ is reduced to that of $f(x)$ and thereby finally to that of $f(1)$.

To prove this theorem, which is also called the *principle of recursion*,[7] we first replace the concept of the function f by that of the set of pairs (x, y) with $y = f(x)$.[8] Then (4) requires the construction of a set P of pairs (x, y) with the properties

$$(5) \quad (1, c) \in P; \qquad \text{from} \quad (x, y) \in P \qquad \text{follows} \quad (x', F(x, y)) \in P.$$

Here it will be prudent to take the smallest such set P, namely the set that is formed from the pair $(1, c)$ by repeated application of the step from (x, y) to $(x', F(x, y))$.[9] In order to define a function f by means of this set, we must prove that for every number $x \in N$ there exists exactly one number y with $(x, y) \in P$. But by complete induction we see from (5) that such a y exists and is unique. For if we use (5) to construct the elements of P, we obtain, apart from the pair $(1, c)$, only pairs of the form (x', z), and thus, since $x' \neq 1$, it follows from $(1, y) \in P$ that $y = c$. If we now assume the desired assertion for x, and if $(x', z_1), (x', z_2) \in P$, then z_1, z_2 must be of the form $F(x, y_1), F(x, y_2)$, with $(x, y_1), (x, y_2) \in P$, since

[7] The same name is given to certain generalizations of Eq. (4), one of which is considered on p. 97.

[8] In IA, §8.4, the functions were directly defined as such sets of pairs. But the concept of a function can also be defined in other ways, independently of the concept of a relation (see, e.g., Lorenzen [1]).

[9] If we wish to proceed here on the basis of set theory, which is not altogether necessary, we will define P as the intersection of all sets P satisfying (5) and must then show, for example, that: if we had $(x', z) \in P$, $z \neq F(x, y)$ for all y, then the deletion of (x', z) from P would produce a set satisfying (5), in contradiction to the definition of P.

otherwise the pairs could not be constructed; then $y_1 = y_2$ by the induction hypothesis and therefore $z_1 = F(x, y_1) = F(x, y_2) = z_2$. Thus there exists a function f satisfying (4).

In order to prove that this function f is unique, we now assume that g is a function satisfying (4), so that $g(1) = c, g(x') = F(x, g(x))$. Then we have $g(1) = f(1)$, and under the hypothesis that $g(x) = f(x)$ we also have $g(x') = F(x, g(x)) = F(x, f(x)) = f(x')$. Thus by complete induction $g(x) = f(x)$ for all $x \in N$, so that $g = f$.

From the proof we see that the theorem is valid under the following weaker hypothesis: the values for the second argument of F need not be numbers but may form an arbitrary set, quite independent of the values for the first argument: this set contains c and the values of F. Of course, we then obtain a function f whose values are no longer necessarily numbers but belong to the arbitrary set.

Our principle of recursion can be made more general if we replace (4) by

$$(4') \qquad f(1) = c, \qquad f(x') = F_x(f(1), ..., f(x)) \qquad \text{for all} \quad x \in N,$$

where F_x is a function of x arguments for every natural number x.[10] But this more general principle can be reduced to (4) by a simple transformation: namely, with the number x we associate the x-tuple[11] $(f(1), ..., f(x))$ and denote this mapping by f^*, so that

$$f^*(x) = (f(1), ..., f(x)).$$

It is clear that the function f is uniquely determined by f^*. Consequently, in order to show the existence and uniqueness of a function f satisfying (4′) we need only transform (4′) into conditions on f^* that are of the form (4) and are therefore satisfied by the mapping f^*. For this purpose, in (4) we replace f by f^* and c by the 1-tuple (c) and define the function F as follows:

$$F(x, y) = (z_1, ..., z_n, F_n(z_1, ..., z_n)), \qquad \text{for} \quad y = (z_1, ..., z_n).$$

Then

$$F(x, f^*(x)) = [f(1), ..., f(x), F_x(f(1), ..., f(x))],$$

so that after these changes the conditions in (4) become identical with those of (4′), in view of the fact that

$$f^*(x') = (f(1), ..., f(x), f(x')).$$

But now we must have recourse to the above-mentioned possibility of weakening the hypotheses in our original principle of recursion: the arbitrary set in question now consists of all n-tuples $(z_1, ..., z_n)$, where n is any natural number and the $z_1, ..., z_n$ are no longer required to be numbers but only members of a set containing the arguments and the values of the functions F_x.

[10] For the concept of the number of elements of a set, see §1.5; in the formulation of (4′) we naturally require the concept of a segment as defined in §1.5.

[11] An x-tuple is a mapping of the segment A_x (see §1.5); thus the x-tuple in question is obtained by restricting the domain of f to A_x.

1.3. *Addition*

By the principle of recursion and the remarks at the beginning of §1.2, there exists exactly one operation, to be denoted by + and called *addition*, which is a function of two arguments, with arguments and values in N such that (1), (2) are satisfied for all $a, b \in N$. In other words, (1), (2) constitute the recursive definition of addition.[12] Addition is *associative*:

$$(a + b) + c = a + (b + c). \tag{6}$$

The proof is based on the argument from c to c': $(a + b) + 1 = (a + b)' = a + (b + 1)$; if (6), then $(a + b) + c' = ((a + b) + c)' = (a + (b + c))' = a + (b + c)' = a + (b + c')$. Addition is also *commutative*:

$$a + b = b + a. \tag{7}$$

For $b = 1$ the proof is by the argument from a to a': $1 + 1 = 1 + 1$; if $1 + a = a + 1$, then $1 + a' = (1 + a)' = (a + 1)' = (a + 1)' = (a + 1) + 1 = a' + 1$. The proof of the general assertion is by the argument from b to b' by means of (6) under the induction hypothesis of (7):

$$a + b' = (a + b)' = (b + a)' = b + a' = b + (a + 1)$$
$$= b + 1 + a) = (b + 1) + a = b' + a.$$

By (6) we may therefore omit the parentheses in a sum with three terms. In order to be able to omit them in sums with more than three terms, we first define the expression $\sum_{i=1}^{n} a_i$ for a given sequence[13] $(a_i)_{i=1,2,\ldots}$ of numbers a_i recursively by setting

$$\sum_{i=1}^{1} a_i = a_1 , \qquad \sum_{i=1}^{n+1} a_i = \sum_{i=1}^{n} a_i + a_{n+1} ; \tag{8}$$

for this purpose we need only set $c = a_1$ and $F(x, y) = y + a_{x+1}$ in (4). In particular, we have $\sum_{i=1}^{3} a_i = (a_1 + a_2) + a_3 = a_1 + a_2 + a_3$ and $\sum_{i=1}^{4} a_i = (a_1 + a_2 + a_3) + a_4$, for which again we naturally write

[12] The recursive definition of addition, in particular (1) and (2), is suitable for instruction at the end of the secondary school, where it could be presented in a course on the axiomatization of the natural numbers. In such a course the proofs given in the present section would be appropriate examples.

[13] A sequence of this sort (infinite) is simply a function $i \to a_i$, defined on N, whose values in this case are also in N.

$a_1 + a_2 + a_3 + a_4$. Moreover, no parentheses are needed to express addition of such sums, as is shown by the equation

$$\sum_{i=1}^{n} a_i + \sum_{i=1}^{m} a_{n+i} = \sum_{i=1}^{n+m} a_i . \tag{9}$$

The proof of this equation is by the argument from m to m': for $m = 1$, Eq. (9) becomes the second of the equations in (8); and from (9) we see by (8) and (6) that

$$\sum_{i=1}^{n} a_i + \sum_{i=1}^{m'} a_{n+i} = \sum_{i=1}^{n} a_i + \left(\sum_{i=1}^{m} a_{n+i} + a_{n+m'} \right)$$

$$= \left(\sum_{i=1}^{n} a_i + \sum_{i=1}^{m} a_{n+i} \right) + a_{n+m'}$$

$$= \sum_{i=1}^{n+m} a_i + a_{(n+m)'} = \sum_{i=1}^{(n+m)'} a_i = \sum_{i=1}^{n+m'} a_i .$$

We note that none of the properties of the numbers (except where they are used as indices) is needed here except property (6). The extension of (7) to sums of more than two terms will be proved in §1.5.

We now prove by means of (9) that any meaningful expression A constructed from numbers $a_1, ..., a_k$ (in this order), and from + signs and parentheses has a value, namely $=\sum_{i=1}^{k} a_i$, which is independent of the distribution of the parentheses (for $k = 1$ the expression A is to be taken equal to a_1). For the proof we make use of induction on k in the altered form of §1.4[14] (with k instead of m and with M as the set of numbers k for which the assertion is true). By the construction of A there must exist natural numbers n, m with $n + m = k$ ($k \neq 1$) such that for expressions B, C formed from $a_1, ..., a_n$ and $a_{n+1}, ..., a_{n+m}$ under appropriate distribution of parentheses, we have the equation $A = B + C$. Since $n, m < k$, the induction hypothesis means that $B = \sum_{i=1}^{n} a_i$, $C = \sum_{i=1}^{m} a_{n+i}$, so that the desired assertion $A = \sum_{i=1}^{k} a_i$, follows from (9). For $k = 1$, the assertion is immediately obvious.

If all $a_i = a$, the sum $\sum_{i=1}^{n} a_i = \sum_{i=1}^{n} a$ is called the *nth multiple na of a*. For this *multiplication*, (9) gives at once the *distributive law*

$$na + ma = (n + m) a. \tag{10}$$

The commutative law for multiplication is dealt with in §1.5, and the

[14] This anticipation of theorems on order relations (which we have already used in speaking of "the numbers $a_1, ..., a_k$") is permissible here, since the present result is not used in §1.4.

associative law in §2.4. From (8), by the argument from n to $n + 1$ it follows directly that $n1 = n$.

Finally, by the argument from c to c' we prove the rule:

(11) $$a = b, \quad \text{if} \quad a + c = b + c.$$

For by (1) the case $c = 1$ is already dealt with by axiom III; and from $a + c' = b + c'$ it follows from (2) that $(a + c)' = (b + c)'$ and therefore $a + c = b + c$; thus, by the induction hypothesis we have $a = b$, as desired.

1.4. *Order*

If for the numbers a, b there exists a number c with $a + c = b$, we write $a < b$ (a isl ess than b), or alternatively $b > a$ (b is greater than a).[15] For the relation $<$ defined in this way we have the following theorems:

(12) *if* $a < b$, *then* $a \neq b$ (*antireflexivity*);

(13) *if* $a < b$ *and* $b < c$, *then* $a < c$ (*transitivity*);

(14) *if* $a < b$, *then* $(a + d) < (b + d)$ (*monotonicity of addition*);

(15) *if* $a \neq b$, *then* $a < b$ *or* $b < a$.

Rule (12), which states that $a + c \neq a$ for all a, c, is proved by complete induction on a, for we have $1 + c \neq 1$ by (1), (7) and axiom IV; and if we had $a' + c = a'$, it would follow that $(a + c)' = a'$, and thus $a + c = a$. For the proof of (13), (14) we set $a + u = b$, $b + v = c$ and thus get $c = (a + u) + v = a + (u + v)$, $b + d = (a + u) + d = a + (u + d) = a + (d + u) = (a + d) + u$. Complete induction on a is again used to prove (15), as follows. The case $a = 1$ is first dealt with by complete induction[16] on b: $1 = 1$; $1 < 1 + b = b + 1 = b'$. Then from (15) (for all b) the same statement with a' instead of a (thus for all b: $a' < b$ or $a' = b$ or $b < a'$) is derived by complete induction on b: $1 < a'$; $a' < b'$ or $a' = b'$ or $b' < a'$ by (15) and (14); here again the induction hypothesis ($a' < b$ or $a' = b$ or $b < a'$) is not used.

From (12), (13) it is easy to see that no two of the statements $a < b$, $a = b$, $b < a$ can be valid at the same time; thus in (15) we can insert the exclusive "either." With $\leqslant$ as an abbreviation for "$<$ or $=$" it follows that $a \leqslant b$ is the negation of $b < a$. From $1 < a'$ we see, by complete induction[16] on a, that

(16) $$1 \leqslant a,$$

[15] Note that by numbers we here mean the natural numbers 1, 2, 3, ..., not including zero.

[16] In this case the induction hypothesis is not used at all, a fact which may make the proof somewhat harder to follow.

and thus $a < 1$ is impossible for any number a. Also

$$(17) \qquad a < b + 1 \quad \text{if and only if} \quad a \leqslant b.$$

For from $a < b$ or $a = b$ it follows by (13) that $a < b + 1$.

On the other hand, if $a + c = b + 1$, it follows from (16) that we need consider only the cases $c = 1$ and $c > 1$ (that is, $c = u + 1$ for a certain u). In the first case we have $a = b$ by (11), and in the second $(a + u)' = b'$ and thus $a + u = b$, or $a < b$.

From the principle of induction we can now derive the following *modified principle of induction*: *If the number m is contained in the number set M whenever $n \in M$ for all numbers $n < m$, then $M = N$.* The induction hypothesis now reads: "$n \in M_1$ for all numbers $n < m$"; and there is no special initial case. For the proof we consider the set M^* of numbers m with $n \in M$ for all $n < m$. Then the hypothesis of our new principle simply states that $M^* \subseteq M$. Since $n < 1$ is not valid for any number n, we get $1 \in M^*$. By (17), any number $n < m'$ is $<m$ or $=m$. If we now assume that $m \in M^*$, then n lies in M not only in the first case but also, since $M^* \in M$, in the second case as well. Thus we have derived $m' \in M^*$ from $m \in M^*$, so that by the argument from m to m' we have $M^* = N$ and thus also $M = N$.

With the new principle of induction, it is very easy to prove the *theorem of well-ordering of the natural numbers*: *every non-empty set of natural numbers contains a smallest number*. For the proof we reformulate the assertion thus: if the set of numbers M contains the number n, then M contains a smallest number. If this statement is assumed for all numbers $n < m$ and if $m \in M$, then M contains a smallest number provided it contains a number $<m$. But otherwise m is itself the smallest number in M. As another method of proving the same theorem, we note that, if we replace M by the set $N - M$ of the numbers not in M, our modified principle of induction can be transformed, by contraposition (see IA, §6.6) and other purely logical operations, into the desired theorem of well-ordering of the natural numbers.

The principle of induction can also be generalized to *complete induction starting from k*, as follows:

If the set M contains the number k and if $n' \in M$ for every number $n \geqslant k$ such that $n \in M$, then M contains all the natural numbers $\geqslant k$.

For the proof we may assume $k > 1$. We set $k = h + 1$ and consider the mapping $x \rightarrow x + h$, which maps the set N into the set of natural numbers $>h$ and thus $\geqslant k$. The inverse mapping [which exists on account of (11)] takes M into a set for which we may prove, by the ordinary principle of induction, that it contains the set N. Thus M does in fact contain all numbers $\geqslant k$. After we have introduced the integers (see the next section), this proof obviously holds for any arbitrary integer k.

1.5. *Segments*

The set of numbers $\leqslant n$ is called the *segment* A_n. By (12) and (13) we see that $A_m \subset A_n$ means the same as $m < n$. A set M (whose elements are not necessarily numbers) is said to be *finite* if it can be mapped one-to-one onto a segment A_n; that is, if there exists a one-to-one (invertible)[17] mapping f of M onto[18] A_n. The number n is then uniquely determined and may therefore be called the *number of elements* in M. In order to prove the uniqueness of n, we consider a one-to-one mapping f of M onto A_n and also a one-to-one mapping g of M onto A_m. By (15) there is no loss of generality in assuming $m \leqslant n$. If we carry out the inversion of f and then the mapping g, we obviously obtain a one-to-one mapping of A_n onto the subset A_m. Our assertion then follows from the theorem:

A one-to-one mapping f of A_n into itself[19] *is a mapping onto A_n.*

We prove this theorem by the argument from n to n'. For $n = 1$ the assertion is clear, since 1 is the only element of A_1. Now let f be a one-to-one mapping of $A_{n'}$ into itself. If $n' \neq f(x)$ for all $x \leqslant n$, then f induces to a one-to-one mapping of A_n into itself, so that by the induction hypothesis $f(A_n) = A_n$. But then we can only have $f(n') = n'$, and consequently $f(A_{n'}) = A_{n'}$. But if $n' = f(k)$ for a number $k \leqslant n$, then by setting

$$g(x) = \begin{cases} f(x) & \text{for } k \neq x \leqslant n \\ f(n') & \text{for } k = x \end{cases},$$

we define a mapping g of A_n into itself, since $f(n') \neq n' = f(k)$ follows from $n' \neq k$. The one-to-one character of g follows easily from that of f, so that by the induction hypothesis we have $g(A_n) = A_n$, and consequently $f(A_{n'}) = A_{n'}$.

If a set M with n elements is mapped onto a set M' with m elements, then $m \leqslant n$, as is easily shown by complete induction on n. But if $m < n$, the mapping cannot be one-to-one; for a one-to-one mapping of M onto M' followed by a one-to-one mapping of M' onto A_m would show that m is the number of elements of M. The application of this fact is often called the *Dirichlet pigeonhole principle*: the "pigeonholes" are the elements of M', into which the "objects" (namely the elements of M) are "inserted"

[17] "One-to-one" or "invertible" means: if $f(x) = f(x^*)$, then $x = x^*$. (Cf. IA, §8.4.)
[18] "Onto" means: for $y \in A_n$ there exists an $x \in M$ with $y = f(x)$. (Cf. IA, §8.4.)
[19] That is, the set $f(A_n)$ of the images $f(x)$ $(x \in A_n)$ is a subset of A_n. (Cf. IA, §8.4.)

by the mapping; if there are more objects than pigeonholes, then at least one pigeonhole must contain two different objects.

In general, two sets are said to be *equivalent* (cf. IA, §7.3) if either of them can be mapped one-to-one onto the other. Thus the finite sets are defined as those sets that are equivalent to the segments A_n. For convenience, the empty set $\emptyset$, which contains no element at all, is also said to be a finite set. In an infinite set M, namely a set which is not finite, it is easy to determine a subset which is equivalent to the set N of all natural numbers: for if f is a mapping which to each non-empty subset X of the set M assigns[20] an element of $f(X)$ of the subset X, the sets $M_1, M_2, \ldots$ can be defined recursively by $M_1 = \{f(M)\}$, $M_{n'} = M_n \cup \{f(M - M_n)\}$, and then the union of the M_n provides us with the desired subset N^*. Thus, since $x \to x + 1$ is a one-to-one mapping of N onto a proper subset of N, there also exists a one-to-one mapping g of $N^*(\subseteq M)$ onto a proper subset of N^*. If each element of $M - N^*$ is assigned to itself, the mapping g is thereby extended to a one-to-one mapping of M onto a proper subset of M. In view of the preceding theorem and the fact that $\emptyset$ has no proper subset, we have the result:[21]

A set M is finite if and only if there exists no proper subset of M equivalent to M.

As a counterpart to the above theorem on the mappings of A_n we prove:

From $f(A_n) = A_n$ it follows that f is one-to-one. For $x \leqslant n$ we determine the smallest number y with $f(y) = x$ and denote by g the mapping of A_n into itself thus defined, so that we have $y = g(x)$. From $g(x) = g(x^*)$ it follows that $x = f(g(x)) = f(g(x^*)) = x^*$, so that g is one-to-one and therefore $g(A_n) = A_n$ (by the first theorem in §1.5). Thus for y with $y^* \leqslant n$ we can always find an x with $x^* \leqslant n$ such that $y = g(x)$, $y^* = g(x^*)$. Then $f(y) = f(y^*)$ implies $x = f(y) = f(y^*) = x^*$ and thus $y = y^*$.

1.6. *Commutativity in Sums with More Than Two Terms*

By making use of segments, we can now prove the commutative law, stated above in §1.3, for sums with more than two terms: for every one-to-one mapping f of A_n onto itself we have

$$\sum_{i=1}^{n} a_{f(i)} = \sum_{i=1}^{n} a_i . \tag{18}$$

[20] The existence of such a mapping follows from the axiom of choice in the theory of sets (see IA, §7.6). Here and below, $M \cup M'$ denotes the union of the sets M, M' (that is, the set of elements which lie in M or M') and $\{a\}$ denotes the set consisting of the element a alone (cf. IA, §7.2).

[21] Taken by Dedekind as the definition of "finite set."

The proof by induction on n will only be indicated here; we confine ourselves to the case[22] $f(k') = n'$, $k + h = n$:

$$\sum_{i=1}^{n'} a_{f(i)} = \sum_{i=1}^{k'} a_{f(i)} + \sum_{i=1}^{h} a_{f(k'+i)} = \sum_{i=1}^{k} a_{f(i)} + a_{n'} + \sum_{i=1}^{h} a_{f(k'+i)}$$

$$= \sum_{i=1}^{k} a_{f(i)} + \sum_{i=1}^{h} a_{f(k'+i)} + a_{n'}$$

$$= \sum_{i=1}^{n} a_{g(i)} + a_{n'} = \sum_{i=1}^{n} a_i + a_{n'} = \sum_{i=1}^{n'} a_i ,$$

where the one-to-one mapping g of A_n onto itself is defined by

$$g(i) = f(i) \quad \text{for} \quad i \leqslant k, \qquad g(k + i) = f(k' + i) \quad \text{for} \quad i \leqslant h.$$

For any finite index set $I \neq \emptyset$, any sequence of numbers[23] $(a_i)_{i \in I}$ and any one-to-one mapping f of A_n onto I, it is easy to see from (18) that $\sum_{i=1}^{n} a_{f(i)}$ does not depend on f but only on the given sequence, so that we can write this sum in the shorter form $\sum_{i \in I} a_i$. From (9) we have in this notation[24]

$$(19) \quad \sum_{i \in I'} a_i + \sum_{i \in I''} a_i = \sum_{i \in I} a_i , \qquad \text{if} \quad I = I' \cup I'', \quad I' \cap I'' = \emptyset.$$

For the case $I' = \emptyset$ we define $\sum_{i \in I'} a_i$ as 0, where 0 is the neutral element of addition (as defined below in §2.3); then it is obvious that (19) still holds.

By complete induction on n we further obtain from (19)

$$(20) \quad \sum_{k=1}^{n} \sum_{i \in I_k} a_i = \sum_{i \in I} a_i , \quad \text{if} \quad I = \bigcup_{k=1}^{n} I_k \quad \text{and} \quad I_k \cap I_h = \emptyset \quad \text{for} \quad k \neq h.$$

[22] The cases $f(1) = n'$ and $f(n') = n'$ require only slight changes.

[23] That is, a mapping of I into N. The indices are not necessarily natural numbers; for example, we could also use pairs of numbers (i, k), in which case (provided there is no danger of misunderstanding) we may write a_{ik} instead of $a_{(i,k)}$, and correspondingly for triples or n-tuples.

In school one often introduces sequences without any mention of their connection with functions. But the concept of a function would be more clearly understood if infinite sequences were presented as mappings of N into N or into some set of numbers.

[24] $M \cap M'$ denotes the intersection of the sets M, M', namely, the set of elements that belong to M and to M'; for the definition of $\cup$ see footnote 20, page 103. (Cf. also IA, §7.2.)

If for I we take the set of pairs (i, k) with $i \leqslant m$, $k \leqslant n$, and for I_k we first take the set of (i, k) with $i \leqslant m$, and then the set of (k, i) with $i \leqslant n$, we have

$$(21) \qquad \sum_{k=1}^{n} \sum_{i=1}^{m} a_{ik} = \sum_{\substack{i \leqslant m \\ k \leqslant n}} a_{ik} = \sum_{k=1}^{m} \sum_{i=1}^{n} a_{ki} \,.$$

Setting $a_{ik} = 1$ gives us (in view of $m1 = m$, $n1 = n$) $nm = mn$, the commutative law of multiplication. Of course, this law could also be proved by complete induction (first on n with $m = 1$ and then on m) but its derivation from the general equation (21) is shorter and corresponds exactly to the usual intuitive argument for the commutative law of multiplication: namely, the nm summands 1 are arranged in m lines and n columns and then added line by line.

2. The Integers

2.1. *Properties Required in an Extension of the Concept of Number*

From now on the symbol a' will no longer be used, as in §1, to denote the successor of a, which will always be written in the form $a + 1$. If a, a' are arbitrary natural numbers, then in the case $a \leqslant a'$ there exists no natural number x with

$$(22) \qquad x + a' = a.$$

But now we wish to proceed to a domain of numbers in which an equation (22) always has a solution.[25] Let us assume that we have already succeeded in finding an extension of the domain of natural numbers in which addition is defined in such a way as to satisfy the laws (6), (7), (11) and, when applied to the natural numbers, to agree with the addition already defined for them. Of course, it will be necessary to prove later that this assumption is justified. But first let us reflect a little on the properties that such an extended domain must have, since in this way we will obtain valuable hints for the construction of the domain.

Whenever we extend a domain of numbers, here and in similar situations below, we shall always require that certain rules of calculation remain valid, a requirement called the *principle of permanence*.[26] But this label should not mislead us into thinking that the principle of permanence justifies once and for all the assumption that such extensions exist. Moreover, it fails to tell us *which* rules of calculation are to be "preserved."

[25] In school it is usual to begin with the requirement that the equation $xa' = a$ has a solution. For this procedure see the end of §3.2.

[26] Often associated with the name of H. Hankel.

It merely confirms the fact of experience that, in making the extensions of the concept of number which from time to time have become necessary, mathematicians have found it convenient to preserve the most important rules of calculation; whether this is possible, and to what extent, must be investigated in each special case. The principle of permanence gives us only a very weak indication of how we ought to proceed, and it by no means deserves the key position it has often been given, without any logical justification.

If we denote by[27] $a - a'$ the solution of (22) in the extended domain, then

$$(a, a') \to a - a' \tag{23}$$

is a mapping of the set of all pairs of natural numbers onto the extended domain, and we must ask: when do two pairs (a, a'), (b, b') have the same image in this mapping? From the equations $(a - a') + a' = a$ and $(b - b') + b' = b$, which characterize $a - a'$ and $b - b'$, it follows that

$$(b - b') + b' + a = b + (a - a') + a',$$

so that

$$a + b' = a' + b \tag{24}$$

means the same as $b - b' = a - a'$. By (22) we must also set

$$c = (c + a') - a'. \tag{25}$$

Finally, we have

$$a' + b' + (a - a') + (b - b') = a + b,$$

and therefore

$$(a - a') + (b - b') = (a + b) - (a' + b'). \tag{26}$$

If now in the set of pairs we define addition by the (clearly associative and commutative) rule

$$(a, a') + (b, b') = (a + b, a' + b')$$

and denote the mapping (23) by f, we can write (26) in the form

$$f(A) + f(B) = f(A + B), \tag{27}$$

where we have used capital letters for the pairs.

[27] This symbol is still completely at our disposal, since we have not used it before.

A mapping f with the property (27) (for all elements A, B of the set of preimages, consisting here of all pairs of natural numbers) is called a *homomorphism* (with respect to addition).

Concerning the homomorphism (23), we have the following fact, which can be expressed in terms of the natural numbers alone: the pairs $A = (a, a')$, $B = (b, b')$ have the same image under f if and only if (24) holds. For (24) we also write $A \doteq B$, since we shall see below that the relation $\doteq$ defined in this way has many properties in common with equality.

2.2 *Construction of the Extended Domain*

After these preliminary remarks we can see that the first part of our task consists of constructing, together with its set of images, a homomorphism f of the set of pairs of natural numbers in such a way that pairs A, B have the same image if and only if $A \doteq B$, with $\doteq$ defined as above. The image of the pair $A = (a, a')$ is created by simply setting between the numbers a, a' a horizontal stroke: $a - a'$. For the moment this stroke, which is now introduced for the first time, does not have the meaning of a minus sign, although it will naturally acquire that meaning later, when we have made the necessary definitions: at present $a - a'$ is nothing but a symbol formed from the two numbers a, a'.[28] But now the real work begins, since the symbols $a - a'$ are completely useless until we have introduced for them the concepts of equality and addition.

These concepts must be introduced in such a way that every statement about $a - a'$, $b - b'$, ..., formed with the symbols $=$ and $+$, is an abbreviation for a statement about the natural numbers a, a', b, b', ..., and in view of the fact that we shall be interested only in statements that could finally be reduced to $=$ and $+$, the new symbols are in principle superfluous, since they could be eliminated from every statement. But they provide us with a much more convenient notation, so that their use is to be recommended on practical grounds.[29] As statements about the new symbols $a - a'$, $b - b'$, ... we shall admit only statements about the natural numbers a, a', b, b', ... that remain unchanged in truth-value (cf. the corresponding remarks in §3.1) when the $a - a'$, $b - b'$, ... are

[28] It makes no difference here whether we regard the natural numbers as complicated logical expressions (sets of equivalent sets) or simply as symbols like ||| (see §1.1).

[29] Of course, we could dispense with these new symbols altogether and work merely with the pairs (a, a'), which would then be called integers and for which we would introduce $\doteq$ as the new relation of equality. But then there is the difficulty that we would like to use the ordinary symbol of equality, introduced before for the natural numbers, for the integers also; for pairs this symbol has already been used in a different sense (see IA, §7.2). The use of classes of pairs of numbers instead of the symbols $a - a'$ is discussed on p. 109.

replaced by any other symbols equal to them (in the sense of equality defined immediately below).

It is to be noted that for the following developments we require only the properties (6), (7), (11) of the natural numbers and their addition. This fact will become important in §3.1, where the procedure described here is applied to multiplication instead of addition (with a/a' instead of $a - a'$). Since in that section we shall be introducing only pairs (a, a') with $a' \neq 0$, we make the further remark here that in what follows (as can easily be seen from the proofs) the rule (11) is needed only for natural numbers c restricted to a proper subset C, provided that the pairs (a, a') are restricted to $a' \in C$ and for $a', b' \in C$ we have $a' + b' \in C$; the only exception is the proof of the existence of the inverse element at the beginning of §2.3, where we must also require $a \in C$.

It is now clear how equality is to be defined, in view of the requirement that if f is the mapping (23), then $f(A) = f(B)$ must mean the same as $A \doteq B$. This requirement is met if we stipulate that $a - a' = b - b'$ means the same as (24). But now, if we wish to calculate with the new concept of equality in the same way as with equality for natural numbers, we must show that the two fundamental rules for equality are satisfied: namely, every expression must be equal to itself (reflexivity); and if each of two expressions is equal to a third, they must be equal to each other (comparativity). For the relation $\doteq$, to which we have reduced our definition of equality, these fundamental rules mean that

$$(28) \qquad A \doteq A;$$

$$(29) \qquad \text{if} \quad A \doteq C \quad \text{and} \quad B \doteq C, \quad \text{then} \quad A \doteq B.$$

But (28) follows immediately from the fact that (24) holds for $a = b$, $a' = b'$. As for the proof of (29), we see that by adding b' to $a + c' = a' + c$ and a' to $b' + c = b + c'$ we obtain the equation $a + b' + c' = a' + b + c'$, so that (24) now follows from (11). Only after (28) has been proved is it clear that (23) is actually a mapping: from $(a, a') = (b, b')$ it follows that $a - a' = b - b'$.

We further note that from the definition of equality we have $(a + d) - (a' + d) = a - a'$.

A relation $\doteq$ which satisfies (28) and (29) is called an *equivalence relation* (see also IA, §8.5). Such a relation is necessarily *symmetric* and *transitive*; for if $C = A$ we see from (29) and (28) that $B \doteq A$ implies $A \doteq B$; and if on the basis of this symmetry we replace $B \doteq C$ by $C \doteq B$ in (29), we obtain the desired transitivity. From $A \doteq B$, $C \doteq A$, and $B \doteq D$ we obtain $C \doteq D$ by a twofold application of transitivity. By symmetry and by the definition of equality this result can also be expressed in the form: A statement $a - a' = b - b'$ is not changed in truth value if $a - a'$ and $b - b'$ are replaced by their equals $c - c'$ and $d - d'$.

If to every A we assign the set $\bar{A}$ of all X with $X \doteq A$, then $A = B$ means the same as $\bar{A} \doteq \bar{B}$; for by (28) we have $A \in \bar{A}$, so that $\bar{A} = \bar{B}$ immediately implies $A \doteq B$; and conversely, if $A \doteq B$, then $X \in \bar{A}$ implies $X \in \bar{B}$ by transitivity and $X \in \bar{B}$ implies $X \in \bar{A}$ by (29). Thus, instead of the $a - a'$ we could simply use the $\bar{A}$, which are called *residue*[30] *classes* with respect to the equivalence relation. A special definition of equality is no longer required, since we have already given a general definition for equality of sets (see IA, §7.2).

Under certain systems of logic the latter possibility appears to be of essential importance, but this fact does not permit us to conclude that the integers *must* be considered as sets of pairs. In fact, our construction of the symbols $a - a'$ corresponds more closely to the way in which the integers are actually used in daily life; when faced with an expression like $2 - 3$, we seldom think of the set of all pairs (x, x') such that $(x, x') \doteq (2, 3)$. Moreover, such a set of pairs of natural numbers is in no sense "more real" than our symbols: for this set of pairs can only be defined by the propositional form $(x, x') \doteq (2, 3)$, where the variables x, x' are quantified by some prefixed symbol (see IA, §7.7); in other words the set of pairs can only be defined by a symbol that is considerably more complicated than $2 - 3$. The question "What is an integer?" has no absolute significance; it can be meaningfully asked only in the framework of a given system for the foundations of mathematics. The unconditionally meaningful question is: "How do we obtain mathematical objects that behave in such and such a way?" And to this question it is possible to give the most varied answers.

In the domain of our new symbols $a - a'$, which henceforth we shall call *integers*, it is now our task to introduce addition in such a way that (27) is valid. At first glance (26) seems to constitute such a definition:

$$(30) \qquad (a - a') + (b - b') = (a + b) - (a' + b').$$

But the sum must actually depend only on the summands, whereas here it seems to depend on a, a', b, b'; in other words: equals added to equals must give equals, or expressed still otherwise: the truth value of (30) must not be altered if the numbers occurring there are replaced by other numbers equal to them. Thus we must prove that $A \doteq C$, $B \doteq D$ always implies $A + B \doteq C + D$;[31] this condition, which alone makes the definition (30) useful, is called *consistency* of $\doteq$ with addition. For the proof it will be sufficient, on account of commutativity, to prove the simpler condition

$$(31) \qquad A + B \doteq C + B, \qquad \text{if} \quad A \doteq C;$$

for then from $B \doteq D$ we will have $C + B \doteq C + D$, and therefore, by transitivity, $A + B \doteq C + D$. But now addition of $b + b'$ to

[30] The name comes from the use of this concept in the theory of divisibility (IB5, §3.6 and IB6, §4.1). For general information on equivalence relations and their sets of residue classes, see IB10, §1.5.

[31] Cf. IA, §8.5, and particularly IA (8.9).

$a + c' = a' + c$ gives $(a + b) + (c' + b') = (a' + b') + (c + b)$, which proves (31). The addition of integers defined by (30) is obviously commutative and associative.

Now we must see to it that certain integers are equal to natural numbers. To do this we define equality between integers and natural numbers, in accordance with (25), by setting:[32]

$$(32) \qquad (c + a') - a' = c, \qquad c = (c + a') - a',$$

with the stipulation that these equations, and only these, are to hold between natural numbers and integers. We must now verify that the two fundamental laws for equality are still valid. Reflexivity remains unaffected by (32), but for the comparativity "if $\alpha = \gamma$ and $\beta = \gamma$, then $\alpha = \beta$," we must distinguish the various cases arising from the fact that each of the letters α, β, γ may represent either a natural number or an integer. Of the eight possible cases we no longer need to examine those in which α, β, γ are of the same kind. In view of the symmetry (32) of equality, we can also strike out those cases that arise from others if α is replaced by β. Thus the following four cases remain:

1. β, γ are the integers $b - b'$, $c - c'$, and α is the natural number a. The assumption $\alpha = \gamma$ can be satisfied only on the basis of (32) and thus implies $c = a + c'$. Consequently, $\beta = \gamma$ implies $b + c' = b' + a + c'$ and also, by (11), $b = b' + a$, which finally, by (32), gives $a = b - b'$, and therefore $\alpha = \beta$.

2. α, β are the integers $a - a'$, $b - b'$, and γ is the natural number c. Then by (32), $\alpha = \gamma$, $\beta = \gamma$ imply $a = c + a'$, $b = c + b'$, from which follows $a + b' = a' + b$, and therefore $\alpha = \beta$.

3. α, β are the natural numbers a, b, and γ is the integer c. Then $\alpha = \gamma$, $\beta = \gamma$ imply by (32) that $c = a + c'$, $c = b + c'$, from which we see that $a + c' = b + c'$, so that finally, by (11), we have $a = b$ and therefore $\alpha = \beta$.

4. α, γ are the natural numbers a, c, and β is the integer $b - b'$. Then by (32), $\beta = \gamma$ implies $b = c + b'$ and therefore, since $\alpha = \gamma$, we have $b = a + b'$, which by (32) gives $a = b - b'$ and therefore $\alpha = \beta$.

Thus we may in fact consider the domain of the integers as an extension of the domain of the natural numbers, since every natural number is actually equal to an integer. But now a new difficulty arises: the sum of two natural numbers a, b can be determined in two different ways, namely, first as $a + b$ and secondly as the integer $((a + c) - c) + ((b + d) - d)$.

[32] In (32) it is necessary to adjoin the second equation in order that equality may be symmetric.

But by (30) the latter number is equal to $(a + b + c + d) - (c + d)$, and therefore again to $a + b$. Thus everything is in order.

In the domain of the integers it is now true that every equation (22) has a solution, namely the integer $a - a'$:

$$(a - a') + a' = (a - a') + ((a' + c) - c) = (a + a' + c) - (a' + c) = a.$$

In §2.3 we shall see that addition of integers actually has the property (11).

2.3. *The Module of the Integers*

From the definition of equality (24) for the integers it follows at once that all integers of the form $a - a$ are equal. If for $1 - 1$ we introduce the abbreviation 0, then

$$a - a = 0, \tag{33}$$

and therefore $0 + a = a$ for every natural number a. But this last result also holds for every integer $a - a'$; for by (30) we have

$$(a - a') + 0 = (a - a') + (1 - 1) = (a + 1) - (a' + 1) = a - a'.$$

Thus we have introduced the number *zero* and have established its most important property. From (30), (33) we also have

$$(a - a') + (a' - a) = (a + a') - (a' + a) = 0.$$

Now a given set, together with an operation defined in it (see IB10, §1.2.2), is called a *module*[33] if (the operation being denoted by +) the following conditions are satisfied:

1. The operation is associative and commutative; i.e., we have the equations $(\alpha + \beta) + \gamma = \alpha + (\beta + \gamma)$ and $\alpha + \beta = \beta + \alpha$ for all elements α, β, γ of the set.

2. There exists in the set a *neutral element* for the operation, namely an element 0 with $\alpha + 0 = \alpha$ for every element α of the set.

3. For every element α of the set there exists an *inverse element*,[34] i.e., an element $-\alpha$ of the set with $\alpha + (-\alpha) = 0$.

The set is said to be a module *with respect to the operation* (which is here written as addition).

[33] Or also a *commutative* (or *Abelian*) *group*; (cf. IB2, §1.1).

[34] Or also simply an *inverse*; the name comes from the fact that the effect of adding $-\alpha$ reverses that of adding α: $(\beta + \alpha) + (-\alpha) = \beta + (\alpha + (-\alpha)) = \beta$. The connection between the notation $-\alpha$ and the use of the "minus" stroke in $a - a'$ will be explained later.

Consequently, the integers form a module under the operation of addition, namely, the *module of integers*. The following remarks are valid for every module;[35] here we may, of course, keep in mind the module of integers that has just been constructed, but we must be careful not to make use of any of its properties that are not common to all modules, since we wish to apply our results to other modules.

For given elements α, β in a module there exists exactly one element ξ with $\xi + \alpha = \beta$.

$$\xi = \xi + 0 = \xi + (\alpha + (-\alpha)) = (\xi + \alpha) + (-\alpha) = \beta + (-\alpha),$$

so that at most one element ξ, namely $\beta + (-\alpha)$, can satisfy the equation; on the other hand, for $\xi = \beta + (-\alpha)$ we have

$$\xi + \alpha = (\beta + (-\alpha)) + \alpha = \beta + ((-\alpha) + \alpha) = \beta + 0 = \beta.$$

Thus rule (11) holds for every module: from $\alpha + \gamma = \beta + \gamma$ it follows that $\alpha = \beta$. For the unique solution $\xi = \beta + (-\alpha)$ of the equation $\xi + \alpha = \beta$ we write $\beta - \alpha$, which agrees with the notation for the integers, since $a - a'$ is the solution of equation (22). Conversely, this abbreviation can be used to define the inverse: $-\alpha = 0 - \alpha\ (= 0 + (-\alpha))$. Thus the minus sign is used in two closely related senses: first in $\beta - \alpha$ as a connective, i.e., as a notation for a function of two arguments; and second in $-\alpha$ for a function of one argument, namely the function which to each element assigns its inverse.

From the uniqueness of the solution of $\xi + \alpha = \beta$ it also follows that $-\alpha$ is already completely determined by α and that the equation $\alpha + (-\alpha) = 0$ [as well as $(-\alpha) + \alpha = 0$] is established. Thus we have at once

$$-(-\alpha) = \alpha \tag{34}$$

and also, since $(\alpha + \beta) + ((-\beta) + (-\alpha)) = \alpha + \beta + (-\beta) + (-\alpha) = \alpha + 0 + (-\alpha) = \alpha + (-\alpha) = 0$, we may write[36]

$$-(\alpha + \beta) = (-\beta) + (-\alpha). \tag{35}$$

Setting $-\beta$ for β we have, by (34),

$$-(\alpha - \beta) = \beta - \alpha. \tag{36}$$

Let us now return to the integers. If $a' < a$, there exists a natural number b with $a = a' + b$, so that by (32) we have $a - a' = b$ and as

[35] In fact, for every group (cf. IB2, §2).

[36] The order of the summands on the right-hand side is so chosen that this result is obviously valid without the assumption of commutativity.

a result we may simply say: $a - a'$ *is* a natural number. If $a' = a$, then $a - a' = 0$, and in the only remaining case, namely $a < a'$ [see (15)], it follows from (36) that $a - a'$ is the inverse $-b$ of a natural number b. For natural numbers b, b' we have $b \neq 0$, $-b \neq 0$ (since $b + 0 \neq 0$), and $b \neq -b'$ (since $b + b' \neq 0$). Consequently, there exist exactly three kinds of integers: the natural numbers, zero, and the inverses of the natural numbers. The latter are called *negative integers* (or *numbers of negative sign*[37]), and then, in contrast, the natural numbers are called *positive integers.*

In view of the above remarks, it would also be possible to extend the set of natural numbers to the module of integers in the following way: For every natural number n we introduce a new symbol $-n$, and also the new symbol 0, for which $-n = -m$ is defined as $n = m$ and there are no other equalities except $0 = 0$. Addition is then defined as follows:

$$0 + 0 = 0, \qquad 0 + m = m + 0 = m, \qquad 0 + (-n) = (-n) + 0 = -n,$$

$$(-m) + (-n) = -(m + n),$$

$$m + (-n) = (-n) + m = k, \qquad \text{with} \qquad n + k = m \qquad \text{for} \qquad n < m,$$

$$m + (-n) = (-n) + m = -k, \qquad \text{with} \qquad m + k = n \qquad \text{for} \qquad m < n,$$

$$m + (-m) = (-m) + m = 0.$$

This procedure is conceptually much simpler but has two serious disadvantages: proofs of the rules for calculation must be divided up into many special cases and thus become much lengthier, and addition must be required to satisfy not only (6), (7), (11) but also (12), (15), which, in contrast to (13), (14), cannot be deduced[38] from (6), (7), (11) alone.

2.4. *Multiplication*

In order to define multiplication, we adopt a plan which may at first sight seem like a detour but has essential advantages over other methods.[39] Our task is to define multiplication of integers in such a way that it satisfies the distributive law and, in the subdomain of the natural numbers, agrees with multiplication as already defined. The distributive law $a(x + y) = ax + ay$ will be regarded as a property of multiplication by a; that is, as a property of the mapping $x \to ax$. So let us first examine mappings with this property [see (38)]. We begin with a discussion of multiplication of the natural numbers from this point of view.

[37] But this terminology readily gives rise to the common error that $-a$ (for an arbitrary integer a) is always a negative integer.

[38] Thus we cannot use this procedure in §3.1.

[39] Two other possible procedures are described on p. 119 (in small print).

By the distributive law (10) and the commutative law as proved in §1.6, the mapping f of the set of natural numbers into itself defined by

$$f(x) = ax \tag{37}$$

has the property

$$f(x + y) = f(x) + f(y) \tag{38}$$

and is therefore a homomorphism with respect to addition. If f and g are two such homomorphisms, it follows from $f(1) = g(1)$ that $f(x) = g(x)$ holds for all natural numbers[40] x, and thus $f = g$; for $f(x) = g(x)$ implies, since (38) holds for g in place of f, that $f(x + 1) = f(x) + f(1) = g(x) + g(1) = g(x + 1)$. Thus

$$f \to f(1) \tag{39}$$

is a one-to-one mapping of the set of homomorphisms in question onto the set of natural numbers, so that in particular each of these homomorphisms has the form (37):

$$f(x) = f(1)\,x. \tag{40}$$

Thus we have obtained a description of multiplication which is very suitable for extension to the domain of integers. In what follows, lowercase italic letters will refer to arbitrary integers or, when the argument is applicable to modules in general, to arbitrary elements of a given module.

Homomorphisms (with respect to addition) of a module M into itself are called endomorphisms of the module. Since we wish to use these endomorphisms, as suggested by (40), in defining multiplication for the module of integers, let us first examine in a general way the set of endomorphisms of a module.

If f and g are endomorphisms of the module M, the mapping $x \to f(x) + g(x)$ is also an endomorphism: for

$$\begin{aligned} f(x + y) + g(x + y) &= f(x) + f(y) + g(x) + g(y) \\ &= (f(x) + g(x)) + (f(y) + f(y)). \end{aligned}$$

This mapping is called the *sum* $f + g$ of the endomorphisms. We now show that with this definition of addition the endomorphisms themselves form a module. To begin with, associativity and commutativity are clear at once. The neutral element is the endomorphism $x \to 0$, which we denote by $\underline{0}(f + \underline{0})(x) = f(x) + \underline{0}(x) = f(x) + 0 = f(x)$.[41] When there

[40] No use is made here of the fact that $f(x)$ and $g(x)$ are natural numbers.

[41] Of course, an expression like $(f + g)(x)$ does not denote any sort of product but rather the value of the function $f + g$ for the argument x.

is no danger of confusion with the number 0, we may write 0 instead of $\underline{0}$. Finally, if f is an endomorphism, then by (35) the mapping $x \to -f(x)$ is also an endomorphism, which is denoted by $-f$ and is the inverse of f: $(f + (-f))(x) = f(x) + (-f)(x) = f(x) + (-f(x)) = 0 = \underline{0}(x)$.[41]

But now, given two endomorphisms, it is possible to define not only their addition but also another operation on them, namely, successive application: the mapping denoted[42] by $f \circ g$ and defined by

$$(f \circ g)(x) = f(g(x)) \qquad \text{for all} \quad x \in M$$

is again an endomorphism for given endomorphisms f and g; for we have

$$\begin{aligned}(f \circ g)(x + y) &= f(g(x + y)) = f(g(x) + g(y)) \\ &= f(g(x)) + f(g(y)) = (f \circ g)(x) \\ &\quad + (f \circ g)(x) + (f \circ g)(y).\end{aligned}$$

The operation $\circ$ will be called *multiplication.* Like every case of successive application of two mappings (cf. IB2, §1.2.5) this multiplication is associative:

$$\begin{aligned}((f \circ g) \circ h)(x) &= (f \circ g)(h(x)) = f[g(h(x))], \\ (f \circ (g \circ h))(x) &= f((g \circ h)(x)) = f[g(h(x))]\end{aligned}$$

for all x and thus

$$(f \circ g) \circ h = f \circ (g \circ h).$$

But it also satisfies the two[43] *distributive laws* with respect to addition

$$f \circ (g + h) = f \circ g + f \circ h, \qquad (f + g) \circ h = f \circ h + g \circ h;$$

since

$$\begin{aligned}(f \circ (g + h))(x) &= f((g + h)(x)) = f(g(x) + h(x)) \\ &= (f \circ g)(x) + (f \circ h)(x) = (f \circ g + f \circ h)(x), \\ ((f + g) \circ h)(x) &= (f + g)(h(x)) = (f \circ h)(x) + (g \circ h)(x) \\ &= (f \circ h + g \circ h)(x).\end{aligned}$$

[42] The symbol $f \circ g$ may be read as "f after g" or also "f times g." The small circle is often omitted but we will retain it here in order to emphasize the difference between this operation and the operation defined by $(fg)(x) = f(x)\,g(x)$ in case a multiplication has been defined in the set of images. If, as is often done, we write the mapping to the right of the object to be mapped, namely, xf or x^f instead of $f(x)$, then the definition of successive application is changed, to the effect that $f \circ g$ denotes the application first of f and then of g [see IB2, §1.2.5 (4)].

[43] Since multiplication of endomorphisms of an arbitrary module is not necessarily commutative (the noncommutative linear mappings in IB3, §2.2 are endomorphisms), these two laws must here be proved separately.

By a *ring* we mean a set of elements with a pair of operations that have the following properties:

1. With respect to the first operation, called addition, the ring is a module.

2. The second operation, called multiplication, obeys the associative law, and also the two distributive laws with respect to addition.

Thus the above results on the endomorphisms of a module can be summarized in words: *the set of endomorphisms of a module forms a ring with respect to addition and multiplication.* This ring is called the *ring of endomorphisms* of the module. It has a neutral element for multiplication, namely the identity *mapping* $x \to x$, which we shall denote by I:[44]

$$(I \circ f)(x) = I(f(x)) = f(x), \qquad (f \circ I)(x) = f(I(x)) = f(x).$$

A neutral element for multiplication in a ring (there exists at most one such element; see IB5, §1.6) is called the *unit element* of the ring.

Let us now return to the module of integers. The mapping (39) is seen to be a one-to-one mapping of the set of endomorphisms onto this module. For, as was proved at the beginning of this section, the equation $f(x) = g(x)$ follows from $f(1) = g(1)$ for all natural numbers x. But from (38) we have $f(0) = f(0) + f(0)$, and thus $f(0) = 0$ and further $f(x) + f(-x) = f(0) = 0$; consequently, $f(-x) = -f(x)$ and also $g(0) = 0$, $g(-x) = -g(x)$, so that $f(x) = g(x)$ holds for all integers. Thus the endomorphism f is already completely determined by the value of $f(1)$: the mapping (39) is one-to-one and can therefore be inverted.

The set of images in (39) or, in other words, the set of numbers $f(1)$ for all endomorphisms f, includes the number 1 (as the image of 1 in I). If it includes $f(1)$, it also includes $f(1) + 1 = (f + I)(1)$; therefore it includes all the natural numbers. In fact, $0 = \underline{0}(1)$ and $-f(1) = (-f)(1)$; this set includes all the integers. Consequently, (39) is in fact a mapping onto the module of integers.

Thus we may use (40) to define multiplication for all integers in such a way that for the domain of natural numbers it agrees with the multiplication already defined at the beginning of this section: *the product ax is defined as the image of x under that endomorphism which takes* 1 *into a.* The rules for calculating with multiplication can now be obtained very simply from the properties of the ring of endomorphisms; since (39) is a homomorphism with respect to addition and multiplication, $f + g$ becomes $(f + g)(1) = f(1) + g(1)$ and $f \circ g$ becomes $(f \circ g)(1) = f(1)\,g(1)$.

[44] If there is no danger of confusion with the number 1, we may also write 1 instead of I.

Thus the associative law and the two distributive laws[45] carry over to the integers:

$$\begin{aligned} f(1)(g(1)\,h(1)) &= f(1)((g \circ h)(1)) = (f \circ (g \circ h))(1) = ((f \circ g) \circ h)(1) \\ &= ((f \circ g)(1))\,h(1) = (f(1)\,g(1))\,h(1), \end{aligned}$$

$$\begin{aligned} (f(1) + g(1))\,h(1) &= ((f + g)(1))\,h(1) = ((f + g) \circ h)(1) \\ &= (f \circ h + g \circ h)(1) \\ &= (f \circ h)(1) + (g \circ h)(1) = f(1)\,h(1) + g(1)\,h(1). \end{aligned}$$

Thus the integers form a ring with respect to addition and multiplication. Since by (39) the endomorphism I corresponds to the number 1, this number 1 is the unit element of the ring of integers:

$$1f(1) = I(1)\,f(1) = (I \circ f)(1) = f(1).$$

$$f(1)1 = f(1)\,I(1) = (f \circ I)(1) = f(1).$$

By an *isomorphism* we mean a one-to-one homomorphism with respect to the operations in question (for a ring, addition and multiplication). Since (39) was shown to be one-to-one, we have the theorem: *the mapping* (39) *is an isomorphism of the ring of endomorphisms of the module of integers onto the ring of integers.*

In order to prove the commutativity of multiplication, we must note that by the second distributive law $x \to xa$ is an endomorphism: for in fact, the image $(x + y)a = xa + ya$ of $x + y$ is the sum of the images of x and y. Application of (40) to this endomorphism gives $xa = (1a)x = ax$, so that the ring of integers is a *commutative ring*.

In an arbitrary ring (for which we denote multiplication in the same way as for the numbers) complete induction on n enables us to generalize the distributive laws to

$$\left(\sum_{i=1}^{n} a_i\right) b = \sum_{i=1}^{n} a_i b, \qquad a \sum_{i=1}^{n} b_i = \sum_{i=1}^{n} a b_i\,,$$

from which we have

$$\sum_{i=1}^{m} a_i \sum_{k=1}^{n} b_k = \sum_{i=1}^{m} \left(a_i \sum_{k=1}^{n} b_k\right) = \sum_{i=1}^{m} \sum_{k=1}^{n} a_i b_k\,,$$

[45] Because of the commutative law (to be proved later) only the second of the two distributive laws needs to be proved for the integers. In any case, the first distributive law is a simple consequence of (30) and (40).

and thus by (21)

$$\sum_{i=1}^{m} a_i \sum_{k=1}^{n} b_k = \sum_{\substack{1 \leqslant i \leqslant m \\ 1 \leqslant k \leqslant n}} a_i b_k . \tag{41}$$

By complete induction we readily obtain[46] from (41)

$$\prod_{i=1}^{m} \sum_{k=1}^{n_i} a_{ik} = \sum_{\substack{1 \leqslant k_i \leqslant n_i \\ (i=1,\dots,m)}} \prod_{i=1}^{m} a_{ik_i} ; \tag{41'}$$

where the index set for the summation on the right-hand side is the set of m-tuples $(k_1, ..., k_m)$, with $1 \leqslant k_i \leqslant n_i$ $(i = 1, ..., m)$.

From the distributive law we further have

$$a(c - d) + ad = a((c - d) + d) = ac,$$

$$(a - b)c + bc = ((a - b) + b)c = ac$$

and thus,

$$a(c - d) = ac - ad, \tag{42}$$

$$(a - b)c = ac - bc. \tag{42'}$$

Replacing a by $a - b$ in (42), we see from (42′) that

$$(a - b)(c - d) = (a - b)c - (a - b)d = ac - bc + (-(ad - bd)),$$

and therefore by (36) and (35)

$$(a - b)(c - d) = ac - bc + bd - ad = (ac + bd) - (ad + bc). \tag{43}$$

Setting $c = d$ and $a = b$ in (42) and (42′) respectively, we obtain $a0 = 0 = 0c$, so that (42), (42′), (43) give

$$a(-d) = -ad, \qquad (-b)c = -bc, \qquad (-b)(-d) = bd. \tag{43'}$$

In particular, we have $-a = (-1)a = a(-1)$ if the ring has a unit element, which for simplicity we have here denoted by 1.

Since the product of two natural numbers is always a natural number, the equations (43′), when applied to the ring of integers, show that the product of a positive with a negative number is a negative number and that the product of two negative numbers is a positive number. But an

[46] The definition of a product of several factors is given at the beginning of §3.3.

integer $\neq 0$ is always either positive or negative; thus $ab \neq 0$, if $a, b \neq 0$; or in other words:

$$\textit{If } ab = 0, \qquad \textit{then} \quad a = 0 \qquad \textit{or} \quad b = 0.$$

A ring with this property is said to have *no divisors of zero*. Thus the ring of integers is recognized as a *commutative ring with unit element that has no divisors of zero.*[47]

Of course, we could also define multiplication of integers by setting, from (43′),

$$(-n)m = m(-n) = -mn, \qquad (-m)(-n) = mn,$$

$$m0 = 0m = 0, \qquad (-n)0 = 0(-n) = 0, \qquad 00 = 0,$$

for the natural numbers m, n. But then the proof of the rules for calculation involves many special cases.

Multiplication for the integers (in the form in which they have been introduced here) could also be defined by setting, from (43),

$$(a, a')(b, b') = (ab + a'b', ab' + a'b)$$

as a multiplication for the pairs of natural numbers and then transferring this multiplication to the integers by means of the mapping $(a, a') \rightarrow a - a'$. Of course, it would then be necessary to show that multiplication of pairs is consistent with the equivalence relation $\doteq$ (of p. 107), but at least we would escape the disadvantage of having many special cases.

In comparison with these two possibilities, the procedure adopted above has the advantage of being independent of the sequence in which the integers are introduced, and secondly of not assuming that multiplication of the natural numbers has already been defined; it is true that we used this multiplication to give us a hint (40) on how to proceed, but the subsequent proofs were independent of it.[48] The endomorphisms by means of which we have introduced multiplication for the integers will be useful to us again in §4.6 for the multiplication of real numbers. Finally, let us remark that from a general point of view the concept of the ring of endomorphisms of a module is of great importance in algebra.

2.5. *Order*

For the time being we denote the set N of natural numbers by P. Then by what has been proved before we have

$$(44_1) \qquad 0 \notin P;$$

[47] A commutative ring without divisors of zero is also called an *integral domain*; cf. also IB5, §1.9.

[48] Except for the fact that the product of two natural numbers is again a natural number; but this fact could easily have been proved by complete induction on the basis of the definition (40) (see p. 116).

(44₂) *if* $0 \neq a \notin P$, *then* $-a \in P$;

(44₃) *if* $a, b \in P$, *then* $a + b \in P$;

(44₄) *if* $a, b \in P$, *then* $ab \in P$.

In general, in a module (with the operation $+$) a subset P with the properties (44_{1-3}) is called a *domain of positivity*; in a ring a domain of positivity must satisfy the additional condition (44_4). From ($44_{1,3}$) it follows at once that $a, -a \in P$ is impossible.

The *ring of integers* has the set N of natural numbers as its only domain of positivity. For $-1 \in P$ would imply from (44_4) that $1 = (-1)(-1) \in P$, which is inconsistent with $-1 \in P$; thus, by (44_2) we have $1 = -(-1) \in P$. Then complete induction shows at once from (44_3) that $N \subseteq P$. Now if there were an $a \in P$ with $a \notin N$, then [since $a \neq 0$ by (44_1)] we would have $-a \in N$, and thus $-a \in P$, which is again impossible; thus we have proved that $N = P$. On the other hand, the *module of integers* has exactly one other domain of positivity; for by what we have just proved, the domain of positivity of the module must be equal to N if it includes 1, and otherwise it must include -1 and therefore all the negative numbers, from which we conclude as before that it must coincide with the set of negative numbers.

The existence of a domain of positivity P enables us to define an ordering in a module: for we may set $a < b$ (or equivalently, $b > a$) if and only if $b - a \in P$. For the module of integers with $P = N$ this order obviously agrees with the order defined for the natural numbers in §1.4. Thus a module or ring with a domain of positivity is called an *ordered module* or *ring*. In an ordered module we can again prove (12)–(15): for (12) follows from (44_1) since $a - a = 0$; and (13) from (44_3) since $c - a = (c - b) + (b - a)$; also (15) follows from (44_2) because of (36); and finally, (14) from $(b + d) - (a + d) = b + d - d - a = b - a$. Since $a < b$ means the same as $b - a \in P$ and $b - a = (-a) - (-b)$, it follows from $a < b$ that $-b < -a$; in other words, in an ordered module the mapping $x \to -x$ [which by (35) is an endomorphism) is monotone decreasing. In an ordered ring we also have, by (42′) and (44_4), the *monotonic law for multiplication*:[49]

(45) $ac < bc$, if $a < b$ and $0 < c$.

Conversely, a module or a ring in which a relation $<$ is defined becomes an ordered module or ring if the conditions (12) to (15) (and for a ring also (45)) are satisfied by the relation. For the proof of this statement we

[49] If the multiplication is not commutative, there is a second law of the same sort, with the factor c on the left.

define P as the set of x with $0 < x$. For then (44_1) follows from (12), and (44_2) is proved as follows: $0 \neq a \notin P$ implies by (15) that $a < 0$, and thus by (14) we have $0 = a + (-a) < -a$, and therefore $-a \in P$. As for (44_3), we see that $0 < a$, $0 < b$ imply by (14) that $a < a + b$ and consequently by (13) that $0 < a + b$. Finally (44_4) is obtained from (45) with $a = 0$. Since by (14) $a < b$ means the same as $0 < b - a$, the relation $<$ actually results in this way from P and is therefore the ordering that corresponds to the domain of positivity P.

The assumption $0 < c$ in (45) is essential; for if $c < 0$, then $0 < -c$, so that $-ac = a(-c) < b(-c) = -bc$ and therefore $bc < ac$; of course for $c = 0$ we have $ac = bc$ $(=0)$. This argument shows that for $c \neq 0$ the endomorphism $x \rightarrow xc$ of the module of integers (by §2.4 every endomorphism is of this form) is a monotone and therefore one-to-one mapping[50] (monotone increasing for $c > 0$ and monotone decreasing for $c < 0$), whereas for $c = 0$ the endomorphism is not one-to-one.

A twofold application of the monotonic law (14) for addition shows that if $a < b$ and $c < d$, then $a + c < b + c$, $b + c < b + d$ and therefore by (13) $a + c < b + d$; thus, inequalities of the same kind may be added. To obtain the same result for multiplication, we must make the additional assumption that $b, c > 0$ or $a, d > 0$; since otherwise (45) would not be applicable.

3. The Rational Numbers

3.1. *Introduction of the Rational Numbers*

But now the integers are incomplete with respect to multiplication in the same way as the natural numbers were incomplete with respect to addition: not every equation $xa' = a$ has a solution. However, multiplication is associative and commutative in exactly the same way as addition, and as a substitute for (11) we at least have: if $ac = bc$ and $c \neq 0$, then $a = b$, as follows from $ac - bc = (a - b)c$ because of the absence of divisors of zero[51] (see §2.4). By the remark in small print at the beginning of §2.2 (for C we take the set of integers $\neq 0$) these properties permit us to follow the construction given there for extending a domain, provided we restrict ourselves to pairs (a, a') with $a' \neq 0$. Instead of $a - a'$ we naturally use the symbol a/a' and replace addition by multiplication. Essential for our present purpose is the following fact: in the product $(a, a')(b, b') = (ab, a'b')$, defined in analogy with (30), the second number $a'b'$ of the pair is also

[50] Of course, unless $c = 1$ or $c = -1$, not every integer will be an image in this mapping; we will have an isomorphism of the module onto a submodule.

[51] Consequently, the procedure we are about to describe may be carried out for any arbitrary commutative ring that has no divisors of zero.

$\neq 0$ because of the absence of divisors of zero. For the *rational numbers* a/a' obtained in this way (they are also called *fractions*) the equation $a/a' = b/b'$ means the same as $ab' = a'b$, so that in particular $ac/a'c = a/a'$ for $c \neq 0$.

Just as for the integers in §2.2, we now admit as statements about the rational numbers a/a', b/b', ... only such statements about the integers a, a', b, b', ... as do not change their truth values when a/a', b/b' are replaced by rational numbers that are equal to them (in the sense of equality just defined). If, as is customary, we call a' the *denominator* of the symbol a/a', then the statement "a/a' has the denominator a'" is not an admissible statement about the rational numbers since, for example, 2/3 has the denominator 3 but 4/6 does not. Of course, there is no serious objection to such statements, even though they are not "equality-invariant," provided we clearly understand their special position. But since it is possible to avoid them altogether in mathematics, we will find it safer and more convenient to exclude them on principle. If we do this, we can still speak about *the* denominator a' of the rational number a/a' if we assume, for example, that $a' > 0$, $a \neq 0$ and a, a' are relatively prime (see IB6, §2.6).

On the other hand, it is customary to speak in school about the *numerator* and the *denominator* of a fraction a/a' even when a and a' have common factors. In this case (on account of the order in which the extensions are usually made in school) a and a' are natural numbers. Moreover, it is a common habit to say that fractions are equal only if they have the same numerator and the same denominator. In this terminology the fractions are actually the pairs of numbers (a, a').[52] Then to obtain the rational numbers one says that the fractions a/a', b/b' are *equal in value* if $ab' = a'b$. This equality of value is our equivalence relation (denoted by $\doteq$ in the analogous developments in §2.2). For the rational numbers one then uses the same symbol a/a' as for fractions, but equality is taken in the sense of equality of value. In contrast to our procedure, in which pairs of numbers are denoted by symbols different from those for rational numbers, the distinction between fractions and rational numbers is now taken into account only in the different concepts of equality. This procedure, which is permissible enough in itself, is obscured by the fact that only equality of value actually appears in the formulas, so that the symbol "=" always means equality of value, whereas the "original" equality of fractions (equality of numerator and denominator) occurs only in informal statements. Here again (in analogy with the use of residue classes or of the equivalence relation $\doteq$ mentioned on p. 109) we may consider a rational number as the set of fractions that are equal in value to a given fraction a/a'; then the rational number is *represented* by the fraction a/a' (or by any other fraction with the same value), and equality of value of fractions means exactly the same as equality of the rational numbers represented by them (in the sense of the definition of equality for sets; see IA, §7.2).

[52] See also Vogel [1].

The equations (32) now become $ca'/a' = c$, $c = ca'/a'$, and we have $(a/a')(b/b') = ab/a'b'$. Instead of the 0 in §2.3 we obtain for any number $a \neq 0$ the fraction a/a as the neutral element for the multiplication of rational numbers; but now, since $a/a = 1$, this element is not a newly adjoined element but simply the number 1, namely the unit element of the ring of integers. If $a, a' \neq 0$, then a'/a is the inverse of a/a'; it is also written $(a/a')^{-1}$ and is called the *reciprocal* of a/a'. Thus the rational numbers $\neq 0$ form a *module*[53] with respect to multiplication, and for given rational numbers α, β with $\alpha \neq 0$ there exists exactly one rational number ξ with $\xi\alpha = \beta$. In agreement with our use up to now of the solidus /, this number is denoted by β/α. In analogy with (34), (35), (36) we have the equations $(\alpha^{-1})^{-1} = \alpha$, $(\alpha\beta)^{-1} = \beta^{-1}\alpha^{-1}$, $(\alpha/\beta)^{-1} = \beta/\alpha$ for all $\alpha, \beta \neq 0$.

It must be noted that the remarks at the end of §2.3 have no analogy in the theory of multiplication: it is not true that every noninteger is the reciprocal of an integer. The explanation is that at that time (end of §2.3) we made use of the properties (12), (13) of the relation $<$. In analogy with $<$ we now define the relation $|$ (to be read: factor of) in the domain of integers: $a \mid b$ if and only if there exists an integer c with $ac = b$. Then the statements analogous to (13), (14) are valid: from $a \mid b$, $b \mid c$ follows $a \mid c$; from $a \mid b$ follows $ad \mid bd$. But in contrast to (12) we always have $a \mid a$, and in contrast to (15) neither $2 \mid 3$ nor $2 = 3$ nor $3 \mid 2$.[54]

3.2. *The Field of Rational Numbers*

But how shall we define addition for the rational numbers introduced above? It is natural to lay down the following three requirements: when applied to the integers, the new addition gives the same results as the old; under the new addition the rational numbers form a module; and finally, multiplication is distributive with respect to addition. If we examine the equations $\xi a' = a$, $\eta b' = b$ (where a, b are integers, and ξ, η are rational numbers), it follows from these requirements that:

$$(\xi + \eta)\, a'b' = ab' + ba'.$$

For the pairs $A = (a, a')$, $B = (b, b')$ it will be natural in the present context (in contrast to §2.2) to define addition as follows:

$$A + B = (ab' + a'b, a'b').$$

[53] When speaking of multiplication, it is customary to use the term "commutative group" rather than "module."

[54] In the terminology introduced in IA, §8.3 the relation $\leqslant$ is an ordering, whereas $|$ is only a partial ordering, and even then only under the restriction to natural numbers (see IB6, §2.2). The relation $|$ is defined in any commutative ring without divisors of zero (cf. IB5, §2.1).

This addition is applicable to all pairs and is commutative. In order to carry it over to the rational numbers by means of the mapping $(a, a') \to a/a'$ we must show, in accordance with the remarks in §2.2, that the equivalence relation $\doteq$ is consistent with it, where $A \doteq B$ now means $ab' = a'b$. To do this, we prove (31) with the new meaning of $\doteq$, namely that from $A \doteq C$, or in other words from $ac' = a'c$, it follows that $(ab' + a'b)\, b'c' = (cb' + c'b)\, a'b'$, so that $A + B \doteq C + B$. After this proof of consistency we may define addition of the rational numbers by

$$a/a' + b/b' = (ab' + a'b)/a'b'. \tag{46}$$

Since $a = a/1$, the new addition agrees with the old for integers. Commutativity of the new addition is clear, and its associativity is easily shown as follows:

$$\begin{aligned}(a/a' + b/b') + c/c' &= (ab' + a'b)/a'b' + c/c' \\ &= (ab'c' + a'bc' + a'b'c)/a'b'c', \\ a/a' + (b/b' + c/c') &= a/a' + (bc' + b'c)/b'c' \\ &= (ab'c' + a'bc' + a'b'c)/a'b'c'.\end{aligned}$$

Since $a/a' + 0 = a/a' + 0/a' = (a + 0)/a' = a/a'$ and $a/a' + (-a)/a' = (a + (-a))/a' = 0/a' = 0$, the rational numbers form a module with respect to addition. Furthermore,

$$\begin{aligned}(a/a')(c/c') + (b/b')(c/c') &= ac/a'c' + bc/b'c' = (ab'c + a'bc)c'/a'b'c'c' \\ &= (ab' + a'b)c/a'b'c' = (a/a' + b/b')(c/c').\end{aligned}$$

Consequently, the distributive law holds and therefore the rational numbers form a commutative ring with respect to addition and multiplication. But this ring has the special property that the nonzero elements in it form a module with respect to multiplication, so that every equation $\xi\alpha = \beta$ $(\alpha \neq 0)$ has a solution.

Now a *field* is defined as a ring with the following properties:

1. For arbitrary elements α, β of the ring with $\alpha \neq 0$, there exists an element ξ in the ring with $\xi\alpha = \beta$.

2. Multiplication is commutative.

Thus by what has just been proved the rational numbers form a field, the *field of rational numbers*. The only property of the integers used in the construction of this field is that they form a commutative ring without divisors of zero.[55] Every such ring R can therefore be extended to a field

[55] The occasional use of the unit element 1 in the ring of integers could easily have been avoided.

that consists, as in the case of the rational numbers, of the symbols a/a' with $a, a' \in R$, $a' \neq 0$; here the a/a' are called *quotients* and the field is called the *quotient field* of R. A field cannot have divisors of zero, since it follows from $\alpha\beta = 0$ and $\alpha \neq 0$ that $\beta = (\alpha^{-1}\alpha)\beta = \alpha^{-1}(\alpha\beta) = \alpha^{-1}0 = 0$.

We could have obtained the field of rational numbers by introducing subtraction and division in the opposite order; beginning with the natural numbers we would then have defined the *positive rational numbers* (cf. §3.4) in the form a/a' (a, a' being natural numbers); in this domain we would introduce addition as before and then apply to it the procedure which led from the natural numbers to the integers. In this case the rational numbers appear in the form $a/a' - b/b'$.

Our chief reason for not adopting this procedure is that then the ring of integers, which is of great importance in algebra, does not appear as an intermediate stage. Of course, such an objection does not mean that the procedure may not be otherwise convenient. For example, it is usually adopted in school.

3.3. *Powers*

In the present section, except where otherwise mentioned, lower-case italicized letters denote the elements of an arbitrary field.[56] With multiplication in place of addition and $\prod$ in place of $\sum$ we can again use the definition (8):

$$\prod_{i=1}^{1} a_i = a_1, \qquad \prod_{i=1}^{n+1} a_i = \left(\prod_{i=1}^{n} a_i\right) a_{n+1} \quad (n \text{ a natural number}).$$

Since the properties of addition used in §1.3 also hold for multiplication, we have the result corresponding to (9):

$$\prod_{i=1}^{n} a_i \prod_{i=1}^{m} a_{n+i} = \prod_{i=1}^{n+m} a_i \quad (n, m \text{ natural numbers}).$$

Now we shall call $\prod_{i=1}^{n} a$ the nth *power* a^n of a. From (10) we have for natural m, n

$$a^n a^m = a^{n+m}. \tag{47}$$

We may also take over (18) and thus introduce $\prod_{i\in I} a_i$. Corresponding to (21) we get the equation

$$\prod_{k=1}^{n} \prod_{i=1}^{m} a_{ik} = \prod_{\substack{i\leqslant m\\ k\leqslant n}} a_{ik} = \prod_{i=1}^{m} \prod_{k=1}^{n} a_{ik} \quad (m, n \text{ natural numbers}). \tag{48}$$

[56] As far as positive exponents are concerned, the developments are valid in an arbitrary commutative ring, or even in an arbitrary (multiplicatively written) Abelian group, where commutativity is needed only for the proof of (48) and (50).

With $a_{ik} = a$ we have

(49) $$(a^m)^n = a^{mn} \qquad (m, n \text{ natural numbers})$$

since from (21), with $a_{ik} = 1$, there exist exactly mn pairs of natural numbers (i, k) with $i \leqslant m$, $k \leqslant n$. For $m = 2$ and $a_{1k} = a$, $a_{2k} = b$ we have from (48)

(50) $$(ab)^n = a^n b^n.$$

This definition of a power is now extended by the convention $a^0 = 1$, where it is often assumed that $a \neq 0$. The value of a^0 for $a \neq 0$ must be defined in exactly this way if (47) is to remain valid: namely, $a^1 a^0 = a^1$ and thus $aa^0 = a$. Then it is easy to show by calculation that not only (47) but also (49), (50) hold for all integers $m, n \geqslant 0$. If we wish to define a^{-n} for every natural number n in such a way that (47) remains valid, we must have $a^n a^{-n} = a^0 = 1$, which shows that $a \neq 0$ is a necessary restriction. Thus $a^{-n} = (a^n)^{-1}$ for every natural number $n > 1$; and of course this equation also holds for $n = 1$. In the proof of (47) for this extended case we may restrict ourselves, on account of the commutativity of multiplication and addition, to replacing n by $-n$. For every nonnegative integer m we then obtain

$$a^{-n}a^m = (a^n)^{-1}\, a^m = \begin{cases} a^{m-n} & \text{for} \quad m > n, \\ 1 & \text{for} \quad m = n, \\ (a^n/a^m)^{-1} = (a^{n-m})^{-1} & \text{for} \quad m < n; \end{cases}$$

for we have $a^{m-n}a^n = a^m$ for $m > n$ and $a^{n-m}a^m = a^n$ for $m < n$. Since for natural numbers m, n we also have $a^{-n}a^{-m} = (a^n)^{-1}\,(a^m)^{-1} = (a^{n+m})^{-1} = a^{-(n+m)} = a^{(-n)+(-m)}$, we see that (47) has now been proved for arbitrary exponents. In order to make the corresponding extension of (49), we need the rule

$$(a^{-1})^n = (a^n)^{-1} \qquad (n \text{ a natural number}),$$

which follows at once from the equation $(a^{-1})^n\, a^n = 1$ [see (50)]. For natural numbers m, n we now obtain

$$(a^{-m})^n = ((a^m)^{-1})^n = ((a^m)^n)^{-1} = (a^{mn})^{-1} = a^{-mn} = a^{(-m)n},$$

$$(a^m)^{-n} = ((a^m)^n)^{-1} = (a^{mn})^{-1} = a^{-mn} = a^{m(-n)},$$

$$(a^{-m})^{-n} = (((a^m)^{-1})^n)^{-1} = (((a^m)^n)^{-1})^{-1} = (a^m)^n = a^{mn} = a^{(-m)(-n)}.$$

Finally, it is also easy to prove (50) for negative exponents:

$$(ab)^{-n} = ((ab)^n)^{-1} = (a^n b^n)^{-1} = (a^n)^{-1}\,(b^n)^{-1} = a^{-n}b^{-n}.$$

3.4. *Order*

With a view to extending the ordering of the integers to the rational numbers, let us first examine the properties which a domain of positivity P of the field of rational numbers must have, in case such a domain exists. The set of integers contained in P is obviously a domain of positivity for the ring of integers, so that in particular $N \subseteq P$. But from $(44_{2,4})$ and $a^2 = (-a)^2$ it follows that a domain of positivity of a ring must contain all squares $\neq 0$. Thus for $a, b \in N$ it follows from $a/b = ab(b^{-1})^2$ that $a/b \in P$. But if P contained a rational number that is not of this form, then by (44_1) it would also contain natural numbers a, b with $(-a)/b \in P$, which contradicts (44_3), since $a/b + (-a)/b = 0 \notin P$ and $a/b \in P$. Thus P consists precisely of the quotients of natural numbers. But these quotients do in fact form a domain of positivity, since a rational number $\neq 0$ which is distinct from these quotients has the form $(-a)/b$ $(a, b \in N)$, which means that (44_2) is valid, while $(44_{1,3,4})$ obviously hold. Thus the field of rational numbers can be ordered in exactly one way. Since

$$b/b' - a/a' = b/b' + (-a)/a' = (ba' + b'(-a))/a'b' = (ba' - b'a)/a'b',$$

we have for integers a, a', b, b': $a/a' < b/b'$ *for* $a', b' > 0$, if and only if $ab' < a'b$. Since $c > 0$ implies $c^{-1} = c(c^{-1})^2 > 0$ for every rational number c, this result can easily be extended by (45) to arbitrary rational numbers a, a', b, b'.

The ordering of the rational numbers is *Archimedean*; that is, for every $a, b > 0$ there exists a natural number n with $na > b$.[57] For the proof we first restrict ourselves to integers a, b. Then from $b + 1 > b$ and $a \geqslant 1$ it follows by the monotonic law that $(b + 1)a > ba \geqslant b1 = b$, so that $na > b$ with $n = b + 1$. But then for the rational numbers $a/a', b/b' > 0$ (a, a', b, b' natural numbers) we have $a/a' = ab'/a'b'$, $b/b' = a'b/a'b'$, so that if we choose a natural number n with $nab' > a'b$, it follows by multiplication with the positive number $(a'b')^{-1}$ that $n(a/a') > b/b'$, as desired. In an arbitrary module, which may not contain the natural numbers, we can always define na as being equal to $\sum_{i=1}^{n} a$, so that the definition of "Archimedean" is applicable to any ordered module. At the end of §4.3 we give an example of an ordered module in which the ordering is not Archimedean.

The ordering of a field[58] is Archimedean if and only if $0 \leqslant a < n^{-1}$, for all natural numbers n implies $a = 0$. For if $a > 0$ in an Archimedean

[57] This statement is often called the "axiom of Archimedes" since it occurs as an axiom in geometry (cf. II2, §1.2).

[58] If the field does not contain the natural numbers, then in the following inequality (and in the proof) the natural number n must be replaced by the nth multiple of the unit element 1 of the field.

ordering, then there exists a natural number n with $na > 1$, so that $a > n^{-1}$. The same argument obviously holds for an ordered ring that contains n^{-1} for every natural number n. Conversely, if $0 \leqslant a \leqslant n^{-1}$ implies $a = 0$, and if $a, b > 0$, so that $a/b > 0$, then the inequality $na \leqslant b$ cannot hold for every natural number n, since it would imply $a/b \leqslant n^{-1}$.

The *absolute value* $|a|$ of the rational number a is defined as follows:

$$|a| = \max(a, -a); \tag{51}$$

where by $\max(a, b)$ we mean the number b if $a < b$ and the number a if $a \geqslant b$. Thus $|a| = a$ or $= -a$ and $|a| \geqslant 0$, and these properties obviously characterize the number $|a|$. Then we can at once derive

$$|ab| = |a||b|. \tag{52}$$

Since $\pm a \leqslant |a|$ for all a, we have $\pm(a + b) \leqslant |a| + |b|$ and therefore

$$|a + b| \leqslant |a| + |b|. \tag{53}$$

Replacing b by $-b$, we see, since $|-b| = |b|$, that $|a - b| \leqslant |a| + |b|$, and if we replace $a - b$ by a, and a by $a + b$, we have

$$|a| \leqslant |a + b| + |b|, \quad \text{and therefore} \quad |a| - |b| \leqslant |a + b|.$$

Since the right-hand side is not altered by the interchange of a and b, it follows that

$$||a| - |b|| \leqslant |a + b|. \tag{54}$$

As in (53), we may replace $a + b$ by $a - b$. Of course, the definition (51) of absolute value, and with it the consequences (52), (53), (54), are valid for any ordered ring.

3.5. *Endomorphisms*

In view of the distributive law, the mapping $x \to cx$ for any rational number c is an endomorphism of the module (with respect to addition) of the rational numbers. As in §2.4, we can show that for two endomorphisms f, g of this module, the equality $f(1) = g(1)$ implies $f(x) = g(x)$ for all integers. From (38) it is easy to prove $f(\sum_{i=1}^{n} x_i) = \sum_{i=1}^{n} f(x_i)$ by complete induction. Thus for a rational a/a' (a, a' integers, $a' > 0$) and two endomorphisms f, g with $f(1) = g(1)$ we have:

$$a'f(a/a') = \sum_{i=1}^{a'} f(a/a') = f(a'(a/a')) = f(a) = g(a) = a'g(a/a').$$

But then it follows that $f(a/a') = g(a/a')$, so that an endomorphism f is completely determined by the value of $f(1)$. Since the image of 1 under the endomorphism $x \to cx$ is the number c, the mapping $f \to f(1)$ is a one-to-one mapping of the ring of endomorphisms onto the field of rational numbers. In fact, this mapping is even an isomorphism, as can be proved in exactly the same way as the corresponding statement in §2.4. As in the corresponding case for the integers, the ring of endomorphisms of the additive module of the rational numbers is isomorphic to the field of rational numbers.

Furthermore, as in §2.5, the endomorphisms $\neq 0$ are monotone mappings: $x \to cx$ is monotone increasing for $c > 0$ and monotone decreasing for $c < 0$.

It is obvious that the endomorphism $x \to cx$ is also a homomorphism with respect to multiplication if and only if $c(x, y) = (cx)(cy)$ for all x, y, or in other words, if and only if $c^2 = c$. But in view of the absence of divisors of zero, the equation $c^2 - c = c(c - 1)$ shows that $c^2 = c$ only for $c = 0$ or $c = 1$. Thus the field of rational numbers has exactly two homomorphisms (with respect to addition and multiplication) into itself, namely the zero mapping $x \to 0$ and the identity mapping $x \to x$.

4. The Real Numbers

4.1. *Decimal Fractions*

In the present section we denote by g a fixed integer $g > 1$, which we call a *base*. For any given positive rational number r we now use g to determine[59] the sequence of integers a_n ($n = 0, 1, 2, \ldots$) by recursion in the following way: a_0 is the greatest integer $\leqslant r$; a_{n+1} ($n \geqslant 0$) is the greatest integer $\leqslant (r - \sum_{i=0}^{n} a_i g^{-i})\, g^{n+1}$. With the abbreviation $r_n = \sum_{i=0}^{n} a_i g^{-i}$, we then have $a_{n+1} \leqslant (r - r_n)\, g^{n+1} < a_{n+1} + 1$, $a_{n+1} g^{-(n+1)} = r_{n+1} - r_n$ and consequently (with n in place of $n + 1$)

$$(55) \qquad r_n \leqslant r < r_n + g^{-n} \qquad (n = 0, 1, 2, \ldots).$$

It follows that $0 \leqslant r - r_n < g^{-n}$ and thus $0 \leqslant a_{n+1} < g$, so that $0 \leqslant a_n < g$ ($n = 1, 2, \ldots$). In view of (55) we can also describe r_n as the greatest integral multiple of g^{-n} which is $\leqslant r$. The sequence of the a_n determines r uniquely; for if (55) holds for r' as well as for r, then

$$-g^{-n} < r - r' < g^{-n} \qquad \text{for all natural numbers } n.$$

[59] The principle of recursion at the end of §1.2 shows that the function $n \to a_n$ is uniquely determined by the above requirements; the fact that in the present case n may take the value 0 represents only an insignificant change from §1.2.

Since $g - 1 > 0$, we have by the binomial theorem (see IB4, §1.3)

$$g^n = (1 + (g - 1))^n > n(g - 1)$$

and thus $g^{-n} < (g - 1)^{-1} n^{-1} < n^{-1}$, so that no positive number can be $\leqslant g^{-n}$ for all natural numbers n. Thus $r' < r$ or $r < r'$ would lead to a contradiction, so that we must have $r' = r$.

In accordance with the usual practice for the base 10 we write the sequence of the a_n in the form $a_0.a_1a_2a_3 \ldots$ and call it an *infinite decimal.*[60] Since this sequence can be regarded as a complete substitute for the number r, we write

$$r = a_0.a_1a_2a_3 \ldots . \tag{56}$$

But we now encounter the following extremely significant fact: although every rational number gives rise in this way to an infinite sequence of nonnegative integers $<g$, it is *not* true that every such sequence can be obtained from some rational number (for an example see §4.8). In order to extend the field of rational numbers, it therefore seems appropriate to consider all infinite sequences of nonnegative integers $a_n (n = 0, 1, 2, \ldots)$ with $a_n < g$ for $n > 0$; these sequences are to be taken as the elements of a domain of numbers which in view of (56) contain[61] the rational numbers $\geqslant 0$. Then for these new numbers we must define equality, order, and addition in such a way that when applied to the rational numbers in accordance with (56) they will yield the same results as the corresponding concepts already defined for rational numbers. For the definition of equality it is natural to set

$$a_0.a_1a_2a_3 \ldots = b_0.b_1b_2b_3 \ldots$$

if and only if $a_n = b_n$ for all $n = 0, 1, 2, \ldots$. As an ordering we take the natural *lexicographic ordering*:[62] $a_0.a_1a_2a_3 \ldots < b_0.b_1b_2b_3 \ldots$ if and only if there exists a nonnegative integer n with $a_i = b_i$ for all $i < n$ and $a_n < b_n$. The definition of addition is necessarily somewhat lengthy,

[60] Of course, from the etymological point of view the word "decimal" ought to be replaced by some other word corresponding to the value of g; for $g = 2$ the phrase *dyadic* fractions is also used.

[61] For brevity we restrict ourselves here to the numbers $\geqslant 0$. For a negative number r we define the a_n as the numbers corresponding to $-r$ in the above procedure and then we set $r = -(a_0.a_1a_2a_3 \ldots)$.

[62] So-called because the words in a lexicon are arranged by this principle; the letters in alphabetic order correspond here to the nonnegative numbers in the order introduced above. Of course, lexicons do not normally contain words with infinitely many letters.

so that we shall content ourselves here with defining the sum $a_0.a_1a_2a_3 \ldots + g^{-k} = b_0.b_1b_2b_3 \ldots$:

$$\text{If} \quad a_k \neq g - 1 \quad \text{or} \quad k = 0, \quad \text{we set}$$
$$b_n = a_n \quad \text{for} \quad n \neq k,$$
$$b_k = a_k + 1,$$

but if $a_n = g - 1$, then for $h < n \leqslant k$ and (if $h > 0$) $a_h \neq g - 1$, we set

$$b_n = a_n \quad \text{for} \quad n < h \quad \text{and} \quad n > k,$$
$$b_n = 0 \quad \text{for} \quad h < n \leqslant k,$$
$$b_h = a_h + 1.$$

One reason for choosing this definition is that for infinite decimals (56) that are equal to rational numbers it is a readily provable rule.

But now there is a difficulty. In the case $a_n = g - 1$ for $n > k$, $a_k \neq g - 1$ or $k = 0$ our definitions (of order and addition) give for all $n > k$:

$$a_0.a_1a_2a_3 \ldots + g^{-n} \sum_{i=0}^{k} a_i g^{-i} + g^{-k} = a_0 , a_1 \ldots a_{k-1}(a_k + 1)000 \ldots .$$

If the monotonic law is to hold for addition and if subtraction is to be possible (for the case when the subtrahend is smaller than the minuend), we have the following inequality (cf. the calculation given above) for the difference $d = \sum_{i=0}^{k} a_i g^{-i} + g^{-k} - a_0 . a_1a_2a_3 \ldots$:

$$d < g^{-n} < (g - 1)^{-1} n^{-1} \qquad \text{for all } n > k.$$

But $d > 0$, so that there exists a positive rational number $< d < (g - 1)^{-n} n^{-1}$ for all $n > k$, in contradiction to the fact that the ordering of the rational numbers is Archimedean. The solution of the difficulty lies, of course, in excluding the sequences with $a_n = g - 1$ for all $n > k$. In fact, such sequences do not occur in the decimal expansions of rational numbers. For if $a_n = g - 1$ for all $n > k$, it follows that

$$r_n - r_k = (g - 1) \sum_{i=k+1}^{n} g^{-i} = (g - 1)\, g^{-n} \sum_{h=0}^{n-k-1} g^h$$
$$= g^{-n}(g^{n-k} - 1) = g^{-k} - g^{-n},$$

so that $r_n + g^{-n} = r_k + g^{-k}$; but then from (55) we would have

$$0 < (r_k + g^{-k}) - r \leqslant g^{-n} < (g - 1)^{-1} n^{-1} \qquad \text{for all} \quad n > k$$

in contradiction to the fact that the ordering of the rational numbers is Archimedean.

In this way the real numbers can be introduced as infinite decimals, and it can be shown that they do in fact form an ordered module (with Archimedean ordering) that includes the module of the rational numbers. Let us now consider the following theorem, which is of basic importance in analysis: every non-empty set of real numbers which is bounded below[63] has an *infimum*, or *greatest lower bound.* This theorem is now very easy to prove. For by the addition of a sufficiently large number (namely $-s$, if s is a negative lower bound) the set M becomes a set of nonnegative numbers, and then r_n is defined, in accordance with (55), as that integral multiple of g^{-n} with $r_n \leqslant x$ for all $x \in M$ such that there exists a number $x \in M$ with $x < r_n + g^{-n}$. As in (55), there then exist integers a_n $(n = 0, 1, 2, ...)$ with $r_n = \sum_{i=0}^{n} a_i g^{-i}$ and $0 \leqslant a_n < g$ $(n = 1, 2, ...)$. From $r_n \leqslant x < r_k + g^{-k}$ it then follows as before that $a_n = g - 1$ for $n > k$ is impossible, so that $a_0.a_1a_2a_3 \ldots$ is actually a real number, which is easily recognized as the greatest lower bound of M.

4.2. *A Survey of Various Possible Procedures*

From the point of view of practical calculation the introduction of the real numbers by means of decimal expansions as described above in §4.1 is very natural and has the advantage that the concepts involved in it are relatively simple. A further advantage is that if we use decimal expansions, we can introduce the real numbers immediately after adjoining zero to the natural numbers without first introducing the integers and then the rational numbers.[64] It is convenient to introduce only the nonnegative real numbers at first and then to apply to them the method of extension described in §2.3.

But these advantages are obtained at high cost; addition can only be defined in a very lengthy way,[65] and the rules for calculation are not very convenient to prove. These disadvantages obviously arise from the special form of the r_n. In order to avoid them, it will be convenient to replace these r_n by more general entities, for which we shall naturally wish to preserve certain properties of the r_n. To do this we may start from either of two facts:

1) $a_0.a_1a_2a_3 \ldots$ *is the least upper bound of the set of* r_n.

2) $a_0.a_1a_2a_3 \ldots$ *is the limit of the sequence of* r_n; for we have

[63] A non-empty set that is bounded above can be reduced to the present case by the mapping $x \rightarrow -x$ and is thus shown to have a *supremum*, or *least upper bound.*

[64] Since the intermediate stages will be important to us later, we have not followed this plan here.

[65] To say nothing of multiplication, which we shall discuss in a separate section. See, for example, F. A. Behrend, A contribution to the theory of magnitudes and the foundations of analysis, Math. Zeitschr. 63, 345–362 (1956).

$0 \leqslant a_0.a_1a_2a_3 \ldots - r_n < g^{-n} < (g-1)^{-1} n^{-1}$, and for every real number $\epsilon > 0$ we can find a natural number n_0 with $n_0(g-1) \geqslant \epsilon^{-1}$, whereupon $| a_0.a_1a_2a_3 \ldots - r_n | < \epsilon$ for all $n \geqslant n_0$.

The first of these two facts suggests that, in a completely general way, we may take as our starting point all non-empty sets of rational numbers bounded from above.[66] This procedure, discussed in §4.3, is essentially the method of Dedekind for defining the real numbers by *Dedekind cuts*. It has the advantage that it defines the real numbers and their order without making any use of addition, so that it can be applied to more general systems in which only an order is defined.[67]

On the other hand, the second of the above listed facts is used as our starting point in §4.4 to introduce the real numbers by means of the *fundamental sequences* of Cantor; that is, sequences of rational numbers which satisfy the Cauchy criterion for convergence. Since the definition of this quite general class of sequences does not require the ordering of the rational numbers but only of their absolute values, it too can be extended to a more general class of modules than the ordered ones.[68] In this case, however, the addition of rational numbers is already employed in the very definition of real numbers.

The introduction of the real numbers by means of nested intervals is to a certain extent a mixed procedure. Here we employ pairs of sequences of rational numbers $(a_n)_{n=1,2,\ldots}$, $(a_n')_{n=1,2,\ldots}$ with $a_n \leqslant a_{n+1} \leqslant a_{m+1}' \leqslant a_m'$ for all m, n and $\lim_{n\to\infty}(a' - a_n) = 0$.[69] Since both order and addition are required here, the usefulness of this procedure for more general systems is considerably reduced, so that we shall not deal with it in detail in the following sections but shall content ourselves with the following remarks. The nests of intervals α and β arising from the pairs of sequences with terms a_n, a_n' and b_n, b_n' are said to be *equal* if and only if for every index pair m, n there exists a rational number x with $a_m \leqslant x \leqslant a_m'$, $b_n \leqslant x \leqslant b_n'$. The number α is set equal to the rational number r if $a_n \leqslant r \leqslant a_n'$ for all n. By the sum $\alpha + \beta$ we mean the nest of intervals defined by the sequences $(a_n + b_n)_{n=1,2,\ldots}$, $(a_n' + b_n')_{n=1,2,\ldots}$, where it remains to be proved that when equality is defined as above, this sum depends only on α, β. If we add to α the nest of intervals defined by the sequences with terms $-a_n'$, $-a_n$ the sum is equal to the rational number 0, which is easily seen to be a neutral element for the addition defined

[66] Of course, we could just as well consider the non-empty sets bounded from below.

[67] To a certain extent (see §4.3) we only require a partial order.

[68] See, for example, van der Waerden [2], §§74, 75. In topology the same procedure is followed with even greater generality for metric spaces.

[69] Here a_n and a_n' are regarded as the endpoints of an interval, which may reduce to a single point. Such intervals are said to be *nested* within one another.

in this way. Since associativity and commutativity are obvious, the real numbers defined as nests of intervals form a module, for which the set of $\alpha \neq 0$ with $a'_n > 0$ is seen to be a domain of positivity. The least upper bound of a non-empty set of real numbers bounded from above is most easily constructed by the *principle of nesting of intervals*: let us first choose an interval with rational endpoints containing an upper bound and at least one number of the set; we then carry out a sequence of bisections, where after each bisection we choose the subinterval as far as possible to the right still containing numbers of the set. The resulting sequence of intervals is then seen to be a nested set which is the desired least upper bound.

We have now indicated various methods for introducing the "real numbers"; for that matter, the decimal procedure already provides us with infinitely many methods, since we have free choice of a base. So it is natural to ask: to what extent do all these methods lead to the same result? Certainly it is true that the entities we have called real numbers are quite different from case to case; for example, an infinite decimal 0.2... cannot occur if 2 is the base. But even if certain objects can occur in several different methods, it is by no means necessary for them to have the same meaning in the different methods; for base 3, for example, the infinite "decimal" 1.111... = 3/2, whereas for base 10 it is =10/9. But in §4.5 we will show that all the domains obtained in this way can be mapped onto one another by isomorphisms that preserve the order (with respect to addition), so that in this sense there is no essential difference among the various systems.

4.3. *Dedekind Cuts*

Our purpose is to extend the module of rational numbers in such a way that every non-empty set of rational numbers bounded from above has a least upper bound. This problem is very similar to the requirement that led us to the integers, namely that every equation (22) should have a solution. In §2.2 we took pairs of numbers as our starting point. Analogously, we now consider all non-empty sets of rational numbers that are bounded from above and to each such set we assign a new symbol fin M.[70] Since the least upper bound of a set M is determined by the set of upper bounds

[70] For the time being "fin" has no meaning whatever; only after we have introduced an ordering will fin M actually turn out to be the least upper bound (finis superior, or supremum) of the set M. The construction of such symbols as fin M is permissible only if we take the constructive or operational attitude toward the foundations of mathematics (see IA, §§1.4 and 10.6), in which the sets themselves are symbols. From other points of view we must proceed somewhat differently; for example, in order to have an entity which is distinct from M but naturally associated with it, we might consider the pairs (0, M), for which we could then introduce the abbreviation fin M.

of M, we will define the equality fin $M =$ fin M' as identity of the sets of upper bounds of M and M'. For abbreviation we let $S(M)$ denote the set of upper bounds of M or, in other words, the set of rational numbers x with $x \geqslant y$ for all $y \in M$. By a *Dedekind cut* we mean the pair consisting of the set $S(M)$ and the set of rational numbers not included in $S(M)$. It is easy to show that a pair of sets M', M'' of rational numbers is a Dedekind cut if and only if it has the following three properties: (i) every rational number x belongs to exactly one of the two sets M', M''; (ii) if $x' \in M'$, $x'' \in M''$, then $x'' < x'$; (iii) the least upper bound of M'', provided it exists (in the set of rational numbers), belongs to M'. These three properties are often taken as the definition of a Dedekind cut, although usually the third one is omitted, since it is quite unimportant. In fact, such a definition may be taken as the starting point for introducing the real numbers, but to us it seems more natural to start from the sets M.[71] An order for Dedekind cuts is immediately available if we note that "pushing up" the least upper bound has the effect of decreasing the set of upper bounds: thus by fin $M <$ fin M' we shall mean simply $S(M') \subset S(M)$. We see at once that this is actually a relation between fin M and fin M' and not only between the sets M and M'; for if fin $M =$ fin M_1 and fin $M' =$ fin M_1', then fin $M <$ fin M' obviously means the same as fin $M_1 <$ fin M_1'. For this relation $<$ the properties (12), (13) are at once clear. Thus in order to prove that we have actually defined an ordering,[72] it only remains to prove (15). To verify (15) for the new symbols (or equivalently, for Dedekind cuts), we now assume that for two sets M, M' neither fin $M' =$ fin M nor fin $M' <$ fin M; that is, it is not true that $S(M) \subseteq S(M')$. Then there exists an upper bound s of M which is not an upper bound of M': that is, there exists $x' \in M'$ with $x' > s$, and for all $x \in M$ we have $x \leqslant s$. Thus $x' > x$ for all $x \in M$. For every upper bound s' of M' it is a fortiori true that $s' > x$ for all $x \in M$, and therefore $s' \in S(M)$. Consequently $S(M') \subseteq S(M)$, so that fin $M <$ fin M', as desired, since fin $M =$ fin M' was excluded.

The symbol fin M (which we may, if we wish, regard as the set of upper bounds of M, or alternatively as the corresponding Dedekind cut) will now be called a *real number*. Relaxing the restriction that M must be bounded from above leads to exactly one new "real number." For if the sets M', M'' are not bounded from above, then $S(M')$, $S(M'')$ are empty

[71] Moreover, the definition of Dedekind cuts as pairs of sets with the above properties requires a total ordering, whereas our procedure can also be used in the case of a partial ordering; see the next footnote.

[72] That is: $\leqslant$ is an order (in the notation of IA, §8.3). Since we have not yet defined addition, (14) requires no attention. Moreover, as long as addition is not yet introduced, the property (15) of the rational numbers is required only here, namely in the proof of (15) for the newly introduced symbols.

and thus equal to each other, so that fin $M' =$ fin M''. If for this *improper real number* we introduce the usual abbreviation ∞, then fin $M < \infty$ for every other real number fin M, since the empty set is a proper subset of any non-empty set. If we also admit the empty set $\emptyset$, we obtain exactly one new improper real number fin $\emptyset$, which is usually written $-\infty$. Since $S(\emptyset)$ contains all the rational numbers,[73] we have $-\infty <$ fin M for every non-empty set M. Thus the improper real numbers ∞, $-\infty$ have some importance in the ordering of the real numbers, but they play no role in addition, as defined below. In what follows we shall disregard them altogether.

It remains to answer the following question: when is a rational number a equal to a real number fin M? Letting $S(a)$ denote the set of $x \geqslant a$ we have the obvious requirement $S(M) = S(a)$, which we shall take as a definition of fin $M = a$ as well as of $a =$ fin M. As in §2.2, we show by simple verification of the four possible cases that comparativity is not destroyed by this definition. It remains to show that the meaning of $a \leqslant a'$ and of $a < a'$ is unchanged if a, a' are replaced by real numbers fin M, fin M' equal to them: but fin $M \leqslant$ fin M' means $S(M') \subseteq S(M)$, and therefore $S(a') \subseteq S(a)$, or $a \leqslant a'$, as desired. Now fin M is actually the least upper bound of the set M. For $x \in M$ we have in every case $S(M) \subseteq S(x)$, so that $x \leqslant$ fin M. On the other hand, if fin $M' \geqslant x$ for all $x \in M$, then $S(M') \subseteq S(x)$ for all $x \in M$. Thus $y \in S(M')$ implies $y \geqslant x$ for all $x \in M$ and therefore $y \in S(M)$. Consequently, $S(M') \subseteq S(M)$ and thus fin $M \leqslant$ fin M', so that fin M is actually the least upper bound.

But so far we have shown only that every non-empty set of rational numbers bounded from above has a least upper bound; now we must demonstrate the same property for any such set of real numbers. In order to describe a set M of real numbers we require a set $\mathfrak{M}$ of non-empty sets of rational numbers that are bounded from above: the real numbers in M are then the fin M with $M \in \mathfrak{M}$. We now let fin M_0 be an upper bound of M, so that fin $M \leqslant$ fin M_0, which means that $S(M_0) \subseteq S(M)$ for all $M \in \mathsf{M}$. Then we form the (non-empty) set[74] $M^* = \bigcup_{M \in \mathfrak{M}} M$ or, in other words, the set of elements x with $x \in M$ for at least one set $M \in \mathfrak{M}$, and

[73] From $x \in \emptyset$ (always false!) it follows that $x \leqslant y$ for every number y.

[74] Here we have a difficulty related to the foundations of mathematics. If we adopt a theory of sets in which the union of any set of sets can always be formed, we must accept the disadvantage that such a theory, at least if it is to satisfy the other demands of mathematics, has not yet been shown to be free of contradictions (see IA, §7.1). On the other hand, we may say that a set of sets is to be formed only in a second language layer, and then, under certain circumstances, we can form a union of sets only in this second layer; in order to preserve our theorem about the least upper bound, we must form a language layer corresponding to every natural number and then distinguish real numbers of the 1st, 2nd, 3rd, ... layer (see Lorenzen [1]). A corresponding difficulty arises in the other methods of introducing real numbers.

prove that fin M^* is the least upper bound of M. In order to show that fin M^* is a real number at all, we must first prove that M^* is bounded from above: but in view of $S(M_0) \subseteq S(M)$, an upper bound for M_0 is also an upper bound for every set $M \in \mathfrak{M}$ and therefore also an upper bound for M^*. We now investigate the upper bounds fin M' of M (where M' is a non-empty set of rational numbers bounded from above, so that fin M' is a real number). The inequality fin $M' \geqslant$ fin M for all $M \in \mathfrak{M}$ means that $S(M') \subseteq S(M)$ for all $M \in \mathfrak{M}$, so that every upper bound of M' is an upper bound of every set $M \in \mathfrak{M}$. But this simply means that every upper bound of M' is an upper bound of M^*; in other words, $S(M') \subseteq S(M^*)$ and therefore fin $M' \geqslant$ fin M^*. Consequently, fin M^* is the least upper bound of M, as desired.

We come now to the introduction of addition. For the sets M of rational numbers it is obvious what we should do: $M + M'$ is to be defined as the set of all $x + x'$ with $x \in M$, $x' \in M'$; this set is non-empty and bounded from above if the sets M, M' have those properties, and addition defined in this way is clearly associative and commutative. But now again we must define addition in the set of real numbers in such a way that $M \to \text{fin } M$ is a homomorphism. To this end we must show (as in §2.2) that $\doteq$ is consistent with addition, where now $M \doteq M'$ simply means $S(M) = S(M')$. So by (31) we must show that $S(M') = S(M'')$ implies $S(M + M') = S(M + M'')$. Or instead, we may show that

$$(57) \qquad S(M + M') \subseteq S(M + M''), \qquad \text{if} \quad S(M') \subseteq S(M'');$$

since the desired result immediately follows from the fact that $\subseteq$ and $\supseteq$ together mean $=$. In order to prove (57) we let z be an upper bound of $M + M'$, so that $z \geqslant x + x'$ for $x \in M$, $x' \in M'$. Then $z - x \geqslant x'$ for all $x' \in M'$, so that $z - x \in S(M')$, and thus, in view of our hypothesis that $S(M') \subseteq S(M'')$, we have $z - x \in S(M'')$, so that $z - x \geqslant x''$ for all $x'' \in M''$. Consequently, $z \geqslant x + x''$ for $x \in M$, $x'' \in M''$ or, in other words, $z \in S(M + M'')$.

The addition of real numbers is now defined by

$$(58) \qquad \text{fin } M + \text{fin } M' = \text{fin}(M + M').$$

When applied to rational numbers, this definition of addition agrees with the former one, since[75] $a = \text{fin}\{a\}$, $a' = \text{fin}\{a'\}$, $a + a = \text{fin}\{a + a'\}$ by (58), and $\{a\} + \{a'\} = \{a + a'\}$. The number 0 is also the neutral element for addition of real numbers, since $\text{fin } M + 0 = \text{fin } M + \text{fin}\{0\} = \text{fin}(M + \{0\}) = \text{fin } M$. Since addition of the sets M is associative and

[75] $\{a\}$ is the set consisting of the single element a.

commutative, as was pointed out before, addition of the real numbers by the definition (58) has the same properties. Thus, in order to prove that the set of real numbers is a module, it only remains to show that every real number fin M has an inverse. For this purpose we consider the set M' of x' with $-x' \in S(M)$ and show that $S(0) \subseteq S(M + M')$ and also $S(M + M') \subseteq S(0)$, so that $S(0) = S(M + M')$, and therefore fin M + fin $M' = 0$, as desired. For the proof of the first inclusion we assume $z \in S(0)$, so that $z \geqslant 0$. From $x \leqslant -x'$ for all $x \in M$, $x' \in M'$ it follows that $x + x' \leqslant z$, so that $z \in S(M + M')$. Thus we have shown that $S(0) \subseteq S(M + M')$. For the proof of the second inclusion we assume $z \in S(M + M')$, so that $z \geqslant x + x'$ for all $x \in M$, $x' \in M'$, from which it follows that $z - x' \geqslant x$, so that $z - x' \in S(M)$. Since $-x'$ is an arbitrary number from $S(M)$, we can prove by complete induction that $(n + 1)\, z - x' = z + (nz - x')$ implies $nz - x' \in S(M)$ for all natural numbers n, and then for $x \in M$ we obtain the result that $n(-z) \leqslant -(x + x')$ for all n. Since the order is Archimedean, it is therefore impossible that $-z > 0$. Thus $z \geqslant 0$, and we have completed the proof that $S(M + M') \subseteq S(0)$.

We have already shown on page 135 that the relation $<$ is an ordering of the real numbers. The monotonic law of addition now follows at once from (57), (58) and the definition of $<$. Consequently, by §2.5 the module of real numbers is an ordered module.

An ordered module in which every non-empty set bounded from above has a least upper bound is called a *complete ordered module.* In such a module every non-empty set bounded from below has a greatest lower bound;[76] as can be seen at once, since the inequality $x < y$, i.e., $0 < y - x$, implies $-y < -x$ in view of the fact that $y - x = (-x) - (-y)$, $-y < -x$, the mapping $x \rightarrow -x$ takes a non-empty set M bounded from below into a non-empty set bounded from above and takes its least upper bound into the greatest lower bound of M.

Our results can now be expressed as follows: *there exists a complete module which includes the module of the rational numbers*, where by inclusion we mean not only that all the rational numbers occur in the new module but also that in it the addition and order for rational numbers are defined in exactly the same way as in the module of rational numbers.[77] The above proofs show that the module of rational numbers, which formed our starting point, could be replaced by any module with an Archimedean

[76] Thus the concept of completeness here is closely related to completeness in lattices (see IB9, §1), with the difference that in a complete lattice *every* set has a greatest lower and a least upper bound; strictly speaking, we ought to say that the module of real numbers is "conditionally complete."

[77] Compare the concept of a subgroup in IB2, §3.2.

ordering.[78] But it is essential that the ordering be Archimedean, as is shown by the theorem:

If an ordered module is complete, its order is Archimedean.

For in a module whose ordering is non-Archimedean there exist elements $a, b > 0$ such that $na \leqslant b$ for all natural numbers n. Thus the set of numbers of the form na is bounded from above, and for every upper bound s of this set there exists a smaller one, namely $s - a$, since $(n + 1)a = na + a \leqslant s$ and therefore $na \leqslant s - a$ for all natural numbers n. Thus the set of numbers na has no least upper bound, so that the module is not complete.

Finally, let us give a simple example of an ordered module in which the ordering is not Archimedean. We form the pairs (a, a') of integers (or of rational or real numbers) and define $(a, a') + (b, b') = (a + b, a' + b')$. Then it is easy to see that the pairs (a, a') for which either $a > 0$ or else $a = 0$ and $a' > 0$ form a domain of positivity. Thus $(a, a') < (b, b')$ if and only if $a < b$ or else $a = b$ and $a' < b'$.[79] Since $n(0, 1) = \sum_{i=1}^{n} (0, 1) = (0, n)$, we have $n(0, 1) < (1, 0)$ for every natural number n. Thus the element $(0, 1)$ in this ordering is said to be *infinitesimal*[80] in comparison with $(1, 0)$. In IB4, §2.5 we will give an example of an ordered field in which the ordering is not Archimedean.

4.4. *Fundamental Sequences*

We consider sequences $(a_n)_{n=1,2,\ldots}$, which for brevity we shall denote[81] simply by a, of rational numbers a_n satisfying the Cauchy criterion for convergence:

For each rational number $\epsilon > 0$ there exists a natural number n_0 such that $| a_n - a_m | < \epsilon$ for all $n, m \geqslant n_0$.

These sequences are called *fundamental sequences* or *Cauchy sequences.* If addition and multiplication are defined by

$$(59) \qquad (a + b)_n = a_n + b_n , \qquad (ab)_n = a_n b_n ,$$

[78] To be sure, we sometimes mention products of the form nz (where n is a natural number and z is an element of the module); but these products could always be considered as sums (of n summands), so that there is no need of a multiplication for the elements of the module.

[79] The ordering here is lexicographic (see page 130, footnote 62).

[80] Of course, the concept defined here has nothing whatever to do with the incorrect use of this expression in analysis.

[81] In $(a_n)_{n=1,2,\ldots}$ the n is a bound variable (cf. IA, §8.4, §2.6), as is indicated by the sign of equality after it; another possible notation is $n \to a_n$, but then we must indicate in some way that in the symbol a_n the n is to be replaced by the natural numbers and by nothing else. The abbreviation a is to be understood as follows: a_n is the value of the function a for the argument n, or in other words a_n is the nth term of the sequence.

the set of fundamental sequences becomes a commutative ring $\mathfrak{R}$ with unit element. Of course, we must first of all show that the sequences $a + b$, ab defined in this way are again fundamental sequences. For $a + b$ this result follows immediately from the inequality [cf. (53)]

$$|(a_n + b_n) - (a_m + b_m)| = |(a_n - a_m) + (b_n - b_m)| \leqslant | a_n - a_m | + | b_n - b_m |.$$

In order to prove the same result for ab, we require the following theorem:

For every fundamental sequence there exists a rational number s with $| a_n | \leqslant s$ *for all n*. To prove this theorem we first determine n_0 in such a way that $| a_n - a_m | < 1$ for all $n, m \geqslant n_0$ and then set $m = n_0$. For $n \geqslant n_0$ it follows from (53) that $| a_n | = | a_{n_0} + (a_n - a_{n_0})| < | a_{n_0} | + 1$. So it is sufficient to take $s \geqslant | a_i |$ $(i = 1, ..., n_0 - 1)$ and $\geqslant | a_{n_0} | + 1$, as is always possible.

Thus we may consider s (>0) to have been so chosen that $| a_n |, | b_n | \leqslant s$ for all n, and then from

$$| a_nb_n - a_mb_m | = | a_n(b_n - b_m) + (a_n - a_m) b_m | \leqslant s | b_n - b_m | + s | a_n - a_m |$$

we readily obtain the desired result that ab is a fundamental sequence.

The ring properties and the commutativity of multiplication follow readily from (59). It is obvious that the sequences $(0)_{n=1,2,...}$ and $(1)_{n=1,2,...}$ are the zero element and the unit element respectively, so that we shall denote them simply by 0 and 1.

Now for every fundamental sequence we construct a new symbol lim a. Since this symbol is to mean the limit of the sequence, we shall set $\lim a = \lim b$ if and only if $c = a - b$ is a *zero sequence*; that is, if for every rational number $\epsilon > 0$ there exists a natural number n_0 with $| c_n | < \epsilon$ for all $n \geqslant n_0$. In other words, if we let $\mathfrak{N}$ denote the set of zero sequences, we are introducing into $\mathfrak{R}$ a relation $\doteq$ by setting $a \doteq b$ if and only if $a - b \in \mathfrak{N}$; in the mapping $a \rightarrow \lim a$ two fundamental sequences have the same image if and only if the relation $\doteq$ holds between them. In order that we may set

(60) $\lim a + \lim b = \lim(a + b), \qquad \lim a \cdot \lim b = \lim ab,$

and thus make the mapping $a \rightarrow \lim a$ into a homomorphism of the ring $\mathfrak{R}$ onto the ring with elements lim a, it merely remains to prove that $\doteq$ is an equivalence relation which is consistent with addition and multiplication. For this purpose we require only the following three properties[82] of $\mathfrak{N}$:

[82] The proof of these properties can be omitted here, since it is exactly the same as for the corresponding theorems in analysis; in (61_3) we require the fact, proved just above, that every fundamental sequence is bounded.

$$(61_1) \qquad 0 \in \mathfrak{N}.$$

$$(61_2) \qquad a - b \in \mathfrak{N}, \quad \text{if} \quad a, b \in \mathfrak{N}.$$

$$(61_3) \qquad ab \in \mathfrak{N}, \quad \text{if} \quad a \in \mathfrak{N}, b \in \mathfrak{R}.$$

For $a \doteq a$ follows from (61_1), and from $a \doteq c$, $b \doteq c$ it follows by (61_2) that $a - b = (a - c) - (b - c) \in \mathfrak{N}$, so that $a \doteq b$, which means that $\doteq$ is an equivalence relation. Since $a \doteq b$ (or in other words, $a - b \in \mathfrak{N}$) implies not only $(a + c) - (b + c) \in \mathfrak{N}$ but by (61_3) also $ac - bc = (a - b)c \in \mathfrak{N}$, the equivalence $\doteq$ is in fact consistent with addition and multiplication.[83]

Under definition (60) the set of symbols $\lim a$ becomes a commutative ring,[84] since the homomorphism $a \to \lim a$ naturally preserves the ring properties of $\mathfrak{R}$. We now wish to make this ring into an extension of the field of rational numbers. Although the problem here is of exactly the same kind as those already solved in §§2.2 and 4.3, we will now solve it in a different way, which can be extended more easily to other cases. We first show[85] that

$$(62) \qquad u \to \lim (u)_{n=1,2,\ldots}$$

is an isomorphism, or in other words, a one-to-one homomorphism of the field of rational numbers: for it follows from $(u)_{n=1,2,\ldots} \doteq (v)_{n=1,2,\ldots}$ that $u = v$, because by the definition of $\doteq$ we have $| u - v | < 1/n$ for every natural number n, so that $| u - v | > 0$ is impossible. The fact that (62) is a homomorphism then follows at once from (59) and (60). Consequently, in the following argument we no longer require the special properties of $\lim a$: we simply set $\lim a = u$, $u = \lim a$ if and only if $\lim a = \lim (u)_{n=1,2,\ldots}$. Since (62) is a one-to-one mapping, the comparativity of equality is thereby preserved, as is easily shown by separate consideration of four cases, as in §2.2. Finally, the fact that (62) is a homomorphism shows that addition and multiplication as defined in (60) are identical, when applied to rational numbers, with the earlier addition and multiplication.

The commutative ring with unit element that has thus been formed as an extension of the field of rational numbers is in fact a field, known as the *field of real numbers*. In §4.6 this assertion will be proved very simply by

[83] This proof is obviously valid for any commutative ring $\mathfrak{R}$ containing a subset $\mathfrak{N}$ with the properties (61) (see the concept of an ideal in IB5, §3).

[84] Namely, the ring of residue classes of $\mathfrak{R}$ mod $\mathfrak{N}$ (see IB5, §3.6).

[85] The u, v here are rational numbers and not sequences. The sequence $(u)_{n=1,2,\ldots}$ has all its elements $= u$; it is obviously a fundamental sequence.

a general argument, but we wish to prove it here also,[86] to which end it only remains to show that every real number $\lim a \neq 0$ has an inverse. But since a is not a zero sequence, there exists an $s > 0$ such that for every natural number n we can find a natural number $m \geqslant n$ with $|a_m| \geqslant s$. On the other hand, since a is a fundamental sequence, there exists a natural number n_0 with $|a_n - a_m| < s/2$ for $n, m \geqslant n_0$. Since $|a_n| = |a_m + (a_n - a_m)| \geqslant |a_m| - |a_n - a_m|$ [see (54)], we thus have $|a_n| \geqslant s/2$, so that $a_n \neq 0$ for $n \geqslant n_0$. The sequence a' defined by $a'_n = 1$ for $n < n_0$, $a'_n = a_n^{-1}$ for $n \geqslant n_0$ is easily seen to be a fundamental sequence, in view of

$$|a_m^{-1} - a_n^{-1}| = |a_m a_n|^{-1} |a_m - a_n| \leqslant 4s^{-2} |a_m - a_n| \quad \text{for } m, n \geqslant n_0,$$

and since $(aa' - 1)_n = 0$ for $n \geqslant n_0$, we have $\lim a \cdot \lim a' = 1$.

A slight extension of the argument shows that for $|a_n| \geqslant s/2$ we can also make the following statement: if $\lim a \neq 0$, then either there exists an n_0 such that $a_n > 0$ for all $n \geqslant n_0$, or else there exists an n_0 such that $a_n < 0$ for all $n \geqslant n_0$. For if there exists an $s > 0$ such that for every natural number n we can find a natural number $m \geqslant n$ with $a_m \geqslant s$, then we have $a_n = a_m + (a_n - a_m) \geqslant s/2 > 0$ for $n \geqslant n_0$; and if not, then for arbitrary $s > 0$ there exists a natural number n_1 with $a_n < s/2$ for all $n \geqslant n_1$. Since for suitable s, n_0 we have already proved that $|a_n| \geqslant s/2$ for all $n \geqslant n_0$, we now have $a_n \leqslant -s/2$ for all $n \geqslant \max(n_0, n_1)$. But the two cases are inconsistent with each other. Thus the argument also shows that in the first case we can choose $c > 0$ such that $a_n \geqslant c$ for all $n \geqslant n_0$, and in the second case we can arrange that $a_n \leqslant -c$ (<0) for $n \geqslant n_0$. So we see that the addition of a zero sequence to a produces no change in these two cases: the decision as to which of the two cases occurs depends only on $\lim a$. The set of elements $\lim a$ for which a satisfies the requirement of the first case is easily seen to form a domain of positivity. In the resulting ordering of the field of real numbers $\lim a < \lim b$ now means that $\lim a \neq \lim b$, and that there exists a natural number n_0 with $a_n \leqslant b_n$ for all $n \geqslant n_0$. To be consistent with the above formulation, we really ought to have written $a_n < b_n$, but the argument shows at once that we can also write $a_n \leqslant b_n$. It is to be noted that $a_n < b_n$ for all $n \geqslant n_0$ does not imply $\lim a < \lim b$ but merely that $\lim a \leqslant \lim b$, a result which follows equally well from $a_n \leqslant b_n$ for all $n \geqslant n_0$.

We now prove that in the field of real numbers every non-empty set M bounded from below has a greatest lower bound, and consequently that every non-empty set bounded from above has a least upper bound. For

[86] Especially because the proof given here requires an ordering for the absolute values only, and thus remains valid for more general systems; cf. van der Waerden [2], §§74, 75.

every real number $\lim a$ there certainly exists a rational number not greater than $\lim a$ and also a rational number not smaller than $\lim a$, for it was proved above that for a suitably chosen rational number s we have $| a_n | \leqslant s$ for all n, so that $-s \leqslant a_n \leqslant s$ for all n and consequently $-s \leqslant \lim a \leqslant s$. Thus the set M has a rational lower bound. After choice of any integer $g > 1$ we now denote by r_n, as in §4.1, the greatest integral multiple of g^{-n} ($n = 0, 1, 2, \ldots$) that is still a lower bound of M. In view of the fact that the ordering of the rational numbers is Archimedean, the existence of r_n implies that every number less than a lower bound of M is again a lower bound and every lower bound lies below some rational number, which may simply be any number that is not smaller than some number in M. For $m > n$ the number g^{-n} is an integral multiple of g^{-m} because $g^{-n} = g^{m-n}g^{-m}$, and thus we have

$$r_n \leqslant r_m < r_n + g^{-n} \qquad \text{for} \quad m \geqslant n. \tag{63}$$

It follows that $| r_m - r_n | g^{-n_0}$ for $m, n \geqslant n_0$, and since §4.1 shows that g^{-n_0} can be made smaller than any given positive number, the sequence $r = (r_n)_{n=1,2,\ldots}$ is a fundamental sequence. Furthermore, it follows from (63) that

$$r_n \leqslant \lim r \leqslant r_n + g^{-n} \qquad \text{for all} \quad n. \tag{64}$$

Now if $\lim a$ is a number such that $\lim a < \lim r$, or in other words if $d = \lim r - \lim a > 0$, let us determine a natural number n with $g^n > d^{-1}$ (as is possible, since there exists a rational number $\geqslant d^{-1}$) and therefore with $\lim a - \lim r < -g^{-n}$. By (64) we then have $\lim a < r_n$, so that $\lim a$ cannot be a member of M and consequently $\lim r$ is a lower bound of M. On the other hand, if $\lim a$ is such that $\lim r < \lim a$, then in the same way $g^{-n} < \lim a - \lim r$. But $r_n + g^{-n}$ is certainly not a lower bound of M; i.e., there exists a real number $\lim b$ with $\lim b \in M$ and $\lim b < r_n + g^{-n}$. Thus we have $\lim b < \lim a + r_n - \lim r$ and therefore, by (64), $\lim b < \lim a$. Consequently, no real number $> \lim r$ is a lower bound of M, so that $\lim r$ is the greatest lower bound of M, as desired.

As in §4.1, we see that $a_n = (r_n - r_{n-1})\, g^n$ ($n = 1, 2, \ldots$) is a nonnegative integer $<g$, that with $r_0 = a_0$ we have $r_n = \sum_{i=0}^{n} a_i g^{-i}$, which shows how our present development is connected with decimal expansions. In general, it is easy to see that for every fundamental sequence a the real number $\lim a$ is in fact the limit of the sequence, so that in our case we may write $\lim r = \lim_{n\to\infty} r_n = \sum_{n=0}^{\infty} a_n g^{-n}$ in the usual notation.

It is also easy to prove directly that the Cauchy criterion for convergence is valid for a sequence of real numbers $\lim a^{(n)}$ ($n = 1, 2, \ldots$). For let us determine, corresponding to each value of n, a natural number n' such that

$| a_{n'}^{(n)} - \lim a^{(n)} | < 1/n$. Then, from the hypothesis of the Cauchy criterion that for $\epsilon > 0$ there exists a natural number n_0 such that $| \lim a^{(n)} - \lim a^{(m)} | < \epsilon$ for $n, m \geqslant n_0$, we see that the sequence b of $b_n = a_{n'}^{(n)}$ is fundamental and that $\lim b$ is the limit of the sequence of the numbers $\lim a^{(n)}$. Since this proof makes use of the ordering only for absolute values, it is more general than the usual theorem for the Cauchy criterion, which is based on the existence of a greatest lower and a least upper bound.

4.5. *Isomorphisms*

In §4.1 and also in §4.3 and §4.4 we have constructed complete ordered modules containing the module for the rational numbers. We now wish to show that any two such modules can be mapped onto each other by an order-preserving isomorphism leaving all the rational numbers fixed, where by an order-preserving mapping f (which may, in particular, be an isomorphism) we mean a mapping that is monotone increasing; that is, if $x < y$, then $f(x) < f(y)$ for all x, y. To do this we first show that the completeness of a module is equivalent in the following sense to a certain maximal property.

A module with Archimedean ordering that contains the rational numbers is complete if and only if it is not contained in a larger module with Archimedean ordering.

For if such a module $\mathfrak{M}$ is not complete, then it can be extended by the procedure of §4.3 to a module with Archimedean ordering. Consequently, if $\mathfrak{M}$ is not contained in a larger module with Archimedean ordering, then $\mathfrak{M}$ is complete. On the other hand, if $\mathfrak{M}$ is complete, and if $\mathfrak{M}'$ is a module with Archimedean ordering that contains $\mathfrak{M}$, then $\mathfrak{M}$ and $\mathfrak{M}'$ coincide, as can be proved in the following way:

Let a' be an element of $\mathfrak{M}'$, let M be the set of rational numbers $<a'$ and let a be the least upper bound of M, which is certainly in $\mathfrak{M}$.

We now require the following lemma: if x, y with $x < y$ are elements of a module with Archimedean ordering that contains the rational numbers, then the module also contains a rational number $r = m/n$ with $x < r < y$. For the proof we simply determine the natural number n and then the natural number m such that $1/n < y - x$ and $(m - 1)/n \leqslant x < m/n$; for then we have

$$m/n = (m - 1)/n + 1/n < x + (y - x) = y.$$

Thus for every element $x < a'$ in $\mathfrak{M}$ we can find a rational number r with $x < r < a'$, so that x is not an upper bound of M and therefore $x \neq a$. On the other hand, if x is an element of $\mathfrak{M}$ with $x > a'$, there exists a rational number r with $a' < r < x$, so that r is a smaller upper bound of M than x, and thus again $x \neq a$. But since $x \neq a'$ implies either $x < a'$ or $a' < x$, we have shown that $a' \neq a$ is impossible; in other words, $a' = a$, and thus every element of $\mathfrak{M}'$ also belongs to $\mathfrak{M}$, which completes the proof of the stated maximal property.

We now let $\mathfrak{M}$ denote any complete ordered module that contains the rational numbers, and we map $\mathfrak{M}$ in the following way onto the complete module, now denoted by $\mathfrak{M}_0$, that was constructed in §4.3:

$$x \to \text{fin } M_x, \tag{65}$$

where M_x is the set of rational numbers $<x$. If $x < y$, then, as already shown, there exists a rational number r with $x < r < y$ and also a rational number s with $r < s < y$, so that r is in $S(M_x)$ but not in $S(M_y)$. On the other hand, since it is obvious that $M_x \subseteq M_y$ and thus $S(M_y) \subseteq S(M_x)$, we see that $S(M_y) \subset S(M_x)$. Consequently, $x < y$ implies fin $M_x <$ fin M_y, so that the mapping (65) is monotone increasing, and therefore one-to-one. That (65) is an isomorphism (with respect to addition) can be seen as follows: Since $r < x, s < y$ implies the inequality $r + s < x + y$, we have $M_x + M_y \subseteq M_{x+y}$. Now in order to show $M_{x+y} \subseteq M_x + M_y$, we choose a rational number $t < x + y$ and then a rational t' with $0 < t' \leqslant x + y - t$ so that $t \leqslant x + y - t'$. But there exist rational numbers r, s with $x - t'/2 \leqslant r < x, y - t'/2 \leqslant s < y$, which implies that $x + y - t' \leqslant r + s$. Taking these inequalities together we see that $t \leqslant r + s$, $r < x$, $s < y$. Thus for $r' = r - (r + s - t)/2$, $s' = s - (r + s - t)/2$ we have $r' \in M_x$, $s' \in M_y$, $t = r' + s'$ and therefore $t \in M_x + M_y$. Consequently, $M_{x+y} \subseteq M_x + M_y$ and thus $M_{x+y} = M_x + M_y$. In view of (58) we have therefore proved that fin M_x + fin M_y = fin M_{x+y}, which means that (65) is actually an isomorphism with respect to addition. Thus the entire set of numbers fin M_x (with $x \in \mathfrak{M}$) also forms a complete ordered module $\mathfrak{M}'$. But this module contains all the rational numbers, since for a rational number r we have $S(M_r) = S(r)$ and therefore fin $M_r = r$. By the maximal property proved at the beginning of this section, it follows that $\mathfrak{M}_0 = \mathfrak{M}_0$, so that the mapping (65) of $\mathfrak{M}$ onto the module $\mathfrak{M}_0$ is an order-preserving isomorphism which leaves the rational numbers fixed. Now if $\mathfrak{M}_1$, $\mathfrak{M}_2$ are two complete ordered modules that contain the rational numbers, we can first map $\mathfrak{M}_1$ by (65) onto $\mathfrak{M}_0$ and then map $\mathfrak{M}_0$ onto $\mathfrak{M}_2$ by the inverse of the mapping (65) from $\mathfrak{M}_2$ into $\mathfrak{M}_0$. Thus we have the following important result:

Two complete ordered modules that contain the rational numbers can be mapped onto each other by a mapping which is isomorphic and order-preserving and leaves the rational numbers fixed.

Moreover, there can be only one such mapping; for (65) is the only order-preserving mapping of $\mathfrak{M}$ onto $\mathfrak{M}_0$ that leaves the rational numbers fixed. To show this we need only note that by §4.3 the set M_x has the least upper bound fin M_x in $\mathfrak{M}_0$, and, on the other hand, has the least upper bound x in $\mathfrak{M}$, since by the lemma at the beginning of this section there exists a rational number $r \in M_x$ such that $y < r$ for every element $y \in \mathfrak{M}$ with $y < x$. But an order-preserving mapping of $\mathfrak{M}$ onto $\mathfrak{M}_0$ must map the least upper bound of M_x in $\mathfrak{M}$ onto the least upper bound of M_x in $\mathfrak{M}_0$, and must therefore map the element x of $\mathfrak{M}$ onto the real number fin M_x.

This result shows that it makes no difference which of the above modules (all of them are ordered and complete and contain the rational numbers) is called the module of the real numbers. For any two of them there is a uniquely determined mapping of one onto the other which leaves the rational numbers fixed and preserves order and addition.

4.6. *Multiplication*

In §4.4 we have already defined multiplication for the real numbers, but we now wish to define multiplication independently of the particular construction chosen there for the module of real numbers. Such a definition is made possible by the existence of monotone endomorphisms. If $c \neq 0$, the mapping $x \to cx$ is a monotone endomorphism (as in §2.5). On the other hand, any monotone endomorphism f is already uniquely determined by $f(1)$; for if g is another monotone endomorphism with $f(1) = g(1)$, then by §3.5 it follows that $f(x) = g(x)$ for all rational numbers.[87] But then $f(x) = g(x)$ for every real number x; for let M_x again denote the set of rational numbers $<x$, so that x is the least upper bound of M_x. Then if the mapping f is monotone increasing, $f(x)$ must be the least upper bound of $f(M_x)$ (= the set of $f(r)$ with $r \in M_x$), and if f is monotone decreasing, then $f(x)$ must be the greatest lower bound of $f(M_x)$. Consequently, just as in §3.5 for the rational numbers, the mappings $x \to cx$ are the only monotone endomorphisms of the module of real numbers.[88] Thus, in exactly the same way as by (40) for the integers, we can define multiplication for the real numbers by setting $f(x) = f(1)x$, where f is a monotone endomorphism or is the zero mapping ($x \to 0$). This definition involves only the concepts of addition (since f is an endomorphism) and order (since f is monotone) and the number 1, so that the isomorphisms of

[87] Of course, we cannot merely cite the above theorem but must prove it anew by the same method, since now $f(x)$, $g(x)$ are no longer required to be rational numbers.

[88] As G. Hamel has shown ("Eine Basis aller Zahlen und die unstetigen Lösungen der Funktionalgleichung $f(x + y) = f(x) + f(y)$," Math., Ann. 60, 459–462, 1905), there exist other (nonmonotone) endomorphisms; these are all discontinuous and even unbounded in every neighborhood of any number, so that in fact they are extremely strange functions.

the complete ordered modules considered in §4.5 remain isomorphisms with respect to multiplication. *The field of real numbers is uniquely determined, up to order-preserving isomorphic mappings, by the requirement that it be a complete ordered field containing the rational numbers*, completeness for an ordered field being defined in exactly the same way as for an ordered module.

By making use of endomorphisms, it is very easy to prove the fact, already proved in §4.4, that every real number $\neq 0$ has an inverse with respect to multiplication; in other words, that we are actually dealing here with a field. Since the endomorphism $x \to cx$ $(c \neq 0)$ is monotone and therefore one-to-one, it maps the module of real numbers isomorphically onto a module $\mathfrak{M}$, which is also ordered and complete. But then the module of the rational numbers is mapped onto an isomorphic module which has all the properties of the module of the rational numbers and therefore differs from it only in the names given to the elements. Thus by the theorem of maximality in §4.5, the set $\mathfrak{M}$ contains all the real numbers, including 1 in particular, so that there exists a number x with $cx = 1$.

Up to now the proof of the existence of a monotone endomorphism f with $f(1) = c$ $(\neq 0)$ has been taken over from §4.4. But we can give an independent proof on the basis of the following definition: for $c, x > 0$ we define $f(x)$ as the least upper bound of the set of all products rs of positive rational numbers with $r < c, s < x$ and then set $f(0) = 0, f(-x) = -f(x)$; then the mapping $-f$ is seen to be a monotone endomorphism with $(-f)(1) = -c$. Since zero and the positive and negative numbers must now be treated as separate cases, the proofs for the rules of calculation will be considerably longer than our earlier proofs. Of course, this difficulty can be avoided if we first construct only the positive rational numbers (see the end of §3.2), proceed from these to the positive real numbers, and then construct the real numbers in the form $a - a'$ as in §2.2. Then the multiplication of positive real numbers is defined as in the present section, and the simplest subsequent procedure is to define the multiplication of real numbers in accordance with (43), by the method indicated at the end of §2.4.

It is obvious that the monotone endomorphism $x \to cx$ is also a homomorphism with respect to multiplication if and only if $c(xy) = cxcy$ for all x, y; that is, if $c^2 = c$ or, since $c \neq 0$, if $c = 1$. Thus there is only one monotone isomorphism of the field of real numbers into itself, namely, the identical mapping. Now any isomorphism of the field of real numbers into itself must be order-preserving, since we shall see in §4.7 that every positive real number is a square (i.e., is of the form x^2), while on the other hand a number $x^2 \neq 0$ must be positive in an ordered ring. Thus the domain of positivity must consist exactly of the elements $x^2 \neq 0$ and will therefore, in view of the equation $f(x^2) = f(x)^2$, be mapped into itself by any isomorphism f of the field of real numbers; but if $x < y$, or in other words $0 < y - x$, then $0 < f(y - x) = f(y) - f(x)$, or $f(x) < f(y)$, so that the isomorphism is order-preserving as stated. An isomorphism of a field onto itself is called an *automorphism* of the field. *Thus, like the field of rational numbers, the field of real numbers admits exactly one automorphism, namely the identical mapping*. In IB7, §6, and IB8, §1.2 we will find examples of fields that admit other automorphisms as well.

4.7. *Roots*

We shall show that after choice of any natural number n in the field of real numbers any number $a \geqslant 0$ is the nth power of exactly one real number $x \geqslant 0$; in other words, there exists exactly one number x with $x^n = a, x \geqslant 0$. For if $0 \leqslant x < y$, complete induction on n shows that $x^n < y^n$ by the monotone law of multiplication, which proves the uniqueness. The existence follows from the mean-value theorem (IB8, §2.1), since on the one hand $0^n \leqslant a$ and on the other $(1 + a/n)^n > a$. The number x, uniquely determined in this way by $x^n = a, x \geqslant 0$, is denoted by $\sqrt[n]{a}$ and is called the *nth root* of a; for $n = 2$ we write $\sqrt{a}$ instead of $\sqrt[2]{a}$. For $n = 2k$ (k a natural number) the assumption $a \geqslant 0$ is necessary, since $x^{2k} = (x^k)^2 \geqslant 0$; and discarding the requirement $x \geqslant 0$ would mean that the uniqueness of the solution of $x^n = a$ is lost, since both $\sqrt[n]{a}$ and $-\sqrt[n]{a}$ satisfy the equation. The fact that solutions of an algebraic equation are also called roots of the equation seems to have led to misunderstanding and to the undesirable practice (for $n = 2k$) of calling both $\sqrt[n]{a}$ and $-\sqrt[n]{a}$ the nth root of a, or even of writing both $\sqrt{4} = 2$ and $\sqrt{4} = -2$, without taking into account the fact that if we are to allow many-valued expressions of this sort, the comparativity of equality is lost (since otherwise $\sqrt{4} = 2$, $\sqrt{4} = -2$ would imply $2 = -2$). Let us again illustrate the language adopted here (which is quite common): *the equation* $x^2 - 4 = 0$ *has the two roots* 2 *and* -2; *but the square root of* 4 *is* 2 *and not* ± 2. Let us note the equation

$$\sqrt[2k]{a^{2k}} = |a|, \tag{66}$$

in which the sign for the absolute value is often quite wrongly omitted; the proof of this equation follows at once from $|a| \geqslant 0$ and $|a|^{2k} = a^{2k}$.

This use of the root sign in the field of real numbers is to be distinguished from its frequent use (as in IB7, §2) in algebraic extensions of fields. *For a rational number a the symbol* $\sqrt[n]{a}$ *denotes an element* α *of an extension of the field of rational numbers which satisfies the equation* $\alpha^n = a$. But even in a prescribed extension the value of α is in general not uniquely determined by the equation $\alpha^n = a$: thus by $\sqrt[n]{a}$ we mean an *arbitrary* one of these elements, which is to remain fixed during a given investigation. The results obtained for $\sqrt[n]{a}$ remain valid if we replace $\sqrt[n]{a}$ by any of the other elements α (from the extension in question) with $\alpha^n = a$. For example, for $\sqrt{-3}$ we may take either the complex number $i\sqrt{3}$ (see IB8, §1) or the complex number $-i\sqrt{3}$, provided the extension in question is contained in the field of complex numbres (which is not necessarily the case).

If $n = 2k - 1$ (k a natural number) we may allow $a < 0$ in the definition of the nth root, provided the restriction $x \geqslant 0$ is also discarded; for $x \to x^n$ now takes positive numbers into positive numbers and negative numbers into negative numbers, and $(-x)^n = -a$ means the same as $x^n = a$. In particular, we have $\sqrt[2k-1]{-1} = -1$, while $\sqrt[2k]{-1}$ is not defined. In contrast to (66), we have

$$\sqrt[2k-1]{a^{2k-1}} = a, \tag{67}$$

the proof of which is an obvious consequence of the identity $a^{2k-1} = a^{2k-1}$.

In view of the general validity (provided a is nonnegative for even n) of the equation $(\sqrt[n]{a}\,|)^n = a$, we are led by (66) and (67) to conjecture that more generally

$$\sqrt[n]{a^m} = \begin{cases} (\sqrt[n]{a}\,|)^m, & \text{for odd } n \\ (\sqrt[n]{|\,a}\,|)^m, & \text{for even } n \text{ with } a^m \geqslant 0 \end{cases} \tag{68}$$

where m is an integer and $a \neq 0$ for $m \leqslant 0$. Since in the second case the expression on the right side is obviously $\geqslant 0$, we need only prove that in both cases the nth power of the right expression is $= a^m$, which is easily proved from (49).

Further rules for calculation with roots are

$$\sqrt[n]{ab} = \sqrt[n]{a}\sqrt[n]{b}, \quad \text{if } n \text{ is odd or } a, b \geqslant 0, \tag{69}$$

$$\sqrt[mn]{a} = \sqrt[m]{\sqrt[n]{a}}, \quad \text{if } m, n \text{ are odd or } a \geqslant 0. \tag{70}$$

Since the right sides are obviously $\geqslant 0$ for even n, we need only show, from (50) and (49), that the nth power of the right side of (69) is $= ab$ and the mnth power of the right side of (70) is $= a$.

For $a > 0$ we also obtain from (70), for natural numbers n, h and arbitrary integer m, that

$$(\sqrt[nh]{a})^{mh} = \sqrt[n]{a})^m. \tag{71}$$

For $m/n = m'/n'$ with integers m, m' and natural numbers n, n' we obtain from (71) that

$$(\sqrt[n]{a})^m = (\sqrt[nn']{a})^{mn'} = (\sqrt[nn']{a})^{nm'} = (\sqrt[n']{a})^{m'}.$$

Thus, the expression $(\sqrt[n]{a})^m$ depends only on a and on the fraction m/n and may therefore be denoted by $a^{m/n}$ (to be read as *a to the m/nth power*); since $\sqrt[1]{a} = a$, this definition agrees with the definition of powers for integral exponents. In particular, for $a > 0$ we can write $\sqrt[n]{a}$ in the form $a^{1/n}$. It is easy to show that the rules (47), (49), (50) hold for arbitrary rational exponents, provided we assume that the real numbers a, b are positive.

The validity of (47) for arbitrary rational exponents r, s, i.e.,

$$a^{r+s} = a^r a^s,$$

means that the mapping $r \to a^r$, defined by the positive real number a (and denoted below by f) is a homomorphic mapping of the group of rational numbers under addition (more concisely, the additive group of the rational numbers) into the group of positive real numbers under multiplication (the multiplicative group of the positive real numbers). The fact that the positive real numbers form a group under multiplication is merely a special case of the more general fact that the domain P of positivity of an ordered field is a group with respect to multiplication: for if $a, b \in P$, then $ab \in P$ by (44_4); and if $a \in P$, then $a^{-1} = a(a^{-1})^2 \in P$ by (44_4), since $(a^{-1})^2 \in P$ by a remark near the beginning of §3.4.

We now assume $a > 1$. For natural numbers n, m we then have $a^{m/n} > 1$; for if $a^{m/n} \leqslant 1$, it would follow from the monotonic law of multiplication that $a^m \leqslant 1$, whereas in fact $a^m > 1$ follows from $a > 1$ by the same monotonic law. Setting $r - s = m/n$, we again obtain from this monotonic law that $a^r = a^{m/n}a^s > a^s$ if $r > s$. Thus if $a^r = a^s$, both $r < s$ and $r > s$ are impossible, so that $r = s$. Consequently the mapping is one-to-one and is even an order-preserving isomorphism: namely, the real numbers a^r are in one-to-one correspondence with the rational numbers r, and act in exactly the same way with respect to order and multiplication as the rational numbers with respect to order and addition. Thus in view of the monotonic law of multiplication and the fact that a set of positive numbers can have only positive upper bounds, the multiplicative group of positive real numbers is a complete ordered module. To be sure, this module does not contain all the rational numbers, but (by what has just been proved) the a^r do constitute a set of numbers which acts exactly like the entire set of rational numbers with respect to order and to the given operation (which is now multiplication instead of addition). Thus the method of proof in §4.5 can be used to extend the isomorphism f to an isomorphism of the additive group of all the real numbers onto the multiplicative group of the positive real numbers: to do this, given any real number x, we merely define $f(x)$ in accordance with (65), namely as the least upper bound of the set of all numbers $f(r) = a^r$ (with rational $r < x$). Writing a^x instead of $f(x)$, we have

$$a^{x+y} = a^x a^y$$

for all real numbers x, y. The function $x \to a^x$ thus defined is called the *exponential function with base a*; as an order-preserving isomorphism it is monotone increasing and thus determines an inverse function on the set of positive real numbers; this inverse function is called the *logarithm*[89] *to the base a* (abbreviated: ${}^a\log$ or also ${}_a\log$). Of course, this function is again an (order-preserving) endomorphism of the multiplicative group of the positive real numbers onto the additive group of the real numbers; that is,

$$^a\log xy = {}^a\log x + {}^a\log y.$$

In making a computation we may therefore replace multiplication by addition, a fact which explains the practical importance of logarithms.

If for a we choose a positive real number < 1, the foregoing results remain valid, except that now the exponential function and the logarithm (to the base a) are monotone decreasing rather than increasing.

4.8. *Uncountability*

A set is said to be *countable* (cf. IA, §7.3) if either it is finite or else is equivalent to the set N of natural numbers[90] (see §1.5); otherwise the set is said to be *uncountable*. A non-empty set is countable if and only if its elements can be represented as the terms of an (infinite) sequence. For if M is infinite, then a one-to-one mapping of N onto M provides such a sequence, and if M is finite and is therefore the image under the mapping $n \to a_n$ of the segment A_m (see §1.5), then after choice of any $a \in M$ the mapping $n \to b_n$, with $b_n = a_n$ for $n \leqslant m$ and $b_n = a$ for $n > m$, has the desired property. On the other hand, if we assume that M consists of all the terms a_n $(n = 1, 2, ...)$ of a sequence and is not finite, then we can define a mapping f of N into M recursively as follows [cf. (4′)!]:

$f(1) = a_1, f(n + 1) = a_k$, where k is the smallest natural number[91] with $a_k \neq f(1), ..., f(n)$.

This mapping is obviously one-to-one, since $f(m) \neq f(n + 1)$ for $m < n + 1$. By complete induction on m we now show that every element a_m in M occurs as an image in the mapping f; for by the definition of f it follows from $a_m = f(n)$ that each of the terms $a_1, ..., a_{m-1}$ has one of the values $f(1), ..., f(n - 1)$ and thus if $a_{m+1} \neq f(1), ..., f(n)$, the equation $a_{m+1} = f(n + 1)$ must hold, which completes the proof that M is equivalent to N.

The set of rational numbers is countable. For the proof we first represent all the positive rational numbers as terms of a sequence by writing them in the order

$$1/1, 1/2, 2/1, 1/3, 2/2, 3/1, 1/4, 2/3, 3/2, 4/1, ...,$$

[89] It is preferable to use the word "logarithm" for the function rather than for the value of the function. The "logarithm of x" is in fact the value of the logarithm function for the argument x.

[90] Thus a set is *countably infinite* if and only if it is equivalent to N.

[91] If there were no such number, M would consist only of the elements $f(1), ..., f(n)$ and would thus be finite.

so that the nth term of the sequence is defined by $a_1 = 1$, $a_n = i/(m - i)$ for $n > 1$, if

$$\sum_{k=1}^{m-2} k < n \leqslant \sum_{k=1}^{m-1} k, \qquad n = \sum_{k=1}^{m-2} k + i.$$

The sequence $n \to a'_n$ with $a'_1 = 0$, $a'_{2k} = a_k$, $a'_{2k+1} = -a_k$ then comprises the entire set of rational numbers, so that by the above argument the set of rational numbers is countable.

But in contrast to the rational numbers, the *set of real numbers is uncountable*;[92] in other words: *given any sequence of real numbers, there exists a real number which is not a term of the sequence.*

To prove this, we represent the nth term ($n = 0, 1, 2, \ldots$) of the given sequence of real numbers as the sum of an integer a_{n0} and a proper decimal fraction with the digits a_{nm} ($<g$, $\geqslant 0$), where $g > 2$ is any chosen base:

$$a_n = a_{n0} + 0.a_{n1}a_{n2}a_{n3} \ldots ; \tag{72}$$

here, as in §4.1, we exclude the case that a number m_0 exists with $a_{nm} = g - 1$ for all $m \geqslant m_0$. Then it is obvious that the a_{nm} are uniquely determined by the a_n. For the sequences $m \to a_{nm}$ we now consider the *diagonal sequence* $n \to a_{nn}$ and form the sequence $n \to b_n$ with

$$b_n = \begin{cases} 0, & \text{if } a_{nn} \neq 0. \\ 1, & \text{if } a_{nn} = 0. \end{cases}$$

Since $b_n < g - 1$, the number $b = b_0 + 0.b_1b_2b_3 \ldots$ is of the same form as (72), so that $b_n \neq a_{nn}$ ($n = 0, 1, 2, \ldots$) implies the inequality $b \neq a_n$ ($n = 0, 1, 2, \ldots$). Thus the real number b is not a term of the given sequence of real numbers.

The procedure by which b is determined from the sequence $n \to a_n$ is called the (second) *Cantor diagonal procedure* (see also IA, §7.3). The uncountability of the set of real numbers proved in this way appears paradoxical, since it is certainly true that every real number must be defined in some way, and such a definition employs only finitely many letters of the alphabet, together with finitely many special symbols (such as |, by repeated use of which we can express all the natural numbers). So if we add the special symbols to the alphabet in any definite arrangement, the definitions of all the real numbers can be arranged in lexicographic order, in contradiction to the fact that the set of real numbers is uncountable. It seems that

[92] It follows that *irrational* numbers (that is, real numbers that are not rational) must exist. It is easy to give some examples directly: from the theorem in IB5, §4.4, it follows at once that $\sqrt{n}$ is irrational for every natural number n which is not the square of a natural number; in particular, $\sqrt{2}$ is irrational.

this paradox, which is really a serious one, can be resolved only by assuming that any natural language is so inexact as to lead inevitably, in extreme cases, to a contradiction. But if we use a formal language, whose formulas (propositions) are constructed according to exact rules prescribed in advance, then any mapping, and in particular a mapping that might possibly map the natural numbers onto all the real numbers, would necessarily be constructed in terms of this formal language. Then the uncountability of the set of real numbers would simply mean that *in this formal language* no such enumeration of them can be constructed. But if we make a suitable extension of the formal language, then all the expressions in it, and in particular all the real numbers that can be described in it, can in fact be enumerated in the new language. The concept of countability is thus dependent on the linguistic expressions[93] at our disposal.

This explanation of the paradox is available only from the constructive (IA, §1.5) or the operational (IA, §10.6) point of view. From the classical point of view (IA, §1.5) the real numbers (or the mathematical entities used to define them, such as the sets of rational numbers in §4.3), exist independently of the way we construct them. But now there is no longer any paradox; we must simply recognize that it is impossible to find any procedure for setting down in succession the definitions of all the real numbers.

Appendix to Chapter 1

Ordinal Numbers

The basic property of the set N of natural numbers, namely that every non-empty subset has a first element, i.e., a smallest element; cf. IB1, §1.3) leads us to consider arbitrary *well-ordered* sets and to interpret them as *ordinal numbers*. The principal purpose of the ordinal numbers is, in fact, to determine the "rank" of each element in a set of elements, and for that purpose the well-ordered sets are exactly suited. After the sequence of finite ordinal numbers $0, 1, 2, ..., n, n + 1, ...$ come the countable infinite ordinal numbers $\omega, \omega + 1, ...$; these ordinal numbers form a well-defined set Ω, which has taken its place in mathematics alongside the set R of real numbers.

[93] In §1.5 we defined the concept of finiteness by means of a mapping; but it can be shown that this concept is independent of the linguistic expressions employed (see Lorenzen [1]).

1. Atomization of the Continuum as an Introductory Example

Let $S = [0, 1]$ denote the set of real numbers in the closed interval 0 to 1. We divide the interval into two subintervals with one point in common, and then divide these subintervals similarly and so forth, each time dividing a closed interval I into two closed subintervals fI. Thus fS denotes the set of the two subintervals of S, $ffS = f^2S$ denotes the set of the four subintervals, f^3S the set of the eight subintervals, and so forth. From the interval S we obtain by the first division the subintervals S_0, S_1 with $S_0 \leqslant S_1$, from S_0 by the second division the intervals S_{00}, S_{01} with $S_{00} \leqslant S_{01}$ and from S_1 the subintervals S_{10} and S_{11}. In general, for any S_i, where the subscript i stands for a dyadic sequence of integers already assigned, we denote by S_{i0} and S_{i1} the left and right subinterval, respectively, of the interval S_i.

For example, we determine in this way the sequence of intervals S_0, S_{01}, S_{010}, S_{0101}, S_{01010}, ... in which 0 and 1 appear alternately. The intersection of the intervals in this sequence, to be denoted by $S_{01010...}$, is either a point or a closed interval. In the latter case we can continue the subdivision and consider the intervals S_{a0} and S_{a1}, where a denotes the infinite sequence of numbers 01010... .

In other words, we have here a "*well-ordered*" set of processes $P_1, P_2, ...$, where P_n leads to the 2^n intervals $S_{i1 \cdots i_n}$ with $i_1, ..., i_n \in \{0, 1\}$. After all these processes P_n there comes a process $P_{1,2,3,...}$, which we shall call P_ν, where ν cannot be a natural number since these have all been used. In P_ν we consider the set of "remainders"

$$S_{i_1 i_2 \ldots i_n \ldots}, \qquad i_1, i_2, ..., i_n, ... \in \{0, 1\}. \tag{1}$$

If this set contains at least one interval consisting of more than a single point, we consider the next process; that is, we subdivide into fI all the intervals I of the form (1) that consist of more than a single point; this is the process $P_{\nu+1}$. Then follow the processes $P_{\nu+2}$, $P_{\nu+3}$ ... . After all these processes $P_{\nu+n}$ with $n \in N$ comes $P_{\nu+\nu}$. In this process we consider all intersections $\bigcap F$, where F is any system of comparable intervals that contains an interval from each of the preceding processes. If the elements of F are denoted by S_{i_0}, $S_{i_0 i_1}$, ..., $S_{i_0 i_1 \ldots i_\nu i_{\nu+1} \ldots}$, it is natural to denote this intersection by

$$S_{i_0 i_1 i_2 \ldots i_{\nu+1} \ldots i_{\nu+n} \ldots},$$

where the index represents a "*double sequence*" (that is, a juxtaposition of two sequences, one after the other). In this way we can continue the subdivision until there is nothing more to divide, namely, until we have arrived at all the isolated points of the reduced continuum [0, 1]. For example, if the subdivision is always a bisection, the process P_ν is the last one, since every set then consists of a single point. It is worth noting that the result is the same if the subdivision fX of X is carried out uniformly; that is, in such a way that the ratio of the lengths of the subintervals is constant.[1] In other cases the "height" of the subdivision can become very great; that is, the necessary number of divisions is "extremely great." If the subdivision of 0, 1 is undertaken in such a way that whenever possible the half-segment $[0, \frac{1}{2}]$ lies in one of the subintervals, then the "height" is at least equal to $\nu + \nu$. At any rate, we see that the set of processes $P_0, P_1, P_2, \ldots$ is *well-ordered* and that the finite ordinal numbers are not sufficient to deal with irregular atomizations.

2. Fundamental Concepts

2.1. A set that is ordered by a relation (cf. IA, §8.3 and §7.4) has a first element x if no element of the set precedes x or, in other words, if there is no element z in the set such that $z < x$; and similarly y is a last element if no element of the set follows y or, in other words, if there is no element z such that $y < z$. An ordered set M is said to be *well-ordered* if the set M itself and each of its non-empty subsets has a first element in the prescribed ordering of M. By definition, the empty set $\emptyset$ is also well-ordered, and the same remark holds for every set consisting of a single element.

The set N of natural numbers in the usual order is well-ordered, which is one of its fundamental properties. Here the ordering is the usual $<$ (smaller than) relation. If we order the set N by putting all the odd natural numbers first and then all the even ones

(2) $$1, 3, 5, \ldots; \qquad 2, 4, 6, \ldots,$$

it is still well-ordered, where by the ordering we mean the relation that holds for the pair of numbers (m, n) if and only if m is odd and n is even or else, if m, n are both even or both odd, then $m < n$. The corresponding remarks hold for the sequences

(3) $$1, 3, 5, \ldots; \qquad 2 \cdot 1, 2 \cdot 3, 2 \cdot 5, \ldots; \qquad 2^2 \cdot 1, 2^2 \cdot 3, 2^2 \cdot 5, \ldots, \qquad 2^n \cdot 1, 2^n \cdot 3, 2^n \cdot 5, \ldots .$$

On the other hand, the set of natural numbers in the order

$$\ldots, n, n-1, \ldots, 3, 2, 1$$

is not well-ordered. The sets Q and R of all rational and all real numbers in order of magnitude are not well-ordered; for example, neither of them

[1] Since $\lim_{n\to\infty} q^n = 0$ for every q with $-1 < q < +1$.

has a first element. However, it is easy to well-order Q.[2] On the other hand, no one has yet succeeded in well-ordering the set R of all real numbers, although this problem is closely related to many other interesting problems; for example, the question *whether for every subset M of R we can define a permutation p_M of M such that $p_M(x) \neq x$ for every $x \in M$.* For if it is possible to well-order R, and therewith every M with $M \subseteq R$, this question can be answered in the affirmative.

2.2. Among the subsets of an ordered set M the *initial* segments are particularly important. A subset A of M is called a *segment* if for every element a of A the set A also contains every element b that precedes the element a in M, or in other words if $a \in A$ and $b < a$ imply $b \in A$. The empty set and the entire set M are also called segments; all other segments are *proper segments* of M.

It is important to note that if A is a proper segment of a well-ordered set M, then the set $M - A$ has a first element.

If we denote by $(\cdot, x)_M$ and $(\cdot, x]_M$ the sets of all $y \in M$ such that $y < x$ and $y \leqslant x$, respectively, then for every $x \in M$ the sets $(\cdot, x)_M$ and $(\cdot, x]_M$ are segments of M; they are sometimes called *initial intervals of M*, or the segments determined by x.

2.3. For ordered sets we define *similarity* or *isomorphism* as follows. An ordered set M is said to be *similar* or *isomorphic* to an ordered set M' if there exists a one-to-one mapping f of M onto M' such that for every pair m_1, m_2 of elements of M we have $m_1 < m_2$ if and only if $fm_1 < fm_2$.[3]

3. Simple Properties of Well-Ordered Sets

3.1. *Let W Be a Well-Ordered Set*

Theorem 1. *Every subset of the well-ordered set W is well-ordered.*

This theorem follows immediately from the definition of a well-ordered set.

[2] If n is a natural number ($n \geqslant 1$), let Q_n be the well-ordered set of rational fractions

$$\frac{1}{n}, -\frac{1}{n}, \frac{2}{n-1}, -\frac{2}{n-1}, \ldots, \frac{n-1}{2}, -\frac{n-1}{2}, \frac{n}{1}, -\frac{n}{1}$$

(note that they are not in the natural order). By forming the sequence $0, Q_1, Q_2 \cdots$ and striking out the terms that have already occurred, we obtain the well-ordered set

$$0; \tfrac{1}{1}, -\tfrac{1}{1}; \tfrac{1}{2}, -\tfrac{1}{2}, \tfrac{2}{1}, -\tfrac{2}{1}; \tfrac{1}{3}, -\tfrac{1}{3}, \tfrac{3}{1}, -\tfrac{3}{1}; \tfrac{1}{4}, -\tfrac{1}{4}, \tfrac{2}{3}, -\tfrac{2}{3}; \tfrac{3}{2}, -\tfrac{3}{2}, \tfrac{4}{1}; -\tfrac{4}{1}, \ldots$$

of all rational numbers. This ordering of Q is obviously different from the natural order, since smaller elements may precede larger ones.

[3] The symbol fa or $f(a)$ denotes the f image of a.

Theorem 2. *A set W is well-ordered if and only if it contains no infinite "decreasing sequences" (regressions).*

Proof: If W has an infinite decreasing sequence $a_0, a_1, a_2, \ldots$ with $a_0 > a_1 > a_2 > \ldots$, then the set of these elements, which is a subset of W, has no first element and therefore W is not well-ordered. Conversely, if W is not well-ordered, we let $A \subseteq W$ denote a non-empty subset with no first element. Then for $a_0 \in A$ there is an $a_1 \in A$ with $a_1 < a_0$, and similarly $a_2 \in A$ with $a_2 < a_1$, and so forth, which leads to the infinite decreasing sequence $a_0, a_1, a_2, \ldots$.

Theorem 3. *The well-ordered set W is not similar to any of its proper segments.*

Proof: Assuming that W is similar to a proper segment A of W, we let f denote a mapping of W onto A. Now let a be the first element of $W - A$; then $fa < a$; but then also $ffa = f^2a < fa, f^3a < f^2a$ and so forth, which leads to the infinite decreasing sequence f^na with n in N, in contradiction to Theorem 2.

3.2. *The Principle of Induction for Well-Ordered Sets*

Theorem 4. *Let W be a non-empty well-ordered set. Let the set M be such that*

(1) *M contains the first element of W.*

(2) *if $x \in W$ and $(\cdot, x)_W \subseteq M$, then $x \in M$* (for the notation here see 2.2). *Then $M \supseteq W$.*

Proof: If the assertion $M \supseteq W$ were false, the set $W - M$ would not be empty. As a *non-empty* subset of W it would have a first element, call it x. Then we would have $(\cdot, x)_W \subseteq M$ and thus by the induction hypothesis (2) also $x \in M$, in contradiction to the fact that $x \in W - M$.

For $W = N$ the principle of induction for well-ordered sets reduces to the ordinary principle of complete induction.

3.3. *Comparison of Well-Ordered Sets*

Theorem 5. *If the well-ordered set A is similar to a segment of the well-ordered set B, then there exists exactly one similar mapping of A such that fA is a segment of B.*

Proof: Let f and g be two distinct similar mappings of A such that fA and gA are segments of B. Let A_0 be the *maximal* segment in which f and g coincide. Then we must show that $A_0 = A$. Otherwise we would have $A_0 \subset A$. Let x be the *first* element of $A - A_0$, so that fx and gx are the first elements of $B - fA_0$ and $B - gA_0$. Now $fA_0 = gA_0$, so that

$fx = gx$, since fx and gx are the first element in one and the same set. Consequently, f and g coincide in a larger segment of A, namely, in $A_0 \cup \{x\}$. But this is a contradiction, so that $A_0 = A$.

Theorem 6. (Fundamental theorem on well-ordered sets.) *If A, B are well-ordered sets, then either A is similar to a segment of B, or B is similar to a proper segment of A.*

Proof: Let us assume that A is not similar to any segment of B, and then prove that B is similar to a proper segment of A. Let b_0 be the first point of B and fb_0 the first point of A. Let us then assume that X is a segment of B which is mapped isomorphically by f_X onto a segment of A; let x be the first element of $B - X$ and $f_x x$ the first element of $A - f_X X$; then we have an isomorphic mapping f_X of $(\,\cdot\,, x]_B$ onto a segment of A. Let M be the set of all elements $x \in B$ with the property that $(\,\cdot\,, x]_B$ is isomorphic to a segment of A. We see at once that the requirements (1) and (2) of the theorem on induction are satisfied for $B = W$ and thus $M \supseteq B$. This means that every segment X of B is isomorphic to a segment $f_X X$ of A, where f_X is the corresponding isomorphic mapping of X. If $X \subset Y \subseteq B$, then $f_X z = f_Y z$ for $z \in X$, since otherwise the restriction of f_Y to X and the mapping f_X would be two distinct isomorphic mappings of X onto a segment of A, in contradiction to Theorem 5. Thus $fz = f_X z$ for every $z \in X$ and every segment X of B defines an isomorphism, and therefore fB is a segment of A.

But the set fB is a proper segment of A; for if we had $fB = A$, then A would be isomorphic to B, in contradiction to the assumption that A is not isomorphic to any segment of B.

4. Definition and Simple Properties of Ordinal Numbers

4.1. *Definition*

By the fundamental theorem on well-ordered sets, two distinct well-ordered sets A and B are *comparable* to each other in the following sense. *Either A and B are similar, or A is similar to a proper segment of B, or B is similar to a proper segment of A.* Thus it is natural to regard the well-ordered sets as ordinal numbers.

Every well-ordered set W represents an ordinal number (OW)[4] *under the following conventions concerning equality and order* (§4.2) *and computation with well-ordered sets* (§4.3).

4.2. *Equality and Order of Ordinal Numbers*

If A and B are two well-ordered sets, their ordinal numbers OA and OB

[4] Following Cantor, many authors write $\overline{\overline{W}}$ instead of OW.

are equal if and only if A and B are similar. Thus any two similar well-ordered sets determine one and the same ordinal number.

The order-relation $OA < OB$ or $OB > OA$ means that A is similar to a proper segment of B, and $OA \leqslant OB$ means either $OA < OB$ or $OA = OB$.

It is easy to prove the transitivity of equality and order: *from* $OA \leqslant OB \leqslant OC$ *follows* $OA \leqslant OC$, where we have $OA < OC$ if the symbol $\leqslant$ actually means $<$ at least once.

4.3. *Computation with Ordinal Numbers*

4.3.1. *Sum of Ordinal Numbers.* If A and B are two *disjoint*[5] well-ordered sets, then $A + B$ denotes the union $A \cup B$ so ordered that the orders of A and B are preserved and all the elements of A precede those of B. Then the well-ordered sum $A + B$ determines the sum of the ordinal numbers $OA + OB$. It is easy to show that the sum is independent of the special representatives of the two ordinal numbers: for if $OA = OA_1$ and $OB = OB_1$ and also $A_1 \cap B_1 = \emptyset$, then we have $OA + OB = OA_1 + OB_1$. More generally, let X be a well-ordered set and for every $x \in X$ let $f(x)$ be a well-ordered set; then if

$$f(x) \cap f(x') = \emptyset \qquad \text{for} \quad x, x' \in X, \qquad x \neq x',$$

let $\sum_{x \in X} f(x)$ be the union $\bigcup_{x \in X} f(x)$, so ordered that $f(x)$ precedes $f(x')$ if and only if $x < x'$ in X.

In case the sets A and B are not disjoint, we proceed as follows. If (A, B) is the ordered pair of well-ordered sets A and B, let $\{1\} \times A$ be the set of all ordered pairs $(1, a)$ with $a \in A$ and let $\{2\} \times B$ be the set of all pairs $(2, b)$ with $b \in B$. If we now let the union $\{1\} \times A \cup \{2\} \times B$ be *lexicographically* ordered, the result is the ordered sum

$$\{1\} \times A + \{2\} \times B.$$

This sum represents a well-ordered set which we may consider as an ordinal number and denote by $OA + OB$.

More generally, if B is a well-ordered set and if to every $b \in B$ there corresponds a well-ordered set $f(b)$, then $\{b\} \times f(b)$ is the set of all ordered pairs (b, x) with $x \in f(b)$. If the union $\bigcup_{b \in B} \{b\} \times f(b)$ is lexicographically ordered, it represents a well-ordered set whose ordinal number is called the sum of the ordinal numbers $Of(b)$ taken over the ordinal number OB; this sum is denoted by $\sum_{b \in B} Of(b)$. Of course, there is no need for the $f(b)$ to be distinct.

[5] That is, $A \cap B = \emptyset$.

4.3.2. *Multiplication of Ordinal Numbers.* If B and C are well-ordered sets with the ordinal numbers OB and OC, then the product $OB \cdot OC$ of their ordinal numbers is defined as the sum $\sum_{b \in B} OC$. In this case $f(b) = C$ for every $b \in B$ in the preceding definition of this sum.

4.4. *Special Symbols*

If we write $0(\emptyset) = 0$, where $\emptyset$ is the empty set, and then set

$$O\{0\} = 1, \qquad O\{0, 1\} = 2, \qquad O\{0, 1, 2\} = 3, \ldots,$$

we obtain the well-ordered set of ordinal numbers

$$0, 1, 2, 3, \ldots, n, \qquad n + 1, \ldots .$$

This set determines the ordinal number $O\{0, 1, 2, \ldots, n, n + 1, \ldots\}$, which is denoted by ω or ω_0 .

4.5. *Arithmetic of the Ordinal Numbers*

4.5.1. *Addition and Multiplication.* For example, $2 + 3 = 5$, and $\alpha + 0 = \alpha$ for every ordinal number α.

Since the well-ordered sets

$$(4) \qquad \begin{gathered} 0, 1, 2, 3, \ldots, n, n + 1, \ldots \\ 1, 2, 3, 4, \ldots, n + 1, n + 2, \ldots \end{gathered}$$

are similar under the mapping $n \to n + 1$, we have $1 + \omega = \omega$. On the other hand $\omega + 1 > \omega$, since the ordinal number $\omega + 1$ may be represented by the well-ordered set

$$(5) \qquad 0, \frac{1}{2}, \frac{2}{3}, \ldots, \frac{n}{n + 1}, \ldots, 1,$$

which is not similar to any segment of (4). For if there existed a similar mapping f of (5) onto a segment of (4), then (4) would contain $f(1)$ in particular, and in front of $f(1)$ would come the infinite set of elements $f[n/(n + 1)]$, which is impossible, since no proper segment of (4) can be infinite, and therefore we must have $\omega + 1 > \omega$.

Consequently, $1 + \omega = \omega < \omega + 1$, so that $1 + \omega < \omega + 1$, which shows that addition of ordinal numbers is not always commutative.

The same remark holds for multiplication; for example, $\omega \cdot 2 = \omega$, $2 \cdot \omega = \omega + \omega > \omega$, so that $\omega \cdot 2 \neq 2 \cdot \omega$. However, it is easy to show that addition and multiplication are associative:

$$(6) \qquad (\alpha + \beta) + \gamma = \alpha + (\beta + \gamma),$$

$$(7) \qquad (\alpha\beta)\,\gamma = \alpha(\beta\gamma).$$

We also have

$$(8) \qquad (\alpha + \beta)\gamma = \alpha\gamma + \beta\gamma;$$

but in some cases

$$\alpha(\beta + \gamma) \neq \alpha\beta + \alpha\gamma,$$

for example, $\omega(1 + 1) \neq \omega \cdot 1 + \omega \cdot 1$, since $\omega \cdot 2 = \omega, \omega + \omega > \omega$. In exactly the same way as for ordinary ordinal numbers $<\omega$, we can prove the following fundamental theorem:

For every ordered pair (α, β), $\beta \neq 0$ *of ordinal numbers there exists a unique ordered pair* (κ, ρ) *of ordinal numbers such that* $\alpha = \kappa\beta + \rho$ with $0 \leqslant \rho < \beta$.

For example, if $\beta = 2$, then every ordinal number is either of the form $\kappa \cdot 2$ (an even ordinal number) or of the form $\kappa 2 + 1$ (an odd ordinal number).

4.5.2. *Subtraction.* If $\alpha \leqslant \beta$, the equation $\alpha + \xi = \beta$ has exactly one solution, which is denoted by $-\alpha + \beta$. The number $-\alpha + \beta$ is the ordinal number $O(B - A)$ of the set $B - A$ if $OB = \beta$ and A is a segment of B with $OA = \alpha$.

For example, $-1 + \omega = \omega$ and in general $-1 + \alpha = \alpha$ for every $\alpha \geqslant \omega$.

Retaining the assumption $\alpha \leqslant \beta$, let us now investigate $\beta - \alpha$; that is, the solution of the equation $\xi + \alpha = \beta$.

For example, $\beta - 0 = \beta$. But consider $\beta - 1$, and in particular $\omega - 1$. The number $\omega - 1$ does not exist, since the equation $\xi + 1 = \omega$ cannot have a solution, in view of the fact that $\xi + 1$ has a last element, whereas ω has no last element.

Thus we have the following result: if $\beta \geqslant \alpha$, then the "*left difference*" $-\alpha + \beta$ is a uniquely determined ordinal number; on the other hand, the "*right difference*" $\beta - \alpha$ does not always exist, and when it does exist it may have more than one value.

For example, $-\omega + \omega = 0$, but $\omega - \omega$ may be any ordinal number $< \omega$.

Definition. An ordinal number β is said to be of the *first kind* (or *to be isolated*) if $\beta - 1$ exists, and to be of the *second kind* if $\beta - 1$ does not exist. An ordinal number of the second kind, other than zero, is also called a *limit number*.

For example, 5 and $\omega + 1$ are isolated, whereas ω and 2ω are limit numbers.

Every number of the second kind is of the form $\kappa\omega$.

4.5.3. *Exponentiation.* We can make the following inductive definitions:

$$\alpha^0 = 1 \quad \text{for every ordinal number} \quad \alpha = 0,$$
$$\alpha^{\beta+1} = \alpha \cdot \alpha^\beta,$$

$\alpha^\lambda = \sup_{\xi<\lambda} \alpha^\xi$ [6] for every non-isolated ordinal number $\alpha \neq 0$.

For example, $2^\omega = \sup_{n<\omega} 2^n = \omega$, and similarly $n^\omega = \omega$ for every n with $0 < n < \omega$.

It is easy to prove that

(9) $$\alpha^\beta \cdot \alpha^\gamma = \alpha^{\gamma+\beta},$$

(10) $$(\alpha^\beta)^\gamma = \alpha^{\gamma\cdot\beta}.$$

But in general it is not true that $(\alpha\beta)^\gamma = \alpha^\gamma\beta^\gamma$.

For example: $(2\omega)^\omega \neq 2^\omega\omega^\omega$, since $(2\omega)^\omega < 2^\omega\omega^\omega$; for we have

$$(2\omega)^\omega = \sup_n (2\omega)^n = \omega^\omega < \omega^{\omega+1} = \omega \cdot \omega^\omega = 2^\omega \cdot \omega^\omega.$$

4.5.4. *Monotonic Laws.* The following monotonic laws (inequalities, cancellations) are valid:

If $\gamma > 0$, then $\alpha + \gamma > \alpha$, and conversely.
If $\alpha < \beta$, then $\gamma + \alpha < \gamma + \beta$, and conversely.
If $\alpha = \beta$, then $\gamma + \alpha = \gamma + \beta$, and conversely.
If $\alpha \leqslant \beta$, then $\alpha + \gamma \leqslant \beta + \gamma$, and conversely,

but the relation $\leqslant$ cannot be replaced by $<$.

For example: although $2 < 3$, nevertheless $2 + \omega = 3 + \omega = \omega$.

If $\alpha = \beta$, then $\alpha\gamma = \beta\gamma$.
If $\alpha\gamma = \beta\gamma, \gamma > 0$, then $\alpha = \beta$.
If $\alpha < \beta, \gamma > 0$, then $\alpha\gamma < \beta\gamma$, and
If $\alpha\gamma < \beta\gamma, \gamma > 0$, then $\alpha < \beta$.
If $\alpha < \beta, \gamma > 0$, then $\gamma\alpha \leqslant \gamma\beta$,

but not necessarily $\gamma\alpha < \gamma\beta$ because, for example, $2 < 3$, $\omega \cdot 2 = \omega \cdot 3$.

5. Enumeration by Means of the Ordinal Numbers

Definition. For each ordinal number α we let $I(\alpha)$ denote the set of all ordinal numbers that are smaller than α; for example, $I(2) = \{0, 1\}$. Then

[6] For a set M of ordinal numbers we denote by $\sup M$ or $\sup_{x\in M} x$ the smallest ordinal number α for which $M \leqslant \alpha$ (that is, $x \leqslant \alpha$ for every $x \in M$).

$I(0)$ is empty and $I(\omega) = \{0, 1, 2 ...\}$, so that the set $I(\omega)$ has no last element.

Theorem 7. *For every ordinal number* α *the set* $I(\alpha)$ *is well-ordered, and* $OI(\alpha) = \alpha$; *that is, the set* $I(\alpha)$ *regarded as an ordinal number is equal to* α. *In other words: every set* A *of type* α *is similar to* $I(\alpha)$.

The proof of this theorem is immediate, since the ordinal numbers $<\alpha$ are represented by the segments $(\cdot, x)_A$ of the set A, and the mapping $x \rightarrow O(\cdot, x)_A$ provides an isomorphism between A and $I(\alpha)$. But instead of this mapping it is often more convenient to consider the inverse mapping $\xi \rightarrow \alpha_\xi$, $\xi < \alpha$, which indicates how the elements α_ξ of A are represented by "*indices*" from $I(\alpha)$. For example, in ftn. 2, p. 156, we have considered an enumeration r_n $(n < \omega)$ of the set Q of all rational numbers.

The rest of the present section is devoted to the following important question: does there exist an ordinal number φ with the property that the entire set R of real numbers can be put into one-to-one correspondence with a sequence a_ξ $(\xi < \varphi)$ of length φ?

Cantor's fundamental theorem (theorem of the uncountability of the set R; cf. IA, §7.3) states that $\varphi \neq \omega$.

We first prove the following theorem.

Theorem 8. *Every set* M *of ordinal numbers arranged in order of magnitude is well-ordered.*

We must show that every non-empty set $X \subseteq M$ has a least element $\inf X$.[7] But if $\beta \in X$, the set $I(\beta)$ is well-ordered, and thus also the set $I(\beta) \cap X$. It is obvious that $\inf (I(\beta) \cap X) = \inf X$.

Now let Ω denote the set of all ordinal numbers α such that the segment $I(\alpha)$ (namely, the set of all ordinal numbers $<\alpha$) is countable.[8] Then Ω is a well-defined set,[9] which by Theorem 8 is well-ordered and thus defines an ordinal number, denoted by ω_1. Then $I(\omega_1) = \Omega$.

Theorem 9. *The set* Ω *has no last element.*

For if $\alpha < \omega_1$, then also $\alpha + 1 < \omega_1$, since the addition of one element to a countable well-ordered set of ordinal numbers produces a set of the same kind.

The special continuum hypothesis of Cantor (cf. IA, §7.6) is that $\varphi = \omega_1$.

In other words, this conjecture states that the cardinal numbers of R and $I(\omega_1)$ are equal to each other; thus there must exist a one-to-one

[7] For a set M of ordinal numbers $\inf M$ or $\inf_{x \in M} x$ denotes the largest number α for which $\alpha \leqslant x$ for every $x \in M$.

[8] Countable means: empty (zero), finite, or equivalent to the set N of all natural numbers.

[9] In contrast, for example, to the "set" of all ordinal numbers, which is meaningless (cf. the Burali-Forti paradox, IA, §7.5).

mapping f of $I(\omega_1)$ onto R. This mapping f enables us to consider, in addition to the natural ordering of the set R, the following well-ordering $<_f$:

$$f(0) <_f f(1) <_f f(2) <_f \dots <_f f(\xi) <_f \dots \quad \text{for every} \quad \xi < \omega_1 .$$

The statement "*Every set can be well-ordered*" (the well-ordering axiom) is equivalent to either of the following two statements:[10]

For every non-empty set S of non-empty sets there exists a set which contains exactly one element from each $X \in S$ (Zermelo axiom of choice).

Every inductive, partially ordered set contains at least one maximal element (Zorn lemma, cf. IB11).

Here we have the following definitions: A set M partially ordered by $<$ is called *inductive* if every subset K of M that is linearly ordered (not only partially ordered) has a least upper bound in M; that is, an element $a \in M$ with $x \leqslant a$ for all $x \in K$ and such that $a \leqslant a'$ for every element $a' \in M$ satisfying the condition $x \leqslant a'$ for all $x \in K$. By a *maximal element* m of M we mean an element for which there is no element $x \in M$ with $m < x$. An ordered set can have at most one maximal element, which in §2.1 we have called its last element.

In order to continue our comparison of the set Ω with the continuum R of real numbers let us prove the following theorem:

Theorem 10. *If $\alpha_n < \omega_1$, $(n \in N)$, then $\sup_n \alpha_n < \omega_1$.*

In particular, if $\alpha_n (n < \omega)$ is a strictly increasing sequence of ordinal numbers $<\omega_1$, then also $\sup_n \alpha_n < \omega_1$.

The corresponding statement is not valid for the linear continuum.

Proof: If A_n is an ordered set of type α_n for every $n < \omega$, then the set $\bigcup_{n<\omega} \{n\} \times A_n$ is countable, since it is the union of countably many countable sets. If we order this set lexicographically, we obtain a well-ordered set of type $\alpha = \alpha_0 + \alpha_1 + \alpha_2 + \dots$, so that $\alpha < \omega_1$ and $\alpha_n < \alpha$. The set X of numbers $\xi \leqslant \alpha$ such that $\xi > \alpha_n$ $(n < \omega)$ is a well-defined subset of the well-ordered set $I(\alpha + 1)$. Thus $\inf X$ exists, and we have $\inf X \leqslant \alpha < \omega_1$ with $\inf X = \sup_n \alpha_n$, as was to be proved.

Theorem 11. *The set Ω is not countable.*

The same statement holds for R.

If $\Omega = I(\omega_1)$ were countable, we would have $\omega_1 \in I(\omega_1)$, so that $\omega_1 < \omega_1$, which is a contradiction.

As mentioned earlier, the famous Cantor continuum hypothesis states

[10] For proofs of the equivalence see, for example, Birkhoff [1], pp. 42–44. The well-ordering axiom has recently been proved to be independent of the other axioms of set theory (See IA, §7.6).

that the sets Ω and R have the same power. This hypothesis represents a postulate independent (see IA, §7.6) of the other axioms of set theory: *The negation of the continuum hypothesis is also a possibility.*

Theorem 12. *For every $a \in R$ the sets $(\cdot, a)_R$, $(a, \cdot)_R$ are isomorphic to each other and to R.*

The mapping $x \to 1/(a - x) + a$ represents an isomorphism between $(\cdot, a)_R$, and $(a, \cdot)_R$.

On the other hand: *for every $\alpha \in \Omega$ the set $(\cdot, \alpha)_\Omega$ is countable, and the set $(\alpha, \cdot)_\Omega$ is not countable.*

Proof: Since $OI(\alpha) = \alpha$ by Theorem 7, we see that $I(\alpha)$ is countable. But if the set $(\alpha, \cdot)_\Omega$ were also countable, Ω itself would be countable, since $\Omega = I(\alpha) \cup \{\alpha\} \cup (\alpha, \cdot)_\Omega$, in contradiction to Theorem 11.

CHAPTER 2

Groups

Introduction

The concept of a group is a creation of modern mathematics. Some notion of it is to be found in the rich ornamentation of classical art and architecture, but its fundamental importance and varied applications were not recognized until the nineteenth century.

The theory of groups originated in the study of algebraic equations, where its central importance was recognized by E. Galois, who introduced the name "group." The work of A. Cauchy, C. Jordan, A. Cayley, L. Sylow, O. Hölder, G. Frobenius, I. Schur, and W. Burnside freed the theory from this subsidiary position and transformed it into an independent branch of mathematics, concerned with algebraic operations on sets of finitely or infinitely many elements.

The late appearance of groups in science shows that a theory based on them could only have resulted from the modern mathematical method of generalization and abstraction, the method of thinking in terms of "systems." With such concepts as "set," "group," "ring," "field," mathematics has reached a stage of great generality. The object of its study is no longer the special character of certain magnitudes but the structure of whole domains. In this way it becomes possible to make statements that are valid for many different fields. For an over-all summary or synthesis of widely varied parts of mathematics, the notion of a group becomes indispensable.

For the theory of groups, as for all branches of modern science, the axiomatic method is characteristic. In this method it becomes unmistakably clear that the axioms and basic theorems are not necessarily "self-evident"; in laying the foundations of a logically constructed science we have the complete "freedom of spirit" that G. Cantor called the very essence of mathematics. Choice of the axioms is restricted by only one condition :

freedom from self-contradiction. Whether we have made a useful choice is determined solely by the applications, which in group theory are especially numerous. Not only does this theory have many applications in other branches of mathematics, for example in Galois theory or in the foundations and development of geometry, but its effectiveness and esthetic appeal make it an important instrument in other branches of science and art as well : in quantum theory, in crystallography, and in the theory of artistic form.

1. Axioms and Examples

1.1. *Axioms*

Let $\mathfrak{G}$ be a non-empty set and v a (binary) *operation* (often also called a *product*) on $\mathfrak{G}$, that is, a function on the set of ordered pairs (G, H) of elements $G, H \in \mathfrak{G}$ (cf. IB10, § 1.2; in particular, 1.2.5). Then $\mathfrak{G}$ is said to be a *group with respect to v* if the following four axioms are satisfied:

(V) The values of v lie in $\mathfrak{G}$:

$$v(G, H) \in \mathfrak{G} \text{ for all } G, H \in \mathfrak{G}.$$

(A) v is *associative*; that is,

$$v(G, v(H, J)) = v(v(G, H), J) \text{ for all } G, H, J \in \mathfrak{G}.$$

(N) There exists a so-called *neutral*, or *unit*, or *identity element* N in $\mathfrak{G}$, with

$$v(N, G) = v(G, N) = G \text{ for all } G \in \mathfrak{G}.\text{[1]}$$

(I) Every element of $\mathfrak{G}$ has an *inverse*; that is, for every element $G \in \mathfrak{G}$ there exists an element $\bar{G}$, such that

$$v(G, \bar{G}) = N.\text{[2]}$$

When there can be no doubt about the operation in question, the phrase "*with respect to* . . ." is ordinarily omitted in the above definition of a group.

From the axioms, as we shall see, it does not follow that

(K) $v(G, H) = v(H, G)$ for all $G, H \in \mathfrak{G}$,

[1] It would be enough to require $v(G, N) = G$ for all $G \in \mathfrak{G}$, since **(N)** could then be derived from the other axioms, but we are not interested here in independence or other refined questions of axiomatic theory.

[2] An element with this property should really be called a *right inverse*; but we shall show later, on the basis of the other axioms, that it is also a *left inverse* [that is, that it satisfies $v(\bar{G}, G) = N$]; and then it is simply called an inverse.

but if a group does satisfy **(K)**, it is said to be *Abelian* or *commutative*. If $v(G, H) = v(H, G)$ for the special elements $G, H \in \mathfrak{G}$, then G and H are said to be *permutable*. For example, by **(N)** the element N is permutable with all the elements of $\mathfrak{G}$.

For simplicity, we shall generally write GH or $G + H$ in place of $v(G, H)$ and shall speak of the operation v as *multiplication* or *addition*. The additive notation is usually restricted to Abelian groups.

The power $|\mathfrak{G}|$ of the set $\mathfrak{G}$ (see IA, §7.3) is called the *order of* $\mathfrak{G}$. If the order is finite, the group $\mathfrak{G}$ is also said to be *finite*, and its order is simply the finite number of elements in $\mathfrak{G}$.

1.2. *Examples*

1.2.1. Let Γ^+ be the set of rational integers $0, \pm 1, \pm 2, \ldots$ and let $v(G, H) = G + H$ be the sum of the numbers G and H in the usual sense. Then **(V)** is certainly satisfied, and **(A)** holds because addition is associative, as is shown in the foundations of the theory of numbers (see IB1, §1.3, §2.2). Thus,

$$(G + H) + J = G + (H + J).$$

The number 0 has the properties of a neutral element: $G + 0 = 0 + G = G$, and $-G$ is inverse to G: $G + (-G) = 0$. Thus Γ^+ is a group, and is Abelian because **(K)** holds. Its order is $\aleph_0$.

1.2.2. Let P^+, R^+, and C^+ be the set of rational, real, and complex numbers, respectively, and again let $v(G, H) = G + H$ be addition in the usual sense. Then the axioms **(V)** through **(I)** hold as in 1.2.1, so that each of these sets is an Abelian group with respect to addition. The order of P^+ is $\aleph_0$, and the order of R^+ and C^+ is the power of the continuum.

1.2.3. Let $P^\times$, $R^\times$, $C^\times$ be the set of nonzero rational, real, and complex numbers, respectively, and let $v(G, H) = GH$ be multiplication in the ordinary sense. Since multiplication of nonzero numbers is associative and the product is also nonzero, axioms **(V)** and **(A)** are satisfied. The number 1 has the properties of a neutral element, and G^{-1} is inverse to G. Since **(K)** holds, the groups $P^\times$, $R^\times$, and $C^\times$ are all Abelian. Their orders coincide with the corresponding orders in example 1.2.2.

1.2.4. Let $\mathfrak{P}$ be the following set of quotients of polynomials in x:

$$\mathfrak{P} = \left\{x, \frac{1}{x}, 1 - x, \frac{x-1}{x}, \frac{1}{1-x}, \frac{x}{x-1}\right\}.$$

For $G, H \in \mathfrak{P}$ let $v(G, H)$ consist of the *substitution* of G into H. In other words, the operation $v(G, H)$ consists of replacing the symbol x in H by the element G. For example, if

$$G = g(x) = \frac{x}{x-1}, \qquad H = h(x) = 1 - x,$$

then

$$v(G, H) = \frac{1}{1-x}.$$

By a finite number of trials we see that the result of this operation is always one of the six functions, so that **(V)** is satisfied. In order to show that the operation is associative, we could in principle test the validity of **(A)** for all the finitely many triples of elements. Less rigorously, we see that in performing the operation $v(G, v(H, K))$ we first replace x in $K = k(x)$ by $H = h(x)$; and then, in the expression $v(H, K) = k(h(x))$ thus obtained, we replace x by $G = g(x)$, with the final result $k[h(g(x))]$. But it is clear that the same result will be obtained if we first construct the expression $v(G, H) = h(g(x))$ and then replace $k(x)$ by x, so as to obtain $v(v(G, H), K)$. In this group x serves as neutral element, the elements x, $1/x$, $1 - x$, $x/(x - 1)$ are their own inverses, $(x - 1)/x$ is inverse to $1/(1 - x)$, and $1/(1 - x)$ is inverse to $(x - 1)/x$, as can easily be verified. Thus $\mathfrak{P}$ is a group of order 6 and is not Abelian; for example,

$$v\left(\frac{1}{x}, 1 - x\right) = \frac{x-1}{x} \neq \frac{1}{1-x} = v\left(1 - x, \frac{1}{x}\right).$$

1.2.5. Let $\mathfrak{R}$ be a set, which we shall now call a *space* in order to distinguish it from other sets to be considered later; and correspondingly, its elements $P, Q, R, \ldots$ will be called *points*. Let $\mathfrak{S}^{\mathfrak{R}}$ be the set of *permutations* on $\mathfrak{R}$; that is, the set of one-to-one mappings of $\mathfrak{R}$ onto itself. If $\sigma \in \mathfrak{S}^{\mathfrak{R}}$, we denote by $P\sigma$ the image of the point $P \in \mathfrak{R}$ under the mapping σ. Then σ has the following properties:

(1) $$P\sigma \in \mathfrak{R} \text{ for all } P \in \mathfrak{R},$$

(2) $$P_1\sigma = P_2\sigma \text{ implies } P_1 = P_2.$$

More generally, if for a subset $\mathfrak{Q} \subseteq \mathfrak{R}$ and a subset $\mathfrak{K} \subseteq \mathfrak{S}^{\mathfrak{R}}$ we denote by $\mathfrak{Q}\mathfrak{K}$ the set of elements $P\sigma$, $P \in \mathfrak{Q}$, $\sigma \in \mathfrak{K}$, then the fact that σ is a mapping onto $\mathfrak{R}$ is equivalent to[3]

(3) $$\mathfrak{R}\sigma = \mathfrak{R}.$$

Thus (1), (2), and (3) characterize the permutations σ among all the mappings of $\mathfrak{R}$. In $\mathfrak{S}^{\mathfrak{R}}$ we now introduce the following product: for $\sigma, \tau \in \mathfrak{S}^{\mathfrak{R}}$ we define the mapping $\sigma\tau$ (see Figure 1) by

(4) $$P(\sigma\tau) = (P\sigma)\tau.$$

[3] No distinction is made here between an element and the set containing it as sole member. Thus σ in (3) in fact represents $\{\sigma\}$, the set consisting only of σ.

Thus $\sigma\tau$ is the mapping of $\mathfrak{R}$ which results from successive applications of the mappings σ and τ. We now show that with this operation $\mathfrak{S}^{\mathfrak{R}}$ is a group.

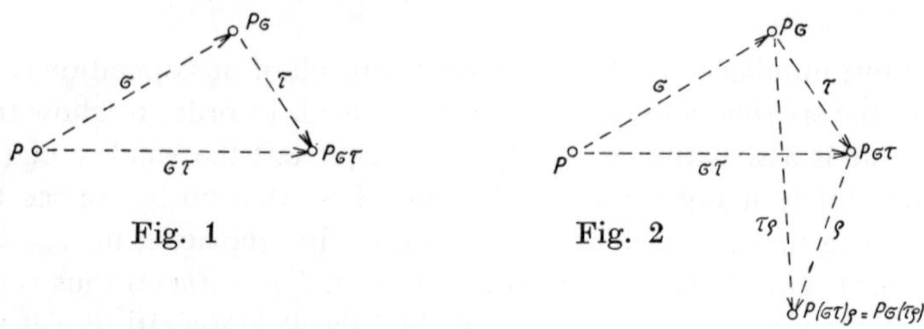

Fig. 1 Fig. 2

(V): We must show that $\sigma\tau$ has the properties (1), (2), (3).

(1): Since $P\sigma \in \mathfrak{R}$ and $P\tau \in \mathfrak{R}$ for all $P \in \mathfrak{R}$, we have $P(\sigma\tau) = (P\sigma)\tau \in \mathfrak{R}$ for all $P \in \mathfrak{R}$.

(2): If $P_1(\sigma\tau) = P_2(\sigma\tau)$, then $(P_1\sigma)\tau = (P_2\sigma)\tau$, and thus $P_1\sigma = P_2\sigma$ and $P_1 = P_2$, since (2) holds for σ and τ.

(3): Since $\mathfrak{R}\sigma = \mathfrak{R}\tau = \mathfrak{R}$, it follows that $\mathfrak{R}(\sigma\tau) = (\mathfrak{R}\sigma)\tau = \mathfrak{R}\tau = \mathfrak{R}$.

(A): On the one hand, we have

$$P[(\sigma\tau)\rho] = [P(\sigma\tau)]\rho = [(P\sigma)\tau]\rho,$$

for all $P \in \mathfrak{R}$ and $\sigma, \tau, \rho \in \mathfrak{S}^{\mathfrak{R}}$, and on the other hand

$$P[\sigma(\tau\rho)] = (P\sigma)(\tau\rho) = [(P\sigma)\tau]\rho,$$

as follows (see Figure 2) from (4).

(N): The mapping 1 defined by $P1 = P$ for all $P \in \mathfrak{R}$ is a permutation, the so-called *identical permutation*. For this permutation we have

$$P(1\sigma) = (P1)\sigma = P\sigma$$

and

$$P(\sigma 1) = (P\sigma)1 = P\sigma,$$

so that $1\sigma = \sigma 1 = \sigma$; and therefore 1 has the properties of a neutral element.

(I): In order to prove the existence of an inverse for $\sigma \in \mathfrak{S}^{\mathfrak{R}}$, it must be remembered that since P runs through all the elements of $\mathfrak{R}$ exactly once, so will $P\sigma$, by (2) and (3). Thus,

$$(P\sigma)\bar{\sigma} = P \tag{5}$$

defines a mapping $\bar{\sigma}$ of $\mathfrak{R}$ for which (1), (2), and (3) are satisfied. From (5) we see at once that $\sigma\bar{\sigma} = \mathbf{1}$, so that $\bar{\sigma}$ is an inverse of σ and therefore **(I)** is satisfied.

The group $\mathfrak{S}^{\mathfrak{R}}$ is called the *symmetric group on* $\mathfrak{R}$. When we examine it more closely, as we shall do below, we see that for finite $\mathfrak{R}$ the order is $|\mathfrak{R}|!$, where $|\mathfrak{R}|$ is the power of $\mathfrak{R}$.

In order to be able to deal more conveniently with multiplication in $\mathfrak{S}^{\mathfrak{R}}$, we identify each of its elements σ with a symbol consisting of two rows: the first row contains every point of $\mathfrak{R}$ exactly once and, directly underneath, the second row contains the images of these points:

$$\sigma = \begin{pmatrix} P & Q & R & \dots \\ P\sigma & Q\sigma & R\sigma & \dots \end{pmatrix}.$$

Two such symbols represent the same permutation if and only if they can be transformed into each other by a permutation of the columns. Since σ is a permutation, the second row also contains every element of $\mathfrak{R}$ exactly once.

If we denote by (n) the set of natural numbers 1, 2, ..., n, then the six elements

$$1 = \begin{pmatrix} 1 & 2 & 3 \\ 1 & 2 & 3 \end{pmatrix}, \quad \alpha = \begin{pmatrix} 1 & 2 & 3 \\ 2 & 3 & 1 \end{pmatrix}, \quad \beta = \begin{pmatrix} 1 & 2 & 3 \\ 3 & 1 & 2 \end{pmatrix},$$

$$\gamma = \begin{pmatrix} 1 & 2 & 3 \\ 1 & 3 & 2 \end{pmatrix}, \quad \delta = \begin{pmatrix} 1 & 2 & 3 \\ 3 & 2 & 1 \end{pmatrix}, \quad \epsilon = \begin{pmatrix} 1 & 2 & 3 \\ 2 & 1 & 3 \end{pmatrix}$$

comprise all the permutations of $\mathfrak{S}^{(3)}$. The multiplication of β and δ, for example, leads to

$$\beta\delta = \begin{pmatrix} 1 & 2 & 3 \\ 3 & 1 & 2 \end{pmatrix}\begin{pmatrix} 1 & 2 & 3 \\ 3 & 2 & 1 \end{pmatrix} = \begin{pmatrix} 1 & 2 & 3 \\ 1 & 3 & 2 \end{pmatrix},$$

which may be read as follows: 1 in β into 3, 3 in δ into 1, therefore 1 in $\beta\delta$ into 1, and so forth. On the other hand, the product $\delta\beta$ produces

$$\delta\beta = \begin{pmatrix} 1 & 2 & 3 \\ 2 & 1 & 3 \end{pmatrix}.$$

Thus the group $\mathfrak{S}^{(3)}$ is not Abelian.

For computation in the space $\mathfrak{R}$ with $|\mathfrak{R}| = n$ this standard model of $\mathfrak{S}^{(n)}$ is very convenient.[4]

1.2.6. Let E_2 be the Euclidean plane and let (P, Q) be the distance between two of its points P, Q. Also, let B_2 denote the set of permutations

[4] The symbol $\mathfrak{S}_n$ is often written in place of $\mathfrak{S}^{(n)}$.

σ on E_2 (regarded as a set of points) which leave invariant the distance between every pair of points:

$$(P\sigma, Q\sigma) = (P, Q).$$

Then B_2 is a group under the same operation as for the permutations in 1.2.5; for now

(V): If $\sigma, \tau \in B_2$, then

$$\begin{aligned}(P(\sigma\tau), Q(\sigma\tau)) &= ((P\sigma)\tau, (Q\sigma)\tau)\\ &= (P\sigma, Q\sigma), \qquad \text{since} \quad \tau \in B_2,\\ &= (P, Q), \qquad \text{since} \quad \sigma \in B_2.\end{aligned}$$

Thus the product $\sigma\tau$ also leaves invariant the distance between every pair of points.

(A) was already proved in §1.2.5 for the product of any two permutations.

(N): The identical permutation 1 (see §1.2.5) is an element of B_2:

$$(P1, Q1) = (P, Q).$$

(I): We shall show that if σ is in B_2, then the permutation $\bar{\sigma}$ defined in §1.2.5 is also in B_2: for let $P, Q \in E_2$ and in accordance with (3) let $P = P^*\sigma$, $Q = Q^*\sigma$ with $P^*, Q^* \in E_2$; then

$$\begin{aligned}(P\bar{\sigma}, Q\bar{\sigma}) &= ((P^*\sigma)\bar{\sigma}, (Q^*\sigma)\bar{\sigma})\\ &= (P^*(\sigma\bar{\sigma}), Q^*(\sigma\bar{\sigma}))\\ &= (P^*, Q^*)\\ &= (P^*\sigma, Q^*\sigma)\\ &= (P, Q),\end{aligned}$$

so that $\bar{\sigma} \in B_2$. But it was shown in §1.2.5 that $\sigma\bar{\sigma} = 1$, so that B_2 is in fact a group. If we think of E_2 as a rigid plate, the elements of B_2 are represented by those motions of the plate which bring it into coincidence with itself without distortion. Thus the elements of B_2 are called the *motions*[5] *of* E_2 and B_2 is the *group of motions* of E_2.

1.2.7. A subset F of the set of points of the Euclidean plane E_2 is called a *figure*. For a given figure F we consider the set $B_{2,F}$ of motions σ in B_2 which map F onto itself; that is, those motions for which, in the notation introduced for permutations in §1.2.5, we have

$$F\sigma = F. \tag{6}$$

[5] They are also called *rigid mappings*.

If the elements of $B_{2,F}$ are combined in the same way as in the preceding examples, then $B_{2,F}$ is a group: for we have

(V): If $F\sigma = F$, $F\tau = F$, then also $F(\sigma\tau) = (F\sigma)\tau = F\tau = F$, so that $\sigma\tau \in B_{2,F}$.

(A) holds for arbitrary permutations, as was shown in §1.2.5.

(N): The mapping 1 defined in §1.2.5 is in $B_{2,F}$ since $F1 = F$. Since $1\sigma = \sigma 1 = 1$ for arbitrary permutations σ on E_2, the mapping 1 is certainly a neutral element for $B_{2,F}$.

(I): If σ belongs to $B_{2,F}$, then so does the element $\bar{\sigma}$ defined in §1.2.5; for if $P \in F$ and, in accordance with (6), $P = P^*\sigma$, $P^* \in F$, then $P\bar{\sigma} = (P^*\sigma)\bar{\sigma} = P^* \in F$, so that every σ has an inverse in $B_{2,F}$.

The group $B_{2,F}$ is called the *group of the figure F.* As an example let us consider the group of the four corners of a square in E_2. It is easy to see that this group is the same as the group $B_{2,Q}$ of the entire square determined by these four corners.

A motion σ of this group is completely characterized by the corresponding permutation of the four corners, since every point of E_2 is determined by its distances from three noncollinear fixed points. If we denote the corners by 1,2,3,4 (see Figure 3), we see that not all permutations of the corners can result from a rigid motion; for example,

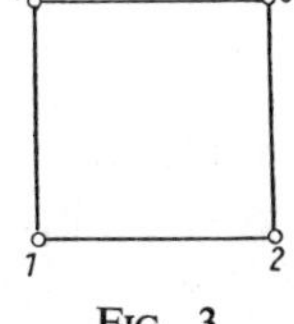

FIG. 3

$$\begin{pmatrix} 1 & 2 & 3 & 4 \\ 1 & 2 & 4 & 3 \end{pmatrix}$$

cannot represent a motion, since the distance is $(1\sigma, 4\sigma) = (1, 3) \neq (1, 4)$. The permutations induced by $B_{2,Q}$ are obviously

$$\begin{pmatrix} 1 & 2 & 3 & 4 \\ 1 & 2 & 3 & 4 \end{pmatrix}, \quad \begin{pmatrix} 1 & 2 & 3 & 4 \\ 2 & 3 & 4 & 1 \end{pmatrix}, \quad \begin{pmatrix} 1 & 2 & 3 & 4 \\ 3 & 4 & 1 & 2 \end{pmatrix}, \quad \begin{pmatrix} 1 & 2 & 3 & 4 \\ 4 & 1 & 2 & 3 \end{pmatrix},$$

$$\begin{pmatrix} 1 & 2 & 3 & 4 \\ 4 & 3 & 2 & 1 \end{pmatrix}, \quad \begin{pmatrix} 1 & 2 & 3 & 4 \\ 2 & 1 & 4 & 3 \end{pmatrix}, \quad \begin{pmatrix} 1 & 2 & 3 & 4 \\ 1 & 4 & 3 & 2 \end{pmatrix}, \quad \begin{pmatrix} 1 & 2 & 3 & 4 \\ 3 & 2 & 1 & 4 \end{pmatrix}.$$

The order of the group $B_{2,Q}$ is therefore 8. This group is not Abelian, as the reader can easily verify.

The developments of §1.2.6 and §1.2.7 are independent of the dimension 2 of E_2. Thus for the n-dimensional Euclidean space E_n, with arbitrary natural number n, we can define the group of motions B_n as the set of permutations on E_n which leave invariant the distance between any two points; and the corresponding remark holds for the group $B_{n,F}$.

1.2.8. Let $\mathfrak{m}$ be a set and let $\mathfrak{B}^{\mathfrak{m}}$ be the set[6] of all its subsets, including the empty set $\emptyset$. For $\mathfrak{a}, \mathfrak{b} \in \mathfrak{B}^{\mathfrak{m}}$ we now define an operation (written additively) as follows (cf. IA, 9.10):

$$\mathfrak{a} \cup \mathfrak{b} - \mathfrak{a} \cap \mathfrak{b} = \mathfrak{a} + \mathfrak{b}.$$

Thus $\mathfrak{a} + \mathfrak{b}$ consists of those elements of $\mathfrak{m}$ which belong to exactly one of the sets $\mathfrak{a}, \mathfrak{b}$ (see Figure 4). Then the set $\mathfrak{B}^{\mathfrak{m}}$ with this multiplication is a group. For we have **(V)**: $\mathfrak{a} + \mathfrak{b}$ is a subset (possibly $\emptyset$) of $\mathfrak{m}$. **(A)**: Let us determine which elements lie in $(\mathfrak{a} + \mathfrak{b}) + \mathfrak{c}$, $\mathfrak{a}, \mathfrak{b}, \mathfrak{c} \in \mathfrak{B}^{\mathfrak{m}}$. For this purpose we think of the elements of $\mathfrak{a}$ as being marked with a cross, and proceed in the same way for the elements of $\mathfrak{b}$. Then $\mathfrak{a} + \mathfrak{b}$ consists of those elements that have been marked with exactly one cross. If we now mark the elements of $\mathfrak{c}$ with a cross, then $(\mathfrak{a} + \mathfrak{b}) + \mathfrak{c}$ contains exactly those elements of $\mathfrak{m}$ which have been marked with either one cross or three crosses. If we construct $\mathfrak{a} + (\mathfrak{b} + \mathfrak{c})$ in the same way, the elements of this subset also receive either one cross or three crosses. But since the number of crosses depends only on the sets $\mathfrak{a}, \mathfrak{b}, \mathfrak{c}$ and not on the parentheses, the result is the same in both cases. Thus $(\mathfrak{a} + \mathfrak{b}) + \mathfrak{c} = \mathfrak{a} + (\mathfrak{b} + \mathfrak{c})$.

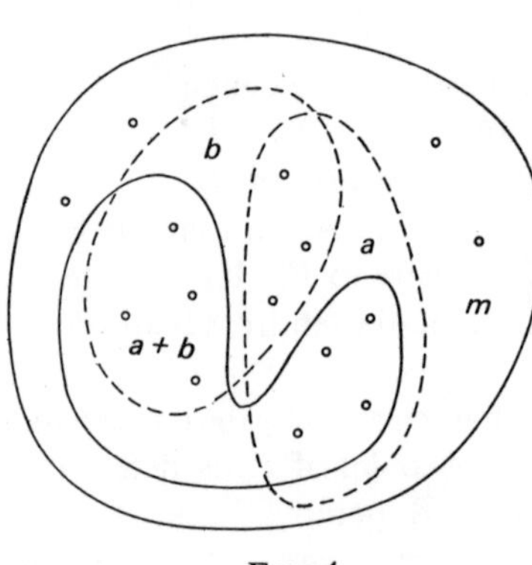

FIG. 4

(N): The empty set $\emptyset$ has the properties of a neutral element.

(I): The element $\mathfrak{a}$ is its own inverse for all $\mathfrak{a} \in \mathfrak{B}^{\mathfrak{m}}$, since $\mathfrak{a} + \mathfrak{a} = \emptyset$. The group $\mathfrak{B}^{\mathfrak{m}}$ is obviously Abelian and has the order $2^{|\mathfrak{m}|}$, where $|\mathfrak{m}|$ is the power of $\mathfrak{m}$.

1.2.9. If n is a natural number, it is shown in the theory of numbers (see IB6, §2.10) that every rational integer g can be represented in the form

$$g = nh + r, \tag{7}$$

where the rational integers h, r are uniquely determined by g, and r is a *reduced remainder for* n; that is, $0 \leqslant r < n$. Such remainders are often said to be *reduced modulo* n.

In the set $\{0, 1, ..., n - 1\}$ of reduced remainders for n, we define an additively written operation as follows. In order to distinguish this

[6] The notation here is intended to suggest the name of G. Boole, who was the first to consider this definition for the multiplication of subsets.

operation from the usual addition $+$ for rational integers, we denote it by $\oplus$ and set

$$s \oplus t = r,$$

where s, t are reduced remainders for n, and r is the reduced remainder uniquely determined by (7), of $g = s + t$ (in the ordinary sense of addition). We now show that the set of reduced remainders for n thus becomes a group:

(V) is obvious.

(A): We must prove that $(s \oplus t) \oplus u = s \oplus (t \oplus u)$. For this purpose we set, as in (7),

$$\begin{aligned} s + t &= nh + r, \\ r + u &= ni + p. \end{aligned} \tag{8}$$

Thus $(s \oplus t) \oplus u = p$. Similarly, if we set

$$\begin{aligned} t + u &= nj + q, \\ s + q &= nk + o, \end{aligned} \tag{9}$$

then $s \oplus (t \oplus u) = o$. From (8) it follows that

$$s + t + u = n(h + i) + p,$$

and thus from (9) that

$$s + t + u = n(j + k) + o.$$

Thus $p = o$, since by (7) these representations are unique.

(N): The number 0 has the properties of a neutral element.

(I): The inverse for r is $n - r$ if $r \neq 0$ and is 0 if $r = 0$.

We denote this group by $\Gamma^{(n)+}$; it is Abelian and its order is n.

1.2.10. In order to construct the corresponding group of reduced remainders with multiplication as the operation instead of addition, we restrict ourselves to those remainders that are prime to n (i.e., have no factor in common with n (see IB6, §2.6)), and then we denote the operation (in order to distinguish it from ordinary multiplication of rational integers) by $\odot$. We now set

$$s \odot t = r,$$

where r is the reduced remainder of $g = s \cdot t$ (under multiplication in the ordinary sense) and is therefore uniquely defined by (7). With this operation the reduced remainders prime to n form a group. For we have

(V): If $(s, n) = (t, n) = 1$ and, by (7),

$$st = nh + r,$$

then also $(r, n) = (st - nh, n) = 1$.

(A): Let s, t, u be reduced remainders and, as in (7), let

$$st = nh + r, \quad ru = ni + p \tag{10}$$

and

$$tu = nj + q, \quad qs = nk + o. \tag{11}$$

Then $(s \odot t) \odot u = p$, $s \odot (t \odot u) = o$. From (10) it then follows that

$$stu = n(hu + i) + p,$$

and from (11) that

$$stu = n(js + k) + o.$$

By the uniqueness of (7) we therefore have $p = o$.

(N): The number 1 has the properties of a neutral element.

(I): In order to show that every element r has an inverse, we determine rational integers $\bar{r}$, $\bar{n}$ (see IB6, §2.9) such that

$$r\bar{r} + n\bar{n} = 1, 0 < \bar{r} < n.$$

Then $(\bar{r}, n) = 1$ and $r\bar{r} = n \cdot (-\bar{n}) + 1$, so that $r \odot \bar{r} = 1$, as desired. We denote this group by $\Gamma^{(n)\times}$; it is Abelian and (see IB6, §4.2, §5) has the order

$$\varphi(n) = p_1^{\alpha_1 - 1} \dots p_r^{\alpha_r - 1}(p_1 - 1) \dots (p_r - 1),$$

where $n = p_1^{\alpha_1} \dots p_r^{\alpha_r}$, with distinct primes $p_1, \dots, p_r$.

1.2.11. We consider the set $\mathfrak{E}^{(3)}$ of real numbers

$$g + h\sqrt{3},$$

where g, h are rational integers and $(g + h\sqrt{3})(g - h\sqrt{3}) = g^2 - 3h^2 = 1$. The set $\mathfrak{E}^{(3)}$ forms a group under ordinary multiplication of real numbers. For we have

(V): If

$$g_1^2 - 3h_1^2 = g_2^2 - 3h_2^2 = 1, \tag{12}$$

then

$$(g_1 + h_1\sqrt{3})(g_2 + h_2\sqrt{3}) = (g_1g_2 + 3h_1h_2) + (g_1h_2 + g_2h_1)\sqrt{3}$$

is a number in $\mathfrak{E}^{(3)}$, since it follows from (12) that

$$(g_1g_2 + 3h_1h_2)^2 - 3(g_1h_2 + g_2h_1)^2 = 1.$$

(A) holds for all real numbers.

(N): The number $1 = 1 + 0\sqrt{3}$ has the properties of a neutral element.

(I): Since

$$(g + h\sqrt{3})(g - h\sqrt{3}) = g^2 - 3h^2 = 1,$$

the number $g - h\sqrt{3}$ is inverse to $g + h\sqrt{3}$. The group $\mathfrak{E}^{(3)}$ is Abelian, and by the theory of sets (see IA, §7.3) it has the order $\aleph_0$, since if $2 + \sqrt{3}$ is in the group, then by **(V)** the numbers $(2 + \sqrt{3})^\gamma, \gamma = 1, 2, \ldots$ are also in $\mathfrak{E}^{(3)}$; but these numbers are all distinct since $2 + \sqrt{3} > 1$ and the one-to-one mapping $(g, h) \to g + h\sqrt{3}$ puts $\mathfrak{E}^{(3)}$ in correspondence with a subset of the countable set of all integral lattice points of a coordinate plane.

In this proof of the properties of a group the number 3 has played no particular role; the reader should consider how the group $\mathfrak{E}^{(n)}$ is to be defined in a corresponding way for every natural number n.

1.2.12. Let K be a field and let $K_n^\times$ be the set of square matrices (see IB3, §2.2, §3.4)

$$A = (a_{ik}), \qquad a_{ik} \in K \tag{13}$$

of order n with nonzero determinant. This set forms a group under the operation of matrix multiplication. For we have

(V): If $A, B \in K_n^\times$, and if we let $|X|$ denote the determinant of a matrix X, then by the rule for the multiplication of determinants

$$|A \cdot B| = |A| \cdot |B|;$$

thus, if $|A|, |B|$ are nonzero, so is $|AB|$.

(A): The associativity of matrix multiplication will be proved in IB3, §2.2.

(N): The unit matrix

$$\begin{pmatrix} 1 & & & 0 \\ & 1 & & \\ & & \ddots & \\ 0 & & & 1 \end{pmatrix}$$

has the properties of a neutral element.

(I): Inverse to A is the matrix $(A_{ki}/|A|)$, where A_{ki} is the algebraic complement of a_{ki}, as is discussed in detail in IB3, §3.5.

1.3. *Examples of Systems That Are Not Groups*

After the numerous examples of groups in the preceding section, the reader may possibly feel that it is difficult to avoid satisfying the axioms for a group. For greater clarity we shall now give some examples of sets with an operation that does not make the set into a group.

1.3.1. Let N be the set of all natural numbers (excluding 0), and for $n, m \in N$ let

$$v(n, m) = nm$$

(with multiplication in the ordinary sense). Then **(V)**, **(A)**, and **(N)** are satisfied but no $n \neq 1$ has an inverse.

1.3.2. For $r, s \in R$, where R is the set of real numbers, let the operation consist of taking the maximum

$$v(r, s) = \max(r, s).$$

Then **(V)** and **(A)** are satisfied, but there is no neutral element and therefore the concept of an inverse remains undefined.

1.3.3. Let $\hat{\Gamma}$ be the set of rational integers excluding the two numbers $+2, -2$. For $g, h \in \hat{\Gamma}$ let

$$v(g, h) = g + h$$

(with addition in the ordinary sense). Then **(A)** is satisfied, 0 has the properties of a neutral element, $-g$ is inverse to g and is contained in $\hat{\Gamma}$, but **(V)** is not satisfied since $v(1, 1) \notin \hat{\Gamma}$.

1.3.4. Let N^* be a set of all natural numbers including 0. For $n, m \in N^*$ set

$$v(n, m) = |n - m|$$

(where the vertical bars denote absolute value). Then **(V)** is satisfied, 0 has the properties of a neutral element, n is its own inverse, but **(A)** is not satisfied, since

$$v(1, v(2, 3)) = |1 - |2 - 3|| = 0,$$
$$v(v(1, 2), 3) = ||1 - 2| - 3| = 2.$$

2. Immediate Consequences of the Axioms for a Group

To simplify the notation, we shall henceforth write the group operation multiplicatively, except where otherwise noted.

2.1. As was proved in IB1, §1.3 for an additively written operation, it follows from the associative law **(A)** that the value of a product of more than three factors depends only on the order of the factors and not on the way in which they are combined in parentheses. For example,

$$(C_1C_2)(C_3C_4) = C_1(C_2(C_3C_4)),$$

and so forth. Thus we can omit the parentheses and for an ordered system of elements $C_1, \ldots, C_r$ simply write

$$C_1C_2 \ldots C_r .$$

Then

$$(C_1 \ldots C_r)(C_{r+1} \ldots C_s) = C_1 \ldots C_rC_{r+1} \ldots C_s .$$

2.2. In **(N)** it was not assumed that a group has only one neutral element. But if N' is also an element with the properties required by **(N)**, then

$$NN' = N,$$
$$NN' = N',$$

and therefore $N = N'$. Thus the neutral element in a group is uniquely defined; if the group is written multiplicatively, we call the neutral element a *unit element* 1; but if additively written, a *zero element* 0.

2.3. Again, in **(I)** it was not required that an element $G \in \mathfrak{G}$ should have only one inverse. But it follows from the other axioms that if $\bar{G}$ is an inverse of G and $\bar{\bar{G}}$ is an inverse of $\bar{G}$, then

$$G\bar{G}\bar{\bar{G}} = 1 \cdot \bar{\bar{G}}$$
$$= G \cdot 1,$$

and so

(1) $$\bar{\bar{G}} = G$$

and

(2) $$\bar{G}G = \bar{G}\bar{\bar{G}} = 1.$$

Now if $\tilde{G}$ is also an inverse of G, we have $G\tilde{G} = 1$, and by multiplication on the left with $\bar{G}$,

$$\bar{G}G\tilde{G} = \bar{G}.$$

From (2) it follows that $\tilde{G} = \bar{G}$, so that the inverse of G is uniquely determined; in the multiplicative notation we denote this inverse by G^{-1}. Then by (1) and (2) we have

$$(3) \qquad \begin{aligned} (G^{-1})^{-1} &= G, \\ GG^{-1} &= G^{-1}G = 1, \end{aligned}$$

so that G is permutable with its inverse. In the additive notation we write $-G$ in place of G^{-1}. Then the rules (3) read

$$\begin{aligned} -(-G) &= G, \\ G + (-G) &= (-G) + G = 0. \end{aligned}$$

In place of $G + (-H)$ we may also write the shorter form $G - H$, but then, in contrast to the situation for the group operation $+$, we must pay attention to parentheses. For example, in Γ^+ (§1.2.1),

$$(3 - 4) - 5 \neq 3 - (4 - 5).$$

It is easy to prove the following important rule for the formation of inverses:

$$(4) \qquad (G_1 \ldots G_r)^{-1} = G_r^{-1} \ldots G_1^{-1}.$$

2.4 From the uniqueness of inverses it follows that a group $\mathfrak{G}$ allows unique two-sided *division*. More precisely: if G, H are arbitrary elements of $\mathfrak{G}$, there exist uniquely determined elements X, $Y \in \mathfrak{G}$, for which

$$(5) \qquad \begin{aligned} GX &= H, \\ YG &= H. \end{aligned}$$

In fact, the elements $X = G^{-1}H$ and $Y = HG^{-1}$ have the desired property:

$$\begin{aligned} GG^{-1}H &= 1 \cdot H = H, \\ HG^{-1}G &= H \cdot 1 = H, \end{aligned}$$

and from $GX_1 = GX_2$, and $Y_1G = Y_2G$, it follows after multiplication by G^{-1} on the left and on the right, respectively, that $G^{-1}GX_1 = G^{-1}GX_2$, and $Y_1G\,G^{-1} = Y_2G\,G^{-1}$, so that $X_1 = X_2$ and $Y_1 = Y_2$, which proves the desired uniqueness.

We now show that in the definition of a group we may replace the axioms **(N)** and **(I)** by two axioms symmetrically constructed with respect to multiplication on the left and on the right:

For every ordered pair (G, H), $G \in \mathfrak{G}$, $H \in \mathfrak{G}$ *we have*

(D_r) *an element* $X \in \mathfrak{G}$ *with* $GX = H$, *and*

(D_l) *an element* $Y \in \mathfrak{G}$ *with* $YG = H$.

Thus we must prove that **(N)** and **(I)** follow from **(V)**, **(A)**, **(D_r)**, and **(D_l)**. To this end, for a fixed $G \in \mathfrak{G}$, we determine R by **(D_r)** from the equation

$$GR = G, \qquad R \in \mathfrak{G}.$$

If H is an arbitrary element of $\mathfrak{G}$ and Y is determined by **(D_l)** from the equation $YG = H$, then

$$HR = (YG)R = Y(GR) = YG = H,$$

and so

$$HR = H \text{ for all } H \in \mathfrak{G}.$$

In the same way, for the element $L \in \mathfrak{G}$ determined by **(D_l)** from $LG = G$, it follows that

$$LH = H \text{ for all } H \in \mathfrak{G}.$$

Thus, in particular, $LR = L$ and $LR = R$, so that $L = R = N$ is the neutral element. Then **(I)** follows at once from **(D_r)**, if we set $H = N$.

Since we have already shown that **(D_r)** and **(D_l)** follow from **(V)**, **(A)**, **(N)**, **(I)**, we see that the two systems of axioms **(V)**, **(A)**, **(N)**, **(I)**, and **(V)**, **(A)**, **(D_r)**, **(D_l)** are equivalent, as desired.

For a finite group $\mathfrak{G}$, the axioms **(D_r)** and **(D_l)** can be replaced by the following *axioms of cancellation*:

(K_r): *If* $GX_1 = GX_2$, *then* $X_1 = X_2$.

(K_l): *If* $Y_1G = Y_2G$, *then* $Y_1 = Y_2$.

To prove this we consider, for an arbitrary but fixed $G \in \mathfrak{G}$, the mappings

$$X \to GX, \quad X \in \mathfrak{G},$$
$$Y \to YG, \quad Y \in \mathfrak{G}$$

of $\mathfrak{G}$ into $\mathfrak{G}$. By **(K_r)** and **(K_l)** these two mappings are one-to-one and thus, since $\mathfrak{G}$ is of finite order, they are mappings onto $\mathfrak{G}$ (see IB1, §1.5). Consequently, every element $H \in \mathfrak{G}$ has the form $H = GX$ and $H = YG$, so that **(D_r)** and **(D_l)** are satisfied. Since we have already proved for arbitrary groups that **(K_r)** and **(K_l)** are satisfied, it is clear that for a finite set $\mathfrak{G}$ the system of axioms **(V)**, **(A)**, **(N)**, **(I)** is equivalent to the system **(V)**, **(A)**, **(K_r)**, **(K_l)**. But, as is shown by the example in §1.3.1, this equivalence does not necessarily hold for infinite sets $\mathfrak{G}$.

3. Methods of Investigating the Structure of Groups

The great importance of group theory is due to the fact that a few simple axioms give rise to a great wealth of theorems. Because of the simplicity and naturalness of its axioms, the theory of groups has penetrated deeply, as the above examples show, into many parts of mathematics, so that its theorems may be interpreted, according to the fields to which they are applied, as theorems about numbers, permutations, motions, residues, and so forth.

Consequently, if we wish to describe these theorems in a natural way, we cannot remain satisfied with the meager vocabulary of the axioms. In the theory of numbers, for example, it is impossible to give any reasonably concise description of the results without introducing such terms as "divisible," "prime number," and so forth; and we must now turn to the construction of a corresponding set of instruments for the analysis of groups.

3.1. *Calculation with Complexes*

Let us consider a fixed group $\mathfrak{G}$ and subsets $\mathfrak{K}$, including the empty set $\emptyset$, of its set of elements. Such subsets will be called *complexes* of $\mathfrak{G}$. As is customary in the theory of sets, we can give a precise description of a set by enclosing its elements in braces { }. For example, in the notation of §1.2.5 we have

$$(n) = \{1, 2, ..., n\}.$$

If the elements of a set are defined by certain properties, these properties are written to the right of a vertical stroke; for example (see §1.2.3)

$$\{a \mid a \in R^{\times}, a^{-1} = a\}$$

is the set of numbers in $R^{\times}$ for which $a^{-1} = a$; that is, the set $\{1, -1\}$.

Since the complexes of a group are subsets of a set, we have already defined for them the set-theoretic concepts (see IA, §7.2) "equal" $=$, "contained in" $\subseteq$, "properly contained in" $\subset$, "intersection" $\cap$ and "union" $\cup$; the last two are applicable to an arbitrary set K of complexes; for them we write $\bigcap_{\mathfrak{K} \in \mathsf{K}} \mathfrak{K}$ and $\bigcup_{\mathfrak{K} \in \mathsf{K}} \mathfrak{K}$.

If $\mathfrak{K}$ and $\mathfrak{L}$ are complexes of $\mathfrak{G}$, we define the *complex-product* $\mathfrak{K}\mathfrak{L}$ as the complex consisting of all elements representable in the form KL with $K \in \mathfrak{K}$, $L \in \mathfrak{L}$:

$$\mathfrak{K}\mathfrak{L} = \{KL \mid K \in \mathfrak{K}, L \in \mathfrak{L}\}.$$

For example, in §1.2.3 for $\mathfrak{G} = P^{\times}$ and $\mathfrak{K} = \{1, \frac{1}{2}, \frac{1}{3}\}$, $\mathfrak{L} = \{-2, \frac{1}{3}, \frac{1}{6}\}$ we have

$$\mathfrak{K}\mathfrak{L} = \{-2, \tfrac{1}{3}, \tfrac{1}{6}, -1, \tfrac{1}{12}, -\tfrac{2}{3}, \tfrac{1}{9}, \tfrac{1}{18}\};$$

and in §1.2.5, for $\mathfrak{G} = \mathfrak{S}^{(3)}$ and $\mathfrak{K} = \{1, \beta\}$, $\mathfrak{L} = \{1, \gamma, \alpha\}$,

$$\mathfrak{K}\mathfrak{L} = \{1, \gamma, \alpha, \beta, \epsilon\}.$$

If $\mathfrak{G}$ is written additively, we write $\mathfrak{K} + \mathfrak{L}$ in place of $\mathfrak{K}\mathfrak{L}$. For $\mathfrak{G} = \Gamma^+$ and $\mathfrak{K} = \{g \mid g \in \Gamma^+, 2 \text{ divides } g\}$, $\mathfrak{L} = \{g \mid g \in \Gamma^+, 4 \text{ divides } g\}$ we thus have $\mathfrak{K} + \mathfrak{L} = \{g \mid g \in \Gamma^+, 2 \text{ divides } g\}$.

This multiplication of the complexes of a fixed group is associative, as follows from the associativity of the operation of the group, and can thus be extended to more than two factors without use of parentheses. In general, multiplication of complexes is not commutative, as can be seen by calculating the elements of $\mathfrak{L}\mathfrak{K}$ in the above example for the group $\mathfrak{S}^{(3)}$.

By $\mathfrak{K}^{-1}$ we denote the complex consisting of the inverses of the elements of the complex $\mathfrak{K}$: thus $\mathfrak{K}^{-1} = \{G^{-1} \mid G \in \mathfrak{K}\}$. Then we have $(\mathfrak{K}\mathfrak{L})^{-1} = \mathfrak{L}^{-1}\mathfrak{K}^{-1}$ by §2.3.(4).

For unions of complexes the following distributive law holds for complex multiplication:

(1) $$\mathfrak{K}(\mathfrak{L} \cup \mathfrak{M}) = \mathfrak{K}\mathfrak{L} \cup \mathfrak{K}\mathfrak{M}, (\mathfrak{L} \cup \mathfrak{M})\mathfrak{K} = \mathfrak{L}\mathfrak{K} \cup \mathfrak{M}\mathfrak{K},$$

but for intersections only a weaker form of distributivity is valid:

(2) $$\mathfrak{K}(\mathfrak{L} \cap \mathfrak{M}) \subseteq \mathfrak{K}\mathfrak{L} \cap \mathfrak{K}\mathfrak{M}, (\mathfrak{L} \cap \mathfrak{M})\mathfrak{K} \subseteq \mathfrak{L}\mathfrak{K} \cap \mathfrak{M}\mathfrak{K}.$$

But if $\mathfrak{K} = G$ contains only one element, then here also we have the equality:

(3) $$G(\mathfrak{L} \cap \mathfrak{M}) = G\mathfrak{L} \cap G\mathfrak{M}, (\mathfrak{L} \cap \mathfrak{M})G = \mathfrak{L}G \cap \mathfrak{M}G.$$

The laws (1) and (2) will not be needed below; the proof of (3) is as follows: since $H \in G(\mathfrak{L} \cap \mathfrak{M})$ is equivalent to $G^{-1}H \in \mathfrak{L} \cap \mathfrak{M}$, it follows that $H \in G\mathfrak{L}$ and $H \in G\mathfrak{M}$.

For a given complex $\mathfrak{K}$ the complexes of the form $G^{-1}\mathfrak{K}G$, $G \in \mathfrak{G}$ are called the *conjugates* or *transforms of* $\mathfrak{K}$ (*under* $\mathfrak{G}$). If $\mathfrak{K}$ is conjugate only to itself, then $\mathfrak{K}$ is said to be *normal* or *invariant* in $\mathfrak{G}$. For example, 1 is normal in $\mathfrak{G}$. Furthermore, the whole group $\mathfrak{G}$ is a normal complex, since for $G \in \mathfrak{G}$ we have:

(4) $$\mathfrak{G}G = G\mathfrak{G} = \mathfrak{G}, \quad \text{or} \quad G^{-1}\mathfrak{G}G = \mathfrak{G},$$

since the equations §2.4(5) have solutions for every pair $G, H \in \mathfrak{G}$. Also, it is clear that

(5) $$\mathfrak{G}\mathfrak{G} = \mathfrak{G},$$

and finally we note that since $(G^{-1})^{-1} = G$, we have:

(6) $$\mathfrak{G}^{-1} = \mathfrak{G}.$$

3.2. *Subgroups*

We now turn our attention to a concept, already used in the examples in §§1.2.6 and 1.2.7, which is of great importance in studying the structure of groups. If $\mathfrak{U}$ is a complex from a group $\mathfrak{G}$, it may happen that the complex $\mathfrak{U}$ forms a group with respect to the operation of group $\mathfrak{G}$. Examples are provided by the set B_2 of motions on E_2 regarded as a complex from the group of all permutations on E_2, and also by the set of motions which map a figure onto itself, regarded as a complex in the set of motions B_2. We describe this situation by saying that $\mathfrak{U}$ is a *subgroup* of $\mathfrak{G}$, by which we mean that under the operation v defined for $\mathfrak{G}$ the set $\mathfrak{U}$ satisfies axioms **(V)**, **(A)**, **(N)**, and **(I)**. Thus B_2 and $B_{2,F}$ are subgroups of $\mathfrak{S}^{E_2}$ and B_2, respectively. The complex $\mathfrak{K} = \{1, -1\}$ is not a subgroup of Γ^+; it is true that $\mathfrak{K}$ is a group with respect to multiplication of numbers, but that is not the operation with respect to which Γ^+ is defined as a group. The sets 1 and $\mathfrak{G}$ are subgroups of every group $\mathfrak{G}$. Subgroups other than these trivial (improper) subgroups 1 and $\mathfrak{G}$ are called *proper* subgroups.

If $\mathfrak{U}$ is a subgroup of $\mathfrak{G}$, the unit element 1 of $\mathfrak{G}$ must be contained in $\mathfrak{U}$ and must also be the unit element of $\mathfrak{U}$. For if $U \in \mathfrak{U}$ and $1_{\mathfrak{U}}$ is the unit element of $\mathfrak{U}$, so that $1_{\mathfrak{U}} U = U$, then this equation for $1_{\mathfrak{U}}$ must also hold in $\mathfrak{G}$ and, by the uniqueness of division in $\mathfrak{G}$, its solution $1_{\mathfrak{U}} = 1$ is unique. Similarly, we can easily show that the inverse of U in $\mathfrak{U}$ is equal to its inverse U^{-1} in $\mathfrak{G}$.

In order to test whether a given complex is actually a subgroup it would be necessary, from a formal point of view, to examine all the four axioms for a group. But as is shown by the examples in §§1.2.6 and 1.2.7, this process can be shortened: for example, if the axiom **(A)** of associativity holds for $\mathfrak{G}$, then it certainly holds for the elements of a subset of $\mathfrak{G}$. We now prove the following *criterion for subgroups*, whereby the process can be still further shortened:

A non-empty complex $\mathfrak{U} \subseteq \mathfrak{G}$ *is a subgroup of* $\mathfrak{G}$ *if and only if*

(7) $$\mathfrak{U}\mathfrak{U}^{-1} \subseteq \mathfrak{U}.$$

The necessity of this condition (7) is clear at once from (5), (6), and the properties of inverses in $\mathfrak{U}$.

In order to prove that the criterion is also sufficient, we let $\mathfrak{U}$ be a complex in $\mathfrak{G}$ which satisfies (7). For arbitrary $U \in \mathfrak{U}$ we then have $1 = UU^{-1} \in \mathfrak{U}$, so that **(N)** is satisfied by the neutral element 1 in $\mathfrak{G}$. Moreover, $1U^{-1} = U^{-1} \in \mathfrak{U}$, so that **(I)** is satisfied. Also, for $U_1, U_2 \in \mathfrak{U}$, we have $U_1(U_2^{-1})^{-1} = U_1U_2 \in \mathfrak{U}$, so that **(V)** is satisfied. Also, we have just seen that

(A) holds in $\mathfrak{U}$ because it holds in $\mathfrak{G}$. Thus $\mathfrak{U}$ is a subgroup of $\mathfrak{G}$, as was to be proved.

3.3. By means of this criterion for subgroups it is easy to show that the following examples are subgroups:

3.3.1. Complex of even numbers in Γ^+.
3.3.2. Complex $\{1, -1\}$ in $P^\times$, $R^\times$, and $C^\times$.
3.3.3. Complex $\mathfrak{U}^{(3)} = \{1, \alpha, \beta\}$ in $\mathfrak{S}^{(3)}$ (§1.2.5).
3.3.4. Complex $\mathfrak{S}_P^{\mathfrak{R}}$ of permutations in $\mathfrak{S}^{\mathfrak{R}}$ (§1.2.5) that leave fixed a given point $P \in \mathfrak{R}$.

The reader will readily verify §3.3.1–3; and in order to show §3.3.4 we need only point out, in view of the criterion for subgroups, that if σ and τ leave the point P fixed, then so does $\sigma\tau^{-1}$:

$$P(\sigma\tau^{-1}) = (P\sigma)\tau^{-1} = P\tau^{-1} = (P\tau)\tau^{-1} = P.$$

3.4. If $\mathfrak{U}$ is a subgroup of $\mathfrak{G}$, the conjugate complexes $\mathfrak{G}^{-1}G$, $G \in \mathfrak{G}$ are also subgroups of $\mathfrak{G}$. For by §2(4), §3(5), (6) we have

$$G^{-1}\mathfrak{U}G(G^{-1}\mathfrak{U}G)^{-1} = G^{-1}\mathfrak{U}GG^{-1}\mathfrak{U}G = G^{-1}\mathfrak{U}G,$$

so that (7) is satisfied with $G^{-1}\mathfrak{U}G$ in place of $\mathfrak{U}$.

If $\mathfrak{U}$ is a subgroup of $\mathfrak{G}$ and $\mathfrak{V}$ is a subgroup of $\mathfrak{U}$, then $\mathfrak{V}$ is also a subgroup of $\mathfrak{G}$, as follows immediately from the criterion for subgroups. Thus the property of being a subgroup is transitive.

If $\mathscr{U}$ is a set of subgroups of $\mathfrak{G}$, the intersection $\mathfrak{D} = \bigcap_{\mathfrak{U}\in\mathscr{U}}\mathfrak{U}$ is also a subgroup of $\mathfrak{G}$; for if U_1, U_2 are in every $\mathfrak{U} \in \mathscr{U}$, then by (7) it follows that $U_1U_2^{-1}$ is in every $\mathfrak{U} \in \mathscr{U}$. On the other hand, the condition (7) is sufficient, and therefore $\mathfrak{D}$ is a subgroup.

Thus for every complex $\mathfrak{K} \subseteq \mathfrak{G}$ there exists, as the intersection of all subgroups $\mathfrak{U} \supseteq \mathfrak{K}$, a smallest subgroup $\langle\mathfrak{K}\rangle$ of $\mathfrak{G}$ containing $\mathfrak{K}$:

$$\langle\mathfrak{K}\rangle = \bigcap_{\mathfrak{K}\subseteq\mathfrak{U}} \mathfrak{U}.$$

This subgroup is said to be *generated* by $\mathfrak{K}$, since $\langle\mathfrak{K}\rangle$ consists of all the finite products $\mathfrak{K}_1^{\epsilon_1} \cdots \mathfrak{K}_r^{\epsilon_r}$, $\epsilon_i = \pm 1$; in other words, of all the products that can be "generated" by the elements $K_1, \ldots, K_r \in \mathfrak{K}$: for we see at once from **(V)** and **(I)** that all such products must lie in $\langle\mathfrak{K}\rangle$ and, on the other hand, the criterion for subgroups shows the complex of these elements is a subgroup of $\mathfrak{G}$ containing $\mathfrak{K}$. A complex $\mathfrak{K}$ for which $\mathfrak{G} = \langle\mathfrak{K}\rangle$ is called a *system of generators* of $\mathfrak{G}$, and $\mathfrak{G}$ is said to be *generated* by $\mathfrak{K}$. If there exists a finite complex $\mathfrak{K} = \{E_1, \ldots, E_n\}$ with this property, then $\mathfrak{G}$ is said to have a *finite system of generators*, or to be *finitely generated*. The minimum number n of such generators is called the *number of generators* of $\mathfrak{G}$.

The set of subgroups $\mathfrak{U}$, $\mathfrak{V}$ of a group forms a lattice (see IB9, 2.1) with respect to the operations $(\mathfrak{U}, \mathfrak{V}) \to \mathfrak{U} \cap \mathfrak{V}$ and $(\mathfrak{U}, \mathfrak{V}) \to \langle \mathfrak{U} \cup \mathfrak{V} \rangle$. This lattice provides us with an easily visualized method for studying the construction of a group. We give the *graphs* (also called *diagrams*; see IA, §9.5) of the lattices of subgroups for some of our examples: $\Gamma^{(6)+}$, §1.2.9 (Figure 5); $\mathfrak{S}^{(3)}$, §1.2.5 (Figure 6); $B_{2,Q}$, §1.2.7 (Figure 7); $\mathfrak{B}^{\{1,2,3\}}$, §1.2.8 (Figure 8). In Figure 7 the elements are written in the form of cycles; see §15.2.1.

For many purposes it is preferable, for easier visualization, to consider only sublattices and their graphs; for example, any two subgroups $\mathfrak{U}$, $\mathfrak{V}$ of a group $\mathfrak{G}$ provide a finite sublattice with the graph represented in Figure 9. In this case several of the subgroups may coincide.

A subgroup $\mathfrak{U} \neq \mathfrak{G}$ is said to be *maximal* in $\mathfrak{G}$ if there exists no subgroup $\mathfrak{V} \neq \mathfrak{G}$ of $\mathfrak{G}$ containing $\mathfrak{U}$ as a proper subgroup. Similarly, a subgroup $\mathfrak{U} \neq 1$ of $\mathfrak{G}$ is *minimal* if $\mathfrak{U}$ has no proper subgroup $\mathfrak{V} \neq 1$.

3.5. *Residue Classes or Cosets*

If $\mathfrak{U}$ is a subgroup of $\mathfrak{G}$, the complexes of the form $\mathfrak{U}G$ and $G\mathfrak{U}$, $G \in \mathfrak{G}$, are called *right residue classes* and *left residue classes* (or *left cosets* and *right cosets*), respectively. In this section we shall consider only right cosets. If $H \in \mathfrak{U}G$, then $\mathfrak{U}G = \mathfrak{U}H$, since H has the form $H = UG$, $U \in \mathfrak{U}$, so that $U^{-1}H = G$ and $\mathfrak{U}G = \mathfrak{U}U^{-1}H = \mathfrak{U}H$, by (4). Thus:

Either $\mathfrak{U}G_1 = \mathfrak{U}G_2$ or $\mathfrak{U}G_1 \cap \mathfrak{U}G_2 = \emptyset$ for arbitrary cosets $\mathfrak{U}G_1$, $\mathfrak{U}G_2$. Since every element of $\mathfrak{G}$ lies in some right coset of $\mathfrak{U}$ (because $G = 1 \cdot G \in \mathfrak{U}G$) the right cosets generate a division into classes (see IA, §8.5) of the set $\mathfrak{G}$. The power of the set of right cosets is called the *index* of $\mathfrak{U}$ (under $\mathfrak{G}$) and is denoted by $\mathfrak{G} : \mathfrak{U}$. Since for $U \in \mathfrak{U}$ the mapping $U \to UG$ of $\mathfrak{U}$ onto $\mathfrak{U}G$ is one-to-one because of the uniqueness of division, all the right cosets of $\mathfrak{U}$ have the same power as $\mathfrak{U}$. Thus,

$$|\mathfrak{G}| = (\mathfrak{G} : \mathfrak{U}) \cdot |\mathfrak{U}|. \tag{8}$$

For finite groups this fact was first proved by Lagrange. As a corollary, the order of a subgroup of a finite group $\mathfrak{G}$ is always a factor of the order of $\mathfrak{G}$.

The right cosets of the subgroup 1 consist of the individual elements of $\mathfrak{G}$. Thus in place of $|\mathfrak{G}|$ we may also write $\mathfrak{G} : 1$. Then (8) takes the form

$$\mathfrak{G} : 1 = (\mathfrak{G} : \mathfrak{U})(\mathfrak{U} : 1).$$

Since the property of being a subgroup is transitive, it follows more generally from (8) that for any sequence of subgroups

$$\mathfrak{U}_1 \supseteq \mathfrak{U}_2 \supseteq \dots \supseteq \mathfrak{U}_r$$

of $\mathfrak{G}$ we have:

$$\mathfrak{U}_1 : \mathfrak{U}_r = (\mathfrak{U}_1 : \mathfrak{U}_2) \cdots (\mathfrak{U}_{r-1} : \mathfrak{U}_r).$$

The reader may verify these results in the graphs of Figures 5 through 9.

A similar discussion of left cosets does not lead to any new results, since $\mathfrak{U}G \to G^{-1}\mathfrak{U}$ is a one-to-one mapping of the set of right cosets onto the set of left cosets: every left coset is an image, and from $G_1^{-1}\mathfrak{U} = G_2^{-1}\mathfrak{U}$ it follows by taking inverses that $\mathfrak{U}G_1 = \mathfrak{U}G_2$. The natural assumption that the set of right cosets is the same as the set of left cosets is false. Consider, for example, $\mathfrak{S}^{(3)}$ (§1.2.5) and

$$\mathfrak{U} = \left\{\begin{pmatrix}1 & 2 & 3\\ 1 & 2 & 3\end{pmatrix}, \begin{pmatrix}1 & 2 & 3\\ 1 & 3 & 2\end{pmatrix}\right\} = \{1, \gamma\}.$$

In this case the set of right cosets is

$$\{\{1, \gamma\}, \{\alpha, \epsilon\}, \{\beta, \delta\}\},$$

and the set of left cosets is

$$\{\{1, \gamma\}, \{\alpha, \delta\}, \{\beta, \epsilon\}\}.$$

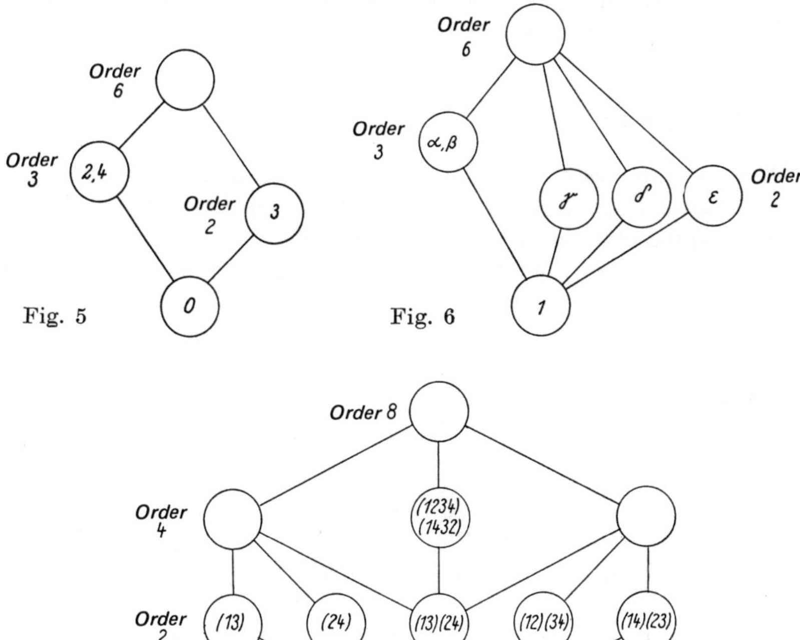

Fig. 5

Fig. 6

Fig. 7

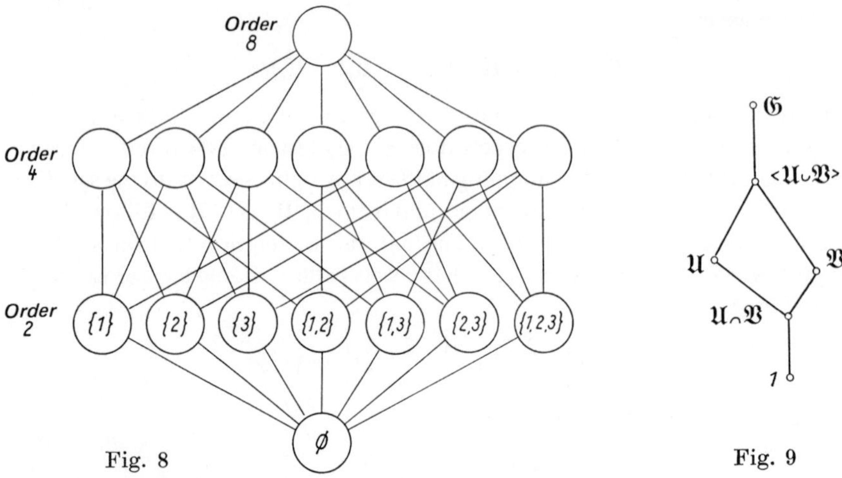

Fig. 8

Fig. 9

Subgroups for which these two sets of cosets are the same (as would always be the case, for example, with Abelian groups) are extremely important, and we shall pay a good deal of attention to them later; they are called *normal subgroups* (see §6).

4. Isomorphisms

If we wish to illustrate the addition of numbers, it makes absolutely no difference what objects (fingers, apples) we use. Obviously, the important element in the situation is not what we add, but how. This phenomenon, which also occurs in group theory and elsewhere, will now be described as clearly as possible and given a mathematical formulation.

4.1. It is obvious that the operation of a group is completely described by the following table:

	G	H	I	..
G	GG	GH	GI	..
H	HG	HH	HI	..
I	IG	IH	II	..
.	.	.	.	
.	.	.	.	..

The elements of 𝔊 are entered in the left-hand column and in the top row and then the product GH is entered at the intersection of the G row and the H column. This table is called the *multiplication* table for 𝔊.

For some of our earlier examples it will have the form:

$\Gamma^{(2)+}$

	0	1
0	0	1
1	1	0

$\mathfrak{B}^{\mathfrak{m}}$, $\mathfrak{m} = \{m\}$

	$\emptyset$	m
$\emptyset$	$\emptyset$	m
m	m	$\emptyset$

$\Gamma^{(8)\times}$

	1	3	5	7
1	1	3	5	7
3	3	1	7	5
5	5	7	1	3
7	7	5	3	1

$\mathfrak{S}^{(3)}$

	1	α	β	γ	δ	ϵ
1	1	α	β	γ	δ	ϵ
α	α	β	1	δ	ϵ	γ
β	β	1	α	ϵ	γ	δ
γ	γ	ϵ	δ	1	β	α
δ	δ	γ	ϵ	α	1	β
ϵ	ϵ	δ	γ	β	α	1

Γ^{+}

	0	1	2	3	⋯	−1	−2	−3	⋯
0	0	1	2	3	⋯	−1	−2	−3	⋯
1	1	2	3	4	⋯	0	−1	−2	⋯
2	2	3	4	5	⋯	1	0	−1	⋯
3	3	4	5	6	⋯	2	1	0	⋯
⋮	.	.	.	.	⋯	.	.	.	⋯
−1	−1	0	1	2	⋯	−2	−3	−4	⋯
−2	−2	−1	0	1	⋯	−3	−4	−5	⋯
−3	−3	−2	1	0	⋯	−4	−5	−6	⋯
⋮	.	.	.	.	⋯	.	.	.	⋯

The two-sidedness and uniqueness of division means that every element of the group will occur exactly once in each column and in each row. The reader should consider how the axioms **(N)** and **(I)**, and such properties as the commutativity of multiplication, are reflected in these tables.

If we look again at the multiplication table for the group $\Gamma^{(8)\times}$, it is obvious that the operation of the group will be equally well described if we replace the Arabic numerals by Roman numerals, so that the table becomes:

	I	III	V	VII
I	I	III	V	VII
III	III	I	VII	V
V	V	VII	I	III
VII	VII	V	III	I

If in the example for $\mathfrak{B}^{\mathfrak{m}}$ we had replaced $\emptyset$ and m by the numbers 0 and 1, we would have obtained the multiplication table for our example $\Gamma^{(2)+}$. The phenomenon arising in this way from the renaming of elements can be described mathematically as follows.

4.2. If $\mathfrak{G}$ is a group and λ is a one-to-one mapping of $\mathfrak{G}$ onto a group $\mathfrak{H}$ such that the image of each product is equal to the product of the images, or in other words if

$$(1) \qquad (G_1G_2)^\lambda = G_1^\lambda G_2^\lambda \qquad \text{for all} \quad G_1, G_2 \in \mathfrak{G},$$

then λ is said to be an *isomorphism* of $\mathfrak{G}$ onto $\mathfrak{H}$. If such an isomorphism exists, we say that $\mathfrak{G}$ and $\mathfrak{H}$ are isomorphic, or that they are of the same type, or the same structure, and write $\mathfrak{G} \simeq \mathfrak{H}$.

As an exercise, the reader may set up an isomorphism between the group $\mathfrak{P}$ (§1.2.4) and $\mathfrak{S}^{(3)}$.

If $\mathfrak{G}$ and $\mathfrak{H}$ are isomorphic, they obviously have the same order, and from the definition it is clear that an isomorphism λ also has the following properties:

4.2.1. The image of the unit element is the unit element.

4.2.2. If G^{-1} is the inverse of $G \in \mathfrak{G}$, then $(G^{-1})^\lambda = (G^\lambda)^{-1}$, which we abbreviate to $G^{-\lambda}$.

4.2.3. The image of a subgroup is a subgroup.

4.2.4. The image of a normal complex is normal.

4.2.5. If $\mathfrak{G}$ is Abelian, then every group isomorphic to $\mathfrak{G}$ is also Abelian.

The reader will readily prove §4.2.2, 3, 5; as examples let us show:

4.2.1. Since $1^\lambda \cdot 1^\lambda = (1 \cdot 1)^\lambda = 1^\lambda$, it follows that 1^λ is the unit element of $\mathfrak{H}$.

4.2.4. Since every element of $\mathfrak{H}$ is an image under the isomorphism, the statement follows from the fact that $G^\lambda \mathfrak{K}^\lambda G^{-\lambda} = (G\mathfrak{K}G^{-1})^\lambda$ for all $H = G^\lambda \in \mathfrak{H}$.

An isomorphism determines a partition into classes; that is, it is a reflexive, symmetric, and transitive relation:[7] the identical mapping of $\mathfrak{G}$ onto $\mathfrak{G}$ is obviously an isomorphism, so that we may write $\mathfrak{G} \simeq \mathfrak{G}$. If λ is an isomorphism of $\mathfrak{G}$ onto $\mathfrak{H}$, the inverse mapping λ^{-1} is an isomorphism of $\mathfrak{H}$ onto $\mathfrak{G}$; for from (1) we have

$$(2) \qquad (G_1^\lambda G_2^\lambda)^{\lambda^{-1}} = [(G_1G_2)^\lambda]^{\lambda^{-1}} = G_1G_2 = (G_1^\lambda)^{\lambda^{-1}}(G_2^\lambda)^{\lambda^{-1}}.$$

Also, if μ is an isomorphism of $\mathfrak{H}$ onto $\mathfrak{J}$, then the mapping $\lambda\mu$ is one-to-one and is an isomorphism of $\mathfrak{G}$ onto $\mathfrak{J}$:

$$(3) \qquad (G_1G_2)^{\lambda\mu} = [(G_1G_2)^\lambda]^\mu = (G_1^\lambda G_2^\lambda)^\mu = G_1^{\lambda\mu}G_2^{\lambda\mu}.$$

[7] The partition into classes also determines the *equivalence relation*; see IA, §8.5.

Thus, $\mathfrak{G} \simeq \mathfrak{H}$ and $\mathfrak{H} \simeq \mathfrak{J}$ imply $\mathfrak{G} \simeq \mathfrak{J}$, so that the set of all groups falls into classes of isomorphic groups such that no group belongs to more than one class.

Now the fundamental problem, the so-called *type* or *structure problem*, of the theory of groups is to select, from each class of isomorphic groups, a representative, or model, which is to be described as precisely as possible. What is meant here by "as precisely as possible" is to a great extent a matter of taste. In general, we shall look for a description in terms of concepts that lie closest to our intuition; for example, numbers, diagrams, and so forth. If we can find a model of this sort in every class, we have given a complete description of all groups, since any group arises from such a model by a mere renaming of the elements. At the present time we are still far from such a goal. Only for certain rather small, special classes of groups, e.g., the finite Abelian groups and a few others, has a satisfactory solution of this problem been found. In the next section we shall carry it through for a very simple class of groups, namely, the cyclic groups.

If we wish to give a complete account of the isomorphisms λ of $\mathfrak{G}$ onto $\mathfrak{H}$, it is enough to consider the case $\mathfrak{H} = \mathfrak{G}$. For if λ_1, λ_2 are two isomorphisms of $\mathfrak{G}$ onto $\mathfrak{H}$, the mapping $\alpha = \lambda_2\lambda_1^{-1}$ is an isomorphism of $\mathfrak{G}$ onto itself. On the other hand, if α is an isomorphism of $\mathfrak{G}$ onto itself, and λ_1 is an isomorphism of $\mathfrak{G}$ onto $\mathfrak{H}$, then $\alpha\lambda_1 = \lambda_2$ is also an isomorphism of $\mathfrak{G}$ onto $\mathfrak{H}$. Thus two isomorphisms of $\mathfrak{G}$ onto $\mathfrak{H}$ differ from each other only by an isomorphism of $\mathfrak{G}$ onto itself. An isomorphism of $\mathfrak{G}$ onto itself is called an *automorphism* of $\mathfrak{G}$. If α, β are two automorphisms of $\mathfrak{G}$, the mapping $\alpha\beta^{-1}$ is also an automorphism of $\mathfrak{G}$, as is easily seen from (2) and (3). Since automorphisms are combined in the same way as permutations, the set of automorphisms of $\mathfrak{G}$ is a subset of $\mathfrak{S}^{\mathfrak{G}}$, the so-called *group of automorphisms* of $\mathfrak{G}$.

For example, $\Gamma^{(3)+}$ has two automorphisms:

$$1 = \begin{pmatrix} 0 & 1 & 2 \\ 0 & 1 & 2 \end{pmatrix}, \qquad \alpha = \begin{pmatrix} 0 & 1 & 2 \\ 0 & 2 & 1 \end{pmatrix}.$$

Thus for any group $\mathfrak{G}$ isomorphic to $\Gamma^{(3)+}$ there exist two isomorphisms of $\Gamma^{(3)+}$ onto $\mathfrak{G}$.

5. Cyclic Groups

Corresponding to any element G of a group there exists the subgroup $\langle G \rangle$ (§3.4). The structure of such groups is particularly simple. In general, a group of the form $\mathfrak{G} = \langle G \rangle$ is called *cyclic*. For example, every group $\mathfrak{G}$ of prime order p is cyclic, since for $G \in \mathfrak{G}$ the theorem of Lagrange §3 (8) shows that $\langle G \rangle : 1$ must be equal either to 1 or to p, so that for $G \neq 1$ we have $\langle G \rangle = \mathfrak{G}$. Let us examine the cyclic groups a little more closely.

5.1. For an arbitrary group $\mathfrak{G}$ and $G \in \mathfrak{G}$ we define the nth power of G as the product

$$G \cdot G \cdot \ldots G, \text{ with } G \text{ written } n \text{ times,}$$

and denote it by G^n. Obviously we have

$$G^n(G^{-1})^n = 1,$$

so that $(G^n)^{-1} = (G^{-1})^n$; and thus we can write G^{-n} for the inverse of G^n. Finally we set $G^0 = 1$. Then for every rational integer g the *gth power* G^g is uniquely defined. Here

$$(G^g)^h = G^{gh}$$

and

$$G^gG^h = G^{g+h}, \tag{1}$$

as follows[8] at once from the definition for the various cases $g \geqslant 0, h \geqslant 0$; $g < 0, h \geqslant 0$; $g \geqslant 0, h < 0$; $g < 0, h < 0$.

By the criterion for subgroups, it follows from (1) that the set of powers of $G \in \mathfrak{G}$ is a subgroup, which is obviously equal to $\langle G \rangle$.

5.2. As a special case of the fundamental problem for the theory of groups we now reduce the description of the structure of cyclic groups to calculation with rational integers. In other words, we prove the following *fundamental theorem for cyclic groups.*

Let the $\mathfrak{G} = \langle G \rangle$ be cyclic. If $\mathfrak{G}$ is of finite order n, then $\mathfrak{G} \cong \Gamma^{(n)+}$. If $\mathfrak{G}$ is of infinite order, then $\mathfrak{G} \cong \Gamma^{+}$.

In particular, we have the following corollaries. *Every cyclic group is Abelian. For every finite order there exists a cyclic group which is uniquely determined up to isomorphism. There are no cyclic groups of power higher than $\aleph_0$.*

For the proof, we consider the various powers of G. By the remark at the end of §5.1 we have $\mathfrak{G} = \{G^g \mid g = 0, \pm 1, \ldots\}$. If $G^{g_1} \neq G^{g_2}$, for all $g_1 \neq g_2$, the mapping $g \to G^g$ is one-to-one and is therefore, by (1), an isomorphism of Γ^+ onto $\mathfrak{G}$; thus in this case the order of $\mathfrak{G}$ is equal to that of Γ^+. On the other hand, if $G^{g_1} = G^{g_2}$, $g_1 < g_2$, so that $G^{g_1 - g_2} = 1$, let $n > 0$ be the smallest positive integer with the property that $G^n = 1$. Let g be a rational integer with $g = nh + r$, $0 \leqslant r < n$, so that $G^g = G^{nh+r} = (G^n)^h G^r = G^r$. From $G^r = G^s$, $0 \leqslant r < s < n$ it follows that $G^{s-r} = 1$, so that $s = r$ by the minimal property of n. Thus $r \to G^r$ is a one-to-one mapping of $\Gamma^{(n)+}$ onto $\mathfrak{G}$. Then by (1) this mapping is an

[8] See also IB1, §3.3.

isomorphism of $\Gamma^{(n)+}$ onto $\mathfrak{G}$, which completes the proof of the fundamental theorem for cyclic groups.

5.3. For an arbitrary group $\mathfrak{G}$ and $G \in \mathfrak{G}$ the number $|\langle G\rangle|$, which has been defined above as the order of the group $\langle G\rangle$, is also called the *order of the element* G. From the proof in §5.2 it follows that the order of G, provided it is finite, can be characterized as the smallest natural number n for which $G^n = 1$. The same proof shows that divisibility of g by $|\langle G\rangle|$ is equivalent to $G^g = 1$. Since the order of G is equal to the order of the subgroup $\langle G\rangle$, the theorem of Lagrange shows that the order of G is a factor of the order of $\mathfrak{G}$. The least common multiple of the orders of all the elements of $\mathfrak{G}$, provided it exists, is called the *exponent* of $\mathfrak{G}$. Thus the exponent of a group $\mathfrak{G}$ is the smallest natural number e for which $G^e = 1$ for all $G \in \mathfrak{G}$.

For example, the exponent of $\mathfrak{S}^{(3)}$ is 6. A group with the exponent 2 is Abelian: for in general we have (§2 (4)) $(G_1G_2)^{-1} = G_2^{-1}G_1^{-1}$, so that if $G^2 = 1$ and therefore $G^{-1} = G$ for all $G \in \mathfrak{G}$, it follows for all such $\mathfrak{G}$ that $G_1G_2 = G_2G_1$.

5.4. We now examine the subgroups $\mathfrak{U}$ of a cyclic group $\langle G\rangle$. To this end we choose G^d in such a way that d is the least positive integer with the property $G^d \in \mathfrak{U}$, $d \geqslant 1$, as is always possible, since the subgroup $\mathfrak{U}$ contains the inverse of every element in $\mathfrak{U}$. Then $\mathfrak{U}$ contains the powers $(G^d)^a$; $a = 0, \pm 1, \dots$. But these powers already exhaust all the elements of $\mathfrak{U}$; for if we had $G^g \in \mathfrak{U}$, $g = hd + r$, $0 < r < d$, it would follow that $G^g(G^{hd})^{-1} = G^r \in \mathfrak{U}$, in contradiction to the choice of d. Thus $\mathfrak{U}$ is cyclic of the form $\langle G^d\rangle$.

It remains to decide when $\langle G^{d_1}\rangle = \langle G^{d_2}\rangle$ with $1 \leqslant d_1, d_2$. A necessary and sufficient condition is obviously the existence of rational integers a, b with

$$G^{d_1} = G^{ad_2},$$

$$G^{d_2} = G^{bd_1};$$

that is,

$$d_1 = ad_2,\ d_2 = bd_1,\ \text{if } |\langle G\rangle| \text{ is infinite,} \tag{2}$$

$$d_1 \equiv ad_2,\ d_2 \equiv bd_1 \bmod n,\ \text{if } |\langle G\rangle| = n \text{ is finite.} \tag{3}$$

Now (2) is equivalent to $d_1 = abd_1$, so that $ab = 1$ and $d_1 = d_2$; and (3) is equivalent, by IB6, §4.1, to

$$n \mid d_1 - ad_2,\ n \mid d_2 - bd_1.$$

But these conditions of divisibility can hold if and only if

$$(n, d_2) \mid d_1 , (n, d_1) \mid d_2$$

or

$$(n, d_1) \mid (n, d_2), (n, d_2) \mid (n, d_1)$$

or

$$(n, d_1) = (n, d_2).$$

We thus obtain the following result:

The subgroups of a cyclic group $\langle G \rangle$ are cyclic of the form $\langle G^d \rangle$, $1 \leqslant d$. Under the mapping $d \to \langle G^d \rangle$ they are in one-to-one correspondence with the natural numbers $d = 1, 2, \ldots$, if $| \langle G \rangle |$ is infinite and to the positive divisors d of n if $| \langle G \rangle | = n$ is finite.

Here d is the index of the subgroup $\mathfrak{U} = \langle G^d \rangle$, since for every $G^d \in \langle G \rangle$ there exists exactly one r, $0 \leqslant r < d$ with $g = hd + r$; thus

$$G^g = (G^d)^h G^r \in \mathfrak{U} G^r,$$

and the complexes $\mathfrak{U} G^r$ generate each coset of $\mathfrak{U}$ exactly once.

As an application of these results the reader may prove the following theorem: *every minimal subgroup is of prime order.*

6. Normal Subgroups and Factor Groups

6.1. In §3.5 we saw that a right coset for a subgroup of $\mathfrak{G}$ is not necessarily a left coset. We now examine those subgroups of a group for which every right coset is also a left coset, a property which, though at first glance it seems insignificant, has several very important consequences. Such subgroups $\mathfrak{N}$ are given the special name of *normal subgroups* or, corresponding to the terminology of §3.1, *invariant subgroups*. By §3.5 they are characterized by the fact that

(1) $\qquad \mathfrak{N} G = G \mathfrak{N} \qquad$ or $\qquad G \mathfrak{N} G^{-1} = \mathfrak{N} \qquad$ for all $\quad G \in \mathfrak{G}$.

The second identity means that $\mathfrak{N}$ coincides with all its conjugates in $\mathfrak{G}$.

6.2. In an Abelian group every subgroup is a normal subgroup, as we have already seen. But the Abelian groups are not the only groups with this property. There exist non-Abelian groups in which every subgroup is invariant. For such groups, called *Hamiltonian groups*, the type problem has been satisfactorily solved. We shall content ourselves here with an example of order 8, the so-called *quaternion group* $\mathfrak{Q}$. Its elements are the following matrices with coefficients from the field of complex numbers, and its operation is matrix multiplication (cf. IB8, §3.1):

$$1 = \begin{pmatrix} 1 & 0 \\ 0 & 1 \end{pmatrix}, \quad -1 = \begin{pmatrix} -1 & 0 \\ 0 & -1 \end{pmatrix}, \quad j = \begin{pmatrix} 0 & i \\ i & 0 \end{pmatrix}, \quad -j = \begin{pmatrix} 0 & -i \\ -i & 0 \end{pmatrix},$$

$$k = \begin{pmatrix} 0 & 1 \\ -1 & 0 \end{pmatrix}, \quad -k = \begin{pmatrix} 0 & -1 \\ 1 & 0 \end{pmatrix}, \quad l = \begin{pmatrix} -i & 0 \\ 0 & i \end{pmatrix}, \quad -l = \begin{pmatrix} i & 0 \\ 0 & -i \end{pmatrix}.$$

The lattice of subgroups of $\mathfrak{Q}$ has the form shown in Figure 10.

If a subgroup $\mathfrak{U}$ of $\mathfrak{G}$ has the index 2, it is a normal subgroup, since in this case the coset distinct from $\mathfrak{U}$ contains exactly those elements of $\mathfrak{G}$ that are not in $\mathfrak{U}$. For example, the subgroup $\{1, \alpha, \beta\}$ (§3.3.3) is a normal subgroup of $\mathfrak{S}^{(3)}$. But the fact that a subgroup of index 3 need not be a normal subgroup is shown by the example $\mathfrak{U} = \{1, \gamma\} \subseteq \mathfrak{G}^{(3)}$.

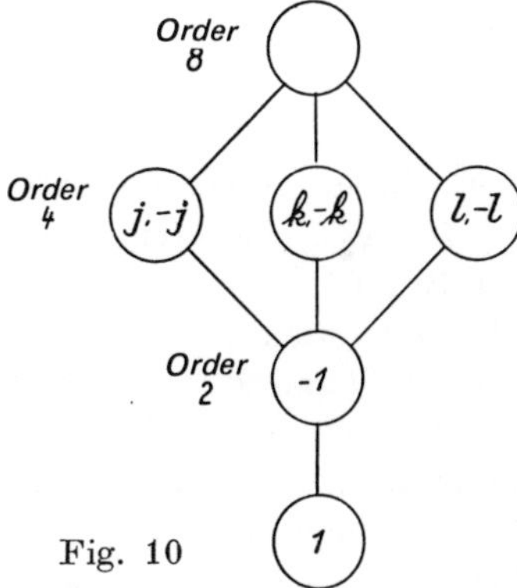

Fig. 10

6.3. The fundamental role of normal subgroups is indicated by the following theorem:

A subgroup of $\mathfrak{G}$ is normal in $\mathfrak{G}$ if and only if the set of its right cosets (or alternatively, of its left cosets) is a group with respect to complex multiplication.

For the proof we first let $\mathfrak{N}$ be a normal subgroup of $\mathfrak{G}$. From (1) it then follows that

$$\text{(2)} \qquad \mathfrak{N}G\mathfrak{N}H = \mathfrak{N}\mathfrak{N}GH = \mathfrak{N}GH,$$

so that **(V)** is satisfied. The associativity of complex multiplication was already shown in §3.1. The neutral element is $\mathfrak{N}$:

$$\mathfrak{N}G\mathfrak{N} = \mathfrak{N}\mathfrak{N}G = \mathfrak{N}G;$$

the inverse of $\mathfrak{N}G$ is $\mathfrak{N}G^{-1}$, since $\mathfrak{N}G\mathfrak{N}G^{-1} = \mathfrak{N}\mathfrak{N} = \mathfrak{N}$. Thus the cosets of $\mathfrak{N}$ form a group.

Conversely, let $\mathfrak{N}$ be a subgroup of $\mathfrak{G}$ whose right cosets form a group, so that in particular the product $\mathfrak{N}G\mathfrak{N}G^{-1}$ is a right coset for all $G \in \mathfrak{G}$. Since $1 = 1G1G^{-1} \in \mathfrak{N}G\mathfrak{N}G^{-1}$, we have by §3.5

$$\mathfrak{N}G\mathfrak{N}G^{-1} = \mathfrak{N} \cdot 1 = \mathfrak{N},$$

and thus

$$G\mathfrak{N}G^{-1} = 1 \cdot G\mathfrak{N}G^{-1} \subseteq \mathfrak{N},$$

and

$$\mathfrak{N} \subseteq G^{-1}\mathfrak{N}G;$$

consequently, $\mathfrak{N} = G\mathfrak{N}G^{-1}$, since if G runs through all elements of $\mathfrak{G}$, so does G^{-1}. Thus $\mathfrak{N}$ is a normal subgroup of $\mathfrak{G}$.

Similarly we could have shown that $\mathfrak{N}$ is a normal subgroup if an only if its left cosets form a group.

The group of cosets of $\mathfrak{N}$ is denoted by $\mathfrak{G}/\mathfrak{N}$ and is called the *factor group of* $\mathfrak{N}$ (*under* $\mathfrak{G}$) or the *factor group* $\mathfrak{G}$ *by* $\mathfrak{N}$. The order of $\mathfrak{G}/\mathfrak{N}$ is equal to the index $\mathfrak{G} : \mathfrak{N}$.

6.4. In an Abelian group every factor group is also Abelian. If $\mathfrak{G} = \langle G \rangle$ is cyclic, every factor group $\langle G \rangle / \langle G^d \rangle$ is also cyclic, namely equal to $\langle \langle G^d \rangle G \rangle$.

If $\mathfrak{N}$ is a normal subgroup with $\mathfrak{N} \neq 1$, the group $\mathfrak{G}$ is split by $\mathfrak{N}$ into two groups $\mathfrak{G}/\mathfrak{N}$ and $\mathfrak{N}$. These groups are usually simpler in structure than $\mathfrak{G}$ itself (for example, in the finite case they are of smaller order) so that by examining them separately we can gain insight into the structure of $\mathfrak{G}$. Especially interesting from this point of view are the groups that do not allow any splitting in this sense; in other words, groups which have no normal subgroups except 1 and $\mathfrak{G}$. Such groups are called *simple*. *Simple Abelian groups are of prime order. Conversely, every group of prime order is simple and Abelian.* As we shall see later, there exist simple groups (sometimes called *properly simple*) that are not Abelian. Their existence offers a serious obstacle to the solution of the structure problem.

It is easy to show that the intersection of a set of normal subgroups is again a normal subgroup.

If $\mathfrak{N}$ is a normal subgroup and $\mathfrak{U}$ is any subgroup of $\mathfrak{G}$, the complex $\mathfrak{N}\mathfrak{U}$ is a subgroup of $\mathfrak{G}$ and is obviously equal to the group $\mathfrak{N} \cup \mathfrak{U}$ generated by $\langle \mathfrak{N} \cup \mathfrak{U} \rangle$. For with $N_1, N_2 \in \mathfrak{N}$; $U_1, U_2 \in \mathfrak{U}$ it follows from (1) that

$$\begin{aligned} N_1U_1(N_2U_2)^{-1} &= N_1U_1U_2^{-1}N_2^{-1} \\ &= N_1(U_1U_2^{-1})\, N_2^{-1}(U_1U_2^{-1})^{-1}U_1U_2^{-1} \\ &\in \mathfrak{N}\mathfrak{U}, \end{aligned}$$

so that the criterion for subgroups is satisfied for $\mathfrak{N}\mathfrak{U}$. In particular, if $\mathfrak{U} = \mathfrak{M}$ is also a normal subgroup of $\mathfrak{G}$, then

$$G\mathfrak{N}\mathfrak{M}G^{-1} = G\mathfrak{N}G^{-1}G\mathfrak{M}G^{-1} = \mathfrak{N}\mathfrak{M},$$

so that $\mathfrak{N}\mathfrak{M}$ is a normal subgroup of $\mathfrak{G}$.

A normal subgroup $\mathfrak{N}$ of $\mathfrak{G}$ is also a normal subgroup of every subgroup $\mathfrak{U} \subseteq \mathfrak{G}$ that is contained in $\mathfrak{N}$. The subgroups of $\mathfrak{U}$ that contain $\mathfrak{N}$ are in one-to-one correspondence under the mapping

$$\mathfrak{U} \to \mathfrak{U}/\mathfrak{N} \tag{3}$$

with the subgroups $\mathfrak{U}/\mathfrak{N}$ of $\mathfrak{G}/\mathfrak{N}$. More precisely, we have here a lattice isomorphism (see IB9, §2.1) of the lattice of subgroups of $\mathfrak{G}$ containing $\mathfrak{N}$ onto the lattice of subgroups of $\mathfrak{G}/\mathfrak{N}$. Moreover, the normal subgroups of the two groups are in one-to-one correspondence.

If $\mathfrak{G}$ is generated by $\mathfrak{K}$, then $\mathfrak{G}/\mathfrak{N}$ is generated by $\{\mathfrak{N}K \mid K \in \mathfrak{K}\}$; for if $\mathfrak{N}G$ is a coset of $\mathfrak{N}$ and $G = K_1^{\epsilon_1} \cdots K_r^{\epsilon_r}$, $K_i \in \mathfrak{K}$, then $\mathfrak{N}G$ can be represented as

$$\begin{aligned} \mathfrak{N}G &= \mathfrak{N}K_1^{\epsilon_1} \ldots K_r^{\epsilon_r} \\ &= (\mathfrak{N}K_1)^{\epsilon_1} \ldots (\mathfrak{N}K_r)^{\epsilon_r}. \end{aligned}$$

7. The Commutator Group

For any given group it is possible, by using the concept of a factor group, to define a certain subgroup which, roughly speaking, measures the extent to which the group departs from being Abelian. For this purpose we consider a set $\mathfrak{N}$ of normal subgroups $\mathfrak{N}$ of $\mathfrak{G}$ whose factor groups are all Abelian. Let $\mathfrak{D}$ be the intersection of all the normal subgroups in $\mathfrak{N}$, so that $\mathfrak{D}$ is a normal subgroup of $\mathfrak{G}$. We will now show that $\mathfrak{G}/\mathfrak{D}$ is also Abelian: for we have $\mathfrak{N}G\mathfrak{N}H = \mathfrak{N}H\mathfrak{N}G$, that is, $\mathfrak{N}GH = \mathfrak{N}HG$ or $\mathfrak{N}GHG^{-1}H^{-1} = \mathfrak{N}$ for all $\mathfrak{N} \in \mathfrak{N}$ and $G, H \in \mathfrak{G}$. Thus every element $GHG^{-1}H^{-1}$ lies in all the $\mathfrak{N} \in \mathfrak{N}$:

$$GHG^{-1}H^{-1} \in \mathfrak{D} \quad \text{and thus} \quad \mathfrak{D}GHG^{-1}H^{-1} = \mathfrak{D}.$$

Consequently,

$$\begin{aligned} \mathfrak{D}GH &= \mathfrak{D}HG, \\ \mathfrak{D}G\mathfrak{D}H &= \mathfrak{D}H\mathfrak{D}G, \end{aligned}$$

so that $\mathfrak{G}/\mathfrak{D}$ is commutative. If $\mathfrak{N}$ is the set of all normal subgroups of $\mathfrak{G}$ with Abelian factor group, it follows that:

In every group there exists a normal subgroup, with Abelian factor group, which is contained in every normal subgroup with Abelian factor group and is the intersection of all such normal subgroups. It is called the *commutator group* of $\mathfrak{G}$ and is denoted by $\mathfrak{G}'$.

It is easy to see that the commutator group is generated by the set of all elements $GHG^{-1}H^{-1}$, the so-called commutators of $\mathfrak{G}$. To say that $\mathfrak{G}' = 1$ is equivalent to saying that $\mathfrak{G}$ is Abelian. In the other extreme case, $\mathfrak{G} = \mathfrak{G}'$, the group $\mathfrak{G}$ is called *perfect*. A properly simple group is perfect, since 1 and $\mathfrak{G}$ are its only normal subgroups and therefore $\mathfrak{G} = \mathfrak{G}'$. The commutator group of $\mathfrak{S}^{(3)}$ is the subgroup $\mathfrak{A}^{(3)}$ (§3.3.3); the factor group $\mathfrak{S}^{(3)}/\mathfrak{A}^{(3)}$ is Abelian and 1 is the only proper subgroup of $\mathfrak{A}^{(3)}$; however, $\mathfrak{S}^{(3)}/1 \cong \mathfrak{S}^{(3)}$ is not Abelian.

8. Direct Products

8.1. Let $\mathfrak{M}$ and $\mathfrak{N}$ be normal subgroups of $\mathfrak{G}$, with

(1) $$\mathfrak{M}\mathfrak{N} = \mathfrak{G}, \qquad \mathfrak{M} \cap \mathfrak{N} = 1.$$

If $M_1N_1 = M_2N_2$, $M_1, M_2 \in \mathfrak{M}$, $N_1, N_2 \in \mathfrak{N}$, then $M_2^{-1}M_1 = N_2N_1^{-1} = 1$ and $M_1 = M_2$, $N_1 = N_2$. Thus if the elements $G \in \mathfrak{G}$ are represented in the form

(2) $$G = MN; \qquad M \in \mathfrak{M}, \quad N \in \mathfrak{N},$$

the M, N are uniquely determined by the G.

Now consider a product $MN = NMM^{-1}N^{-1}MN$. Since $\mathfrak{M}$ and $\mathfrak{N}$ are normal subgroups, we have

$$(M^{-1}N^{-1}M)\,N = M^{-1}(N^{-1}MN) \in \mathfrak{M} \cap \mathfrak{N} = 1$$

and thus

$$M^{-1}N^{-1}M \in \mathfrak{N}, \qquad N^{-1}MN \in \mathfrak{M},$$

so that every element of $\mathfrak{M}$ permutes with every element of $\mathfrak{N}$.

Thus the operation of the group $\mathfrak{G}$ is completely determined by the products of elements from $\mathfrak{M}$ and $\mathfrak{N}$:

(3) $$(M_1N_1)(M_2N_2) = M_1M_2 \cdot N_1N_2\,.$$

We describe this special case by saying that $\mathfrak{G}$ is *decomposable* into the *direct product of* $\mathfrak{M}$ *and* $\mathfrak{N}$. It is customary to write (1) in the shorter form: $\mathfrak{G} = \mathfrak{M} \times \mathfrak{N}$.

The mapping σ, with $(\mathfrak{M}N)^\sigma = N$, is an isomorphism of $\mathfrak{G}/\mathfrak{M}$ onto $\mathfrak{N}$:

$$\mathfrak{G}/\mathfrak{M} \cong \mathfrak{N}.$$

For by (1) the requirement §4(1) is satisfied, and the mapping σ is onto $\mathfrak{N}$ and is one-to-one, since every coset of $\mathfrak{M}$ contains its image.

8.2. More generally, we say that $\mathfrak{G}$ is the *direct product of* $\mathfrak{M}_i$; $i = 1, \ldots, r$, and write

$$\mathfrak{G} = \mathfrak{M}_1 \times \ldots \times \mathfrak{M}_r\,,$$

if the $\mathfrak{M}_i$ are normal subgroups of $\mathfrak{G}$, with $M_i M_k = M_k M_i$ for $M_i \in \mathfrak{M}_i$, $M_k \in \mathfrak{M}_k$, $i \neq k$, and if every element $G \in \mathfrak{G}$ can be written as the product

$$G = M_1 \dots M_r \tag{4}$$

with uniquely determined $M_i \in \mathfrak{M}_i$. This extension of our notation is justified by the fact that the direct product as now defined can be obtained by iteration from the earlier definition. For if

$$\mathfrak{G} = \mathfrak{M} \times \mathfrak{N}, \qquad \mathfrak{N} = \mathfrak{L} \times \mathfrak{K},$$

we have

$$\mathfrak{G} = \mathfrak{M} \times \mathfrak{L} \times \mathfrak{K}$$

in the generalized notation.

From the uniqueness of the representation (4) it follows that the order of a direct product is given by

$$| \mathfrak{G} | = | \mathfrak{M}_1 | \dots | \mathfrak{M}_r |.$$

If $\mathfrak{G}$ is the direct product of Abelian groups, then $\mathfrak{G}$ itself is Abelian, as follows directly from (3). Every group can be decomposed into the direct product $\mathfrak{G} = 1 \times \mathfrak{G}$. A group which allows no other decomposition into a direct product is said to be *indecomposable.* For example, the group $\mathfrak{S}^{(3)}$ is indecomposable, since it has only one normal subgroup $\neq 1$.

8.3. An exact knowledge of the indecomposable groups would be very desirable; in a certain sense they represent an analogue in the theory of groups to prime numbers in the theory of numbers. For a very extensive class of groups, which includes for example the finite groups and the finitely generated Abelian groups, we have the following theorem, in analogy with unique decomposition into prime numbers.

If

$$\mathfrak{M}_1 \times \dots \times \mathfrak{M}_r \cong \mathfrak{N}_1 \times \dots \times \mathfrak{N}_s;$$

with $\mathfrak{M}_i \mathfrak{N}_k$ *indecomposable*; $i = 1, \dots, r$; $k = 1, \dots, s$, *then* $r = s$, *and there exists a permutation* σ *of the indices such that* $\mathfrak{M}_i \cong \mathfrak{N}_{i\sigma}$.

In the following section we shall determine the indecomposable groups in the class of all finite Abelian groups, and the direct product will provide us with a suitable solution of the type problem for this class of groups.

9. Abelian Groups

In §9.1 we state certain elementary but basic properties of Abelian groups which will be useful in what follows.

9.1. As mentioned before, every subgroup of an Abelian group $\mathfrak{G}$ is a normal subgroup, and for two such groups $\langle \mathfrak{U} \cup \mathfrak{V} \rangle = \mathfrak{U}\mathfrak{V}$. Also, in Abelian groups we have the following *power rule:*

$$(GH)^n = GH \dots GH = G^n H^n; \qquad G,\ H \in \mathfrak{G}.$$

From this rule it follows that for every g the complex $\mathfrak{G}^g = \{G^g \mid G \in \mathfrak{G}\}$ is a subgroup of $\mathfrak{G}$: $G^g H^{-g} = (GH^{-1})^g$; and dually, for fixed g, the set of elements $G \in \mathfrak{G}$ for which $G^g = 1$ is a subgroup $\mathfrak{G}_g$: for if $G_1, G_2 \in \mathfrak{G}$, then $(G_1 G_2^{-g} = G_1^g G_2^{-g} = 1$.

If $G, H \in \mathfrak{G}$ are of finite order and $G^g = 1$, $H^h = 1$, then $(GH^{-1})^{gh} = 1$. Thus the elements of finite order in $\mathfrak{G}$ form a subgroup $\mathfrak{T}$, the so-called *torsion group* of $\mathfrak{G}$. For example, the torsion group of $P^{\times}$ (§1.2.3) is the subgroup $\{1, -1\}$. The torsion group $\mathfrak{T}$ itself is not necessarily of finite order, as is shown by the example P^+/Γ^+ (§1.2.2): for if $g/h \in P^+$, $g, h \in \Gamma^+$, then for h summands we have

$$\left(\Gamma^+ + \frac{g}{h}\right) + \dots + \left(\Gamma^+ + \frac{g}{h}\right) = \Gamma^+ + g = \Gamma^+.\ P^+/\Gamma^+,$$

so that P^+/Γ^+ is identical with its torsion group. But P^+/Γ^+ is not of finite order, since the elements $1, \frac{1}{2}, \frac{1}{3}, \dots$ lie in distinct cosets of Γ^+.

Abelian groups in which only the unit element is of finite order are said to be *torsion-free*. Thus the cyclic group of infinite order is torsion-free, and so are all direct products of torsion-free groups. Furthermore, *for an arbitrary Abelian group the factor group by its torsion group is torsion-free:* for if $(\mathfrak{T}G)^m = \mathfrak{T}$, then $G^m = T \in \mathfrak{T}$; and if $T^t = 1$, then $G^{mt} = 1$, so that $G \in \mathfrak{T}$. Thus $\mathfrak{T}$ is the only element of $\mathfrak{G}/\mathfrak{T}$ that is of finite order.

9.2. *Finite and Finitely Generated Abelian Groups*

The purpose of the present section is as follows: by making use of the direct product we reduce the type problem for finite Abelian groups to the same problem for cyclic groups, for which it was already solved in §5.2. In speaking of direct products we include the case that the product has only one factor.

9.2.1. We now prove the fundamental theorem on finite Abelian groups: *a finite Abelian group $\mathfrak{G}$ is the direct product of cyclic groups of prime power order:*

(1) $\qquad \mathfrak{G} = \langle G_1 \rangle \times \dots \times \langle G_r \rangle,\ |\langle G_i \rangle| = p_i^{\alpha_i}, \qquad p_i = \text{prime}.$

To prove this theorem we first prove the following two lemmas:

9.2.2. *If the exponent of a finite Abelian group is not a prime power, the group is decomposable.*

9.2.3. *If the exponent of a finite Abelian group* $\mathfrak{G}$ *is a prime power and if M is an element of maximal order in* $\mathfrak{G}$, *then there exists a direct decomposition*

$$\mathfrak{G} = \langle \mathfrak{M} \rangle \times \tilde{\mathfrak{G}}$$

with a suitable subgroup $\tilde{\mathfrak{G}} \subseteq \mathfrak{G}$.

From these two lemmas it is clear that an Abelian group which is not cyclic of prime power exponent must allow a proper decomposition

$$\mathfrak{G} = \mathfrak{H} \times \mathfrak{J}; \qquad |\,\mathfrak{H}\,|, \mathfrak{J}\,| < |\,\mathfrak{G}\,|.$$

Since by the theorems on cyclic groups the exponent of such a group is equal to the order of the group, the fundamental theorem now follows immediately by complete induction on the order of the group and iterated construction of the direct product.

Proof of §9.2.2. We assume that the exponent e of $\mathfrak{G}$ is the product of mutually prime positive integers $\neq 1$:

$$e = ab, \qquad (a, b) = 1,$$

and then show that

$$\mathfrak{G} = \mathfrak{G}^a \times \mathfrak{G}^b. \tag{2}$$

For this purpose we determine $\bar{a}$, $\bar{b}$ such that $a\bar{a} + b\bar{b} = 1$. Then $G = (G^a)^{\bar{a}}(G^b)^{\bar{b}}$ for all $G \in \mathfrak{G}$, and thus

$$\mathfrak{G} = \mathfrak{G}^a\mathfrak{G}^b.$$

Let $G_1^a = G_2^b \in \mathfrak{G}^a \cap \mathfrak{G}^b$; $G_1, G_2 \in \mathfrak{G}$. Then

$$(G_1^a)^{\bar{a}} = (G_2^b)^{\bar{a}},$$

$$G_1^{1-b\bar{b}} = G_2^{b\bar{a}},$$

$$G_1 = (G_2^{\bar{a}}G_1^{\bar{b}})^b.$$

Since $e = ab$, it follows that $G_1^a = (G_2^{\bar{a}}G_1^{\bar{b}})^{ab} = 1$ and $\mathfrak{G}^a \cap \mathfrak{G}^b = 1$, so that we are dealing here with a direct product (2). But a and b are $\neq 1$, so that by definition of the exponent $\mathfrak{G}^a$, $\mathfrak{G}^b \neq 1$.

Proof of §9.2.3. If $\mathfrak{G}$ is cyclic, there is nothing to prove. If not, we first construct a subgroup $\mathfrak{H} \neq 1$ whose intersection with $\langle M \rangle$ is 1. To do this, we let the exponent of $\mathfrak{G}$ be a power of the prime p. The theorems on the subgroups of cyclic groups show that for $G \notin \langle M \rangle$,

$$\langle G \rangle \cap \langle M \rangle = \langle G^{p^s} \rangle = \langle M^{p^r} \rangle,$$

so that

$$p^s \cdot |\langle G^{p^s} \rangle| = |\langle G \rangle|,$$

$$p^r \cdot |\langle M^{p^r} \rangle| = |\langle M \rangle|$$

with, let us say, $G^{p^s} = M^{ap^r}$. By the maximality of M we then have $s \leqslant r$. For the element $H = GM^{-ap^{r-s}}$, which is different from 1, it follows that $H^n \in \langle M \rangle$ is equivalent to $G^n \in M$, so that

$$\langle H \rangle \cap \langle M \rangle = \langle H^{p^s} \rangle.$$

But

$$H^{p^s} = G^{p^s} M^{-ap} = 1,$$

and therefore

$$\langle H \rangle \cap \langle M \rangle = 1.$$

Consequently, $\mathfrak{H} = \langle H \rangle$ has the desired property.

The factor group $\mathfrak{G}/\mathfrak{H}$ is also of prime power exponent, so that each of its elements is of prime power order. Also, $\mathfrak{H}M$ is of maximal order in $\mathfrak{G}/\mathfrak{H}$; for if $(\mathfrak{H}M)^{p^t} = \mathfrak{H}$, $(\mathfrak{H}M^*)^{p^t} \neq \mathfrak{H}$, $M^* \in \mathfrak{G}$, it would follow that $M^{p^t} \in \mathfrak{H}$, so that by the construction of $\mathfrak{H}$

$$M^{p^t} = 1$$

and

$$M^{*p^t} \notin \mathfrak{H},$$

so that $M^{*p} \neq 1$ and M^* would be of greater order than M, in contradiction to the choice of M. But now the fundamental theorem follows immediately by complete induction on the order of $\mathfrak{G}$, since the theorem holds trivially for $|\,\mathfrak{G}\,| = 1$, and since $\mathfrak{G}/\mathfrak{H}$ is of smaller order than $\mathfrak{G}$, there exists a direct decomposition

$$\mathfrak{G}/\mathfrak{H} = \langle \mathfrak{H}M \rangle/\mathfrak{H} \times \tilde{\mathfrak{G}}/\mathfrak{H}$$

with a subgroup $\tilde{\mathfrak{G}}$ of $\mathfrak{G}$. In other words, $\langle \mathfrak{H}M \rangle \tilde{\mathfrak{G}} = \mathfrak{G}$, $\langle \mathfrak{H}M \rangle \cap \tilde{\mathfrak{G}} = \mathfrak{H}$. But from $\langle \mathfrak{H}M \rangle = \mathfrak{H}\langle M \rangle$ and $\mathfrak{H} \subseteq \tilde{\mathfrak{G}}$ we then have $\langle M \rangle \tilde{\mathfrak{G}} = \mathfrak{G}$ and $\langle M \rangle \cap \tilde{\mathfrak{G}} \subseteq \mathfrak{H}$. Thus it follows from the choice of $\mathfrak{H}$ that $\langle M \rangle \cap \tilde{\mathfrak{G}} = 1$.

Thus the proof of the fundamental theorem is complete.

9.2.4. The fundamental theorem provides the complete decomposition of a given group, as can be seen from the following fact: every cyclic group $\langle G \rangle$ of prime power order p^α is indecomposable. For if $\langle G \rangle$ were decomposable, each of the factors would contain a minimal subgroup of order p, in contradiction to the fact that a cyclic group cannot have more than one subgroup of a given order.

9.2.5. The cyclic groups allow us to solve the type problem for a more general class of groups, in view of the theorem:

If $\mathfrak{G}$ *is a finitely generated Abelian group, then* $\mathfrak{G}$ *is a direct product of finite cyclic groups, whose orders are prime powers, and of infinite cyclic groups:*

$$(3) \qquad \mathfrak{G} = \langle G_1 \rangle \times ... \times \langle G_r \rangle \times \langle H_1 \rangle \times ... \times \langle H_s \rangle;$$

$$|\langle G_i \rangle| = p_i^{\alpha_i},\ p_i \text{ prime}, \quad i = 1, ..., r; \quad |\langle H_k \rangle| = \aleph_0\,, \quad k = 1, ..., s.$$

It is obvious that the torsion group of $\mathfrak{G}$ is given by

$$\mathfrak{T} = \langle \mathfrak{G}_1 \rangle \times ... \times \langle G_r \rangle,$$

which is finite.

We shall omit the proof of this more general theorem, since it does not require the introduction of any essentially new ideas. The chief difficulty lies in replacing the complete induction on the order of the group by complete induction on the number of generators, which may become laborious, since it is not always easy to see whether a proper subgroup has a smaller number of generators than the group itself.

9.2.6. By means of the theorem on the subgroups of a cyclic group it is easy to prove that an infinite cyclic group is indecomposable. Then the theorem in §8.3 on the uniqueness of direct decompositions states that for finitely generated Abelian groups the numbers $p_1^{\alpha_1}, ..., p_r^{\alpha_r}$ and s in (3) are uniquely determined by $\mathfrak{G}$. On account of their importance in combinatorial topology, these numbers $p_1^{\alpha_1}, ..., p_r^{\alpha_r}$ are called the *torsion numbers* of $\mathfrak{G}$, and s is the *Betti number* of $\mathfrak{G}$.

9.2.7. From the decomposition (1) it is clear that we can determine the exponent e of $\mathfrak{G}$ as follows: for every prime p that divides $\mathfrak{G} : 1$ we determine the group $\langle G_i \rangle$ of highest p-power order p^{α}; then

$$(4) \qquad e = \prod_{p | \mathfrak{G}:1} p^{\alpha}.$$

9.2.8. *Examples.* The reader may consider for himself how the general finite cyclic groups are included in the preceding theorems and how they are to be decomposed into direct products of cyclic groups of prime power order.

The expression (4) for the exponent enables us to give a complete description of the structure of finite groups with exponent 2. As pointed out in §5.3, these groups are Abelian, so that they must be direct products of cyclic groups of order 2. Thus, for example, the type of all groups $\mathfrak{B}^{\mathfrak{m}}$ with finite $|\ \mathfrak{m}\ |$ has been satisfactorily described.

As a further example of the preceding theorems we shall show how they apply to the group §1.2.11. The torsion group $\mathfrak{T}$ of $\mathfrak{E}^{(3)}$ consists of the elements $1, -1$, since these are the only real numbers for which some power is equal to 1. If we denote by $\tilde{\mathfrak{E}}^{(3)}$ the set of positive numbers in $\mathfrak{E}^{(3)}$, then $\tilde{\mathfrak{E}}^{(3)}$ is obviously a subgroup of $\mathfrak{E}^{(3)}$. Since every number in $\mathfrak{E}^{(3)}$ can be written in the form $1 \cdot \epsilon$ or $(-1) \cdot \epsilon$ with an element $\epsilon \in \tilde{\mathfrak{E}}^{(3)}$ and since $\mathfrak{T} \cap \tilde{\mathfrak{E}}^{(3)} = 1$, we have the direct product decomposition $\mathfrak{E}^{(3)} = \mathfrak{T} \times \tilde{\mathfrak{E}}^{(3)}$.

Let us investigate the group $\tilde{\mathfrak{E}}^{(3)}$. To every real number $g + h\sqrt{3}$ with integral g, h we assign the lattice point (g, h) in a Cartesian coordinate system. Then the elements of $\mathfrak{E}^{(3)}$ correspond to the lattice points of the hyperbola $x^2 - 3y^2 = 1$. We now consider the set $\mathfrak{g}$ of parallel lines

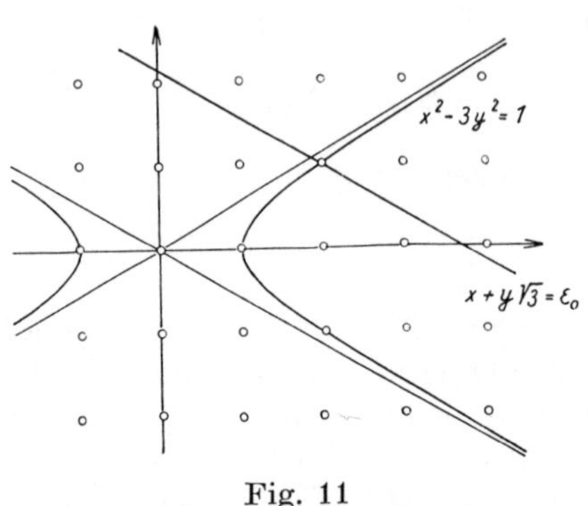

Fig. 11

$$x + y\sqrt{3} = \rho, \qquad \rho \text{ real} \geqslant 1.$$

As can be seen at once from Figure 11, $\mathfrak{g}$ consists of the lines that are parallel to the line $x + y\sqrt{3} = 1$ and lie above it to the right. Thus there exists a line $x + y\sqrt{3} = \epsilon_0$ in $\mathfrak{g}$, with $\epsilon_0 > 1$, that passes through a lattice point of the hyperbola and has the minimal ϵ_0 for all such lines. Then ϵ_0 is an element of $\tilde{\mathfrak{E}}^{(3)}$, and we can show that the elements $\epsilon \in \tilde{\mathfrak{E}}^{(3)}$ are the powers of ϵ_0. For let us choose an integer m such that

$$\epsilon_0^m \leqslant \epsilon < \epsilon_0^{m+1}.$$

Then

$$1 \leqslant \frac{\epsilon}{\epsilon_0^m} < \epsilon_0 .$$

Now, $\epsilon/\epsilon_0^m = \bar{\epsilon}$ is an element of $\tilde{\mathfrak{E}}^{(3)}$, so that the line $x + y\sqrt{3} = \bar{\epsilon}$ passes through a lattice point of the hyperbola. From the minimal property of ϵ_0 it follows that $\bar{\epsilon} = 1$ and $\epsilon_0^m = \epsilon$. Since the powers ϵ^n form an infinite set of numbers, the group $\tilde{\mathfrak{E}}^{(3)}$ is a cyclic group of infinite order.

9.3. *Group Properties of Ornaments*

We shall now give an example to show how the symmetry of ornaments can be analyzed by means of group concepts. The characteristic feature of the ornaments of interest to us is "infinite repetition," by which we mean the repetition of a definite pattern, the "elementary ornament," at equal distances. This repetition, or translation, can take place in one, two, or three noncoplanar directions, with the corresponding classification of ornaments into linear (that is, occurring on a band or strip), planar, and spatial types, the last two of which play an important role in crystallography. Whether there exist other symmetry operations, in addition to these pure translations, depends on the nature of the given ornament.

In illustration of these remarks let us consider the classical decoration shown in Figure 12. If we interpret this ornamental strip as a sequence of partially overlapping discs situated in three-dimensional space, then the

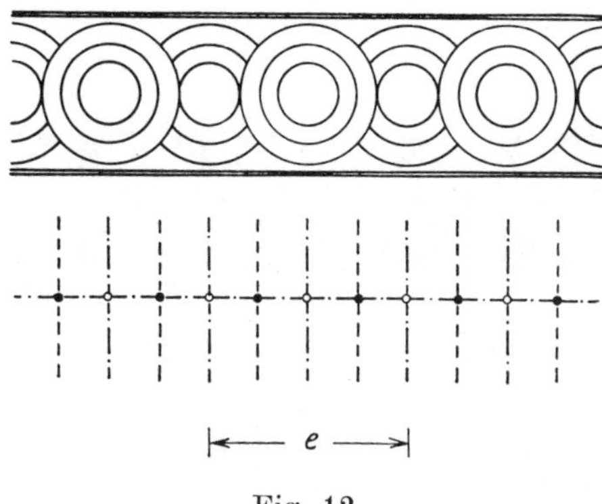

Fig. 12

strip can be mapped onto itself by the following, simple or composite, operations:

Translations by distances that are multiples of a certain elementary vector $\vec{e}$ in the direction of the axis of the ornament.

Reflection in the plane perpendicular to the strip and through the axis, with or without simultaneous translation by a multiple of the elementary vector (longitudinal reflections).

Reflections in planes that are perpendicular to the axis of the ornament and are at a distance from one another of half the elementary distance e (transverse reflections; in the figure the traces of these planes are marked by dots and dashes).

Rotations by 180° about transverse axes in the plane of the strip, where the axes are again at a distance from one another of half the elementary distance (half-turns around a transverse axis; in the figure the axes are marked by dashes).

Rotations around centers at a distance from one another of half the elementary distance (in the figure these centers are indicated by small circles).

Translations by an odd multiple of half the elementary vector with simultaneous reflection in the plane of the ornament, so that the overlappings of the circles are reversed (planar glide reflections).

Translations by an odd multiple of half the elementary vector with simultaneous rotation of 180° around the longitudinal axis (spiral motions).

Rotations around points separated by half the elementary distance (marked in the figure by small black circles) with simultaneous reflection in the plane of the ornament (rotatory reflections).

The schematic representation in Figure 13 shows the structure of the ornament.

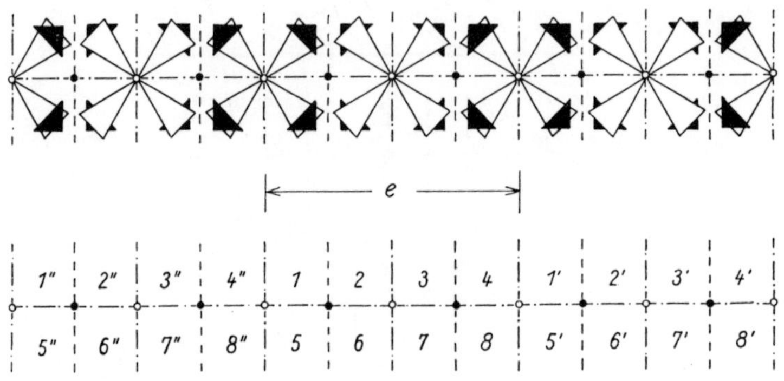

FIG. 13

It is obvious that these mappings are the elements of an infinite non-Abelian group. We divide them into the complexes listed in the above classification; among these complexes the translations, including the identity, clearly form a normal subgroup. We can gain insight into the structure of the ornament by forming the factor group with respect to translations.

In order to investigate this factor group, we make use again of the above schema, consisting of overlapping triangles with the same structure as the ornament. We consider each elementary part of the ornament as being divided into eight fields, each of which contains a figure consisting of two overlapping triangles; this figure can be mapped onto the other figures by an operation of the group, indicated below by primes.

1. The translations

$B_1^{(0)} : 1 \to 1$ (the identity mapping),

$B_1^{(1)} : 1 \to 1'$ (translation by $\vec{e}$),

$B_1^{(2)} : 1 \to 1''$ (translation by $-\vec{e}$), and so forth

form the normal subgroup and the unit element of the corresponding factor group is $\mathfrak{B}_1 = \{B_1^{(n)} \mid n = 0, 1, 2, \ldots\}$.

2. The mappings

$B_2^{(0)} : 1 \to 2,$

$B_2^{(1)} : 1 \to 2',$

$B_2^{(2)} : 1 \to 2'',$ and so forth

represent half-turns. All these motions arise from a single B_2 by complex-multiplication with the group of translations: $\mathfrak{B}_2 = \{B_2^{(n)}\} = B_2\mathfrak{B}_1$.

3. The mappings

$B_3^{(0)} : 1 \to 3,$

$B_3^{(1)} : 1 \to 3',$

$B_3^{(2)} : 1 \to 3'',$ and so forth

characterize the complex of planar glide reflections. This complex arises from one element B_3 as follows: $\mathfrak{B}_3 = \{B_3^{(n)}\} = B_3\mathfrak{B}_1$.

4. The mappings

$B_4^{(0)} : 1 \to 4,$

$B_4^{(1)} : 1 \to 4',$

$B_4^{(2)} : 1 \to 4'',$ and so forth

are reflections in transverse axes; they may be written as cosets of $\mathfrak{B}_1$ with arbitrary B_4: $\mathfrak{B}_4 = \{B_4^{(n)}\} = B_4\mathfrak{B}_1$.

5. Reflections and glide reflections in the longitudinal axis

$B_5^{(0)} : 1 \to 5,$

$B_5^{(1)} : 1 \to 5',$

$B_5^{(2)} : 1 \to 5'',$ and so forth

are elements of a complex which may be represented by $\mathfrak{B}_5 = \{B_5^{(n)}\} = B_5\mathfrak{B}_1$.

6. From

$B_6^{(0)} : 1 \to 6,$

$B_6^{(1)} : 1 \to 6',$

$B_6^{(2)} : 1 \to 6'',$ and so forth

we obtain the rotatory reflections. Here $\mathfrak{B}_6 = \{B_6^{(n)}\} = B_6\mathfrak{B}_1$.

7. The helical motions

$B_7^{(0)} : 1 \to 7,$

$B_7^{(1)} : 1 \to 7',$

$B_7^{(2)} : 1 \to 7'',$ and so forth

form the complex $\mathfrak{B}_7 = \{B_7^{(n)}\} = B_7\mathfrak{B}_1$.

8. Finally, the simple rotations around centers

$B_8^{(0)} : 1 \to 8,$

$B_8^{(1)} : 1 \to 8',$

$B_8^{(2)} : 1 \to 8'',$ and so forth

may be represented in the form $\mathfrak{B}_8 = \{B_8^{(n)}\} = B_8\mathfrak{B}_1$.

The symmetry of the ornament is now characterized by the structure of the factor group in the following way: if $\mathfrak{G}$ is an infinite non-Abelian group of mappings of the ornamental strip onto itself, then the factor group with respect to the translations

$$\mathfrak{G}/\mathfrak{B}_1 = \{\mathfrak{B}_1, \mathfrak{B}_2, \mathfrak{B}_3, \mathfrak{B}_4, \mathfrak{B}_5, \mathfrak{B}_6, \mathfrak{B}_7, \mathfrak{B}_8\}$$

is Abelian. The element $\mathfrak{B}_1$ is the unit element, and every element is inverse to itself (involution). The multiplication table is as follows:

	$\mathfrak{B}_1$	$\mathfrak{B}_2$	$\mathfrak{B}_3$	$\mathfrak{B}_4$	$\mathfrak{B}_5$	$\mathfrak{B}_6$	$\mathfrak{B}_7$	$\mathfrak{B}_8$
$\mathfrak{B}_1$	$\mathfrak{B}_1$	$\mathfrak{B}_2$	$\mathfrak{B}_3$	$\mathfrak{B}_4$	$\mathfrak{B}_5$	$\mathfrak{B}_6$	$\mathfrak{B}_7$	$\mathfrak{B}_8$
$\mathfrak{B}_2$	$\mathfrak{B}_2$	$\mathfrak{B}_1$	$\mathfrak{B}_4$	$\mathfrak{B}_3$	$\mathfrak{B}_6$	$\mathfrak{B}_5$	$\mathfrak{B}_8$	$\mathfrak{B}_7$
$\mathfrak{B}_3$	$\mathfrak{B}_3$	$\mathfrak{B}_4$	$\mathfrak{B}_1$	$\mathfrak{B}_2$	$\mathfrak{B}_7$	$\mathfrak{B}_8$	$\mathfrak{B}_5$	$\mathfrak{B}_6$
$\mathfrak{B}_4$	$\mathfrak{B}_4$	$\mathfrak{B}_3$	$\mathfrak{B}_2$	$\mathfrak{B}_1$	$\mathfrak{B}_8$	$\mathfrak{B}_7$	$\mathfrak{B}_6$	$\mathfrak{B}_5$
$\mathfrak{B}_5$	$\mathfrak{B}_5$	$\mathfrak{B}_6$	$\mathfrak{B}_7$	$\mathfrak{B}_8$	$\mathfrak{B}_1$	$\mathfrak{B}_2$	$\mathfrak{B}_3$	$\mathfrak{B}_4$
$\mathfrak{B}_6$	$\mathfrak{B}_6$	$\mathfrak{B}_5$	$\mathfrak{B}_8$	$\mathfrak{B}_7$	$\mathfrak{B}_2$	$\mathfrak{B}_1$	$\mathfrak{B}_4$	$\mathfrak{B}_3$
$\mathfrak{B}_7$	$\mathfrak{B}_7$	$\mathfrak{B}_8$	$\mathfrak{B}_5$	$\mathfrak{B}_6$	$\mathfrak{B}_3$	$\mathfrak{B}_4$	$\mathfrak{B}_1$	$\mathfrak{B}_2$
$\mathfrak{B}_8$	$\mathfrak{B}_8$	$\mathfrak{B}_7$	$\mathfrak{B}_6$	$\mathfrak{B}_5$	$\mathfrak{B}_4$	$\mathfrak{B}_3$	$\mathfrak{B}_2$	$\mathfrak{B}_1$

The group belongs to the type dealt with in the second paragraph of §9.2.8. Since it is of order 8, the theorem of Lagrange suggests that we look for subgroups of order 4 and 2. Such subgroups are to be found in the table, or more simply by a glance at the schematic figure (Figure 14).

The full ornament (occupying eight fields) has the structure of the group $\mathfrak{G}/\mathfrak{B}_1$ (holohedrism).

If we reduce the number of fields to four, we obtain seven different possibilities, corresponding to suitable choices of the occupied fields, for ornaments with the structure of the subgroups of order 4 (hemihedrism).

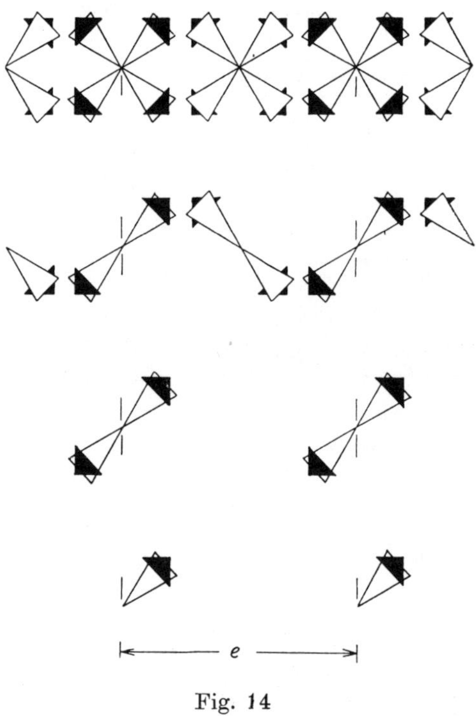

Fig. 14

If we restrict ourselves to two fields, we can find seven further arrangements, which have the structure of the subgroups $\mathfrak{B}_i$ of order 2 (tetartohedrism).

Finally, if only one field is occupied, we have the unit element of the factor group.

In Figures 15 through 29 we give examples of ornaments with the structure of the various subgroups.

In addition to the eight symmetry operations described above, there exist three others that can appear in ornaments. The number of possible holohedrisms is thereby increased to four; they contain sixteen hemihedrisms and ten tetartohedrisms, the unit element in each case being the group of translations.

Groups of the same order are isomorphic to one another. Figure 30 shows the lattice of subgroups of the holohedrism.

FIG. 15
Group $\mathfrak{U}_1 = \{\mathfrak{B}_1, \mathfrak{B}_2, \mathfrak{B}_3, \mathfrak{B}_4\}$

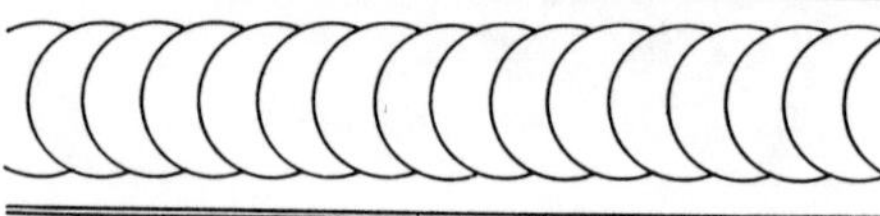

FIG. 16
Group $\mathfrak{U}_2 = \{\mathfrak{B}_1, \mathfrak{B}_2, \mathfrak{B}_5, \mathfrak{B}_6\}$

FIG. 17
Group $\mathfrak{U}_3 = \{\mathfrak{B}_1, \mathfrak{B}_2, \mathfrak{B}_7, \mathfrak{B}_8\}$

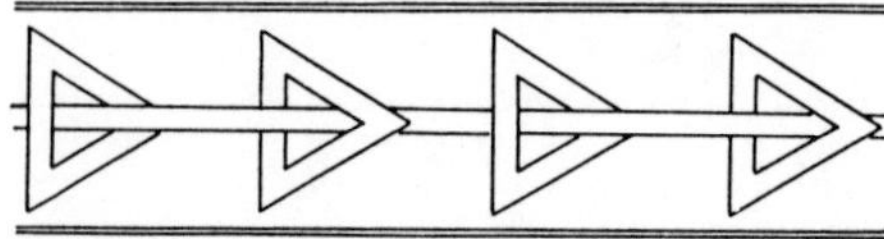

FIG. 18
Group $\mathfrak{U}_4 = \{\mathfrak{B}_1, \mathfrak{B}_3, \mathfrak{B}_5, \mathfrak{B}_7\}$

FIG. 19
Group $\mathfrak{U}_5 = \{\mathfrak{B}_1, \mathfrak{B}_3, \mathfrak{B}_6, \mathfrak{B}_8\}$

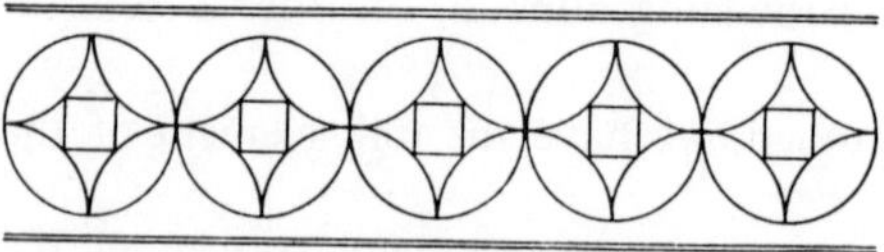

FIG. 20
Group $\mathfrak{U}_6 = \{\mathfrak{B}_1, \mathfrak{B}_4, \mathfrak{B}_5, \mathfrak{B}_8\}$

FIG. 21
Group $\mathfrak{U}_7 = \{\mathfrak{B}_1, \mathfrak{B}_4, \mathfrak{B}_6, \mathfrak{B}_7\}$

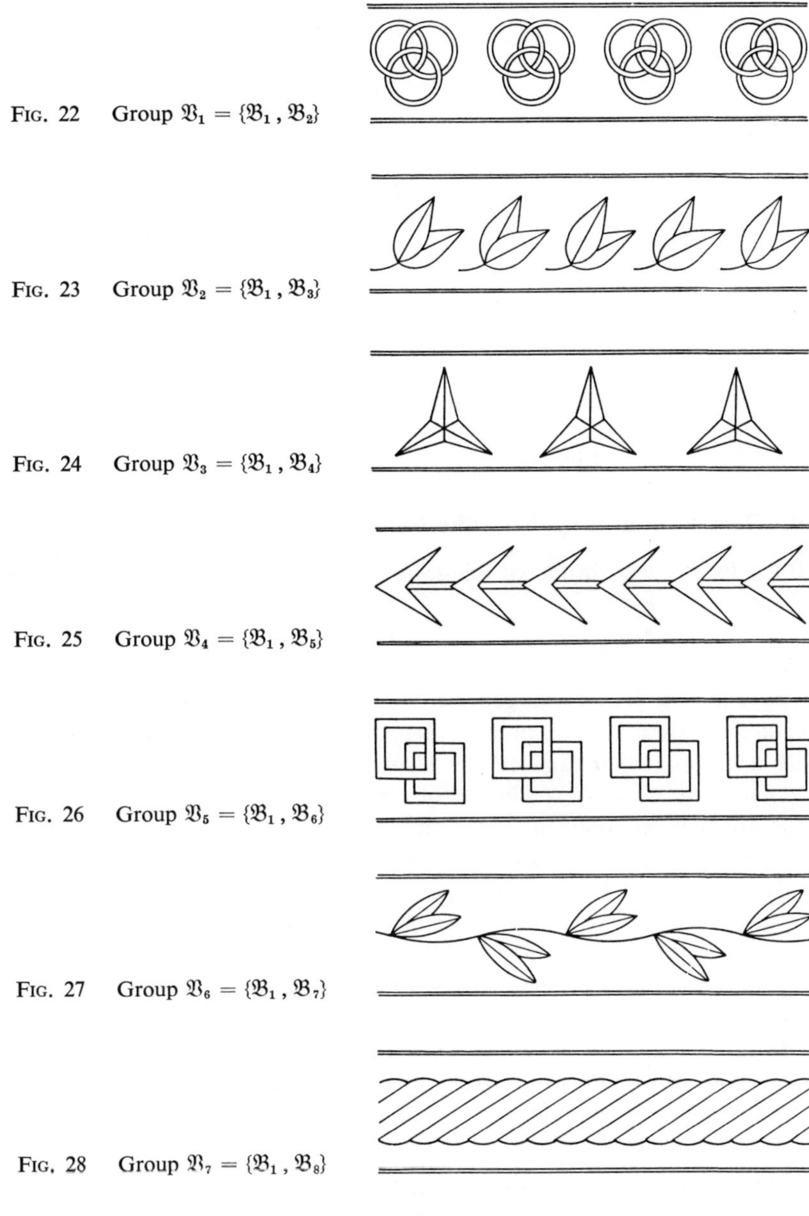

Fig. 22 Group $\mathfrak{B}_1 = \{\mathfrak{B}_1, \mathfrak{B}_2\}$

Fig. 23 Group $\mathfrak{B}_2 = \{\mathfrak{B}_1, \mathfrak{B}_3\}$

Fig. 24 Group $\mathfrak{B}_3 = \{\mathfrak{B}_1, \mathfrak{B}_4\}$

Fig. 25 Group $\mathfrak{B}_4 = \{\mathfrak{B}_1, \mathfrak{B}_5\}$

Fig. 26 Group $\mathfrak{B}_5 = \{\mathfrak{B}_1, \mathfrak{B}_6\}$

Fig. 27 Group $\mathfrak{B}_6 = \{\mathfrak{B}_1, \mathfrak{B}_7\}$

Fig. 28 Group $\mathfrak{B}_7 = \{\mathfrak{B}_1, \mathfrak{B}_8\}$

Fig. 29 Group $\mathfrak{B}_1$

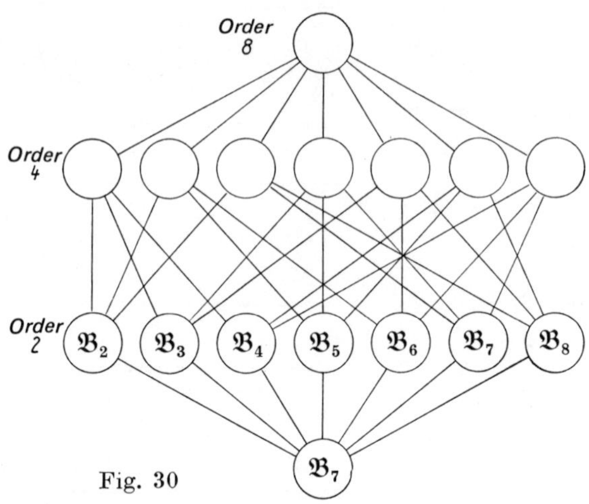

Fig. 30

10. The Homomorphism Theorem

10.1. In §4.2 we considered one-to-one mappings λ of a group $\mathfrak{G}$ for which

(1) $$(G_1G_2)^\lambda = G_1^\lambda G_2^\lambda; \qquad G_1, \quad G_2 \in \mathfrak{G}.$$

The following important generalization is obtained by dropping the requirement that the mapping be one-to-one. A mapping λ of a group $\mathfrak{G}$ into a group $\mathfrak{H}$ that satisfies (1) is called a *homomorphism of* $\mathfrak{G}$ *into* $\mathfrak{H}$. The set of elements of $\mathfrak{G}$ that are mapped onto the unit element $1_{\mathfrak{H}}$ of $\mathfrak{H}$ is denoted by $\mathfrak{G}_\lambda$ and is called the *kernel of* λ: $\mathfrak{G}_\lambda = \{G \mid G \in \mathfrak{G}, G^\lambda = 1_{\mathfrak{H}}\}$; the set $\mathfrak{G}^\lambda = \{G^\lambda \mid G \in \mathfrak{G}\}$ is called the *image* of $\mathfrak{G}$ under the mapping λ, or simply the image under λ (see Figure 31).

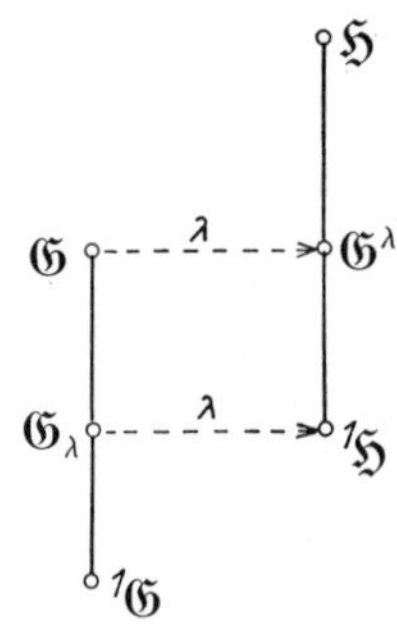

Fig. 31

The rules §4.2.1–5 are proved for homomorphisms in the same way as for isomorphisms.

The following *homomorphism theorem* shows that homomorphisms offer a very useful means of finding normal subgroups in a group and of investigating their factor groups:

The kernel of a homormorphism λ *of* $\mathfrak{G}$ *into* $\mathfrak{H}$ *is a normal subgroup of* $\mathfrak{G}$, *and the image of* $\mathfrak{G}$ *under* λ *is a subgroup of* $\mathfrak{H}$. *Moreover*, $\mathfrak{G}/\mathfrak{G}_\lambda \cong \mathfrak{G}^\lambda$.

Proof. From $G_1^\lambda = \mathfrak{G}_2^\lambda = 1_{\mathfrak{H}}$ it follows that $1_{\mathfrak{H}} = G_1^\lambda G_2^{-\lambda} = (G_1G_2^{-1})^\lambda$, so that $\mathfrak{G}_\lambda$ is a subgroup of $\mathfrak{G}$. Also, for every $X \in \mathfrak{G}$ and $G \in \mathfrak{G}_\lambda$

$$(XGX^{-1})^\lambda = X^\lambda G^\lambda X^{-\lambda} = X^\lambda X^{-\lambda} = 1_{\mathfrak{H}},$$

so that $\mathfrak{G}_\lambda$ is a normal subgroup of $\mathfrak{G}$. For arbitrary $X, Y \in \mathfrak{G}$ we have $X^\lambda(Y^\lambda)^{-1} = (XY^{-1})^\lambda$, so that $\mathfrak{G}^\lambda$ is a subgroup of $\mathfrak{H}$. In order to show that $\mathfrak{G}/\mathfrak{G}_\lambda \cong \mathfrak{G}^\lambda$ we consider the mapping induced by λ on the set of cosets of $\mathfrak{G}/\mathfrak{G}_\lambda$. This mapping is in fact a mapping *onto* $\mathfrak{G}^\lambda$, since

$$(\mathfrak{G}_\lambda X)^\lambda = \mathfrak{G}_\lambda^\lambda X^\lambda = X^\lambda \qquad \text{for} \qquad X \in \mathfrak{G}.$$

The mapping is one-to-one, since $(\mathfrak{G}_\lambda X_1)^\lambda = (\mathfrak{G}_\lambda X_2)^\lambda$ implies $X_1^\lambda = X_2^\lambda$, which means that $(X_1 X_2^{-1})^\lambda = 1_{\mathfrak{H}}$, $X_1 X_2^{-1} \in \mathfrak{G}_\lambda$ and $\mathfrak{G}_\lambda X_1 = \mathfrak{G}_\lambda X_2$. Finally,

$$\begin{aligned}(\mathfrak{G}_\lambda X_1 \mathfrak{G}_\lambda X_2)^\lambda &= (\mathfrak{G}_\lambda X_1 X_2)^\lambda \\ &= (X_1 X_2)^\lambda \\ &= X_1^\lambda X_2^\lambda \\ &= (\mathfrak{G} X_1)^\lambda (\mathfrak{G} X_2)^\lambda.\end{aligned}$$

Thus we have shown that the induced mapping is an isomorphism of $\mathfrak{G}/\mathfrak{G}_\lambda$ onto $\mathfrak{G}^\lambda$.

In order to complete the homomorphism theorem we note that every normal subgroup $\mathfrak{N}$ of a group $\mathfrak{G}$ is the kernel of a homomorphism of $\mathfrak{G}$, since the mapping $G \to G\mathfrak{N}$ is a homomorphism onto $\mathfrak{G}/\mathfrak{N}$, as is easily seen from §6(2), and its kernel is obviously $\mathfrak{N}$.

10.2. Let us consider two applications of the homomorphism theorem. For the first example let $\mathfrak{G} = K_n^\times$ (i.e., the set of $n \times n$ matrices over a field K; see §1.2.12) and let a mapping λ from K_n into the multiplication group $K^\times$ of the field K be defined as follows: to every matrix the mapping λ assigns the value of its determinant:

$$A^\lambda = |A|, \qquad A \in K_n^\times.$$

Then, by the rule for multiplication of determinants, λ is a homomorphism and its kernel is the group of matrices with determinant 1. Thus the latter group is a normal subgroup of $K_n^\times$. But every element of $K^\times$ is an image under λ, since $k \in K^\times$ is the image of

$$\begin{pmatrix} k & & & 0 \\ & 1 & & \\ & & \ddots & \\ 0 & & & 1 \end{pmatrix}$$

Thus the factor group of this normal subgroup is isomorphic to $K^\times$.

For the next example we take $\mathfrak{G} = B_2$ (§1.2.6). If for $\sigma \in B_2$ we set $\epsilon_\sigma = +1$ or $\epsilon_\sigma = -1$, according to whether the orientation of a triangle in E_2 is preserved or changed by the motion σ (that is, whether σ involves or does not involve a half-turn of the plane), then obviously $\epsilon_{\sigma_1}\epsilon_{\sigma_2} = \epsilon_{\sigma_1\sigma_2}$. Thus the mapping $\sigma \to \epsilon_\sigma$ is a homomorphism of B_2 onto the cyclic group of order 2. By the homomorphism theorem its kernel is a normal subgroup of B_2 whose cyclic factor group is of order 2. This group is called the group of *proper motions in* E_2 and is denoted by B_2^+.

In B_2 we can also construct a normal subgroup. For this purpose we first note that in a proper motion the angle between a line and its image is the same for all lines. If to each element $\sigma \in B_2^+$ we assign this angle ω_σ in radian measure, the product of two proper motions $\sigma_1\sigma_2$ corresponds to the angle $\omega_{\sigma_1} + \omega_{\sigma_2}$, as follows from the theorem on the exterior angles of a triangle. Since two angles are equal if and only if their radian measure differs by a multiple of 2π, the mapping $\sigma \to \omega_\sigma$ is a homomorphism of B_2^+ onto the factor group $R^+/\langle 2\pi\rangle$, the so-called *planar rotation group.* The kernel of this homomorphism is the group T_2 of *translations;* that is, of proper motions in which the image $g\sigma$ of every line g is parallel to g. By the homomorphism theorem the factor group B^+/T_2 is isomorphic to $R^+/\langle 2\pi\rangle$.

Our examination of the structure of B_2 can now be brought to an end with the remark that T_2 is isomorphic to the direct product of two groups of type R^+ (§1.2.2), a fact whose proof we leave to the reader. The structure of B_1 may be examined in the same way.

11. The Isomorphism Theorem

11.1. In our discussion of direct products $\mathfrak{G} = \mathfrak{M} \times \mathfrak{N}$ we noted that $\mathfrak{G}/\mathfrak{N} \cong \mathfrak{M}$ and $\mathfrak{G}/\mathfrak{M} \cong \mathfrak{N}$. We now prove an important generalization of this fact.

Isomorphism theorem: if $\mathfrak{N}$ *is a normal subgroup and* $\mathfrak{U}$ *is a subgroup of* $\mathfrak{G}$, *then* $\mathfrak{N} \cap \mathfrak{U}$ *is a normal subgroup of* $\mathfrak{U}$, *and*

$$\mathfrak{N}\mathfrak{U}/\mathfrak{N} \cong \mathfrak{U}/\mathfrak{N} \cap \mathfrak{U}.$$

The above statement about direct products corresponds to the special case that $\mathfrak{U}$ is also a normal subgroup of $\mathfrak{G}$ and $\mathfrak{G} = \mathfrak{N}\mathfrak{U}$, $\mathfrak{N} \cap \mathfrak{U} = 1$.

Proof. Let $X \in \mathfrak{U}$, $U \in \mathfrak{N} \cap \mathfrak{U}$; then $XUX^{-1} \in \mathfrak{U}$ since $\mathfrak{U}$ is a group, and $XUX^{-1} \in \mathfrak{N}$ since $\mathfrak{N}$ is a normal subgroup, so that the subgroup $\mathfrak{N} \cap \mathfrak{U}$ is a normal subgroup of $\mathfrak{U}$. The elements of $\mathfrak{N}\mathfrak{U}/\mathfrak{N}$ are the cosets of the form $\mathfrak{N}U$, $U \in \mathfrak{U}$. For U_1, $U_2 \in \mathfrak{U}$ the two statements $\mathfrak{N}U_1 = \mathfrak{N}U_2$ and $(\mathfrak{N} \cap \mathfrak{U})\, \mathfrak{U}_1 = (\mathfrak{N} \cap \mathfrak{U})\, U_2$ are equivalent to each other. Thus $(\mathfrak{N}U)^\lambda = (\mathfrak{N} \cap \mathfrak{U})U$ defines a one-to-one mapping λ of $\mathfrak{N}\mathfrak{U}/\mathfrak{N}$ onto $\mathfrak{U}/\mathfrak{N} \cap \mathfrak{U}$. Since

$$\begin{aligned}(\mathfrak{N}U_1\mathfrak{N}U_2)^\lambda &= (\mathfrak{N}U_1U_2)^\lambda\\ &= (\mathfrak{N} \cap \mathfrak{U})\, U_1U_2\\ &= (\mathfrak{N} \cap \mathfrak{U})\, U_1(\mathfrak{N} \cap \mathfrak{U})\, U_2\\ &= (\mathfrak{N}U_1)^\lambda(\mathfrak{N}U_2)^\lambda,\end{aligned}$$

the mapping λ is an isomorphism of $\mathfrak{N}\mathfrak{U}/\mathfrak{N}$ onto $\mathfrak{U}/\mathfrak{N} \cap \mathfrak{U}$.

The various interrelations involved in the isomorphism theorem can be represented very clearly in a diagram if in the graph of the corresponding lattice we draw the line segments $\mathfrak{N}\mathfrak{U} - \mathfrak{N}$ and $\mathfrak{U} - \mathfrak{N} \cap \mathfrak{U}$ parallel to each other and of equal length (see Figure 32).

Then the geometric fact that the segments $\mathfrak{N}\mathfrak{U} - \mathfrak{N}$ and $\mathfrak{U} - \mathfrak{N} \cap \mathfrak{U}$ are necessarily parallel and equal finds its group-theoretic expression in the fact that

$$\mathfrak{N} : \mathfrak{N} \cap \mathfrak{U} = \mathfrak{N}\mathfrak{U} : \mathfrak{U},$$

which follows at once from $\mathfrak{N}\mathfrak{U} : \mathfrak{N} \cap \mathfrak{U} = (\mathfrak{N}\mathfrak{U} : \mathfrak{N})(\mathfrak{N} : \mathfrak{N} \cap \mathfrak{U}) = (\mathfrak{N}\mathfrak{U} : \mathfrak{U})(\mathfrak{U} : \mathfrak{N} \cap \mathfrak{U})$ and the isomorphism theorem.

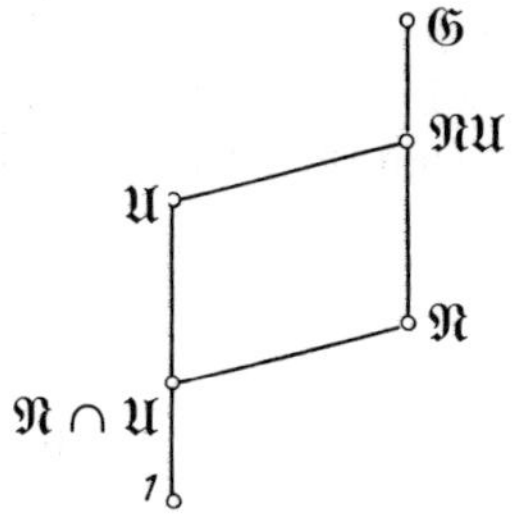

Fig. 32

11.2. In the next section we shall use the isomorphism theorem to prove an important result in the general theory of groups, but first let us illustrate the theorem for the above group B_2^+ (§10.2) of proper motions in E_2. It is easy to show, as in §3.3.4, that the set $B_{2,P}^+$ of elements of B_2 leaving a point P fixed is a subgroup of B_2^+. Obviously, this subgroup consists of the planar rotations around the point P, so that like the factor group B_2^+/T_2 it is isomorphic to the group $R^+/\langle 2\pi \rangle$. This fact also follows from the isomorphism theorem, since every proper motion can be obtained as the result of a translation followed by a rotation around P, which means in group-theoretic language that $B_2^+ = T_2 B_{2,P}^+$. Since $B_{2,P}^+$ and T_2 have only the mapping 1 in common, it follows from the isomorphism theorem that

$$B_2^+/T_2 \cong B_{2,P}^+ \cong R^+/\langle 2\pi \rangle.$$

12. Composition Series, Jordan-Hölder Theorem

In dealing with the type problem we naturally try to divide up every group into its simplest possible components, as was done above for the case of Abelian groups. In the section on direct products we remarked that although the theorem on unique decomposition into a direct product is valid for an extensive class of groups, the indecomposable groups themselves are still too complicated to provide an acceptable survey of all types of groups. In the present section we undertake an analysis leading to a simpler class of basic components, though now there is the disadvantage, not found in the direct product, that the given group is no longer uniquely determined by its components.

12.1. We consider a group $\mathfrak{G} \neq 1$ for which there exists a finite sequence of subgroups

(1) K: $$\mathfrak{G} = \mathfrak{N}_0 \supset \mathfrak{N}_1 \supset \ldots \supset \mathfrak{N}_l = 1$$

such that $\mathfrak{N}_i$ is a maximal normal subgroup of $\mathfrak{N}_{i-1}$ for $i = 1, ..., l$. Here l is called the *length of the composition series K*. The minimum of the lengths of all composition series of $\mathfrak{G}$ is called the *length of* $\mathfrak{G}$. There exist groups, for example Γ^+, that have no composition series, but such groups are always of infinite order. If $\mathfrak{G}$ is simple, then $l = 1$, since $\mathfrak{G} \supset 1$ is the only composition series of $\mathfrak{G}$.

From the lattice isomorphism §6(3) it follows that if

$$(2)\quad K_1: \qquad \mathfrak{G}/\mathfrak{N} = \mathfrak{N}_0/\mathfrak{N} \supset \mathfrak{N}_1/\mathfrak{N} \supset ... \supset \mathfrak{N}_s/\mathfrak{N} = \mathfrak{N}/\mathfrak{N}$$

and

$$(3)\quad K_2: \qquad \mathfrak{N} = \mathfrak{M}_0 \supset \mathfrak{M}_1 \supset ... \supset \mathfrak{M}_t = 1$$

are composition series of $\mathfrak{G}/\mathfrak{N}$ and $\mathfrak{N}$, respectively, then

$$(4)\quad K_3: \qquad \mathfrak{G} = \mathfrak{N}_0 \supset \mathfrak{N}_1 \supset ... \supset \mathfrak{N}_s \supset \mathfrak{M}_1 \supset ... \mathfrak{M}_t = 1$$

is a composition series of $\mathfrak{G}$. Thus for any subgroup of $\mathfrak{N}$ it is possible to construct a composition series of $\mathfrak{G}$ that includes the subgroup $\mathfrak{N}$, provided $\mathfrak{G}/\mathfrak{N}$ and $\mathfrak{N}$ themselves have composition series. The part of K that lies in $\mathfrak{N}_i$ is a composition series for $\mathfrak{N}_i$.

The factor groups $\mathfrak{N}_{i-1}/\mathfrak{N}_i$ occurring in (1) are called *composition factors of K*. Since $\mathfrak{N}_i$ is a maximal normal subgroup of $\mathfrak{N}_{i-1}$, these composition factors are simple groups. If all the composition factors of a composition series are Abelian or if they are all simple groups of prime power order, the group $\mathfrak{G}$ is said to be *solvable*, a term which arises from the applications of groups to the theory of fields (see IB7, §10). A solvable group $\mathfrak{G}$ is necessarily finite.

Now it is natural to ask whether a solvable group may not have other composition series with non-Abelian composition factors. The answer to this question is part of the following, more general, theorem.

Jordan-Hölder theorem: if l is the length of $\mathfrak{G}$ *and*

$$K: \qquad \mathfrak{G} = \mathfrak{N}_0 \supset ... \supset \mathfrak{N}_l = 1$$

and

$$L: \qquad \mathfrak{G} = \mathfrak{M}_0 \supset ... \supset \mathfrak{M}_s = 1$$

are composition series of $\mathfrak{G}$, *then* $s = 1$, *and there exists a permutation* σ *of the indices such that* $\mathfrak{N}_{i-1}/\mathfrak{N}_i \cong \mathfrak{M}_{i\sigma-1}/\mathfrak{M}_{i\sigma}$.

Proof. From each class of isomorphic simple groups we choose a fixed representative $\mathfrak{E}$ and denote by $n_{\mathfrak{E}}^K$ and $n_{\mathfrak{E}}^L$ the number of composition factors of K and L that are isomorphic to $\mathfrak{E}$. In (4) it is clear that

$$n_{\mathfrak{E}}^{K_3} = n_{\mathfrak{E}}^{K_1} + n_{\mathfrak{E}}^{K_2}.$$

Our theorem is proved if $n_{\mathfrak{E}}^{K} = n_{\mathfrak{E}}^{L}$ for every $\mathfrak{E}$; that is, if $n_{\mathfrak{E}}^{K} = n_{\mathfrak{E}}^{L} = n_{\mathfrak{E}}^{\mathfrak{G}}$ depends only on $\mathfrak{G}$ and not on the particular composition series. If $\mathfrak{G}$ is simple, the theorem is obvious. Arguing by complete induction on the length l of $\mathfrak{G}$, we now assume that $l \neq 1$ and that the theorem is already proved for all groups of smaller length than $\mathfrak{G}$. Then, if $\mathfrak{N}_1 = \mathfrak{M}_1$, we have

$$n_{\mathfrak{E}}^{K} = n_{\mathfrak{E}}^{\mathfrak{G}/\mathfrak{N}_1} + n_{\mathfrak{E}}^{\mathfrak{N}_1} = n_{\mathfrak{E}}^{\mathfrak{G}/\mathfrak{M}_1} + n_{\mathfrak{E}}^{\mathfrak{M}_1} = n_{\mathfrak{E}}^{L},$$

since $\mathfrak{G}/\mathfrak{N}_1 = \mathfrak{G}/\mathfrak{M}_1$ and $\mathfrak{N}_1 = \mathfrak{M}_1$ are of smaller length than $\mathfrak{G}$. Otherwise we must have $\mathfrak{G} = \mathfrak{N}_1\mathfrak{M}_1$, since $\mathfrak{N}_1$ and $\mathfrak{M}_1$ are maximal normal subgroups of $\mathfrak{G}$. By the isomorphism theorem we then have $\mathfrak{G}/\mathfrak{N}_1 \cong \mathfrak{M}_1/\mathfrak{N}_1 \cap \mathfrak{M}_1$ and $\mathfrak{G}/\mathfrak{M}_1 \cong \mathfrak{N}_1/\mathfrak{M}_1 \cap \mathfrak{N}_1$. But $\mathfrak{G}/\mathfrak{N}_1$, $\mathfrak{G}/\mathfrak{M}_1$, and $\mathfrak{D} = \mathfrak{M}_1 \cap \mathfrak{N}_1$ are of smaller length than $\mathfrak{G}$, so that

$$n_{\mathfrak{E}}^{K} = n_{\mathfrak{E}}^{\mathfrak{G}/\mathfrak{N}_1} + n_{\mathfrak{E}}^{\mathfrak{N}_1/\mathfrak{D}} + n_{\mathfrak{E}}^{\mathfrak{D}} = n_{\mathfrak{E}}^{\mathfrak{M}_1/\mathfrak{D}} + n_{\mathfrak{E}}^{\mathfrak{G}/\mathfrak{M}_1} + n_{\mathfrak{E}}^{\mathfrak{D}} = n_{\mathfrak{E}}^{L}.$$

Thus $n_{\mathfrak{E}}^{K} = n_{\mathfrak{E}}^{L} = n_{\mathfrak{E}}^{\mathfrak{G}}$ depend only on $\mathfrak{G}$, as was to be proved.

12.2. The great importance of simple groups in the general theory of groups becomes even clearer from the Jordan-Hölder theorem than from the remark in §6.4. Thus it is natural to ask whether we can find a satisfactory solution of the type problem for simple groups. But here the situation is as follows. In addition to the simple groups of prime order, research has uncovered many non-Abelian simple groups of finite and infinite order, but even the proof of such a basic statement as *every finite non-Abelian simple group is of even order* (conjectured by W. Burnside in the nineteenth century) seems to require almost all the immense apparatus of the present-day theory of finite groups.[9] So we shall content ourselves here with the proof, given in §15.4.3, that there exist infinitely many of these finite non-Abelian simple groups.

13. Normalizer, Centralizer, Center

In order to acquire insight into the structure of non-Abelian (i.e., noncommutative) groups, we must first introduce certain concepts, such as the commutator group, which measure the extent to which a group departs from being commutative.

13.1. For a given complex $\mathfrak{K}$ of the group $\mathfrak{G}$ let us enumerate the complexes conjugate to $\mathfrak{K}$. For this purpose we note that the elements $G \in \mathfrak{G}$ for which $G^{-1}\mathfrak{K}G = \mathfrak{K}$ form a subgroup, since from $G_1^{-1}\mathfrak{K}G_1 = G_2^{-1}\mathfrak{K}G_2 = \mathfrak{K}$ it follows that $G_2\mathfrak{K}G_2^{-1} = \mathfrak{K}$ and $(G_1G_2^{-1})^{-1}\mathfrak{K}(G_1G_2^{-1}) = G_2G_1^{-1}\mathfrak{K}G_1G_2^{-1} = \mathfrak{K}$.

[9] For the proof of this theorem, see: Feit, W., and Thompson, J. G., *Solvability of groups of odd order*, Pacific J. of Math. 13 (1963), pp. 775–1029.

This subgroup is denoted by $N_{\mathfrak{K}}$ and is called the *normalizer of* $\mathfrak{K}$ (in $\mathfrak{G}$). Then the statement

$$G^{-1}\mathfrak{K}G = H^{-1}\mathfrak{K}H$$

or

$$HG^{-1}\mathfrak{K}GH^{-1} = \mathfrak{K}$$

is equivalent to

$$HG^{-1} \in N_{\mathfrak{K}}$$

or

$$H \in N_{\mathfrak{K}}G.$$

Thus the number of complexes conjugate to $\mathfrak{K}$ is equal to $\mathfrak{G}: N_{\mathfrak{K}}$.

If $\mathfrak{U}$ is a subgroup of $\mathfrak{G}$, it follows from the definition of $N_{\mathfrak{U}}$ that $N_{\mathfrak{U}}$ is the largest subgroup of $\mathfrak{G}$ in which $\mathfrak{U}$ is a normal subgroup. If the complex in question consists of the single element G, then N_G is called the *centralizer of* G, and more generally

$$Z_{\mathfrak{K}} = \bigcap_{G \in \mathfrak{K}} N_G$$

is the *centralizer of the complex* $\mathfrak{K}$. This centralizer consists of all the elements of $\mathfrak{G}$ that commute with every element of $\mathfrak{K}$. If $\mathfrak{G}$ is Abelian, then $Z_{\mathfrak{K}} = \mathfrak{G}$ for every complex $\mathfrak{K} \subseteq \mathfrak{G}$. The complex $Z_{\mathfrak{K}}$ is a subgroup of $\mathfrak{G}$, since it is the intersection of certain subgroups. The centralizer $Z_{\mathfrak{G}}$ of the whole group $\mathfrak{G}$ is called the *center of* $\mathfrak{G}$. The center $Z_{\mathfrak{G}}$ is Abelian, and every subgroup of it is a normal subgroup of $\mathfrak{G}$.

13.2. *If the factor group* $\mathfrak{G}/\mathfrak{Z}$ *of a subgroup* $\mathfrak{Z}$ *in the center of* $\mathfrak{G}$ *is cyclic, then* $\mathfrak{G}$ *is Abelian:* for if $\mathfrak{G}/\mathfrak{Z} = \langle \mathfrak{Z}\mathfrak{G} \rangle$, then every element $\mathfrak{G} \in \mathfrak{G}$ can be written in the form

$$G = ZG^r,$$

with $Z \in \mathfrak{Z}$. But for two elements $G_1 = Z_1G^{r_1}$, $G_2 = Z_2G^{r_2}$, $Z_1, Z_2 \in \mathfrak{Z}$ we then have

$$G_1G_2 = Z_1G^{r_1}Z_2G^{r_2} = Z_2Z_1G^{r_2}G^{r_1} = Z_2G^{r_2}Z_1G^{r_1} = G_2G_1 .$$

13.3. The relation of conjugacy is easily seen to produce a partition into classes for the elements of $\mathfrak{G}$. The single-element classes are exactly the elements of the center of $\mathfrak{G}$. If $K_1, ..., K_r$ are representatives of the remaining classes, the result of §13.1 shows that the following equation must hold for the number of elements in

$$(1) \qquad \mathfrak{G} : 1 = Z_{\mathfrak{G}} : 1 + \mathfrak{G} : Z_{K_1} + ... + \mathfrak{G} : Z_{K_r},$$

with $Z_{K_i} \neq \mathfrak{G}$; $i = 1, ..., r$. This equation is called the *class equation of* $\mathfrak{G}$.

For $\mathfrak{S}^{(3)}$ there exist three classes of conjugate elements:

$$\{1\}, \quad \{\alpha, \beta\}, \quad \{\gamma, \delta, \epsilon\}.$$

Thus the class equation is

$$6 = 1 + 2 + 3.$$

14. p-Groups

14.1. A finite group $\mathfrak{G}$ of prime power order p^α is called a *p-group*. From the results of the preceding section we can show that p-groups are solvable. For this purpose we require the following lemma:

If $\mathfrak{G}$ is a p-group, then $Z_{\mathfrak{G}} \neq 1$.

Proof. The indices $\mathfrak{G} : Z_i$ on the right side of the class equation for $\mathfrak{G}$ are $\neq 1$, and since they are factors of the order of the group, they must be powers of p. The left side is p^α; therefore p must be a factor of $Z_{\mathfrak{G}} : 1$.

From this lemma we see at once that *every p-group is solvable.*

Proof. The theorem is true if $\mathfrak{G}$ is Abelian. For $\mathfrak{G} : 1 \neq p$ it then follows by complete induction on the order, first for $\mathfrak{G}/Z_{\mathfrak{G}}$ and $Z_{\mathfrak{G}}$ and then for $\mathfrak{G}$, if we construct a composition series for $\mathfrak{G}$ that passes through $Z_{\mathfrak{G}}$.

We leave to the reader the proof that every maximal subgroup of a p-group is a normal subgroup, as well as the verification of these theorems for the examples $B_{2,Q}$ and $\mathfrak{Q}$ (§1.2.7, §6.2).

14.2. Since the foregoing theorems show that the order of the center of a group of order p^2 is either p or p^2, the factor group with respect to the center of such a group is either of order p or of order 1, and thus is cyclic in every case. As was shown in §13.2, it follows that a group of order p^2 must be Abelian. The existence of non-Abelian groups of order p^3 is shown by the examples $B_{2,Q}$ and $\mathfrak{Q}$ (§§1.2.7, 6.1).

14.3. The great importance of p-groups for the general theory of finite groups rests on the fact that the order of a subgroup is a factor of the order of the group (§3.5). But the example $\mathfrak{S}^{(4)}$ shows that the converse is not necessarily true; i.e., there exists a factor d of $\mathfrak{S}^{(4)} : 1$ for which there is no subgroup of order d. However, we do have the *theorem of Sylow: if p^α is the highest power of the prime p which is a factor of $\mathfrak{G} : 1$, then there exists in $\mathfrak{G}$ a subgroup of order p^α.* Thus every group contains at least one subgroup whose order is the highest possible power of p permitted by the theorem of Lagrange (§3.3(8)). The proof proceeds by induction on the order of $\mathfrak{G}$, as follows. For $\mathfrak{G} = 1$ the theorem is obvious. Now let us first suppose that there exists a proper subgroup $\mathfrak{U}$ of $\mathfrak{G}$, whose index $\mathfrak{G} : \mathfrak{U}$ is not divisible by p. Then p^α is a factor of $\mathfrak{U} : 1$. If we now assume that the theorem has already been proved for all groups of order smaller than $\mathfrak{G} : 1$, then $\mathfrak{U}$ contains a subgroup of order p^α, which as a subgroup of $\mathfrak{G}$ provides the desired result. If p is a factor of the indices of all the subgroups, then p must be a factor of $Z_{\mathfrak{G}} : 1$ in the class equation §13(1) for $\mathfrak{G}$, since $\mathfrak{G} : 1$ and all the $\mathfrak{G} : Z_{K_i}$ are divisible by p. By the fundamental theorem on finite Abelian groups, there exists in $Z_{\mathfrak{G}}$ a subgroup $\mathfrak{P}_0 \neq 1$ whose order is a power of p, say p^{α_0}. This subgroup is a normal subgroup of $\mathfrak{G}$ and the group $\mathfrak{G}/\mathfrak{P}$ is of smaller order than $\mathfrak{G}$. The greatest power of p that divides $\mathfrak{G} : \mathfrak{P}_0$ is $p^{\alpha-\alpha_0}$. Thus by the induction hypothesis there exists in $\mathfrak{G}/\mathfrak{P}_0$ a subgroup $\mathfrak{P}/\mathfrak{P}_0$ of order $p^{\alpha-\alpha_0}$. Then the subgroup $\mathfrak{P}$ of $\mathfrak{G}$ has the order p^α, as was to be proved.

The p-subgroups whose existence has just been proved are called *Sylow p-groups* of $\mathfrak{G}$, in honor of their discoverer. The theorem does not state that there exists only one Sylow p-group for every prime p, but Sylow did prove that every p-subgroup of $\mathfrak{G}$ has a conjugate in an arbitrarily preassigned Sylow p-group of $\mathfrak{G}$, so that, in particular, all Sylow p-groups of $\mathfrak{G}$ for a fixed prime p are conjugate to one another.

The p-groups are included in a larger class of groups, the so-called *nilpotent groups*, which are defined as the direct products of groups of prime power order. They are of particular interest because for them we can prove the converses of the theorems given above for p-groups. For example: *a finite group is nilpotent if and only if every factor group has the property that its center is not merely the unit element.* Or: *a finite group is nilpotent if and only if every maximal subgroup is a normal subgroup.* For lack of space the proofs must be omitted.

15. Permutation Groups

15.1. *Representations*

In the example of groups of motions we have seen that a group may be much easier to investigate if it is not defined abstractly, say by its multiplication table, but in some geometric way. One simple but effective method (see the examples below) for getting a clearer picture of the concept of a group is to investigate the possibilities of representing the group as a permutation group, that is as a subgroup of $\mathfrak{S}^{\mathfrak{R}}$ for a suitable space $\mathfrak{R}$. Let us examine these possibilities.

15.1.1. For an arbitrary group $\mathfrak{G}$ let σ be a homomorphism of $\mathfrak{G}$ into $\mathfrak{S}^{\mathfrak{R}}$. Then σ is called a *permutation representation of* $\mathfrak{G}$ *in* $\mathfrak{R}$ or simply a *representation.* The number $|\mathfrak{R}|$ of elements in $\mathfrak{R}$ is called the *degree* of the representation.

If σ_1, σ_2 are representations of $\mathfrak{G}$ in $\mathfrak{R}_1$ and $\mathfrak{R}_2$, respectively, and if th.

two permutations differ only by a renaming of the permuted points, i.e., if there exists a one-to-one mapping τ of $\mathfrak{R}_1$ onto $\mathfrak{R}_2$ such that

$$(1) \qquad (PG^{\sigma_1})^{\tau} = P^{\tau}G^{\sigma_2} \qquad \text{for all} \qquad G \in \mathfrak{G} \quad \text{and} \quad P \in \mathfrak{R}_1,$$

then we say that σ_1 and σ_2 are *similar*. The relation of similarity obviously produces a partition into classes and, as we have done up to now for isomorphic groups, we shall consider similar representations as essentially not distinct. For example, the two representations σ_1, σ_2 of the cyclic group $\mathfrak{G} = \{1, G\}$ of order 2 in $\mathfrak{R}_1 = \{a, b, c\}$ and $\mathfrak{R}_2 = \{A, B, C\}$ are similar:

$$1^{\sigma_1} = \begin{pmatrix} a & b & c \\ a & b & c \end{pmatrix}, \qquad 1^{\sigma_2} = \begin{pmatrix} A & B & C \\ A & B & C \end{pmatrix},$$

$$G^{\sigma_1} = \begin{pmatrix} a & b & c \\ b & a & c \end{pmatrix}, \qquad G^{\sigma_2} = \begin{pmatrix} A & B & C \\ B & A & C \end{pmatrix}.$$

Here the relation (1) is established by the mapping τ with

$$a^{\tau} = A, \qquad b^{\tau} = B, \qquad c^{\tau} = C.$$

15.1.2. When a fixed representation σ of $\mathfrak{G}$ in $\mathfrak{R}$ is being considered, we shall for brevity set

$$PG^{\sigma} = PG, \qquad P \in \mathfrak{R}.$$

For every $P \in \mathfrak{R}$ the elements $G \in \mathfrak{G}$ leaving P fixed form a subgroup $\mathfrak{G}_P = \{G \mid PG = P\}$, as we have seen in §3.3.4. We shall call it the *fix-group* of P.

Subspaces of $\mathfrak{R}$ of the form $P\mathfrak{G}$ are called *domains of transitivity* of σ. Every element $P \in \mathfrak{R}$ is contained in exactly one domain of transitivity, since

$$Q \in P_1\mathfrak{G} \cap P_2\mathfrak{G},$$

i.e.,

$$Q = P_1G_1 = P_2G_2, \qquad G_1, G_2 \in \mathfrak{G},$$

implies

$$P_1G_1\mathfrak{G} = P_2G_2\mathfrak{G}$$

and

$$P_1\mathfrak{G} = P_2\mathfrak{G}.$$

Thus the domains of transitivity $\mathfrak{R}^{(i)}$ of $\mathfrak{R}$ produce a partition into classes of $\mathfrak{R}$:

$$\mathfrak{R} = \mathfrak{R}^{(1)} \cup \ldots \cup \mathfrak{R}^{(r)},$$

$$\mathfrak{R}^{(i)} \cap \mathfrak{R}^{(k)} = \emptyset \qquad \text{for} \quad i \neq k.$$

For every $i = 1, ..., r$ the representation σ induces a representation σ_i of $\mathfrak{G}$ in $\mathfrak{R}^{(i)}$, which is related to σ in the following way:

$$P_i G^\sigma = P_i G^{\sigma_i} \qquad \text{for} \quad P_i \in \mathfrak{R}^{(i)}.$$

Thus it is sufficient to consider representations σ of $\mathfrak{G}$ in $\mathfrak{R}$ for which $\mathfrak{R}$ itself is a domain of transitivity. Such representations σ are said to be *transitive*.

Transitive representations of $\mathfrak{G}$ can be obtained in the following way: choose a subgroup $\mathfrak{U}$ of $\mathfrak{G}$ and then in the set $\mathfrak{R}$ of cosets $\mathfrak{U}H$ of $\mathfrak{U}$ define the permutation $G^{\sigma_\mathfrak{U}}$ by

$$(\mathfrak{U}H)G^{\sigma_\mathfrak{U}} = \mathfrak{U}HG.$$

Since

$$\begin{aligned}(\mathfrak{U}H)(G_1G_2)^{\sigma_\mathfrak{U}} &= \mathfrak{U}HG_1G_2 \\ &= (\mathfrak{U}HG_1)\, G_2 \\ &= ((\mathfrak{U}H)\, G_1^{\sigma_\mathfrak{U}})\, G_2^{\sigma_\mathfrak{U}} \\ &= (\mathfrak{U}H)\, G_1^{\sigma_\mathfrak{U}} G_2^{\sigma_\mathfrak{U}},\end{aligned}$$

it follows that $\sigma_\mathfrak{U}$ is a representation of $\mathfrak{G}$. Every coset has the form $\mathfrak{U}G$, $G \in \mathfrak{G}$, so that the representation is transitive. We call it the *representation of* $\mathfrak{G}$ *induced by* $\mathfrak{U}$. It is clear that $\mathfrak{U}$ is the fix-group of $\mathfrak{U}$ (as a point of $\mathfrak{R}$) and $\mathfrak{G} : \mathfrak{U}$ is the degree of $\sigma_\mathfrak{U}$.

Then the following theorem shows that, up to similarity, we have thus obtained all the transitive representations of $\mathfrak{G}$: if σ is any transitive permutation representation of $\mathfrak{G}$ in $\mathfrak{R}$ and if $P \in \mathfrak{R}$, then σ is similar to the representation $\sigma_\mathfrak{U}$ of $\mathfrak{G}$ induced by $\mathfrak{U} = \mathfrak{G}_P$.

Proof. Since

$$PH_1^\sigma = PH_2^\sigma$$

is equivalent to

$$PH_1^\sigma H_2^{\sigma-1} = P$$

and to

$$H_1 \in \mathfrak{G}_P H_2 \qquad \text{or} \qquad \mathfrak{G}_P H_1 = \mathfrak{G}_P H_2,$$

the mapping τ defined by

$$(PH^\sigma)^\tau = \mathfrak{G}_P H$$

is a one-to-one mapping of $\mathfrak{R}$ into the set of right cosets of $\mathfrak{G}_P$, and the transitivity means that every coset of $\mathfrak{G}_P$ is an image under τ. Finally, for $Q = PH^\sigma$ we have

$$Q^\tau G^{\sigma_\mathfrak{U}} = \mathfrak{G}_P HG = (P(HG)^\sigma)^\tau = (QG^\sigma)^\tau,$$

so that (1) is satisfied for $\sigma_1 = \sigma_\mathfrak{U}$, $\sigma_2 = \sigma$.

For transitive representations of $\mathfrak{G}$ in $\mathfrak{R}$ the preceding theorem implies in particular that

(2) $$\mathfrak{G} : 1 = (\mathfrak{G}_P : 1) \cdot |\,\mathfrak{R}\,|.$$

Consequently, *the degree of a transitive representation is a factor of the order of the group.*

We remark without proof that the representations induced by two subgroups $\mathfrak{U}$, $\mathfrak{B}$ are similar if and only if the subgroups are conjugate under $\mathfrak{G}$.

15.1.3. Of particular importance is the representation of $\mathfrak{G}$ induced by the subgroup 1, the so-called *regular* representation of $\mathfrak{G}$. It plays an especial role in the history of the subject, since it was used by Cayley to show that every abstractly defined finite group can also be defined "concretely," namely, as a permutation group.

15.1.4. We now wish to determine the kernel $\mathfrak{G}_\sigma$ for a transitive representation σ of $\mathfrak{G}$. For this purpose we first note that if $PG = Q$, $G \in \mathfrak{G}$, $P \in \mathfrak{R}$, and $Q \in \mathfrak{R}$, then $\mathfrak{G}_Q = G^{-1}\mathfrak{G}_P G$, as is clear from the fact that $Q = QH$ is equivalent to

$$PG = PGH$$

and

$$P = PGHG^{-1},$$

so that

$$GHG^{-1} \in \mathfrak{G}_P,$$
$$H = G^{-1}\mathfrak{G}_P G.$$

Thus it follows from the transitivity of σ that for an arbitrary but fixed point $P \in \mathfrak{R}$ the subgroup

$$\mathfrak{G}_\sigma = \bigcap_{G \in \mathfrak{G}} G^{-1}\mathfrak{G}_P G$$

leaves every point of $\mathfrak{R}$ fixed and is the kernel of σ. It is the greatest normal subgroup of $\mathfrak{G}$ in $\mathfrak{G}_P$, as is easily shown.

15.2. *The Symmetric Group of Permutations*

In the preceding section we have discussed group representations as subgroups of the symmetric group. This discussion can be made more useful for the general theory of groups through a closer study of the structure of $\mathfrak{S}^{\mathfrak{R}}$, to which we shall devote the next two sections. By regarding $\mathfrak{S}^{\mathfrak{R}}$ and its subgroups as being represented by the identical representation of themselves we can make use of the results of the preceding section, for which purpose it is convenient to apply the concept of transitivity not to the representation but to the group itself. We assume throughout that $\mathfrak{R}$ is finite.

15.2.1. The group $\mathfrak{S}^{\mathfrak{R}}$ is obviously transitive, and for $P \in \mathfrak{R}$ we have $\mathfrak{S}^{\mathfrak{R}-P} \cong \mathfrak{S}^{\mathfrak{R}}_P$, since $\mathfrak{S}^{\mathfrak{R}}_P$ contains all permutations of the points in $\mathfrak{R} - P$. Thus by (2)

$$\mathfrak{S}^{\mathfrak{R}} : 1 = (\mathfrak{S}^{\mathfrak{R}-P} : 1) \cdot |\mathfrak{R}|.$$

Since $\mathfrak{S}^{\mathfrak{R}} : 1 = 1$ for $|\mathfrak{R}| = 1$, it follows by complete induction on the power $|\mathfrak{R}|$ that

$$\mathfrak{S}^{\mathfrak{R}} : 1 = |\mathfrak{R}|!$$

Now let α be an element of $\mathfrak{S}^{\mathfrak{R}}$. The domains of transitivity of $\langle\alpha\rangle$ have the form

$$\{P, P\alpha, ..., P\alpha^{n-1}\},$$

where n is the smallest positive integer with the property $P\alpha^n = P$. Then α can be written as

$$\text{(3)} \qquad \alpha = \begin{pmatrix} P & P\alpha & \dots & P\alpha^{n-1} & Q & \dots \\ P\alpha & P\alpha^2 & \dots & P & Q\alpha & \dots \end{pmatrix},$$

where $P, Q, ...$ are representatives of the various domains of transitivity of $\langle\alpha\rangle$. If the points of $\mathfrak{R}$, except for $P\langle\alpha\rangle$, are left unchanged by α, we write simply

$$\alpha = (P, P\alpha, ..., P\alpha^{n-1})$$

and call α a *cycle of length n*. For example,

$$\alpha = \begin{pmatrix} 1 & 2 & 3 & 4 & 5 & 6 \\ 1 & 6 & 3 & 2 & 4 & 5 \end{pmatrix} = (2\ 6\ 5\ 4)$$

is a cycle in $\mathfrak{S}^{(6)}$.

Cycles are of special importance, since every element $\alpha \in \mathfrak{S}^{\mathfrak{R}}$ is the product of pointwise disjoint cycles. As can be calculated at once, the element (3) may be written in the form

$$\text{(4)} \qquad \alpha = (P, P\alpha, ..., P\alpha^{n-1})(Q, Q\alpha, ...)(...) ...,$$

which is called the *canonical decomposition of* α into cycles. Thus for

$$\alpha = \begin{pmatrix} 1 & 2 & 3 & 4 & 5 & 6 & 7 \\ 3 & 4 & 1 & 5 & 6 & 7 & 2 \end{pmatrix}$$

this decomposition is

$$\alpha = (1 \;\; 3)(2 \;\; 4 \;\; 5 \;\; 6 \;\; 7). \tag{5}$$

In order to include the identical permutation **1** in this form of writing we set

$$\mathbf{1} = (P)$$

with an arbitrary $P \in \mathfrak{S}^{\mathfrak{R}}$.

15.2.2. The advantage of the cycle notation lies not only in its greater conciseness but also in its convenience for determining the order and conjugacy of permutations.

Thus we see at once that for a single cycle the order is equal to the degree:

$$(P_1 P_2 \dots P_n)^n = \mathbf{1}, \qquad (P_1 \dots P_n)^m \neq \mathbf{1} \qquad \text{for} \quad 1 \leqslant m < n.$$

Since pointwise disjoint cycles are permutable, we have

$$\begin{aligned} \alpha &= (P_{11} \dots P_{1n_1})(P_{21} \dots P_{2n_2}) \dots (P_{s1} \dots P_{sn_s}): \\ \alpha^m &= (P_{11} \dots P_{1n_1})^m (P_{21} \dots P_{2n_2})^m \dots (P_{s1} \dots P_{sn_s})^m, \end{aligned} \tag{6}$$

so that α^m is equal to the identical permutation if and only if

$$m \mid n_1, \dots, m \mid n_s.$$

Thus the order of α is equal to the least common multiple of the length of the cycles in the canonical decomposition of α.

For example, the element α in (5) has the order 10.

We now wish to determine which elements are conjugate in $\mathfrak{S}^{\mathfrak{R}}$ to the permutation whose canonical decomposition is given by (6). Let $\beta \in \mathfrak{S}^{\mathfrak{R}}$. We see that

$$\beta^{-1}\alpha\beta = (P_{11}\beta, P_{12}\beta, \dots P_{1n_1}\beta) \dots (P_{s1}\beta \dots P_{sn_s}\beta)$$

is the canonical decomposition of $\beta^{-1}\alpha\beta$, since

$$(P_{11}\beta)\,\beta^{-1} = P_{11}, \qquad P_{11}\alpha = P_{12},$$

so that

$$(P_{11}\beta)\,\beta^{-1}\alpha\beta = P_{12}\beta,$$

and so forth.

For example,

$$\beta = \begin{pmatrix} 1 & 2 & 3 \\ 2 & 1 & 3 \end{pmatrix}, \qquad \alpha = (1\ \ 3)$$

$$\beta^{-1}\alpha\beta = (2\ \ 3).$$

We shall say that α and $\alpha^* = \beta^{-1}\alpha\beta$ are *similar*, by which we mean that the cycles in the canonical representation of α and α^* can be put into one-to-one correspondence with each other in such a way that corresponding cycles have the same length. Then our problem is already solved: *Two elements of* $\mathfrak{S}^{\mathfrak{R}}$ *are conjugate in* $\mathfrak{S}^{\mathfrak{R}}$ *if and only if they are similar.*

For the proof we need only show that two similar permutations

$$\alpha = (P_{11} \ldots P_{1n_1}) \ldots (P_{s1} \ldots P_{sn_s}),$$
$$\alpha^* = (P_{11}^* \ldots P_{1n_1}^*) \ldots (P_{s1}^* \ldots P_{sn_s}^*)$$

are conjugate to each other. But with

$$\beta = \begin{pmatrix} P_{11} & \ldots & P_{sn_s} \\ P_{11}^* & \ldots & P_{sn_s}^* \end{pmatrix},$$

it is easy to calculate that $\beta^{-1}\alpha\beta = \alpha^*$, as required.

15.2.3. From this result a permutation $\xi \in \mathfrak{S}^{\mathfrak{R}}$ in which $\xi\alpha = \alpha\xi$ for all $\alpha \in \mathfrak{S}^{\mathfrak{R}}$ cannot be similar to any other permutation in $\mathfrak{S}^{\mathfrak{R}}$. For if $\xi = 1$, the statement is certainly true for an arbitrary $\mathfrak{R}$, and for $|\mathfrak{R}| = 2$ it is true for the permutations distinct from 1. For $|\mathfrak{R}| \geqslant 3$ we see that

$$\alpha = (P_{11}P_{12} \ldots) \ldots \neq 1$$

and

$$\alpha' = (P_{11}P_{12}' \ldots) \ldots, \qquad P_{12}' \neq P_{12}$$

are similar and that $\alpha \neq \alpha'$, since $P_{11}\alpha \neq P_{11}\alpha'$. Thus for $|\mathfrak{R}| \geqslant 3$ *the center of* $\mathfrak{S}^{\mathfrak{R}}$ *consists only of the identical permutation* 1. For $|\mathfrak{R}| = 1, 2$ the group $\mathfrak{S}^{\mathfrak{R}}$ is obviously Abelian.

15.3. *The Alternating Group*

15.3.1. A cycle of degree 2 is called a *transposition*. Every finite cycle is a product of transpositions:

$$(P_1 \ldots P_n) = (P_{n-1}P_n)(P_{n-2}P_{n-1}) \ldots (P_1P_2).$$

Since every permutation can be written as a product of cycles, it follows that all permutations are also products of transpositions. However, the transpositions are not necessarily pointwise disjoint. For example,

$$\begin{pmatrix} 1 & 2 & 3 & 4 & 5 \\ 2 & 3 & 1 & 5 & 4 \end{pmatrix} = (1\ 2\ 3)(4\ 5) = (2\ 3)(1\ 2)(4\ 5).$$

Thus the group $\mathfrak{S}^{\mathfrak{R}}$ is generated by the transpositions in $\mathfrak{R}$.

15.3.2. The elements of $\mathfrak{S}^{\mathfrak{R}}$ that can be written as the product of an even number of transpositions are called *even transpositions*. They form a normal subgroup of $\mathfrak{S}^{\mathfrak{R}}$, since for

$$\alpha = (P_1Q_1) \ldots (P_{2k}Q_{2k}),$$
$$\beta = (R_1S_1) \ldots (R_{2l}S_{2l})$$

the element $\alpha\beta^{-1} = (P_1Q_1) \ldots (P_{2k}Q_{2k})(R_{2l}S_{2l}) \ldots (R_1S_1)$ is also the product of an even number of transpositions, and conjugate permutations are similar to each other. This normal subgroup is called the *alternating group* in $\mathfrak{R}$ and is denoted by $\mathfrak{A}^{\mathfrak{R}}$. We now wish to prove: *for* $|\mathfrak{R}| \neq 1$, *the alternating group* $\mathfrak{A}^{\mathfrak{R}}$ *has index* 2 *in* $\mathfrak{S}^{\mathfrak{R}}$, *so that* $\mathfrak{A}^{\mathfrak{R}}$ *has the order* $|\mathfrak{R}|!/2$.

For the proof we take $\mathfrak{R} = (n)$. Then for an arbitrary polynomial $f(x_1, \ldots, x_n)$ in the indeterminates $x_1, \ldots, x_n$ $\alpha \in \mathfrak{S}^{\mathfrak{R}}$ we define

$$f^{\alpha}(x_1, \ldots, x_n) = f(x_{1\alpha}, \ldots, x_{n\alpha}),$$

so that

$$(f^{\alpha})^{\beta} = f^{\alpha\beta}.$$

As a particular polynomial let us consider the difference product

$$\varDelta(x_1, \ldots, x_n) = \prod_{\substack{i,k=1,\ldots,n \\ i<k}} (x_i - x_k).$$

Apart from a change of sign, the application of α to $\varDelta$ merely permutes the factors of the right-hand side:

$$\varDelta^{\alpha} = \epsilon_{\alpha}\varDelta, \qquad \epsilon_{\alpha} = \pm 1,$$

so that

$$\epsilon_{\alpha\beta}\varDelta = \varDelta^{\alpha\beta} = \epsilon_{\alpha}\varDelta^{\beta} = \epsilon_{\alpha}\epsilon_{\beta}\varDelta.$$

The mapping $\epsilon : \alpha \to \epsilon_{\alpha}$ is therefore a homomorphism of $\mathfrak{S}^{\mathfrak{R}}$ into the cyclic group of order 2. The desired theorem now follows from the homomorphism theorem if we show that ϵ is a homomorphism onto this group and that its kernel is $\mathfrak{A}^{\mathfrak{R}}$. But it is easy to show that

$$\varDelta^{(12)} = -\varDelta,$$

so that $\epsilon_{(12)} = -1$. Also, since every transposition (ab) in $\mathfrak{S}^{(n)}$ is conjugate to (12), say

$$\beta^{-1}(1\ 2)\,\beta = (a\ b),$$

we have

$$\epsilon_{(ab)} = \epsilon_{\beta^{-1}}\epsilon_{(12)}\epsilon_\beta = (\epsilon_\beta)^{-1}\epsilon_{(12)}\epsilon_\beta = \epsilon_{(12)}\,.$$

The product of an even number of transpositions is thus mapped by ϵ onto 1, and the product of an odd number onto -1, so that the proof of the theorem is complete.

15.3.3. For $|\,\mathfrak{N}\,| = 1$ we have $\mathfrak{A}^{(1)} = \mathfrak{S}^{(1)}$, and for $n = 2, 3$ the order of $\mathfrak{A}^{(n)}$ is 1, 3, respectively. For $n = 4$ the permutations

$$(1\ 2)(3\ 4), \qquad (1\ 3)(2\ 4), \qquad (1\ 4)(2\ 3),$$

together with the unit element **1** form a normal subgroup of $\mathfrak{A}^{(4)}$ of order 4, the so-called four-group. The lattice of subgroups of $\mathfrak{A}^{(4)}$ is given in Figure 33. In all other cases $\mathfrak{A}^{(n)}$ is simple (and non-Abelian), as we shall show in the next section for $n = 5$.

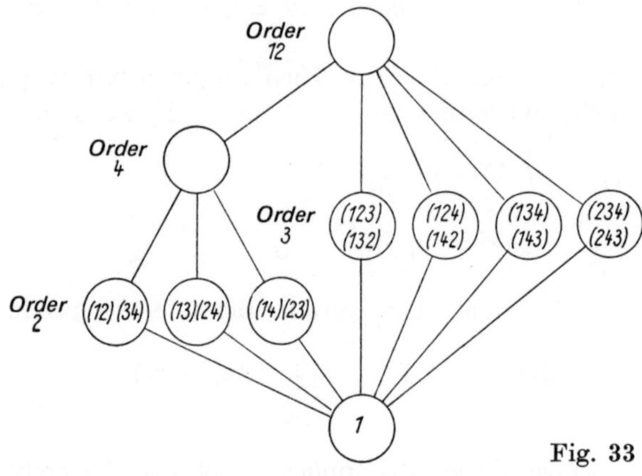

Fig. 33

15.4. *Applications of the Theory of Permutation Groups to the General Theory of Groups*

Let us now give some examples to show how our results on permutation groups can be applied to the general theory of groups. They are all concerned with the important question of finding conditions under which a group is simple.

15.4.1. For a representation σ of $\mathfrak{G}$ in $\mathfrak{N}$ the index $\mathfrak{G} : \mathfrak{G}_\sigma$ is always a factor of $|\,\mathfrak{N}\,|!$, since the homomorphism theorem shows that $\mathfrak{G}/\mathfrak{G}_\sigma$ is isomorphic to a subgroup of $\mathfrak{G}^{\mathfrak{N}}$. Thus for the representation induced by a subgroup $\mathfrak{U}$ of $\mathfrak{G}$, the index of the greatest normal subgroup of $\mathfrak{G}$

contained in $\mathfrak{U}$ is a factor of $(\mathfrak{G} : \mathfrak{U})!$. *If $\mathfrak{G}$ is simple, there is no subgroup $\mathfrak{U}$ in $\mathfrak{G}$ for which* $(\mathfrak{G} : \mathfrak{U})! < \mathfrak{G} : 1$. Since the group $\mathfrak{A}^{(5)}$ is shown below to be simple, it contains no subgroup with index 2, 3, or 4.

15.4.2. *If $\mathfrak{G}$ is a group of order $2u$ with odd u, then $\mathfrak{G}$ contains a normal subgroup $\mathfrak{N}$ with index* 2. For let σ be the regular representation of $\mathfrak{G}$ in $\mathfrak{R}$ $(=\mathfrak{G})$. Then $G \in \mathfrak{G}$, $G \neq 1$ leaves no point of $\mathfrak{R}$ fixed, since 1 is the fix-group for all points of $\mathfrak{R}$. Now let G be an element of order 2, which must exist by the theorem of Sylow for $p = 2$. The canonical decomposition of G^σ into cycles consists of u transpositions. Thus G^σ is not an element of $\mathfrak{A}^{\mathfrak{R}}$, so that the intersection $\mathfrak{G}^\sigma \cap \mathfrak{A}^{\mathfrak{R}}$ is $\neq \mathfrak{G}^\sigma$. Also, since $\mathfrak{A}^{\mathfrak{R}}$ is a maximal subgroup of $\mathfrak{S}^{\mathfrak{R}}$, we have $\mathfrak{A}^{\mathfrak{R}}\mathfrak{G}^\sigma = \mathfrak{S}^{\mathfrak{R}}$. By the isomorphism theorem it follows that $\mathfrak{N}^\sigma = \mathfrak{A}^{\mathfrak{R}} \cap \mathfrak{G}^\sigma$ is a normal subgroup of $\mathfrak{G}^\sigma$ and

$$\mathfrak{G}^\sigma/\mathfrak{N}^\sigma \cong \mathfrak{S}^{\mathfrak{R}}/\mathfrak{A}^{\mathfrak{R}}.$$

Consequently, $\mathfrak{N}$ is a normal subgroup of $\mathfrak{G}$. Since the Burnside conjecture is now known to be true (§12.2), we have the result:

The order of every non-Abelian simple group is divisible by 4.

15.4.3. From the results of the preceding sections we now prove the existence, as stated above, of infinitely many non-Abelian simple groups, by showing that *if $\mathfrak{G}$ is a permutation group in $\mathfrak{R}$ with $|\mathfrak{R}| = r$ a prime, and if $\mathfrak{G}$ is generated by cycles of length r, then $\mathfrak{G}$ is simple.*

Proof. The group $\mathfrak{G}$ is transitive, since $P\langle Z\rangle = \mathfrak{R}$ for each of the generating cycles Z and $P \in \mathfrak{R}$. Let $\mathfrak{N}$ be a normal subgroup $\neq 1$ of $\mathfrak{G}$. By (2) the fix-group $\mathfrak{G}_P$ has prime index r in $\mathfrak{G}$ and is thus a maximal subgroup of $\mathfrak{G}$. By §15.1.5 the normal subgroup $\mathfrak{N}$ is not in $\mathfrak{G}_P$, since 1 is the only permutation in $\mathfrak{G}$ that leaves fixed all points of $\mathfrak{R}$. Thus $\mathfrak{N}\mathfrak{G}_P = \mathfrak{G}$ and by the isomorphism theorem

$$\mathfrak{G}/\mathfrak{N} \cong \mathfrak{G}_P/\mathfrak{N} \cap \mathfrak{G}_P .$$

Now only the first power of the prime r is a factor of $\mathfrak{G} : 1$, since $\mathfrak{G}$ is a subgroup of $\mathfrak{S}^{\mathfrak{R}}$ and $\mathfrak{S}^{\mathfrak{R}}$ has the order $r!$. Thus $\mathfrak{G}_P : 1$ and $\mathfrak{G} : \mathfrak{N}$ are prime to r. If one of the generating cycles Z were not in $\mathfrak{N}$, then $\mathfrak{N}\langle Z\rangle$ would be greater than $\mathfrak{N}$, and on the other hand it would follow from the theorem of isomorphism that

$$\mathfrak{N}\langle Z\rangle : \mathfrak{N} = \langle Z\rangle : \mathfrak{N} \cap \langle Z\rangle,$$

so that $\mathfrak{N}\langle Z\rangle : \mathfrak{N}$ would be a factor of $\langle Z\rangle : 1 = r$ and would thus be $=r$, in contradiction to the fact that $\mathfrak{G} : \mathfrak{N}$ is not divisible by r. Consequently, all the generating cycles lie in $\mathfrak{N}$, so that $\mathfrak{G} = \mathfrak{N}$, and $\mathfrak{G}$ is therefore simple.

If in this theorem we choose the generating cycles in such a way that they are not all powers of a fixed element, as is always possible for $r \geqslant 5$, then the group generated by them is simple (and non-Abelian). Thus there exist infinitely many (nonisomorphic) non-Abelian simple groups.

15.4.4. For example, the (non-Abelian) group generated by the two cycles

$$(1\ 2\ 3\ 4\ 5\ 6\ 7), \qquad (2\ 1\ 3\ 4\ 5\ 6\ 7)$$

is simple.

For $r = 5$ there are $4! = 24$ distinct cycles of degree 5. By §15.3.1 they are all products of $r - 1$ transpositions and thus lie in $\mathfrak{A}^{(5)}$. Since

$$(a\ b\ c\ d\ e)(c\ b\ a\ d\ e) = (c\ e\ d),$$

the 10 cycles of degree 3 also lie in the subgroup of $\mathfrak{A}^{(5)}$ generated by the 24 cycles of degree 5. But since $\mathfrak{A}^{(5)} : 1 = 60$ has no factor $\geqslant 24 + 10$, the theorem of Lagrange shows that this subgroup must be equal to $\mathfrak{A}^{(5)}$. Thus the above theorem shows that $\mathfrak{A}^{(5)}$ is simple. Let us state here without proof that there are no non-Abelian simple groups of order less than 60.

16. Some Remarks on More General Infinite Groups

The nature of the present work has restricted us to a discussion of the simplest and most general results concerning groups, although we have given some special theorems about groups of finite order. This preference for finite groups is justified by the fact that, both historically and in their importance for the whole of mathematics, finite groups have formed the backbone of group theory. Moreover, the general theory of infinite groups is not yet very well developed. Of greatest importance here are the finitely generated groups and the topological groups. To conclude the present chapter, we give one striking example for each of these two classes of groups.

16.1. *Free Groups with Finitely Many Generators*

Let $\{E_1, ..., E_n\}$ be a finite set of distinct elements, $E_i \neq E_k$ for $i \neq k$. We consider *words* W of the form

(1)
$$W = E_{i_1}^{\epsilon_1} \dots E_{i_r}^{\epsilon_r},$$
$$1 \leqslant i_1, ..., i_r \leqslant n,$$
$$\epsilon_\nu = \pm 1,$$
$$\epsilon_{\nu+1} \neq -\epsilon_\nu \qquad \text{for} \qquad i_\nu = i_{\nu+1}.$$

A word of this sort is to be combined with another such word

$$V = E_{k_1}^{\delta_1} \dots E_{k_s}^{\delta_s}$$

by the following rule:

$$WV = E_{i_1}^{\epsilon_1} \dots E_{i_{r-n-1}}^{\epsilon_{r-n-1}} E_{k_{n+2}}^{\delta_{n+2}} \dots E_{k_s}^{\delta_s},$$

$$\begin{array}{llcl} \text{if} & i_r = k_1 & \text{and} & \epsilon_r = -\delta_1 \\ \text{and} & i_{r-1} = k_2 & '' & \epsilon_{r-1} = -\delta_2 \\ & & \vdots & \\ \text{and} & i_{r-n} = k_{n+1} & '' & \epsilon_{r-n} = -\delta_{n+1}, \\ \text{but} & i_{r-n-1} \neq k_{n+2} & \text{or} & \epsilon_{r-n-1} \neq -\delta_{n+2}. \end{array}$$

Then WV is again a word of the form (1) if we consider the empty set $\emptyset$ as a word (the so-called *empty word*). If we now define $W\emptyset = \emptyset W = W$ for all words W, the set of words with the operation defined in this way forms a group $\mathfrak{F}_n$, called the *free group of rank n*. Its unit element is $\emptyset = 1$, and $W^{-1} = E_{i_r}^{-\epsilon_r} \dots E_{i_1}^{-\epsilon_1}$ is the inverse of (1). Clearly the structure of this group depends only on the number n and not on the special set $\{E_1, \dots, E_n\}$; moreover, $\mathfrak{F}_n = \langle E_1, \dots, E_n \rangle$. The group $\mathfrak{F}_n$ may be considered as the most general group that can be generated by n elements. More precisely, every group $\mathfrak{G}$ that can be generated by n elements $G_1, \dots, G_n$ is isomorphic to a factor group of $\mathfrak{F}_n$. For example, an isomorphism of this sort results, by the homomorphism theorem, from the mapping

$$W = E_{i_1}^{\epsilon_1} \dots E_{i_r}^{\epsilon_r} \to G_{i_1}^{\epsilon_1} \dots G_{i_r}^{\epsilon_r},$$

which is easily seen to be an homomorphism. On the other hand, it is clear from §6.4 that every factor group of $\mathfrak{F}_n$ can be generated by n elements.

If a system $R_1, \dots, R_t$ of elements of $\mathfrak{F}_n$, taken together with their conjugates under $\mathfrak{F}_n$, generates the kernel of a homomorphism λ of $\mathfrak{F}_n$ onto $\mathfrak{G}$, then the set of elements

$$R_1, \dots, R_t$$

is said to *define* $\mathfrak{G}$. This term comes from the fact that if a product of elements in the system of generators $\{E_1^\lambda, \dots, E_n^\lambda\}$ of $\mathfrak{G}$ is equal to 1, then it can be obtained by constructing the inverses, transforms, and products of the elements R_i, $i = 1, \dots, t$, and then modifying them by replacing E_i by E_i^λ, $i = 1, \dots, n$.

For example,

$$E_1^3\,, \qquad E_2^2\,, \qquad E_1E_2E_1E_2$$

or

$$E_1^2\,, \qquad E_2^2\,, \qquad (E_1E_2)^2$$

define the group $\mathfrak{S}^{(3)}$.

For a given $R_1, ..., R_t \in \mathfrak{F}_n$, the question whether an element $W \in \mathfrak{F}_n$ lies in the normal subgroup generated by the R_i and their conjugates is usually very difficult. It is the so-called *word problem*; cf. IA, §5.3.

16.2. *Topological Groups*

If a topology is defined on a group $\mathfrak{G}$ regarded as a set of elements the group $\mathfrak{G}$ is said to be a *topological group*, provided the topology is consistent with the operation of the group; more precisely, if the mapping $(G, H^{-1}) \to GH^{-1}$, $G, H \in \mathfrak{G}$, of the product space $\mathfrak{G} \times \mathfrak{G}$ is continuous on $\mathfrak{G}$.

For example, the group R^+ is a topological group if for every $\epsilon > 0$ the set of numbers $U_{a,\epsilon} = \{a^* \mid a^* \in R^+, \mid a - a^* \mid < \epsilon\}$ is taken as a neighborhood of a. The continuity of addition is seen as follows: if $U_{a-b,\epsilon}$ is a neighborhood of $a - b$, then $a^* - b^* \in U_{a-b,\epsilon}$ for $a^* \in U_{a,\epsilon/2}$, $b^* \in U_{b,\epsilon/2}$, as is easily shown. Thus the image of the neighborhood $V = U_{a,\epsilon/2} \times U_{b,\epsilon/2}$ of (a, b) under the mapping $(a, b) \to a - b$ lies in $U_{a+b,\epsilon}$, so that the mapping is continuous.

More picturesquely, though perhaps somewhat less precisely, we may say that a topological group is defined if the product G^*H^{*-1} "approaches" GH^{-1} as G^* and H^* "approach" G and H. Then we may exploit the resources of topology, in much the same way as permutations were used above for finite groups, to investigate the structure of a group in which a topology is defined.

CHAPTER 3

Linear Algebra*

Outline

The statements in this outline are not intended to be complete or rigorous and they sometimes refer to concepts that are not explained until later in the chapter.

The solutions of a system of linear equations consist of n-tuples of numbers: $\mathfrak{x} = (x^1, ..., x^n)$;[1] it may be that there is no solution, exactly one solution, or infinitely many. We are interested here in the existence of solutions, and particularly in how many of them there are. It turns out that the set of solutions (of a homogeneous system of equations, to which a nonhomogeneous system can be reduced) has a definite algebraic structure: in it we can perform the operations of addition and of multiplication by a number; in short, we can construct linear combinations. Such a structure is called a *vector space* and its elements are *vectors*. The name is to be explained by the historical association with geometry and physics, but here we are discussing a purely algebraic situation. *The theory of solutions of systems of linear equations is part of the more inclusive theory of a certain algebraic structure, a "vector space."*

Starting from the n-tuples of numbers, we arrive at this structure by defining the two operations mentioned above and then working out the rules of computation that hold for these operations, whereupon the rules of computation become axioms for the "vector space" thus defined. To

* The authors of this chapter are deeply indebted to G. Pickert for his valuable advice and assistance.

[1] As is customary in tensor analysis, we make use of subscripts and superscripts in such a way as to indicate the behavior of the magnitude in question under a transformation of coordinates. Except for -1 and for the exponents in (3) on page 286 there are no exponents in the present chapter.

begin with, the n-tuples of numbers are merely examples or models of vector spaces, but it turns out that the vectors of any vector space can be represented by n-tuples of numbers (coordinates), though only after a (largely arbitrary) choice of n vectors as *basis vectors.*

Thus, in the theory of vector spaces there are two points of view to be distinguished: either we base the development solely on the rules of computation (the axioms of the vector space) and produce a coordinate-free theory, or else we introduce coordinates, in which case we must subsequently make our theory independent of the special choice of basis; that is, we must investigate the invariants of a transformation of basis or at least ask what happens under such a transformation. In any case, any concrete representation of vectors will usually be in the form of coordinates.

The two points of view will be presented here side by side. In §1 the foundations of the theory are developed. In §2 we investigate *linear mappings* of a vector space V_n of dimension n into a vector space V_k of dimension k. Generally speaking, these mappings will themselves form a vector space of dimension $n \cdot k$. In the coordinates of the vector spaces V_n and V_k the mappings or transformations are represented by *matrices.* For the mappings of V_n into itself ($V_k = V_n$) it is possible, beside the operations of the vector space, to introduce a multiplication, namely successive application of mappings. The resulting structure is a ring, and we obtain the foundations of the theory of matrices.

A change of basis in V_n and V_k results in a certain transformation of the matrix of the mapping. In this way a given matrix can be transformed into a diagonal matrix, and one of the applications of this particular transformation is the solution of a system of linear equations, the problem from which we started out.

Linear mappings of the vectors of V_n onto numbers (i.e., those linear mappings for which V_k is a vector space of dimension 1; in particular, the domain of scalars) are *linear forms.* They constitute the vector space *dual* to V_n, which is important in applications to geometry and elsewhere (cf. II8, §4).

In §3 we set ourselves the problem of introducing *products of vectors.* It turns out that we cannot satisfy all the rules for computation with numbers. After making a suitable choice of these rules we can state the requirements for such a product as follows: the taking of a product is to be a *bilinear mapping* of the pairs of vectors of V onto the vectors of a new vector space W. The various kinds of products result from various choices of W. If W is the domain of scalars, we obtain the *inner* or *scalar product*; then it is natural to investigate those changes of basis that leave the *factors* of the inner product invariant (of course, the inner product itself is invariant by definition); and thus we are led to the orthogonal transformations. In the applications an important role is played by the reduction of symmetric matrices to diagonal form by means of orthogonal transformations. We shall deal with this problem in the complex plane; i.e., we investigate unitary transformations of Hermitian forms, since the proofs are the same as in the real field; but we must first develop the theory of determinants.

If for W we choose a vector space of dimension $n \cdot n$, we obtain the *tensor product*, and as its alternating part, so to speak, the *outer* or *alternating product*. The outer product of several factors leads to the *determinant*. The vanishing of the outer product characterizes the linear dependence of vectors. Since a system of (homogeneous) linear equations can be regarded as a query concerning the linear dependence of certain vectors (the column vectors of the coefficient matrix), we again obtain an insight into the theory of such systems of equations.

1. The Concept of a Vector Space

1.1. *Introduction*

We start with the problem of finding the solutions of a system of linear equations

$$\text{(I)} \qquad \sum_{\nu=1}^{n} a_\nu{}^\kappa x^\nu = b^\kappa, \qquad (\kappa = 1, ..., k).$$

In the applications the $a_\nu{}^\kappa$, b^κ are generally real or complex numbers. For the present we need only assume that they are elements of a field, which we call the *domain of scalars S*. Up to §3.2 we may even dispense with the commutative law. A system of elements that satisfies all the axioms for a field (cf. IB1, §3.2) except commutativity of multiplication, is called a *skew field*. Thus fields are themselves skew fields; an example of a noncommutative skew field, i.e., of a skew field that is not a field, is given by the quaternions (see IB8, §3).

Thus we assume that the domain of scalars S is a skew field, and we call its elements *scalars*. The set of equations

$$\text{(II)} \qquad \sum_{\nu=1}^{n} a_\nu{}^\kappa x^\nu = 0, \qquad (\kappa = 1, ..., k)$$

is called the *homogeneous system* corresponding to (I). We shall nowhere need to assume that (I) is strictly nonhomogeneous, i.e., that at least one $b^i \neq 0$.

The solutions of such a system of equations are n-tuples of elements of S, which we write in the form:

$$\mathfrak{x} = (x^1, \ldots, x^n) = (x^\nu)^{\nu=1,\ldots,n},$$

or more briefly, if n is known, as (x^ν).

Now our chief questions are *whether* such a system of equations has solutions and, if so, *how many*. Methods for numerical calculation of the solutions will turn up incidentally, but they are not the object of our investigation.

We see at once that if (I) has two solutions

$$\mathfrak{x} = (x^1, \ldots, x^n), \qquad \mathfrak{y} = (y^1, \ldots, y^n),$$

then $\mathfrak{z} = (x^1 - y^1, \ldots, x^n - y^n)$ is also a solution of (II).

If (II) has several solutions $\mathfrak{x}_\lambda = (x_\lambda{}^1, \ldots, x_\lambda{}^n)$, $\lambda = 1, \ldots, l$, then all linear combinations

$$\mathfrak{z} = \sum_{\lambda=1}^{l} \mathfrak{x}_\lambda c^\lambda = \left(\sum_\lambda x_\lambda{}^1 c^\lambda, \ldots, \sum_\lambda x_\lambda{}^n c^\lambda\right)$$

with arbitrary c^λ in S are solutions of (II). Thus it is natural to introduce an addition for n-tuples and a multiplication by scalars, and then to investigate the algebraic structure of the resulting configuration (for the definition of this word see IB10).

1.2. *Calculation with n-tuples*

Two n-tuples are *equal* if the corresponding scalars are equal; in other words,

(G) $\qquad (x^\nu) = (y^\nu)$ if and only if $x^\nu = y^\nu$ for every ν.

Of course, this definition of equality is part of the definition of an n-tuple as a mapping of the numbers $1, \ldots, n$ into the set of scalars.

It is easy to show that equality as thus defined is reflexive, symmetric, and transitive and is therefore an equivalence (see IA, §8.3 and 5).

Addition can be introduced as follows: to the n-tuples $\mathfrak{x} = (x^\nu)$, $\mathfrak{y} = (y^\nu)$ we assign the n-tuple $(x^\nu + y^\nu)$ as their *sum*. The sum defined in this way clearly has the following properties:

(A1) Existence and uniqueness: two n-tuples $\mathfrak{x}, \mathfrak{y}$, have exactly one n-tuple as their sum, which we denote by $\mathfrak{x} + \mathfrak{y}$;

(A2) Consistency with equality: from $\mathfrak{x} = \mathfrak{u}$ and $\mathfrak{y} = \mathfrak{v}$ it follows that $\mathfrak{x} + \mathfrak{y} = \mathfrak{u} + \mathfrak{v}$.

(A2) is a special case of the general principle of equality in logic: if $A(\mathfrak{x}, \mathfrak{y}, \mathfrak{z})$ is a statement about $\mathfrak{x}, \mathfrak{y}, \mathfrak{z}$, then $A(\mathfrak{x}, \mathfrak{y}, \mathfrak{z})$ and $\mathfrak{x} = \mathfrak{u}, \mathfrak{y} = \mathfrak{v}, \mathfrak{z} = \mathfrak{w}$ imply $A(\mathfrak{u}, \mathfrak{v}, \mathfrak{w})$. In our case $A(\mathfrak{x}, \mathfrak{y}, \mathfrak{z})$ is the statement $\mathfrak{x} + \mathfrak{y} = \mathfrak{z}$. Only this special case, and the corresponding case (M2) for multiplication, will be needed below.

On the basis of (A1) and (A2) our addition is an *operation* defined on the set of n-tuples, and the corresponding statement holds below for (M1), (M2). We write

(A) $$\mathfrak{x} + \mathfrak{y} = (x^\nu) + (y^\nu) = (x^\nu + y^\nu).$$

By (G), (A), and the rules for a skew field we have the

(A3) associative law: $\mathfrak{x} + (\mathfrak{y} + \mathfrak{z}) = (\mathfrak{x} + \mathfrak{u}) + \mathfrak{z}$;

(A4) commutative law: $\mathfrak{x} + \mathfrak{y} = \mathfrak{y} + \mathfrak{x}$;

(A5) neutral element: there exists an n-tuple $\mathfrak{o}$ such that $\mathfrak{x} + \mathfrak{o} = \mathfrak{x}$ for every $\mathfrak{x}$, namely $\mathfrak{o} = (0, ..., 0)$. It follows that there can be *only one* neutral element, for if $\mathfrak{o}$ and $\mathfrak{o}'$ were two such elements, we would have $\mathfrak{o}' + \mathfrak{o} = \mathfrak{o}'$ and $\mathfrak{o} + \mathfrak{o}' = \mathfrak{o}$, so that by (A4) and the transitivity of equality $\mathfrak{o}' = \mathfrak{o}$.

(A6) inverse elements: for every $\mathfrak{x} = (x^\nu)$ there exists an $\mathfrak{x}'$ with $\mathfrak{x} + \mathfrak{x}' = \mathfrak{o}$, namely $\mathfrak{x}' = (-x^\nu)$. We write $\mathfrak{x}' = -\mathfrak{x}$ and $\mathfrak{y} + (-\mathfrak{x}) = \mathfrak{y} - \mathfrak{x}$.

From (A1) to (A6) follows (A5′): for every pair of n-tuples $\mathfrak{u}, \mathfrak{v}$ there exists exactly one n-tuple with $\mathfrak{u} + \mathfrak{x} = \mathfrak{v}$, namely $\mathfrak{x} = \mathfrak{v} - \mathfrak{u}$.

On the other hand, (A5) and (A6) follow from (A1) to (A4) and(A5′), if we postulate the existence of at least one element. (Cf. also IB1, §2.3 and IB2, §2.4.)

Multiplication by scalars is introduced as follows. Since we shall naturally wish to be able, for example, to write $\mathfrak{x} + \mathfrak{x} = \mathfrak{x} \cdot 2$, i.e.,

$$(x^\nu) \cdot 2 = (x^\nu) + (x^\nu) = (x^\nu + x^\nu) = (x^\nu \cdot 2),$$

we define right-multiplication of an n-tuple (x^ν) by an element s of S (*S-multiplication* on the right; the scalar s is written to the right of the vector $\mathfrak{x}$) by setting

(S–M) $$\mathfrak{x} \cdot s = (x^1, ..., x^n) \cdot s = (x^1 \cdot s, ..., x^n \cdot s),$$

or, more concisely,

$$(x^\nu) \cdot s = (x^\nu \cdot s).$$

This scalar product, as well as the product in S and later the product of two matrices, will be denoted by writing one factor after the other with or without an intervening dot, depending on whether or not the dot seems conducive to clarity. Since these three kinds of products will be distinguished from one another by the notation for their factors, they can all be written in the same way.

It must be pointed out that $s \cdot \mathfrak{x}$ is not yet defined, even when S is commutative. For the time being we shall not require this left-multiplication; when it is needed later, we shall define it by setting $s \cdot (x^\nu) = (s \cdot x^\nu)$. If S is commutative, then $s \cdot \mathfrak{x} = \mathfrak{x} \cdot s$.

This definition (of scalar multiplication on the right) implies

(M1) the existence and uniqueness of the S product: for every n-tuple $\mathfrak{x}$ and every s in S there exists exactly one n-tuple $\mathfrak{y} = \mathfrak{x} \cdot s$.

In the same way as for addition, it is easy to show that

(M2) the S-multiplication is consistent with the equality: $\mathfrak{x} = \mathfrak{y}$ and $s = t$ implies $\mathfrak{x} \cdot s = \mathfrak{y} \cdot t$;

(M3) the associative law: $\mathfrak{x} \cdot (s \cdot t) = (\mathfrak{x} \cdot s) \cdot t$; the distributive laws

(M4) $(\mathfrak{x} + \mathfrak{y}) \cdot s = \mathfrak{x} \cdot s + \mathfrak{y} \cdot s$,

(M5) $\mathfrak{x} \cdot (s + t) = \mathfrak{x} \cdot s + \mathfrak{x} \cdot t$;

and further,

(M6) if 1 is the unit element of the skew field S, then $\mathfrak{x} \cdot 1 = \mathfrak{x}$.

A set of elements (here n-tuples) in which these rules for computation are defined, is called a *vector space*. Abstractly we make the following definition:

Let V be a set in which there is defined an addition satisfying the laws (axioms) (A1) *to* (A6); *let S be a skew field, and let there be defined an S-multiplication satisfying the laws* (M1) *to* (M6). *Then V is called a vector space over S, and its elements are called vectors.*

Thus, V is a commutative group with respect to addition. The skew field S is a skew subfield (up to isomorphism) of the ring of endomorphisms of this group. For if to each element s of S we assign a mapping $\sigma : \mathfrak{x} \to \mathfrak{x} \cdot s$, then (M4) means that σ is an endomorphism, (M3) states that the product of two endomorphisms is defined by successive application of the mapping, and (M5) is the usual definition of the sum of two endomorphisms. (See IB1, §2.4 or B. L. van der Waerden [2], page 148). Consequently, from a given commutative group we can construct a vector space by selecting a skew field from the ring of endomorphisms of the group.

Consequences from the axioms of a vector space are:

1) If 0 is the zero element of S, then $\mathfrak{x} \cdot 0 = \mathfrak{o}$ for every $\mathfrak{x}$.

Proof: Since

$$\mathfrak{x} \cdot 0 = \mathfrak{x} \cdot (0 + 0) = \mathfrak{x} \cdot 0 + \mathfrak{x} \cdot 0,$$

therefore, by (A5′),

$$\mathfrak{x} \cdot 0 = \mathfrak{x} \cdot 0 - \mathfrak{x} \cdot 0 = \mathfrak{o}.$$

2) For every s in S we have $\mathfrak{o} \cdot s = \mathfrak{o}$. The proof is analogous to that of (1).

3) Conversely, if $\mathfrak{x} \cdot s = \mathfrak{o}$, then $s = 0$ or $\mathfrak{x} = \mathfrak{o}$.

Proof. Let $\mathfrak{x} \cdot s = \mathfrak{o}$ and $s \neq 0$. Then

$$\mathfrak{x} \cdot 1 = \mathfrak{x} \cdot s \cdot (1/s) = \mathfrak{o} \cdot (1/s) = \mathfrak{o}.$$

4) $-\mathfrak{x} = \mathfrak{x} \cdot (-1)$.

Proof. $\mathfrak{x} + \mathfrak{x} \cdot (-1) = \mathfrak{x} \cdot (1 - 1) = \mathfrak{o}$.

Often (for example, in IB6, §8) we speak of a vector space over S even when S is not a skew field but only a ring (with unit element). But then, besides (A1 to A6) and (M1 to M6) it is necessary to postulate the existence of a basis (see §1.3); for if S is not a skew field, it is no longer possible (as in §1.3) to deduce the existence of a basis from (A1 to A6) and (M1 to M6).

1.3. *Linear Dependence. Basis*

Addition and S-multiplication can be combined into the concept of a "linear combination." A vector $\mathfrak{c}$ is said to be a *linear combination* of the vectors $\mathfrak{a}_1, ..., \mathfrak{a}_n$, or to be *linearly dependent* on these vectors, if there exist $c^1, ..., c^n$ in S such that

$$\mathfrak{c} = \mathfrak{a}_1 c^1 + \cdots + \mathfrak{a}_n c^n = \sum_{\nu=1}^{n} \mathfrak{a}_\nu c^\nu.$$

With regard to the notation, we shall usually omit the sign $\sum$ by agreeing once and for all (with Einstein) that summation is to be taken over equal Greek superscripts or Greek subscripts. Of course, it must be clear from the context what values the indices are to assume.

We use Greek letters (as variables) for the indices when we mean that all possible values of these indices are to be assumed successively (in the language of logic, a Greek letter denotes a bound index variable; in our notation for n-tuples the binding can be expressed by $(x^\nu)^{\nu=1,...,n}$). If we are referring to a definite value of the index, we use a Latin letter. Consequently, summation is not to be taken over Latin indices.

Thus, a vector space over S is a set in which it is always possible to form linear combinations of its elements with the elements of S. It is to this characteristic feature that the concept "vector space" owes its importance.

Preliminary discussion. If $\mathfrak{e}_1, ..., \mathfrak{e}_n$ are given vectors, then the set of linear combinations $\mathfrak{x} = \mathfrak{e}_\nu x^\nu$, $x^\nu \in S$, forms a vector space, as is easily shown. Conversely, we shall obtain a good picture of a given vector space V if we succeed in finding a set (presumably finite) of vectors $\mathfrak{e}_1, ..., \mathfrak{e}_n$ such that every vector $\mathfrak{x}$ in V is representable (as far as possible, uniquely) as a linear combination $\mathfrak{x} = \mathfrak{e}_\nu x^\nu$, and we shall see that in fact this is always possible. But then the vector is determined by the n-tuple (x^ν), so that the vector space V is characterized by the n-tuples of S. Thus a set of n-tuples is not only a particular example of a vector space, but *every* vector space is representable as the set of n-tuples in S; the various vector spaces over S are distinguished from one another only by the number n.

Development. To carry out these ideas precisely we shall need the following concepts: the vectors $\mathfrak{a}_1, ..., \mathfrak{a}_n$ are said to be *linearly dependent* if there exist scalars $c^1, ..., c^n$, not all 0, such that $\mathfrak{a}_\nu c^\nu = \mathfrak{o}$; but if $\mathfrak{a}_\nu c^\nu = \mathfrak{o}$ implies that all $c^\nu = 0$, the given vectors are said to be *linearly independent.*

The following theorems are immediate consequences of the definition:

Theorem 1*a*: *if for some* i $(1 \leqslant i \leqslant n)$ *we have* $\mathfrak{a}_i = \mathfrak{o}$, *then* $\mathfrak{a}_1, ..., \mathfrak{a}_n$ *are linearly dependent.*

Theorem 1*b*: *if the vectors* $\mathfrak{a}_1, ..., \mathfrak{a}_n$ *are linearly dependent, then so are the vectors* $\mathfrak{a}_1, ..., \mathfrak{a}_n, \mathfrak{a}_{n+1}, ..., \mathfrak{a}_{n+p}$.

Theorem 1*c*: *if the vectors* $\mathfrak{a}_1, ..., \mathfrak{a}_n, \mathfrak{a}_{n+1}, ..., \mathfrak{a}_{n+p}$ *are linearly independent, then so are the vectors* $\mathfrak{a}_1, ..., \mathfrak{a}_n$.

Note. The order of the vectors has no effect on linear dependence.

The system $\mathfrak{e}_1, ..., \mathfrak{e}_n$ is called a *basis* of V if for every vector $\mathfrak{a}$ in V there exist n elements a^ν in S such that

1) $\mathfrak{a} = \mathfrak{e}_\nu a^\nu$ (reminder to the reader: summation over ν is from 1 to n),

2) $\mathfrak{a} = \mathfrak{e}_\nu a^\nu = \mathfrak{e}_\nu b^\nu$ implies $a^\nu = b^\nu$ for all ν; in other words, the representation is unique.

The a^ν are called the *coordinates* of $\mathfrak{a}$ with respect to the basis $(\mathfrak{e}_\nu)$. Condition (2) is equivalent to

2′) $\mathfrak{e}_1, ..., \mathfrak{e}_n$ are linearly independent.

Proof. (*a*) Assume (2) and $\mathfrak{e}_\nu c^\nu = \mathfrak{o}$. Since $\sum_\nu \mathfrak{e}_\nu \cdot 0 = \mathfrak{o}$, it follows from (2) that $c^\nu = 0$ for all ν.

(*b*) assume (2′) and $\mathfrak{e}_\nu a^\nu = \mathfrak{e}_\nu b^\nu$.

Then it follows that $\mathfrak{e}_\nu(a^\nu - b^\nu) = \mathfrak{o}$, so that by (2′) we have $a^\nu - b^\nu = 0$ for all ν.

A vector space can have various bases; but we shall see that the number (n) of basis elements is always the same. This number is called the *dimension* of the vector space, which we then denote by V_n.

The above assertion follows easily from the next theorem.

Theorem 2: *If the* $n+1$ *vectors* $\mathfrak{a}_1, \ldots, \mathfrak{a}_{n+1}$ *are linear combinations of* n *vectors* $\mathfrak{e}_1, \ldots, \mathfrak{e}_n$,

$$\mathfrak{a}_\kappa = \mathfrak{e}_\nu a_\kappa{}^\nu, \qquad \nu = 1, \ldots, n, \qquad \kappa = 1, \ldots, n+1,$$

then $\mathfrak{a}_1, \ldots, \mathfrak{a}_{n+1}$ *are linearly dependent.*

The proof is by complete induction.

Initial step: for $n = 1$ we have $\mathfrak{a}_1 = \mathfrak{e}_1 a_1{}^1$, $\mathfrak{a}_2 = \mathfrak{e}_1 a_2{}^1$. If $a_1{}^1 = 0$ or $a_2{}^1 = 0$, then the assertion is correct by Theorem 1*a*. Otherwise,

$$\mathfrak{a}_1 \cdot (a_1{}^1)^{-1} - \mathfrak{a}_2 (a_2{}^1)^{-1} = \mathfrak{o}.$$

Completion of the induction: let

$$\begin{aligned} \mathfrak{a}_1 &= \mathfrak{e}_1 a_1{}^1 + \mathfrak{e}_2 a_1{}^2 + \cdots + \mathfrak{e}_n a_1{}^n \\ \mathfrak{a}_2 &= \mathfrak{e}_1 a_2{}^1 + \mathfrak{e}_2 a_2{}^2 + \cdots + \mathfrak{e}_n a_2{}^n \\ &\ldots\ldots\ldots\ldots \\ \mathfrak{a}_{n+1} &= \mathfrak{e}_1 a^1_{n+1} + \mathfrak{e}_2 a^2_{n+1} + \cdots + \mathfrak{e}_n a^n_{n+1}. \end{aligned}$$

If all $a_1{}^\nu = 0$, the assertion is correct by Theorem 1*a*. We may assume $a_1{}^1 \neq 0$, since for $a_1{}^1 = 0$, $a_1{}^i \neq 0$ the proof is exactly the same. Then the n vectors

$$\begin{aligned} \mathfrak{b}_2 &= \mathfrak{a}_2 - \mathfrak{a}_1 \cdot (a_1{}^1)^{-1} \cdot a_2{}^1 \\ &\ldots\ldots\ldots\ldots \\ \mathfrak{b}_{n+1} &= \mathfrak{a}_{n+1} - \mathfrak{a}_1 \cdot (a_1{}^1)^{-1} \cdot a^1_{n+1} \end{aligned}$$

are linear combinations of the $n-1$ vectors $\mathfrak{e}_1, \ldots, \mathfrak{e}_n$, so that they are linearly independent by the induction hypothesis; that is, there exist $x^2, \ldots, x^{n+1} \in S$, not all $= 0$, such that

$$\mathfrak{b}_2 x^2 + \cdots + \mathfrak{b}_{n+1} x^{n+1} = \mathfrak{o}.$$

Substitution of the above expressions for the $\mathfrak{b}_\nu$ gives

$$\mathfrak{a}_2 x^2 + \cdots + \mathfrak{a}_{n+1} x^{n+1} - \mathfrak{a}_1 (a_1{}^1)^{-1} (a_2{}^1 x^2 + \cdots + a^1_{n+1} x^{n+1}) = \mathfrak{o}$$

where at least one of the coefficients of $\mathfrak{a}_2, \ldots, \mathfrak{a}_{n+1}$ is $\neq 0$.

From Theorem 2 follows

Theorem 3: *if* V *has a basis of* n *elements* $\mathfrak{e}_1, \ldots, \mathfrak{e}_n$, then:

a) *Every* $n+1$ *elements of* V *are linearly dependent.*

b) *Consequently, no basis of* V *has more than* n *elements.*

c) *No basis can consist of fewer than* n *elements*; for otherwise $\mathfrak{e}_1, \ldots, \mathfrak{e}_n$ would be linearly dependent.

d) Every n linearly independent vectors $\mathfrak{v}_1, \ldots, \mathfrak{v}_n$ *of V form a basis.* For if $\mathfrak{a}$ is an arbitrary element of V, then by (*a*) the vectors $\mathfrak{v}_1, \ldots, \mathfrak{v}_n, \mathfrak{a}$ are linearly dependent; that is, there exist a^ν ($\nu = 0, 1, \ldots, n$), not all $= 0$, such that

$$\mathfrak{a}a^0 + \mathfrak{v}_1 a^1 + \cdots + \mathfrak{v}_n a^n = \mathfrak{o}.$$

Here $a^0 \neq 0$, since otherwise $\mathfrak{v}_1, \ldots, \mathfrak{v}_n$ would be linearly independent, and therefore $\mathfrak{a}$ is representable as a linear combination of $\mathfrak{v}_1, \ldots, \mathfrak{v}_n$.

Thus the dimension n of V can also be characterized by the fact that *there exist n linearly independent vectors in V and every n + 1 vectors are linearly dependent.* If and only if it has this property, does the vector space V have a basis of n elements. But this characterization of dimension is *independent of the choice of basis.*

The concept of a "basis" was first introduced by Dedekind for modules (supplement XI to Dirichlet's *Zahlentheorie*, 3rd ed., 1879, §165) and was used for ideals. He did not require that the representation be unique, and he allowed the domain of coefficients to be a ring. The concept of a "basis" in this sense occurs in Chapter 5, §3.2.

The dimension of a vector space over S determines the vector space uniquely up to isomorphism; in other words, we have the isomorphism theorem: *two vector spaces V, V′ over the same domain of scalars S are isomorphic if they have the same dimension.*

The proof is almost trivial when we consider *what* is to be proved. Vector spaces V and V' are said to be *isomorphic* if there exists a one-to-one mapping of V onto V' ($\mathfrak{x} \to \mathfrak{x}'$) with the property that

$$(\mathfrak{x} + \mathfrak{y})' = \mathfrak{x}' + \mathfrak{y}' \qquad \text{and} \qquad (\mathfrak{x} \cdot s)' = \mathfrak{x}' \cdot s.$$

But if $\mathfrak{e}_1, \ldots, \mathfrak{e}_n$ is a basis of V, and $\mathfrak{e}'_1, \ldots, \mathfrak{e}'_n$ is a basis of V', then $\mathfrak{e}_\nu x^\nu \to \mathfrak{e}'_\nu x^\nu$ is a mapping with the required properties.

In analysis, an important role is played by vector spaces of infinitely many dimensions; for example, the real functions that can be represented as a Fourier polynomial in the interval $(-\pi, +\pi)$ form a vector space over the field of real numbers; for this vector space the functions 1, $\cos(\nu x)$, $\sin(\nu x)$, ($\nu = 1, \ldots$) are a basis. More generally, the representation of functions in an orthogonal series may be regarded as a representation of the vector space formed by the functions (cf. III, 11). *But in the present chapter we consider only vector spaces of finite dimension.* (For the existence of a basis in a vector space that is not necessarily finite-dimensional, see IB11, §3, Theorem 2.)

When we consider the sets of n-tuples of elements of S as a vector space, as we did in §1.2, we are representing the space in terms of the special basis $\mathfrak{e}_1 = (1, 0, \ldots, 0)$, $\mathfrak{e}_2 = (0, 1, 0, \ldots, 0) \ldots$, $\mathfrak{e}_n = (0, 0, \ldots, 0, 1)$.

It is easy to see that S itself is a vector space of dimension 1 over S.

1.4. *Vector Subspaces*

A subset U of a vector space V over S is called a *vector subspace* if U itself is a vector space over S and if the addition and the S-product in U are the same as in V.

If U is a vector subspace of V, we have:

1) if $\mathfrak{a} \in U$ and $\mathfrak{b} \in U$, then $\mathfrak{a} + \mathfrak{b} \in U$;

2) if $\mathfrak{a} \in U$ and $s \in S$, then $\mathfrak{a} \cdot s \in U$.

These two conditions are also sufficient for a non-empty subset U to be a vector subspace of V. To prove the sufficiency, we must verify the laws (A), (M) for U. The associative law

$$(\mathfrak{x} + \mathfrak{y}) + \mathfrak{z} = \mathfrak{x} + (\mathfrak{y} + \mathfrak{z})$$

holds for arbitrary elements of V, and thus in particular when $\mathfrak{x}, \mathfrak{y}, \mathfrak{z}$ are elements of U. The same remark holds for all those laws in which the words "there exists" do not occur. The existence of the sum and the S-product in U is guaranteed by (1) and (2). It remains to verify (A5) and (A6). As for (*A*5), for an arbitrary $\mathfrak{x} \in U$ we have $\mathfrak{x} \cdot 0 \in U$ and from §1.2, corollary 1, it follows that $\mathfrak{x} \cdot 0 = \mathfrak{o}$. As for (A6), if $\mathfrak{x}$ is an element of U, then so is $\mathfrak{x} \cdot (-1)$ (see §1.2, corollary 4).

The intersection of two or arbitrarily many vector subspaces is again a vector subspace. For the proof we need only verify (1), (2). The necessary definitions are:

$$\mathfrak{x} \in U_1 \cap U_2 \quad \text{if and only if} \quad \mathfrak{x} \in U_1 \quad \text{and} \quad \mathfrak{x} \in U_2 .$$

If $\mathfrak{M}$ is a non-empty set of vector subspaces, then $\mathfrak{x} \in \bigcap_{U \in \mathfrak{M}} U$ if and only if $\mathfrak{x} \in U$ for all $U \in \mathfrak{M}$.

If $\mathfrak{a}_1, \ldots, \mathfrak{a}_k$ are arbitrarily given elements of V_n, the smallest subspace U of V_n containing $\mathfrak{a}_1, \ldots, \mathfrak{a}_k$ is called the subspace *spanned* or *generated* by $\mathfrak{a}_1, \ldots, \mathfrak{a}_k$. The word "smallest" here means that U is the intersection of all subspaces containing $\mathfrak{a}_1, \ldots, \mathfrak{a}_k$, so that the existence of U is guaranteed. Of course, it may happen that $U = V$.

A basis for the subspace spanned by the vectors $\mathfrak{a}_1, \ldots, \mathfrak{a}_k$ can be found by writing the $\mathfrak{a}_1, \ldots, \mathfrak{a}_k$ in any order and then striking out every $\mathfrak{a}_k$ that is linearly dependent on its predecessors.

To prove this statement we enumerate the vectors in such a way that each of the first l vectors $\mathfrak{a}_1, \ldots, \mathfrak{a}_l$ is linearly independent of its predecessors but the vectors $\mathfrak{a}_{l+1}, \ldots, \mathfrak{a}_k$ are linearly dependent on $\mathfrak{a}_1, \ldots, \mathfrak{a}_l$, and then show that

1) $\mathfrak{a}_1, \ldots, \mathfrak{a}_l$ are linearly independent.

The proof is by induction on l. For $l = 1$ the statement is correct, since we may assume that $\mathfrak{a}_1 \neq \mathfrak{o}$. We now assume that $\mathfrak{a}_1, \ldots, \mathfrak{a}_{l-1}$ are linearly independent. If $\mathfrak{a}_1, ..., \mathfrak{a}_l$ were linearly dependent, there would exist scalars $x^1, ..., x^l$, not all 0, such that $\mathfrak{a}_1 x^1 + \cdots + \mathfrak{a}_l x^l = 0$. If $x^l = 0$, then $\mathfrak{a}_1, ..., \mathfrak{a}_{l-1}$ would be linearly dependent, and if $x^l \neq 0$, then $\mathfrak{a}_l$ would be linearly dependent on $\mathfrak{a}_1, ..., \mathfrak{a}_{l-1}$.

2) Every subspace U, even a smallest one, which contains $\mathfrak{a}_1, ..., \mathfrak{a}_k$, certainly contains $\mathfrak{a}_1, ..., \mathfrak{a}_l$ and all linear combinations $\mathfrak{a}_1 x^1 + \cdots + \mathfrak{a}_l x^l$. But the totality of these linear combinations is a subspace containing $\mathfrak{a}_1, ..., \mathfrak{a}_k$, and is thus the desired subspace.

The basis $\mathfrak{a}_1, ..., \mathfrak{a}_l$ of a subspace can be extended to a basis of V_n. For, either the $\mathfrak{a}_1, ..., \mathfrak{a}_l$ already span V_n or else there exists a vector $\mathfrak{a}_{l+1}$, linearly independent of them, that can then be adjoined as a further basis vector for V_n. Since the dimension of V is finite, this procedure comes to an end after finitely many steps.

1.5. *Change to a New Basis*

After choice of a basis the vectors in V_n can be represented as n-tuples from S. In the applications where it is natural to choose some special basis, the vectors are often given in this way.

Thus it is important to determine how the coordinates of a vector behave under a change of basis or, as is often said, of coordinate system.

Let $\mathfrak{e}_1, ..., \mathfrak{e}_n$ and $\mathfrak{e}_{1'}, ..., \mathfrak{e}_{n'}$ be two bases of the vector space V_n, where for greater uniformity in our subsequent notation the primes have been written not on the $\mathfrak{e}$ but on the indices. Thus $1', ..., n'$ are simply marks to identify the elements of the second (the "primed") basis; of course, there are n of them.

Every vector of one basis can be expressed as a linear combination of the vectors of the other basis. The coordinates of the basis vectors will be denoted by $t^{\kappa}_{\nu'}$ and $t^{\mu'}_{\nu}$, respectively (later also, cf. §2.3, by $s^{\kappa}_{\nu'}$ and $s^{\mu'}_{\nu}$ respectively):

$$\mathfrak{e}_{\nu'} = \mathfrak{e}_{\kappa} t^{\kappa}_{\nu'}, \tag{1a}$$

$$\mathfrak{e}_{\lambda} = \mathfrak{e}_{\mu'} t^{\mu'}_{\lambda}. \tag{1b}$$

The t with primed superscripts and the t with primed subscripts are to be carefully distinguished. They are defined in completely different ways.

The purpose of using the same letter t in these two cases is to make the equations for transformation of coordinates easy to remember. One only needs to recall that summation is taken over any index appearing both as superscript and subscript and that the indices over which summation is not taken occur either only as subscripts or only as superscripts. It was for this reason that we put the primes on the indices rather than on the $\mathfrak{e}$.

The relations between the $t^{\kappa}_{\nu'}$ and the $t^{\mu'}_{\lambda}$ are found by substituting (1*a*) into (1*b*):

$$\mathfrak{e}_{\lambda} = \mathfrak{e}_{\kappa} t^{\kappa}_{\mu'} t^{\mu'}_{\lambda}$$

(summation for κ is from 1 to n, and for μ' it is from $1'$ to n').

Since the $\mathfrak{e}_{\lambda}$ are linearly independent, it follows that

(2*a*) $$t^{\kappa}_{\mu'} t^{\mu'}_{\lambda} = \delta_{\lambda}^{\kappa} = \begin{cases} 1 & \text{for} \quad \kappa = \lambda, \\ 0 & \text{for} \quad \kappa \neq \lambda. \end{cases}$$

In the same way, by substitution of (1*b*) into (1*a*) we obtain

(2*b*) $$t^{\mu'}_{\kappa} t^{\kappa}_{\nu'} = \delta^{\mu'}_{\nu'} = \begin{cases} 1 & \text{for} \quad \mu' = \nu', \\ 0 & \text{for} \quad \mu' \neq \nu'. \end{cases}$$

The Kronecker symbols δ *with two indices, which may also occur either both as superscripts or both as subscripts, will always be used in this sense.*

The equations (2) show that not every arbitrary system of $n \cdot n$ elements $t^{\kappa}_{\nu'}$ of S can occur as the schema of coefficients for a transformation of basis: corresponding to a system $t^{\kappa}_{\nu'}$ there must exist a second system $t^{\mu'}_{\lambda}$ such that equations (2) are satisfied. But this condition is also sufficient. For we assert that, if for the system $t^{\kappa}_{\nu'}$ there exists a system $t^{\mu'}_{\lambda}$ such that the equations (2) are satisfied, and if $\mathfrak{e}_1, ..., \mathfrak{e}_n$ is a basis of V_n, then the vectors $\mathfrak{e}_{\nu'} = \mathfrak{e}_{\kappa} t^{\kappa}_{\nu'}$ also form a basis of V_n.

For the proof of this assertion, it is only necessary, since vectors of V_n form a basis, to show that the $\mathfrak{e}_{\nu'}$ are linearly independent; that is, that $\mathfrak{e}_{\nu'} c^{\nu'} = \mathfrak{o}$ implies $c_{\nu'} = 0$ for all ν'.

Now we have $\mathfrak{e}_{\nu'} c^{\nu'} = \mathfrak{e}_{\kappa} t^{\kappa}_{\nu'} c^{\nu'}$. Since the $\mathfrak{e}_{\kappa}$ are linearly independent, the right-hand can be $= \mathfrak{o}$ only if $t^{\kappa}_{\nu'} c^{\nu'} = 0$ for every κ. Multiplying the κth of these equations with $t^{\mu'}_{\kappa}$ (where μ' is arbitrary) and summing, we obtain:

$$t^{\mu'}_{\kappa} t^{\kappa}_{\nu'} c^{\nu'} = 0.$$

From (2*b*) it follows that $c^{\mu'} = 0$ and, of course, this procedure can be carried out for every μ'.

The reader is advised to make the computation for a few simple examples (say for $n = 2$). In the case $n = 2$ the determination of the $t^{\mu'}_{\lambda}$ for a given $t^{\kappa}_{\nu'}$ requires the solution of four linear equations which for commutative S can be solved if and only if $t^{1}_{1'} t^{2}_{2'} - t^{1}_{2'} t^{2}_{1'} \neq 0$.

Let us now ask how the coordinates of a vector $\mathfrak{x}$ are transformed under the change of basis (1). Let the coordinates of $\mathfrak{x}$ with respect to the basis $(\mathfrak{e}_{\lambda})$ be x^{λ}, and with respect to the basis $(\mathfrak{e}_{\nu'})$ let them be $x^{\nu'}$, so that we have

(3) $$\mathfrak{x} = \mathfrak{e}_{\lambda} x^{\lambda} = \mathfrak{e}_{\nu'} x^{\nu'}.$$

Substitution of (1*b*) in (3) gives

$$\mathfrak{c}_{\mu'} t^{\mu'}_{\lambda} x^{\lambda} = \mathfrak{c}_{\nu'} x^{\nu'},$$

so that, since the $\mathfrak{c}_{\mu'}$ are linearly independent,

$$x^{\nu'} = t^{\nu'}_{\lambda} x^{\lambda}. \tag{4a}$$

The solution of this system of equations for the x^{λ} is obtained either by substituting (1*a*) in (3) or by multiplying (4*a*) with t and using (2):

$$x^{\kappa} = t^{\kappa}_{\nu'} x^{\nu'}. \tag{4b}$$

Here the x^{κ} and $x^{\nu'}$ are coordinates of the same vector with respect to different bases. The "kernel letter" x denotes the (fixed) vector, and the change of basis is expressed in the index. This "kernel-index notation" is due to Schouten.

2. Linear Transformations of Vector Spaces

2.1. *General Properties of Linear Transformations*

In the equations §1.5 (4) we could also interpret x^{λ} and $x^{\nu'}$ as coordinates of different vectors with respect to the same basis. In the kernel-index notation such a transformation has the form

$$u^{\kappa} = a_{\nu}{}^{\kappa} x^{\nu}; \tag{1}$$

that is, we use different kernel letters for the two vectors and do not use any primed indices.

By (1) there is assigned to each vector $\mathfrak{x}$ exactly one vector $\mathfrak{u}$, which we denote by $\mathfrak{u} = \mathsf{A}\mathfrak{x}$ (A is to be read: alpha) and

$$\text{from} \quad \mathfrak{x} = \mathfrak{y} \quad \text{follows} \quad \mathsf{A}\mathfrak{x} = \mathsf{A}\mathfrak{y}. \tag{Lv}$$

Thus we have a *mapping* or *transformation of* V_n *into itself*, with the following properties:

$$\mathsf{A}(\mathfrak{x} + \mathfrak{y}) = \mathsf{A}\mathfrak{x} + \mathsf{A}\mathfrak{y} \tag{La}$$

$$\mathsf{A}(\mathfrak{x} \cdot s) = (\mathsf{A}\mathfrak{x}) \cdot s. \tag{Lm}$$

Transformations with these properties are said to be *linear*. They are simply the homorphisms of V_n. Since it will be necessary for us to consider them in various forms, we will in general take $\mathsf{A}, \mathsf{B}, \ldots$ to be linear transformations of $V_n(S)$ into a $\tilde{V}_k(S)$. Thus the dimensions of the two vector spaces are not required to be the same. We shall be particularly interested in the cases $\tilde{V}_k = V_n$ and $\tilde{V}_k = V_1 = S$, and when it is desirable to indicate the dimensions, we shall speak of an $n \times k$ transformation.

We do not assume that every vector in V_k is the image of a vector in V_n. The set of vectors in $\tilde{V}_k$ which are images of vectors in V_n is denoted by $\mathsf{A}(V_n)$ and is called the *image domain* (cf. IA, §8.4) or the *image space* of the transformation A. We first note that $\mathsf{A}(V_n)$ is a vector space over S and is thus a vector subspace of $\tilde{V}_k$.

To prove this we must show that $\tilde{\mathfrak{x}} \in \mathsf{A}(V_n)$, $\tilde{\mathfrak{y}} \in \mathsf{A}(V_n)$, $s \in S$ implies

$$(1)\quad \tilde{\mathfrak{x}} + \tilde{\mathfrak{y}} \in \mathsf{A}(V_n), \qquad (2)\quad \tilde{\mathfrak{x}}s \in \mathsf{A}(V_n).$$

We outline the proof for (2) and leave the proof of (1) to the reader. Since $\tilde{\mathfrak{x}} \in \mathsf{A}(V_n)$ means that there exists an $\mathfrak{x} \in V_n$ such that $\mathsf{A}\mathfrak{x} = \tilde{\mathfrak{x}}$, it follows from (Lm) that $\tilde{\mathfrak{x}}s = (\mathsf{A}\mathfrak{x})\, s = \mathsf{A}(\mathfrak{x}s) \in \mathsf{A}(V_n)$.

The dimension r of the image space of $\mathsf{A}(V_n) = \tilde{V}_r$ is called the *rank of the transformation* A. It is obvious that $r \leqslant n$.

We now turn our attention to the set of $n \times k$ transformations themselves. In this set we can introduce an algebraic structure as follows: two transformations are said to be *equal*, $\mathsf{A} = \mathsf{B}$, if the equation $\mathsf{A}\mathfrak{x} = \mathsf{B}\mathfrak{x}$ holds (cf. IA, §8.4) for all $\mathfrak{x}$ in V_n. This equality is reflexive, symmetric, and transitive.

An *addition* is defined by

$$(2)\qquad (\mathsf{A} + \mathsf{B})\, \mathfrak{x} = \mathsf{A}\mathfrak{x} + \mathsf{B}\mathfrak{x},$$

and an *S-multiplication* on the left by

$$(3)\qquad (s \cdot \mathsf{A})\, \mathfrak{x} = s \cdot (\mathsf{A}\mathfrak{x}).$$

For this purpose it is necessary that $\tilde{V}_k$ be not only a right-space but also a left-space over S. This situation certainly holds if $\tilde{V}_k = S$; and if S is commutative (cf. §1.2), it can easily be brought about by defining $s\mathfrak{x} = \mathfrak{x}s$. Our applications will be confined to these two cases.

Theorem 1: *With respect to these operations the $n \times k$ transformations form a vector space L_n^k* of dimension $n \cdot k$.

What we must prove is:

1) $\mathsf{A} + \mathsf{B}$ is a linear transformation. To show this we must verify (Lv), (La), and (Lm). As an example, we shall give the proof for (La), leaving the other proofs here and below to the reader.

$$\begin{aligned}(\mathsf{A} + \mathsf{B})(\mathfrak{x} + \mathfrak{y}) &= \mathsf{A}(\mathfrak{x} + \mathfrak{y}) + \mathsf{B}(\mathfrak{x} + \mathfrak{y}) && \text{by (2)}\\ &= \mathsf{A}\mathfrak{x} + \mathsf{A}\mathfrak{y} + \mathsf{B}\mathfrak{x} + \mathsf{B}\mathfrak{y} && \text{by (La), applied to A and B,}\\ &= (\mathsf{A} + \mathsf{B})\, \mathfrak{x} + (\mathsf{A} + \mathsf{B})\, \mathfrak{x}\end{aligned}$$

on the basis of the associative and commutative law for addition in $\tilde{V}_k$ and (2).

2) The sum $\mathsf{A} + \mathsf{B}$ satisfies the condition (A1–A6). As an example, let us prove (A2). The assertion is that if for every $\mathfrak{x}$ we have $\mathsf{A}\mathfrak{x} = \mathsf{B}\mathfrak{x}$ and $\Gamma\mathfrak{x} = \Delta\mathfrak{x}$, then for every $\mathfrak{x}$ it follows that

$$(\mathsf{A} + \Gamma)\,\mathfrak{x} = (\mathsf{B} + \Delta)\,\mathfrak{x}.$$

But

$$\begin{aligned}(\mathsf{A} + \Gamma)\,\mathfrak{x} &= \mathsf{A}\mathfrak{x} + \Gamma\mathfrak{x} && \text{by (2)}\\ &= \mathsf{B}\mathfrak{x} + \Delta\mathfrak{x} && \text{by (A2), applied in } \tilde{V}_k,\\ &= (\mathsf{B} + \Delta)\,\mathfrak{x} && \text{by (2).}\end{aligned}$$

The zero element is the transformation $\mathsf{O}\mathfrak{x} = \tilde{\mathfrak{o}}$, where $\tilde{\mathfrak{o}}$ is the zero element of $\tilde{V}_k$.

3) $s\mathsf{A}$ is a linear transformation.

4) The S-multiplication satisfies (M1–M6), but with left s-factors.

The theorem further states that the dimension of L depends on the dimensions of V and $\tilde{V}$. To prove this statement we shall require a representation of the transformations in terms of a basis $(\mathfrak{e}_\nu)$ of V_n and $(\tilde{\mathfrak{e}}_\kappa)$ of $\tilde{V}_k$. Therefore we shall postpone this proof until we have completed our discussion of coordinate-free theorems.

The *multiplication of one transformation by another* is defined in the usual way as successive application of the two transformations:

$$(\mathsf{AB})\,\mathfrak{x} = \mathsf{A}(\mathsf{B}\mathfrak{x}). \tag{4}$$

Here B is a transformation of V_n into $\tilde{V}_k$, and A is a transformation of $\tilde{V}_k$ into a new vector space W_l, and in the product of several transformations additional vector spaces are introduced in the same way. It is permitted, but not required, that these vector spaces be distinct from one another.

The product of two transformations is defined only if the image space of the first transformation is contained in the preimage space of the second. Moreover we have defined the sum of two transformations only for transformations from the same V_n into the same $\tilde{V}_k$. But so far as the sums and products exist, they satisfy the conditions for a ring. Beside those already stated for addition, these conditions are:

5) Consistency of multiplication with equality: if for all $\mathfrak{x}$ we have $\mathsf{A}\mathfrak{x} = \mathsf{B}\mathfrak{x}$ and $\Gamma\mathfrak{x} = \Delta\mathfrak{x}$, then for all $\mathfrak{x}$ we also have $\mathsf{A}\Gamma\mathfrak{x} = \mathsf{B}\Delta\mathfrak{x}$.

Proof. By hypothesis $\Gamma\mathfrak{x} = \Delta\mathfrak{x}$. Thus it follows from (Lv) that $\mathsf{A}(\Gamma\mathfrak{x}) = \mathsf{A}(\Delta\mathfrak{x})$ and then from $\mathsf{A}\mathfrak{x} = \mathsf{B}\mathfrak{x}$ that $\mathsf{A}(\Delta\mathfrak{x}) = \mathsf{B}(\Delta\mathfrak{x})$ for all $\mathfrak{x}$, so that the assertion is true by transitivity of equality.

6) The associative law:

$$\mathsf{A}(\mathsf{B}\Gamma) = (\mathsf{A}\mathsf{B})\,\Gamma.$$

This law holds generally for transformations with successive application as the rule for multiplication (see IB2, §1.2.5.).

7) The distributive laws:

$$\mathsf{A}(\mathsf{B} + \Gamma) = \mathsf{A}\mathsf{B} + \mathsf{A}\Gamma; \qquad (\mathsf{A} + \mathsf{B})\,\Gamma = \mathsf{A}\Gamma + \mathsf{B}\Gamma.$$

The proofs always depend on the same fundamental idea.

The existence of sum and product is in every case guaranteed if we consider only the set of linear transformations of V_n into itself:

Theorem 2: *The linear transformations of a vector space V_n into itself form a ring with respect to the addition* (2) *and the multiplication* (4).

Corollary: This ring has a unit element (for rings, also called unity element), namely the transformation E with $\mathsf{E}\mathfrak{x} = \mathfrak{x}$. (For the concept of a ring see IB1, §2.4, and IB5, §1.2.)

Under what circumstances does there exist, for a transformation with $\mathsf{A}(V_n) = \tilde{V}_r \subseteq \tilde{V}_k$, an inverse transformation $\bar{\mathsf{A}}$ such that $\bar{\mathsf{A}}(\tilde{V}_r) = V_n$? Since the dimension of $\bar{\mathsf{A}}(\tilde{V}_r)$ is smaller than or equal to r, and on the other hand $r \leqslant n$, we must have $r = n$.

But this necessary condition is also sufficient, as can be shown, for example, in the following way: Let $(\mathfrak{e}_\nu)$ be a basis of V_n, so that the images $\mathsf{A}\mathfrak{e}_\nu = \tilde{\mathfrak{a}}_\nu$ form a basis of $\tilde{V}_n$. The desired inverse transformation is then given by $\bar{\mathsf{A}}\tilde{\mathfrak{a}}_\nu = \mathfrak{e}_\nu$, as the reader may easily show. (The argument is similar to the one at the beginning of §2.2; cf. also the end of §2.3.)

By the definition of $\bar{\mathsf{A}}$ it follows that $\bar{\mathsf{A}}\mathsf{A} = \mathsf{E}$, so that $\bar{\mathsf{A}}$ is a left inverse of A. But if we start from the basis $(\tilde{\mathfrak{a}}_\nu)$ in $\tilde{V}_n$, we see that $\mathsf{A}\bar{\mathsf{A}} = \mathsf{E}$, so that $\bar{\mathsf{A}}$ is also a right inverse of A.

But $\bar{\mathsf{A}}$ is uniquely determined by A; for if $\bar{\mathsf{A}}\mathsf{A} = \mathsf{E}$ and $\bar{\mathsf{B}}\mathsf{A} = \mathsf{E}$, then by multiplying the second equation on the right with $\bar{\mathsf{A}}$ we see that $\bar{\mathsf{A}} = \bar{\mathsf{B}}$. Thus we may call $\bar{\mathsf{A}}$ *the* inverse mapping for A, and denote it by A^{-1}. The transformation A^{-1} itself has an inverse (since it is of rank n) and in fact $(\mathsf{A}^{-1})^{-1} = \mathsf{A}$.

Thus, *a transformation* A *is invertible* (*i.e., has an inverse*) *if and only if its rank, i.e., the dimension of the image space, is equal to the dimension of the original space.*

2.2. *Matrices*

Let us now consider the representation of linear transformations in terms of a basis $(\mathfrak{e}_\nu)$ of V_n and a basis $(\tilde{\mathfrak{e}}_\kappa)$ of $\tilde{V}_k$. From (La) and (Lm) it follows that

$$\mathsf{A}\mathfrak{x} = \mathsf{A}(\mathfrak{e}_\nu x^\nu) = (\mathsf{A}\mathfrak{e}_\nu)\,x^\nu.$$

Consequently, *after choice of a basis* $(\mathfrak{e}_\nu)$ *of* V_n *the transformation* A *is completely determined if the images of basis elements*

$$\mathsf{A}\mathfrak{e}_\nu = \tilde{\mathfrak{a}}_\nu$$

are given.

In terms of a basis of $\tilde{V}_k$ we have $\tilde{\mathfrak{a}}_\nu = \tilde{\mathfrak{e}}_\kappa a_\nu{}^\kappa$.

After choice of a basis $(\mathfrak{e}_\nu)$ *of* V_n *and a basis* $(\tilde{\mathfrak{e}}_\kappa)$ *of* $\tilde{V}_k$ *the transformation* A *is completely determined by the "rectangular array" of* $n \cdot k$ *elements of* S

$$\mathfrak{A} = \begin{pmatrix} a_1{}^1, \ldots, a_n{}^1 \\ \cdot\;, \ldots, \;\cdot \\ a_1{}^k, \ldots, a_n{}^k \end{pmatrix} = (a_\nu{}^\kappa)_{\substack{\kappa=1,\ldots,k \\ \nu=1,\ldots,n}}.$$

An array of this sort is called an $n \times k$ (n by k) *matrix*.

Here the superscript outside the bracket indicates the row, and the subscript indicates the column. The notation $(a_{\kappa\nu})_{\kappa=1,\ldots,k;\nu=1,\ldots,n}$ for matrices is also common; in this case the first index indicates the row. If the position and range of values of κ, ν are clear, we write $\mathfrak{A} = (a_\nu{}^\kappa)$.

The expression "rectangular array" means simply that to every pair of numbers (κ, ν) there is assigned an element $a_\nu{}^\kappa$ of the domain of scalars. Thus the matrix is a mapping of the pairs of numbers into the domain of scalars. By the general definition for equality of mappings (see IA, §8.4) we thus have

(1) $$\mathfrak{A} = \mathfrak{B} \quad \text{if and only if} \quad a_\nu{}^\kappa = b_\nu{}^\kappa \quad \text{for all} \quad \kappa, \nu.$$

If we set $\mathsf{A}\mathfrak{x} = \mathfrak{y} = \tilde{\mathfrak{e}}_k y^k$, then

$$\tilde{\mathfrak{e}}_k y^k = \mathsf{A}\mathfrak{x} = \tilde{\mathfrak{a}}_\nu x^\nu = \tilde{\mathfrak{e}}_k a_\nu{}^k x^\nu.$$

Since the $\tilde{\mathfrak{e}}_k$ are linearly independent, the transformation can also be represented by the system of equations

(2) $$y^\kappa = a_\nu{}^\kappa x^\nu.$$

It was from such a system that we started out in the first place, and now we have shown that every linear transformation of V_n can be represented in this way.

Let us now examine the question of uniqueness. Let $\mathfrak{A} = (a_\nu{}^\kappa)$ and $\mathfrak{B} = (b_\nu{}^\kappa)$ be the matrices assigned in a given coordinate system to the transformations A, B. It is obvious that if $a_\nu{}^\kappa = b_\nu{}^\kappa$ for all κ, ν, then $\mathsf{A}\mathfrak{x} = \mathsf{B}\mathfrak{x}$ for all $\mathfrak{x}$; that is, $\mathsf{A} = \mathsf{B}$.

Conversely, for a pair of indices i, j assume $a_j{}^i \neq b_j{}^i$; then there certainly exists a vector $\mathfrak{x}$ for which $\mathsf{A}\mathfrak{x} \neq \mathsf{B}\mathfrak{x}$, for example, the vector $\mathfrak{e}_j$ with coordinates δ_j^{ν}. By (2) its image vectors have the coordinates

$$y^{\kappa} = a_{\nu}{}^{\kappa}\delta_j{}^{\nu} = a_j{}^{\kappa} \qquad \text{and} \quad z^{\kappa} = b_{\nu}{}^{\kappa}\delta_j{}^{\nu} = b_j{}^{\kappa},$$

which differ from one another in the ith coordinate.

Thus $\mathsf{A} = \mathsf{B}$ if and only if $a_{\nu}{}^{\kappa} = b_{\nu}{}^{\kappa}$ for all κ, ν. Then (1) shows that $\mathsf{A} = \mathsf{B}$ if and only if $\mathfrak{A} = \mathfrak{B}$.

For matrices we now wish to introduce the rules of computation which will correspond to those already introduced for transformations in such a way that the resulting "configurations" (cf. IB10, §1.2) for the matrices are isomorphic to the corresponding configurations for the transformations.

So we ask: which matrix will correspond to the sum $\mathsf{A} + \mathsf{B} = \Gamma$? By §2.1 (2) we have $\Gamma\mathfrak{x} = \mathsf{A}\mathfrak{x} + \mathsf{B}\mathfrak{x}$, so that, if matrices are denoted by the corresponding letters:

$$c_{\nu}{}^{\kappa}x^{\nu} = a_{\nu}{}^{\kappa}x^{\nu} + b_{\nu}{}^{\kappa}x^{\nu} = (a_{\nu}{}^{\kappa} + b_{\nu}{}^{\kappa})\, x^{\nu}.$$

This system of equations is satisfied for all possible vectors (x^{ν}) if and only if

$$c_{\nu}{}^{\kappa} = a_{\nu}{}^{\kappa} + b_{\nu}{}^{\kappa} \qquad \text{for all} \quad \kappa, \nu.$$

Thus we define the *addition of two matrices* by

$$(3) \qquad (a_{\nu}{}^{\kappa}) + (b_{\nu}{}^{\kappa}) = (a_{\nu}{}^{\kappa} + b_{\nu}{}^{\kappa}).$$

By an analogous argument we are led to define *left-multiplication of a matrix by an element of S* by setting

$$(4) \qquad s \cdot (a_{\nu}{}^{\kappa}) = (s \cdot a_{\nu}{}^{\kappa}).$$

Then the $n \times k$ matrices form a vector space isomorphic to $L_n{}^k$, whose zero element is given by the matrix $\mathfrak{O}$ with $a_{\nu}{}^{\kappa} = 0$ for all κ, ν. For this vector space it is easy to assign a basis, namely, the matrices $\mathfrak{E}_j{}^i$ with 1 in the position $x_i{}^j$ (note that the indices are interchanged) and zero elsewhere. Then every matrix $\mathfrak{A}$ can be represented in the form

$$\mathfrak{A} = (a_{\nu}{}^{\kappa}) = a_{\nu}{}^{\kappa}\mathfrak{E}_{\kappa}{}^{\nu},$$

where the $\mathfrak{E}_j{}^i$ are linearly independent, since (1) implies that $a_{\nu}{}^{\kappa}\mathfrak{E}_{\kappa}{}^{\nu} = (a_{\nu}{}^{\kappa}) = \mathfrak{O}$ if and only if $a_{\nu}{}^{\kappa} = 0$. Thus the remaining part of Theorem 1, *$L_n{}^k$ has dimension $n \cdot k$*, is proved if we show that isomorphic vector spaces have the same dimension, a detail which we leave to the reader.

It is also possible to show without the use of matrices that the transformations $\mathsf{E}_j{}^i$ (corresponding to the matrices $\mathfrak{E}_j{}^i$) form a basis, where the $\mathsf{E}_j{}^i$ are defined by

$$\mathsf{E}_j{}^i\mathfrak{e}_i = \tilde{\mathfrak{e}}_j\,, \qquad \mathsf{E}_j{}^i\mathfrak{e}_l = \tilde{\mathfrak{o}} \qquad \text{for} \quad l \neq i;$$

note that summation is not taken over the Latin index i.

Matrix multiplication: the transformation $\mathsf{A}(\mathsf{B}\mathfrak{x})$ is represented in the matrix notation by $y^\lambda = a_\kappa{}^\lambda(b_\nu{}^\kappa x^\nu)$. Thus the matrix product $\mathfrak{A}\mathfrak{B} = \mathfrak{C}$ is to be defined by

$$(5) \qquad (c_\nu{}^\lambda) = (a_\kappa{}^\lambda b_\nu{}^\kappa).$$

The element $c_\nu{}^\lambda$ is formed by multiplying the elements of the λth row of $\mathfrak{A}$ (as left factors) with those of the νth column of $\mathfrak{B}$ and adding the products. Thus the product exists only if the number of elements in a row (i.e., the number of columns) of $\mathfrak{A}$ is the same as the number of elements in a column (i.e., the number of rows) of $\mathfrak{B}$.

The isomorphism between matrices and transformations shows that the same rules of calculation hold here as for transformations.

The following example shows that multiplication of matrices, and therefore multiplication of transformations, is not commutative:

$$\begin{pmatrix} 0, & 1 \\ -1, & 0 \end{pmatrix}\begin{pmatrix} 0, & -1 \\ -1, & 0 \end{pmatrix} = \begin{pmatrix} -1, & 0 \\ 0, & 1 \end{pmatrix}$$

$$\begin{pmatrix} 0, & -1 \\ -1, & 0 \end{pmatrix}\begin{pmatrix} 0, & 1 \\ -1, & 0 \end{pmatrix} = \begin{pmatrix} 1, & 0 \\ 0, & -1 \end{pmatrix}.$$

But calculation with matrices can be also defined by (1), (3), (4), (5) independently of the linear transformations, whereupon the rules for calculation can be verified by direct computation. Then the right of (2) can be interpreted as a product of matrices, where $(x^\nu) = \mathfrak{x}$ and $(y^\kappa) = \tilde{\mathfrak{y}}$ are matrices consisting of a single column. In place of (2) we then write

$$(2') \qquad \tilde{\mathfrak{y}} = \mathfrak{A}\mathfrak{x}.$$

Thus the equation §1.5, (4*a*) would be written in the form $\mathfrak{x}' = \mathfrak{T}\mathfrak{x}$. Here of course the kernel-index notation must be given up.

Since the matrices corresponding to transformations of V_n into itself are the square $n \times n$ matrices, they form a ring, a fact which again can be verified by actual computation without any reference to the theory of linear transformations. The unit element in this ring is the unit matrix

$$\mathfrak{E} = (\delta_\nu{}^\kappa) = \begin{pmatrix} 1, & 0, \ldots, & 0 \\ 0, & 1, \ldots, & 0 \\ \cdot & \cdot \quad \cdot & \cdot \\ 0, \ldots, & 0, & 1 \end{pmatrix}.$$

If the transformation A has an inverse transformation $\bar{\mathsf{A}}$, then the matrix $\mathfrak{A}$ has an inverse matrix $\bar{\mathfrak{A}}$, which by the remarks at the end of §2.1 is seen to be both a left and a right inverse. A matrix $\mathfrak{A}$ of this sort, for which there exists an $\bar{\mathfrak{A}}$ with $\mathfrak{A}\bar{\mathfrak{A}} = \bar{\mathfrak{A}}\mathfrak{A} = \mathfrak{E}$, is called *regular*. Comparison with the equations §1.5, (2) shows that a given matrix can be the matrix of coefficients of a transformation of basis if and only if it is regular.

Not every matrix is regular; for example,

$$\begin{pmatrix} 1, & 0 \\ 0, & 0 \end{pmatrix}\begin{pmatrix} u, & v \\ x, & y \end{pmatrix} = \begin{pmatrix} u, & v \\ 0, & 0 \end{pmatrix}$$

is certainly different from the unit matrix, no matter how the u, v, x, y are chosen.

If we set $u = v = 0$ and say $x = y = 1$, we see that the matrices $\begin{pmatrix} 1, 0 \\ 0, 0 \end{pmatrix}$ and $\begin{pmatrix} 0, 0 \\ 1, 1 \end{pmatrix}$ are divisors of zero. (Their product is zero, but neither factor is equal to zero.)

2.3. *Rank and Transformation of Basis for Matrices*

Now that we have laid the foundations for calculation with matrices, let us discuss the following question:

1) How is the rank of the transformation indicated in the matrix? We return to the beginning of §2.2. Every vector of V_n is a linear combination of the $\mathfrak{e}_\nu$, so that every vector of $\tilde{V}_r$ is a linear combination of the $\tilde{\mathfrak{a}}_\nu$, which means that $\tilde{V}_r$ is spanned by the $\tilde{\mathfrak{a}}_\nu$. Thus by §1.4 a suitably chosen set of r of the vectors $\tilde{\mathfrak{a}}_\nu$ forms a basis of $\tilde{V}_r$; in other words: among the $\tilde{\mathfrak{a}}_\nu$ there exist r linearly independent vectors and every $r + 1$ of these vectors are linearly dependent. Now the coordinates of the $\tilde{\mathfrak{a}}_\nu$ form the columns of the matrix $\mathfrak{A}$. The maximal number of linearly independent column vectors is called the *column rank* of $\mathfrak{A}$, so that we have the theorem: *the column rank of the matrix* $\mathfrak{A}$ *is equal to the rank of the transformation* A.

2) How does the matrix $\mathfrak{A}$ representing a transformation A change with a change of basis?

Let the transformation A be represented with respect to the bases $(\mathfrak{e}_\nu)$, $(\tilde{\mathfrak{e}}_\kappa)$ by the matrix $\mathfrak{A} = (a_\nu{}^\kappa)$, and let the image of the vector $\mathfrak{x}$ be

$$\mathsf{A}\mathfrak{x} = \mathfrak{y}.$$

(We write $\mathfrak{y}$ here instead of $\tilde{\mathfrak{x}}$ in order to avoid having too many diacritical marks on the same letter.)

By §2.1, (1) the coordinates of this image vector are

(1) $$y^\kappa = a_\nu{}^\kappa x^\nu, \qquad \text{or in matrix notation:} \quad \mathfrak{y} = \mathfrak{A}\mathfrak{x}.$$

Here we consider a vector as a matrix with one column, whose elements are the coordinates of the vector with respect to the given basis.

We now make a transformation of basis in the two vector spaces:

$$x^{\nu} = t^{\nu}_{\mu'} x^{\mu'} \qquad \text{or} \quad \mathfrak{x} = \mathfrak{T}\mathfrak{x}',$$

$$y^{\kappa} = s^{\kappa}_{\lambda'} y^{\lambda'} \qquad \text{or} \quad \mathfrak{y} = \mathfrak{S}\mathfrak{y}.$$

Then from (1), by multiplication with $(s^{\rho'}_{\kappa}) = \mathfrak{S}^{-1}$, we obtain

$$y^{\rho'} = s^{\rho'}_{\kappa} a_{\nu}{}^{\kappa} t^{\nu}_{\mu'} x^{\mu'} \qquad \text{or} \quad \mathfrak{y}' = \mathfrak{S}^{-1}\mathfrak{A}\mathfrak{T}\mathfrak{x}'.$$

Thus, the transformation A *is now represented, in terms of the new bases, by the matrix*

$$\mathfrak{A}' = \mathfrak{S}^{-1}\mathfrak{A}\mathfrak{T}, \tag{2}$$

where $\mathfrak{S}$ *and* $\mathfrak{T}$ *are regular matrices.*

Definition: two matrices $\mathfrak{A}$, $\mathfrak{A}'$ that stand in the relation (2) to each other are said to be *equivalent*: $\mathfrak{A} \sim \mathfrak{A}'$.

Thus we have the result: if two matrices represent the same transformation, they are equivalent to each other. The converse is also true; let us state it in detail: if $\mathfrak{A}$ and $\mathfrak{A}'$ are equivalent matrices and if $\mathfrak{A}$ represents the transformation A with respect to the bases $(\mathfrak{e}_{\nu})$, $(\tilde{\mathfrak{e}}_{\kappa})$, then there exist bases $(\mathfrak{e}_{\mu})$, $(\tilde{\mathfrak{e}}_{\lambda})$ with respect to which the matrix $\mathfrak{A}'$ represents the transformation A.

It follows that equivalence of matrices is reflexive, symmetric and transitive, and is thus an equivalence relation (IA, §8.3 and 5), as can also be shown by actual computation from (2).

3) Is it possible, by a suitable choice of bases, to represent a given transformation in a particularly illuminating way?

We had $\mathsf{A}\mathfrak{e}_{\nu} = \tilde{\mathfrak{a}}_{\nu}$. By reindexing (if necessary) we can arrange that the $\tilde{\mathfrak{a}}_1, \ldots, \tilde{\mathfrak{a}}_r$ are linearly independent and that $\tilde{\mathfrak{a}}_{r+1}, \ldots, \tilde{\mathfrak{a}}_n$ are linearly dependent on these first r vectors. Thus by §1.4 we can choose a basis of $\tilde{V}_k$ such that $\tilde{\mathfrak{e}}_1 = \tilde{\mathfrak{a}}_1, \ldots, \tilde{\mathfrak{e}}_r = \tilde{\mathfrak{a}}_r$ (and of course $\tilde{\mathfrak{e}}_{r+1}, \ldots, \tilde{\mathfrak{e}}_k$ are linearly dependent on them). Then for all $\tilde{\mathfrak{a}}_{\nu}$ the coordinates $a_{\nu}{}^{\kappa} = 0$ for $\kappa > r$; thus $\mathfrak{A}$ has the form

$$\begin{array}{cccc} \tilde{\mathfrak{a}}_1, \ldots, & \tilde{\mathfrak{a}}_r, & \tilde{\mathfrak{a}}_{r+1}, \ldots, & \tilde{\mathfrak{a}}_n \\ \hline \end{array}$$

$$\begin{pmatrix} 1 & 0 & a^1_{r+1} & a_n{}^1 \\ \cdots & \cdots & \cdots & \cdots \\ 0 & 1 & a^r_{r+1} & a_n{}^r \\ 0 & 0 & 0 & 0 \\ \cdots & \cdots & \cdots & \cdots \\ 0 & 0 & 0 & 0 \end{pmatrix}$$

We now introduce a new basis for V_n as follows: if $a^1_{r+1} \neq 0$, we set

$$\mathfrak{e}_{(r+1)'} = \mathfrak{e}_{r+1} - \mathfrak{e}_1 \cdot (a^1_{r+1})^{-1}, \qquad \mathfrak{e}_{\nu'} = \mathfrak{e}_\nu \quad \text{for} \quad \nu \neq r+1.$$

Then

$$\tilde{\mathfrak{a}}_{(r+1)'} = \mathsf{A}\mathfrak{e}_{(r+1)'} = \tilde{\mathfrak{a}}_{r+1} - \tilde{\mathfrak{a}}_1 (a^1_{r+1})^{-1}$$

and thus

$$a^1_{(r+1)'} = 0.$$

It is clear that repetition of this procedure must lead to a new basis in V_n with respect to which the transformation A is represented by the following matrix:

$$\mathfrak{E}_r = \left(\begin{array}{ccc|c} 1 & & \overset{r}{\overline{0}} & \\ & \ddots & & 0 \\ 0 & & 1 & \\ \hline & 0 & & 0 \end{array}\right) r.$$

Thus we have obtained the theorem: *every matrix of rank r is equivalent* to the matrix $\mathfrak{E}_r$. Consequently, if two matrices have the same rank, they are equivalent. Conversely, it is obvious that equivalent matrices have the same rank. Thus equivalent matrices are characterized by their rank alone.

In saying "rank" here instead of "column rank" we are anticipating a result at the end of §2.4.

With respect to the new bases the transformation A is represented by the following equations:

(3) $$y^1 = x^1, \ldots, y^r = x^r, \qquad y^{r+1} = 0, \ldots, y^n = 0.$$

If necessary, we may also consider a transformation of $\tilde{V}_r$ into V_n by means of $\tilde{\mathfrak{e}}_1 \to \mathfrak{e}_1, \ldots, \tilde{\mathfrak{e}}_r \to \mathfrak{e}_r$ and then say (for the moment, simply for the sake of visualization) that the transformation represented by (3) is a *projection.* Then *every linear transformation of a vector space V_n into a vector space is a projection of V_n onto an r-dimensional vector subspace, where r is the rank of* A. Conversely, it can be shown immediately that every projection is a linear transformation.

From (3) we see at once what was already proved in §2.1: if $r = n$, then the transformation is invertible.

2.4. *Systems of Linear Equations*

In the system of equations

(I) $$\sum_{\nu=1}^{n} a_\nu{}^\kappa x^\nu = b^\kappa \qquad (\kappa = 1, \ldots, k)$$

we consider, for every ν, the k-tuples $(a_\nu^1, ..., a_\nu^k) = \mathfrak{a}_\nu$ and $(b^1, ..., b^k) = \mathfrak{b}$ as vectors in a space V_k. In other words, for these k-tuples we introduce addition and S-multiplication as in §1.2. Then (I) can be written:

$$\text{(I}'\text{)} \qquad \mathfrak{a}_\nu x^\nu = \mathfrak{b},$$

and the question of solutions becomes simply whether, and in how many ways, the vector $\mathfrak{b}$ can be represented as a linear combination of the vectors $\mathfrak{a}_\nu$. The solutions, i.e., the n-tuples (x^ν), can be regarded in their turn as vectors of a (different) linear vector space W_n.

The following preliminary discussion is expressed in geometric terms, the $\mathfrak{a}_1, ..., \mathfrak{a}_n, \mathfrak{b}$ being thought of as vectors in a k-dimensional affine space. In this case, however, the W_n cannot be interpreted geometrically.

If the $\mathfrak{a}_1, ..., \mathfrak{a}_n$ already span the whole V_k, then every $\mathfrak{b} \in V_k$ can be represented as a linear combination of the $\mathfrak{a}_\nu$. If the $\mathfrak{a}_\nu$ form a basis, i.e., if $n = k$, this representation is unique, but if $n > k$, then it may be possible to choose a basis for V_k in various ways from the $\mathfrak{a}_1, ..., \mathfrak{a}_n$, so that the representation of $\mathfrak{b}$ will no longer be unique.

If the $\mathfrak{a}_1, ..., \mathfrak{a}_n$ span a proper subspace $V_l < V_k$, $l < k$, then only those $\mathfrak{b}$ are representable that belong to this subspace. Thus the possibility of a solution and the total number of solutions will depend on the dimension l of the subspace V_l' spanned by the $\mathfrak{a}_1, ..., \mathfrak{a}_n$.

We now repeat the theorems in §1:

If $\mathfrak{x} = (x^\nu)$, $\mathfrak{y} = (y^\nu)$ *are solutions of* (I), *then* $\mathfrak{x} - \mathfrak{y}$ *is a solution of*

$$\text{(II)} \qquad \sum_{\nu=1}^{n} a_\nu^\kappa x^\nu = 0 \qquad \text{or} \quad \mathfrak{a}_\nu x^\nu = \mathfrak{o}.$$

Thus we can find all solutions of (I) *by adding to a particular solution of* (I) *all solutions of* (II).

If $\mathfrak{x}, \mathfrak{y}$ *are solutions of* (II), *then* $\mathfrak{x} + \mathfrak{y}$ *and* $\mathfrak{x} \cdot s$ $(s \in S)$, *are also solutions of* (II).

Thus the solutions of (II) form a vector space W' which is a subspace of W_n. The question of the number of solutions of (II), and thus of (I), becomes a question of the dimension of W'. By the above preliminary discussion this dimension will depend on the dimension l of V_l', so that we must now bring this latter space into play.

We assume that the $a_1, ..., a_l$ are linearly independent and that $\mathfrak{a}_{l+1}, ..., \mathfrak{a}_n$ are linearly dependent on them. Of course, this assumption requires a reindexing of the $\mathfrak{a}_\nu$ and the x^ν, which can easily be reversed at the end of the calculation. Thus we assume that

$$\begin{aligned} \mathfrak{a}_{l+1} &= \mathfrak{a}_1 c_{l+1}^1 + \cdots + \mathfrak{a}_l c_{l+1}^l, \\ &\cdots\cdots\cdots \\ \mathfrak{a}_n &= \mathfrak{a}_1 c_n{}^1 + \cdots + \mathfrak{a}_l c_n{}^l. \end{aligned}$$

If these equations are substituted into (II), we obtain

$$\sum_{\lambda=1}^{l} \mathfrak{a}_\lambda \left(x^\lambda + \sum_{\mu=l+1}^{n} c_\mu{}^\lambda x^\mu\right) = \mathfrak{o}. \tag{1}$$

Since the $\mathfrak{a}_\lambda$ $(\lambda = 1, \ldots, l)$ are linearly independent, the system (1), and with it (II), is satisfied if and only if for every λ the system

$$x^\lambda + \sum_{\mu=l+1}^{n} c_\mu{}^\lambda x^\mu = 0 \tag{2}$$

is satisfied. But from the latter system we can at once read off all possible solutions as follows. If for $x^{l+1}, \ldots, x^n$ we insert arbitrary elements from S, the corresponding $x^1, \ldots, x^l$ can be calculated uniquely from (2). We thus obtain as a solution $n - l$ linearly independent vectors $\mathfrak{x}_1, \ldots, \mathfrak{x}_{n-l}$ and therewith a basis for W', if we set

$$(x_1^{l+1}, x_1^{l+2}, \ldots, x_1{}^n) = (1, 0, \ldots, 0)$$

$$\cdots\cdots\cdots\cdots$$

$$(x_{n-l}^{l+1}, x_{n-l}^{l+2}, \ldots, x_{n-l}^{n}) = (0, \ldots, 0, 1)$$

and then in each case calculate the corresponding $x_\rho{}^1, \ldots, x_\rho{}^n$ from (2). A solution with arbitrary values $x^{l+1}, \ldots, x^n$ is obtained as a linear combination $\mathfrak{x}_1 x^{l+1} + \mathfrak{x}_2 x^{l+2} + \cdots + \mathfrak{x}_{n-l} x^n$.

In this way we obtain the theorem: *if the vectors* $\mathfrak{a}_1, \ldots, \mathfrak{a}_n$ *span an* l*-dimensional subspace* $V_l' \subseteq V_k$, *then the solutions of* (II) *form an* $(n - l)$*-dimensional vector space* W_{n-l}'.

Thus the concept of a vector space is seen to be well adapted to the theory of systems of linear equations.

A basis of W_{n-l}' is called a *system of fundamental solutions*, or more briefly a *fundamental system*. The number l of linearly independent vectors among the $\mathfrak{a}_1, \ldots, \mathfrak{a}_n$ is the column rank of the matrix $(a_\nu{}^\kappa)$.

But how do we determine the l and the $c_\mu{}^\lambda$? By §1.4 we must decide which of the $\mathfrak{a}_\nu$ are linearly dependent on their predecessors. Of course, this actually means that we must solve systems of linear equations; thus we would simply be going around in circles, if it were not possible to determine the rank of a system of vectors in some other way. We shall return to this question in §3.6, but at present we follow another path: we recast the system of equations (I); i.e., we set up another system whose solutions are the same as those of (I) but are much easier to perceive.

What is wanted, of course, is a system of the form

$$\begin{aligned} x^1 \qquad\qquad &= q^1 \\ x^2 \qquad &= q^2 \\ \cdots\cdots \\ x^r &= q^r; \end{aligned}$$

that is, a system whose coefficient matrix has the form $\mathfrak{E}_r$. In §2.2 we have already spoken about transforming a matrix into such a form; let us now examine this question somewhat more closely from our present point of view. We must see what such transformations mean for our system of equations.

If $a_1{}^1 \neq 0$, let us subtract a suitable multiple of the first equation from the other equations, so that x^1 no longer occurs in them, repeating the same process with the second equation and with x^2, if $a_2{}^2 \neq 0$, and so forth. In order to have $a_1{}^1 \neq 0$ we change the order of the equations, if necessary, or else change the numbering of the x^ν. In order to obtain $a_i{}^i \neq 0$ at a later stage, it may sometimes be necessary to take both these steps.

We now wish to interpret these operations in our vector space. It is a matter of showing that they do not change the linear dependence or independence of the vectors $\mathfrak{a}_\nu$, $\mathfrak{b}$.

1) Renumbering the x_ν means renumbering the $\mathfrak{a}_\nu$, which produces no change in the relation of linear dependence.

2) Moreover, this relation is not changed by a change of basis. For example, the transformation of basis (for arbitrary fixed i, j)

$$\mathfrak{e}_{i'} = \mathfrak{e}_j, \qquad \mathfrak{e}_{j'} = \mathfrak{e}_i, \qquad \mathfrak{e}_{\nu'} = \mathfrak{e}_\nu \qquad \text{for} \quad \nu \neq i, j$$

merely interchanges the ith and jth equations.

3) Let us now suppose that $a_1{}^1 \neq 0$ has been brought about by (1) or (2).

Actually we should here write $a_1^{1'}$ or $a_{1'}^{1'}$, and in each of the successive steps we should adjoin one more prime, but the reader will allow us in each case to write only the prime arising from the next step and then, in the final equations, to write only one prime on the superscript and on the subscript.

By the transformation

$$\begin{aligned} \mathfrak{e}_{1'} &= \mathfrak{e}_1 + \mathfrak{e}_2 a_1{}^2 (a_1{}^1)^{-1} \\ \mathfrak{e}_{2'} &= \mathfrak{e}_2, \ldots, \mathfrak{e}_{n'} = \mathfrak{e}_n \end{aligned}$$

we obtain

$$\mathfrak{a}_\nu = \mathfrak{e}_{1'} a_\nu{}^1 + \mathfrak{e}_{2'}(a_\nu{}^2 - a_1{}^2(a_1{}^1)^{-1}\, a_\nu{}^1) + \mathfrak{e}_{3'} a_\nu{}^3 + \cdots;$$

i.e.,

$$a_\nu^{2'} = a_\nu{}^2 - a_1{}^2(a_1{}^1)^{-1}\, a_\nu{}^1, \qquad a_\nu^{\kappa'} = a_\nu{}^\kappa \qquad \text{for} \quad \kappa \neq 2,$$

so that in particular $a_1^{2'} = 0$.

With respect to the *coefficient matrix* $\mathfrak{A} = (a_\nu^\kappa)$, or to the *extended matrix*

$$\mathfrak{B} = \begin{pmatrix} a_1^1, \ldots, a_n^1, b^1 \\ \cdot\ \cdot\ \cdot\ \cdot\ \cdot\ \cdot\ \cdot \\ a_1^k, \ldots, a_n^k, b^k \end{pmatrix},$$

the operations admitted up to now consist of an interchange of columns, an interchange of rows, and the addition of a multiple of one row to another row. By repeated application of these rules we can, as a first step, bring the matrix $\mathfrak{A}$ into a matrix of the form

$$\mathfrak{A}' = \begin{pmatrix} a_{1'}^{1'}, & \cdot & \cdot & \cdot & \cdot & \cdot & \cdot \\ 0, & a_{2'}^{2'}, & \cdot & \cdot & \cdot & \cdot & \cdot \\ \cdot & \cdot & \cdot & \cdot & \cdot & \cdot & \cdot \\ 0, & \cdot & 0, & a_{l'}^{l'}, & \cdot & \cdot & \cdot \\ 0, & \cdot & \cdot & \cdot & \cdot & \cdot & 0, \\ 0, & \cdot & \cdot & \cdot & \cdot & \cdot & 0, \end{pmatrix},$$

called an *echelon matrix*. It is characterized by the fact that in the main diagonal $a_{1'}^{1'}, \ldots, a_{l'}^{l'} \neq 0$, whereas to the left of this diagonal and in the rows beginning with the $(l+1)$th there occur only zeros. The other elements may be arbitrary.

From this matrix, or the corresponding system of equations

$$\begin{aligned} a_{1'}^{1'}x^{1'} + a_{2'}^{1'}x^{2'} + \cdots + a_{n'}^{1'}x^{n'} &= 0 \\ a_{2'}^{2'}x^{2'} + \cdots + a_{n'}^{2'}x^{n'} &= 0 \\ a_{l'}^{l'}x^{l'} + \cdot + a_{n'}^{l'}x^{n'} &= 0, \end{aligned} \tag{3}$$

it is easy to see that the solutions can be calculated for an arbitrary choice of $x^{(l+1)'}, \ldots, x^{n'}$. Also, we can at once read off the column rank of the transformed matrix $(a_{\nu'}^{\kappa'}) = \mathfrak{A}'$, since it is equal to the number of nonzero elements in the main diagonal. Since our transformations have made no change in linear dependence, this number is also the column rank of $\mathfrak{A}$.

If we carry out the same operations on the extended matrix $\mathfrak{B}$, leaving the column (b^κ) unchanged, then instead of (3) we have a system of equations of the form

$$\begin{aligned} a_1^{1'}x^1 + \cdots + a_n^{1'}x^n &= b^{1'} \\ \cdot\ \cdot\ \cdot\ \cdot\ \cdot\ \cdot\ \cdot\ \cdot\ \cdot\ &\cdot\ \cdot\ \cdot \\ a_l^{l'}x^l + \cdots + a_n^{l'}x^n &= b^{l'} \\ 0 &= b^{(l+1)'} \\ \cdot\ \cdot\ &\cdot\ \cdot \\ 0 &= b^{k'}, \end{aligned}$$

which is solvable if and only if $b^{(l+1)'} = \cdots = b^{k'} = 0$. In this case the rank of $\mathfrak{B}'$, and thus also of $\mathfrak{B}$, is equal to the rank of $\mathfrak{A}$, since otherwise the column of the $b^{\kappa'}$ would provide, after an interchange of columns, a nonzero element in the main diagonal. Thus we have the theorem: *the system* (I) *is solvable if and only if the rank of the extended matrix is equal to the rank of the coefficient matrix.*

The above proof of this theorem contains at the same time a procedure for the numerical solution of a given system of equations which is more convenient in practice than, for example, the method of solution by means of determinants.

It remains to justify our use here of the word "rank" instead of "column rank" of a matrix. Among the admissible operations for the transformation of a matrix we have made no mention of the addition of a multiple of one column to another column. But here also the number of linearly independent column vectors remains unchanged, as is shown by the theorem: *if* $\mathfrak{a}_1, \ldots, \mathfrak{a}_n$ *are linearly dependent or linearly independent, then the same is true for* $\mathfrak{a}_1, \mathfrak{a}_2 + \mathfrak{a}_1 s, \mathfrak{a}_3, \ldots, \mathfrak{a}_n$. The proof is left to the reader.

Since it is clear that rows and columns of a matrix are on an equal footing with each other, we can also regard the rows as coordinates of vectors (in a space other than that of the column vectors), in which case the S-multiplication will be on the left. To determine the number of linearly independent row vectors we make use of the same operations as for the column vectors. Thus the resulting echelon matrix has the same number of nonzero elements in the main diagonal. This method shows that the row rank of a matrix is equal to the column rank, so that we may speak simply of the *rank* of the matrix.

2.5. *Transformation of a Matrix into Diagonal Form*

By the addition of multiples of columns to other columns we can bring the echelon matrix $\mathfrak{A}'$ into the diagonal form

$$\mathfrak{D} = \begin{pmatrix} d_1 & & & & & & & 0 \\ & d_2 & & & & & & \\ & & \cdot & & & & & \\ & & & \cdot & & & & \\ & & & & d_r & & & \\ & & & & & 0 & & \\ 0 & & & & & & \cdot & \\ & & & & & & & \cdot \\ & & & & & & & 0 \end{pmatrix}$$

(with zeros everywhere except on the main diagonal). In fact, exactly this process is carried out, in a somewhat indirect way, in the ordinary method of solution of a system of linear equations corresponding to an echelon matrix. A more important interpretation of the process is described in §2.2.

Our first remark is that the operations used above to transform the matrix $\mathfrak{A}$, namely:

interchange of two rows or columns,

addition of a multiple of a row or column to another row, or column,

can be effected by multiplication of $\mathfrak{A}$ with suitably chosen regular matrices.

For let $\mathfrak{U}_{i,j}$ be the matrix that arises from $\mathfrak{E}$ by interchange of the ith and jth rows (or what amounts to the same thing, of the ith and jth columns). Then the matrix $\mathfrak{A}\mathfrak{U}_{i,j}$ arises from $\mathfrak{A}$ by interchange of the ith and jth column, and $\mathfrak{U}_{i,j}\mathfrak{A}$ arises by interchange of the ith and jth rows.

If $\mathfrak{B}_j{}^i(q)$ is the matrix that arises from $\mathfrak{E}$ by adding the element q in the position $x_j{}^i$, then $\mathfrak{A}\mathfrak{B}_j{}^i(q)$ arises from $\mathfrak{A}$ by addition of the qth multiple of the ith column to the jth column, and $\mathfrak{B}_j{}^i(q)\,\mathfrak{A}$ arises by addition of the qth multiple of the jth row to the ith row.

The matrices $\mathfrak{U}$, $\mathfrak{B}$ are square, whereas $\mathfrak{A}$ may be rectangular, in which case the matrices used for right multiplication will not have the same number of rows as those used for left multiplication.

The matrices $\mathfrak{U}$, $\mathfrak{B}$ are regular, since they represent transformations of basis. Since the product of regular matrices is regular, we obtain the theorem: *a matrix $\mathfrak{A}$ can be transformed by multiplication with suitably chosen regular matrices $\mathfrak{S}$, $\mathfrak{T}$ into a diagonal matrix*

$$\mathfrak{D} = \mathfrak{S}\mathfrak{A}\mathfrak{T}.$$

Then by multiplication with regular matrices $\mathfrak{D}$ can be further transformed into $\mathfrak{E}_r$, by left-multiplication with $\mathfrak{E}$ and right-multiplication with the matrix that arises from $\mathfrak{E}$ through replacement of the first r diagonal element by $1/d_1, \ldots, 1/d_r$.

Since the inverse $\mathfrak{S}^{-1}$ of an arbitrary regular matrix $\mathfrak{S}$ is regular, we have obtained another proof for the theorem: every matrix of rank r is equivalent to the matrix $\mathfrak{E}_r$.

This partition into equivalence classes is rather coarse, since the possibilities for transforming matrices into one another are still extremely numerous. In what follows we shall restrict the transformations in various ways, examining only the simplest case in detail.

One restriction consists of regarding the vectors $\mathfrak{x}$ and $\mathfrak{y}$ as elements of the same space, so that the same transformation is applied to both of them. Matrices that are in the relation

$$\mathfrak{A}' = \mathfrak{T}^{-1}\mathfrak{A}\mathfrak{T} \tag{1}$$

are said to be *similar*, in which case the matrix $\mathfrak{A}$ must be square. In §3.7 we shall examine some invariants under similarity.

A further specialization refers to the admissible transformations. For example, the orthogonal transformations (see the following section and II.7) are important for geometry. We shall discuss a question of this sort in §3.7.

2.6. *Linear Forms*

In our discussion of systems of linear equations we have encountered three vector spaces: the space of solutions, the space of column vectors, and the space of row vectors. Multiplication with elements of S was on the left for row vectors, and on the right for column and solution vectors. Let us now examine the relationship between solution vectors and row vectors in the case of a single equation,

$$\sum_{\nu=1}^{n} a_\nu x^\nu = b.$$

Here we regard the x^ν as coordinates of a vector $\mathfrak{x}$ in a space V_n. If $a_1, ..., a_n$ are given elements of S, the mapping

$$\mathfrak{x} \to \langle \alpha, \mathfrak{x} \rangle = a_1 x^1 + \cdots + a_n x^n = a_\nu x^\nu \tag{1}$$

assigns to every element $\mathfrak{x}$ of V_n an element $\langle \alpha, \mathfrak{x} \rangle$ of S. This mapping α has the properties:

(Lv) From $\mathfrak{x} = \mathfrak{y}$ follows $\langle \alpha, \mathfrak{x} \rangle = \langle \alpha, \mathfrak{y} \rangle$,

(La) $\langle \alpha, \mathfrak{x} + \mathfrak{y} \rangle = \langle \alpha, \mathfrak{x} \rangle + \langle \alpha, \mathfrak{y} \rangle$,

(Lm) $\langle \alpha, \mathfrak{x}s \rangle = \langle \alpha, \mathfrak{x} \rangle \cdot s$,

and is thus a linear mapping of V_n into S, a special case of the mappings considered in §2.1. Here $k = 1$ and 1 is regarded as a basis of S (as a V_1). The matrix corresponding to the mapping α consists of a single row $(a_\nu{}^1) = (a_\nu)$; this matrix will also be denoted by the letter α.

The sign of equality in $\alpha = (a_\nu)$ and in $\mathfrak{x} = (x^\nu)$ refers to the representation of α and $\mathfrak{x}$ with respect to a given basis. As a sign of equality that is valid only with respect to a given basis, Schouten uses the symbol $\stackrel{*}{=}$. We consider it unnecessary to introduce any special symbol in our present context.

By §2.1 every linear mapping of V_n into S can be represented in the form (1). Such mappings are called *linear forms*.[2] They constitute an

[2] In IB5, §3.9, on the other hand, a linear form will mean a linear homogeneous polynomial.

n-dimensional vector space L_n with S as the domain of left multipliers, namely, the *space of linear forms* or the *dual space* of V_n, also called a *module of linear forms.*

For vectors as "one-dimensional" matrices we have used lower-case letters; so now in the same way we use lower-case (instead of capital) letters for the linear mappings of V_n into S. The notation $\langle \alpha, \mathfrak{x} \rangle$ is intended to emphasize the symmetry of the two "factors."

After choice of a basis $(\mathfrak{e}_\nu)$ for V_n and $\tilde{\mathfrak{e}}_1 = 1$ for S, the linear forms ϵ^ρ defined by $\langle \epsilon^\rho, \mathfrak{e}_\nu \rangle = \delta^\rho_\nu$ constitute a basis for L_n. These linear forms correspond to the mappings E^i_j (for $k = 1$).

Since the bases of V_n and L_n have been put in correspondence with each other in this way, a transformation of basis in V_n corresponds to a transformation in L_n. Let us examine the effect of such a transformation on the coordinates.

Let the image of $\mathfrak{x}$ under the linear mapping α be represented in terms of one basis of V_n by $\langle \alpha, \mathfrak{x} \rangle = a_\nu x^\nu$, and in terms of another by $\langle \alpha, \mathfrak{x} \rangle = a_{\nu'} x^{\nu'}$. Then, since the linear transformation is independent of the choice of basis, it follows that for a transformation which takes x^ν into

(2) $$x^{\nu'} = t^{\nu'}_{\mu} x^\mu,$$

the a_ν must be transformed in such a way that

$$a_{\nu'} x^{\nu'} = a_\nu x^\nu.$$

By substituting the inverse transformation $x^\nu = t^\nu_{\mu'} x^{\mu'}$ of (2) into this equation we obtain:

(3) $$a_{\mu'} x^{\mu'} = a_\nu t^\nu_{\mu'} x^{\mu'}.$$

Thus the transformation

(4) $$a_{\mu'} = a_\nu t^\nu_{\mu'}$$

produces the desired result, and no other transformation can do so, as is clear from the fact that (3) must hold for every vector $\mathfrak{x}$ and therefore in particular for $x^{1'} = 1$, $x^{2'} = \cdots = x^{n'} = 0$, and so forth.

In order to obtain the transformation matrix in (4) from the matrix in (2), we must first form the inverse matrix $(t^\nu_{\mu'})$ and then sum over the superscripts on the t's rather than over the subscripts.

The equations of the transformation (4) are the same as for the basis vectors:

(5) $$\mathfrak{e}_{\mu'} = \mathfrak{e}_\nu t^\nu_{\mu'},$$

so that the a_ν are said to transform *cogrediently* with the basis vectors, whereas the x^ν transform *contragrediently*; thus the linear forms $\alpha = (a_\nu)$ are called *covariant* vectors and the $\mathfrak{x} = (x^\nu)$ *contravariant*.

The advantage of using superscripts and subscripts is now clear, and the convention of summing over equal superscripts and subscripts is convenient because it takes an expression involving a covariant and a contravariant vector into a magnitude that is invariant (under transformation of coordinates).

The mapping $\alpha \to \langle \alpha, \mathfrak{x} \rangle$ may be regarded as a mapping, defined by $\mathfrak{x}$, of the dual vector space into the domain of scalars. More generally, we also consider mappings of $V_n \times V_n \times \cdots$ or $L_n \times L_n \times \cdots$, and so forth, into the domain of scalars. Let us describe the next simplest case. A mapping of the set of pairs of contravariant vectors into the domain of scalars

$$\mathfrak{x}, \mathfrak{y} \to c = \Gamma(\mathfrak{x}, \mathfrak{y})$$

is called a *bilinear form* (cf. page 268) if it is linear in each variable; in other words,

$$\text{(La)} \qquad \begin{aligned} \Gamma(\mathfrak{x}_1 + \mathfrak{x}_2, \mathfrak{y}) &= \Gamma(\mathfrak{x}_1, \mathfrak{y}) + \Gamma(\mathfrak{x}_2, \mathfrak{y}); \\ \Gamma(\mathfrak{x}, \mathfrak{y}_1 + \mathfrak{y}_2) &= \Gamma(\mathfrak{x}, \mathfrak{y}_1) + \Gamma(\mathfrak{x}, \mathfrak{y}_2), \end{aligned}$$

$$\text{(Lm)} \qquad \Gamma(s\mathfrak{x}, \mathfrak{y}) = s \cdot \Gamma(\mathfrak{x}, \mathfrak{y}); \qquad \Gamma(\mathfrak{x}, \mathfrak{y}s) = \Gamma(\mathfrak{x}, \mathfrak{y}) \cdot s.$$

Here we assume that $s\mathfrak{x} = \mathfrak{x}s$ and that S is commutative, although for the time being it would be sufficient to assume that V is both a left and right vector space. From (La), (Lm) it follows that for $\mathfrak{x} = x^\kappa \mathfrak{e}_\kappa$, $\mathfrak{y} = \mathfrak{e}_\lambda y^\lambda$:

$$\text{(6)} \qquad \Gamma(\mathfrak{x}, \mathfrak{y}) = x^\kappa \cdot \Gamma(\mathfrak{e}_\kappa, \mathfrak{e}_\lambda) \cdot y^\lambda,$$

and if we set $\Gamma(\mathfrak{e}_\kappa, \mathfrak{e}_\lambda) = g_{\kappa\lambda}$,

$$\text{(7)} \qquad \Gamma(\mathfrak{x}, \mathfrak{y}) = x^\kappa g_{\kappa\lambda} y^\kappa.$$

Under a transformation of coordinates §1.5 (1) the transformed values of the $g_{\kappa\lambda}$ will be such that

$$g_{\mu'\nu'} = \Gamma(\mathfrak{e}_{\mu'}, \mathfrak{e}_{\nu'}) = t^\kappa_{\mu'} g_{\kappa\lambda} t^\lambda_{\nu'}.$$

The bilinear form Γ is also called a *covariant tensor of second order* and the $g_{\kappa\lambda}$ are its coordinates.

In general, tensors can be defined as *multilinear forms*, i.e., as mappings of systems of covariant and contravariant vectors into the domain of scalars (cf. page 268) linear in each of the variables. For example,

$$\Delta(\mathfrak{x}, \alpha, \mathfrak{y}) = d_{\kappa.\mu}^{\ \lambda} x^\kappa a_\lambda y^\mu$$

defines a tensor of the third order which is covariant with respect to two indices and contravariant with respect to one. The coordinates of a tensor are the coefficients of a multilinear form; for example, under a transformation of coordinates we have

$$d_{\kappa'\cdot\mu'}^{\;\;\lambda'} = t^{\kappa}_{\kappa'} t^{\lambda'}_{\lambda} t^{\mu}_{\mu'} d_{\kappa\cdot\mu}^{\;\;\lambda}.$$

Let us now raise the question: do there exist transformations under which the coordinates of a covariant vector are transformed in the same way as those of a contravariant vector? Under such a transformation we must have for every vector $\mathfrak{x} = (\mathfrak{x}^{\kappa})$ not only

(8) $$x^{\kappa'} = t^{\kappa'}_{\lambda} x^{\lambda}$$

but also

(9) $$x^{\kappa'} = \sum_{\lambda} t^{\lambda}_{\kappa'} x^{\lambda}.$$

If we solve (8) for the x^{μ} and substitute in (9), we obtain

$$x^{\mu} = t^{\mu}_{\kappa'} x^{\kappa'} = \sum_{\kappa'} \sum_{\lambda} t^{\mu}_{\kappa'} t^{\lambda}_{\kappa'} x^{\lambda}.$$

By setting $(x^{\lambda}) = (1, 0, \ldots, 0)$, $(1, 1, 0, \ldots, 0)$, and so forth we obtain the *conditions of orthogonality*

$$\sum_{\kappa'} t^{\mu}_{\kappa'} t^{\lambda}_{\kappa'} = 0 \qquad \text{for} \quad \mu \neq \lambda$$

and of *normality*

$$\sum_{\kappa'} t^{\mu}_{\kappa'} t^{\mu}_{\kappa'} = 1,$$

or taken together

(O1) $$\sum_{\kappa'} t^{\mu}_{\kappa'} t^{\lambda}_{\kappa'} = \delta^{\mu\lambda}.$$

In the same way

(O2) $$\sum_{\lambda} t^{\lambda}_{\kappa'} t^{\lambda}_{\mu'} = \delta_{\kappa'\mu'}.$$

Conversely, let us assume (O) and (8). Then

$$x^{\lambda} = t^{\lambda}_{\mu'} x^{\mu'} \quad \text{[solution of (8)]},$$

$$\sum_{\lambda} t^{\lambda}_{\kappa'} x^{\lambda} = \sum_{\lambda} \sum_{\mu'} t^{\lambda}_{\kappa'} t^{\lambda}_{\mu'} x^{\mu'} = x^{\kappa'} \qquad \text{by (O), so that (9) holds.}$$

Thus we have the theorem: *for transformations satisfying the conditions* (O), *and only for such transformations, it is unnecessary to distinguish between covariant and contravariant vectors.*

These transformations are called *orthogonal.*

Let us repeat the argument in matrix notation. For this purpose we must first define the *transpose* $\mathfrak{A}^T = (\overset{T}{a_\kappa{}^\nu})$ of a given matrix $\mathfrak{A} = (a_\nu{}^\kappa)$, which is formed by interchanging rows and columns $\overset{T}{a_\kappa{}^\nu} = a_\nu{}^\kappa$ in $\mathfrak{A}$. In more detailed notation the transpose of the matrix $(a_\nu{}^\kappa)_{\nu=1,\ldots,n}^{\kappa=1,\ldots,k}$ is represented by $(a_\nu{}^\kappa)_{\kappa=1,\ldots,k}^{\nu=1,\ldots,n}$.

We leave to the reader the proof that

$$(10)\quad \mathfrak{A}^{TT} = \mathfrak{A}, \qquad \text{and if } S \text{ is commutative,} \qquad (\mathfrak{A}\mathfrak{B})^T = \mathfrak{B}^T\mathfrak{A}^T.$$

Then we interpret $a_\nu x^\nu$ as the matrix product $\mathfrak{a}^T \cdot \mathfrak{x}$, where $\mathfrak{x}$ and $\mathfrak{a}$ each consist of one column, so that $\mathfrak{a}^T$ consists of one row.

Now if $\mathfrak{x}$ is transformed into $\mathfrak{x}' = \mathfrak{T}\mathfrak{x}$, and if we are to have

$$\mathfrak{a}^T\mathfrak{x} = (\mathfrak{a}^T)'\mathfrak{x}' = \mathfrak{a}'^T\mathfrak{x}' = \mathfrak{a}'^T\mathfrak{T}\mathfrak{x},$$

then we must have $\mathfrak{a}^T = \mathfrak{a}'^T\mathfrak{T}$, so that we must set $\mathfrak{a}' = (\mathfrak{T}^{-1})^T\,\mathfrak{a}$; that is, $\mathfrak{a}$ must be transformed by the inverse transposed matrix, corresponding to the passage from (2) to (4) on page 263.

In order that the transformation for $\mathfrak{a}$ be the same as for $\mathfrak{x}$, we must have $\mathfrak{a}' = \mathfrak{T}\mathfrak{a}$, so that $(\mathfrak{T}^{-1})^T = \mathfrak{T}$, or $\mathfrak{T}^{-1} = \mathfrak{T}^T$. The equations

$$\mathfrak{T}^T\mathfrak{T} = \mathfrak{T}\mathfrak{T}^T = \mathfrak{E}$$

are the same as (O). Thus the conditions of orthogonality and normality state that the inverse matrix is the same as the transpose.

3. Products of Vectors

3.1. *General requirements*

The reader is already acquainted with several different products of vectors. We define, with respect to a given basis,

a) the *inner* or *scalar product* by

$$\mathfrak{x} \cdot \mathfrak{y} = x^1y^1 + \cdots + x^ny^n,$$

b) in a V_3 the *vector product* of two vectors by

$$\mathfrak{x} \times \mathfrak{y} = (x^2y^3 - x^3y^2,\, x^3y^1 - x^1y^3,\, x^1y^2 - x^2y^1),$$

c) in a V_2 the *complex product* by (cf. IB8, §1)

$$\mathfrak{x} \circ \mathfrak{y} = (x^1y^1 - x^2y^2,\, x^1y^2 + x^2y^1);$$

that is, if $\mathfrak{x} = x^1 + ix^2$, $\mathfrak{y} = y^1 + iy^2$, then

$$\mathfrak{x} \circ \mathfrak{y} = x^1y^1 - x^2y^2 + i(x^1y^2 + x^2y^1).$$

It is not customary to use any special symbol, like the $\circ$ here, since this product does not usually occur in the same context with the others.

Only the complex product satisfies all the rules for computation familiar from the multiplication of real numbers; but this complex product is possible only in a two-dimensional vector space.

In (*a*) the product of two vectors is not itself a vector; in (*b*) it is true that for every system of coordinates $\mathfrak{x} \times \mathfrak{y}$ is defined as a vector, but this product is not independent of the coordinate system. The desired independence can be attained only if we restrict ourselves to coordinate systems that arise from the original system by orthogonal transformations (see §2.6 and II, 7), or if we consider the product not as a vector but as a tensor of second order with the 9 coordinates $x^\mu y^\nu - x^\nu y^\mu$ $(\mu, \nu = 1, 2, 3)$. Neither (*a*) nor (*b*) satisfies the associative law, and in both cases a product can be equal to zero even though neither of the factors is equal to zero.

We now consider the problem of defining for vectors one or several operations that can reasonably be called "multiplication." What must be required of such an operation?

1) Multiplication will assign to two vectors a "product." We cannot require that the product belong to the same vector space as the factors, but we will require that it belong to some vector space over the same domain of scalars S; this requirement means only that for the objects which turn up as products of two vectors there is defined, or can be defined, an addition and an S-multiplication.

On the other hand, the "factors" may come from different vector spaces, although they must be over the same domain of scalars.

(P_1) By multiplication we shall mean a procedure which to each vector $\hat{\mathfrak{a}} \in \hat{V}_k(S)$ and each vector $\mathfrak{b} \in V_l(S)$ assigns exactly one vector $\mathfrak{c} = \Pi(\hat{\mathfrak{a}}, \mathfrak{b}) \in W_r(S)$, provided certain further requirements are met. Thus, a multiplication is a mapping of the set $\hat{V} \times V$ into W, where $\hat{V} \times V$ is the set of pairs $(\hat{\mathfrak{a}}, \mathfrak{b})$ with $\hat{\mathfrak{a}} \in \hat{V}_k(S)$, $\mathfrak{b} \in V_l(S)$. Included in this statement is the consistency of multiplication with equality:

(P_2) From $\hat{\mathfrak{a}} = \hat{\mathfrak{a}}'$ and $\mathfrak{b} = \mathfrak{b}'$ it follows that $\Pi(\hat{\mathfrak{a}}, \mathfrak{b}) = \Pi(\hat{\mathfrak{a}}', \mathfrak{b}')$.

2) If the product does not belong to the same vector space as the factors, we can hardly expect a straightforward associative law. Next in importance come the distributive laws:

$$\text{(La)} \quad \begin{aligned} \Pi(\hat{\mathfrak{a}}_1 + \hat{\mathfrak{a}}_2, \mathfrak{b}) &= \Pi(\hat{\mathfrak{a}}_1, \mathfrak{b}) + \Pi(\hat{\mathfrak{a}}_2, \mathfrak{b}), \\ \Pi(\hat{\mathfrak{a}}, \mathfrak{b}_1 + \mathfrak{b}_2) &= \Pi(\hat{\mathfrak{a}}, \mathfrak{b}_1) + \Pi(\hat{\mathfrak{a}}, \mathfrak{b}_2), \end{aligned}$$

which we shall require from a multiplication.

3) If in (La) we set $\hat{\mathfrak{a}}_1 = \hat{\mathfrak{a}}_2$ and $\mathfrak{b}_1 = \mathfrak{b}_2$, we obtain

$$\Pi(2\hat{\mathfrak{a}}, \mathfrak{b}) = 2\Pi(\hat{\mathfrak{a}}, \mathfrak{b}),$$
$$\Pi(\hat{\mathfrak{a}}, \mathfrak{b}2) = \Pi(\hat{\mathfrak{a}}, \mathfrak{b})\,2,$$

provided we assume that $\hat{V}$ is a left vector space, V a right vector space, and W both a left and right vector space over S. This assumption is satisfied if, for example, S is commutative and $s\mathfrak{a} = \mathfrak{a}s$ is defined in all three vector spaces; the assumption with respect to W is also satisfied for noncommutative S if $r = 1$ and $W = S$, which is sufficient for our present purposes. Thus we demand from a "multiplication" that the two rules written above for (2) shall hold for arbitrary s in S:

$$\text{(Lm)} \qquad \begin{aligned} \Pi(s\hat{\mathfrak{a}}, \mathfrak{b}) &= s\Pi(\hat{\mathfrak{a}}, \mathfrak{b}), \\ \Pi(\hat{\mathfrak{a}}, \mathfrak{b}s) &= \Pi(\hat{\mathfrak{a}}, \mathfrak{b})\, s. \end{aligned}$$

Then by (*P*), (La), (Lm) the multiplication is a mapping of $\hat{V} \times V$ into W which is linear with respect to each of the two factors; such a mapping is called *bilinear*, or in the case of several factors *multilinear*, and if $W = S$ is the domain of scalars, it is also called a *bilinear form* or a *multilinear form*.

For later use let us note that if $\hat{V} = V$, then to the bilinear form Π we can assign a *quadratic form*, namely the mapping $\mathfrak{a} \to \Pi(\mathfrak{a}, \mathfrak{a})$.

Thus we have set up the requirements that must be satisfied by an operation if it is to be called a "multiplication." But how are we to give a concrete definition of such a multiplication? We must state some rule for assigning a product to every pair of vectors. Now, a given vector can, on the one hand, be defined geometrically, and in this case we must give a geometric definition of multiplication. This problem will be dealt with in II,7.

On the other hand, a vector can also be defined by its coordinates, after choice of a basis. Then the rule for multiplication will determine the coordinates of the product from those of the factors and we must insure that the result is independent of the special choice of basis.

If we let $(\hat{\mathfrak{e}}_\kappa)$, $\kappa = 1, \ldots, k$ be a basis of $\hat{V}_k$ and $(\mathfrak{e}_\lambda)$, $\lambda = 1, \ldots, l$ a basis of V_l, it follows from (La) and (Lm), exactly as for linear mappings, that

$$\Pi(\hat{\mathfrak{a}}, \mathfrak{b}) = \Pi(a^\kappa \hat{\mathfrak{e}}_\kappa, \mathfrak{e}_\lambda b^\lambda) = a^\kappa \Pi(\hat{\mathfrak{e}}_\kappa, \mathfrak{e}_\lambda)\, b^\lambda.$$

Thus, a multiplication is completely defined by the products of the basis vectors. For abbreviation we write

$$\Pi(\hat{\mathfrak{e}}_\kappa, \mathfrak{e}_\lambda) = \Pi_{\kappa\lambda}.$$

These products belong to W, but this fact is not very helpful, at least not yet, because up to now we have said nothing about the space W. In fact, it will be necessary, at least to some extent, to construct W. But to do this in a suitable way we must first investigate the behavior of products under a transformation of basis.

For a given transformation

$$(1) \qquad \hat{\mathfrak{e}}_\kappa = \hat{t}^{\mu'}_\kappa \hat{\mathfrak{e}}_{\mu'} \qquad \mathfrak{e}_\lambda = \mathfrak{e}_{\nu'} t^{\nu'}_\lambda,$$

in which the coordinates a^κ, b^λ become

$$(2) \qquad a^{\mu'} = a^\kappa \hat{t}^{\mu'}_\kappa \qquad b^{\nu'} = t^{\nu'}_\lambda b^\lambda,$$

it will be necessary to define a transformation of the $\Pi_{\kappa\lambda}$ into $\Pi_{\mu'\nu'}$ in such a way that the product remains invariant; that is, for all vectors (a^κ), (b^λ) we must have

$$(3) \qquad a^\kappa \Pi_{\kappa\lambda} b^\lambda = a^{\mu'} \Pi_{\mu'\nu'} b^{\nu'}.$$

From (2) and (3) it follows that

$$(\mathrm{T}) \qquad \Pi_{\mu'\nu'} = \hat{t}^\kappa_{\mu'} \Pi_{\kappa\lambda} t^\lambda_{\nu'} .$$

In order for the product to remain invariant under transformation of basis, we must assign to the transformations (1) of $\hat{V}$ and V a transformation of the form (T) in W. In the following two sections we shall describe two possibilities, and in each case the existence of a "multiplication" with the required properties will be proved by our giving an explicit statement (in coordinates) of what the products are.

3.2. *The Inner or Scalar Product*

Let $\hat{V} = V$, and $W = S$, so that $r = 1$. If we write $g_{\kappa\lambda}$ in place of $\Pi_{\kappa\lambda}$, then (T) states that the $g_{\kappa\lambda}$ must be the coordinates of a covariant tensor of second order. Thus to define this multiplication we must first choose a basis $(\mathfrak{e}_\nu)$ in V and then choose arbitrary numbers $g_{\kappa\lambda}$. With respect to this basis the *inner* or *scalar* product is defined by

$$(1) \qquad \mathfrak{a}\mathfrak{b} = a^\kappa g_{\kappa\lambda} b^\lambda.$$

If we make a transformation of basis, then instead of the $g_{\kappa\lambda}$ we must use the numbers

$$(2) \qquad g_{\mu'\nu'} = t^\kappa_{\mu'} g_{\kappa\lambda} t^\lambda_{\nu'}$$

in order to form the product; that is, we must have

$$\mathfrak{a}\mathfrak{b} = a^{\mu'} g_{\mu'\nu'} b^{\nu'}.$$

It is obvious that the requirements (P), (La), (Lm) are satisfied by the product defined in (1).

If in a given vector space we have defined a covariant tensor of second order, namely a bilinear form, as the "fundamental tensor" or "fundamental form" and have thereby defined an inner product, we say that the vector space has a *metric structure*, or that it is a *metric space*, a name that is explained by the fact that the inner product can be used for the introduction of a metric. For if S is the field of real numbers and if the quadratic form $x^\kappa g_{\kappa\lambda} x^\lambda$ is positive definite, then the mapping d, defined by

$$d(\mathfrak{a}, \mathfrak{b}) = \sqrt{(b^\kappa - a^\kappa)\, g_{\kappa\lambda}(b^\lambda - a^\lambda)} = \sqrt{(\mathfrak{b} - \mathfrak{a})(\mathfrak{b} - \mathfrak{a})},$$

satisfies the requirements for a distance function.

Let us now suppose that for a given basis an inner product has been defined by the coordinates $g_{\kappa\lambda}$ of the fundamental tensor. We ask whether it is possible to choose a new basis in such a way that the representation of this product will become especially simple; for example, the matrix $(g_{\mu'\nu'})$ resulting from the transformation (2) will be a diagonal matrix, or if possible the unit matrix.

In matrix notation, (2) becomes

$$\mathfrak{G}' = \mathfrak{T}^T \mathfrak{G} \mathfrak{T}. \qquad (2')$$

Two matrices $\mathfrak{G}$, $\mathfrak{G}'$ related to each other in this way are said to be *congruent*. The difference between similarity (§2.5 (1)) and congruence consists in the fact that for similarity the matrix $\mathfrak{A}$ represents a mixed tensor, covariant with respect to one index and contravariant with respect to the other, whereas for congruence the matrix represents a doubly covariant tensor.

If $\mathfrak{G}'$ is to be a diagonal matrix, then it must at least be symmetric; i.e., $\mathfrak{G}'^T = \mathfrak{G}'$. *If S is commutative, as we shall assume from now on, then*

$$\mathfrak{G}'^T = \mathfrak{T}^T \mathfrak{G}^T \mathfrak{T} \qquad \text{(cf. §2.6 (10))}.$$

By multiplication with $(\mathfrak{T}^T)^{-1}$ and $\mathfrak{T}^{-1}$ we see that $\mathfrak{G}'^T = \mathfrak{G}'$ if and only if $\mathfrak{G}^T = \mathfrak{G}$; in other words, the symmetry of a matrix remains unchanged by transformation to a congruent matrix.

We now assume that $\mathfrak{G}$ is *symmetric*. Then the transformation to a diagonal matrix can be effected, exactly as in §2.5, by multiplication with matrices $\mathfrak{U}$, $\mathfrak{B}$; for by right-multiplication with $\mathfrak{U}$ and $\mathfrak{B}$, or by left-multiplication with $\mathfrak{U}^T$ and $\mathfrak{B}^T$, we perform exactly the same operations on the rows as on the columns, as the reader may easily verify.

If S is the *field of complex numbers*, it is appropriate to introduce another concept: for the matrix $\mathfrak{G} = (g_{\kappa\lambda})$ we define the *conjugate transposed matrix* $\mathfrak{G}^* = (g^*_{\kappa\lambda})$ by

$$g^*_{\kappa\lambda} = \overline{g_{\lambda\kappa}}; \qquad \mathfrak{G}^* = \overline{\mathfrak{G}}^T = \overline{\mathfrak{G}^T},$$

where the bars denote complex conjugates. From the one-column matrix $\mathfrak{x}$ we obtain the one-row matrix $\mathfrak{x}^*$ with the elements $x^{*\lambda} = \overline{x^\lambda}$.

A matrix with the property

$$\mathfrak{G}^* = \mathfrak{G}, \qquad \text{i.e.,} \quad g_{\kappa\lambda} = \overline{g_{\lambda\kappa}}$$

is called a Hermitian matrix. The diagonal elements of a Hermitian matrix are real ($g_{\kappa\kappa} = \overline{g_{\kappa\kappa}}$).

If $\mathfrak{G}$ is a Hermitian matrix, the mapping $(\mathfrak{x}, \mathfrak{y}) \to \mathfrak{x}^*\mathfrak{G}\mathfrak{y} = \overline{x^\kappa} g_{\kappa\lambda} y^\lambda$ is called a *Hermitian bilinear form*, and the mapping $\mathfrak{x} \to \mathfrak{x}^*\mathfrak{G}\mathfrak{x}$ is a *Hermitian form.*

Under the coordinate transformation $\mathfrak{x} = \mathfrak{T}\mathfrak{x}'$, the matrix $\mathfrak{G}$ becomes $\mathfrak{G}' = \mathfrak{T}^*\mathfrak{G}\mathfrak{T}$, so that a Hermitian matrix goes into a Hermitian matrix ($\mathfrak{G}'^* = \mathfrak{G}'$ if and only if $\mathfrak{G}^* = \mathfrak{G}$). Thus the property of being Hermitian is independent of the choice of basis.

If in a Hermitian bilinear form we take the basis vectors as arguments, we obtain, in view of $\mathfrak{e}_i = (\delta_i{}^\nu)$,

$$\mathfrak{e}_i^*\mathfrak{G}\mathfrak{e}_k = g_{ik} ,$$

or in other words exactly the coefficients of the bilinear form. This result holds for any basis.

The values of a Hermitian form are real; for if $\bar{\mathfrak{x}}^T\mathfrak{G}\mathfrak{x} = w$, then $\bar{w} = \mathfrak{x}^T\overline{\mathfrak{G}}\bar{\mathfrak{x}}$. Now $\bar{w}$ is a number, i.e., a one-row matrix, so that $\bar{w} = \bar{w}^T = \bar{\mathfrak{x}}^T\overline{\mathfrak{G}}^T\mathfrak{x}$, and therefore, since $\overline{\mathfrak{G}}^T = \mathfrak{G}$, we have $\bar{w} = w$.

This statement does not hold for quadratic forms with complex argument, a fact which explains why Hermitian forms are the appropriate ones in our present discussion.

We now define an inner product by

(3) $$\Pi(\mathfrak{a}, \mathfrak{b}) = \mathfrak{a}^*\mathfrak{G}\mathfrak{b} = \overline{a^\kappa} g_{\kappa\lambda} b^\lambda.$$

In order that this definition may be independent of the choice of basis, the matrix $\mathfrak{G}$ must be transformed according to the rule

(4) $$\mathfrak{G}' = \mathfrak{T}^*\mathfrak{G}\mathfrak{T} \qquad \text{or} \quad g_{\mu'\nu'} = \overline{t^\kappa_{\mu'}} g_{\kappa\lambda} t^\lambda_{\nu'} .$$

If we restrict S to the field of real numbers, we obtain our earlier result: a real Hermitian matrix is a symmetric matrix, and a real Hermitian form is a quadratic form, so that (3), (4) become (1), (2).

A Hermitian matrix is taken into a Hermitian matrix by the transformation (4), and can be transformed, in the same way as a real symmetric matrix, into a (real) diagonal matrix. We omit the proof here, since the next section proves a sharper result.

Moreover, by transformations of the form

$$\mathfrak{D}' = \mathfrak{S}^*\mathfrak{D}\mathfrak{S} \tag{5}$$

with regular matrix $\mathfrak{S}$, the diagonal elements of a real diagonal matrix can be transformed into ± 1; for example:

$$\begin{bmatrix} \frac{1}{\sqrt{|d|}} & & & \\ & 1 & & \\ & & \ddots & \\ & & & 1 \end{bmatrix} \begin{bmatrix} d & & & \\ & \cdot & & \\ & & \ddots & \\ & & & \cdot \end{bmatrix} \begin{bmatrix} \frac{1}{\sqrt{|d|}} & & & \\ & 1 & & \\ & & \ddots & \\ & & & 1 \end{bmatrix} = \begin{bmatrix} \frac{d}{|d|} & & & \\ & \cdot & & \\ & & \ddots & \\ & & & \cdot \end{bmatrix}$$

But there are other possibilities; for instance, we could also make the transformation

$$x^1 = x^{2'}, \qquad x^2 = x^{1'}.$$

It is an important fact that for all such transformations the number of positive and negative terms remains the same (Sylvester's law of inertia).

The proof of this law is as follows. Transformation (5) does not change the rank r of the matrix (in view of this invariance, r is also called the rank of the form represented by the matrix), so that the number of nonzero diagonal elements remains the same in every diagonal representation. So let us assume that the Hermitian form has been brought by two transformations (5) into the forms

$$\begin{aligned} x^1x^1 + \cdots + x^px^p - x^{p+1}x^{p+1} - \cdots - x^rx^r \\ = y^1y^1 + \cdots + y^qy^q - y^{q+1}y^{q+1} - \cdots - y^ry^r, \end{aligned} \tag{6}$$

where the x^λ and y^ν are related by $y^\nu = s_\lambda{}^\nu x^\lambda$, and we shall suppose that $p > q$. Then the system of equations

$$\begin{aligned} y^i = s_\lambda{}^i x^\lambda &= 0 \qquad & i &= 1, \ldots, q \\ x^j &= 0 & j &= p+1, \ldots, r \end{aligned}$$

has a nontrivial solution. If we substitute this solution in (6), then, since not all x^i are 0, the left side is > 0, and the right side is $\leqslant 0$. Since a corresponding contradiction can be derived from $p < q$, it follows that $p = q$, as desired.

For real quadratic forms the theorem and the proof are the same with restriction to real transformations $(s_\lambda{}^i)$.

The difference between the number of positive and the number of negative terms is called the *signature* of the form. We have shown that the *rank and signature of a Hermitian form or of a real quadratic form are invariant under the transformations* (5). They are also the only such invariants, since every such form can be transformed by (5) into a form with diagonal matrix and with diagonal terms ± 1.

If and only if the signature is equal to the rank, will all the values of the Hermitian form $\mathfrak{x}^*\mathfrak{G}\mathfrak{x}$ be nonnegative; and if, in addition, the rank is equal to the dimension of V, then the form will assume the value 0 only for $\mathfrak{x} = \mathfrak{o}$. Such a form is said to be *positive definite.*

Our discussion has now led to the result: *if an inner product with respect to a given basis is represented by a positive definite Hermitian form, then with respect to a suitably chosen basis it can be represented in the form*

$$\mathfrak{x}^*\mathfrak{y} = \sum_{\nu} \bar{x}^{\nu} y^{\nu},$$

i.e., by the *unit matrix.*

We now ask which transformations will leave this form of the representation unchanged. By (4) it will be those transformations for which $\mathfrak{T}^*\mathfrak{E}\mathfrak{T} = \mathfrak{E}$, or

$$\mathfrak{T}^*\mathfrak{T} = \mathfrak{E};$$

such transformations are called *unitary.* If S is restricted to the field of real numbers, we obtain $\mathfrak{T}^T\mathfrak{T} = \mathfrak{E}$, or in other words the *orthogonal* transformations. The corresponding *matrices* are also called *unitary,* or *orthogonal.*

The set of unitary (or if real, orthogonal) transformations is already sufficient to transform any Hermitian (or if real, symmetric) matrix into a diagonal matrix. We shall prove this statement in §3.7, after we have introduced the concept of a determinant.

3.3. *The Tensor Product and the Outer Product*

A second possibility for introducing a multiplication with the properties (P), (La), (Lm), (T) consists of regarding $\Pi_{\kappa\lambda}$ as the basis vectors of a vector space W_r, whose dimension is therefore $r = kl$. More precisely: let W_r be a vector space of dimension $r = kl$ with the basis $\epsilon_{\kappa\lambda}$. Then we define the *tensor product* by

(1) $$\Pi(\hat{\mathfrak{e}}_{\kappa}, \mathfrak{e}_{\lambda}) = \epsilon_{\kappa\lambda},$$

where it must be remembered that S is assumed to be commutative; moreover, we assume that $\mathfrak{a}s = s\mathfrak{a}$ is defined in $\hat{V}$, V and W.

The mapping defined by (1) is in general not a mapping of $\hat{V} \times V$ *onto* W; for the images of the elements of $\hat{V} \times V$ are $a^{\kappa}b^{\lambda}\epsilon_{\kappa\lambda}$, but the elements of W are the $q^{\kappa\lambda}\epsilon_{\kappa\lambda}$.

The product will be invariant under basis transformations $(\hat{t}^{\kappa}_{\mu'})$ in $\hat{V}$ and $(t_{\nu}{}^{\lambda})$ in V, if in W we make the corresponding transformation

$$\text{(T)} \qquad \epsilon_{\mu'\nu'} = \hat{t}^{\kappa}_{\mu'}\epsilon_{\kappa\lambda}t^{\lambda}_{\nu'} = \hat{t}^{\kappa}_{\mu'}t^{\lambda}_{\nu'}\epsilon_{\kappa\lambda}\,.$$

We must still verify that (T) is a basis transformation, i.e., that the $\epsilon_{\mu'\nu'}$ are linearly independent. But if we assume that

$$c^{\mu'\nu'}\epsilon_{\mu'\nu'} = \mathfrak{o},$$

then from (T) and from the fact that the $\epsilon_{\kappa\lambda}$ are linearly independent it follows that

$$\hat{t}^{\kappa}_{\mu'}c^{\mu'\nu'}t^{\lambda}_{\nu'} = 0 \qquad \text{for all } \kappa,\, \lambda.$$

If (with arbitrary ρ', σ') we multiply by $\hat{t}^{\rho'}_{\kappa}, t^{\sigma'}_{\lambda}$ and add, we obtain

$$0 = \hat{t}^{\rho'}_{\kappa}\hat{t}^{\kappa}_{\mu'}c^{\mu'\nu'}t^{\lambda}_{\nu'}t^{\sigma'}_{\lambda} = c^{\rho'\sigma'}.$$

By a transformation (T), the coordinates $q^{\kappa\lambda}$ of an element of W are transformed according to

$$q^{\mu'\nu'} = \hat{t}^{\mu'}_{\kappa}t^{\nu'}_{\lambda}q^{\kappa\lambda}.$$

If $\hat{V} = V$, so that $\hat{t}^{\mu'}_{\kappa} = t^{\mu'}_{\kappa}$, comparison with §2.6 shows that the elements of W (if $\hat{V} = V$) are contravariant tensors of second order with respect to V. If $\hat{V}$ and V are dual to each other, then the elements of W are mixed tensors.

The tensor product is denoted by

$$\hat{\mathfrak{a}} \otimes \mathfrak{b} = a^{\kappa}b^{\lambda}(\hat{\mathfrak{e}}_{\kappa} \otimes \mathfrak{e}_{\lambda}).$$

We have retained the distinction between the vector spaces $\hat{V}$ and V in order to make it easier to define the tensor product of several factors, since we need only use W in place of $\hat{V}$ or V, although then, of course, the vector space to which the products belong is a new one.

We obtain

$$(\hat{\mathfrak{a}} \otimes \mathfrak{b}) \otimes \mathfrak{c} = a^{\kappa}b^{\lambda}c^{\mu}((\hat{\mathfrak{e}}_{\kappa} \otimes \mathfrak{e}_{\lambda}) \otimes \mathfrak{e}_{\mu}).$$

The product of arbitrarily many factors is now defined by induction.

However, we are in fact chiefly interested in the case in which all the factors come from the same vector space $\hat{V} = V$, when we may write

$$(\mathfrak{a} \otimes \mathfrak{b}) \otimes \mathfrak{c} = a^{\kappa}b^{\lambda}c^{\mu}((\mathfrak{e}_{\kappa} \otimes \mathfrak{e}_{\lambda}) \otimes \mathfrak{e}_{\mu}),$$
$$\mathfrak{a} \otimes (\mathfrak{b} \otimes \mathfrak{c}) = a^{\kappa}b^{\lambda}c^{\mu}(\mathfrak{e}_{\kappa} \otimes (\mathfrak{e}_{\lambda} \otimes \mathfrak{e}_{\mu})).$$

To begin with, the vectors $(\mathfrak{e}_\kappa \otimes \mathfrak{e}_\lambda) \otimes \mathfrak{e}_\mu$ and $\mathfrak{e}_\kappa \otimes (\mathfrak{e}_\lambda \otimes \mathfrak{e}_\mu)$ are to be regarded as basis elements of two distinct vector spaces. But since these spaces have the same dimension l^3, we can set up an isomorphism between them by an arbitrary one-to-one correspondence of their basis elements, whereupon we regard the basis elements assigned to each other by $(\mathfrak{e}_\kappa \otimes \mathfrak{e}_\lambda) \otimes \mathfrak{e}_\mu \to \mathfrak{e}_\kappa \otimes (\mathfrak{e}_\lambda \otimes \mathfrak{e}_\mu)$ as being "the same." But now the tensor product is obviously associative, and we can write

(2) $$\mathfrak{a} \otimes \mathfrak{b} \otimes \mathfrak{c} = a^\kappa b^\lambda c^\mu (\mathfrak{e}_\kappa \otimes \mathfrak{e}_\lambda \otimes \mathfrak{e}_\mu).$$

The tensor product is not the only bilinear mapping of $\hat{V} \times V$ in W_r. Like the linear mappings (see above), the bilinear and multilinear mappings form a vector space under appropriately defined S-multiplication. We shall consider only the following special case:

If $(\mathfrak{a}, \mathfrak{b}) \to \mathfrak{a} \otimes \mathfrak{b}$, and consequently also $(\mathfrak{a}, \mathfrak{b}) \to \mathfrak{b} \otimes \mathfrak{a}$, are bilinear mappings, then so are

(3) $$(\mathfrak{a}, \mathfrak{b}) \to \mathfrak{a} \vee \mathfrak{b} = \mathfrak{a} \otimes \mathfrak{b} + \mathfrak{b} \otimes \mathfrak{a},$$

(4) $$(\mathfrak{a}, \mathfrak{b}) \to \mathfrak{a} \wedge \mathfrak{b} = \mathfrak{a} \otimes \mathfrak{b} - \mathfrak{b} \otimes \mathfrak{a},$$ [3]

so that these two mappings may be regarded as products. Since

$$\mathfrak{a} \vee \mathfrak{b} = \mathfrak{b} \vee \mathfrak{a},$$

the first of these products could be called the symmetric product. More important is the second one, which because of

$$\mathfrak{a} \wedge \mathfrak{b} = -\mathfrak{b} \wedge \mathfrak{a}$$

is called the *alternating product* or, by Grassmann, the *outer product*. In the present section we consider only this product.

As an element of W the alternating product is a contravariant tensor of second order. After choice of a basis $(\mathfrak{e}_\kappa)$ for V and a corresponding basis for W, the coordinates of this product are to be obtained from (4). If we set

$$\mathfrak{a} \wedge \mathfrak{b} = p^{\kappa\lambda}(\mathfrak{e}_\kappa \otimes \mathfrak{e}_\lambda),$$

we obtain from (4)

(5) $$p^{\kappa\lambda} = a^\kappa b^\lambda - a^\lambda b^\kappa.$$

From (5) follows $p^{\kappa\lambda} = -p^{\lambda\kappa}$, and in particular $p^{\kappa\kappa} = 0$. Thus the k^2 coordinates $p^{\kappa\lambda}$ are completely determined by the values of a suitably

[3] In the present context the symbols $\wedge$, $\vee$ are defined by (3) and (4), and do *not* mean "and" and "or."

chosen set of $k(k-1)/2$ of these coordinates. In this sense we can say that $\mathfrak{a} \wedge \mathfrak{b}$ has only $k(k-1)/2$ essentially distinct coordinates.

Furthermore, we can show that every skew-symmetric tensor $\mathfrak{p} = (p^{\kappa\lambda})$ with $p^{\lambda\kappa} = -p^{\kappa\lambda}$, has this property; i.e., for the representation of such a tensor it is sufficient to know $k(k-1)/2$ basis vectors, namely

$$\mathfrak{e}_\kappa \wedge \mathfrak{e}_\lambda = \mathfrak{e}_\kappa \otimes \mathfrak{e}_\lambda - \mathfrak{e}_\lambda \otimes \mathfrak{e}_\kappa, \qquad (\kappa < \lambda).$$

For if in

$$\mathfrak{p} = \sum_{\kappa<\lambda} p^{\kappa\lambda}(\mathfrak{e}_\kappa \otimes \mathfrak{e}_\lambda) + \sum_{\lambda<\kappa} p^{\kappa\lambda}(\mathfrak{e}_\kappa \otimes \mathfrak{e}_\lambda)$$

we make a change of summation indices in the second sum, we obtain

$$\mathfrak{p} = \sum_{\kappa<\lambda} p^{\kappa\lambda}(\mathfrak{e}_\kappa \otimes \mathfrak{e}_\lambda) + \sum_{\kappa<\lambda} p^{\lambda\kappa}(\mathfrak{e}_\lambda \otimes \mathfrak{e}_\kappa),$$

from which, since $p^{\lambda\kappa} = -p^{\kappa\lambda}$, it follows that

$$\mathfrak{p} = \sum p^{\kappa\lambda}(\mathfrak{e}_\kappa \otimes \mathfrak{e}_\lambda - \mathfrak{e}_\lambda \otimes \mathfrak{e}_\kappa) = \sum_{\kappa<\lambda} p^{\kappa\lambda}(\mathfrak{e}_\kappa \wedge \mathfrak{e}_\lambda).$$

It is to be noted that for the alternating product we have

$$\mathfrak{a} \wedge \mathfrak{b} = \sum_{\kappa=1}^{k} \sum_{\lambda=1}^{k} a^\kappa b^\lambda(\mathfrak{e}_\kappa \wedge \mathfrak{e}_\lambda) = \sum_{\kappa<\lambda} (a^\kappa b^\lambda - a^\lambda b^\kappa)(\mathfrak{e}_\kappa \wedge \mathfrak{e}_\lambda).$$

Another important property is that the alternating product is equal to 0 if and only if $\mathfrak{a}, \mathfrak{b}$ are linearly dependent.

Proof. (*a*) If $\mathfrak{a}, \mathfrak{b}$ are linearly dependent, so that $p\mathfrak{a} + q\mathfrak{b} = \mathfrak{o}$ and say $q \neq 0$, then $\mathfrak{b} = -(p/q)\,\mathfrak{a}$, and from (4) or (5) it follows that $\mathfrak{a} \wedge \mathfrak{b} = \mathfrak{o}$.

(*b*) If $\mathfrak{a} \wedge \mathfrak{b} = \mathfrak{o}$, then

(6) $$a^\kappa b^\lambda - a^\lambda b^\kappa = 0 \qquad \text{for all} \quad \kappa, \lambda;$$

so we wish to find two numbers p and q, not both $= 0$, such that

$$pa^\kappa + qb^\kappa = 0 \qquad \text{for all } \kappa.$$

But we may assume $a^i \neq 0$ (since if $\mathfrak{a} = \mathfrak{o}$, then $\mathfrak{a}, \mathfrak{b}$ are linearly dependent) and therefore, by (6) we may take $p = -b^i$, $q = a^i$.

The alternating product of more than two factors could also be regarded as defined by (4). Then we would have

(7) $$\begin{aligned}(\mathfrak{a} \wedge \mathfrak{b}) \wedge \mathfrak{c} &= (\mathfrak{a} \otimes \mathfrak{b} - \mathfrak{b} \otimes \mathfrak{a}) \otimes \mathfrak{c} - \mathfrak{c} \otimes (\mathfrak{a} \otimes \mathfrak{b} - \mathfrak{b} \otimes \mathfrak{a}) \\ &= \mathfrak{a} \otimes \mathfrak{b} \otimes \mathfrak{c} - \mathfrak{b} \otimes \mathfrak{a} \otimes \mathfrak{c} - \mathfrak{c} \otimes \mathfrak{a} \otimes \mathfrak{b} + \mathfrak{c} \otimes \mathfrak{b} \otimes \mathfrak{a}.\end{aligned}$$

On the other hand,

$$\mathfrak{a} \wedge (\mathfrak{b} \wedge \mathfrak{c}) = \mathfrak{a} \otimes \mathfrak{b} \otimes \mathfrak{c} - \mathfrak{a} \otimes \mathfrak{c} \otimes \mathfrak{b} - \mathfrak{b} \otimes \mathfrak{c} \otimes \mathfrak{a} + \mathfrak{c} \otimes \mathfrak{b} \otimes \mathfrak{a};$$

so that the operation $\wedge$ would not be associative. Of course, it would satisfy the following law:

$$(\mathfrak{a} \wedge \mathfrak{b}) \wedge \mathfrak{c} + (\mathfrak{b} \wedge \mathfrak{c}) \wedge \mathfrak{a} + (\mathfrak{c} \wedge \mathfrak{a}) \wedge \mathfrak{b} = 0.$$

This law holds for the vector product in a three-dimensional orthogonal space (cf. II, 7, §1.9):

$$(\mathfrak{a} \times \mathfrak{b}) \times \mathfrak{c} + (\mathfrak{b} \times \mathfrak{c}) \times \mathfrak{a} + (\mathfrak{c} \times \mathfrak{a}) \times \mathfrak{b} = \mathfrak{o}.$$

Thus the definition by (7) is suitable if we wish to interpret the alternating product as a vector product. But this interpretation is restricted to the three-dimensional orthogonal space, and we also wish to preserve the associative law. With this in mind, we argue as follows: on the right-hand side of (7) the three vectors $\mathfrak{a}$, $\mathfrak{b}$, $\mathfrak{c}$ are not all on the same footing, since only four of the six permutations of $\mathfrak{a}$, $\mathfrak{b}$, $\mathfrak{c}$ actually appear. Thus we simply write down *all* the permutations, with a plus sign for the even permutations and a minus sign for the odd ones, and replace (7) by the new definition:

$$\begin{aligned} \mathfrak{a} \wedge \mathfrak{b} \wedge \mathfrak{c} &= (\mathfrak{a} \wedge \mathfrak{b}) \wedge \mathfrak{c} = \mathfrak{a} \wedge (\mathfrak{b} \wedge \mathfrak{c}) \\ &= \mathfrak{a} \otimes \mathfrak{b} \otimes \mathfrak{c} + \mathfrak{b} \otimes \mathfrak{c} \otimes \mathfrak{a} + \mathfrak{c} \otimes \mathfrak{a} \otimes \mathfrak{b} \\ &\quad - \mathfrak{a} \otimes \mathfrak{c} \otimes \mathfrak{b} - \mathfrak{c} \otimes \mathfrak{b} \otimes \mathfrak{a} - \mathfrak{b} \otimes \mathfrak{a} \otimes \mathfrak{c}. \end{aligned}$$

More generally: let $\pi 1, \ldots, \pi n$ be a permutation of the numbers $1, \ldots, n$, and let $(-1)^\pi = \pm 1$ according as π is an even or an odd permutation (cf. IB2, §15.3.2), then we define

$$(8) \qquad \mathfrak{a}_1 \wedge \mathfrak{a}_2 \wedge \cdots \wedge \mathfrak{a}_n = \sum_\pi (-1)^\pi\, \mathfrak{a}_{\pi 1} \otimes \mathfrak{a}_{\pi 2} \otimes \cdots \otimes \mathfrak{a}_{\pi n}.$$

This sum is to be taken over all permutations of the numbers $1, \ldots, n$.

In the case $n = 2$ we again obtain the definition (4), but (4) must be restricted to the case that the two vectors $\mathfrak{a}$, $\mathfrak{b}$ belong to the vector space V; then and only then do we have the freedom to define

$$(\mathfrak{a} \wedge \mathfrak{b}) \wedge \mathfrak{c} = \mathfrak{a} \wedge (\mathfrak{b} \wedge \mathfrak{c}) = \mathfrak{a} \wedge \mathfrak{b} \wedge \mathfrak{c}.$$

The alternating product is also denoted by square brackets: $\mathfrak{a}_1 \wedge \mathfrak{a}_2 \wedge \cdots \wedge \mathfrak{a}_n = [\mathfrak{a}_1, \mathfrak{a}_2, \ldots, \mathfrak{a}_n]$.

If we interpret $\mathfrak{a} \wedge \mathfrak{b}$ as a vector product in a three-dimensional orthogonal space, the expression $\mathfrak{a} \wedge \mathfrak{b} \wedge \mathfrak{c}$ now corresponds to $(\mathfrak{a} \times \mathfrak{b})\, \mathfrak{c}$.

3.4. *The determinant*

On the basis of the definition §3.3 (7) the alternating product has the following properties:

(La) $$[..., \mathfrak{a}_i' + \mathfrak{a}_i'' ...] = [..., \mathfrak{a}_i', ...] + [..., \mathfrak{a}_i'', ...].$$

(The vectors represented here by the dots remain unchanged in each of the successive steps of the summation.)

(Lm) $$[..., \mathfrak{a}_i s, ...] = [..., \mathfrak{a}_i, ...]\, s$$

(a) $$[... \mathfrak{a}_i ... \mathfrak{a}_j ...] = \mathfrak{o}, \quad \text{if} \quad \mathfrak{a}_i = \mathfrak{a}_j \quad \text{and} \quad i \neq j.$$

From (a) and (La) it follows that

(a') $$[\cdots \mathfrak{a}_i \cdots \mathfrak{a}_j \cdots] = -[\cdots \mathfrak{a}_j \cdots \mathfrak{a}_i \cdots],$$

since

$$[\cdots \mathfrak{a}_i \cdots \mathfrak{a}_j \cdots] + [\cdots \mathfrak{a}_j \cdots \mathfrak{a}_i \cdots] = [\cdots \mathfrak{a}_i + \mathfrak{a}_j \cdots \mathfrak{a}_j + \mathfrak{a}_i \cdots] = \mathfrak{o}.$$

If the characteristic of S (cf. IB5, §1.11) is not equal to 2, so that $1 + 1 \neq 0$, then (a) also follows from (a').

From these properties it follows that *if* $\mathfrak{a}_1, ..., \mathfrak{a}_n$ *are linearly dependent, then* $[\mathfrak{a}_1 ... \mathfrak{a}_n] = \mathfrak{o}$. For then one of the vectors, say $\mathfrak{a}_n$, is a linear combination of the others, and we have

$$\begin{aligned}[\mathfrak{a}_1 ... \mathfrak{a}_{n-1}, \mathfrak{a}_1 c^1 + \cdots + \mathfrak{a}_{n-1} c^{n-1}]& \\ = [\mathfrak{a}_1 \cdots \mathfrak{a}_{n-1}, \mathfrak{a}_1]\, c^1 + \cdots + [\mathfrak{a}_1 ... \mathfrak{a}_{n-1}, \mathfrak{a}_{n-1}]\, c^{n-1}& \\ = \mathfrak{o} \quad \text{by} \quad (a).&\end{aligned}$$

We shall see later that $[\mathfrak{a}_1 ... \mathfrak{a}_n] = \mathfrak{o}$ *only* if the $\mathfrak{a}_1, ..., \mathfrak{a}_n$ are linearly dependent, a fact closely related to the solvability of the system of homogeneous equations

$$a_\nu^\kappa x^\nu = 0 \qquad (\kappa = 1, ..., n, \nu = 1, ..., n).$$

It is an important fact that the alternating product is determined in an essentially unique way by the properties (La), (Lm), (a). We shall prove this statement here only for the case $k = n$, by actually computing the coordinates of $[\mathfrak{a}_1, ..., \mathfrak{a}_n]$ on the basis of these properties alone.

Let $\mathfrak{e}_1, ..., \mathfrak{e}_n$ be a basis of V_n, although it would be sufficient that all $\mathfrak{a}_\nu$ are expressible as linear combinations

$$\mathfrak{a}_\nu = \mathfrak{e}_\kappa a_\nu^\kappa;$$

in other words the $\mathfrak{e}_\lambda$ need not be linearly independent.

Then by (La), (Lm) we have

$$[\mathfrak{a}_1 \ldots, \mathfrak{a}_n] = \sum_{\nu_1}\sum_{\nu_2} \cdots \sum_{\nu_n} [\mathfrak{e}_{\nu_1}, \ldots, \mathfrak{e}_{\nu_n}]\, a_1^{\nu_1} \cdots a_n^{\nu_n}.$$

From (a) it follows that the product $[\mathfrak{e}_{\nu_1} \ldots e_{\nu_n}] = \mathfrak{o}$ if two indices are equal. Thus for the $\nu_1, \ldots, \nu_n$ we need to consider only the permutations $\pi 1, \ldots, \pi n$ of the numbers $1, \ldots, n$:

$$[\mathfrak{a}_1, \ldots, \mathfrak{a}_n] = \sum_{\pi} [\mathfrak{e}_{\pi 1}, \ldots, \mathfrak{e}_{\pi n}]\, a_1^{\pi 1} a_2^{\pi 2} \cdots a_n^{\pi n}.$$

By (a) we have $[\mathfrak{e}_{\pi 1}, \ldots, \mathfrak{e}_{\pi n}] = (-1)^{\pi} [\mathfrak{e}_1, \ldots, \mathfrak{e}_n]$, so that

(D) $$[\mathfrak{a}_1, \ldots, \mathfrak{a}_n] = [\mathfrak{e}_1, \ldots, \mathfrak{e}_n] \sum_{\pi} (-1)^{\pi}\, a_1^{\pi 1} \cdots a_n^{\pi n}.$$

Thus $[\mathfrak{a}_1, \ldots, \mathfrak{a}_n]$ is determined up to the "factor" $[\mathfrak{e}_1, \ldots, \mathfrak{e}_n]$. This factor is a vector in W_r ($r = n^n$).

To a great extent, the present discussion can be made independent of the preceding section. Given a system of vectors $\mathfrak{a}_1, \ldots, \mathfrak{a}_n$ in a space V_k, let us set ourselves the problem of assigning to it a vector $[\mathfrak{a}_1 \ldots \mathfrak{a}_n]$, in another space W, which is to be equal to zero if and only if the $\mathfrak{a}_1, \ldots, \mathfrak{a}_n$ are linearly dependent. Such a mapping must in any case satisfy condition (a). Furthermore, it is reasonable, though unnecessary, to demand (La) and (Lm). For if, for example, both $\mathfrak{a}_1'$ and $\mathfrak{a}_1''$ are linearly dependent on $\mathfrak{a}_2, \ldots, \mathfrak{a}_n$, then $\mathfrak{a}_1' + \mathfrak{a}_1'', \mathfrak{a}_2, \ldots, \mathfrak{a}_n$ are linearly dependent. But then $[\mathfrak{a}_1', \mathfrak{a}_2, \ldots, \mathfrak{a}_n] = [\mathfrak{a}_1'', \mathfrak{a}_2, \ldots, \mathfrak{a}_n] = \mathfrak{o}$ implies $[\mathfrak{a}_1' + \mathfrak{a}_1'', \mathfrak{a}_2, \ldots, \mathfrak{a}_n] = \mathfrak{o}$, which is exactly the case if (La) holds.

Furthermore, if $\mathfrak{a}_1 \ldots \mathfrak{a}_n$ are linearly dependent, so are $\mathfrak{a}_1 s, \mathfrak{a}_2, \ldots, \mathfrak{a}_n$, and then $[\mathfrak{a}_1, \ldots, \mathfrak{a}_n] = \mathfrak{o}$ implies $[\mathfrak{a}_1 s, \mathfrak{a}_2, \ldots, \mathfrak{a}_n] = \mathfrak{o}$, which is exactly the case if (Lm) holds. Thus (La) and (Lm) are not proved (that would be impossible) but they are motivated.

From the definition of a mapping we have the requirement:

(Lv) From $\mathfrak{a}_i' = \mathfrak{a}_i''$ follows $[\ldots, \mathfrak{a}_i', \ldots] = [\ldots, \mathfrak{a}_i'', \ldots]$.

Now by the argument of the present section alone we see that if a mapping with the desired properties exists at all, then for $k = n$ it can only be the mapping represented by (D). The numerical factor will be denoted by

(1) $$\sum_{\pi} (-1)^{\pi}\, a_1^{\pi 1} \cdots a_n^{\pi n} = |(a_\nu^{\kappa})| = |\, a_\nu^{\kappa}\,| = |\,\mathfrak{A}\,| = A$$

and will be called the *determinant of the matrix* $\mathfrak{A}$.

If we wish, we may *normalize* the mapping (D) by choosing a basis $(\mathfrak{e}_\kappa)$ such that

$$(n) \qquad [\mathfrak{e}_1, \ldots, \mathfrak{e}_n] = 1.$$

In this sense we also speak of the determinant of the vector system $\mathfrak{a}_1, \ldots, \mathfrak{a}_n$ with respect to the given basis. The determinant is then a number, but under a change of basis it is transformed like the system of coordinates of an n-order tensor whose coordinates either all vanish or are all alike except for sign. Because of the alternation, we may simplify the transformation as follows:

Let $\mathfrak{a}_\nu = \mathfrak{e}_\kappa a_\nu{}^\kappa = \mathfrak{e}_{\lambda'} a_\nu^{\lambda'}$ with

$$(2) \qquad \mathfrak{e}_{\lambda'} = \mathfrak{e}_\kappa t^\kappa_{\lambda'}, \qquad \text{so that} \quad a_\nu{}^\kappa = t^\kappa_{\lambda'} a_\nu^{\lambda'}.$$

Then

$$\begin{aligned}[\mathfrak{a}_1, \ldots, \mathfrak{a}_n] &= [\mathfrak{e}_1, \ldots, \mathfrak{e}_n] \cdot |\, a_\nu{}^\kappa \,| \\ &= [\mathfrak{e}_{1'}, \ldots, \mathfrak{e}_{n'}] \cdot |\, a_\kappa^{\lambda'} \,|.\end{aligned}$$

But $[\mathfrak{e}_{1'}, \ldots, \mathfrak{e}_{n'}] = [\mathfrak{e}_1, \ldots, \mathfrak{e}_n] \cdot |\, t^\kappa_{\lambda'} \,|$, so that

$$(3) \qquad |\, a_\nu{}^\kappa \,| = |\, t^\kappa_{\lambda'} \,|\, |\, a_\nu^{\lambda'} \,|;$$

in other words, under a transformation of basis the given determinant is multiplied by the determinant of the matrix of the transformation.

From (3) we can draw an important conclusion: (2) states (among other things) that the matrix $\mathfrak{A} = (a_\nu{}^\kappa)$ is the product of the matrices $\mathfrak{T} = (t^\kappa_{\lambda'})$ and $\mathfrak{A}' = (a^\lambda.)$, and therefore (3) means that

$$(4) \qquad |\, \mathfrak{T} \cdot \mathfrak{A}' \,| = |\, \mathfrak{T} \,| \cdot |\, \mathfrak{A}' \,|.$$

This relation is called the *law of multiplication of determinants.* We seem to have proved it here only under the assumption that $\mathfrak{T}$ is a regular matrix, but in (2) we may in fact consider an arbitrary matrix. In this case the $\mathfrak{e}_{\lambda'}$ will not necessarily be linearly independent, but it was pointed out at the time that the equation (D), the only equation used here, does not require the vectors $\mathfrak{e}_{\lambda'}$ to be linearly independent. Thus (4) is valid for arbitrary matrices $\mathfrak{T}, \mathfrak{A}'$.

Our argument shows that if there exists a mapping which is multilinear (Lv, *a*, *m*) and alternating (*a*) and has the normalization (*n*), then it can only be represented by the determinant (1). But does A actually have the property of vanishing if and only if $\mathfrak{a}_1, \ldots, \mathfrak{a}_n$ are linearly dependent?

At the beginning of the present section we proved:

If $\mathfrak{a}_1, \ldots, \mathfrak{a}_n$ are linearly dependent, then $[\mathfrak{a}_1, \ldots, \mathfrak{a}_n] = 0$.

On the other hand, if the $\mathfrak{a}_1, \ldots, \mathfrak{a}_n$ are linearly independent, they can be chosen as a basis, and the original basis vectors can be represented in the form $\mathfrak{e}_\kappa = \mathfrak{a}_\nu e_\kappa{}^\nu$. But then by (D) we have

$$1 = [\mathfrak{e}_1, \ldots, \mathfrak{e}_n] = [\mathfrak{a}_1, \ldots, \mathfrak{a}_n] \mid e_\kappa{}^\nu \mid,$$

so that

$$[\mathfrak{a}_1, \ldots, \mathfrak{a}_n] \neq 0.$$

So the vanishing or nonvanishing of the determinant provides a criterion for the linear dependence of a given system of n vectors in V_n; in other words, it determines whether a system of n homogeneous linear equations in n unknowns has a nontrivial solution or only the solution $x^1 = \cdots = x^n = 0$. Thus the most important step in the theory of systems of linear equations has been taken. In order to provide a complete answer to the question whether an arbitrary system of k linear equations in n unknowns has a solution and, if so, how many solutions, we need only introduce certain refinements, to be described in the next two sections. For this purpose we require from the present section only the definition of the determinant of a matrix given by (1) above. The preceding discussion has motivated this definition, but if we are willing to adopt it without motivation, the theory of systems of linear equations can be developed independently of the theory of vector spaces.

3.5. *Rules for Calculation with Determinants*

1. In order to emphasize that we are taking over almost nothing from the foregoing discussion, we repeat the definition of a determinant: the *determinant* of the quadratic matrix $\mathfrak{A} = (a_\nu{}^\kappa)_{\nu=1,\ldots,n}^{\kappa=1,\ldots,n}$ is the number

(D1) $$A = \mid \mathfrak{A} \mid = |(a_\nu{}^\kappa)| = \sum_\pi (-1)^\pi \, a_1^{\pi 1} \cdots a_n^{\pi n}.$$

2. *Interchange of rows and columns.* The transpose $\mathfrak{A}^T = (\overset{T}{a}{}_\kappa^\nu)$ of the matrix $\mathfrak{A}$ is defined by $\overset{T}{a}{}_\kappa^\nu = a_\nu^\kappa$. Its determinant is

$$\mid \mathfrak{A}^T \mid = \sum_\pi (-1)^\pi \, a_{\pi 1}^1 \cdots a_{\pi n}^n .$$

In each of these summands let us make the permutation π^{-1} in the factors. Then

$$\mid \mathfrak{A}^T \mid = \sum_\pi (-1)^\pi \, a_1^{\pi^{-1} 1} \cdots a_n^{\pi^{-1} n}.$$

But since π^{-1} is even or odd together with π, and since π^{-1} also runs through all possible permutations, we have the result

(D2) $$\mid \mathfrak{A}^T \mid = \mid \mathfrak{A} \mid.$$

Consequently, any rule for calculation that concerns the columns of a matrix is also valid for the rows, and conversely.

3. *Expansion of a determinant by the elements of a column (or of a row).* If in (D1) we combine all the summands containing a_1^1 and then all those containing a_1^2, and so forth, we obtain the determinant in the form

$$A = a_1^1 A_1^1 + a_1^2 A_2^1 + \cdots + a_1^n A_n^1.$$

Here, for example,

$$A_1^1 = \sum_{\pi'} (-1)^{\pi'} a_2^{\pi'2} \cdots a_n^{\pi'n},$$

where π' runs through all permutations of the numbers 2, ..., n. But this expression is exactly the determinant of the matrix obtained from $\mathfrak{A}$ by deleting the first row and the first column. We speak here of an $(n - 1)$-rowed subdeterminant of $\mathfrak{A}$ or of A, and we shall later use the expression "r-rowed subdeterminant" in the corresponding sense for rectangular matrices. If we denote by U_κ^ν the subdeterminant obtained from A by striking out the κth row and the νth column, we have: $A_\kappa^\nu = (-1)^{\nu+\kappa} U_\kappa^\nu$. We leave it to the reader to verify this rule, and the following equations, by actual calculation:

$$A = \sum_{\kappa=1}^{n} a_i^\kappa A_\kappa^i \qquad \text{(for every } i\text{; no summation over } i\text{)}$$

$$= \sum_{\nu=1}^{n} a_\nu^j A_j^\nu \qquad \text{(for every } j\text{; no summation over } j\text{).}$$

By forming the sum $\sum_\kappa a_i^\kappa A_\kappa^j$, we obtain the determinant of the matrix formed from $\mathfrak{A}$ by deleting the jth column and replacing it by the ith column. But this matrix contains two equal columns, so that its determinant is 0, a fact which we can either take from §3.4 (a) or derive directly from (1). We thus obtain, together with the above equations,

$$\text{(D3)} \qquad a_i^\kappa A_\kappa^j = \delta_i^j \cdot A; \qquad a_\nu^j A_i^\nu = \delta_i^j \cdot A.$$

Here again we take summation over equal indices.

4. *The Laplace expansion.* The result (D3) can be generalized in the following way: instead of the 1-rowed subdeterminant (a_i^κ) consisting of the elements of a fixed column, we can consider the q-rowed subdeterminants formed from q fixed columns. Let these columns be numbered $i_1, ..., i_q$, let the rows of such a subdeterminant be numbered $\kappa_1, ..., \kappa_q$, and denote the subdeterminant itself by

$$a_{i_1,\ldots,i_q}^{\kappa_1,\ldots,\kappa_q}.$$

By the *algebraic complement* of this subdeterminant we mean the subdeterminant (with suitable sign) formed from A by deleting the columns $i_1, \ldots, i_q$ and the rows $\kappa_1, \ldots, \kappa_q$. If we denote its columns by $i_{q+1}, \ldots, i_n$ and its rows by $\kappa_{q+1}, \ldots, \kappa_n$, then $i_1, \ldots, i_q, i_{q+1}, \ldots, i_n$ and $\kappa_1, \ldots, \kappa_q, \kappa_{q+1}, \ldots, \kappa_n$ are permutations of 1, ..., n, and the desired algebraic complement is given by

$$A^{i_1, \ldots, i_q}_{\kappa_1, \ldots, \kappa_q} = (-1)^i(-1)^\kappa \, a^{\kappa_{q+1}, \ldots, \kappa_n}_{i_{q+1}, \ldots, i_n}.$$

The generalization of (D3) is now given by the *Laplace expansion*: if we choose q arbitrary columns (or rows) and hold them fixed, and then multiply every q-rowed subdeterminant that can be formed from them by its algebraic complement and sum up, we obtain the determinant A:

$$\text{(D4)} \quad \sum_{(\kappa)} a^{\kappa_1, \ldots, \kappa_q}_{i_1, \ldots, i_q} \cdot A^{i_1, \ldots, i_q}_{\kappa_1, \ldots, \kappa_q} = \sum_{(\kappa)} (-1)^i(-1)^\kappa \, a^{\kappa_1, \ldots, \kappa_q}_{i_1, \ldots, i_q} \cdot a^{\kappa_{q+1}, \ldots, \kappa_n}_{i_{q+1}, \ldots, i_n} = A.$$

For a fixed permutation i the summation here is to be taken over all possible choices of q numbers $\kappa_1, \ldots, \kappa_q$ from among the numbers 1, ..., n.

For the proof we may either write out the subdeterminants in full and verify that exactly the same products $a_1^{\pi 1} \cdots a_n^{\pi n}$ occur as in A, or we may verify (La), (Lm), (a), (n) and make use of the fact that the determinant is the only function with these properties.

5. *The inverse matrix.* From (D3) it follows, if $A \neq 0$, that the matrix $\overline{\mathfrak{A}} = (A_\lambda{}^\nu / A)$ satisfies the equations

$$\mathfrak{A}\overline{\mathfrak{A}} = \overline{\mathfrak{A}}\mathfrak{A} = \mathfrak{E},$$

and is thus both a right and a left inverse.

If $A = 0$, then $\mathfrak{A}$ has no inverse, since it follows from $\mathfrak{A}\overline{\mathfrak{A}} = \mathfrak{E}$ by the theorem for multiplication of determinants that

$$|\,\mathfrak{A}\,|\,|\,\overline{\mathfrak{A}}\,| = 1, \qquad \text{so that} \quad |\,\mathfrak{A}\,| \neq 0.$$

The multiplication theorem for determinants can also be proved by direct calculation without use of the earlier theory.

3.6. *Applications of Determinants to Systems of Linear Equations*

1. If in

$$\text{(I)} \qquad \sum_{\nu=1}^{n} a_\nu{}^\kappa \kappa^\nu = b^\kappa \qquad (\kappa = 1 \ldots k)$$

we have $k = n$, then for each $\lambda = 1, \ldots, n$ we multiply the κth equation by $A_\kappa{}^\lambda$ (see §3.5.3) and add:

$$A_\kappa{}^\lambda a_\nu{}^\kappa x^\nu = A_\kappa{}^\lambda b^\kappa.$$

From (D3) it follows that

(1) $$Ax^\lambda = A_\kappa{}^\lambda b^\kappa.$$

Every solution of (I) is a solution of (1). If $A \neq 0$, then (1) can have only the solution

(2) $$x^\lambda = A_\kappa{}^\lambda b^\kappa / A \qquad (\lambda = 1 \ldots n).$$

By actual substitution it is easy to see that this system (x^λ) is actually a solution of (I). Thus (I) has the unique solution (2). The same remark holds for the case with all $b^\kappa = 0$, when the unique solution is the so-called trivial solution: all $x^\lambda = 0$.

2. If $A = 0$, and also if $k \neq n$, let us find the "largest possible" subdeterminant $\neq 0$ that can be formed by striking out rows and columns. Let there exist an r-rowed subdeterminant $\neq 0$, whereas all $(r + 1)$-rowed subdeterminants $= 0$. From §3.5 (D4) it follows that all subdeterminants with more than $r + 1$ rows are equal to zero.

By renumbering, if necessary, the equations and the x's, we may assume that

$$\hat{A} = \begin{vmatrix} a_1{}^1 & \cdots & a_r{}^1 \\ \vdots & & \\ a_1{}^r & \cdots & a_r{}^r \end{vmatrix} \neq 0.$$

Then we must solve the first r equations

$$\begin{aligned} a_1{}^1 x^1 + \cdots + a_r{}^1 x^r &= -a_{r+1}^1 x^{r+1} - \cdots - a_n{}^1 x^n + b^1, \\ &\cdots\cdots\cdots\cdots\cdots \\ a_1{}^r x^1 + \cdots + a_r{}^r x^r &= -a_{r+1}^r x^{r+1} - \cdots - a_n{}^r x^n + b^r \end{aligned}$$

for the $x^1, \ldots, x^r$, where the solution will contain $x^{r+1}, \ldots, x^n$ as "parameters" whose values may be chosen arbitrarily. We can then show that these solutions, for arbitrary values of $x^{r+1}, \ldots, x^n$, are also solutions of the remaining equations of the original system, provided this system has any solutions at all. The details are to be found in any textbook.

From this result it follows that the *rank of a matrix* (cf. §2.3) is the number r characterized by the following property: *there exists at least one r-rowed subdeterminant* $\neq 0$, *but all* $(r + 1)$*-rowed subdeterminants are* $= 0$.

3.7. *Unitary Transformations of Hermitian Forms*

We now come to the proof of the assertion at the end of §3.2.

In the present context S is the field of complex numbers, a basis $(\mathfrak{e}_\nu)$

is given in V_n, and only unitary transformations are admissible. For the given basis, and consequently for all admissible bases, i.e., all bases arising from unitary transformations, let there be an inner product defined by

$$\mathfrak{x}^*\mathfrak{y} \qquad \text{(in the real case: } \mathfrak{x}^T\mathfrak{y}).$$

We call $\mathfrak{x}$ a *unit vector* if $\mathfrak{x}^*\mathfrak{x} = 1$, and we say that the vectors $\mathfrak{x}$, $\mathfrak{y}$ are *orthogonal* if $\mathfrak{x}^*\mathfrak{y} = 0$. This definition satisfies all the customary requirements for orthogonality (cf. II, 7, §2.5): namely, (1) $\mathfrak{o}$ is orthogonal to every vector; (2) if $\mathfrak{a}$ is orthogonal to $\mathfrak{b}$, then $\mathfrak{b}$ is orthogonal to $\mathfrak{a}$; (3) the vectors orthogonal to a given vector $\mathfrak{a}$ form an $(n - 1)$-dimensional vector space. With this definition the given basis and all other admissible bases consist of orthogonal unit vectors.

We wish to show that a Hermitian form represented for the given basis by $\mathfrak{x}^*\mathfrak{A}\mathfrak{x}$ (i.e., by the Hermitian matrix $\mathfrak{A}$) can be reduced by unitary transformations to the form $\mathfrak{y}^*\mathfrak{D}\mathfrak{y}$ with diagonal matrix $\mathfrak{D}$; in other words, there exists a unitary matrix $\mathfrak{T}$ such that

$$\mathfrak{T}^*\mathfrak{A}\mathfrak{T} = \mathfrak{D} \tag{1}$$

is a diagonal matrix.

In the real field this matrix produces the orthogonal transformation of a quadratic form into a sum of squares; but the proofs in the complex field are exactly the same, so that here we discuss the more general case.

By §1.5 the columns of $\mathfrak{T}$ are the coordinates of the new basis vectors with respect to the old basis. So these coordinates are precisely what we are looking for; i.e., we seek a suitable system of orthogonal unit vectors.

Since $\mathfrak{T}^* = \mathfrak{T}^{-1}$, it follows from (1) that $\mathfrak{A}\mathfrak{T} = \mathfrak{T}\mathfrak{D}$. In this matrix equation let us consider the ith column. If $\mathfrak{e}_{i'}$ is the ith column of $\mathfrak{T}$, and d_i is the ith diagonal element of $\mathfrak{D}$, we obtain

$$\mathfrak{A}\mathfrak{e}_{i'} = d_i\mathfrak{e}_{i'} \qquad \text{(no summation over } i).$$

Thus the desired vectors of the new basis are the solutions of the equation

$$\mathfrak{A}\mathfrak{v} = d\mathfrak{v}. \tag{2}$$

This analysis of the problem shows that we must find n mutually orthogonal vectors which are solutions of (2), where d may have various values still to be suitably determined.

If $\mathfrak{v}$ is a solution of (2), then so is $s\mathfrak{v}$, so that for each solution we may arrange that $\mathfrak{v}^*\mathfrak{v} = 1$.

It is easy to verify that the matrix $\mathfrak{T}$ formed from n such vectors provides the desired transformation of $\mathfrak{A}$.

The trivial solution of (2) is not a solution of our problem. A nontrivial solution exists only if

(Ch)
$$| \mathfrak{A} - d\mathfrak{E} | = \begin{vmatrix} a_1{}^1 - d, a_1{}^2, \ldots, a_n{}^1 \\ \cdot \quad \cdot \quad \cdot \quad \cdot \quad \cdot \quad \cdot \quad \cdot \quad \cdot \quad \cdot \quad \cdot \\ a_n{}^1, \ldots, a_n{}^n - d \end{vmatrix} = 0.$$

This equation is called the *characteristic equation* of the matrix $\mathfrak{A}$, and its solutions are the *eigenvalues* of $\mathfrak{A}$.

In some contexts it is customary to define the eigenvalues as the solutions of $| d\mathfrak{A} - \mathfrak{E} | = 0$.

To each eigenvalue there correspond nontrivial solutions of (2), which are called *eigenvectors* of the matrix $\mathfrak{A}$. The equation (Ch) is of the nth degree in d:

(3)
$$A_0 - A_1 d + \cdots + (-1)^{n-1} A_{n-1} d^{n-1} + (-1)^n d^n = 0.$$

The coefficients are sums of principal subdeterminants, formed by striking out rows and columns with the same indices. In particular,

$$A_0 = A = | \mathfrak{A} |$$

$$A_{n-1} = \sum_{i=1}^{n} a_{ii} = \text{tr } \mathfrak{A}, \text{ the } \textit{trace} \text{ of } \mathfrak{A}.$$

The characteristic equation has the following invariant property. If $\mathfrak{T}$ is an arbitrary regular matrix, then

$$\begin{aligned} | \mathfrak{A} - d\mathfrak{E} | &= | \mathfrak{T}^{-1} | \, | \mathfrak{A} - d\mathfrak{E} | \, | \mathfrak{T} | = | \mathfrak{T}^{-1} (\mathfrak{A} - d\mathfrak{E}) \mathfrak{T} | \\ &= | \mathfrak{T}^{-1} \mathfrak{A} \mathfrak{T} - d\mathfrak{E} |. \end{aligned}$$

Consequently, *similar matrices have* the same characteristic equation and thus also *the same eigenvalues*. Since the A_ν are the elementary symmetric functions of the eigenvalues, they also, and in particular the determinant and the trace, are invariant under similarity transformations (in the sense of equation (1) in §2.5).

For the problem of finding n mutually orthogonal eigenvectors of the Hermitian matrix $\mathfrak{A}$ the following two theorems are fundamental:

1) *The eigenvalues of a Hermitian matrix, and thus of a real symmetric matrix, are real.*

Proof: Let d be a complex eigenvalue and $\mathfrak{v}$ a corresponding eigenvector, so that $\mathfrak{A}\mathfrak{v} = d\mathfrak{v}$ and thus also $\overline{\mathfrak{A}}\overline{\mathfrak{v}} = \bar{d}\overline{\mathfrak{v}}$. Multiplying by $\overline{\mathfrak{v}}^T$ and $\mathfrak{v}^T$ respectively, we obtain

(4)
$$\overline{\mathfrak{v}}^T \mathfrak{A} \mathfrak{v} = d \overline{\mathfrak{v}}^T \mathfrak{v},$$

(5)
$$\mathfrak{v}^T \overline{\mathfrak{A}} \overline{\mathfrak{v}} = \bar{d} \mathfrak{v}^T \overline{\mathfrak{v}}.$$

In (5) we form the transposes:

$$\bar{\mathfrak{v}}^T \bar{\mathfrak{A}}^T \mathfrak{v} = \bar{d} \bar{\mathfrak{v}}^T \mathfrak{v}. \tag{6}$$

Since $\bar{\mathfrak{A}}^T = \mathfrak{A}$, it follows from (4) and (6) that

$$(d - \bar{d}) \bar{\mathfrak{v}}^T \mathfrak{v} = 0.$$

But

$$\bar{\mathfrak{v}}^T \mathfrak{v} = \sum_{\nu=1}^{n} \bar{v}^\nu v^\nu \neq 0, \qquad \text{so that} \quad d = \bar{d},$$

as desired.

2) *Eigenvectors belonging to distinct eigenvalues are mutually orthogonal.*

Proof: let $\mathfrak{A}\mathfrak{v}_1 = d_1 \mathfrak{v}_1$, and $\mathfrak{A}\mathfrak{v}_2 = d_2 \mathfrak{v}_2$. Then

$$\mathfrak{v}_2^* \mathfrak{A} \mathfrak{v}_1 = d_1 \mathfrak{v}_2^* \mathfrak{v}_1, \tag{7.1}$$

$$\mathfrak{v}_1^* \mathfrak{A} \mathfrak{v}_2 = d_2 \mathfrak{v}_1^* \mathfrak{v}_2. \tag{7.2}$$

In (7.2) we form the conjugate transposes. Since d_2 is real and $\mathfrak{A}^* = \mathfrak{A}$, we obtain

$$\mathfrak{v}_2^* \mathfrak{A} \mathfrak{v}_1 = d_2 \mathfrak{v}_2^* \mathfrak{v}_1.$$

Comparison with (7.1) gives

$$(d_1 - d_2)\, \mathfrak{v}_2^* \mathfrak{v}_1 = 0,$$

so that if $d_1 \neq d_2$, then $\mathfrak{v}_2^* \mathfrak{v}_1 = 0$, as desired.

If the equation (Ch) has n distinct zeros, these theorems show that our problem is completely solved. The zeros of (Ch) are themselves the elements of the desired diagonal matrix. If we are interested only in this matrix or, in other words, in the result of the transformation, the eigenvectors do not need to be calculated at all.

Let us give a simple example to show what may happen if (Ch) has multiple roots. Consider the real quadratic form

$$x^1 x^1 + x^2 x^2 + c x^3 x^3. \tag{8}$$

The fact that the matrix is already in diagonal form will make the calculation shorter. The characteristic equation is

$$(1 - d)^2 (c - d) = 0. \tag{9}$$

The system of equations (2) becomes

$$\begin{aligned} (1 - d)\, x^1 \qquad\qquad\qquad\qquad &= 0, \\ (1 - d)\, x^2 \qquad\qquad &= 0, \\ (c - d)\, x^3 &= 0. \end{aligned} \tag{10}$$

For the double root $d = 1$ of (9) the matrix in (10) has the rank $n - 2$, so that the solutions of (10) form a two-dimensional vector space, in which we can find two mutually orthogonal solution vectors (as eigenvectors), one of which may be chosen arbitrarily. In our case the eigenvectors (normalized by $\mathfrak{v}^*\mathfrak{v} = 1$) comprise all the following vectors with arbitrary φ:

$$\begin{aligned} \mathfrak{v}_1 &= (\cos\varphi, \sin\varphi, 0), \\ \mathfrak{v}_2 &= (-\epsilon\sin\varphi, \epsilon\cos\varphi, 0), \qquad \epsilon = \pm 1, \\ \mathfrak{v}_3 &= (0, 0, 1). \end{aligned}$$

In geometric language this result means that in an ellipsoid of rotation the principal axes are not all uniquely determined.

In general, we can prove the existence of n orthogonal eigenvectors in a totally different manner, which depends on the property of a quadric that its principal axes are of extremal (more precisely: of stationary) length. A quadric is defined by

$$\mathfrak{x}^*\mathfrak{A}\mathfrak{x} = 1,$$

and the length of a vector is measured by $\sqrt{\mathfrak{x}^*\mathfrak{x}}$. Let us ask the question: When does $\mathfrak{x}^*\mathfrak{x}$ assume a stationary value under the subsidiary condition $\mathfrak{x}^*\mathfrak{A}\mathfrak{x} = 1$ or, what amounts to the same thing, when does $\mathfrak{x}^*\mathfrak{A}\mathfrak{x}$ assume a stationary value under the subsidiary condition $\mathfrak{x}^*\mathfrak{x} = 1$? Introducing the Lagrange multiplier k, we see that the partial derivatives of

$$\mathfrak{x}^*\mathfrak{A}\mathfrak{x} - k\mathfrak{x}^*\mathfrak{x}$$

must be set equal to zero. The calculation is slightly different for the real and the complex case. In the real field we must form the equations (for the meaning of the $\delta_{\kappa\lambda}$ see §1.5)

$$\frac{\partial}{\partial x^i}\,[x^\kappa(a_{\kappa\lambda} - k\delta_{\kappa\lambda})\,x^\lambda] = (a_{i\lambda} - k\delta_{i\lambda})\,x^\lambda + x^\kappa(a_{\kappa i} - k\delta_{\kappa i}) = 0.$$

Since $a_{i\lambda} = a_{\lambda i}$, it follows that

$$(a_{i\lambda} - k\delta_{i\lambda})\,x^\lambda = 0 \qquad \text{or} \qquad (\mathfrak{A} - k\mathfrak{E})\,\mathfrak{x} = 0.$$

In the complex case we may consider x^i and $\bar{x}^i$ as independent variables and then construct the equations

$$(a) \qquad \frac{\partial}{\partial x^i}\,[\bar{x}^\kappa(a_{\kappa\lambda} - k\delta_{\kappa\lambda})\,x^\lambda] = \bar{x}^\kappa(a_{\kappa i} - k\delta_{\kappa i}) = 0,$$

$$(b) \qquad \frac{\partial}{\partial \bar{x}^i}\,[\bar{x}^\kappa(a_{\kappa\lambda} - k\delta_{\kappa\lambda})\,x^\lambda] = (a_{i\lambda} - k\delta_{i\lambda})\,x^\lambda = 0.$$

In matrix notation we have

$$(a)\quad \mathfrak{x}^*(\mathfrak{A} - k\mathfrak{E}) = 0, \qquad (b)\quad (\mathfrak{A} - k\mathfrak{E})\,\mathfrak{x} = 0.$$

Since $\mathfrak{A}^* = \mathfrak{A}$, these two equations have exactly the same meaning, namely (2). The existence of the desired solution-vectors now follows from a theorem of analysis. Since the "points" for which $\mathfrak{x}^*\mathfrak{x} = 1$ form a closed set, and since $\mathfrak{x}^*\mathfrak{A}\mathfrak{x}$ is continuous, there exists a vector $\mathfrak{x} = \mathfrak{v}_1$, for which $\mathfrak{x}^*\mathfrak{A}\mathfrak{x}$ assumes an extreme value with $\mathfrak{x}^*\mathfrak{x} = 1$. Next we determine a vector $\mathfrak{v}_2$ such that $\mathfrak{v}_2^*\mathfrak{A}\mathfrak{v}_2$ assumes an extreme value under the subsidiary conditions $\mathfrak{v}_2^*\mathfrak{v}_2 = 1, \mathfrak{v}_2^*\mathfrak{v}_1 = 0$. We proceed in this way until the equations $\mathfrak{v}_k^*\mathfrak{v}_i = 0$ for all $i < k$ can no longer be satisfied, i.e., until we have n such vectors.

We shall omit the proof that the matrix $\mathfrak{T} = (v_\nu{}^\kappa)$ constructed in this way actually has the desired properties.

Thus we have reached the desired result that every Hermitian matrix can be transformed into a diagonal matrix by a unitary matrix $\mathfrak{T}$ with

$$\mathfrak{T}^*\mathfrak{A}\mathfrak{T} = \mathfrak{D}.$$

The elements of $\mathfrak{D}$ are the eigenvalues of the matrix $\mathfrak{A}$; i.e., they are the solutions of the equation

$$|\,\mathfrak{A} - d\mathfrak{E}\,| = 0.$$

The number of nonzero d_i is equal to the rank of $\mathfrak{D}$, and thus also to the rank of $\mathfrak{A}$.

If in addition to transformations with unitary matrices we allow transformations

$$\mathfrak{A}' = \mathfrak{S}^*\mathfrak{A}\mathfrak{S}$$

with arbitrary regular matrices $\mathfrak{S}$, then, as was shown in §3.2, we can reduce the terms of a real diagonal matrix to ± 1.

List of Formulas

I	System of linear equations §1.1 §2.4, §3.6
II	System of homogeneous linear equations §1.1, §2.4
G	Equality of n-tuples §1.2
A	Addition of n-tuples §1.2
A1—A6	Laws of addition §1.2
S-M	Multiplication of an n-tuple with a scalar §1.2
M1—M6	Laws of S-multiplication §1.2

Lv, La, Lm	Properties of linear transformations §2.1 §2.6 §3.1 §3.4
1, 2	Orthogonality and normalization of matrices §2.6
P1, P2	Postulates for products of vectors §3.1
T	Transformation of $\Pi_{\kappa\lambda}$ §3.1
a	Alternation (for the outer product) §3.4
D	Determinant §3.4
n	Normalization of the determinant §3.4
D1–D4	Rules for calculation with determinants §3.5
Ch	Characteristic equation §3.7

CHAPTER 4

Polynomials

1. Entire Rational Functions

1.1. *Definition and Standard Notation*

By an entire rational function we mean a function (defined, let us say, for all real numbers and assuming real values) that can be constructed by addition and multiplication alone. Of course, we must make this remark more precise, and in doing so we shall free ourselves from any definite domain of numbers, considering instead an arbitrary commutative ring R with unit[1] element 1 (see IB1, §2.4). Thus in what follows we may take R to be the field of real numbers, or of rational numbers, or also the ring of integers, but in each case we must take care to use only the properties implied by the definition of a "commutative ring with unit element." We now consider a function of one argument, defined in R and with values in R; in other words, we consider mappings of R into itself. Two of these mappings have a particularly simple character; namely, for every $c \in R$ the *constant function* $x \to c$ (which to every argument $x \in R$ assigns the value c) and the *identical* function $x \to x$. For every pair f, g of mappings of the ring R into itself we can form the further mappings

$$x \to f(x) + g(x), \qquad x \to f(x)\,g(x).$$

We denote these by $f + g$ and fg, so that

$$(f + g)(x) = f(x) + g(x), \qquad (fg)(x) = f(x)\,g(x).$$

It is obvious that these mappings again take R into itself. In the set of mappings of R into itself we have thus defined an addition and a multi-

[1] For this element we use the symbol 1 from "force of habit" without implying thereby that the natural numbers are contained in R.

plication.[2] By an *entire rational function of one argument in R* we now mean the constant functions, the identity function, and every mapping of R into itself that can be formed from these functions by repeated application of addition and multiplication (a finite number of times). Entire rational functions of several arguments can be defined in the same way, but we shall introduce them in a different manner in §2.3, so that for the present we restrict ourselves, without special mention of the fact, to functions of one argument.

As a result of the rules for computation in R, the set of entire rational functions can be characterized very simply:

A mapping f of R into itself is an entire rational function in R if and only if there exist elements $a_0, \ldots, a_n \in R$ such that for all $x \in R$ we have the equation[3]

$$(1) \qquad f(x) = \sum_{i=0}^{n} a_i x^i.$$

To show that every function f defined by (1) (for all $x \in R$) is an entire rational function, we introduce the notation $\underline{c}$ for the constant function $x \to c$ and I for the identical function $x \to x$. Then (if we set $I^0 = \underline{1}$), it is obvious that $x \to a_i x^i$ is the function $\underline{a}_i I^i$ and thus from (1) it follows that $f = \sum_{i=0}^{n} \underline{a}_i I^i$; that is, f can be formed from I and the $\underline{a}_i$ by addition and multiplication. In order to show, on the other hand, that every entire rational function f can be represented in the form (1), it is only necessary to prove that this statement holds for $\underline{c}$ and I and that it holds for $f + g$ and fg if it holds for f and g. For $\underline{c}$ we need only set $a_0 = c$, $n = 0$ in (1), and for I we set $a_0 = 0$, $a_1 = 1$, $n = 1$; if, besides the representation (1), we also have $g(x) = \sum_{i=0}^{m} b_i x^i$, then we define $a_i = 0$, $b_k = 0$ for $i > n$, $k > m$ and obtain,[4] with $l = \max(m, n)$,

$$(2) \qquad (f + g)(x) = f(x) + g(x) = \sum_{i=0}^{l} (a_i + b_i)\, x^i,$$

$$(3) \quad (fg)(x) = f(x)\, g(x) = \sum_{0 \leqslant i \leqslant n, 0 \leqslant k \leqslant m} a_i b_k x^{i+k} = \sum_{h=0}^{n+m} \left(\sum_{i=0}^{h} a_i b_{h-i} \right) x^h,$$

which completes the proof.

If $f \neq \underline{0}$, then in (1) we may obviously assume $a_n \neq 0$. But are n and the a_i already uniquely determined by f? Since in (2) we may obviously replace $+$ by $-$ on both sides, we see that the answer to this question is affirmative

[2] It is easy to see that with these operations the mappings form a commutative ring with unit element.

[3] In order that this notation may be applicable to the case $x = 0$ we must define $0^0 = 1$, as we shall do throughout the present chapter.

[4] In (3) we must apply IB1, (41) (20).

if and only if $f = \underset{\sim}{0}$ in (1) implies $a_0 = \cdots = a_n = 0$. In §1.2 we shall see that this is certainly the case if R has no divisors of zero and contains infinitely many elements. But in the theory of numbers it will often be necessary to consider rings with only finitely many elements. One example is the ring containing only two elements, namely the zero element 0 and the unit element 1, with $1 + 1 = 0$ (the residue class ring $G/2$ in IB5, §3.7), which has $x^2 + x = 0$ for all x; in other words, the entire rational function $x \to x^2 + x$ is the constant function $\underset{\sim}{0}$, even though it has the representation (1) with $a_0 = 0$, $a_1 = a_2 = 1$, $n = 2$.

1.2. *Zeros*

In this subsection we let R be a commutative ring with unit element and without divisors of zero. By a *zero* of the entire rational function f in R we mean an element $\alpha \in R$ such that $f(\alpha) = 0$. If not every element of R is a zero of f, then f has only finitely many zeros.[5] More precisely, we prove the following theorem:

If f admits a representation (1) *with $a_n \neq 0$, then the number of zeros of f is at most n.*

In the first place, we deduce from (1), and from the fact that

$$(x - \alpha) \sum_{k=0}^{i-1} x^k \alpha^{i-1-k} = \sum_{k=0}^{i-1} x^{k+1} \alpha^{i-1-k} - \sum_{k=0}^{i-1} x^k \alpha^{i-k}$$

$$= \sum_{k=1}^{i} x^k \alpha^{i-k} - \sum_{k=0}^{i-1} x^k \alpha^{i-k} = x^i - \alpha^i$$

for $n > 0$, the equation

$$f(x) - f(\alpha) = (x - \alpha) f_1(x), \tag{4}$$

with

$$f_1(x) = \sum_{k=0}^{n-1} a'_k x^k, \qquad a'_k = \sum_{i=k+1}^{n} a_i \alpha^{i-1-k}. \tag{5}$$

The theorem can now be proved by complete induction on n. For $n = 0$ the function f has no zeros, since $a_0 \neq 0$, so that the assertion is true. For $n > 0$, the induction hypothesis can be applied to f_1, since $a'_{n-1} = a_n$; in other words, f_1 has at most $n - 1$ zeros. Now if α is a zero of f, then (4) gives us the equation $f(x) = (x - \alpha) f_1(x)$. For a zero $x \neq \alpha$ of f we thus have $(x - \alpha) f_1(x) = 0$, and therefore $f_1(x) = 0$, since $x - \alpha \neq 0$ and R has no divisors of zero. Thus f has at most[6] one zero more than f_1 and therefore at most n zeros.

[5] Since by IB1, §1.5, the empty set is to be considered as a finite set, the case where f has no zero is included here.

[6] It may happen that α is also a zero of f_1; cf. §2.2.

From this theorem it follows that (1) holds with $a_n \neq 0$ and $f(x) = 0$ for all $x \in R$ only if R has at most n elements. It follows at once that if R has infinitely many elements and if $f(x) = 0$ for all $x \in R$, then (1) implies $a_0 = \cdots = a_n = 0$.

1.3. *Horner's Rule*

From the definition of the a_k in (5) it follows at once that $a'_{n-1} = a_n$, $a'_{k-1} = a'_k\alpha + a_k$ $(k = 1, \dots, n-1)$, $f(\alpha) = \alpha'_0\alpha + a_0$. These equations lead to the following simple rule, due to Horner, for calculating[7] $f(\alpha)$:

$$\begin{array}{cccccccc}
a_n & & a_{n-1} & & \cdots & a_1 & & a_0 \\
 & \big| & a'_{n-1}\alpha & \big| & \cdots & a'_1\alpha & \big| & a'_0\alpha \; \big| \\
\hline
\downarrow & \nearrow & \downarrow & \nearrow & \nearrow & \downarrow & \nearrow & \downarrow \\
a'_{n-1} & & a'_{n-2} & & \cdots & a'_0 & \mid & f(\alpha)
\end{array}$$

The vertical arrows denote addition, and the diagonal ones multiplication with α. This procedure produces not only $f(\alpha)$ but also the a'_k and therefore f_1. If we apply the the same procedure to f_1 in the case $n > 1$, we obtain, corresponding to (4), (5), the values of $f_1(\alpha)$, the a''_k with $f_2(x) = \sum_{k=0}^{n-2} a''_k x^k$, and $f_1(x) = f_1(\alpha) + (x - \alpha) f_2(x)$, so that after substituting these values in (4) we have

$$f(x) = f(\alpha) + f_1(\alpha)(x - \alpha) + (x - \alpha)^2 f_2(x).$$

Continuing this way, we obtain recursively f_h, the a_k^h $(k = 0, \dots, n-h)$ for $h = 1, \dots, n$ with $f_h(x) = \sum_{k=0}^{n-h} a_k^h\, x^k$, and $f_{h-1}(x) = f_{h-1}(\alpha) + (x - \alpha) f_h(x)$, where we have set $f_0 = f$, $a_k^{(0)} = a_k$. By complete induction on h we have

$$(6) \qquad f(x) = \sum_{i=0}^{h-1} f_i(\alpha)(x - \alpha)^i + (x - \alpha)^h f_h(x) \qquad \text{for} \quad h \leqslant n.$$

Since f_n is obviously a constant, so that $f_n(x) = f_n(\alpha)$, we further obtain from (6)

$$(7) \qquad f(x) = \sum_{i=0}^{n} f_i(\alpha)(x - \alpha)^i,$$

which converts the representation (1) from x to $x - \alpha$. By complete induction on h it is easy to show that $a_{n-h}^{(h)} = a_n$ and thus[8] $f_n(\alpha) = a_0^{(n)} = a_n$.

[7] The rule is particularly convenient for use with a slide rule, since the only multiplications are with the constant factor α.

Applied to $f(x) = 2x^4 - 3x^3 + x^2 - 1$ and $\alpha = 1$, Horner's rule gives:

$$\begin{array}{rrrrr}
2 & -3 & 1 & 0 & -1 \\
 & 2 & -1 & 0 & -0 \\
\hline
2 & -1 & 0 & 0 & |\ -1 \\
 & 2 & 1 & 1 & | \\
\hline
2 & 1 & 1 & |\ 1 & \\
 & 2 & 3 & | & \\
\hline
2 & 3 & |\ 4 & & \\
 & 2 & | & & \\
\hline
2 & |\ 5 & & & \\
2 & & & &
\end{array}
\qquad
\begin{array}{l}
f(x) \\
= 2(x-1)^4 + 5(x-1)^3 + 4(x-1)^2 + (x-1) - 1
\end{array}$$

We can also use the rule to transform the expression $(1 + z)^n$ by setting $f(x) = x^n$, $\alpha = 1$, and $z = x - 1$; if, for simplicity, we omit the intermediate rows (with the $a_k^{(h)}\alpha$), we obtain:

$$\begin{array}{ccccccc}
1 & 0 & 0 & \cdot & \cdot & \cdot & 0 \\
1 & 1 & 1 & \cdot & \cdot & \cdot & \boxed{1} \\
1 & 2 & 3 & \cdot & \cdot & \boxed{n} & \\
1 & 3 \rightarrow & \downarrow 6 & \cdot & | & & \\
\cdot & \cdot & \cdot & \cdot & & &
\end{array}$$

If we strike out the first row, number the remaining rows and columns from 0 to n, and denote by c_{ik} the number in the ith row and kth column, then by Horner's rule (as indicated by the two small arrows) we have

(8) $c_{0k} = c_{i0} = 1, \quad c_{i+1,k+1} = c_{i+1,k} + c_{i,k+1}$ for $0 \leqslant i + k \leqslant n - 2$

and

$$(1 + z)^n = \sum_{k=0}^{n} c_{i,n-i} z^i. \tag{9}$$

The c_{ik} thus recursively determined by (8) are called the *binomial coefficients*. By complete induction on n it is easy to show from (8) that $c_{ik} = \binom{i+k}{i}$, where as usual we have set

$$\binom{n}{i} = \Big(\prod_{k=n-i+1}^{n} k\Big) \Big/ \prod_{k=1}^{i} k = \frac{n!}{i!(n-i)!}. \tag{10}$$

If we multiply (9) by a^n and set $az = b$, we obtain the *binomial theorem*

$$(a + b)^n = \sum_{i=0}^{n} \binom{n}{i} a^{n-i} b^i. \tag{11}$$

[8] Since the method of comparison of coefficients (see §2.1) is not yet at our disposal, we cannot simply deduce this fact from (7).

The table for the c_{ik} is usually turned through 45° and called the *Pascal triangle*:

$$\begin{array}{ccccccccccc} & & & & & 1 & & & & & \\ & & & & 1 & & 1 & & & & \\ & & & 1 & & 2 & & 1 & & & \\ & & 1 & & 3 & & 3 & & 1 & & \\ & 1 & & 4 & & 6 & & 4 & & 1 & \\ . & & . & & . & & . & & . & & . \end{array}$$

Since we are working here in an arbitrary commutative ring R with unit element 1, the c_{ik} are not to be considered as natural numbers but rather as elements $\sum_{j=1}^{m} 1$ of R. But then we could have $1 + 1 = 0$, for example (cf. §1.1), so that the quotients in (10) would cause trouble, which may be avoided by regarding the c_{ik} as natural numbers and interpreting an expression like mr (m a natural number, $r \in R$) in (9) and (11) as $\sum_{j=1}^{m} r$. With this precaution the result (11) holds in any commutative ring with unit element.[9]

2. Polynomials

2.1. *Formation of a Ring of Polynomials*

In §1.1 we have already seen that in general the function $x \to \sum_{i=0}^{n} a_i x^i$ does not uniquely determine the a_i. But for calculation with such expressions as $\sum_{i=0}^{n} a_i x^i$ it would be very convenient to be able to assume that the coefficients a_i are uniquely determined by the values of the expression. This will unquestionably be the case (for an element x with certain properties) if in R or in a suitable extension of R we can find an element x such that an equation $\sum_{i=0}^{n} a_i x^i = 0$ always implies $a_0 = \cdots = a_n = 0$; for then we can recognize, as in §1.1, that $\sum_{i=0}^{n} a_i x^i = \sum_{i=0}^{n} b_i x^i$ implies (*comparison of coefficients*) the equations $a_i = b_i$ ($i = 0, ..., n$). An element x with this property will be called a *transcendent* over R. If R is the field of rational numbers, then in agreement with the definition in IB6, §8.1, any transcendental number may be chosen as a transcendent over R in the present sense. Since a transcendent x cannot satisfy any algebraic equation $\sum_{i=0}^{n} a_i x^i = 0$ with $a_n \neq 0$, it cannot be characterized (i.e., determined) by statements involving only x, elements of R, and equality, addition, and multiplication in R. Thus

[9] Of course, the proof could have been carried out independently of Horner's rule. We can also dispense with the existence of a unit element in R if we agree that in (11) $a^n b^0$, $a^0 b^n$ are to be interpreted simply as a^n, b^n.

the transcendents are also called *indeterminates*.[10] But a name of this sort must not be allowed to conceal the fact that a transcendent must be a definite element (of an extension ring of R) and that the existence of such elements must in every case be proved. As an indeterminate over the field of rational numbers we may, as remarked above, choose any transcendental number (such as e or π).

Thus it becomes our task to extend the commutative ring R with unit element 1 to a commutative ring R' containing a transcendent x over R. By saying that R' arises from extension of R and may thus be called an *extension ring* we mean that the elements of R are all contained in R', and that addition and multiplication of these elements leads to the same result in R' as in R; we express the same idea by saying that R is a *subring* of R'.[11] Throughout the present chapter we shall make the tacit assumption that the *unit element of R is also the unit element of R'*, and we shall also assume that all the rings in question are *commutative*.

Such a ring R' certainly contains all the expressions $\sum_{i=0}^{m} a_i x^i$. But by the definition of a transcendent these expressions are in one-to-one correspondence with the sequences $(a_n)_{n=0,1,\ldots}$, if from the sequence $a_0, \ldots, a_m$ we construct an infinite sequence by setting $a_n = 0$ for $n > m$. So let us see what will happen if for R' we simply take the set of sequences[12] $a = (a_n)_{n=0,1,\ldots}$ with the property that there exists a natural number m, such that $a_n = 0$ for $n > m$. Motivated by (2) and (3), we now define addition and multiplication in R' by

$$(12) \qquad (a + b)_n = a_n + b_n, \qquad (ab)_n = \sum_{i=0}^{n} a_i b_{n-i},$$

from which it is easy to see that the sequences $a + b$, ab are again contained in R'. With respect to this addition the set R' is obviously a module, and we see at once that the multiplication is commutative and distributive. Finally, associativity is shown thus:

$$((ab)\, c_n) = \sum_{k=0}^{n} \sum_{i=0}^{k} a_i b_{k-i} c_{n-k} = \sum_{i=0}^{n} a_i \sum_{k=i}^{n} b_{k-i} c_{n-k}$$

$$= \sum_{i=0}^{n} a_i \sum_{h=0}^{n-i} b_h c_{n-i-h} = (a(bc))_n .$$

[10] In §§2 and 3 the symbol x will almost always denote an indeterminate; more precisely, x is a variable for which only indeterminates can be substituted. On the other hand, in §1 the variable x (provided it is not bound) may be replaced by any of the elements of a ring.

[11] If R' or R is a field, we speak of an *extension field* or *subfield*, respectively.

[12] For the notation see IB1, §4.4. Instead of simply writing a we shall sometimes use the more complete symbol $(a_0, \ldots, a_n, 0, \ldots)$.

Consequently, R' is a commutative ring. Also, it is easy to show that $a \to (a, 0, ...)$ is an isomorphism of the ring R into the ring R'. Thus we may extend the equality (cf. IB1, §4.4), which up to now has been defined only between elements of R and elements of R', by setting

$$(13) \qquad a = (a, 0, ...) \quad \text{and} \quad (a, 0, ...) = a \qquad \text{for} \quad a \in R.$$

Then R is a subring of R', and the unit element 1 of R is also the unit element of R'; moreover, the zero element of R is also the zero element of R', as follows at once from (12), (13).

Then R' certainly contains a transcendent over R, namely $z = (0, 1, 0, ...)$. To prove this we derive the equation

$$(14) \qquad (a_0, ..., a_n, 0, ...) = \sum_{i=0}^{n} a_i z^i$$

by complete induction on n. For $n = 0$ the equation follows from (13). From (14) for an integer $n \geqslant 0$ we have

$$(a_0, ..., a_{n+1}, 0, ...) = \sum_{i=0}^{n} a_i z^i + (b_0, ..., b_{n+1}, 0, ...),$$

if $b_i = 0$ for $i = 0, ..., n$ and $b_{n+1} = a_{n+1}$. For the case $a_0 = \cdots = a_{n-1} = 0$, $a_n = a_{n+1}$, we further obtain from (14)

$$a_{n+1} z^n = (c_0, ..., c_n, 0, ...),$$

with $c_i = 0$ for $i = 0, ..., n - 1$ and $c_n = a_{n+1}$. From the definition of b_i, c_i, z it now follows at once from (12) that

$$(a_{n+1} z^n)\, z = (b_0, ..., b_{n+1}, 0, ...),$$

and thus

$$(a_0, ..., a_{n+1}, 0, ...) = \sum_{i=0}^{n} a_i z^i + a_{n+1} z^{n+1} = \sum_{i=0}^{n+1} a_i z^i,$$

which completes the proof by induction. If we now have $\sum_{i=0} a_i z^i = 0$, it follows from (14) that $(a_0, ..., a_n, 0, ...) = 0$; but since (13) are the only equations holding between an element of R and an element of R', we thus have $a_0 = \cdots = a_n = 0$, so that z is a transcendent. As a generalization of the concepts in IB3, §1.3, we may now state the content of (14) in the following way: the x^i $(i = 0, 1, ...)$ form a basis of R'; that is, R' is a vector space (of infinite dimension) over a domain of scalars that is not necessarily a skew field but only a commutative[13] ring.

[13] The commutativity is required only for the multiplication. Compare the multiplication in a vector space R' with the multiplication in an algebra of finite order (IB5, §3.9).

Any commutative ring which, like the ring R' just constructed, contains the ring R and also a transcendent x with respect to R, and also consists only of elements of the form $\sum_{i=0}^{n} a_i x^i$ ($a_i \in R$), is called a *polynomial ring in the indeterminate* (or also in the *generator*) *x over R* and its elements are called *polynomials in x over R*. The above discussion shows that a polynomial $y = \sum_{i=0}^{n} a_i x^i$ determines the sequence $(a_0, ..., a_n, 0, ...)$ uniquely. The terms of this sequence are called the *coefficients of y* (more precisely: a_i = *coefficient of* x^i *in y*). For $y \neq 0$ it is obvious that there exists exactly one greatest integer $n \geqslant 0$ with $y = \sum_{i=0}^{n} a_i x^i$, $a_i \in R$ and $a_n \neq 0$. This number is called the *degree of y*, and a_n is the *leading coefficient*. To the polynomial 0 we shall assign the degree 0, although this is not usually done. Then the set of polynomials of degree $\leqslant n$ (or also $< n$ in case $n > 0$) is a module with respect to addition, as is easily shown. In particular, the set of polynomials of degree 0 is equal to R itself.

The ring R' thus defined is not the only polynomial ring in an indeterminate x over R, although it is the easiest to construct; but every such polynomial ring is mapped isomorphically onto R' by the correspondence

$$\sum_{i=0}^{n} a_i x^i \to (a_0, ..., a_n, 0, ...),$$

where the elements of R remain fixed and x is mapped onto $z = (0, 1, 0, ...)$. The calculations given above in (2), (3) remain valid here, so that with $1 = \max(n, m)$, $a_i = 0 = b_k$ for $i > n$, $k > m$ we have

$$(15)\quad \begin{cases} \sum_{i=0}^{n} a_i x^i + \sum_{i=0}^{m} b_i x^i = \sum_{i=0}^{l} (a_i + b_i)\, x^i \\ \left(\sum_{i=0}^{n} a_i x^i\right)\left(\sum_{i=0}^{m} b_i x^i\right) = \sum_{i=0}^{n+m} \left(\sum_{h=0}^{i} a_h b_{i-h}\right) x^i \end{cases}$$

Moreover, it is easy to show that this is the only isomorphism with the desired properties. The various methods of construction are therefore completely equivalent to one another, so that we may speak of *the* polynomial ring in x over R. To denote this ring we shall use the symbol $R[x]$.

If S is an arbitrary extension ring of R (the simplest case would be $S = R$), then every polynomial $\sum_{i=0}^{n} a_i x^i \in R[x]$ defines an entire rational function

$$(16)\quad u \to \sum_{i=0}^{n} a_i u^i$$

in S; for it follows from $\sum_{i=0}^n a_i x^i = \sum_{i=0}^m b_i x^i$ that $(a_0, ..., a_n, 0, ...) = (b_0, ..., b_m, 0, ...)$ and thus $\sum_{i=0}^n a_i u^i = \sum_{i=0}^m b_i u^i$, so that in fact the mapping (16) is uniquely defined by the polynomial alone. Moreover, the definition of addition and multiplication of polynomials is such that for every $u \in S$ the mapping $\sum_{i=0}^n a_i x^i \to \sum_{i=0}^n a_i u^i$ is a homomorphism of the ring $R[x]$ into the ring S, as is easily shown. The homomorphism determined by u in this way is often referred to as *substitution of* u (for the indeterminate x). In particular, if S is an extension ring of $R[x]$ and if the function (16) is denoted by f, we have $\sum_{i=0}^n a_i x^i = f(x)$: the polynomial is the value, for the argument x, of the function corresponding to it, and therefore it is uniquely determined by f. In this way we obtain a one-to-one correspondence between the polynomials and all functions defined[14] in an extension ring of $R[x]$ (which may be $R[x]$ itself) by the $a_0, ..., a_n \in R$ as in (16). Consequently, polynomials in x will be written below in the form $f(x)$, where f is the function so defined. Then the results of §1.2, §1.3 also hold for polynomials. The value $f(u)$ of the function f (i.e., of the function corresponding to the polynomial $f(x)$ for an argument u) in an extension ring of $R[x]$ will be called, concisely though inexactly, the *value of the polynomial* $f(x)$ at the point u. But this abbreviated way of speaking must not be allowed to conceal the fact that a polynomial over R is not necessarily a function defined in R (or in an extension ring of R). In any case, the above polynomials $(a_0, ..., a_n, 0, ...)$ by means of which we demonstrated the existence of polynomial rings, are not functions of this sort. Of course, it is *possible* that polynomials over R may also be functions in R. For example, in an infinite ring R without divisors of zero we may, by §1.1, §1.2, define the polynomial ring over R as the ring of entire rational functions in R with the identity function I as generator, provided we set the constant function $\underline{c}$ equal to c. But in many cases (though not in the case just mentioned) we must even then distinguish between the polynomial as a function in R and the function (16) corresponding to it. Under the assumptions just mentioned for R and the $\underline{c}$, not only I but also I^2 is a transcendent over R in $R[I]$, so that we can form the ring $R[I^2]$ of polynomials in I^2 over R, and then of course the elements of this ring are entire rational functions in R. The polynomial $1 + I^2$ is then the function $u \to 1 + u^2$, whereas (16) assigns to it the function $u \to 1 + u$, since we may set $a_0 = a_1 = 1, n = 1$.

This sharp distinction between a polynomial and an entire rational function is a necessary one from the logical point of view, but it is often

[14] If the domain of definition of the function (16) is restricted to R, then in general the uniqueness of this correspondence is lost and can be restored only under certain special assumptions, e.g., that R has infinitely many elements and no divisors of zero (see §1.1, §1.2).

disregarded in the various branches of mathematics; in many cases only one of the two concepts is actually needed but both names are used for it. There are historical reasons for this practice. Originally the word "polynomial" denoted any expression with several terms, and then more particularly an expression of the form $a_0 + a_1x + \cdots + a_nx^n$ in powers of a variable x. Now it is common practice in analysis to use the word "function" not only for the function (= mapping) but also for its value at the point x (and similarly for the case of several arguments). Thus it became customary in analysis to use "polynomial" and "entire rational function" as synonyms, and this practice can be justified on the basis of our definitions, provided we make a strict distinction between a function and its value; for in analysis the coefficients are taken either from the field of real numbers or of complex numbers, and thus, since each of these fields contains infinitely many elements, every entire rational function can be interpreted, as remarked above, as a polynomial in I, where I is the identical mapping $x \to x$.

It was Steinitz, in his fundamental work [1a] of the year 1910, who first introduced the precise concept of an indeterminate as an element that is transcendental over the domain of coefficients. What we call a "polynomial" is called by him an "entire rational function of the transcendent x"; in our present language it would have been more precise to call it the "value of an entire rational function at the point x." In a textbook [1] published in 1926, H. Hasse distinguishes between an "entire rational function in the sense of analysis" and an "entire rational function in the sense of algebra" (i.e., polynomials in our nomenclature). In the later textbooks on algebra (e.g., van der Waerden [1], Haupt [1]) the word "polynomial" is used exclusively for expressions $a_0 + \cdots + a_nx^n$ in a transcendent x (over the ring containing the a_i), but with the remark, in concession to the older usage, that the words "entire rational function" are also used. Finally, in Bourbaki [3] a polynomial is clearly distinguished from an entire rational function by the notation itself, although a function of this sort is given the name *fonction polynome* which emphasizes the close connection between the two concepts.

2.2. *Zeros*

In this section R is a *commutative ring with unit element* 1 and without divisors of zero. Then for $a_n \neq 0$, $b_m \neq 0$ it follows at once from (15) that $a_nb_m(\neq 0)$ is the coefficient of x^{n+m} and is at the same time the leading coefficient of $\sum_{i=0}^n a_ix^i \sum_{i=0}^m b_ix^i$. Consequently, *the product of two polynomials* $\neq 0$ *is also* $\neq 0$, *and its degree is the sum of the degrees of the two polynomials. In particular,* $R[x]$ *has no divisors of zero.* By complete induction on the number of factors it is easy to extend this theorem to products of several polynomials.

By a *zero*[15] of the polynomial $f(x) \in R[x]$ we mean a zero of f in any extension ring of $R[x]$.[16] For a zero α of $f(x)$ it follows from (4) that $f(x) = (x - \alpha) f_1(x)$. This equation naturally raises the question: for which natural numbers m do we have

$$f(x) = (x - \alpha)^m g(x) \tag{17}$$

for a $g(x)$ (dependent on m) with $g(x) \in R[x]$? If $f(x) = 0$, then m may of course have any value, but if $f(x) \neq 0$, our discussion shows that the degree of $f(x)$ must be equal to the sum of m and the degree of $g(x)$, so that $m \leqslant$ degree of $f(x)$. Thus, for $f(x) \neq 0$ there exists a greatest natural number m with an equation (17); this number is called the *multiplicity of the zero* α. It can also be characterized by (17) and $g(\alpha) \neq 0$. For on the one hand, if $g(\alpha) = 0$, it follows from (4) that $g(x) = (x - \alpha) g_1(x)$ and thus by (17) that $f(x) = (x - \alpha)^{m+1} g_1(x)$; and on the other hand, from (17) and $f(x) = (x - \alpha)^{m'} h(x)$ with $m' > m$ we have

$$(x - \alpha)^m (g(x) - (x - \alpha)^{m'-m} h(x)) = 0,$$

and also from $x - \alpha \neq 0$ and the absence of divisors of zero,[17] $g(x) = (x - \alpha)^{m'-m} h(x)$ and consequently $g(\alpha) = 0$. Thus:

If the nonzero polynomial $f(x)$ has s distinct zeros $\alpha_1, ..., \alpha_s$ with multiplicities $m_1, ..., m_s$, then there exists a polynomial $h(x)$ with

$$f(x) = h(x) \prod_{i=1}^{s} (x - \alpha_i)^{m_i}. \tag{18}$$

The proof is by complete induction on s. For $s = 1$ the result (18) follows at once from the definition of m_1. Now let us assume (18) and let α be another zero ($\neq \alpha_1, ..., \alpha_s$) of $f(x)$. Since R has no divisors of zero, it follows that α is then a zero of $h(x)$, so that $h(x) = (x - \alpha)^m g(x)$, $g(\alpha) \neq 0$, $m \geqslant 1$. Setting $g^*(x) = g(x) \prod_{i=1}^{s} (x - \alpha_i)^{m_i}$, we then have $f(x) = (x - \alpha)^m g^*(x)$, $g^*(\alpha) \neq 0$, so that m is the multiplicity of α as a zero of $f(x)$. Substitution of $(x - \alpha)^m g(x)$ for $h(x)$ in (18) then provides the statement necessary for the induction. By comparing the degrees on both sides of (18) we obtain a sharpening of the theorem in §1.2:

[15] Instead of "zero" the word "root" is often used. This meaning of "root" is of course different from the concept of an nth root in the field of real numbers (IB1, §4.7). The connection between the two concepts lies in the fact that the nth root of a is a root of the polynomial $x^n - a$.

[16] We could write R here in place of $R[x]$, but then we would have to change our notation, since then the function (16) formed for a polynomial in an extension ring of R no longer determines the polynomial uniquely in every case.

[17] Here and in the preceding discussion we could dispense with the postulate of absence of divisors of zero, since the highest coefficient of $x - a$ is equal to 1.

The sum of the multiplicities of the zeros of a nonzero polynomial is not greater than its degree.

In the notation of §1.3 it follows readily from (7) that the multiplicity m of a zero α is characterized by the equations $f_i(\alpha) = 0$ for $0 \leqslant i < m$, $f_m(\alpha) \neq 0$. In particular, α is a *multiple zero* (i.e., $m > 1$), if and only if $f(\alpha) = f_1(\alpha) = 0$. Instead of $f_1(x)$ it is often more convenient to make use of the *derivative* $f'(x)$ of $f(x)$, defined as follows:[18]

$$(19) \qquad f'(x) = \sum_{i=1}^{n} ia_i x^{i-1};$$

of course (since R does not necessarily contain the natural numbers), ia_i is to be interpreted here as $\sum_{j=1}^{i} a_i$ (i.e., as the sum of i summands, each of which is $= a_i$). If in the definition (5) of $f_1(x)$ we now replace α by x, then $f_1(x)$ becomes $f'(x)$, as is easily shown by changing the order of summation in (5); in particular, we thus have $f_1(\alpha) = f'(\alpha)$. It is obvious from (19) that the derivative of a sum of polynomials is equal to the sum of their derivatives. The rule for the derivative of a product, namely

$$(20) \qquad h'(x) = f'(x)\,g(x) + f(x)\,g'(x), \qquad \text{if} \quad h(x) = f(x)\,g(x),$$

can of course be proved from (19), but we shall give a simpler proof in §3.1. By complete induction on n we then obtain from (20)

$$(20') \qquad f'(x) = \sum_{k=1}^{n} (g_k'(x) \prod_{i \neq k} g_i(x)\,, \qquad \text{if} \quad f(x) = \prod_{i=1}^{n} g_i(x).$$

In particular, if we set $g_i(x) = x - \alpha_i$, then from (20′) we have

$$f'(x) = \sum_{k=1}^{n} \prod_{i \neq k} (x - \alpha_i).$$

If R, and thus also $R[x]$, has no divisors of zero, we can form $(x - \alpha_k)^{-1}$ in the quotient field $R(x)$ (see §2.3) and then obtain

$$f'(x) = f(x) \sum_{k=1}^{n} (x - \alpha_k)^{-1}.$$

2.3. *Polynomials in Several Indeterminates*

We again let R be a commutative ring with unit element 1. As a generalization of the concepts in §2.1, the elements $x_1, \ldots, x_n$ of an

[18] In this definition of the derivative no use is made of the concept of a limit, but if R is the field of real numbers, then the definition of $f'(x)$ by the usual limiting process produces exactly the same result as here.

extension ring R' of R are now called *independent transcendents*, or also *indeterminates*, if an equation[19] of the form

$$(21) \qquad \sum_{0 \leqslant i_k \leqslant m} a_{i_1 \dots i_n} x_1^{i_1} \cdots x_n^{i_n} = 0 \qquad (a_{i_1 \dots i_n} \in R)$$

implies $a_{i_1 \dots i_n} = 0$ for all indices. Every subring of R' which contains R and all the x_i also contains all expressions of the form on the left-hand side of (21), and it is easy to show that these expressions again form a ring. This ring is called the *polynomial ring* $R[x_1, \dots, x_n]$ *in the indeterminates* $x_1, \dots, x_n$ *over* R and its elements are called *polynomials*[20] in the $x_1, \dots, x_n$ over R. For $n = 1$ this is obviously the definition in §2.1. For $n > 1$ we have

$$(22) \qquad R[x_1, \dots, x_n] = R[x_1, \dots, x_{n-1}][x_n].$$

To prove this statement we denote $R[x_1, \dots, x_{n-1}]$ by S, and $R[x_1, \dots, x_n]$ by T. Then obviously $S \subseteq T$. From $\sum_{i=0}^{m} u_i x_n^i = 0$ $(u_i \in S)$, if we express the u_i in terms of the $x_1, \dots, x_{n-1}$, we obtain an equation of the form (21), whose coefficients are thus all $= 0$, which means that $u_i = 0$. Consequently, x_n is an indeterminate over S, so that in T we can form the polynomial ring $S[x_n]$. Conversely, for

$$u_i = \sum_{0 \leqslant i_k \leqslant m} a_{i_1 \cdots i_{n-1} i} x_1^{i_1} \cdots x_{n-1}^{i_{n-1}} \qquad (i = 0, \dots, m)$$

we at once obtain

$$\sum_{0 \leqslant i_k \leqslant m} a_{i_1 \cdots i_n} x_1^{i_1} \cdots x_n^{i_n} = \sum_{i=0}^{n} u_i x_n{}^i, \qquad u_i \in S$$

and therefore $T \subseteq S[x_n]$, which completes the proof of (22).

By (22) we have reduced the construction of a polynomial ring in several indeterminates to the successive construction of polynomial rings in one indeterminate. Thus for every R and for every natural number n there exists a polynomial ring over R in n indeterminates.

The $a_{i_1 \dots i_n}$ in

$$(23) \qquad y = \sum_{0 \leqslant i_k \leqslant m} a_{i_1 \cdots i_n} x_1^{i_1} \cdots x_n^{i_n} \qquad (\in R[x_1, \dots, x_n])$$

are called the *coefficients* of the polynomial y. They are uniquely determined by y, since a second representation (23) would lead, when subtracted

[19] Strictly speaking, we should also write $k = 1, \dots, n$ below the sign of summation; this summation is taken over all n-tuples $(i_1, \dots, i_n)$ with $0 \leqslant i_k \leqslant m$ $(k = 1, \dots, n)$.

[20] The same term is sometimes used (see, e.g., IB5, §3.9) even when the x_i are not independent transcendents.

from the first one, to an equation of the form (21). For each coefficient $a_{i_1 \ldots i_n} \neq 0$ in y the number $i_1 + \cdots + i_n$ is called the *degree of the term* $a_{i_1 \ldots i_n} x_1^{i_1} \cdots x_n^{i_n}$. For $y \neq 0$ the maximum of the degrees of the individual terms (with coefficient $\neq 0$) is the *degree of* y *in the* $x_1, ..., x_n$. This degree is to be distinguished from the *degree of* y *in* x_i, which is defined as the degree of y as a polynomial in x_i over the ring of polynomials in the other indeterminates. Thus $x_1 x_3^2 + x_2$ is of degree 3 in x_1, x_2, x_3, of degree 2 in x_3, and of degree 1 in x_1 and x_2. If all the terms of y (with coefficient $\neq 0$) have the same degree, then y is said to be *homogeneous*. Thus the homogeneous polynomials of degree 1 have the form $\sum_{i=1}^{n} a_i x_i$.

As in §2.1 for the case $n = 1$, we now assign to each polynomial $y \in R[x_1, ..., x_n]$ a function f of n arguments in an extension S of $R[x_1, ..., x_n]$ by defining $f(u_1, ..., u_n)$ for $u_1, ..., u_n \in S$ as the element that arises on the right side of (23) by substitution[21] of u_i for x_i. In particular, we then have $y = f(x_1, ..., x_n)$, so that this correspondence is one-to-one. If $u_1, ..., u_n \in R$, then $f(u_1, ..., u_n)$ is also in R, so that if we restrict the domain of definition, f becomes a function of n arguments in R. The functions defined in this way are called the *entire rational functions of n arguments in R*.

By complete induction on n it follows from (22), in view of the theorem at the beginning of §2.2, that if R has no divisors of zero, then $R[x_1, ..., x_n]$ has none either. Thus if R has no divisors of zero, then by IB1, §3.2 we can form the quotient field of $R[x_1, ..., x_n]$, whose elements are therefore of the form y/z with $y, z \in R[x_1, ..., x_n]$, $z \neq 0$. A quotient field of this sort is denoted by $R(x_1, ..., x_n)$. In view of the close connection between polynomials and entire rational functions, this field is usually called the *field of rational functions over* R, although its elements, being defined as quotients of polynomials, are in general not functions. But of course, to every element $f(x_1, ..., x_n)/g(x_1, ..., x_n)$ of $R(x_1, ..., x_n)$ there corresponds a rational function R,

$$(u_1, ..., u_n) \to f(u_1, ..., u_n)/g(u_1, ..., u_n);$$

its domain of definition consists of the n-tuples $(u_1, ..., u_n)$ with $u_1, ..., u_n \in R$, $g(u_1, ..., u_n) \neq 0$, and its values are in the quotient field of R.

The remarks at the end of §2.1 about the terminology "polynomial" and "entire rational function" apply equally well here to the case of several transcendents and to the terms "polynomial quotient" and "rational function," which again are often used synonymously, as is clear from the choice of name for the quotient field of a polynomial ring. A

[21] As in the case $n = 1$, this substitution is possible even if S is only an extension of R and not necessarily of $R[x_1, ..., x_n]$.

striking result of this practice is the fact that the elements of an algebraic extension field (see IB7, §2) over $R(x_1, ..., x_n)$, where R is a field, are called *algebraic functions* over R. Of course, these functions in the sense of algebra are not necessarily functions at all; in particular, they are certainly not algebraic functions in the sense of the theory of functions of a complex variable (see III6, §5). On the other hand, the latter functions can always be regarded as algebraic functions in the sense of algebra.

2.4. *Symmetric Polynomials*

An important special case of (18) arises when $h(x) \in R$. Comparison of degrees then shows that $\sum_{i=1}^{s} m_i = n$, where n is the degree of $f(x)$. Since multiplicities will play no role in what follows, we index the zeros from 1 to n in such a way that the number of times each zero appears is equal to its multiplicity; we then have

$$f(x) = c(x - \alpha_1) \cdots (x - \alpha_n). \tag{24}$$

By multiplying out on the right-hand side [see IB1, (41′)] we see that the coefficient a_{n-i} of x^{n-i} in $f(x)$, with $a_n = c$, satisfies the equation

$$a_{n-i} = (-1)^i c \sum_{0<k_1<\dots<k_i \leqslant n} \alpha_{k_1} \cdots \alpha_{k_i} \qquad (i = 1, ..., n). \tag{25}$$

The notation under the summation sign indicates that the summation is to be taken over the set of all i-tuples $(k_1, ..., k_i)$ with positive integers $k_h \leqslant n$ $(h = 1, ..., i)$ and $k_h < k_{h+1}$ $(h = 1, ..., i - 1)$. This relationship between the coefficients and the zeros of a polynomial in the case (24) suggests that in the polynomial ring $R[x_1, ..., x_n]$ (where R is an arbitrary commutative ring with unit element) we should pay special attention to the polynomials

$$\sigma_i(x_1, ..., x_n) = \sum_{0<k_1<\dots<k_i \leqslant n} x_{k_1} \cdots x_{k_i} \qquad (i = 1, ..., n), \tag{26}$$

where in particular

$$\sigma_1(x_1, ..., x_n) = x_1 + \cdots + x_n, \qquad \sigma_n(x_1, ..., x_n) = x_1 \cdots x_n.$$

Then (25) can be written in the form $a_{n-i} = (-1)^i \sigma_i(\alpha_1, ..., \alpha_n)$. Obviously $\sigma_i(x_1, ..., x_n)$ is homogeneous of degree i, and has the further property, immediately obvious from (26), that it is left unchanged by an arbitrary permutation of the $x_1, ..., x_n$. Polynomials with this property are called *symmetric*. From the uniqueness of the coefficients it follows that a polynomial is symmetric if and only if each coefficient is left unchanged by an arbitrary permutation of the indices. Now the polynomials

$\sigma_i(x_1, ..., x_n)$ are of basic importance for all symmetric polynomials, in the following sense:

For every symmetric polynomial $f(x_1, ..., x_n)$ from $R[x_1, ..., x_n]$ there exists[22] *a polynomial $F(x_1, ..., x_n)$ in $R[x_1, ..., x_n]$, such that*

$$f(x_1, ..., x_n) = F(\sigma_1(x_1, ..., x_n), ..., \sigma_n(x_1, ..., x_n)).$$

For this reason the $\sigma_i(x_1, ..., x_n)$ are called the *elementary symmetric polynomials*[23] in the $x_1, ..., x_n$, and the above theorem is called the *fundamental theorem of the elementary symmetric polynomials;* this theorem states that every symmetric polynomial can be expressed as an entire rational expression in the elementary symmetric polynomials.

For the proof we choose a natural number $g > 1$ and confine our attention to the symmetric polynomials $f(x_1, ..., x_n) \neq 0$ of degree $< g$. As indices for the coefficients we will then have only n-tuples $(i_1, ..., i_n)$ with $0 \leqslant i_k < g$ $(k = 1, ..., n)$. By the mapping

$$(i_1, ..., i_n) \rightarrow \sum_{k=1}^{n} i_k g^{n-k}$$

this set of n-tuples is put into one-to-one correspondence with the set of non-negative integers $< g^n$; the image of an n-tuple will be called its numeral.[24] Let the greatest numeral of an n-tuple $(i_1, ..., i_n)$ with $a_{i_1 \dots i_n} \neq 0$ in $f(x_1, ..., x_n)$ be denoted by h. By the principle of complete induction (see IB1, §1.4) it is sufficient to prove the assertion for $f(x_1, ..., x_n)$ under the assumption that it is already known to be correct for all symmetric polynomials whose nonzero coefficients have an index numeral $< h$. If $(i_1, ..., i_n)$ is the n-tuple with numeral h, it follows that $i_k \geqslant i_{k+1}$ $(k = 1, ..., n - 1)$, for if $i_k < i_{k+1}$, then the n-tuple arising from $(i_1, ..., i_n)$ by interchange of i_k with i_{k+1} would have a numeral $> h$, so that its coefficient would necessarily be zero, whereas in view of the symmetry of $f(x_1, ..., x_n)$ this coefficient is $= a_{i_1 \dots i_n} \neq 0$. Abbreviating $\sigma_i(x_1, ..., x_n)$ to σ_i, we now write down the obviously symmetric polynomial

$$(27) \qquad f^*(x_1, ..., x_n) = f(x_1, ..., x_n) - a_{i_1 \dots i_n} \prod_{k=1}^{n-1} \sigma_k^{i_k - i_{k+1}} \sigma^{i_n}.$$

[22] It can be proved that there is "exactly one" such polynomial. See, e.g., van der Waerden [2], §29.

[23] The entire functions corresponding to them (see the end of §2.3) are called the *elementary symmetric functions.*

[24] It is obvious that in this enumeration the n-tuples are arranged in lexicographic order (see IB1, §4.1).

Since σ_k is of degree k and

$$\sum_{k=1}^{n-1} k(i_k - i_{k+1}) + ni_n = \sum_{k=1}^{n} i_k < g,$$

the polynomial $f^*(x_1, ..., x_n)$ is also of degree $< g$. In a product of σ_i-factors we can find the term $\neq 0$ with greatest index numeral by choosing, for each factor σ_i in (26), the summand with least indices $k_1, ..., k_i$ (in other words, $x_1 \cdots x_i$) and then multiplying these summands together. For the product subtracted in (27) this rule gives

$$a_{i_1 \ldots i_n} \prod_{k=1}^{n-1} (x_1 \cdots x_k)^{i_k - i_{k+1}} (x_1 \cdots x_n)^{i_n};$$

or, in other words, in view of $\sum_{k=j}^{n-1} (i_k - i_{k+1})$, precisely the term $a_{i_1 \ldots i_n} x_1^{i_1} \cdots x_n^{i_n}$. Thus every coefficient $\neq 0$ in $f^*(x_1, ..., x_n)$ has an index numeral $< h$, so that the induction hypothesis can be applied to this polynomial: $f^*(x_1, ..., x_n) = F^*(\sigma_i, ..., \sigma_n)$. Then from (27) the desired result follows at once for $f(x_1, ..., x_n)$. The proof provides a method for actually calculating the F, for example:[25]

$$\begin{aligned}
&y = x_1^3 + x_2^3 + x_3^3,\\
&y - \sigma_1^3\\
&= -3x_1^2x_2 - 3x_2^2x_3 - 3x_3^2x_1 - 3x_1x_2^2 - 3x_2x_3^2 - 3x_3x_1^2 - 6x_1x_2x_3 = z,\\
&z + 3\sigma_1\sigma_2 = 3x_1x_2x_3 = 3\sigma_3,\\
&y = \sigma_1^3 - 3\sigma_1\sigma_2 + 3\sigma_3 .
\end{aligned}$$

This procedure also enables us to solve the following problem: under the assumption (24) with $c = 1$ and with a given entire rational function g of one argument in R it is required to calculate the coefficients of the polynomial $\prod_{i=1}^{n} (x - g(\alpha_i))$ in terms of those of $f(x)$. For this purpose we represent the symmetric polynomial $\sigma_i(g(x_1), ..., g(x_n))$ in the form $F_i(\sigma_1, ..., \sigma_n)$ and then obtain the desired coefficient of x^{n-i} in the form $(-1)^i F_i(-a_{n-1}, ..., (-1)^n a_0)$.

2.5. *Power Series*

In analysis, an important role is played by power series (or sums of power series) in the variable x; that is, by expressions of the form $\sum_{k=0}^{\infty} a_k x^k$.

[25] A simpler method for this case is given at the end of §2.5.

Their addition and multiplication proceeds, provided x lies inside the circle of convergence, according to the formulas:

$$\sum_{k=0}^{\infty} a_k x^k + \sum_{k=0}^{\infty} b_k x^k = \sum_{k=0}^{\infty} (a_k + b_k)\, x^k, \tag{28}$$

$$\sum_{k=0}^{\infty} a_k x^k \sum_{k=0}^{\infty} b_k x^k = \sum_{k=0}^{\infty} \Big(\sum_{i=0}^{k} a_i b_{k-i} \Big)\, x^k.$$

A power series is determined by its sequence of coefficients, i.e., by the mapping $k \to a_k$, and conversely the sequence of coefficients is determined by the power series (more precisely, by the corresponding function) if the radius of convergence is $\neq 0$ (cf. III7, §1.3). But in algebraic applications of power series we are often interested, not in the numerical values obtained by inserting some x-value (from the circle of convergence), but only in the sequences of coefficients and their combinations according to (28). Then the restriction to real or complex numbers and the consideration of questions of convergence is not only superfluous but even troublesome, since it is frequently convenient to deal with a sequence as though it were the sequence of coefficients of a power series in a ring that is not a subring of the field of complex numbers. Just as in the construction of the polynomial ring in §2.1, it is desirable here to calculate with the sequences themselves. In order to be able to divide by power series, we must also consider series of the form $\sum_{k=h}^{\infty} a_k x^k$ (with negative integer h), for which the equations (28) must be slightly generalized.

These remarks suggest the following definitions. Let R be a commutative ring with unit element 1. We consider the mappings a of the set of integers into R (with a_k as the image of k in R[26]) and confine our attention to mappings a with the property that there exists an integer h with $a_k = 0$ for $k < h$. In the set R^* of these mappings we can define an addition and multiplication, corresponding to (12), as follows:

$$(a + b)_k = a_k + b_k\,; \qquad (ab)_k = \sum_{i=h}^{k-h} a_i b_{k-i}\,, \tag{29}$$

if $a_l = b_l = 0$ for all $l < h$. Then (cf. IB1, §1.6) for $h > k - h$, or $k > 2h$ the sum $= 0$, and, in general, this sum is independent of h, provided only $a_l = b_l = 0$ for all $l < h$. Now, exactly as for R' in §2.1, it is easy to see that R^* is a commutative ring. To each $a \in R$ we assign the mapping $\bar{a}$ with $\bar{a}_0 = a$, $\bar{a}_k = 0$ for $k \neq 0$, so that R is mapped isomorphically into R^*. Thus we may set $a = \bar{a}$ and $\bar{a} = a$, whereby R^* becomes an

[26] We shall use this notation below, even when the mapping is not denoted by a single letter.

extension ring of R. Letting x denote the mapping which sends 1 into 1 and all other integers into 0, we see that x^i $(i > 0)$ is the mapping which sends i into 1 and all other integers into 0. Then it is clear that $\sum_{i=0}^n a_i x^i)_k$, (i.e., the image of k under the mapping $\sum_{i=0}^n a_i x^i$ where it is to be noted that $a_i = \bar{a}_i$) is equal to a_k for $k = 0, ..., n$ and is otherwise 0. In particular, $\sum_{i=0}^n a_i x^i = 0$ implies $a_i = 0$, so that x is a transcendent over R. Thus we can construct the polynomial ring $R[x]$ as a subring of R^*.

In the particular case that R is a field, the extension ring R^* also contains the quotient field $R(x)$ and is thus itself a field. For the proof of this assertion we first note that the mapping which sends -1 into 1 and other integers into 0 is the inverse x^{-1} of x with respect to multiplication. Now for $a \neq 0$ in R^* there exists by assumption an integer h such that $a_h \neq 0$, $a_k = 0$ for $k < h$. With $b = ax^{-h}$ we then have $b_0 = a_h$, $b_k = 0$ for $k < 0$. Thus it remains only to construct an inverse c for b. But by (29) we obtain such an inverse if we set $c_k = 0$ for $k < 0$ and calculate the remaining c_k from the equations

$$\sum_{i=0}^{k} b_i c_{k-i} = \begin{cases} 1 & \text{for } k = 0 \\ 0 & \text{for } k > 0 \end{cases},$$

i.e., for $k = 0, 1, 2, ...$

$$\begin{aligned} &b_0 c_0 = 1, \\ &b_0 c_1 + b_1 c_0 = 0, \\ &b_0 c_2 + b_1 c_1 + b_2 c_0 = 0, \\ &\cdots\cdots\cdots\cdots\cdots \end{aligned}$$

Since $b_0 \neq 0$, it is obvious that this system of equations can be satisfied with elements $c_0, c_1, ... \in R$, which completes the proof that R^* is a field.

To every element a of R^* we now assign a rational number, which we call[27] the *value* $|a|$ of a: $|0| = 0$, $|a| = 2^{-h}$, if $a_h \neq 0$ and $a_k = 0$ for all $k < h$. Since

$$\left(a - \sum_{i=h}^{n} a_i x^i\right)_k = a_k - \left(\sum_{i=h}^{n} a_i x^i\right)_k = 0$$

for all $k \leqslant n$ and for every $a \in R^*$, we have

$$\left| a - \sum_{i=h}^{n} a_i x^i \right| < 2^{-n}.$$

[27] This valuation has nothing to do with an ordering, as is the case, say, for the absolute value of the rational and the real numbers (cf. IB1, §3.4). We are only interested in the fact that the value here has the properties IB1, (52), (53), as is easily shown; see also van der Waerden [2], §§74, 75.

If now by means of this valuation we define the concept of a limit in the usual way, then $a = \lim_{n\to\infty} \sum_{i=h}^{n} a_i x^i$. Thus, as is customary in the theory of infinite series, we write

$$a = \sum_{n=h}^{\infty} a_n x^n$$

and call R^* the *power-series ring in x over R*, or the *power-series field*, if R (and therefore R^*) is a field.

As a result of this construction of R^*, the purely algebraic properties of power series (addition, multiplication, formation of inverses) can be investigated in a purely algebraic way, i.e., independently of the special properties of the field of real (or complex) numbers of analysis. But it must be pointed out that the above concept of a limit for R^* does not coincide with the concept of a limit for real (or complex) numbers if the a_i are such numbers and x is replaced by such a number.

If R is ordered, the power series ring R^* can be ordered in a very simple way. As domain of positivity we take the set of all power series $a \neq 0$ for which the leading coefficient a_h (i.e., $a_h \neq 0$, $a_k = 0$ for $k < h$) is positive.[28] The properties of a domain of positivity [IB1, (44)] are easily verified. Since all the positive elements of R obviously belong to the domain of positivity just defined, the order in R^* determines the same order for the elements of R as they are assumed to have in the first place. In view of the fact that for every natural number n the element $x^i - nx^k$, with $i < k$, has the leading coefficient 1, so that $x^i > nx^k$, we see that the ordering is non-Archimedean (cf. IB1, §4.3), since x^k is infinitesimal with respect to x^i.

As an example of the usefulness of these power series which we have just introduced in a purely algebraic way, we shall consider the problem (see §2.4) of expressing the power sums $s_i = \sum_{k=1}^{n} x_k^i$ $(i = 1, 2, ...)$ as entire rational functions of the elementary symmetric polynomials $\sigma_i(x_1, ..., x_n)$, for which we again write simply σ_i. For an arbitrary commutative ring R with unit element 1, we construct the power-series ring in x over the polynomial ring $R[x_1, ..., x_n]$. By the definition of the σ_i the polynomial $f(x) = \prod_{i=1}^{n} (1 - x_i x)$ over $R[x_1, ..., x_n]$ can be written in the form (with $\sigma_0 = 1$):

$$(30) \qquad f(x) = 1 + \sum_{i=1}^{n} (-1)^i \sigma_i(xx_1, ..., xx_n) = \sum_{i=0}^{n} (-1)^i \sigma_i x^i.$$

In the power-series ring we can now prove

$$(31) \qquad -f'(x) = f(x) \sum_{i=0}^{\infty} s_{i+1} x^i.$$

[28] Of course, this ordering does not lead (by IB1, (51) to the valuation introduced above.

For by the definition of $f(x)$, of the s_{i+1} and of addition as in (29), the right side of (31) is

$$= \sum_{k=1}^{n} \left(x_k(1 - x_k x) \sum_{i=0}^{\infty} x_k^i x^i \prod_{j \neq k} (1 - x_j x)\right).$$

By (29) we have

$$(1 - x_k x) \sum_{i=0}^{\infty} x_k^i x^i = \sum_{i=0}^{\infty} x_k^i x^i - x_k x \sum_{i=0}^{\infty} x_k^i x^i = 1,$$

so that for the right side of (31) we obtain the expression $\sum_{k=1}^{n} x_k \prod_{j \neq k} (1 - x_j x)$ which is seen at once from (20′) to be equal to $-f'(x)$. Then comparison of coefficients in (30) and (31) leads to

$$(32) \qquad \begin{cases} m\sigma_m + \sum_{k=1}^{m} (-1)^k \sigma_{m-k} s_k = 0 & \text{for } m = 1, \ldots, n, \\ \sum_{k=m-n}^{m} (-1)^k \sigma_{m-k} s_k = 0 & \text{for } m > n. \end{cases}$$

In particular, for $m = 1, 2, 3$ and $n \geqslant 3$:

$$\sigma_1 - s_1 = 0,$$
$$2\sigma_2 - \sigma_1 s_1 + s_2 = 0,$$
$$3\sigma_3 - \sigma_2 s_1 + \sigma_1 s_2 - s_3 = 0,$$

so that $s_1 = \sigma_1$, $s_2 = \sigma_1^2 - 2\sigma_2$, $s_3 = \sigma_1^3 - 3\sigma_1\sigma_2 + 3\sigma_3$.

For $n = 2$, $m = 3$ the equation (32) becomes

$$-\sigma_2 s_1 + \sigma_1 s_2 - s_3 = 0,$$

so that $s_3 = \sigma_1^3 - 3\sigma_1\sigma_2$.

3. The Use of Indeterminates as a Method of Proof

3.1. *The Derivative of a Product*

The rules for the derivative of a product (20) can be proved most conveniently in the following way. Over the polynomial ring $R[x]$ we construct the polynomial ring $R[x]\,[u]\,(=R[x, u])$. Then by the calculations leading to (4), (5) we can obviously arrive at an equation

$$(33) \qquad f(x) - f(u) = (x - u)\, F(x, u),$$

with $F(x, u) \in R[x, u]$, such that

$$(34) \qquad f'(x) = F(x, x).$$

Now $F(x, u)$ is uniquely determined by (33) (even without the assumption that R has no divisors of zero). For if $F^*(x, u)$ in $R[x, u]$, considered as a polynomial in u over $R[x]$, has the leading coefficient c, then $-c$ is the leading coefficient of $(x - u) F^*(x, u)$, so that the latter polynomial is $\neq 0$, if $F^*(x, u) \neq 0$. Thus $f'(x)$ is characterized by (33) and (34). From the equation

$$g(x) - g(u) = (x - u) G(x, u) \qquad \text{with} \quad g'(x) = G(x, x)$$

corresponding to (33) a short calculation now leads to

$$h(x) - h(u) = (x - u)(F(x, u) g(x) + f(u) G(x, u)).$$

For $h(x) = f(x) g(x)$, this is the equation for $h(x)$ corresponding to (33), so that the equation corresponding to (34) gives us the desired result

$$h'(x) = F(x, x) g(x) + f(x) G(x, x) = f'(x) g(x) + f(x) g'(x).$$

It is to be noted that by use of the indeterminate u we have created an exact proof out of the well-known faulty argument in which the difference quotient $(f(x) - f(a))/(x - a)$ is first calculated as an entire rational expression in x and a under the assumption that $x \neq a$ and then $x = a$ is substituted into this expression. For the field of real numbers our definition of the derivative of an entire rational function leads to the same result as the usual definition of analysis by means of the limiting value of the difference quotient, a fact which follows at once from the continuity of the entire rational function F in (33).

3.2. *Determinant of a Skew-Symmetric Matrix*

Let

$$A = (a_{ik})_{i,k=1,\ldots,n}$$

be a *skew-symmetric matrix*, i.e.,

$$(35) \qquad a_{ik} = -a_{ki} \qquad (i, k = 1, \ldots, n).$$

For the *transpose*[29] $A^T = (a_{ki})_{i,k=1,\ldots,n}$ we thus have $A^T = -A$. Formation of determinants then leads to $| A^T | = (-1)^n | A |$. For odd n we have $| A^T | = - | A |$, so that $2 | A | = 0$ in view of the fact that $| A^T | = | A |$.

[29] Cf. IB3, §2.6.

Thus, if the a_{ik} are numbers, it follows that $|A| = 0$. But in an arbitrary ring we may have $2a = 0$ even for $a \neq 0$; for example, in §1.1 we have noted that $1 + 1 = 0 \neq 1$. Nevertheless, we can prove that even in an arbitrary commutative ring the determinant of a skew-symmetric matrix with an odd number of rows and columns is $= 0$, provided that in the definition of *skew-symmetric* we further require[30] that

$$(36) \qquad a_{ii} = 0 \qquad (i = 1, ..., n).$$

For the proof we construct over the ring of integers the polynomial ring in the $n(n-1)/2$ indeterminates x_{ik} $(1 \leqslant i < k \leqslant n)$ for an odd number $n > 1$. With

$$(37) \qquad x_{ii} = 0\ (i = 1, ..., n), \qquad x_{ik} = -x_{ki}\ (1 \leqslant k < i \leqslant n)$$

the matrix $X = (x_{ik})_{i,k=1,...,n}$ then satisfies the condition corresponding to (35), so that $2\,|X| = 0$. From the fact that the polynomial ring has no divisors of zero and that in the ring of integers $1 + 1 \neq 0$, it follows that $|X| = 0$. But by the definition of a determinant $|X|$ is a polynomial in the x_{ik} $(1 \leqslant i < k \leqslant n)$. Thus the coefficients of this polynomial are all $= 0$. If in an arbitrary ring R we have a matrix A satisfying (35), (36) and if we replace the indeterminates x_{ik} $(1 \leqslant i < k \leqslant n)$ by the a_{ik} with the same indices, then X is transformed into A, in view of (35), (36), (37).

But what happens to the polynomial $|X|$? In order to answer this question, we consider a polynomial $f(x_1, ..., x_n) = \sum_{0 \leqslant i_k \leqslant m} c_{i_1 \ldots i_n} x_1^{i_1} \cdots x_n^{i_n}$ over the ring of integers and elements $a_1, ..., a_n$ from an arbitrary commutative ring R. For any natural number c and any $a \in R$ we shall let $c \cdot a$ denote the sum $\sum_{i=1}^{c} a$ (i.e., the c-fold multiple of a); also we set $0 \cdot a = 0$ and $(-c) \cdot a = -c \cdot a$.[31] We then define

$$f(a_1, ..., a_n) = \sum_{0 \leqslant i_k \leqslant m} c_{i_1 \ldots i_n} \cdot a_1^{i_1} \cdots a_n^{i_n}.$$

Now $f(x_1, ..., x_n) \to f(a_1, ..., a_n)$ (for fixed $a_1, ..., a_n$) is a homomorphism of the polynomial ring into the ring R; that is, $f(x_1, ..., x_n) + g(x_1, ..., x_n)$ goes into $f(a_1, ..., a_n) + g(a_1, ..., a_n)$ [and therefore $f(x_1, ..., x_n) - g(x_1, ..., x_n)$ into $f(a_1, ..., a_n) - g(a_1, ..., a_n)$], and $f(x_1, ..., x_n)\, g(x_1, ..., x_n)$

[30] In Bourbaki [2] a skew-symmetric matrix with this further property is said to be "alternating," although etymologically speaking this word again refers only to the change of sign under interchange of indices. If R is such that $2a = 0$ implies $a = 0$, then (36) is obviously a consequence of (35), since by (35) we have $a_{ii} = -a_{ii}$, so that $2a_{ii} = 0$.

[31] If R contains the ring of integers, then obviously $c \cdot a = ca$. Otherwise ca is not defined and as a substitute for it we simply take $c \cdot a$. The only purpose of the dot is to call attention to this distinction.

goes into $f(a_1, ..., a_n)\, g(a_1, ..., a_n)$, as is easily proved from the formulas $c \cdot a + c' \cdot a = (c + c') \cdot a$, $(c \cdot a)(c' \cdot a') = cc' \cdot aa'$. The proof of the first formula, exactly as in IB1, (47), is based on the associativity of addition. The second formula is easily reduced to the case $c, c' > 0$, when it can proved by simply multiplying out the factors [IB1, (41)].

If we now apply this homomorphism to $|X|$, we obtain $|A|$ (since all additive-multiplicative relations remain unchanged), and on the other hand we also obtain 0, since the coefficients of $|X|$ are all $=0$. Thus $|A| = 0$, as desired.

3.3. *Determinant of the Adjoint Matrix*

For the matrix $A = (a_{ik})_{i,k=1,...,n}$ $(n > 1)$ we denote the subdeterminant belonging to a_{ik} by A_{ik} or $-A_{ik}$, according as $i + k$ is even or odd (this is the usual notation; cf. IB3, §3.5.3). Then the *adjoint* $A^* = (A_{ik})_{i,k=1,...,n}$ of A is such that $A^T A^*$ is a diagonal matrix with every diagonal element $=|A|$. Formation of determinants leads to

(38) $$|A|\,|A^*| = |A|^n.$$

If the a_{ik} are the elements of a commutative[32] ring without divisors of zero (e.g., numbers) and if $|A| \neq 0$, then by (38) we have

(39) $$|A^*| = |A|^{n-1}.$$

But we can also prove this equation even when the a_{ik} come from an arbitrary commutative ring and the case $|A| = 0$ is not excluded. For just as in §3.2, let us construct the polynomial ring in the n^2 indeterminates x_{ik} $(i, k = 1, ..., n)$ over the ring of integers. For $X = (x_{ik})_{i,k=1,...,n}$ the determinant $|X|$ is then a polynomial whose coefficients are not all $= 0$. Thus $|X| \neq 0$. But because the polynomial ring has no divisors of zero, the above discussion shows that $|X^*| = |X|^{n-1}$, which means that all the coefficients of the polynomial $|X^*| - |X|^{n-1}$ are $= 0$. As in §3.2, if we replace X by A, we obtain the equation (39) for an arbitrary matrix A with elements from an arbitrary commutative ring.

[32] For noncommutative rings the concept of a determinant is of very restricted usefulness.

CHAPTER 5

Rings and Ideals

1. Rings, Integral Domains, Fields

1.1. The simplest example of a ring is the set of rational integers (IB1, §2)

$$(G) \qquad 0, \quad \pm 1, \quad \pm 2, \ldots .$$

This set is closed with respect to addition, subtraction, and multiplication, by which we mean that the sum, difference, and product of two rational integers is always a rational integer. Furthermore, there are certain rules for calculation with these numbers, e.g., the familiar rules for the removal of parentheses and for the sign of a product.

On the other hand, this ring is not closed with respect to division, since the quotient of two rational integers is not always a rational integer.

There are many other examples of rings, e.g., the ring $G[i]$ of *Gaussian integers* (IB6) consisting of all numbers $a + bi$, where i is the imaginary unit (see IB8, §1) and a, b are rational integers. The set of Gaussian integers is easily seen to be closed with respect to addition, subtraction, and multiplication, and the general rules for calculation with them are the same as for the rational numbers (IB1).

Another important ring is the ring $G[x]$ of the polynomials

$$p(x) = a_0 + a_1 x + a_2 x^2 + \cdots + a_n x^n$$

of all possible degrees $n = 0, 1, 2, \ldots$, where the coefficients $a_0, a_1, \ldots, a_n$ are rational integers (IB4, §2.1). Since we shall be dealing below with many other examples of rings, let us first give an exact definition.

1.2. *Definition of a Commutative*[1] *Ring*

A (commutative) ring $\mathfrak{R}$ *is a non-empty set of elements* $a, b, c, d, \ldots$, *for which two operations, namely an addition and a multiplication, are defined; in other words, for any two elements* $a, b \in \mathfrak{R}$ *there exists a uniquely determined element* $c \in \mathfrak{R}$ *which is the result of the addition*

$$a + b = c,$$

and also a uniquely determined element $d \in \mathfrak{R}$ *which is the result of the multiplication*[2]

$$ab = d.$$

These operations must satisfy the following laws (also called "axioms" or "postulates"):

I. *The associative law for addition and multiplication: for arbitrary elements* $a, b, c \in \mathfrak{R}$ *we have the equations*

$$(a + b) + c = a + (b + c), \tag{Ia}$$

$$(ab)\,c = a(bc); \tag{Ib}$$

II. *The commutative law for addition and multiplication: for arbitrary elements* $a, b \in \mathfrak{R}$ *we have the equations*

$$a + b = b + a, \tag{IIa}$$

$$ab = ba; \tag{IIb}$$

III. *Invertibility of addition; i.e., subtraction is always possible: if* a, b *are any two elements of the ring* $\mathfrak{R}$, *there exists a uniquely determined solution* x *of the equation*

$$a + x = b. \tag{III}$$

This operation, inverse to addition, is usually called *subtraction*, and the solution of (III) is written in the form

$$x = b - a, \tag{1}$$

[1] In the present chapter we confine ourselves to *commutative* rings, usually without explicit mention of the fact. If all the axioms except (IIb) are satisfied, the ring is said to be "*noncommutative*." Important examples of noncommutative rings are the *quaternions* (IB8, §3) and the general matrix rings (IB3, §2.2). The definition of a ring in IB1, §2.4 is formulated somewhat differently from the one given here, yet it is easy to see from the following discussion that the two definitions (apart from the assumption here that multiplication is commutative) are equivalent to each other (see also IB2, §2.4).

[2] Multiplication is usually denoted, not by any special sign such as a dot, but merely by juxtaposition of the two elements.

so that for arbitrary elements a, b we have the identity

$$a + (b - a) = b; \tag{2}$$

IV. *The distributive law: for arbitrary elements $a, b, c \in \mathfrak{R}$ we have*[3]

$$a(b + c) = ab + ac. \tag{IV}$$

It is easy to verify that all these postulates are satisfied by the examples given above. On the other hand, we must also show that these laws, which have been reduced to the simplest possible form, imply all the usual rules for calculation (IB1, §2.4).

1.3. *Remark on the Associative Law*

Up to now we have defined the sum of two elements only, so that if three or more elements $a, b, c, \ldots, d$ are to be added, we may, for example, begin by adding the first two elements, and then the third and the fourth, and so on, and the order in which these operations are to be performed may be indicated by parentheses. But (Ia) states that in a sum of three terms it makes no difference how we place the parentheses, and thus we may simply omit them. The same result may be deduced for a sum of any number of terms, the proof being the same as in IB1, §1.3. Thus for such sums it is customary to omit the parentheses and simply to write

$$a + b + c + \cdots + d.$$

Furthermore, the commutative law (IIa) shows that in a sum of this sort we may permute the terms at will, without affecting the result (IB1, §1.4).

From (Ib) and (IIb) it follows that the same remarks may be made for products of three or more factors.

The distributive law can also be extended by induction to more than two summands and to products of factors in parentheses (cf. IB1, §2 (41)):

$$a(b + c + \cdots + d) = ab + ac + \cdots + ad, \tag{3}$$

$$(a + b)(c + d) = a(c + d) + b(c + d) = ac + ad + bc + bd. \tag{4}$$

1.4. The postulates (Ia), (IIa) and (III) state that *the element of a ring $\mathfrak{R}$ form an Abelian group with respect to addition; this is the so-called "additive group" of the ring* (IB2, §2.4 and IB1, §2.3).[4]

[3] In noncommutative rings, for which (IIb) is not postulated, we must make the separate postulate

$$(b + c)a = ba + ca.$$

[4] With respect to addition the ring is thus a *module*, by which we mean an additively written Abelian group (IB1, §2.3).

In view of the uniqueness (1) of its solution the equation

$$a + b = a + c \tag{5}$$

implies $b = c$; in other words, we may *cancel* equal summands on both sides of an equation (*first rule of cancellation*).

By III the equation

$$a + x = a, \tag{6}$$

where a is any element of the ring, also has a unique solution

$$x = a - a. \tag{6'}$$

Now if b is any other element of the same ring, we may add the element $b - a$ to both sides of (6), which in view of the identity (2) gives

$$b + x = b, \tag{6''}$$

so that the same element x is the solution of (6) and of (6″). This element

$$a - a = b - b = 0 \tag{7}$$

is called the *zero element* of the ring $\mathfrak{R}$. For the time being we shall denote it by the letter o, but later we shall also use[5] the customary symbol 0. Then we have the identities

$$a + o = o + a = a, \tag{8}$$

$$a - o = a, \tag{8'}$$

where (8′) follows from (2) and (8), if in (2) we put o for a and a for b. *Thus the zero element is the "identity element" or "neutral element" of the additive group of the ring.*

By (III) the equation

$$a + x = o$$

also has a uniquely determined solution $o - a$, which we abbreviate to $-a$:

$$x = o - a = -a.$$

Thus every ring element a has an *inverse* element $-a$ satisfying the identity

$$a + (-a) = o; \tag{9}$$

[5] The set consisting of the zero element alone satisfies all the postulates for a ring, but in general we shall assume that besides the zero element, which is always present, every ring contains at least one further element.

but since $a - a = o$, we may abbreviate $a + (-a)$ to $a - a$. More generally,

$$b + (-a) = b - a, \tag{10}$$

since it follows from (Ia), (IIa), (9) and (8) that

$$a + [b + (-a)] = a + (-a) + b = o + b = b,$$

which implies the assertion (10) on account of the uniqueness of subtraction.[6]

If to both sides of the equation (III) we add the element $-x$, it follows from (9) and (8) that

$$b + (-x) = a,$$

so that in view of (1) and the uniqueness of subtraction:

$$-(b - a) = a - b. \tag{11}$$

1.5. *We have*

$$b - a = d - c \tag{12a}$$

if and only if

$$a + d = b + c. \tag{12b}$$

It follows from the identity (2) and the equation (12a) that

$$a + d = a + [c + (d - c)] = a + c + (b - a) = b + c;$$

and conversely from (12b) and again from (2) that

$$a + c + (b - a) = b + c = a + d,$$

so that by the rule of cancellation (5)

$$c + (b - a) = d,$$

from which (12a) follows by the uniqueness of subtraction.

We note the easily proved formulas

$$(b - a) + (d - c) = (b + d) - (a + c), \tag{13}$$

$$(b - a) - (d - c) = (b - a) + (c - d) = (b + c) - (a + d), \tag{14}$$

[6] In IB1, §2.3, the equation (9) is taken as the definition of $-a$ and (10) as the definition of $b - a$, and equation III, which is here taken as the definition of $b - a\,(=x)$, is proved there as a theorem. The present equation (11) is proved there as equation (36).

which contain the usual rules for a change of sign under the removal of parentheses; since these rules are logical consequences of the postulates (I) through (III), they are valid in every ring.[7] Furthermore, if we multiply the equation

$$c + (b - c) = b$$

on both sides by a and apply the distributive law (IV), we have

$$ac + a(b - c) = ab,$$

from which by subtraction we obtain the following complement to (IV):

$$a(b - c) = ab - ac; \tag{15}$$

finally, by repeated application of this formula

$$(b - a)(d - c) = (ac + bd) - (ad + bc). \tag{15'}$$

1.6. *Unity Element*

An element e of the ring $\mathfrak{R}$, satisfying the relation

$$ea = ae = a \tag{16}$$

for every element $a \in \mathfrak{R}$, is called a *unity element* (or a *unit element*, or an *identity*). In particular,

$$e^2 = ee = e.$$

Not every ring contains a unity element; e.g., the ring of even integers $0, \pm 2, \pm 4, \ldots$ satisfies all the postulates for a ring but contains no unity element. But if a ring does contain a unity element, it cannot contain more than one; for if e' were a second unity element, then by (16) we would have $ee' = e$ and also $ee' = e'$, so that $e = e'$. In most cases we shall denote the unity element by the usual symbol 1.

1.7. *Divisors of Zero*

If o denotes the (always present) zero element of the ring $\mathfrak{R}$, we have

$$oa = ao = o \tag{17}$$

for every element $a \in \mathfrak{R}$, as follows immediately from (7) and (15). But in

[7] In IB1, §2.1, the equation (13) for the natural numbers a, b, c, d occurs in the definition of addition of integers in the form (26), and (12b) occurs in the form (24) in the definition of equality of integers; in the present context we are chiefly interested in showing that these rules can be deduced from the axioms for a ring.

certain rings it can happen that a product is equal to zero even though neither factor is zero:

$$ab = o, \quad \text{with} \quad a \neq o \quad \text{and} \quad b \neq o.$$

In this case the two elements a, b are called *divisors of zero*.

Examples of rings with divisors of zero are provided by the residue class rings G/n, with composite n (cf. §3.7 and IB6, §4.1). Another example is the set of two-rowed matrices (IB3, §2.2):

$$\begin{pmatrix} a & b \\ b & a \end{pmatrix} \quad \text{with} \quad a, b \in G.$$

By the rules for calculation with matrices, these matrices form a commutative ring whose zero element is the zero matrix

$$\begin{pmatrix} 0 & 0 \\ 0 & 0 \end{pmatrix}$$

and whose unity element is the unit matrix

$$\begin{pmatrix} 1 & 0 \\ 0 & 1 \end{pmatrix}.$$

This ring contains divisors of zero, e.g., the two matrices

$$\begin{pmatrix} 1 & -1 \\ -1 & 1 \end{pmatrix} \quad \text{and} \quad \begin{pmatrix} 1 & 1 \\ 1 & 1 \end{pmatrix},$$

whose product is the zero matrix.

1.8. A subset $\mathfrak{S}$ of a ring $\mathfrak{R}$ that is closed with respect to addition, subtraction, and multiplication satisfies all the ring postulates as a part of $\mathfrak{R}$ and is thus called a *subring* of $\mathfrak{R}$; e.g., the set of all integers divisible by by 3, 0, ± 3, ± 6, ..., forms a subring of the ring G of all the rational integers.

As the following example shows, it can happen that the unity element of the subring $\mathfrak{S}$ is different from that of $\mathfrak{R}$. Let $\mathfrak{R}$ be the ring of all two-rowed diagonal matrices $\begin{pmatrix} a & 0 \\ 0 & b \end{pmatrix}$ with $a, b \in G$, and let $\mathfrak{S}$ consist of all such matrices with $b = 0$. The unity element of $\mathfrak{R}$ is $\begin{pmatrix} 1 & 0 \\ 0 & 1 \end{pmatrix}$, and that of $\mathfrak{S}$ is $\begin{pmatrix} 1 & 0 \\ 0 & 0 \end{pmatrix}$. Like G, the subring $\mathfrak{S}$ has no divisors of zero, but $\mathfrak{R}$ does have such divisors.

1.9. *Integral Domains*

The feature of greatest importance for the structure of a ring is the presence or absence of divisors of zero; in order to emphasize this feature with a special name, (commutative) rings without divisors of zero are also called *integral domains*[8] (*or domains of integrity*). Most of the rings in the present chapter have no divisors of zero and are therefore integral domains, e.g., all the polynomial rings $G[x]$, $G[x, y]$, $G[x, y, z]$, ... (cf. IB4, §2).

In an integral domain we may cancel any nonzero factor that appears on both sides of an equation; that is, we have the *second rule of cancellation*:

(18) $$ab = ac \quad \text{and} \quad a \neq o \quad \text{imply} \quad b = c.$$[9]

The above rule can also be stated: An integral domain $\mathfrak{J}$ is a ring in which the solution of an equation

(19) $$ax = b, \qquad a, b \in \mathfrak{J}, \qquad a \neq o,$$

is unique, provided it exists at all. For if there were two distinct solutions x_1 and x_2, we would have $ax_1 = ax_2 = b$, so that $a(x_1 - x_2) = o$, which would imply the existence of two divisors of zero $a \neq o$ and $x_1 - x_2 \neq o$. On the other hand, if a ring has divisors of zero: $ab = o$, $a \neq o$, $b \neq o$, then $ax = a(x + b)$ for every element x.

1.10. *Fields*

It may happen that in the given ring $\mathfrak{J}$ every equation (19) is uniquely solvable *without exception*; such a ring is called a *field* (see also IB1, §3.2). In a field, division (with the exception of division by zero) is unique and always possible. In other words, we may define a field as a set of elements in which besides the above listed postulates for a ring (1.2), the following postulate holds:

Postulate for a field: *In a field $\mathfrak{K}$ every equation $ax = b$ with $a \neq o$*[10] *but with otherwise arbitrary elements $a, b \in \mathfrak{K}$ has a uniquely determined solution.*

It follows that *every field is free of divisors of zero and has a unity element.* The freedom from divisors of zero is proved in exactly the same way as for integral domains (§1.9), and the existence of a unity element follows from the solvability of the equation

$$ax = a.$$

For if $x = e$ is a solution for some definite element $a \neq o$, so that $ae = a$, and if b is any other element of the field, then by the postulate for a field there exists an element c satisfying the equation $ca = b$. If we multiply

[8] In analysis they are usually called *integral rings*, since the word "domain" could easily lead to ambiguity.

[9] Since $a(b - c) = o$ and a is neither zero nor a divisor of zero, it follows that $b - c = o$, so that $b = c$.

[10] If $a = o$, it follows from (17) that $b = o$, and then *every* element x is a solution of the equation $ax = b$.

by c on both sides of the equation $ae = a$, we obtain $be = b$, for *every* element $b \in \mathfrak{K}$; in other words e is the unity element (1.6).

By postulates (I*b*), (II*b*) and the field postulate, the elements of a field, excluding the zero element, form an Abelian group with respect to multiplication; this group is called the *multiplicative group* of the field (for the concept of a group see IB2). Conversely, the field postulate is a consequence of this property. Thus we have the following important result:[11] *Every integral domain with finitely many elements is a field*; for example, every residue class ring G/p modulo a prime (p) is a field (IB6, §4.3).

For the proof let $x_1, x_2, ..., x_n$ be the finitely many elements of the integral domain $\mathfrak{J}$ and multiply them one after another by the element $a (a \neq 0)$. By the rule of cancellation (18), the products $ax_1, ax_2, ..., ax_n$ are distinct and therefore represent all the elements of $\mathfrak{J}$, including the element b; thus we have a solution for the equation $ax = b$.

If we do not postulate the commutativity of multiplication, then in order to obtain the analogue of a field, we must postulate that every equation $ax = b$ and also every equation $ya = b$ is solvable, provided only $a \neq o$. (It is sufficient to postulate the existence of the solutions, since their uniqueness can then be proved.) The resulting system of elements is called a *skew field*. From the postulates it follows again that the elements of a skew field, excluding the zero element, form a *group*; so that the existence of an identity element and inverse elements a^{-1}, the absence of divisors of zero in a skew field, and the uniqueness of the postulated solutions can then be proved exactly as in group theory (IB2, §2.4). We can also show that *every finite skew field* (i.e., every skew field containing only finitely many elements) is *necessarily commutative and is therefore a field.*

1.11. *Prime Fields and the Characteristic of a Field*

The simplest and best known field is the field R of all rational numbers (IB1, §3.2). Next in simplicity are the above-mentioned residue class rings G/p, i.e., fields with finitely many (namely p) elements. Since these fields contain no proper subfields, we call them *prime fields*. The residue class rings are prime fields "*of characteristic p*" (§3.7), and the field R of rational numbers is a prime field of "*characteristic* 0."

We already know that every field $\mathfrak{K}$ contains both a zero element o and a unity element e; but then $\mathfrak{K}$ also contains the element $e + e$, $e + e + e$, and so forth, which we abbreviate to $2e$, $3e$ ... (ne is the sum of n summands e). Now it may happen that these elements are not all distinct but that some of the me being the same as some earlier ne:

$$ne = me, \qquad \text{so that} \quad (n - m)\, e = o.$$

[11] Cf. IB2, §2.4, where it is proved that a *finite* set with an associative operation (V), (A) and rules for cancellation (K_r), (K_l) is a group.

Then if p is the smallest natural number for which $pe = o$, it follows that p must be prime[12] and the elements $o, e, 2e, \ldots, (p-1)e$ are distinct. In this case we say that the field $\mathfrak{K}$ has "*characteristic* p"; and otherwise, if all the elements $e, 2e, \ldots, ne, \ldots$ are distinct, we assign to $\mathfrak{K}$ the "*characteristic* 0."

The *characteristic of an integral domain* $\mathfrak{J}$ with unity element e is defined in exactly the same way.

It can be shown that every field $\mathfrak{K}$ which is not a prime field contains as its smallest subfield a prime field of the above type with the same characteristic as $\mathfrak{K}$.

1.12 *Quotient Fields*

For every integral domain $\mathfrak{J}$ that is not a field we can construct a field, the *quotient field*, that contains $\mathfrak{J}$ and is constructed from $\mathfrak{J}$ in exactly the same way as the field R of rational numbers is constructed from the integral domain G of rational integers IB1, §3.2). We first construct the set of all formal *fractions* a/b, where a, b are arbitrary elements of $\mathfrak{J}$ with $b \neq o$. In this set of fractions we introduce a partition into classes by means of the *equality*:

$$a/b = c/d \qquad \text{if and only if} \quad ad = bc.$$

An element a in $\mathfrak{J}$ is identified with the class of fractions ab/b. Computation with these fractions follows the well-known rules. The proof of these statements is to be found in IB1, §3.1–3.2; the argument given there, as was remarked at that time, can be transferred verbatim to the present case, since the only necessary postulate is that the domain $\mathfrak{J}$ be a commutative ring without divisors of zero, i.e., an integral domain.

The domain $\mathfrak{K}$ constructed in this way is a field, the quotient field[13] of $\mathfrak{J}$; this field contains $\mathfrak{J}$ as a subdomain, and the results of any computation in $\mathfrak{J}$ remain valid for $\mathfrak{K}$.

The most important quotient fields are the field R of rational numbers, as quotient field of the integral domain G of rational integers, and the field $R(x)$ of all polynomial quotients in one indeterminate x with rational coefficients. The field $R(x)$ is a quotient field not only of $G[x]$ but also of $R[x]$.

[12] For if p were 6, say, there would be divisors of zero, since we would then have

$$(2e)(3e) = (e + e)(e + e + e) = 6e = o,$$

and the minimal property of p would mean that $2e \neq o$, $3e \neq o$.

[13] Two different integral domains may belong to the same (or isomorphic) quotient fields; e.g., the rational field R can also be obtained as the quotient field of the integral domain of all even numbers.

1.13. *Isomorphism*

The concepts of *isomorphism* and *homomorphism* are defined (see also IB1, §2.4) for rings in the same way as for groups (IB2, §4.2 and §10.1). Two rings $\mathfrak{R}$ and $\mathfrak{R}^*$ are said to be *isomorphic*, in symbols $\mathfrak{R} \rightleftarrows \mathfrak{R}^*$, if the elements $a, b, c, \ldots$ of $\mathfrak{R}$ are in *one-to-one* correspondence with the elements $a^*, b^*, c^*, \ldots$ of $\mathfrak{R}^*$:

$$a \leftrightarrow a^*, \qquad b \leftrightarrow b^*, \qquad c \leftrightarrow c^*$$

in such a way that this correspondence is *consistent* with addition and multiplication; i.e., if for all a, b we have

$$a + b \leftrightarrow a^* + b^*, \qquad ab \leftrightarrow a^*b^*.$$

Since the postulates (§1.2) are satisfied in both rings, it follows that the correspondence is also consistent for subtraction and for more complicated expressions, e.g.,

$$a - b \leftrightarrow a^* - b^*, \qquad a(b + c) \leftrightarrow a^*(b^* + c^*), \quad \text{etc.}$$

Thus, the zero element o of $\mathfrak{R}$ corresponds to the zero element o^* of $\mathfrak{R}^*$, the inverse element $-a$ corresponds to the inverse element $-a^*$, a divisor of zero in $\mathfrak{R}$ corresponds to a divisor of zero in $\mathfrak{R}^*$, the unity element of $\mathfrak{R}$ (if its exists) corresponds to the unity element of $\mathfrak{R}^*$. If $\mathfrak{R}$ contains a unity element, $\mathfrak{R}^*$ also necessarily contains one, and if $\mathfrak{R}$ contains divisors of zero, then so does $\mathfrak{R}^*$; if $\mathfrak{R}$ is an integral domain or a field, then so is $\mathfrak{R}^*$, and conversely.

If $\mathfrak{R}_1$ is a *subring* (that is, a subset closed with respect to addition and multiplication; see also IB4, §2.1) of $\mathfrak{R}$, then the corresponding elements in $\mathfrak{R}^*$ form a subring $\mathfrak{R}_1^*$ isomorphic to $\mathfrak{R}_1 : \mathfrak{R}_1 \rightleftarrows \mathfrak{R}_1^*$.

Isomorphism between rings is an *equivalence relation* (cf. IA, §8.5 and IB1, §2.2); every ring is trivially isomorphic to itself, and if two rings are isomorphic to a third, then they are isomorphic to each other. As far as their algebraic structure as rings is concerned, two isomorphic rings can differ only in the notation, so that we can identify them by disregarding all the properties of their elements that do not affect their structure as rings.[14]

If the rings $\mathfrak{R}$ and $\mathfrak{R}^*$ contain a common subring $\mathfrak{R}_0$, then $\mathfrak{R}$ is said to be "isomorphic to $\mathfrak{R}^*$ with respect to $\mathfrak{R}_0$" if every element of the subring $\mathfrak{R}_0$ corresponds to itself.

If the mapping of the ring $\mathfrak{R}$ onto the ring $\mathfrak{R}^*$ is (possibly) many-to-one, so that each of the elements $a, b, \ldots$ of $\mathfrak{R}$ has a unique image $a^*, b^*, \ldots$ in $\mathfrak{R}^*$:

$$a \rightarrow a^*, \qquad b \rightarrow b^*, \ldots,$$

[14] In the example of §1.8 the two rings $\mathfrak{S}$ and G are isomorphic.

and if *every* element d^* in $\mathfrak{R}^*$ has at least one (and perhaps more than one) preimage d in $\mathfrak{R}$: $d \to d^*$, and if further the mapping is consistent with the operations of addition and multiplication:

$$a + b \to a^* + b^*, \qquad ab \to a^*b^*,$$

then $\mathfrak{R}^*$ is said to be a *homomorphic image* of $\mathfrak{R}$, or in symbols $\mathfrak{R} \simeq\!\!\!\!\to \mathfrak{R}^*$ (cf. §3.6).

2. Divisibility in Integral Domains

2.1 In the present section we consider integral domains $\mathfrak{J}$ with a unity element 1; in particular, the integral domain G of all rational integers, the polynomial rings $G[x]$ and $R[x]$ of all polynomials in x with coefficients in G or R, and the field of rational numbers (IB4, §2). In an integral domain $\mathfrak{J}$ the field postulate (§1.10) is not satisfied, in general; i.e., the equation $ax = b$ for given a, b generally has solution in $\mathfrak{J}$; but in the special cases in which this equation does have a solution[15] we say that b is *divisible* by a and write

(19′) $$a|b, \qquad \text{or: "}a \text{ divides } b.\text{"}$$

We also say that a is a *divisor*, or a *factor*, of b.

The relation of divisibility is *transitive*; i.e., from (19′) and $b|c$ it follows that $a|c$. For by hypothesis there exist in $\mathfrak{J}$ elements x and y such that

$$ax = b \qquad \text{and} \qquad by = c,$$

but then $z = xy$ is a solution of the equation $az = c$, so that $a|c$.

Since $1a = a$, every element a in $\mathfrak{J}$ has the so-called "trivial" divisors 1 and a.[16]

2.2. *Units*

In the theory of divisibility in an integral domain $\mathfrak{J}$ an important role is played by the so-called *units*, defined as divisors of the unity element; for example, the elements e_1 and e_2 are units if

(20) $$e_1e_2 = 1.$$

In G, and also in $G[x]$, the only units are 1 and -1; in $R[x]$ on the other hand, every polynomial of zero degree, i.e., all the rational numbers, are

[15] If the solution exists, it is *unique* (§1.9). By (17) every element $a \in \mathfrak{J}$ is a divisor of the zero element: $a \mid o$. But by a "divisor of zero" we mean only the elements defined in §1.7. Divisors of zero do not occur in integral domains.

[16] Concerning the properties of divisibility in an integral domain compare the discussion in IB6, §1.

units.[17] In the integral domain $G[i]$ of the Gaussian numbers there are four units: $1, -1, i, -i$, and similar remarks hold for other rings of algebraic numbers (IB6, §8).

Thus the units are characterized by the property that they have a *reciprocal* element in $\mathfrak{J}$; for it follows from (20), in the usual notation, that

$$e_1^{-1} = \frac{1}{e_1} = e_2 \in \mathfrak{J}$$

and conversely. The product of two or more units is again a unit; for it follows from (20) and $e_3e_4 = 1$ that $(e_1e_3)(e_2e_4) = 1$, so that e_1e_3 and e_2e_4 are also units. Thus the units of an integral domain form an Abelian group (IB2, §1.1) with respect to multiplication.

Two elements a and $a' = ae$ of an integral domain $\mathfrak{J}$ are called *associates*[18] if each is the product of the other by a unit of $\mathfrak{J}$. Associate elements a, a' are characterized by the divisibility relations

$$(21) \qquad a|a' \quad \text{and} \quad a'|a,$$

the second of which follows from $a'e^{-1} = a$. Thus each of two associates is divisible by the other, and conversely. For by (21) there exist two elements x, y in $\mathfrak{J}$ such that

$$ax = a' \quad \text{and} \quad a'y = a,$$

from which we have $a(xy) = a$, and by the rule (18) of cancellation $xy = 1$; thus x and $y = x^{-1}$ are units, so that a and a' are associates, as was to be proved.

Every element $a \neq 0$ in $\mathfrak{J}$ has as its *trivial* factors all the units and all its associates; by a *proper* factor of a we mean a factor which is not an associate of a.[19]

2.3. *Irreducible and Prime Elements*

An element $a(\neq 0)$ in $\mathfrak{J}$ *irreducible* if it has only trivial factors, i.e., if a factorization $a = bc$ into two factors is possible only when one of the factors is a unit and the other is an associate of a. Otherwise a is said to be *reducible.*

Examples of irreducible elements are: all the prime numbers in G, and all the *irreducible* polynomials in $G[x]$. But we must not confuse "*irreducible*" with "*prime*," even though in certain integral domains,

[17] In a *field* every element is divisible by every other element (except 0). Since this statement is true in particular for 1, *every nonzero element of a field is a unit.*

[18] In particular, every element a is an associate of itself.

[19] It is convenient to include the units among the *proper* factors.

including the most important ones (namely G, $G[x]$, and $R[x]$), every irreducible element is prime and conversely, as will be proved below.

An element p in $\mathfrak{J}$ is said to be *prime if* $p|ab$ *implies at least one of the two relations* $p|a$, $p|b$;[20] i.e., a product is divisible by a prime element p if and only if at least one of its factors is divisible by p.[21] We now prove the following theorem.

2.4. *Every prime element is irreducible, but the converse is not necessarily true.*

For let $p = ab$ be a factorization of the prime element p, so that $p \mid ab$; then one of the factors, say a, must be divisible by p. Writing $a = pc$, we obtain $p = pbc$, or $1 = bc$, by cancellation (18). Therefore b is a unit and a is an associate of p, so that $p = ab$ is a trivial factorization of p. Thus p has no nontrivial factorization, which means that p is irreducible.

On the other hand, it may happen in certain integral domains that an element is irreducible but not prime. For example, in the integral domain $G[\sqrt{-3}]$ of all numbers $a + b\sqrt{-3}$ with $a, b \in G$[22] the number 2 is *irreducible*, because the Diophantine equation (IB6, §7)

$$2 = (a + b\sqrt{-3})(a - b\sqrt{-3}) = a^2 + 3b^2$$

clearly has no solution in integers.

On the other hand, the number 2 is not *prime* in $G[\sqrt{-3}]$, since the product

$$(1 + \sqrt{-3})(1 - \sqrt{-3}) = 4$$

is divisible by 2, although neither of the factors $1 + \sqrt{-3}$ and $1 - \sqrt{-3}$ has this property.

However, we shall prove below that in the integral domains of greatest importance to us, in particular in G, $G[x]$ and $R[x]$, *every irreducible element is also prime*, so that in these domains the two concepts may be identified.

[20] In the language of the theory of ideals (§3) we may say: *an element p is prime if the ideal* (p) *generated by p is a prime ideal* (§3.6).

[21] The zero element of an integral domain $\mathfrak{J}$ is not regarded as a prime element although, formally speaking, it is both irreducible and prime.

[22] We can construct a simpler example in an integral domain without unity element. For example, let G_2 be the integral domain of all even numbers; then the number 30 is irreducible, because every product of two even numbers is divisible by 4. On the other hand, 30 is not prime in G_2, since the product $6 \cdot 10 = 60$ is divisible by 30, although neither factor has this property.

2.5. *Divisor Chain Theorem*

This important theorem, valid for all the integral domains considered here, will be needed in several places below. It states that a *proper divisor chain*

$$(22) \qquad a_1, a_2, \ldots, a_i, \ldots \qquad (a_i \neq 0, a_{i+1} \mid a_i, i = 1, 2, \ldots),$$

i.e., a sequence of elements a_i in the given integral domain such that each a_i is a *proper* divisor of its predecessor a_{i-1}, contains[23] only finitely many terms.

The divisor chain condition holds in G, because the sequence of absolute values $|a_1|, |a_2|, \ldots$ corresponding to a divisor chain (22) consists of monotonically decreasing natural numbers, so that the smallest number 1 must be reached after a finite number of steps.

The same result can be proved for a divisor chain (22) in the polynomial ring $R[x]$ by considering, instead of the absolute value, the degree of the successive polynomials in x. To prove the result for $G[x]$, we may consider the sum of the degree of the polynomial and the absolute value of its first coefficient.[24]

If the divisor chain condition holds in an integral domain $\mathfrak{I}$, every element $a(\neq 0)$ in $\mathfrak{I}$ can be written in at least one way as the product of (finitely many) irreducible elements u_i:

$$(23) \qquad a = u_1 u_2 \cdots u_3 .$$

For if $a = u$ is irreducible, we already have a representation (23) with $s = 1$; but if a is reducible, we can split up each factor into irreducible terms, which must be reached after *finitely* many steps, since otherwise we would have a nonterminating proper divisor chain, in contradiction to the hypothesis.

In general, the representation (23) is not unique, even apart from associate elements and the order of the factors. For example, in $G[\sqrt{-3}]$ (see also §2.4) the number 4 can be represented in two essentially different ways as the product of irreducible numbers (in $G[\sqrt{-3}]$):

$$4 = 2 \cdot 2 = (1 + \sqrt{-3})(1 - \sqrt{-3}).$$

[23] Thus G contains, among others, the following divisor chains beginning with 100:

100, 50, 10, 2, 1 and 100, 20, 4, 2, 1.

[24] It is not easy to construct examples of integral domains without the divisor chain condition; one example is the integral domain of *all algebraic integers*, containing the infinite proper divisor chain:

$$\sqrt{2}, \sqrt[4]{2}, \sqrt[8]{2}, \ldots, \sqrt[2^n]{2}, \ldots .$$

But if in a given integral domain $\mathfrak{J}$ every representation (23) is unique in the above sense, then every irreducible element in $\mathfrak{J}$ is prime, so that the concepts "irreducible" and "prime" coincide. For if $\mathfrak{J}$ were to contain an irreducible element u that is not prime (§2.3), we could find an equation $ua = bc$ with elements a, b, c in $\mathfrak{J}$ such that u is not a divisor of b nor of c. But now we could derive a contradiction by splitting up the elements a, b, c into their irreducible factors and noting that both sides of the equation must yield the same factorization, for then the irreducible element u must be an associate of an irreducible factor in the right-hand side, arising either from b or from c.

2.6. *Unique Factorization (u.f.) Rings*

Of great importance are the integral domains $\mathfrak{J}$ in which every factorization (23) is unique. Such rings are called *u.f. rings*; in them we have the *u.f. theorem* (unique factorization theorem):[25] *Every element $a(\neq 0)$ of a u.f. ring $\mathfrak{J}$ can be expressed uniquely as the product of prime elements p_i*:[26]

$$a = p_1 p_2 \cdots p_s , \tag{24}$$

where uniqueness means that the prime elements p_i in (24) *are uniquely determined up to order and unit factors.*

We now prove the theorem: *an integral domain $\mathfrak{J}$ with unity element satisfying the divisor chain condition and the condition that every irreducible element is prime is a u.f. ring.*

Since the existence of at least one factorization (24) has just been shown to follow from the divisor chain condition, only the uniqueness remains to be proved. Let us assume that

$$a = q_1 q_2 \cdots q_t \tag{24$'$}$$

is a second factorization with prime elements q_i; since the product $q_1 q_2 \cdots q_t$ is divisible by the prime element p_1, at least one factor must be divisible by p_1. We may assume that the order of factors in (24′) is such that $p_1 \mid q_1$; since q_1 is also prime, it follows that p_1 and q_1 are associates: $q_1 = ep_1$ with a suitably chosen unit e. By setting (24) and (24′) equal to each other and cancelling with p_1 we obtain the result: $p_2 p_3 \cdots p_s = e q_2 q_3 \cdots q_t$. Proceeding in this way we find successively that q_2 is an associate of p_2, q_3 of p_3, ..., q_s of p_s, and also that $t = s$, which completes the proof.

In the quotient field of a u.f. ring, every fraction a/b can be written "in lowest terms," i.e., in such a way that the numerator and the denominator have no prime factor in common; the reduction to lowest terms is

[25] If the factorization (23) is unique every irreducible element is also prime, as has just been shown, so that here the two concepts coincide.

[26] Of course, two or more of the prime factors p_i may be equal.

essentially unique, since it can be altered only by the adjunction of the same unit factor to numerator and denominator: $a/b = ea/eb$.

2.7. Let a, b be two arbitrary elements in the integral domain $\mathfrak{J}$. An element $d \in \mathfrak{J}$ which is a divisor of a and also of b is called a *common divisor* of a, b. We say that a, b are *coprime* if they have only the units as common divisors.

An element d is called the *greatest common divisor* of a, b, or in symbols

$$\text{(25)} \qquad \mathrm{GCD}(a, b) = d,$$

if d is a common divisor of a, b and every other common divisor d' of a, b is a divisor of d: $d' | d$. The $\mathrm{GCD}(a, b, c, ..., d)$ of the elements $a, b, c, ..., d$ is defined analogously.

It is obvious that if the greatest common divisor of two or more elements exists, *it is determined only up to a unit factor*.

2.8. *Euclidean Rings*

In a u.f. ring it is easy (cf. IB6, §2.6) to read off the $\mathrm{GCD}(a, b)$ from the "canonical factorization" (24) of a and b; but often these factorizations are not immediately known, and generally it is a very time-consuming task to determine them. Thus it is advantageous to have another procedure for finding the GCD, namely, the *Euclidean algorithm*, which can be carried out in certain rings, the so-called *Euclidean rings*, and leads very quickly to the $\mathrm{GCD}(a, b)$. Let us now discuss these concepts.

An integral domain $\mathfrak{J}$[27] is called a *Euclidean ring* if the division algorithm is available, i.e., if to every element $a(\neq 0)$ in $\mathfrak{J}$ it is possible to assign a nonnegative integer $H(a)$ such that

$$\text{(26')} \qquad H(ab) \geqslant H(a) \qquad \text{for all} \quad a, b(\neq 0) \text{ in } \mathfrak{J}$$

and for any two given elements $a, b(\neq 0)$ in $\mathfrak{J}$ we can find elements q and r in $\mathfrak{J}$ such that

$$\text{(26'')} \qquad a = bq + r \quad \text{with} \quad H(b) > H(r) \quad \text{or} \quad r = 0.$$

The integral domains G and R[x] are Euclidean rings, since in G the correspondence $H(a) = |a|$, and in $R[x]$ the correspondence $H(a) =$ "the degree of the polynomial a," satisfy the conditions (26′) and (26″) (see IB6, §2.10); and for the same reason every polynomial ring $\mathfrak{K}[x]$ over a field $\mathfrak{K}$ of coefficients is a Euclidean ring.[28]

[27] We need not assume the existence of a unity element in $\mathfrak{J}$, because its existence, as we shall show in §2.10, follows from (26′) and (26″).

[28] Certain rings of algebraic numbers, e.g., the rings $G[i]$, $G[\sqrt{-2}]$, $G[\zeta]$ with $\zeta = (-1 + \sqrt{-3})/2$, are Euclidean (see IB6, §2.10).

2.9. *In a Euclidean ring it is possible to carry out the Euclidean algorithm,* which consists of a certain repetition of the division algorithm (26″). Applied to two elements $a, b(\neq 0)$ of a Euclidean ring $\mathfrak{J}$ this algorithm leads, as will be shown in detail in IB6, §2.10, to the result:

(I) *The last nonzero remainder is the* GCD(a, b);

(II) *There exist in* $\mathfrak{J}$ *two (coprime) elements* x, y *satisfying the equation*

$$ax + by = \text{GCD}\,(a, b).[29] \tag{27}$$

In particular, if the elements a, b are *coprime*, then GCD $(a, b) = 1$, and there exist two elements $x, y \in \mathfrak{J}$ satisfying the equation

$$ax + by = 1; \tag{28}$$

conversely, it follows from such an equation that a, b are coprime. The elements x, y can be calculated by the Euclidean algorithm, and every pair of the form

$$x' = x + bc, \qquad y' = y - ac$$

with arbitrary $c \in \mathfrak{J}$ is then a solution of (27) or (28).

2.10 (I) *Every Euclidean ring contains a unity element.*

(II) *In a Euclidean ring the divisor chain condition* (§2.5) *is satisfied.*

(III) *In a Euclidean ring irreducible elements and prime elements coincide* (§2.3).

Finally, it follows (§2.6) that *every Euclidean ring is a u.f. ring.*[30]

Proof of (I). Let $a(\neq 0)$ be an element of the Euclidean ring $\mathfrak{J}$ (§2.8) for which $H(a)$ has the smallest possible value, and let $b \in \mathfrak{J}$ be arbitrary. Then the division algorithm (26″) applied to the pair b, a gives:

$$b = aq + r;$$

but here we must have $r = 0$, since otherwise $H(r) < H(a)$, contrary to assumption. Thus $b = aq$; that is, every element $b \in \mathfrak{J}$ is divisible by a. In particular, for $b = a$ we have the equation $a = ae$ or, after multiplication by q, also $b = be$, so that e is the unity element in $\mathfrak{J}$ (§1.6).

Proof of (II). The fact that $\mathfrak{J}$ cannot contain a proper divisor chain (§2.5) with infinitely many terms follows immediately from the lemma: *if b is a proper divisor of a, then* $H(b) < H(a)$.

[29] If we expand a/b in a *regular continued fraction* (IB6, §3.1), the numerator and the denominator of the second-to-last approximating fraction, taken with suitable signs, form a solution y, x of equation (27).

[30] This theorem is also a consequence of the idealtheoretic theorems §3.3 and 3.4.

To prove the lemma, let $a = bc$ and let the division algorithm (26″), when applied to b, a, give

$$b = aq + r,$$

where we must have $r \neq 0$, since otherwise a and b would be divisible by each other and would therefore be associates, contrary to assumption. Thus $H(r) < H(a)$; on the other hand, it follows from

$$r = b - aq = b(1 - cq)$$

and from $1 - cq \neq 0$ (since by assumption c is not a unit) that $H(b) \leqslant H(r)$, by (26′) and therefore $H(b) < H(a)$, as was to be proved.

Proof of (III). Now let p denote an *irreducible* element of $\mathfrak{J}$ and assume that the product ab is divisible by p, so that $pq = ab$. If $p|a$, then the criterion in §2.3 for p to be a prime is already satisfied; thus we assume that a is not divisible by p and then prove that we must have $p \mid b$. But p, a are now coprime, since the irreducible element p has only associates and units as its divisors, and therefore by (28) there exist elements x, y in $\mathfrak{J}$ such that

$$px + ay = 1.$$

If we multiply this equation by b and the equation

$$pq - ab = 0$$

by y and add, we obtain $p(bx + qy) = b$, so that $p \mid b$, as was to be proved.

Since the important integral domains G and $R[x]$ have already been proved to be Euclidean rings (§2.8),*it now follows that they are also u.f. rings. The same result follows for all polynomial rings $\mathfrak{K}[x]$ over a field $\mathfrak{K}$ of coefficients.*

But the converse is not true, since there exist u.f. rings that are not Euclidean. An example is the ring $G[x]$ of all integral polynomials in x which, as we shall prove in §2.14, is a u.f. ring but is not Euclidean, for if it were, we could apply the Euclidean algorithm to the coprime elements x and 2 and obtain an equation (28):

$$xh(x) + 2k(x) = 1 \quad \text{with} \quad h(x), k(x) \in G[x],$$

which leads to a contradiction, as is clear at once if we set $x = 0$.

2.11. *The Polynomial Ring $\mathfrak{J}[x]$*

Let us now discuss, somewhat more generally, the polynomial ring $\mathfrak{J}[x]$ consisting of all the polynomials

$$f(x) = a_0 + a_1x + a_2x^2 + \cdots + a_nx^n \tag{29}$$

of degree $n = 0, 1, 2, \ldots$, with coefficients $a_i \in \mathfrak{J}$,[31] where we shall assume that $\mathfrak{J}$ is a u.f. ring. For example, $G[x]$ is a polynomial ring of this kind.

Since $\mathfrak{J}$ is a u.f. ring, the greatest common divisor

$$\text{GCD}(a_0, a_1, \ldots, a_n) = d \tag{30}$$

is uniquely determined up to a unit. To determine its value we carry out the *canonical factorization* (24) of the coefficients a_i in the usual way and select all the common prime factors. If we then set

$$a_i = da_i^*, \qquad a_i^* \in \mathfrak{J}, \qquad i = 0, 1, \ldots, n,$$

we have

$$f(x) = df^*(x) \qquad \text{with} \quad f^*(x) = a_0^* + a_1^*x + a_2^*x^2 + \cdots + a_n^*x^n \tag{31}$$

and

$$\text{GCD}(a_0^*, a_1^*, \ldots, a_n^*) = 1. \tag{31$'$}$$

A polynomial in $\mathfrak{J}[x]$ whose coefficients, like those of $f^*(x)$, have no common factors except the units is said to be *primitive*. Thus we have the result that *every polynomial $f(x)$ in $\mathfrak{J}[x]$ can be written uniquely (apart from units) as the product* (31) *of the GCD of its coefficients* (30) *and a primitive polynomial $f^*(x)$.*[32]

Along with $\mathfrak{J}[x]$ we now consider the polynomial ring $\mathfrak{K}[x]$ over the quotient field (cf. §1.12 and §2.6) $\mathfrak{K}$ of the u.f. ring $\mathfrak{J}$. By the analogous argument we see that *every polynomial $f(x)$ in $\mathfrak{K}[x]$ can be written uniquely (apart from units) as the product*

$$f(x) = \frac{a}{b} f^*(x), \tag{32}$$

where a, b are coprime elements of $\mathfrak{J}$ and $f^*(x)$ is a primitive polynomial in $\mathfrak{J}[x]$.

[31] For degree 0 we thus obtain all the elements of $\mathfrak{J}$, so that $\mathfrak{J}$ is a *subring* of $\mathfrak{J}[x]$. Under our present convention the zero element of the ring $\mathfrak{J}$, like every other element in the same ring, has the degree 0. In certain other contexts, which we shall not discuss here, it is convenient to leave the degree of the zero element undetermined. If we wish to preserve the formula

$$\text{degree } f(x) \cdot g(x) = \text{degree } f(x) + \text{degree } g(x)$$

for identically vanishing factors, we may set degree $0 = -\infty$ (cf., p. 361, ftn. 7).

[32] If $n = 0$, we set $f^*(x) = 1$.

2.12. *Theorem of Gauss for Primitive Polynomials*

The product of two primitive polynomials is again a primitive polynomial.

For if the polynomial $f(x)$ in (29) is primitive, and if $g(x)$ is another primitive polynomial

$$g(x) = b_0 + b_1x + b_2x^2 + \cdots + b_mx^m,$$

then their product is a polynomial

$$h(x) = f(x)\,g(x) = c_0 + c_1x + c_2x^2 + \cdots + c_{n+m}x^{n+m}$$

with coefficients formed as follows:[33]

$$\begin{aligned} c_0 &= a_0b_0\,, \\ c_1 &= a_0b_1 + a_1b_0\,, \\ &\ldots\ldots\ldots\ldots\ldots\ldots \\ c_i &= a_0b_i + a_1b_{i-1} + a_2b_{i-2} + \cdots + a_ib_0\,, \\ &\ldots\ldots\ldots\ldots\ldots\ldots\ldots\ldots \end{aligned}$$

Now let it be assumed that π is a prime divisor of all the coefficients $c_0, c_1, \ldots, c_{n+m}$; then since the polynomials $f(x)$ and $g(x)$ are both primitive, there exist indices j and $k (0 \leqslant j \leqslant n, 0 \leqslant k \leqslant m)$ such that π is not a divisor of a_j or b_k but is a divisor of all the preceding $a_0, \ldots, a_{j-1}$ and $b_0, \ldots, b_{k-1}$. Then, in contradiction to our assumption, the coefficient

$$c_{j+k} = a_0b_{j+k} + a_1b_{j+k-1} + \cdots + a_jb_k + \cdots + a_{j+k-1}b_1 + a_{j+k}b_0\,,$$

is certainly not divisible by π, since all the terms in this sum are divisible by π, with the single exception of the term a_jb_k. Thus $h(x)$ is primitive.

2.13. *The Polynomial Ring* $\mathfrak{J}[x]$ *Satisfies the Divisor Chain Condition* (§2.5)

For if an element $f(x) = df^*(x)$ written in the form (31) has a proper divisor $g(x) = bg^*(x)$, then $b \mid d$ and $g^*(x) \mid f^*(x)$, where the polynomials $g^*(x)$ and $f^*(x)$ are primitive and at least one of the divisions is proper. But the element d of the u.f. ring $\mathfrak{J}$ has only finitely many divisors b, and the proper divisor $g^*(x)$ must be of lower degree than $f^*(x)$. Thus every proper divisor chain (22) in $\mathfrak{J}[x]$ is finite.

2.14. *A polynomial ring* $\mathfrak{J}[x]$ *is a u.f. ring if and only if the domain of coefficients* $\mathfrak{J}$ *is a u.f. ring. In particular,* $G[x]$ *is a u.f. ring.*[34]

The condition is obviously necessary; in order to show that it is also sufficient, we need only prove, in view of §2.6 and §2.13, that *every irreducible element in* $\mathfrak{J}[x]$ *is prime.* Now an irreducible element in $\mathfrak{J}[x]$ is

[33] In order to avoid a troublesome listing of various cases, we shall assume that all a_j with index $j > n$ and all b_k with index $k > m$ are set equal to zero.

[34] Note that $G[x]$ is *not a Euclidean ring* (§2.10).

either *independent* of x (i.e., the representation (29) has the degree $n = 0$), when it is an irreducible and therefore prime element π of the u.f. ring $\mathfrak{J}$, or else it is an irreducible and therefore primitive polynomial $p^*(x)$.

In the first case, if the product $f(x)\,g(x)$ of two polynomials in $\mathfrak{J}[x]$ is divisible by π, i.e., if in the notation of §2.13

$$\pi \mid f(x)\,g(x) \qquad \text{or also} \qquad \pi \mid bdf^*(x)\,g^*(x),$$

it follows that $\pi | bd$. But π is prime in $\mathfrak{J}$, so that one of the two factors b, d, say d, is divisible by π: $\pi \mid d \mid f(x)$. Thus π is also prime in $\mathfrak{J}[x]$.

In the second case it follows from

$$p^*(x) \mid f(x)\,g(x) \qquad \text{or} \qquad p^*(x) \mid bdf^*(x)\,g^*(x),$$

in view of the fact that $p^*(x)$ and also the product $f^*(x)\,g^*(x)$ are primitive (§2.12), that

$$p^*(x) \mid f^*(x)\,g^*(x).$$

If we examine these divisibility relations (see §2.11) in $\mathfrak{R}[x]$, which is a u.f. ring (§2.10), we see that one of the two factors, say $f^*(x)$, is divisible by the prime element[35] $p^*(x)$:

$$f^*(x) = p^*(x)\,h(x) \qquad \text{with} \quad h(x) \in \mathfrak{R}[x].$$

By §2.11 we can write $h(x) = a/b\, h^*(x)$, with $h^*(x)$ primitive in $\mathfrak{J}[x]$, so that the above equation, after multiplication by b, gives:

$$bf^*(x) = ap^*(x)\,h^*(x);$$

thus we see from the theorem of Gauss (§2.12) that $a = eb$ and $f^*(x) = ep^*(x)\,h^*(x)$, where e is a unit, so that $f(x) = df^*(x)$ is in fact divisible by $p^*(x)$ with a quotient in $\mathfrak{J}[x]$, as was to be proved.

A slight generalization of this theorem leads to the result: *not only the polynomial rings $R[x]$ and $G[x]$ in one indeterminate, but also the rings*

[35] The polynomial $p^*(x)$ is *prime*, since it too is irreducible in $\mathfrak{R}[x]$. For if $p^*(x)$ were reducible in $\mathfrak{R}[x]$:

$$p^*(x) = p_1(x)\,p_2(x), \qquad p_1(x) = \frac{a}{b}\,p_1^*(x), \qquad p_2(x) = \frac{c}{d}\,p_2^*(x)$$

with $a, b, c, d \in \mathfrak{J}$ and primitive polynomials $p_1^*(x)$, $p_2^*(x)$ in $\mathfrak{J}[x]$ of degree less than the degree of $p^*(x)$, we would have

$$bdp^*(x) = acp_1^*(x)\,p_2^*(x);$$

but then by the theorem of Gauss (§2.12) we could write $ac = ebd$ with a unit $e \in \mathfrak{J}$ and $p^*(x) = ep_1^*(x)\,p_2^*(x)$, in contradiction to the assumed irreducibility of $p^*(x)$ in $\mathfrak{J}[x]$.

$R[x, y]$, $G[x, y]$, $R[x, y, z]$, $G[x, y, z]$ ... *in several indeterminates are u.f. rings. The same result holds for all polynomial rings* $\mathfrak{J}[x, y]$, $\mathfrak{J}[x, y, z]$ ... , *provided* $\mathfrak{J}$ *is a u.f. ring.* The proof follows at once from the remark that the polynomial ring $\mathfrak{J}[x, y]$ can also be regarded as a polynomial $\mathfrak{J}^*[y]$ if we set $\mathfrak{J}^* = \mathfrak{J}[x]$. But $\mathfrak{J}^*$ is a u.f. ring, so that $\mathfrak{J}^*[y] = \mathfrak{J}[x, y]$ is also such a ring, and so forth.

3. Ideals in Commutative Rings, Principal Ideal Rings, Residue Class Rings

3.1. In IB2, §3 we saw that the inner structure of a group can best be determined by a study of its subgroups, particularly of its invariant subgroups; thus it is natural, in investigating the *inner structure*[36] of a ring $\mathfrak{R}$, to examine its subrings. Here also subrings of a certain type will be distinguished, namely the *ideals.*

Definition of an ideal. A non-empty subset $\mathfrak{a}$ *of a ring* $\mathfrak{R}$ *is called an "ideal" if it has the following properties:*

(I) *the module property*

$$a, b \in \mathfrak{a} \quad \text{implies} \quad a - b \in \mathfrak{a}; \tag{33}$$

(II) *the ideal property*

$$a \in \mathfrak{a} \quad \text{and} \quad r \in \mathfrak{R} \quad \text{imply} \quad ra \in \mathfrak{a}. \tag{34}$$

The module property (I) implies that every ideal of a ring is a *subgroup of the additive group* of the ring (§1.4). To see this, we note that every ideal $\mathfrak{a}$ contains the zero element $a - a = 0$ and thus by (33) contains the *inverse* $-a = 0 - a$ of every element a in the ideal. But then, from $a - (-b) = a + b$ it follows[37] by (33) that

$$a, b \in \mathfrak{a} \quad \text{implies} \quad a + b \in \mathfrak{a}. \tag{33'}$$

Similarly the ideal property (II) implies that $\mathfrak{a}$ is a *subring* of $\mathfrak{R}$; that is, $\mathfrak{a}$ is closed not only with respect to addition (33′) and subtraction (33), but also under multiplication; for from $a, b \in \mathfrak{a}$ it follows by (34) that $ab \in \mathfrak{a}$. But (34) makes a stronger demand, namely that the product ab shall still lie in $\mathfrak{a}$ even though only one of the two factors lies in $\mathfrak{a}$. Our reason for confining attention to the subrings distinguished in this way will become clear in our discussion of congruences and residue class rings (§3.6).

The zero element by itself forms an ideal, the *zero ideal,* denoted by (0); in the same way the ring $\mathfrak{R}$ is itself an ideal, the *unit ideal.* These two

[36] By the *inner structure* of a ring we chiefly mean the divisibility relations in the ring.

[37] Thus we can derive (33′) from (33), but not conversely.

ideals occur in every ring, and in a field they are the only ideals; for in a field $\mathfrak{K}$ any ideal $\mathfrak{a}$ containing a nonzero element must contain all the elements of $\mathfrak{K}$, by the field postulate (§1.10) and the ideal property (34).

From (33′) and (34) we can now derive the following important property of an ideal $\mathfrak{a}$ in a ring $\mathfrak{R}$:

$$(35) \qquad a, b \in \mathfrak{a} \quad \text{and} \quad r, s \in \mathfrak{R} \quad \text{imply} \quad ra + sb \in \mathfrak{a}.$$

3.2. More generally we can say: If an ideal $\mathfrak{a}$ contains the elements $a_1, a_2, \ldots, a_s$, then it contains all the *linear combinations*

$$(36) \qquad r_1 a_1 + r_2 a_2 + \cdots + r_s a_s \quad \text{with} \quad r_1, r_2, \ldots, r_s \in \mathfrak{R}.$$

Conversely, the set of all elements (36) forms an ideal, call it $\mathfrak{b}$, as can be verified at once from the conditions (33) and (34). The ideal $\mathfrak{b}$[38] is said to be *generated by the elements* $a_1, a_2, \ldots, a_s$ and is written:

$$(37) \qquad \mathfrak{b} = (a_1, a_2, \ldots, a_s).$$

The elements $a_1, a_2, \ldots, a_s$ are said to form a *basis for the ideal* $\mathfrak{b}$, although it is not thereby asserted that the basis cannot be shortened, i.e., that the same ideal cannot be generated by fewer *basis elements.*

A priori it is not clear whether every ideal in an arbitrary ring has a *finite* basis or not; nevertheless, in all the rings considered here it is true that *every ideal has a finite basis.* Such rings are said to satisfy the *basis condition* and are called *Noetherian rings.*

Particularly important are the rings in which every ideal $\mathfrak{a}$ has a basis consisting of a single element $\mathfrak{a} = (a)$. In this case $\mathfrak{a}$ is the set of all multiples $ra (r \in \mathfrak{R})$ of a; in other words, the ideal comprizes the set of all elements in the ring $\mathfrak{R}$ that are *divisible* by a. Such ideals are called *principal ideals*, and if all the ideals of a ring are principal, the ring is called a *principal ideal ring.*

3.3. *Every Euclidean Ring is a Principal Ideal Ring*

For if $\mathfrak{J}$ is a Euclidean ring (§2.8) and $\mathfrak{a}$ is an arbitrary ideal ($\neq (0)$) in $\mathfrak{J}$, let $a (\neq 0)$ be an element in $\mathfrak{a}$ such that the function $H(a)$ defined in §2.8 assumes its smallest value. Then for any b in $\mathfrak{a}$ the division algorithm (26″), applied to b and a, produces an equation $b = aq + r$. By (35) the element r is contained in $\mathfrak{a}$ and, if it is not zero, satisfies the condition $H(r) < H(a)$, in contradiction to the assumption. Thus $r = 0$ and $b = aq$ with $q \in \mathfrak{J}$. Consequently, $\mathfrak{a} = (a)$ is a principal ideal, as was to be proved.

[38] It is assumed here that the ring $\mathfrak{R}$ contains a unity element. The ideal $\mathfrak{b}$ is a *subideal* of $\mathfrak{a}$, which may coincide with $\mathfrak{a}$; in symbols:

$$\mathfrak{a} \supseteq \mathfrak{b} \quad \text{or} \quad \mathfrak{b} \subseteq \mathfrak{a}.$$

In particular, the rings G and $R[x]$, *which are Euclidean rings by* §2.8, *are principal ideal rings.*[39]

We already know (§2.10) that every Euclidean ring is a u.f. ring; somewhat more generally, we have the following theorem.

3.4. *Every Principal Ideal Ring Is a U.F. Ring*

For let $\mathfrak{J}$ be a principal ideal ring, i.e., an integral domain with unity element, in which every ideal has a basis consisting of a single element; then by §2.6 we must prove that every irreducible element in $\mathfrak{J}$ is prime and that $\mathfrak{J}$ satisfies the divisor chain condition.

The greatest common divisor (§2.7)

$$\text{GCD}(a_1, a_2, ..., a_s) = d \tag{38}$$

of two or more elements is always a well-defined[40] element of $\mathfrak{J}$, and d is the basis element of the principal ideal generated by the a_i,

$$(a_1, a_2, ..., a_s) = (d). \tag{38'}$$

For on the one hand all the a_i are divisible by d, and on the other d is a linear combination (36) of the a_i and is thus divisible by each of their common factors; consequently, d is their greatest common factor.

In particular, if two elements a, b in $\mathfrak{J}$ are coprime, i.e., if $\text{GCD}(a, b) = 1$, then the principal ideal $(a, b) = (1)$ generated by them is the unit ideal, so that there exist elements x, y in $\mathfrak{J}$ satisfying the equation

$$ax + by = 1.$$

It follows, exactly as in §2.10 (III), that *in a principal ideal ring* $\mathfrak{J}$ *every irreducible element is prime.*

Now let $a_1, a_2, ..., a_i, ...$ be a divisor chain (22) in $\mathfrak{J}$, and form the set $\mathfrak{m}$ of all elements of $\mathfrak{J}$ that are divisible by any of the a_i.[41] This set $\mathfrak{m}$ is an *ideal* in $\mathfrak{J}$; for if a, b are any two elements of a set $\mathfrak{m}$, then there exists a first a_i in the divisor chain that is a divisor of a, and also a first a_j that is a divisor of b. Let us assume that $j \geqslant i$; then a_j is a common divisor of a and b and thus a divisor of $a - b$, so that $a - b \in \mathfrak{m}$, and $\mathfrak{m}$ has the module property (33). The condition (34) is obviously satisfied, since together with a all elements $ra (r \in \mathfrak{J})$ are divisible by a_i and therefore are contained in $\mathfrak{m}$.

[39] On the other hand, the integral polynomial ring $G[x]$ is a Noetherian ring but not a principal ideal ring.

[40] Defined up to a unit factor; in (38) and (38′) d can be replaced by any of its associates.

[41] An element that is divisible by a_i is also, of course, divisible by each of the subsequent $a_{i+1}, ...$.

By hypothesis this ideal $\mathfrak{m}$ has a one-element basis: $\mathfrak{m} = (m)$. Since the basis element m is an element of $\mathfrak{m}$, it must be divisible by a first $a_k : a_k | m$. If $l > k$, then $a_l \mid a_k$, and since $a_l \in \mathfrak{m} = (m)$, we also have $m \mid a_l$, so that $a_k | m | a_l$, or $a_k \mid a_l$; in other words, a_k is an associate of every subsequent element of the divisor chain. Thus a proper divisor chain in $\mathfrak{J}$ cannot have infinitely many terms; in other words, *the divisor chain condition holds in* $\mathfrak{J}$, which completes the proof that $\mathfrak{J}$ is a u.f. ring.

But the converse of this theorem does not hold; for example, the integral polynomial ring $G[x]$ is a u.f. ring (§2.14) but not a principal ideal ring; for it is easy to see (cf. §2.10) that the ideal $(x, 2)$ in $G[x]$ is not a principal ideal.

3.5. *Congruences Modulo* $\mathfrak{a}$

Two elements a and a' of a ring $\mathfrak{R}$ are said to be "*congruent modulo* $\mathfrak{a}$," or in symbols: $a \equiv a'(\mathfrak{a})$, if their difference $a - a'$ is contained in the ideal $\mathfrak{a}$:

(39) $\qquad a \equiv a'(\mathfrak{a}) \qquad$ means the same as $\quad a - a' \in \mathfrak{a}$.

In particular, the congruence $a \equiv 0(\mathfrak{a})$ means that the element a itself is in the ideal $\mathfrak{a}$.

With these congruences we can compute in exactly the same way as with equations: for it follows from $a \equiv a'(\mathfrak{a})$ and $b \equiv b'(\mathfrak{a})$ that

(40) $\qquad a + b \equiv a' + b', \qquad a - b \equiv a' - b', \qquad ab \equiv a'b'(\mathfrak{a})$.

To prove the first of these three congruences we note (cf. (13)) that:

$$(a + b) - (a' + b') = (a - a') + (b - b') \in \mathfrak{a}.$$

The proof of the second is analogous, and by (35) the third follows from:

$$ab - a'b' = (a - a')\, b + a'(b - b') \in \mathfrak{a}.$$

The fact that this last conclusion requires the ideal property (34) explains the peculiar importance of ideals in the class of all subrings (cf. §3.1).

3.6. *Residue Classes*

The congruence relation just defined is an *equivalence relation*[42] (IA, §8.5, and IB1, §2.2) for the elements of the ring $\mathfrak{R}$ and therefore generates a partition of these elements into classes, called "*residue classes*" modulo $\mathfrak{a}$. Every residue class is uniquely determined by any element a contained in it, since together with a it contains all the elements a' satisfying (39).[43] In

[42] It is easily shown that congruence is *reflexive*, *symmetric*, and *transitive* (cf. IA, §8.3 and IB1, §2.2).

[43] Every element of the ring lies in exactly one residue class; two distinct residue classes have no element in common.

particular, the residue class containing the zero element coincides with the ideal $\mathfrak{a}$.

By $[a]$ we denote the residue class modulo $\mathfrak{a}$ containing the element a and say that a is a *representative* of this class; any other element a' of the same class will serve equally well as a representative.

We now consider these residue classes $[a]$, $[b]$, $[c]$, ... as new elements, for which addition and multiplication can be defined in accordance with the postulates in §1.2 for a ring, with the result that the residue classes form a new ring, the *residue class ring* $\mathfrak{R}/\mathfrak{a}$.[44] The sum and the product of two residue classes $[a]$ and $[b]$ are naturally defined by

$$(41) \qquad [a] + [b] = [a + b], \qquad [a]\,[b] = [ab].$$

This definition is unique, since the result remains the same (cf. (40)) for all other choices a' and b' of the representatives for the two classes.

Computation with residue classes modulo $\mathfrak{a}$ is essentially the same as with the congruences modulo $\mathfrak{a}$ described in §3.5, where two elements congruent modulo $\mathfrak{a}$ are regarded as equal to each other. Consequently, all the ring postulates (§1.2) are satisfied in the residue class ring, since they are valid for the elements of the orginal ring $\mathfrak{R}$.

If to every element a in $\mathfrak{R}$ we assign the residue class $[a]$ of the residue class ring $\mathfrak{R}/\mathfrak{a}$, we have a *homomorphism* $\mathfrak{R} \simeq \mathfrak{R}/\mathfrak{a}$ in the sense of §1.13, in which all the elements of the ideal $\mathfrak{a}$ are mapped onto the zero element of the residue class ring $\mathfrak{R}/\mathfrak{a}$. Thus *every residue class ring* $\mathfrak{R}/\mathfrak{a}$ *is a homomorphic image of the ring* $\mathfrak{R}$.

Conversely, if a ring $\mathfrak{R}^*$ is the homomorphic image of $\mathfrak{R} : \mathfrak{R} \simeq \mathfrak{R}^*$ *there exists an ideal* $\mathfrak{a}$ *in* $\mathfrak{R}$ *such that the residue class ring* $\mathfrak{R}/\mathfrak{a}$ *is isomorphic to* $\mathfrak{R}^* : \mathfrak{R}/\mathfrak{a} \cong \mathfrak{R}^*$ (*the homomorphism theorem for rings*). The ideal $\mathfrak{a}$ consists of the set of all elements of $\mathfrak{R}$ mapped by the homomorphism $\mathfrak{R} \simeq \mathfrak{R}^*$ onto the zero element of $\mathfrak{R}^*$, where it is clear that this set is actually an ideal, since addition and multiplication are preserved under a homomorphism; i.e., if a and b are mapped onto the zero element in $\mathfrak{R}^*$, then so is $ra + sb$, $r, s \in \mathfrak{R}$. We see that the elements of the residue class ring $\mathfrak{R}/\mathfrak{a}$ are in one-to-one correspondence with the elements of $\mathfrak{R}^*$.

An ideal $\mathfrak{p}$ whose residue class ring $\mathfrak{R}/\mathfrak{p}$ has no divisors of zero is called a *prime ideal*; thus a prime ideal is characterized by the property that a product ab is contained in $\mathfrak{p}$ if and only if at least one factor a, b lies in $\mathfrak{p}$ (cf. p. 340, ftn. 2).

A *primary ideal* $\mathfrak{q}$ is an ideal whose residue class ring $\mathfrak{R}/\mathfrak{q}$ contains only *nilpotent* divisors of zero, i.e., elements which become equal to zero when raised

[44] The residue class ring corresponds to the concept in group theory (IB2, §3.3) of a *factor group* with respect to a normal subgroup.

to a certain power. Thus we can characterize a primary ideal $\mathfrak{q}$ by the following condition:

(42) $$ab \in \mathfrak{q} \quad \text{and} \quad a \notin \mathfrak{q} \quad \text{imply} \quad b^\rho \in \mathfrak{q}$$

for a natural number ρ. Setting $\rho = 1$, we have the condition for a prime ideal, so that prime ideals are special cases of primary ideals. To every primary ideal $\mathfrak{q}$ there corresponds a prime ideal $\mathfrak{p}$ consisting of all the elements of $\mathfrak{R}$ that lie in nilpotent residue classes modulo $\mathfrak{q}$.

In *Noetherian rings* (§3.2) we have the *general factorization theorem*: *every ideal can be represented (in an essentially unique way) as the intersection of finitely many primary ideals* corresponding to distinct prime ideals. To a certain extent, this factorization theorem for ideals takes the place of the (no longer valid) u.f. theorem. Somewhat more special are the rings occurring in the theory of algebraic numbers; in these rings of the "classical ideal theory" we have the following *factorization theorem*: *every ideal can be represented (uniquely, apart from the order of the factors) as a product of prime ideals* (cf. also IB6, §8.2).

The reader should compare the discussion in IB11, §3.

3.7. *Residue Class Rings* G/n[45]

The integral domain G of rational integers is a principal ideal ring (§3.3); every ideal (in this context often called a "module") can thus be written as the principal ideal $\mathfrak{a} = (n)$ consisting of all the multiples in G of the natural number n. In every residue class there is exactly one number between 0 and $n - 1$ (inclusive); this number may be chosen as the representative of the class. Thus there are exactly n residue classes

$$[0], [1], [2], \ldots, [n-1].$$

With these classes we compute exactly as with integers, except that the result of a computation modulo n must be reduced to the smallest nonnegative remainder (cf. IB6, §4.1).

The residue class ring G/n is finite[46] and has no divisors of zero if and only if $n = p$ is a prime; as a finite integral domain (§1.10) it is then a *field*, namely the *prime field of characteristic* p (§1.11).

3.8. *The Integral Domain of the Gaussian Numbers* $G[i]$ *as Residue Class Ring* $G[x]/\mathfrak{p}$

In the polynomial ring $G[x]$ we now consider the principal ideal[47]

(43) $$\mathfrak{p} = (x^2 + 1);$$

[45] The notation G/n is often used in place of the more exact $G/(n)$.

[46] A "finite" ring is a ring with only finitely many elements.

[47] This ideal $\mathfrak{p}$ is a *prime ideal*, since the basis polynomial $x^2 + 1$ is *irreducible* in $G[x]$ and is therefore *prime* (§2.14). In other words, a product $f(x)\,g(x)$ of two polynomials in $G[x]$ is contained in $\mathfrak{p}$ (i.e., is divisible by $x^2 + 1$) if and only if at least one of the factors is already contained in $\mathfrak{p}$.

if $p(x)$ is any polynomial in $G[x]$, division of $p(x)$ by $x^2 + 1$ gives the equation

$$(43') \quad p(x) = (x^2 + 1)\, q(x) + (ax + b), \quad \text{with} \quad q(x) \in G[x], \quad a, b \in G,$$

so that

$$(43'') \qquad p(x) \equiv ax + b(\mathfrak{p}).$$

So in every residue class modulo $\mathfrak{p}$ there exists exactly one integral polynomial $ax + b$ of degree $\leqslant 1$; and, of course, we may choose this particular polynomial as representative of its class. Computation with these classes, i.e., with congruences modulo $\mathfrak{p}$ follows the rules:

$$(ax + b) \pm (a'x + b') \equiv (a \pm a')\, x + (b \pm b')(\mathfrak{p}),$$
$$(ax + b)(a'x + b') \equiv (ab' + a'b)\, x + (bb' - aa')(\mathfrak{p});$$

but exactly the same rules must be followed if we replace x by the imaginary unit i and write ordinary equations instead of the congruences modulo $(\mathfrak{p})$. Consequently, except for the somewhat different notation, the residue class ring $G[x]/\mathfrak{p}$ is identical with the integral domain $G[i]$ of Gaussian integers (§1.1); in other words we have the *isomorphism*:

$$(44) \qquad G[i] \rightleftarrows G[x]/(x^2 + 1).$$

In exactly the same way we can show that the *field C of complex numbers* is isomorphic to the residue class ring $K[x]/(x^2 + 1)$, where K is the field of real numbers (see also IB8, §1.2).

3.9. *Residue Class Rings* $\mathfrak{K}[x]/f(x)$

Now let $f(x)$ be any polynomial of degree $\geqslant 1$ in the polynomial ring $\mathfrak{K}[x]$ over an arbitrary base field $\mathfrak{K}$, and consider the residue class ring $\mathfrak{K}[x]/\mathfrak{a}$ of $\mathfrak{K}[x]$ with respect to the principal ideal $\mathfrak{a} = (f(x))$, for which we also write $\mathfrak{K}[x]/f(x)$.

If $f(x)$ is of degree n in x, every polynomial $p(x)$ in $\mathfrak{K}[x]$ can be reduced, if necessary by the division algorithm (26″), to a polynomial of degree $\leqslant n - 1$:

$$(45) \qquad p(x) = f(x)\, q(x) + r(x),$$

so that

$$(45') \qquad p(x) \equiv r(x)(\mathfrak{a}),$$

with $r(x) = c_1 + c_2 x + c_3 x^2 + \cdots + c_n x^{n-1}$. So in every residue class modulo $\mathfrak{a}$ there is exactly one polynomial $r(x)$ of degree $\leqslant n - 1$ with coefficients in $\mathfrak{K}$. The residue classes can be put in one-to-one correspondence with these polynomials.

Computation with the residue classes is the same as ordinary computation with the polynomials $r(x)$, except that the result must be reduced modulo $f(x)$ whenever the degree exceeds $n - 1$. The calculations become simpler if we introduce the following notation for the residue classes

$$(46) \qquad [1] = e_1\,, \qquad [x] = e_2\,, \qquad [x^2] = e_3\,, \ldots, [x^{n-1}] = e_n\,.$$

Then the residue class represented by the polynomial

$$r(x) = c_1 + c_2x + \cdots + c_nx^{n-1}$$

can be written as a *linear form*[48] in the e_i

$$(46') \qquad c_1e_1 + c_2e_2 + \cdots + c_ne_n\,.$$

It is clear how such linear forms are to be added, but for multiplication we need a *multiplication table*, which must be constructed by computing congruences modulo $f(x)$. It is sufficient to calculate the result[49] for all products e_ie_j:

$$(46'') \qquad e_ie_j = \gamma_{ij}^1e_1 + \gamma_{ij}^2e_2 + \cdots + \gamma_{ij}^ne_n\,, \qquad i, j = 1, \ldots, n.$$

When the coefficients γ_{ij}^k have been determined, it is easy to carry out all the operations (except division) on the linear forms (46′), and in each case the result is a linear form denoting the same residue class as would result from the same operations applied to residue classes.

A ring consisting of linear forms (46′) provided with a multiplication table (46″) is called an *algebra* or a *hypercomplex system*. Thus we can now say: *every residue class ring* $\mathfrak{K}[x]/f(x)$ *is a (commutative) algebra over the base field* $\mathfrak{K}$.

3.10. *Residue Class Rings as Field Extensions*

If

$$(47) \qquad f(x) = a_0 + a_1x + a_2x^2 + \cdots + a_nx^n \qquad (n \geqslant 1)$$

is an irreducible polynomial in $\mathfrak{K}[x]$, *the residue class ring* $\mathfrak{K}[x]/f(x)$ *is a field*, which we shall denote by Ω.

[48] A *form* is a homogeneous polynomial, a *linear form* is a linear homogeneous polynomial.

[49] By (46) the residue class e_ie_j contains x^{i+j-2}; if $i + j - 2 \leqslant n - 1$, we have simply $e_ie_j = e_{i+j-1}$; otherwise, we must reduce x^{i+j-2} by (45) and express the resulting $r(x)$ by (46) as a linear form.

For if the polynomial $p(x)$ is not divisible by $f(x)$, i.e., if the irreducibility of $f(x)$ implies that $p(x)$ and $f(x)$ are coprime, then by (28) the Euclidean ring $\mathfrak{K}[x]$ contains two polynomials $h(x)$ and $k(x)$ with

$$p(x)\,h(x) + f(x)\,k(x) = 1;$$

and therefore, since $[f(x)] = [0]$,

$$[p(x)][h(x)] = [1],$$

which means that every nonzero element in Ω is a unit (§2.2), so that Ω is a *field.*

The field Ω contains a subfield $\mathfrak{K}^*$ isomorphic to $\mathfrak{K}$, where $\mathfrak{K}^*$ consists of all the residue classes that contain an element of $\mathfrak{K}$. Letting the element $a \in \mathfrak{K}$[50] represent its residue class $[a]$, we see that the correspondence $a \leftrightarrow [a]$ is an isomorphism (§1.13), so that we can identify the isomorphic fields $\mathfrak{K}$ and $\mathfrak{K}^*$ by setting their elements equal to each other. For the residue classes $[a]$ in $\mathfrak{K}^*$ we write simply a, as may be done without fear of ambiguity.

If we now denote by $f(X)$ the polynomial (47) in the new indeterminate X

$$f(X) = a_0 + a_1X + a_2X^2 + \cdots + a_nX^n, \tag{47'}$$

it is easy to see that *in the field Ω this polynomial has the zero* $[x]$, since in the residue class ring Ω

$$\begin{aligned} f([x]) &= a_0 + a_1[x] + \cdots + a_n[x]^n = [a_0 + a_1x + \cdots + a_nx^n] \\ &= [f(x)] = [0]. \end{aligned}$$

To sum up: *a polynomial $f(X)$ irreducible*[51] *over the field $\mathfrak{K}$ has at least one zero in the residue class field $\mathfrak{K}[x]/f(x)$, which may be regarded as an extension field of $\mathfrak{K}$.*

4. Divisibility in Polynomial Rings Elimination

4.1. The process of deciding whether a given element in a u.f. ring is reducible or irreducible is usually very time-consuming, if it is possible at all. Even in the very simple u.f. ring G of rational integers the only practical way of deciding whether a given number is prime or not is to consult a table of primes (provided the given number lies within the range of the table).[52] So we must expect that in complicated u.f. rings, partic-

[50] Two distinct elements in $\mathfrak{K}$ cannot be contained in the same residue class modulo $f(x)$.

[51] "Irreducible over $\mathfrak{K}$" means "irreducible in $\mathfrak{K}[x]$."

[52] A famous example for the difficulty of recognizing a prime is the number $2^{32} + 1 = 4294967297$, which Fermat (1601–1665) considered to be prime; but Euler (1707–1783) discovered that in fact it is composite: $641 \cdot 6700417$.

ularly in the polynomial rings $G[x]$, $R[x]$, $G[x, y]$, ..., this question will not be a simple one. Of course, we can show, just as for the ring G, that in principle the question can be decided in *finitely many steps*, which it may be possible to simplify by more or less ingenious devices; but even then the actual process will usually require far more time than can be devoted to it. So we must content ourselves here with some useful lemmas and a few special criteria for irreducibility.

4.2. *A polynomial* $p(x)$ *is irreducible in* $R[x]$ (*"irreducible over* R") *if and only if the corresponding primitive polynomial* $p^*(x)$ (*see* §2.11) *is irreducible in* $G[x]$.

For if the primitive polynomial $p^*(x)$ is reducible in $G[x]$, then certainly it remains so in $R[x]$; if it is prime in $G[x]$, then it remains prime in $R[x]$, as was shown in §2.14 (ftn. 35).

4.3. If $\mathfrak{J}$ is an integral domain with unity element ($\mathfrak{J}$ may be a field), then *a polynomial* $p(x)$ *in* $\mathfrak{J}[x]$ *is divisible* by *a linear polynomial* $x - \alpha$ ($\alpha \in \mathfrak{J}$) *if and only if* α *is a zero of* $p(x)$, i.e., ($p(\alpha) = 0$).

For if $p(x) = (x - \alpha)\, q(x)$, $q(x) \in \mathfrak{J}[x]$, it follows at once that $p(\alpha) = 0$. Conversely, if $p(\alpha) = 0$, then by the division algorithm[53] we can set up the identity

$$p(x) = (x - \alpha)\, q(x) + r, \qquad q(x) \in \mathfrak{J}[x], \qquad r \in \mathfrak{J}$$

and replace the indeterminate x by α; since $p(\alpha) = 0$, it follows that $r = 0$, so that $(x - \alpha) | p(x)$, as was to be proved.

4.4. For a *primitive* polynomial in $G[x]$:

$$p(x) = c_0 + c_1 x + c_2 x^2 + \cdots + c_n x^n \tag{48}$$

it is clear that a linear polynomial $a_0 + a_1 x$ can be a divisor of $p(x)$ only if $\mathrm{GCD}(a_0, a_1) = 1$, $a_0 | c_0$, $a_1 | c_n$. In particular, if $c_n = 1$, then necessarily $a_1 = \pm 1$: *a rational zero of an integral polynomial with* 1 *as highest coefficient is necessarily an integer.*

4.5. *The Irreducibility Criterion of Eisenstein*

A polynomial $p(x) \in G[x]$ *is irreducible in* $G[x]$ *if there exists a prime number* π *such that all the coefficients* c_i ($i = 0, 1, ..., n - 1$) *with the exception of* c_n *are divisible by* π *and and the first coefficient* c_0 *is not divisible by* π^2.

[53] Cf. IB4, §1 (4); the division algorithm (26″) can also be carried out in the non-Euclidean ring $\mathfrak{J}[x]$ if the "divisor" has a unit for its highest coefficient.

For if we examine the factorization $p(x) = f(x)\,g(x)$ we see that $\pi \mid c_0$, $\pi^2 \nmid c_0$, and $c_0 = a_0 b_0$ imply $\pi \mid a_0$, $\pi \nmid b_0$,[54] where the notation is the same as in §2.12. From the remaining conditions $\pi \mid c_1$, $\pi \mid c_2$, and so forth, it follows that $\pi \mid a_1$, $\pi \mid a_2$, ..., up to $\pi \mid a_{n-1}$; from $\pi \nmid c_n$ it follows that $\pi \nmid a_n$, so that $a_n \neq 0$. Thus $f(x)$ has the same degree as $p(x)$, so that $f(x)$ is not a proper divisor.[55]

4.6. The greatest divisor GCD($f(x)$, $g(x)$), of two polynomials $f(x)$ and $g(x)$ in a polynomial ring $\mathfrak{R}[x]$[56] can be calculated by the Euclidean algorithm (§2.9). But we now give another criterion, usually much easier to apply, for deciding whether two polynomials are coprime or not. Here it is convenient to write the polynomials in the following way:

$$(49) \qquad \begin{aligned} f(x) &= a_0 x^m + a_1 x^{m-1} + \cdots + a_m, \qquad a_0 \neq 0, \\ g(x) &= b_0 x^n + b_1 x^{n-1} + \cdots + b_n, \qquad b_0 \neq 0; \qquad a_i, b_j \in \mathfrak{R}. \end{aligned}$$

The polynomials $f(x)$ and $g(x)$ have a nontrivial common divisor GCD($f(x)$, $g(x)$) $= d(x)$ if and only if there exist in $\mathfrak{R}[x]$ a polynomial $h(x)$ of degree $\leqslant n - 1$ and a polynomial $k(x)$ of degree $\leqslant m - 1$ satisfying the identity

$$(50) \qquad h(x)\,f(x) + k(x)\,g(x) = 0 \qquad (h(x), k(x) \neq 0).$$

For if $f(x)$ and $g(x)$ are coprime, an equation of the form (50) would imply (since $\mathfrak{R}[x]$ is a u.f. ring) that $f(x) \mid k(x)$ and $g(x) \mid h(x)$, which is impossible since $k(x)$ is of lower degree than $f(x)$, and $h(x)$ is of lower degree than $g(x)$. Conversely, if the GCD($f(x)$, $g(x)$) $= d(x)$ is a polynomial of positive degree, the polynomials $h(x) = g(x)/d(x)$ and $k(x) = -f(x)/d(x)$ satisfy all the conditions of the theorem.

4.7. *A Criterion Based on the Resultant*

From the preceding section we can at once deduce that *the polynomials $f(x)$ and $g(x)$ in $\mathfrak{R}[x]$ are coprime if and only if the following $m + n$ polynomials*

$$(51) \qquad x^{n-1}f(x), \quad x^{n-2}f(x), \ldots, f(x), \quad x^{m-1}g(x), \quad x^{m-2}g(x), \ldots, g(x)$$

are linearly independent[57] *over* $\mathfrak{R}$.

[54] Of course there is no loss of generality in assuming $\pi \mid a_0$ rather than $\pi \mid b_0$. The notation $\pi \nmid b_0$ means "π is not a divisor of b_0."

[55] In this proof we have used only the u.f. theorem; so the Eisenstein criterion is valid in any polynomial ring $\mathfrak{I}[x]$ over a u.f. ring $\mathfrak{I}$.

[56] Here $\mathfrak{R}$ can be an entirely arbitrary field; in particular, one of the polynomial quotient fields $R(y)$, $R(y, z)$, ...

[57] For linear dependence see IB3, §1.3.

For equation (50) simply expresses the linear dependence over $\mathfrak{K}$ of the polynomials (51); and conversely, this linear dependence implies the existence in $\mathfrak{K}$ of elements α_i and β_j (not all zero) such that

$$\begin{aligned}\alpha_0 x^{n-1} f(x) + \alpha_1 x^{n-2} f(x) + \cdots + \alpha_{n-1} f(x) \\ + \beta_0 x^{m-1} g(x) + \beta_1 x^{m-2} g(x) + \cdots + \beta_{m-1} g(x) = 0\end{aligned}$$

is an identity in x, and this identity is of the form (50) with

$$\begin{aligned}h(x) &= \alpha_0 x^{n-1} + \alpha_1 x^{n-2} + \cdots + \alpha_{n-1}\,,\\ k(x) &= \beta_0 x^{m-1} + \beta_1 x^{m-2} + \cdots + \beta_{m-1}\,.\end{aligned}$$

If we now consider the polynomials (51) as linear forms in the $m + n$ magnitudes x^{m+n-1}, x^{m+n-2}, ..., x, 1, the question of their linear dependence or independence can be decided (IB3, §3.4) by constructing the determinant of their coefficient matrix. For the polynomials (49) this determinant, which is called the *Sylvester determinant*, has the following form:

$$\left|\begin{array}{ccccccccccc} a_0 & a_1 & \cdot & \cdot & \cdot & \cdot & \cdot & \cdot & a_m & & \\ & a_0 & a_1 & \cdot & \cdot & \cdot & \cdot & \cdot & \cdot & a_m & \\ & & \cdot & \cdot & \cdot & \cdot & \cdot & \cdot & \cdot & \cdot & \cdot \\ & & & & a_0 & a_1 & \cdot & \cdot & \cdot & \cdot & a_m \\ b_0 & b_1 & \cdot & \cdot & \cdot & b_n & & & & & \\ & b_0 & b_1 & \cdot & \cdot & \cdot & b_n & & & & \\ & & \cdot & \cdot & \cdot & \cdot & \cdot & \cdot & \cdot & \cdot & \cdot \\ & & & & & & b_0 & b_1 & \cdot & \cdot & b_n \end{array}\right| \begin{array}{l} \left.\vphantom{\begin{array}{c}a\\a\\a\\a\end{array}}\right\} n \text{ rows} \\ \left.\vphantom{\begin{array}{c}a\\a\\a\\a\end{array}}\right\} m \text{ rows.} \end{array}$$

The first row contains the coefficients a_0, a_1, ..., a_m of $f(x)$ followed by zeros; the second row begins with a zero and is otherwise equal to the first row shifted one place to the right, and so on; and the second half of the determinant is constructed analogously. It is easy to see that in this way we will obtain exactly $m + n$ columns.

The Sylvester determinant is called the resultant $R(f, g)$ *of the polynomials* (49). *The vanishing of the resultant is a necessary and sufficient condition for the polynomials* $f(x)$ *and* $g(x)$ *to have a nontrivial common factor.*

For in fact the vanishing of $R(f, g)$ is a necessary and sufficient condition for the linear dependence of the polynomials (51).

4.8. *The resultant* $R f, g)$ *is homogeneous of degree n in the* a_i *and of degree m in the* b_j ; *and it is isobaric of weight mn in the two sets of coefficients; its leading term is* $a_0^n b_n^m$ *(with the coefficient* $+1$*)*. Also

(53) $$R(g, f) = (-1)^{mn} R(f, g).$$

For it is easy to see,[58] by developing the determinant (52) (see IB3, §3.4), that

[58] The indices i_1, ..., i_n and j_1, ..., j_m in (54) are not necessarily distinct.

the individual terms of the resultant consist of n factors a_i and m factors b_j:

$$(54) \qquad a_{i_1}a_{i_2} \cdots a_{i_n}b_{j_1}b_{j_2} \cdots b_{j_m},$$

with rational integers for coefficients. The sum of all the indices in (54) is mn (i.e., the resultant is "isobaric of weight mn"):

$$(54') \qquad i_1 + i_2 + \cdots + i_n + j_1 + j_2 + \cdots + j_m = mn.$$

The proof of (54′) is as follows. In the determinant (52) we make the substitution

$$a_i \to \rho^i a_i, \qquad b_j \to \rho^j b_j;$$

then we multiply the successive rows of (52) by $1, \rho, \rho^2, \ldots, \rho^{n-1}, 1, \rho, \rho^2, \ldots, \rho^{m-1}$, and divide the successive columns by the factors $1, \rho, \rho^2, \ldots, \rho^{m+n-1}$. We thus obtain the original determinant[59] multiplied by the factor ρ^{mn}; but in the above substitution each individual term (54) is multiplied by the factor

$$\rho^{i_1+i_2+\cdots+i_n+j_1+j_2+\cdots+j_m} = \rho^{mn},$$

which completes the proof of (54′).

The formula (53) is a simple consequence of a well-known theorem on determinants (IB3, §3.4(a′) and §3.5. 2). Note that the determinant on the right arises from the one on the left by mn interchanges of rows (each of the m lower rows is interchanged with each of the n upper rows), which produce the factor $(-1)^{mn}$.

The leading term $a_0^n b_n^m$, by means of which the resultant may be normalized, is the product of the elements in the leading diagonal in (52).

For example, if $m = n = 2$, the resultant is given by

$$R(f, g) = \begin{vmatrix} a_0 & a_1 & a_2 & 0 \\ 0 & a_0 & a_1 & a_2 \\ b_0 & b_1 & b_2 & 0 \\ 0 & b_0 & b_1 & b_2 \end{vmatrix}$$

$$= a_0{}^2b_2{}^2 + a_2{}^2b_0{}^2 - a_0a_1b_1b_2 - a_1a_2b_0b_1 + a_0a_2b_1{}^2 + a_1{}^2b_0b_2 - 2a_0a_2b_0b_2$$

$$= (a_0b_2 - a_2b_0)^2 - (a_0b_1 - a_1b_0)(a_1b_2 - a_2b_1).$$

4.9. *The Resultant as a Function of the Zeros or Roots*

We shall denote the zeros[60] of the polynomial $f(x)$ in (49) by $\alpha_1, \alpha_2, \ldots, \alpha_m$, and the zeros of $g(x)$ by $\beta_1, \beta_2, \ldots, \beta_n$, and consider these zeros as independent transcendents in the sense of IB4, §2.3, which

[59] By the well-known formula for the sum of an arithmetic progression,

$$1 + 2 + 3 + \cdots + (n-1) = \tfrac{1}{2}n(n-1),$$

we here obtain

$$1 + 2 + \cdots + (m+n-1) - 1 - 2 - \cdots - (m-1) - 1 - 2 - \cdots - (n-1) = mn.$$

[60] Cf. IB4, §2.2.

must be adjoined [61] to the base field $\mathfrak{K}$ of the polynomial ring $\mathfrak{K}[x]$. Then $f(x)$ and $g(x)$ can be factored into linear factors (IB4, §2.2):

$$(55') \qquad f(x) = a_0(x - \alpha_1)(x - \alpha_2) \cdots (x - \alpha_m),$$

$$(55'') \qquad g(x) = b_0(x - \beta_1)(x - \beta_2) \cdots (x - \beta_n).$$

Apart from sign, the quotients a_i/a_0, b_j/b_0 of the polynomial coefficients (49) are the *elementary symmetric polynomials* (IB4, §2.4) of the α_i and β_j; so $a_0^{-n}b_0^{-m}R(f, g)$ is an entire rational function of the elementary symmetric polynomials and is therefore a *symmetric polynomial* in the indeterminates α_i and in the β_j:

$$a_0^{-n}b_0^{-m}R(f, g) = P(\alpha_1, \ldots, \alpha_m, \beta_1, \ldots, \beta_n).$$

Considered as a polynomial in the α_i, the expression P has the zeros β_j $(j = 1, \ldots, n)$, since substitution of β_j for α_i is necessary and sufficient for the polynomials $f(x)$ and $g(x)$ to acquire a common factor $x - \beta_j$, whereupon the resultant $R(f, g)$ becomes equal to zero. By §4.3, it follows that P is divisible by every linear factor $\alpha_i - \beta_j$, so that we may write

$$a_0^{-n}b_0^{-m}R(f, g) = C \prod_{i=1}^{m} \prod_{j=1}^{n} (\alpha_i - \beta_j),$$

with a factor C still to be determined. From (55″) we have

$$a_0^{-n}b_0^{-m}R(f, g) = Cb_0^{-m} \prod_{i=1}^{m} g(\alpha_i) = Cb_0^{-m} \prod_{i=1}^{m} (b_0\alpha_i^n + \cdots + b_n),$$

so that the *leading term* $a_0{}^n b_n{}^m$ in $R(f, g)$ (see §4.8) corresponds to $Cb_0^{-m}b_n{}^m$, which gives $C = 1$ and finally:

$$(56) \qquad R(f, g) = a_0{}^n b_0{}^m \prod_{i=1}^{m} \prod_{j=1}^{n} (\alpha_i - \beta_j).$$

This relation is an identity in the indeterminates α_i and β_j if the a_i/a_0, b_j/b_0 are replaced by the elementary symmetric polynomials; thus the relation continues to hold if the zeros α_i, β_j are no longer indeterminates but are arbitrary elements of the field $\mathfrak{K}$ or of an extension field of $\mathfrak{K}$.

From (56) we can at once read off the characteristic property of the resultant: namely, *the resultant of two polynomials vanishes if and only if the two polynomials have a common root.*

[61] The original base field $\mathfrak{K}$ is thus replaced by the transcendental extension $\mathfrak{K}(\alpha_1, \ldots, \alpha_m, \beta_1, \ldots, \beta_n)$.

The above discussion also leads at once to the two expressions for the resultant:

(56′) $$R(f, g) = a_0^n \prod_{i=1}^{m} g(\alpha_i) = (-1)^{mn} b_0^m \prod_{j=1}^{n} f(\beta_j).$$

Furthermore if $g(x) = g_1(x)\, g_2(x)$ is the product of two polynomials of (positive) degree n_1, $n_2 (n_1 + n_2 = n)$, it is easy to show that

$$R(f, g) = a_0^n \prod_{i=1}^{m} g_1(\alpha_i)\, g_2(\alpha_i) = \left\{a_0^{n_1} \prod_{i=1}^{m} g_1(\alpha_i)\right\} \left\{\alpha_0^{n_2} \prod_{i=1}^{m} g_2(\alpha_i)\right\},$$

which gives the important formula

(56″) $$R(f, g_1 g_2) = R(f, g_1)\, R(f, g_2).$$

4.10. The *discriminant* $D(f)$ of a polynomial $f(x)$ in $\mathfrak{K}[x]$ is defined, up to a numerical factor, as the resultant of the polynomial $f(x)$ and its derivative[62] $f'(x)$:

(57) $$D(f) = \frac{(-1)^{m(m-1)/2}}{a_0} R(f, f').$$

From (56′) and (58)[62] we find

$$D(f) = (-1)^{m(m-1)/2}\, a_0^{m-2} \prod_{i=1}^{m} f'(\alpha_i) = (-1)^{m(m-1)/2}\, a_0^{2m-2} \prod_{i=1}^{m} \prod_{j=1}^{m}{}' (\alpha_i - \alpha_j).$$

In the double product on the right every factor $(\alpha_i - \alpha_j)$ occurs twice, the second time with opposite sign; taking these factors together and noting the sign, we have

(59) $$D(f) = a_0^{2m-2} \prod_{i<j} (\alpha_i - \alpha_j)^2.$$

The vanishing of the discriminant $D(f)$ of a polynomial $f(x)$ in $\mathfrak{K}[x]$ is a necessary and sufficient condition for $f(x)$ and $f'(x)$ to have a common zero, or in other words for $f(x)$ to have a multiple zero (cf. IB4, §2.2).

[62] For the derivative of a polynomial see IB4, §2.2. From the formula proved there,

$$f'(x) = \frac{f(x)}{x - \alpha_1} + \frac{f(x)}{x - \alpha_2} + \cdots + \frac{f(x)}{x - \alpha_m},$$

it follows that

(58) $$f'(\alpha_i) = a_0 \prod_{j=1}^{m}{}' (\alpha_i - \alpha_j),$$

where the prime on the product means that the term with $j = i$ is to be omitted.

4.11. *Elimination Theory*

The problem of solving a system of algebraic equations in several unknowns, i.e., of finding their common zeros, is part of the theory of elimination. The case of systems of linear equations has already been handled in IB3, §§2.4 and 3.6; the solutions there can be found by means of determinants. For nonlinear systems of equations we adopt the method of *elimination*; i.e., from the given system of equations we deduce another system containing one fewer unknowns (from which one unknown has been eliminated), and we repeat this elimination until we reach a single equation with one unknown.

Provided we have taken certain precautions, every solution of the system of equations obtained by elimination can be extended, in at least one way, to a solution of the original system. Here we take for granted that we know how to solve an algebraic equation in one unknown.

We must be content with illustrating the method for the case of two equations[63] in two unknowns

$$f(x, y) = 0, \qquad g(x, y) = 0. \tag{60}$$

If $f(x, y)$ and $g(x, y)$ are polynomials in the polynomial ring $\mathfrak{K}[x, y]$, we first consider them as polynomials in $\mathfrak{K}^*[x]$ with $\mathfrak{K}^* = \mathfrak{K}(y)$. Then the polynomials f and g have the form (49), where the coefficients a_i, b_j are now polynomials in y. In order to apply the following theory we must first insure that the leading coefficients a_0 and b_0 (which may depend on y) are elements $(\neq 0)$ in the base field $\mathfrak{K}$. This condition can always be satisfied by means of a sufficiently general linear transformation.

Then the necessary and sufficient condition for the polynomials (60) to have a common zero is the vanishing of the resultant $R(f, g)$, which is here (cf. (52)) a polynomial in y. For the complete solution of the system (60) we must determine[64] the zeros $\beta_1, \ldots, \beta_t$ and substitute them into (60); the resulting polynomials $f(x, \beta_j)$ and $g(x, \beta_j)$ have a GCD of degree $\geqslant 1$, which can be calculated by the Euclidean algorithm; let its zeros be denoted by $\alpha_{j1}, \ldots, \alpha_{j,s_j}$. Then the common solutions of (62) are given by

$$x = \alpha_{jk}, \qquad y = \beta_j, \qquad j = 1, \ldots, t; \qquad k = 1, \ldots, s_j.$$

A completely satisfactory theory of elimination can be given only in terms of the theory of ideals in polynomial rings $P_n = \mathfrak{K}[x_1, \ldots, x_n]$. The left-hand sides of a given system of algebraic equations are polynomials $p_1, \ldots, p_s$ in P_n; they generate an ideal $a = (p_1, \ldots, p_s)$ in P_n, and our task is to determine the

[63] Strictly speaking, these are not *equations* but *problems*, namely, to find all the common zeros of the polynomials on the left-hand sides.

[64] These zeros are finite in number, unless $R(f, g)$ vanishes identically, which would mean that the polynomials $f(x, y)$ and $g(x, y)$ are not coprime.

manifold of the zeros of this ideal. We eliminate x_n by forming the *elimination ideal* $\mathfrak{a}_1 = \mathfrak{a} \cap P_{n-1}$ in[65] $P_{n-1} = \mathfrak{K}[x_1, ..., x_{n-1}]$, and then form the second *elimination ideal* $\mathfrak{a}_2 = \mathfrak{a} \cap P_{n-2}$ in $P_{n-2} = \mathfrak{K}[x_1, ..., x_{n-2}]$, and so forth. The last nonzero elimination ideal is a principal ideal. If it is the unit ideal (the entire ring P_n), then there are no zeros at all, otherwise its manifold of zeros is a sum of *algebraic manifolds*.

The investigation becomes even more difficult if the *multiplicity* of the zeros is to be taken into account. The concept of multiplicity plays an important role in theorems of enumeration, modeled after the theorem that a polynomial $f(x)$ of degree n has exactly n zeros, counting multiplicities. The most important theorem of this kind is the *theorem of Bézout*:

If multiplicities are taken into account, n homogeneous polynomials in $\mathfrak{K}[x_0, x_1, ..., x_n]$ *have either infinitely many zeros or a number of zeros equal to the product of their degrees.*

In view of the homogeneity here, the trivial zero $\{0, 0, ..., 0\}$ is excluded and two zeros $\{\xi_0, \xi_1, ..., \xi_n\}$ and $\{\rho\xi_0, \rho\xi_1, ..., \rho\xi_n\}$, with $\rho \neq 0$ and ξ_i in $\mathfrak{K}$ or in an algebraic extension of $\mathfrak{K}$, are regarded as identical. The multiplicity of an individual zero can also be defined, in terms of the theory of ideals, as the length (see below) of a corresponding primary ideal, which arises as the intersection of primary ideals (§3.6) in the factorization of the ideal generated by the n forms.

The length l of a primary ideal $\mathfrak{q}$ is defined as the length of a *composition series* (cf. the corresponding concept for groups in IB2, §12.1), extending from the primary ideal $\mathfrak{q}$ to the *associated* (§3.6, p. 355) prime ideal $\mathfrak{p}$:

$$\mathfrak{q} = \mathfrak{q}_1 \subset \mathfrak{q}_2 \subset \cdots \subset \mathfrak{q}_l = \mathfrak{p}.$$

Here it is assumed that all the terms $\mathfrak{q}_j (j = 1, ..., l)$ are primary ideals associated with the same prime ideal $\mathfrak{p}$, that $\mathfrak{q}_j$ is a *proper subideal* (i.e., a *proper subset*) of $\mathfrak{q}_{j+1}$, and that the series cannot be made longer by the insertion of further terms. In particular, every prime ideal is of length 1. The theorems here are similar to those for the composition series of a group: e.g., in a Noetherian ring (§3.2) every primary ideal $\mathfrak{q}$ has at least one composition series of *finite* length l; and every other composition series for the same primary ideal $\mathfrak{q}$ has the same length l (Jordan-Hölder theorem, IB2, §12.1).

In algebraic geometry still other definitions, some of them quite complicated, have been introduced for the multiplicity of points of intersection, but as long as we are dealing with applications of the concept as it occurs in the simple theorem of Bézout, the various definitions of multiplicity are all equivalent to the idealtheoretic one given here.

[65] Thus $\mathfrak{a}_1$ contains all those polynomials in $\mathfrak{a}$ which are independent of x_n.

CHAPTER 6

Theory of Numbers

1. Introduction

The theory of the natural numbers may be regarded as number theory in the narrower sense (see IB1, §1), but no matter how far we may wish to set the boundaries, it remains one of the most attractive parts of mathematics; for the most part its problems can be understood without extensive preparation, and they range from questions that can easily be answered to famous unsolved conjectures.

The modern theory of numbers includes the study of so-called algebraic numbers, i.e., the roots (zeros) of polynomials with coefficients that are integers in the ordinary sense (rational integers). Under the influence of the structure-theoretic methods of present-day mathematics, certain parts of number theory have become more abstract. The advantages of such a treatment of the subject are particularly clear in the theory of divisibility, described in §2. The concepts and theorems of that section are equally valid for the ring of Gaussian integers (see IB5, §1.1), i.e., the numbers $a + bi$ with rational integers a and b, for the polynomial ring in one indeterminate (see IB4, §2.1) with coefficients from a field, and for many other rings. Thus it is unnecessary to begin the argument afresh for each new application.

2. Divisibility Theory

2.1. For the time being we consider an arbitrary commutative[1] ring $\mathfrak{R}$. Let a and b be two elements of $\mathfrak{R}$; then a is said to be a *divisor of* b, in symbols $a \mid b$, if $b = ac$ with $c \in \mathfrak{R}$. To denote the opposite, we write

[1] Throughout Chapter 6, except where otherwise noted, we shall be dealing with commutative rings, so that the word "commutative" will ordinarily be omitted.

$a \nmid b$. This divisibility relation is obviously transitive: from $a \mid b$ and $b \mid c$ follows $a \mid c$. For all $x \in \mathfrak{R}$ we have $x \mid 0$, and thus in particular $0 \mid 0$. If $\mathfrak{R}$ has a unity element, we have the reflexive law: $x \mid x$ for all $x \in \mathfrak{R}$, and also $1 \mid x$. In the relation $b = ac$ the element c is said to be *complementary* to the divisor a.

2.2. The relation of divisibility defined in this way can be regarded as a weakening of the order relation (see IA, §8.3). In an order relation "$\leqslant$" it follows from $a \leqslant b$ and $b \leqslant a$ that $a = b$, but in the ring $\mathfrak{C}$ of integers[2] the fact that $(-2)|(+2)$ and $(+2)|(-2)$ are both true shows that the divisibility relation is certainly not an ordering, not even a partial ordering (see IA, §8.3). If $1 \in \mathfrak{R}$, such a relation is called a quasi-ordering, i.e., a reflexive and transitive relation (to be denoted, say, by "$\leqslant$" for which there may exist unrelated elements, i.e., elements a and b such that neither $a \leqslant b$ nor $a = b$ nor $b \leqslant a$.

2.3. Now let $\mathfrak{R}$ be a ring with unity element. An element $\epsilon \in \mathfrak{R}$ is called a *unit* if there exist an $\eta \in \mathfrak{R}$, such that $\epsilon\eta = 1$. Then η is an inverse of ϵ in sense of IB1, §3.1, so that $\eta = \epsilon^{-1}$ is also a unit. The unity element is obviously a unit, and so is the product of two units. Thus the units form a group with respect to multiplication, so that the inverse of a unit is uniquely determined (see IB2, §2.3).

For example, in the ring of Gaussian integers the numbers 1, -1, i and $-i$ are the only units, as is easily seen. In a field all the nonzero elements are obviously units.

If $c = ab$ is a factorization of c, then so is $c = (\epsilon a) \cdot (\epsilon^{-1}b)$, where ϵ is a unit; thus it is clear that, as far as factorization is concerned, the elements a and $a\epsilon$ are not essentially distinct. Such elements a and $a\epsilon$ are said to be *associates:* $a \sim a\epsilon$. The relation of "associate" is obviously reflexive, symmetric, and transitive and is thus an equivalence (see IA, §8.5). The corresponding equivalence classes are the classes of associated elements.

2.4. In general, in a quasi-ordered set two elements a and b are said to be associates if $a \leqslant b$ and also $b \leqslant a$. It is easy to show that *the quasi-ordering induces* (see IA, §8.3) *a partial ordering in the set of corresponding equivalence classes.*

2.5. Now let a, b and t be elements of a ring $\mathfrak{R}$ (with unity element) such that $t \mid a$ and $t \mid b$. Then it is clear that t is also a divisor of any *linear combination* $xa + yb(\{x, y\} \subseteq \mathfrak{R})$. The set $\mathfrak{B}(\mathfrak{a}, \mathfrak{b}) = \mathfrak{B}$ of all linear combinations of a and b has certain remarkable properties:

I From $\{v_1, v_2\} \subseteq \mathfrak{B}$ it follows that $v_1 \pm v_2 \in \mathfrak{B}$.

II If $v \in \mathfrak{B}$ and $r \in \mathfrak{R}$, then $rv \in \mathfrak{B}$.

[2] The symbol $\mathfrak{C}$ will be used throughout Chapter 6 to denote the ring of rational integers.

The first property states that $\mathfrak{B}$ is a module (i.e., an Abelian group with additively written operation; see also IB2, §1.1); more precisely: $\mathfrak{B}$ is a submodule of the additive group $\mathfrak{R}^+$ of $\mathfrak{R}$.[3] If we take $r \in \mathfrak{B}$ in (II), we see at once that $\mathfrak{B}$ is a subring of $\mathfrak{R}$; but the condition (II) is much stronger, since we may take $r \in \mathfrak{R}$. By IB5, §3.1 the conditions (I) and (II) are precisely the definition of an *ideal* in $\mathfrak{R}$, so that in the notation of IB5, §3.2 we see that $\mathfrak{B}$ is the ideal (a, b).

If $\mathfrak{R}$ is a not necessarily commutative ring, we have *left ideals* and *right ideals*, and also *two-sided ideals*, according to whether in II we may set r on the left, on the right, or on both sides.[4]

From $a \sim b$ it obviously follows that the principal ideals (a) and (b) coincide, and from $a \mid b$ follows $(b) \subseteq (a)$ and conversely.

2.6. Any element of a ring is obviously divisible not only by its associates but also by every unit in the ring. The associates and the units are said to be *trivial divisors* of a. A divisor d of a which is not an associate of a is said to be a *proper divisor* of a, which we shall occasionally denote by $d \mid_{\mathrm{pr}} a$. An element $a \in \mathfrak{R}$ is said to be reducible if there exists at least one factorization $a = a_1 a_2 \cdots a_n$ in which all the a_i are proper divisors of a; otherwise a is *irreducible*. If a is irreducible and $a = ab$, then either $b = 1$ or $1 - b$ is a nontrivial divisor of zero (see IB5, §1.7), since $a(1 - b) = 0$. *In an integral domain* (i.e., a commutative ring without divisors of zero; see IB5, §1.9) *the nonzero irreducible elements are identical with the elements that have only trivial divisors of zero.*

In the ring $\mathfrak{C}$ the positive irreducible numbers are called prime numbers.[5]

We now say that a ring $\mathfrak{R}$ with unity element admits a theory of divisibility if it satisfies the following condition:

[3] With respect to addition alone, every ring $\mathfrak{R}$ is obviously an Abelian group, the so-called *additive group of the ring*, in symbols $\mathfrak{R}^+$. With respect to multiplication the set $\mathfrak{R} - \{0\}$ is a semigroup, the so-called *multiplicative semigroup of* $\mathfrak{R}$, in symbols $\mathfrak{R}^\times$. If $\mathfrak{R}$ is a field, or only a skew field, then $\mathfrak{R}^\times$ is obviously a group, the *multiplicative group of the "skew" field.* In this case the multiplicative group is obviously identical with the group of units. It was merely to preserve this group property for the special case of a field that we excluded the zero element from $\mathfrak{R}^\times$.

[4] If $\mathfrak{R}$ has no unity element, the conditions (I) and (II) still define an ideal. By (I) the left ideal $(a_1, a_2, \ldots, a_n)$ consists of all expressions of the form

$$x_1 a_1 + x_2 a_2 + \cdots + x_n a_n + g_1 a_1 + \cdots + g_n a_n \quad \text{with} \quad \begin{array}{l}\{g_1, g_2, \ldots, g_n\} \subset \mathfrak{C}, \\ \{x_1, x_2, \ldots, x_n\} \subseteq \mathfrak{R},\end{array}$$

where for positive integer g the product ga is defined by $\sum_{\nu=1}^{g} a$, and $(-g)a = -(ga)$. In a ring with unity element we have $g \cdot a = g \cdot ea = (ge) \cdot a (g \in \mathfrak{C})$, so that in view of $ge \in \mathfrak{R}$ we may omit the expression $g_1 a_1 + \cdots + g_n a_n$.

[5] This definition is not universally accepted. For many authors the zero element and the unity element are not prime numbers. By the above definition the zero element and all the units in an integral domain are irreducible.

Fundamental condition of the theory of divisibility: let $\mathfrak{P}$ be a system of representatives of the classes of associated irreducible elements, excluding the class of units. Then for every $a \neq 0$ there exists a representation of the form

$$a = \epsilon \prod_{\lambda=1}^{s} p_{\lambda}^{\alpha_\lambda}, \qquad s \geqslant 0,^{6} \qquad \alpha_\lambda > 0 \quad (\epsilon \text{ a } unit), \tag{1}$$

$$p_\kappa \neq p_\lambda \quad \text{for} \quad \kappa \neq \lambda, \qquad p_\lambda \in \mathfrak{P} \qquad (\lambda = 1, \ldots, s),$$

which is unique apart from the order of the factors, i.e., for two representations of the form (1)

$$\epsilon_1 \prod_{\lambda=1}^{s} p_{\lambda}^{\alpha_\lambda} = \epsilon_2 \prod_{\rho=1}^{t} q_{\rho}^{\beta_\rho}$$

it follows that $s = t \geqslant 0$ and with a suitable arrangement of the factors, $p_\lambda = q_\lambda$, $\alpha_\lambda = \beta_\lambda$ for all $\lambda = 1, \ldots, s$ and $\epsilon_1 = \epsilon_2$.

In rings satisfying this condition the irreducible elements are also called *prime elements* (for the general definition of a prime element in rings see §8.2 and IB5, §2.3), the rings themselves are called *u.f. rings* (unique factorization rings), the above fundamental condition is called the *u.f. condition* and the factorization (1) is said to be *canonical.*

If $\mathfrak{R}$ has nontrivial divisors of zero, so that $ab = 0$, $a \neq 0$, $b \neq 0$, and if for $a = a + ab$ we construct the factorization (1), then for $a = a(1 + b)$ we obtain a second factorization (since $b \neq 0$) by factoring $1 + b$ canonically: for if $1 + b$ is not a unit, then the exponents are not identical, but if $1 + b = \epsilon$ is a unit, it follows from the assumed u.f. condition that $\epsilon_1 = \epsilon_1 \epsilon$, and thus, since the units form a group, we have at once $\epsilon = 1$, or in other words $b = 0$, in contradiction to our assumption concerning b. From this contradiction of the theorem of unique factorization we have:

Every u.f. ring is an integral domain.

The above definition of a u.f. ring is identical with the definition in IB5, §2.6, as follows at once from the discussion of prime elements in §8.2.

It is convenient to introduce the following definitions: a prime element p is called a *prime divisor* of a if $p \mid a$, $p \not\sim 1$ and $p \neq 0$. If a and b are arbitrary elements of the ring and $d \mid a$, $d \mid b$, so that d is a *common divisor*, then the *greatest common divisor* (GCD), which we denote by (a, b), is defined, provided it exists, as a common divisor g such that $d \mid g$ holds for all common divisors d. Similarly, a v with $a \mid v$ and $b \mid v$ is a *common*

[6] Here we are adopting the convention that an empty product (e.g., $\prod_{\nu=1}^{0} a_\nu$) always has the value 1. Similarly, an empty sum has the value 0.

multiple of a and b, and a k with $a \mid k$, $b \mid k$, $k \mid v$ for all common multiples v is the *least common multiple* (LCM), in symbols $[a, b]$. For the general case $n \geqslant 1$, the GCD $(a_1, a_2, ..., a_n)$ and the LCM $[a_1, a_2, ..., a_n]$ are defined correspondingly. If all $a_\nu \neq 0$ and if we set $b_\nu = a_1 a_2 \cdots a_n / a_\nu$ $(\nu = 1,2, ..., n)$, it is easy to show that

$$(a_1, a_2, ..., a_n) \cdot [b_1, b_2, ..., b_n] = a_1 a_2 \cdots a_n .$$

In the special case $n = 2$ this relation becomes the simpler $(a, b) \cdot [a, b] = ab$.

For $a \neq 0$ and $b \neq 0$, if in (1) we allow zero as an exponent, we may write $a = \epsilon_1 \prod_{\lambda=1}^{s} p_\lambda^{\alpha_\lambda}$ and $b = \epsilon_2 \prod_{\lambda=1}^{s} p_\lambda^{\beta_\lambda}$; so that in u.f. rings we have the formulas

$$(2) \qquad (a, b) = \prod_{\lambda=1}^{s} p_\lambda^{\min(\alpha_\lambda, \beta_\lambda)} \qquad \text{and} \qquad [a, b] = \prod_{\lambda=1}^{s} p_\lambda^{\max(\alpha_\lambda, \beta_\lambda)}.$$

Thus the (a, b) and $[a, b]$ necessarily exist, but as long as we make no convention about normalization, they are determined only up to associates, so that it would be more correct to regard the symbols as denoting the corresponding equivalence classes in the sense of §2.3.

If $(a, b) = 1$, we say that a and b are *coprime*.

Finally, from the u.f. condition we obtain the so-called *fundamental lemma of the theory of divisibility:*

If p is irreducible, it follows from $p \mid ab$ that $p \mid a$ or $p \mid b$.

As examples of u.f. rings that do not fall under the special headings of §2.9 and §2.10 let us mention the polynomial rings $\mathfrak{R}[x_1, x_2, ..., x_n]$ (see IB4, §2.3) in n indeterminates, where $\mathfrak{R}$ itself is assumed to be a u.f. ring (see also §2.10, next-to-last paragraph).

The prime elements in polynomial rings are called *irreducible polynomials*, and the other polynomials are said to be *reducible*.

Concerning the number of prime elements in a ring we have the following generalization of the classical theorem on the infinitude of primes.

Theorem of Euclid: If $\mathfrak{R}$ is a u.f. ring, which is not a field and which has the property that for every nonunit $a \neq 0$ there exists a *unit ϵ such that $a + \epsilon \not\sim 1$, there exist infinitely many prime elements, no two of which are associates.* (See also the last paragraph of §2.10.)

Proof: No prime divisor of $a + \epsilon$ is an associate of a prime divisor of a. Since $\mathfrak{R}$ is not a field, there exist non-unit elements a, and by hypothesis there also exist prime divisors of $a + \epsilon$; consequently, in the canonical factorization of a there cannot appear a complete system of representatives of the classes of associated prime elements.

For u.f. rings with only finitely many nonassociated prime elements see the end of §2.10.

2.7. On the basis of §2.2 and §2.4 we now regard the classes of associated elements in a u.f. ring as a partially ordered set. If we let $\hat{a}$ and $\hat{b}$ denote the equivalence classes defined by a and b, then the GCD class $\widehat{(a, b)}$ is the greatest common predecessor of the classes $\hat{a}$ and $\hat{b}$ with the property that every common predecessor $\hat{d}$ of $\hat{a}$ and $\hat{b}$ is also a predecessor of $\widehat{(a, b)}$, from which it follows that there exists exactly one greatest common predecessor. Similarly, the LCM class $\widehat{[a, b]}$ is the unique least common successor of $\hat{a}$ and $\hat{b}$ in the sense that is precedes every common successor. A partially ordered set of this sort, in which every two elements have a greatest common predecessor and a least common successor in the above sense, is called a *lattice* (see IB9, §1). *Thus the classes of associated elements in a u.f. ring form a lattice.*

If the construction of $\widehat{(a, b)}$ and $\widehat{[a, b]}$ is regarded as two operations, in symbols $\widehat{(a, b)} = \hat{a} \cup \hat{b}$, $\widehat{[a, b]} = \hat{a} \cap \hat{b}$, it is easy to verify the associative law; and the commutative law is trivial.

2.8. Since every u.f. ring $\mathfrak{R}$ is an integral domain, it can be embedded, by IB1, §3.2, in a quotient field $\mathfrak{Q}(\mathfrak{R})$. If we write the elements $\kappa \in \mathfrak{Q}(\mathfrak{R})$ in the form $\kappa = a/b$ with $\{a, b\} \in \mathfrak{R}$ and apply (1) to a and b, we obtain, for all $\kappa \neq 0$, a representation of the form (1) (allowing negative exponents) with the corresponding uniqueness properties. We thus arrive at a theory of divisibility for $\mathfrak{Q}(\mathfrak{R})$ relative to $\mathfrak{R}$ by defining $\kappa_1 \mid \kappa_2$ with $\{\kappa_1, \kappa_2\} \subseteq \mathfrak{Q}(\mathfrak{R})$, to mean that $\kappa_2/\kappa_1 \in \mathfrak{R}$. Then by (2) we can define the GCD and LCM for all $\kappa \in \mathfrak{Q}(\mathfrak{R})$. If $(a, b) = 1$, the fraction a/b for κ is said to be *in lowest terms*, and it follows from the u.f. condition applied to $\mathfrak{Q}(\mathfrak{R})$ that this representation is unique up to associates. The significance of *cancellation in fractions* becomes clear from this discussion.

By the *lowest common denominator of* $a_1/b_1, a_2/b_2, ..., a_n/b_n$ we mean the LCM $[b_1, b_2, ..., b_n]$.

2.9. *Principal Ideal Rings*

It is natural now to ask for sufficient conditions that a given ring should be a u.f. ring. Here we may refer to the result already proved in IB5, §3.4 on principal ideal rings:

Theorem 1: *Every principal ideal ring is a u.f. ring.*

The part of the proof that there exists at least one factorization into prime elements depended on the so-called *divisor-chain condition* (IB5, §2.5). For later use let us state an equivalent condition in ideal-theoretic terms. Let $a_2 \mid a_1, a_3 \mid a_2, ..., a_{n+1} \mid a_n, ...$ be a divisor chain. By §2.5 the condition $a_{n+1} \mid a_n$ is equivalent to $(a_{n+1}) \supseteq (a_n)$. Now, for an arbitrary ring we say that the *maximal condition* is satisfied if in every ascending sequence of ideals: $\mathfrak{a}_1 \subseteq \mathfrak{a}_2 \subseteq \cdots \subseteq \mathfrak{a}_n \subseteq \mathfrak{a}_{n+1} \subseteq ...$, from some place onward all the ideals are equal to one another. The theorem proved in IB5, §3.4

to the effect that the divisor-chain condition holds for principal ideal rings can thus be formulated as follows:

Theorem 2: *The maximal condition is satisfied in every principal ideal ring.*

In §8.2 we shall return to rings with the maximal condition that are not necessarily principal ideal rings.

In theorem 1 we now have an important sufficient condition for a ring to be a u.f. ring. Our next aim is to find sufficient conditions for a ring to be a principal ideal ring.

2.10. *Euclidean Rings*

As models for the following concepts we may consider the ring $\mathfrak{C}$ of rational integers and the polynomial ring $\mathfrak{K}[x]$ in one indeterminate (where $\mathfrak{K}$ is a field) (see IB4, §2.1),[7] since in each of these two rings there exists a division algorithm (see below).

Definition: *A ring $\mathfrak{E}$ without divisors of zero is said to be Euclidean if the following conditions are satisfied:*

(I) *In $\mathfrak{E} - \{0\}$ there is defined a nonnegative integer-valued function $w(x)$, the so-called (absolute) value function, or valuation.*

(II) *For every pair of elements a and b in $\mathfrak{E}$ with $b \neq 0$ there exist elements q and r in $\mathfrak{E}$ such that $a = qb + r$ and $w(r) < w(b)$ or $r = 0$ (division algorithm).*

Lemma: In a Euclidean ring $\mathfrak{E}$ every ideal $\mathfrak{a}$ is a principal ideal (a) in the sense that all $x \in \mathfrak{a}$ are multiples qa of a.[8]

Proof: Let a be an element with the smallest possible value $w(a)$ in $\mathfrak{a}$; by (I) there exists at least one such element, if $\mathfrak{E} \neq \{0\}$. Then for arbitrary $b \in \mathfrak{a}$ there exists by (II) a representation in the form $b = qa + r$ ($r = 0$ or $w(r) < w(a)$). By the module property of an ideal it follows that $r = b - qa \in \mathfrak{a}$. Thus the minimal property of $\mathfrak{a}$ implies $r = 0$.

If we apply the lemma to the unit ideal (IB5, §3.1) $\mathfrak{E}$, it follows that $\mathfrak{E} = (\epsilon)$, so that for every $x \in \mathfrak{E}$ there exists a $q(=q(x))$, with $x = \epsilon q$; in particular, for $x = \epsilon$ let $\epsilon = \epsilon e$. Then

$$x = q\epsilon = q\epsilon e = q\epsilon \cdot e = xe \qquad \text{for all} \quad x \in \mathfrak{E}; \tag{3}$$

that is, e is the unity element in $\mathfrak{E}$. To sum up, we have

[7] Since in general the degree $(a_0 + a_1x + \cdots + a_nx^n = n$ if $a_n \neq 0$, it is customary not to assign any degree, or possibly the degree $-\infty$ to the "0" (the zero polynomial) (in IB4, §2.1 and IB5, §2.11, on the other hand, we set degree $0 = 0$). From the definition of a unit it also follows that all $a \in \mathfrak{K}$, $a \neq 0$, and only these are units of $\mathfrak{K}[x]$.

[8] Note that the existence of a unity element in $\mathfrak{E}$ is not postulated. Cf. §2.5, footnote 4.

Theorem 3: *Every Euclidean ring is a principal ideal ring and thus a u.f. ring.*

In order to show that the ring $\mathfrak{C}$ of of integers is Euclidean, we set $w(x) = |x|$, so that (I) is satisfied. As for (II), let us first assume $0 \leqslant a < |b|$. Then $a = 0 \cdot b + a$ is a division formula (II). Now let $a \geqslant |b| > 0$; then a and $|b|$ are natural numbers with the Archimedean property (see IB1, §3.4) that there exists a natural number n such that $n|b| > a$. But the set of natural numbers is well-ordered by the "$<$" relation (see IB1, §1.4), so that the subset of all n with $n|b| > a$ contains a smallest number, say $q_1 + 1$; then $(q_1 + 1) \cdot |b| > a \geqslant q_1 \cdot |b|$. Setting $q = q_1 \cdot \operatorname{sgn} b$ and $r = a - q_1 b$ and subtracting $q_1 | b$, we obtain

$$(4) \qquad 0 \leqslant a - q_1 |b| = r = |r| < |b|, \quad \text{d.h.} \quad a = (q_1 \cdot \operatorname{sgn} b)\, b + r = qb + r,$$

so that (II) is again satisfied. Finally, if $a < 0$, then by (4) $(-a) = q_1 |b| + r$, $0 \leqslant r < |b|$, so that $a = (-q_1 \cdot \operatorname{sgn} b)\, b - r$, and

$$(5) \qquad a = qb + (-r), \qquad |-r| < |b| \qquad (q = -q_1 \cdot \operatorname{sgn} b),$$

and therefore II holds in every case. In (4) the remainder is nonnegative, so that (4) represents a *division with smallest positive remainder.* We can also bring (5) into the same form: $a = -q_1 |b| - r = -(q_1 + 1)|b| + |b| - r$, where $0 \leqslant |b| - r < |b|$ for $r \neq 0$.

On the other hand, we could have put (4) in a form with a negative remainder, and then by choosing the remainder, positive or negative, with smaller absolute value we obtain the *division with smallest absolute remainder*. Except when "$2 \mid b$ and $|r| = |b/2|$", where the two possibilities provide the same absolute value for the remainder, it is easy to show that the q and r are uniquely determined in every case.

In the polynomial ring $\mathfrak{K}[x]$ in one indeterminate over a field $\mathfrak{K}$ it is obvious that $w(f(x)) = \operatorname{degree} f(x)$ $(f(x) \in \mathfrak{K}[x],\ f(x) \neq 0)$ is a valuation satisfying (I). But the division algorithm (II) also holds, so that $\mathfrak{K}[x]$ *is a Euclidean ring*. In order to prove (II), we must first show that for two polynomials $f(x) = a_0 x^n + \cdots + a_n$ and $g(x) = b_0 x^k + \cdots + b_k$, $b_0 \neq 0$ with degree $f(x) \geqslant$ degree $g(x)$ there exists a $q(x) \in \mathfrak{K}[x]$, such that degree $(f(x) - q(x)\, g(x)) <$ degree $f(x)$; but it is at once clear that $q(x) = (a_0/b_0)\, x^{n-k}$ is satisfactory for the purpose. Now let $f(x)$ and $g(x)$ be arbitrary with $g(x) \neq 0$; then in the case degree $f(x) <$ degree $g(x)$ we can at once satisfy (II) with $q(x) = 0$, $r(x) = f(x)$, and if degree $f(x) \geqslant$ degree $g(x)$, then let $q(x)$ be so chosen that degree $(f(x) - q(x) g(x))$ is minimal, provided we do not already have the trivial case $g(x) \mid f(x)$. If we set $r(x) = f(x) - q(x)\, g(x)$ and assume that degree $r(x) \geqslant$ degree

$g(x)$, we have already shown that there exists a $q_1(x) \in \Re[x]$, such that degree $(r(x) - q_1(x)\, g(x)) <$ degree $r(x)$. But then

$$\begin{aligned} \text{degree}\, r(x) &> \text{degree}\, (r(x) - q_1(x)\, g(x)) \\ &= \text{degree}\, \big(f(x) - (q(x) + q_1(x))\, g(x)\big), \end{aligned}$$

in contradiction with our having chosen $q(x)$ so as to minimize degree $r(x)$. Thus for $f(x) = q(x)\, g(x) + r(x)$ we have degree $r(x) <$ degree $g(x)$.

The ring $\Re[x]$ has the further property that in the *division formula* $f(x) = q(x)\, g(x) + r(x)$ in (II) *the polynomials* $q(x)$ *and* $r(x)$ *are uniquely determined*, as can easily be shown by an indirect proof based on their degrees.

The ring $\mathfrak{C}[i]$ of Gaussian integers (IB5, §1.1) is also Euclidean. For $w(\alpha + \beta i) = (\alpha + \beta i)(\alpha - \beta i) = \alpha^2 + \beta^2$ obviously satisfies (I) and if the *norm* (see §8.1 and IB8, §1.2) $N(z) = z\bar{z}$ of a complex number $z = \alpha + \beta i$ is chosen as its absolute value (so that the distributivity $w(z_1 z_2) = w(z_1) \cdot w(z_2)$ (IB8, (10)) is immediately clear), then (II) is proved as follows. In order to find, for given z_1 and $z_2 \neq 0$, the q and r demanded by (II), we first determine a (perhaps fractional) complex number q', such that $z_1 - z_2 q' = 0$, and then in $q' = \gamma' + \delta' i$ we replace the rational numbers γ' and δ' by the nearest integers, say γ and δ. With $q = \gamma + \delta i$ and $z_1 = q z_2 + r$ we then have

$$\begin{aligned} N(r) &= N(z_1 - q z_2) = N(z_1 - q' z_2 + (q' - q)\, z_2) = N((q' - q)\, z_2) \\ &= N(q' - q)\, N(z_2), \end{aligned}$$

$$N(q' - q) = (\gamma' - \gamma)^2 + (\delta' - \delta)^2 \leqslant (\tfrac{1}{2})^2 + (\tfrac{1}{2})^2 < 1,$$

so that $N(r) = N(q' - q)\, N(z_2) < N(z_2)$, which satisfies (II).

In the same way it can be shown that the set of numbers $\alpha + \beta\sqrt{2}$, $\{\alpha, \beta\} \subseteq \mathfrak{C}$, is a Euclidean ring if we put

$$w(\alpha + \beta\sqrt{2}) = |(\alpha + \beta\sqrt{2})(\alpha - \beta\sqrt{2})| = |\, \alpha^2 - 2\beta^2 \,|.$$

In general, the set of numbers $\alpha + \beta\sqrt{\delta}$, $\{\alpha, \beta\} \subseteq \mathfrak{C}$, where $\delta \in \mathfrak{C}$ is not a perfect square, forms a ring, as is easily proved; but in general this ring is not Euclidean, as may be shown by the examples $\delta = -5, -3, +10$, and so forth. For $\delta = -5$, for example,

$$21 = (4 + \sqrt{-5})(4 - \sqrt{-5}) = 3 \cdot 7$$

shows two essentially different factorizations into irreducible factors.[9]

[9] For details see, e.g., Hasse [3], §16.

A subring of a Euclidean ring is not necessarily even a u.f. ring, as may be seen from the ring of even numbers; it is clear that all numbers of the form $2u$ with odd u, and only these, are irreducible, and for 60 we have the two factorizations, $2 \cdot 30$ and $6 \cdot 10$.

The valuation $w(x)$ is not uniquely determined; in fact, for every fixed integer $\lambda > 0$ the function $\lambda w(x)$ is easily seen to be another valuation. In general, it is possible to construct valuations that are not connected with one another in such a simple way, and for some of them it is necessary, in order to preserve Axiom II, to use other magnitudes in place of q and r in the division formula $a = qb + r$. Thus it is natural to ask how we can normalize the valuations so as to restrict them to convenient forms. In this direction we have[10] the following theorem.

Theorem 4: *For a Euclidean ring $\mathfrak{E}$ there exist valuations $w(x)$ such that associate elements have the same values; and this property is equivalent to the property that $w(x, y) \geqslant w(x)$ for $x \neq 0$, $y \neq 0$.*

Proof: Let $x^*(x)$ be any valuation for which $\mathfrak{E}$ is Euclidean. Let $\hat{a}$ denote the class of all associates of a in $\mathfrak{E}$. Set $w(\hat{a}) = w(a) = \min_{x \in \hat{a}} w^*(x)$. Since $w^*(x)$ is an integer, there exists an $a_m \in \hat{a}$ such that $w(a) = w^*(a_m)$, and therefore

$$(6) \qquad w(a) = w(a_m) = w^*(a_m) \leqslant w^*(a) \qquad \text{for all} \quad a \sim a_m .$$

Here Axiom I and the additional condition are obviously satisfied. Now let $a \neq 0$ and let b be arbitrary, define $a_m = \epsilon a$ (where ϵ is a unit) as before and let $b = qa_m + r$ be a division formula with respect to $w^*(x)$. If $r = 0$, then Axiom II with respect to a and b is satisfied for all valuations, since $b = qa_m = q\epsilon \cdot a$. For $r \neq 0$ it follows from (6) that $w(b - qa_m) \leqslant w^*(b - qa_m) = w^*(r) < w^*(a_m) = w(a)$; thus $w(a) > w(b - qa_m) = w(b - q\epsilon\epsilon^{-1}a_m) = w(b - q_1 a)(q_1 = q\epsilon)$, so that $b = q_1 a + r$, and Axiom II is again satisfied. Only the last part of the theorem now remains to be proved. Let us first assume that $w(x, y) \geqslant w(x)$. If ϵ is a unit, we have

$$w(a\epsilon) \geqslant w(a) = w(a\epsilon \cdot \epsilon^{-1}) \geqslant w(a\epsilon), \quad \text{and therefore} \quad w(a\epsilon) = w(a).$$

For the proof of the converse it is sufficient to show that $w(ab) > w(a)$ for nonunits. Let us assume to the contrary that $w(ab) \leqslant w(a)$. In the division formula $a = q \cdot ab + r$ we then have $r \neq 0$, since $b \not\sim 1$, and therefore $w(a) \geqslant w(ab) > w(r) = w((1 - qb)\, a)$; but then the strict inequality shows that $1 - qb = b_1$ is not a unit. The same procedure

[10] H. J. Claus, Über die Partialbruchzerlegung in nicht notwendig kommutativen Ringen. Journ. f. reine u. angew. Math. (Crelle) 194, (1955), 88–100.

applied to ab_1 and a in place of ab and a leads to a nonunit element b_2, and thus to the inequality $w(a) \geqslant w(ab) > w(ab_1) > w(ab_2)$. Continuation of the procedure produces a nonterminating, strictly monotone decreasing series of values $w(ab_\nu)$, in contradiction to Axiom I.

On the basis of this theorem we may now adjoin to the Axioms I and II the following axiom:

(III) *From $a \sim b$ it follows that $w(a) = w(b)$*, from which we may also assume $w(ab) \geqslant w(a)$ for all $a \neq 0$, $b \neq 0$, whereby we have reached agreement with IB5, §2.8.

We can state the further result: *if a is a proper factor of b*, i.e., $a \mid_{\text{pr}} b$, then $w(a) < w(b)$, and if $w(a) = w(1)$, then $a \sim 1$ and conversely.

Proof: Since $b \nmid a$, we have $r \neq 0$ in every division formula $a = qb + r$. If we set $b = ac$, then

$$w(b) > w(r) = w(a - qb) = w(a(1 - qc)) \geqslant w(a),$$

as desired. The second statement follows immediately.

The above development of the theory of divisibility is based on the theory of principal ideal rings and may thus be regarded as an ideal-theoretic method. If we begin with a Euclidean ring $\mathfrak{E}$ in the first place, we can reach the same results by a different method, which is more elementary and has the advantage of being constructive, namely, by explicitly calculating the GCD rather than by proving its existence from the properties of a principal ideal. For this calculation we use the *Euclidean algorithm* in the following way. Let a and b, $b \neq 0$, be arbitrary elements of $\mathfrak{E}$; then Axiom II allows us to set up in succession the division formulas:

$$\begin{aligned} a &= q_1 b + r_1, & w(r_1) &< w(b), \\ b &= q_2 r_1 + r_2, & w(r_2) &< w(r_1), \\ r_1 &= q_3 r_2 + r_3, & w(r_3) &< w(r_2), \\ &\vdots \\ r_{n-2} &= q_n r_{n-1} + r_n, & w(r_n) &< w(r_{n-1}), \\ r_{n-1} &= q_{n+1} r_n + r_{n+1}, & w(r_{n+1}) &< w(r_n) \quad \text{or} \quad r_{n+1} = 0. \end{aligned} \tag{7}$$

The sequence is to be regarded as terminating as soon as a zero remainder occurs. Since $w(b) > w(r_1) > w(r_2) > \cdots$, it follows from Axiom I that such a remainder must eventually occur. If we run through the algorithm (7) from the first line down to the last, we see that every common divisor of a and b is a divisor of all the r_ν $(1 \leqslant \nu \leqslant n + 1)$, and on the other hand, if we run through (7) from the last line up to the first, assuming $r_{n+1} = 0$, we have $r_n \mid r_{n-1}$, $r_n \mid r_{n-2}$, ..., $r_n \mid b$, $r_n \mid a$. Taken together, these results show that the last nonzero remainder r_n is the common divisor of greatest absolute value, a property which, on the basis of Axiom III, can be used as a definition of the GCD, for which we now have the following theorem.

GCD Theorem: *Every common divisor is a divisor of the GCD.*

It is easy to prove that $(a_1, a_2, ..., a_n) = ((a_1, a_2, ..., a_{n-1}), a_n)$ and also that the GCD theorem holds for arbitrary $n \geqslant 1$ (cf. the last paragraph of §2.7).

Finally, if we begin at the next-to-last equation in (7) and work backwards, we obtain a representation of the form $(a, b) = r_n = ax_0 + by_0$. Then the fundamental lemma, and with it the u.f. theorem, can be proved in exactly the same way as for principal ideal rings.

In analogy with the GCD, we can now define an LCM $[a_1, a_2, ..., a_n]$ as a common nonzero multiple of least absolute value, for which we obtain the following theorem:

Theorem of the LCM: *For every common multiple v_1 we have $v = [a_1, a_2, ..., a_n] \mid v_1$. In particular, all the LCM's are associates.*

Proof: We apply the Euclidean algorithm to v, v_1 and obtain the GCD $(v, v_1) = d$. From the minimal property of $w(v)$ and from $d \mid v$ it follows that $w(d) = w(v)$, so that $d \nmid_{\text{pr}} v$; that is, $d \sim v$, and thus, in view of $d \mid v_1$, we have at once $v \mid v_1$.

Finally, we must mention a third way of constructing the theory of divisibility in Euclidean rings, namely by first proving the u.f. theorem, i.e., without using the concept of the GCD, and then defining the GCD and LCM by (2). But in order to obtain the important representation of the GCD $(a_1, a_2, ..., a_n)$ as a sum of multiples $a_1x_1 + a_2x_2 + \cdots + a_nx_n$, we must then proceed either by way of the principal-ideal-property (if we are satisfied with proving the existence of the desired representation) or else by way of the Euclidean algorithm.

In order to prove the u.f. theorem directly (i.e, without using the GCD) we require a sharpening of the above Axiom III,[11] which will also be necessary for the discussion of partial fractions in 2.11 below. In place of Axiom III we now require the following axiom:

(III′) *From $w(a) < w(b)$ it follows that $w(ac) < w(bc)$ for all $c \neq 0$, and conversely.*

Corollary 1: *from $w(a) = w(b)$ it follows that $w(ac) = w(bc)$ for all $c \neq 0$, and conversely.*

[11] No immediate proof of the u.f. theorem is known at present without this sharpening of Axiom III. It is possible that all Euclidean rings are no longer included; however, up to the present no known Euclidean rings fail to satisfy the new requirement. Thus it would be of interest to know whether a theorem analogous to Theorem 4 is valid.

Corollary 2: *from* III′ *follows* III.

Corollary 3: *from* $w(a) < w(b)$ *and* $w(c) < w(d)$ *follows* $w(ac) < w(bd)$.

Proof: Corollary 1 is easy to prove indirectly. As for corollary 2, it is sufficient by theorem 4 to prove that $w(ab) \geqslant w(a)$. But if we had $w(ab) < w(a) = w(a \cdot 1)$, it would follow that $w(1) > w(b)$ and thus $w(b) = w(1 \cdot b) > w(b^2)$, and then $w(b^2) > w(b^3)$, and so forth; but the chain $w(1) > w(b) > w(b^2) > \ldots$ would be in contradiction to I. Corollary 3 follows from $w(ac) < w(bc) < w(bd)$.

Proof of the u.f. theorem in Euclidean rings under the assumptions I, II *and* III.[12] We make the induction hypothesis that the theorem is true for all x with $w(x) < w(a)$ and assume the existence of an a contradicting the assertion. Now let $p \not\sim 1$ be a divisor of a with the smallest possible value, from which it follows that p is irreducible. Let $a = bp$. Since $p \not\sim 1$, we have $b \,|_{\text{pr}} a$, and thus $w(b) < w(a)$, so that b has a canonical factorization, and a has at least one factorization (1). Let $q \not\sim 1$ be an irreducible factor of the second (assumed) factorization (1) of a, with $a = qc$. The two factorizations cannot have associated irreducible factors, since by cancellation of such factors we would obtain an element of smaller value than $w(a)$, which would therefore, by the induction hypothesis, have a unique factorization (1). Consequently, the original factorizations of a cannot, after all, be different from each other. In $a = pb = qc$ we now insert the division formulas $q = q_1 p + r_1$ and $c = q_2 p + r_2$, with $r_1 \neq 0$ and $r_2 \neq 0$, since $p \not\sim q$ and $p \nmid c$. We thus obtain:

$$(8) \quad a = pb = (q_1 p + r_1)(q_2 p + r_2) = p(q_1 q_2 p + r_1 q_2 + r_2 q_1) + r_1 r_2 .$$

From the minimal property of p it follows for the division remainders that

$$w(r_1) < w(p) \leqslant w(q),$$
$$w(r_2) < w(p) \leqslant w(c),$$

and thus by Corollary 3:

$$w(r_1 r_2) < w(qc) = w(a),$$

so that $r_1 r_2$ has a unique factorization. Since $w(r_i) < w(p)$ $(i = 1, 2)$, the factor p cannot occur in this factorization, but by (8) we have $p \mid r_1 r_2$, which provides the desired contradiction.[13]

[12] See H. Klappauf, Beweis des Fundamentalsatzes der Zahlentheorie. Jahresbericht DMV 45, 130 Kursiv. The first proof was given by Zermelo.

[13] In this form of proof by induction it is unnecessary to prove the initial statement (although its correctness for all $a \sim 1$, i.e., $w(a) = w(1)$, is obvious, since units are

If $\mathfrak{R}$ is merely a u.f. ring, then on the basis of the fact that $\mathfrak{Q}(\mathfrak{R})[x]$ is Euclidean and is thus a u.f. ring, it can be shown (see IB5, §2.14) that $\mathfrak{R}[x_1] = \mathfrak{R}_1$ is also a u.f. ring (Gauss). By successive application of this theorem, the same result follows for $\mathfrak{R}_1[x_2] = \mathfrak{R}[x_1, x_2]$, $\mathfrak{R}[x_1, x_2, x_3]$, ..., as was mentioned at the end of §2.6 (see also IB5, §2.14).

If now in the field P of rational numbers we consider the set $\mathfrak{M}$ of all fractions s/t, $\{s, t\} \subseteq \mathfrak{C}$, $3 \nmid t$, it is easy to show that $\mathfrak{M}$ is a ring, a so-called *quotient ring*. All s/t with $(s, 3) = (t, 3) = 1$ form the group of units. Apart from associates, the only prime element is 3, and it is clear that every number in $\mathfrak{M}$ can be represented uniquely in the form $\epsilon 3^\alpha$ (where ϵ is a unit) and $\alpha \geqslant 0$ is an integer. With the definition $w(\epsilon 3^\alpha) = \alpha$, it is easy to show that $\mathfrak{M}$ is Euclidean, and to generalize to the case of more than one prime element. On the other hand, it is clear that the assumptions for the Euclidean theorem (see §2.6) hold for the rings $\mathfrak{C}$, $\mathfrak{C}[i]$, $\mathfrak{K}[x]$ (where $\mathfrak{K}$ is a field).

2.11. *Decomposition into Partial Fractions in Euclidean Rings*

We now assume that $\mathfrak{E}$ satisfies the Axioms I, II and III′. Let $\mathfrak{Q} = \mathfrak{Q}(\mathfrak{E})$ be the quotient field of $\mathfrak{E}$. Those elements $u \in \mathfrak{Q}$ that are also in $\mathfrak{E}$ are called *integers*. Also, $u = a/b$ ($\{a, b\} \subseteq \mathfrak{E}$) is said to be a *proper fraction* if $a = 0$ or $w(a) < w(b)$. It is a consequence of III′ that *the property of being a "proper fraction" is invariant*[14] *under cancellation or under multiplication of numerator and denominator by the same number.*

If a and b are integers with $(a, b) = 1$, then a/b is called a *partial fraction* if and only if $b = 1$ or $w(a) < w(b)$.

If s and t are integers with $(s, t) = 1$, and if $t = \prod_{\rho=0}^{r} q_\rho$, where the q_ρ are coprime integers, and $q_0 = 1$, then

$$\frac{s}{t} = \sum_{\rho=0}^{r} \frac{a_\rho}{q_\rho} \qquad (a_\rho \text{ integers}) \tag{9}$$

is called a *decomposition into partial fractions* (abbreviated DPF) *of the*

always irreducible): for if the correctness of the statement $A(x)$ for all $x < n$ implies the correctness of $A(n)$ ($n \geqslant 1$), then $A(n)$ holds for all natural numbers n, since the correctness of $A(1)$ is now included in the proof: i.e., the induction hypothesis is true because it consists of the (empty) statement "$A(x)$ holds for all $a < 1$," which is certainly not false (cf. IB1, §1.4). But if for the argument by induction it is necessary that there should exist an $x_0 < n$ for which $A(x_0)$ is correct, then the above form of proof by induction cannot be applied.

[14] Under Axioms I, II, III alone it is easy to construct valuations in $\mathfrak{C}$ for which this invariance no longer holds (see Ostmann, Euklidische Ringe mit eindeutiger Partialbruchzerlegung, Journ. f. reine u. angew. Math. (Crelle) 188 (1950) 150–161.

first form, if all the a_ρ/q_ρ are partial fractions. If $t = \prod_{\rho=1}^{r} p_\rho^{\lambda_\rho}$ is a canonical factorization, then

$$(10) \qquad \frac{s}{t} = a_0 + \sum_{\rho=1}^{r} \sum_{\kappa=1}^{\rho} \frac{a_{\rho\kappa}}{p_\rho^{\kappa}}, \; a_0$$

with all the $a_{\rho\kappa}$ integers, $w(a_{\rho\kappa}) < w(p_\rho)$ or $a_{\rho\kappa} = 0$, is called a DPF of the second form.

Theorem: *Every $\kappa \in \mathfrak{Q}(\mathfrak{E})$ has DPF's of both forms.*

Proof: As for (9), it is sufficient to consider the case $r = 2$. Since $(q_1, q_2) = 1$, there exist in $\mathfrak{E}$ two elements x_0 and y_0 such that $1 = q_1x_0 + q_2y_0$; thus

$$\frac{s}{t} = \frac{sq_1x_0 + sq_2y_0}{q_1q_2} = \frac{sx_0}{q_2} + \frac{sy_0}{q_1}.$$

But now from the division formulas $sy_0 = a_0'q_1 + a_1$, $sx_0 = a_0''q^2 + a_2$, with $a_0' + a_0'' = a_0$, we deduce (9) as follows: from $(s, t) = 1$ we have $(sx_0, q_2) = (sy_0, q_1) = 1$, since otherwise $q_1q_2 = t$ would not be the least common denominator; thus we must also have $(a_i, q_i) = 1$ $(i = 1, 2)$. To obtain (10) we start from (9) with $q_\rho = p_\rho^{\lambda_\rho}$. If in $a_\rho/p_\rho^{\lambda_\rho}$ we insert the division formula $a_\rho = b_\rho p_\rho + a_{\rho\lambda_\rho}$, we have

$$\frac{a_\rho}{p_\rho^{\lambda_\rho}} = \frac{b_\rho}{p_\rho^{\lambda_\rho - 1}} + \frac{a_{\rho\lambda_\rho}}{p_\rho^{\lambda_\rho}},$$

where $a_{\rho\lambda_\rho}/p_\rho^{\lambda_\rho}$ is already a partial fraction. By successive application of this procedure to the first summands on the right, we finally obtain (10).

For example, $1/12 = 1/3 \cdot 4$ has four DPF's of the first form:

$$\tfrac{1}{12} = \tfrac{1}{3} - \tfrac{1}{4} = -1 + \tfrac{1}{3} + \tfrac{3}{4} = 1 - \tfrac{2}{3} - \tfrac{1}{4} = -\tfrac{2}{3} + \tfrac{3}{4};$$

on the other hand, it is easy to show that for $\mathfrak{E} = \mathfrak{K}[x]$ both these DPF's are unique in $\mathfrak{Q} = \mathfrak{K}(x)$, if we take $w(f(x)) = \text{degree } f(x)$.[15]

The DPF's in $\mathfrak{K}(x)$ have a well-known application to the integration of rational functions. As a geometric application of the DPF's in the field $\mathfrak{Q}(\mathfrak{C}) = \mathrm{P}$ of the rational numbers, let us mention the construction of a regular n-polygon with composite n of the form $n = 2^\alpha q_1q_2 \cdots q_s$, $\alpha \geqslant 0$, where the q_i are odd and pairwise coprime, and the regular q_i-polygon is assumed to be

[15] On the existence of a division algorithm in $\mathfrak{K}[x]$ with respect to other valuations and for DPF's in general, see the references in footnotes 10 and 14.

constructible with rules and compass (e.g., the 15-polygon). In order to construct the angle $2\pi/n = 2\pi[1/(2^\alpha q_1 \cdots q_s)]$ we construct the DPF of the first form $1/n$:

$$(11) \qquad \frac{2\pi}{n} = 2\pi\left(\frac{a_0}{2^\alpha} + \frac{a_1}{q_1} + \cdots + \frac{a_s}{q_s}\right) = \frac{2\pi a_0}{2^\alpha} + \frac{2\pi a_1}{q_1} + \cdots + \frac{2\pi a_s}{q_s}.$$

Since the a_σ ($\sigma = 0, 1, ..., s$) are integers, the angles $2\pi/q_\sigma \mid a_\sigma \mid$ ($\sigma = 0, 1, ..., s$; $q_0 = 2^\alpha$) are constructible by hypothesis, and thus we can also construct the angle $2\pi/n$ (see also the last paragraph of §2.12.).

2.12. *Number of Divisors, Sum of Divisors; Certain Special Types of Numbers*

Let $\mathfrak{R}$ be a u.f. ring and let $\mathfrak{P}$ be a system of representatives of the classes of associate prime elements, excluding the class of units. If $n \in \mathfrak{R}$ has the canonical factorization $n = \epsilon\prod_{i=1}^s p_i^{\alpha_i}$, $p_i \in \mathfrak{P}$, then the divisors of n obviously have the form $\eta \prod_{i=1}^s p_i^{\beta_i}$, $0 \leqslant \beta_i \leqslant \alpha_i$, where η is a unit. By a *normalized* factor d of n with respect to $\mathfrak{P}$ we mean a $d = \prod_{i=1}^s p_i^{\beta_i}$, $0 \leqslant \beta_i \leqslant \alpha_i$, and the symbol $\sum_{d/n}$ means that d runs only through the normalized factors. In the ring $\mathfrak{C}$, unless otherwise noted, the set of prime numbers >1 (see §2.6) will always be taken for $\mathfrak{P}$, so that in this case d runs through all the positive divisors of n.

We now define the function

$$(12) \qquad \sigma_k(n) = \sum_{d|n} d^k \qquad (k \text{ real}).$$

For $k = 0$ we obtain the *number of divisors* $\sigma_0(n) = \tau(n)$, and for $k = 1$ the *sum of the divisors* $\sigma_1(n) = \sigma(n)$. The values of these functions are given by

$$(13) \quad \tau(n) = \prod_{i=1}^s (\alpha_i + 1), \qquad \sigma_k(n) = \prod_{i=1}^s \frac{p_i^{k(\alpha_i+1)} - 1}{p_i^k - 1} \qquad (k \neq 0).$$

Proof: Since there are exactly $\alpha_i + 1$ possibilities for the β_i in $d = \prod_{i=1}^s p_i^{\beta_i}$ the stated value of $\tau(n)$ follows at once by complete induction. Furthermore,

$$\sum_{d|n} d^k = \sum_{\substack{\beta_1=0,\dots,\alpha_1\\ \vdots \\ \beta_s=0,\dots,\alpha_s}} p_1^{k\beta_1} p_2^{k\beta_2} \cdots p_s^{k\beta_s} = \sum_{\beta_1=0}^{\alpha_1} \sum_{\beta_2=0}^{\alpha_2} \cdots \sum_{\beta_s=0}^{\alpha_s} p_1^{k\beta_1} \cdots p_s^{k\beta_s}$$

$$= \prod_{i=1}^s \sum_{\beta_i=0}^{\alpha_i} p_i^{k\beta_i} = \prod_{i=1}^s (1 + p_i^k + (p_i^k)^2 + \cdots + (p_i^k)^{\alpha_i})$$

$$= \prod_{i=1}^s \frac{(p_i^k)^{\alpha_i+1} - 1}{p_i^k - 1},$$

as the sum of a geometric progression if $k \neq 0$.

By (13) we have at once:

(14) *From* $(u_1, u_2) = 1$ *follows* $\sigma_k(u_1 u_2) = \sigma_k(u_1) \cdot \sigma_k(u_2)$ *(k real).*

In general, a function $f(x)$ is called *multiplicative* if $(x, y) = 1$ implies $f(xy) = f(x) f(y)$, and *distributive* if this functional equation holds for arbitrary x, y. Thus the function $\sigma_k(n)$ is multiplicative.

Also, a function $F(x)$ is called the *summatory function* of $f(x)$ if $F(x) = \sum_{d|x} f(d)$.

It is easy to see that if $f(x)$ is multiplicative, then so is $F(x)$.

In $\mathfrak{C}$ a positive number is said to be *κ-perfect* if $\sigma(n) = \kappa n$, *κ-deficient* if $\sigma(n) < \kappa n$, and *κ-abundant* if $\sigma(n) \geqslant \kappa n$; and for $\kappa = 2$ the given number is simply called *perfect*, *deficient*, and *abundant*, respectively. For example, 6, 28, 996, and 8128 are perfect, 4 is deficient, and 12 is abundant. For κ-perfect numbers κ is necessarily rational.

Since $\sigma(1) = 1 < 2$ and $\sigma(p) = p + 1 < 2p$, the prime numbers $p (p > 1)$ are deficient; and since 1 is the sole 1-perfect number, only the case $\kappa > 1$ is of interest. It is easy to show that *every multiple of a κ-abundant number is κ-abundant.*

Theorem 5 (Euclid-Euler): *If n is even and perfect, then* $n = 2^\rho(2^{\rho+1} - 1)$, $\rho > 0$, *and* $2^{\rho+1} - 1 = p$ *is a prime number (Euler), and conversely every such number is an even perfect number.*

Proof: Since n is even, we may set $n = 2^\rho u$ (with u odd). Then if n is perfect,

$$\sigma(n) = \sigma(2^\rho u) = \sigma(2^\rho)\, \sigma(u) = (2^{\rho+1} - 1)\, \sigma(u) = 2n = 2^{\rho+1} u,$$

so that $2^{\rho+1} \mid \sigma(u)$, and thus $\sigma(u) = 2^{\rho+1}\lambda$, $\lambda \geqslant 1$. Consequently,

$$2^{\rho+1} u = (2^{\rho+1} - 1)\, \sigma(u) = (2^{\rho+1} - 1)\, 2^{\rho+1}\lambda, \quad \text{i.e.,} \quad u = (2^{\rho+1} - 1)\, \lambda,$$

and thus, since $\rho > 0$, we have $\sigma(u) \geqslant (2^{\rho+1} - 1)\, \lambda + \lambda = 2^{\rho+1}\lambda = \sigma(u)$, so that the equality sign must hold, and u has exactly two factors; therefore u is prime and $\lambda = 1$; i.e., $u = 2^{\rho+1} - 1$. The converse follows at once from

$$\sigma(n) = \sigma(2^\rho p) = \sigma(2^\rho)\, \sigma(p) = (2^{\rho+1} - 1)(p + 1) \qquad (\text{since } \rho > 0)$$
$$= p \cdot 2^{\rho+1} = 2 \cdot 2^\rho \cdot p = 2n.$$

Numbers of the form $2^n - 1$ are called *Mersenne numbers*. Since $2^{ab} - 1 = (2^a)^b - 1 = (2^a - 1)(1 + 2^a + \cdots + 2^{a(b-1)})$, we have the following theorem:

Theorem 6: *A number* $2^p - 1$ *can be a Mersenne prime only if p is prime.*

Thus the search for even perfect numbers is reduced to the search for Mersenne primes. It is still an open question whether odd perfect numbers exist and also whether there are infinitely many Mersenne primes.

Still less is known about *amicable numbers*,[16] i.e., pairs a, b with $\sigma(a) = \sigma(b) = a + b$, or in other words $\sum_{d|a,d<a} d = b$, $\sum_{d|b,d<b} d = a$. Example: 220, 284.

From the factorization

$$a^u + b^u = (a + b)(a^{u-1} - a^{u-2}b + \cdots - ab^{u-2} + b^{u-1})$$

for odd u we have at once the following theorem:

Theorem 7: *The number* $2^n + 1$ *can be prime only if* $n = 2^\nu$ $(\nu \geqslant 0)$.

Numbers of the form $2^{2^\nu} + 1$ *are called Fermat numbers or Gauss numbers or, if prime, Fermat* (*or Gauss*) *primes*. For $\nu = 0, 1, 2, 3, 4$ we have the primes 3, 5, 17, 257, 65537, but no further primes of this form are known. At least $\nu = 5$ is not a prime, in view of the fact that $641 \mid 2^{2^5} + 1$ (Euler). In any case, $(2^{2^\nu} + 1, 2^{2^\mu} + 1) = 1$ for $\nu \neq \mu$.

As a supplement to the last paragraph in §2.11, let us mention the following theorem of Gauss: *a regular p-polygon, where p is a prime number, is constructible with ruler and compass if and only if p is a Gauss prime; and all constructible n-gons are obtained by replacing the* q_i *in* (11) *with Gauss primes.*

3. Continued Fractions

3.1. By the Euclidean algorithm §2 (7) the fraction a/b admits the representation:

$$\text{(15)} \qquad \frac{a}{b} = q_1 + \frac{r_1}{b} = q_1 + \cfrac{1}{\cfrac{b}{r_1}} = q_1 + \cfrac{1}{q_2 + \cfrac{r_2}{r_1}}$$

$$= q_1 + \cfrac{1}{q_2 + \cfrac{1}{q_3 + \ddots + \cfrac{1}{q_n + \cfrac{1}{q_{n+1}}}}}.$$

[16] For a detailed account of amicable numbers, see A. Wulf: Die befreundeten Zahlen nebst einem Ausblick auf die vollkommenen und aliquoten Zahlen. Göttingen, 1950, hectographed.

In general, a fraction of the form

$$(16)\qquad b_0 + \cfrac{a_1}{b_1 + \cfrac{a_2}{b_2 + \cfrac{a_3}{b_3 + \ddots}}} = b_0 + \frac{a_1 |}{| b_1} + \frac{a_2 |}{| b_2} + \cdots$$

is called a *continued fraction*, the a_i are the partial numerators, the b_i the partial denominators, and b_0 is the first term. If all $a_i = 1$, all b_i are rational integers, and $b_i > 0$ for $i > 0$, then (16) is called a *regular continued fraction*. Such a fraction can obviously be normalized so as to make the last partial denominator greater than unity. In the present section we shall always assume that this has been done. The notation

$$b_0 + \cfrac{1}{b_1 + \cfrac{}{\ddots + \cfrac{1}{b_n}}} = [b_0 , b_1 , ..., b_n]$$

is in common use. For convenience in the statement of proofs, it is often desirable to arrange that the b_ν are real and positive and to include this property in the definition of the symbol. Then we can show at once that for $n \geqslant 1$

$$(17)\quad [b_0 , b_1 , ..., b_n] = b_0 + \frac{1}{[b_1 , ..., b_n]} = [b_0 , [b_1 , ..., b_n]],$$

$$[b_0 , b_1 , ..., b_n] = \left[b_0 , ..., b_{n-1} + \frac{1}{b_n}\right] \quad \text{(recursion formula).}$$

From the fact that

$$[b_0 + b_0' , b_1 , ..., b_n] = b_0' + [b_0 , b_1 , ..., b_n]$$

we see that without loss of generality we can confine our attention to continued fractions for which $b_0 \geqslant 0$, as will be assumed throughout the present section. The reduced fraction

$$(18)\qquad \frac{A_\nu}{B_\nu} = [b_0 , b_1 , ..., b_\nu], \qquad \nu = 0, 1, 2, ..., n,$$

is called the νth *convergent*, A_ν is the νth partial numerator, and B_ν is the νth partial denominator. Obviously $A_0 = b_0$, $B_0 = 1$. For convenience we also introduce

$$(19)\qquad A_{-2} = 0, \qquad A_{-1} = 1, \qquad B_{-2} = 1, \qquad B_{-1} = 0.$$

Then we have the fundamental linear recursion formulas:

$$(20) \qquad A_\nu = b_\nu A_{\nu-1} + A_{\nu-2}\,, \qquad B_\nu = b_\nu B_{\nu-1} + B_{\nu-2} \qquad (\nu \geqslant 0)$$

and the identity

$$(21) \qquad A_\nu B_{\nu-1} - A_{\nu-1} B_\nu = (-1)^{\nu-1} \qquad (\nu \geqslant -1).$$

Proof: For $\nu = 0$ the formulas (20) follow from (19). If (20) holds for $\nu - 1$ ($\nu \geqslant 1$), then by the second equation (17), again assuming that the b_i are real and positive,

$$\frac{A_\nu}{B_\nu} = \frac{\left(b_{\nu-1} + \dfrac{1}{b_\nu}\right) A_{\nu-2} + A_{\nu-3}}{\left(b_{\nu-1} + \dfrac{1}{b_\nu}\right) B_{\nu-2} + B_{\nu-3}} \left(= \left[b_0\,, b_1\,, ..., b_{\nu-1} + \frac{1}{b_\nu}\right]\right)$$

$$= \frac{b_\nu(b_{\nu-1}A_{\nu-2} + A_{\nu-3}) + A_{\nu-2}}{b_\nu(b_{\nu-1}B_{\nu-2} + B_{\nu-3}) + B_{\nu-2}} = \frac{b_\nu A_{\nu-1} + A_{\nu-2}}{b_\nu B_{\nu-1} + B_{\nu-2}},$$

where the last equation follows from the induction hypothesis. In order to show that for integral b_i the last fraction is already reduced, we consider the A_ν, B_ν as defined recursively by (20) and (19). Again the last equation holds, and (18) follows from it by induction; so it remains to prove (21). But for $\nu = -1$, (21) follows from (19), and by complete induction

$$\begin{aligned} A_\nu B_{\nu-1} - A_{\nu-1} B_\nu &= (b_\nu A_{\nu-1} + A_{\nu-2})\, B_{\nu-1} - A_{\nu-1}(b_\nu B_{\nu-1} + B_{\nu-2}) \\ &= -(A_{\nu-1} B_{\nu-2} - A_{\nu-2} B_{\nu-1}) = -(-1)^{\nu-2} = (-1)^{\nu-1}, \end{aligned}$$

which is (21), as desired. For integral b_i we at once have $(A_\nu, B_\nu) = 1$, so that the fractions (A_ν/B_ν) are reduced.

As an estimate for the B_ν, it follows from $B_0 > 0$, $B_1 = b_1 \geqslant 1$, $B_2 = b_2 b_1 + 1 \geqslant 2$, by complete induction for $\nu \geqslant 3$, that

$$B_\nu = b_\nu B_{\nu-1} + B_{\nu-2} \geqslant \nu - 1 + \nu - 2 = \nu + (\nu - 3) \geqslant \nu,$$

and thus

$$(22) \qquad B_\nu \geqslant \nu \qquad \text{for all} \quad \nu \geqslant 0.$$

If we now define the nonterminating regular continued fraction $[b_0\,, b_1\,, b_2\,, ...]$ as the sequence of partial quotients $(A_n/B_n) = [b_0, b_1, ..., b_n]$, we have the following theorem.

Convergence theorem: *Every nonterminating regular continued fraction is convergent.*

Proof: Since

$$0 \leqslant \frac{A_n}{B_n} = \frac{A_0}{B_0} + \sum_{\nu=1}^{n} \left(\frac{A_\nu}{B_\nu} - \frac{A_{\nu-1}}{B_{\nu-1}}\right) \leqslant \frac{A_0}{B_0} + \sum_{\nu=1}^{n} \left| \frac{A_\nu B_{\nu-1} - A_{\nu-1} B_\nu}{B_\nu B_{\nu-1}} \right|$$

$$= b_0 + \sum_{\nu=1}^{n} \left| \frac{(-1)^{\nu-1}}{B_\nu B_{\nu-1}} \right| \leqslant b_0 + 1 + \sum_{\nu=2}^{n} \frac{1}{\nu(\nu-1)}$$

$$\leqslant b_0 + 1 + \sum_{\nu=1}^{\infty} \frac{1}{\nu^2} \left(= b_0 + 1 + \frac{\pi^2}{6}\right),$$

it follows that $\sum_{\nu=1}^{\infty} [(A_\nu/B_\nu) - (A_{\nu-1}/B_{\nu-1})]$ is absolutely convergent, and thus also convergent.

For the proof of the following theorem we need a slight generalization of the Euclidean algorithm for real numbers. For every real number ρ we let $[\rho]$ denote the *greatest integer* $g \leqslant \rho$, $g = [\rho] \leqslant \rho < [\rho] + 1$. Such a g always exists, since the field of real numbers has an Archimedean ordering (see IB1, §4.1), which means that for every real $\rho \geqslant 0$ there exists a natural number n such that $n \cdot 1 > \rho$. In the set of all such n there is a smallest, which we put equal to $g + 1$. Then obviously $g = [\rho]$. For a negative nonintegral ρ we have $[\rho] = -([-\rho] + 1)$ with the desired properties; for an integer ρ it is clear that $[\rho] = \rho$. In the division with smallest positive remainder for the ring $\mathfrak{C}$, say $a = qb + r$ with $\rho = a/b = q + r/b$, we have $q = [\rho]$, since $0 \leqslant r/b = \delta < 1$, and therefore $\rho = [\rho] + \delta$, $0 \leqslant \delta < 1$. From the definition of $[\rho]$ it is clear that every real ρ admits such a formula, which is the desired generalization.

Expansion theorem: *Every real number can be expanded in exactly one way as a regular continued fraction.*

Proof: Set $b_0 = [\rho]$ and $\rho = b_0 + 1/\eta_1$; thus $\eta_1 > 1$ if ρ is not already an integer. Furthermore, let $\eta_0 = \rho$, $b_1 = [\eta_1]$, $\eta_1 = b_1 + 1/\eta_2$, so that

$$\rho = b_0 + \cfrac{1}{b_1 + \cfrac{1}{\eta_2}} = [b_0, b_1, \eta_2];$$

proceeding in this way we obtain a sequence, terminating or nonterminating, of integers b_ν such that

$$\rho = [b_0, b_1, \ldots, b_{\nu-1}, \eta_\nu]. \tag{23}$$

From (20) and (21) for $\nu > 1$ it follows in the nonterminating case

$$(24)\quad \left|\rho - \frac{A_{\nu-1}}{B_{\nu-1}}\right| = \left|\frac{\eta_\nu A_{\nu-1} + A_{\nu-2}}{\eta_\nu B_{\nu-1} + B_{\nu-2}} - \frac{A_{\nu-1}}{B_{\nu-1}}\right| = \left|\frac{A_{\nu-2}B_{\nu-1} - A_{\nu-1}B_{\nu-2}}{B_{\nu-1}(\eta_\nu B_{\nu-1} + B_{\nu-2})}\right|$$

$$= \frac{1}{B_{\nu-1}(\eta_\nu B_{\nu-1} + B_{\nu-2})} < \frac{1}{B_{\nu-1}} \leqslant \frac{1}{\nu - 1},$$

from which we see at once that $\lim_{\nu\to\infty} A_{\nu-1}/B_{\nu-1} = \rho$, so that $\rho = [b_0, b_1, ...]$. With respect to the uniqueness, we note that for (23) we have in the terminating case

$$\eta_\nu = [b_\nu, b_{\nu+1}, ..., b_n], \qquad 0 \leqslant \nu \leqslant n,$$

and in the nonterminating case $\eta_\nu = [b_\nu, b_{\nu+1}, ...]$, $\nu \geqslant 0$, so that $\eta_\nu > 1$ for all $\nu \geqslant 1$ (where in the terminating case we must take account of the normalization $b_n > 1$). Thus in $\eta_\nu = [b_\nu, \eta_{\nu+1}] = b_\nu + 1/\eta_{\nu+1} (\nu \geqslant 0)$ we necessarily have $b_\nu = [\eta_\nu]$, so that b_ν is uniquely determined by η_ν; in particular, b_0 is uniquely determined by the value $\eta_0 = \rho$. If we now assume the uniqueness of the numbers $b_0, b_1, ..., b_{\nu-1}$, then by (23) η_ν is also uniquely determined and thus, as we have just seen, so is b_ν.

Since by (15) rational numbers have a terminating continued fraction, the preceding theorem gives us this result: *terminating regular continued fractions represent rational numbers, and nonterminating continued fractions represent irrational numbers.*

From (21) it follows from division by $B_{\nu-1}B$ for $\nu \geqslant 1$ that

$$\frac{A_\nu}{B_\nu} - \frac{A_{\nu-1}}{B_{\nu-1}} = \frac{(-1)^{\nu-1}}{B_\nu B_{\nu-1}},$$

so that the sequence of first differences of the sequence A_ν/B_ν is alternating; taking into account the inequalities

$$B_\nu = b_\nu B_{\nu-1} + B_{\nu-2} \begin{cases} > B_{\nu-1}, & \text{for } \nu \geqslant 2 \quad \text{and} \quad \nu = 0, \\ \geqslant B_{\nu-1}, & \text{for all } \nu \geqslant 0, \end{cases}$$

which show that $1/B_{\nu-1}B_\nu$ is strictly decreasing, we see from

$$\frac{A_0}{B_0} = b_0 < [b_0, b_1, ...] = k$$

that

$$\frac{A_0}{B_0} < \frac{A_2}{B_2} < \cdots < \frac{A_{2n}}{B_{2n}} < \cdots < k < \cdots < \frac{A_{2n+1}}{B_{2n+1}} < \cdots < \frac{A_3}{B_3} < \frac{A_1}{B_1}.$$

The recursion formulas (20) provide us with a simple setup for practical calculation of the partial quotients (A_ν/B_ν) of $[b_0, b_1, ...]$:

ν	-2	-1	0	1	$\cdots$	$\nu-2$	$\nu-1$	ν
b_ν	—	—	b_0	b_0	$\cdots$	$b_{\nu-2}$	$b_{\nu-1}$	b_ν
A_ν	0	1	b_0	$b_1 b_0 + 1$	$\cdots$	$A_{\nu-2}$	$A_{\nu-1}$	$A_\nu = b_\nu A_{\nu-1} + A_{\nu-2}$
B_ν	1	0	1	$b_1 1 + 0$	$\cdots$	$B_{\nu-2}$	$B_{\nu-1}$	$B_\nu = b_\nu B_{\nu-1} + B_{\nu-2}$

The last two lines are calculated, independently of each other, by the same rule (see the last column). For example, if we wish to calculate $\pi \approx 3.14159265358$, the Euclidean algorithm gives $\pi = [3, 7, 15, 1, 292, 1, 1, 1, ...]$; and the corresponding setup is:

ν	-2	-1	0	1	2	3	4
b_ν	—	—	3	7	15	1	292
A_ν	0	1	3	22	333	355	103993
B_ν	1	0	1	7	106	113	33102
Error: ϵ_ν	—	—	$+1{,}4 \cdot 10^{-1}$	-10^{-3}	$+0{,}8 \cdot 10^{-4}$	$-2{,}7 \cdot 10^{-7}$	$+0{,}6 \cdot 10^{-9}$

Here the well-known approximations 3, $\frac{22}{7}$ ... stand one under the other. It is remarkable that the slight increase from A_2 to A_3 and B_2 to B_3 produces a considerable improvement in the accuracy, whereas the far greater increase involved in passing from $\nu = 1$ to $\nu = 2$ and from $\nu = 3$ to $\nu = 4$ fails to produce any correspondingly great improvement in the approximation. This phenomenon finds its general explanation in the formula (24). With the increase of η_ν and therefore of b_ν, the approximation $A_{\nu-1}/B_{\nu-1}$ is improved.

3.2. One of the essential properties of the above partial quotients of a number k rests on the fact that among all rational numbers these partial quotients *best approximate* the number k, in the sense of the following definition: *p/q is called a best approximation to k if from*

$$\left|\frac{a}{b} - k\right| \leqslant \left|\frac{p}{q} - k\right| \quad \textit{and} \quad \frac{a}{b} \neq \frac{p}{q} \quad \textit{it follows that } b > q.$$

In other words, in order to make a better approximation than the above-defined best approximation, we must have recourse to larger denominators.[17] Now we can show that *all the partial quotients A_ν/B_ν are best approximations to k*. In order to find *all* the best approximations to k, we

[17] The problem of finding best approximations is of practical importance in the technology of power machinery, where the gears should approximate a given transmission ratio as closely as possible, but with the smallest possible number of cogs.

must also consider the so-called *subsidiary partial quotients* (the A_ν/B_ν are then called the A_ν/B_ν *principal partial quotients*)

$$N_{\lambda,\rho} = \frac{A_{\lambda-2} + \rho A_{\lambda-1}}{B_{\lambda-2} + \rho B_{\lambda-1}}, \qquad \rho = 1, 2, \ldots, b_{\lambda-1}.$$

It is easy to show that the $N_{\lambda,\rho}$ lie in the open interval $[(A_{\lambda-2}/B_{\lambda-2}), (A_\lambda/B_\lambda)]$ and change monotonically for $\rho = 1, 2, \ldots, b_{\lambda-1}$ $(\lambda \geqslant 1)$. Then those $N_{\lambda,\rho}$ for which

$$\left| k - \frac{A_{\lambda-2}}{B_{\lambda-2}} \right| > | k - N_{\lambda,\rho} | > \left| k - \frac{A_{\lambda-1}}{B_{\lambda-1}} \right|$$

are certainly not best approximations, since the $A_{\lambda-1}/B_{\lambda-1}$ are better approximations and have a smaller denominator. More precisely, it can be shown (see Perron [3]) that we obtain

for $\rho < \frac{1}{2}b_\lambda$ *no best approximations,*
for $\rho > \frac{1}{2}b_\lambda$ *best approximations,*
and for $\rho = \frac{1}{2}b_\lambda$ *a best approximation if and only if*

$$[b_\lambda, b_{\lambda-1}, \ldots, b_1] > [b_\lambda, b_{\lambda+1}, \ldots].$$

3.3. A continued fraction $k = [b_0, b_1, \ldots]$ is said to be *periodic* if there exist two numbers n and p such that $b_{n+\lambda p+\kappa} = b_{n+\kappa}$ for all $\lambda \geqslant 0$ and all $\kappa = 0, 1, p - 1$; in analogy with the notation for periodic decimals we then write $k = [b_0, \ldots, b_{n-1}, \overline{b_n, \ldots, b_{n+p-1}}]$. If the choice of p and n is minimal, p is called the (*primitive*) *period* and $[b_0, b_1, \ldots, b_{n-1}]$ the *preperiodic part*; if this latter part is missing, we speak of a *purely periodic* (or simply *periodic*) *continued fraction.* The following theorem is due to Euler:

If the continued fraction expansion of a number k is periodic, then k is the solution of a quadratic equation which is irreducible in P$[x]$[18] *and has integral coefficients (in other words, k is a quadratic irrationality, i.e., an algebraic number of second degree).*

Proof: Every real quadratic irrationality has the form $(a + \sqrt{b})/c$ (a, b, c integers, $b > 0$, b not a perfect square), and conversely every number of this form is a quadratic irrationality. Thus it is sufficient to show that a purely periodic continued fraction $\eta_n = \overline{[b_n, \ldots, b_{n+p-1}]}$ is a quadratic irrationality; in other words, we assume $n = 0$. But then

$$\begin{aligned} \eta_p &= \overline{[b_p, \ldots, b_{2p-1}]} = \overline{[b_0, \ldots, b_{p-1}]} = \eta_0 = k \\ &= [b_0, \ldots, b_{p-1}, \eta_p] = \frac{\eta_p A_{p-1} + A_{p-2}}{\eta_p B_{p-1} + B_{p-2}} = \frac{kA_{p-1} + A_{p-2}}{kB_{p-1} + B_{p-2}}, \end{aligned}$$

[18] P is the field of rational numbers.

so that k is the solution of the equation

$$B_{p-1}x^2 + (B_{p-2} - A_{p-1})\,x - A_{p-2} = 0.$$

Of great importance is the converse of this theorem, due to Lagrange (see Perron [2], [3]):

Every quadratic irrationality has a periodic continued fraction expansion.

3.4. A generalization of periodic partial fractions is due to Hurwitz. Let each of the l rows

$$\begin{array}{c} a_1^{(1)}, a_2^{(1)}, \dots \\ a_1^{(2)}, a_2^{(2)}, \dots \\ \cdot\ \cdot\ \cdot\ \cdot\ \cdot\ \cdot \\ \cdot\ \cdot\ \cdot\ \cdot\ \cdot\ \cdot \\ \cdot\ \cdot\ \cdot\ \cdot\ \cdot\ \cdot \\ a_1^{(l)}, a_2^{(l)}, \dots \end{array} \tag{25}$$

be an arithmetic progression of arbitrary order. Then

$$\begin{aligned}[b_0, b_1, \dots, b_{n-1}, &\overline{a_\rho^{(1)}, a_\rho^{(2)}, \dots, a_\rho^{(l)}}]_{\rho=1}^{\infty} \\ &= [b_0, \dots, b_{n-1}, a_1^{(1)}, \dots, a_1^{(l)}, a_2^{(1)}, \dots a_2^{(l)}, \dots]\end{aligned}$$

is called a *Hurwitz continued fraction.* If all the sequences in (25) have the order zero, in other words, if they are sequences of constants, the result is obviously a periodic partial fraction. The following theorem is due to Hurwitz (see Perron [2], [3]):

For numbers η and ξ related by

$$\eta = \frac{a\xi + b}{c\xi + d}, \qquad ad - bc \neq 0; \qquad a, b, c, d \text{ rational},^{19}$$

if ξ is a Hurwitz partial fraction, so also is η and conversely; moreover, the number of arithmetic progressions of nth order for every $n > 0$ is the same in both cases.[20]

As an example of a Hurwitz continued fraction let us consider

$$k = [(2n + 1)\,b]_{n=1}^{\infty} = [b, 3b, 5b, \dots];$$

here we have

$$k = \frac{e^{\frac{1}{b}} + e^{-\frac{1}{b}}}{e^{\frac{1}{b}} - e^{-\frac{1}{b}}};$$

[19] By taking the fractions over a least common denominator, we may confine ourselves to rational integers a, b, c, d.

[20] Here it may be necessary to break up a sequence $a_1, a_2, \dots$, say into $a_1, a_3, a_5, \dots$ and $a_2, a_4, a_6, \dots$ and so forth. Constant sequences may be inserted or removed.

taking $b = 2$, we see from the preceding theorem that e also has an expansion in Hurwitz continued fractions; we have

$$e = [2, \overline{1, 2n, 1}]_{n=1}^{\infty} = [2, 1, 2, 1, 1, 4, 1, 1, 6, 1, ...].$$

The regular continued fraction expansion for π is unknown. On the other hand,

$$\pi = \frac{4|}{|1} + \frac{1|}{|2} + \frac{3^2|}{|2} + \frac{5^2|}{|2} + \frac{7^2|}{|2} + \cdots$$

or

$$\pi = 2 - \frac{2|}{|3} - \frac{2\cdot 3|}{|1} - \frac{1\cdot 2|}{|3} - \frac{4\cdot 5|}{|1} - \frac{3\cdot 4|}{|3} - \frac{6\cdot 7|}{|1} - \frac{5\cdot 6|}{|3} - \cdots$$

are expansions in the form (16) (see Perron [3]).

4. Congruences

4.1. Let $\mathfrak{G}$ be an Abelian group (see IB2, §1.1), and let $\mathfrak{U}$ be a subgroup of $\mathfrak{G}$. As was shown in IB2, §3.5, every group $\mathfrak{G}$ can be represented as the union of disjoint cosets (residue classes) $g\mathfrak{U}$. The property of two elements g_1, g_2 of belonging to the same coset is obviously an equivalence relation.[21] This relation is called a congruence and is written

$$(26) \qquad g_1 \equiv g_2 (\text{mod } \mathfrak{U}) \qquad \text{or} \quad g_1 \equiv g_2(\mathfrak{U});$$

in words: g_1 *is congruent to* g_2 *modulo* $\mathfrak{U}$. From $g_1 \in g_2\mathfrak{U}$ follows the existence of a $u \in \mathfrak{U}$ such that $g_1 = g_2 u$, and therefore $g_2^{-1} g_1 = u$, or in other words $g_2^{-1} g_1 \in \mathfrak{U}$; and conversely, $g_2^{-1} g_1 \in \mathfrak{U}$ implies that $g_1 \equiv g_2(\mathfrak{U})$ or, in other words, if e is the unity of $\mathfrak{G}$, we have $g_2^{-1} g_1 \equiv e(\mathfrak{U})$. Since $\mathfrak{G}$ is Abelian, it follows from $g_2^{-1} g_1 \in \mathfrak{U}$ and $g_4^{-1} g_3 \in \mathfrak{U}$ that $g_2^{-1} g_1 \cdot g_4^{-1} g_3 = (g_2 g_4)^{-1} g_1 g_3 \in \mathfrak{U}$, i.e., $g_1 g_3 \equiv g_2 g_4(\mathfrak{U})$; in other words: *congruences* $g_1 \equiv g_2(\mathfrak{U})$ *and* $g_3 \equiv g_4(\mathfrak{U})$ *may be multiplied, and naturally also divided.* If the operation of the group is written as addition, i.e., if $\mathfrak{G}$ is a module, then (26) is obviously equivalent to $g_1 - g_2 \in \mathfrak{U}$ or $g_1 - g_2 \equiv 0(\mathfrak{U})$, and then the congruences can be added and subtracted.

If $\mathfrak{G}$ is the additive group of a ring $\mathfrak{R}$, $\mathfrak{G} = \mathfrak{R}^+$, and if the submodule $\mathfrak{U}$ is an ideal $\mathfrak{a}$ in $\mathfrak{R}^+$, we have the following theorem:

Theorem 1: *Congruences* mod $\mathfrak{a}$ *may be added, subtracted, and multiplied.*

[21] That is, it is reflexive, symmetric, and transitive (cf. IA, §8.3 and §8.5).

Proof: Let $a_1 \equiv a_2(\mathfrak{a})$ and $a_3 \equiv a_4(\mathfrak{a})$; from $a_1 - a_2 \in \mathfrak{a}$ and $a_3 \in \mathfrak{R}$ it follows [see §2.5. (II)] that $(a_1 - a_2) a_3 \in \mathfrak{a}$, so that $a_1a_3 \equiv a_2a_3(\mathfrak{a})$; similarly, $a_2a_3 \equiv a_2a_4(\mathfrak{a})$ and therefore, by the transitivity, $a_1a_3 \equiv a_2a_4(\mathfrak{a})$.

If $\mathfrak{a} = (m)$ is a principal ideal, we also write $a \equiv b\ (\text{mod}\ m)$, or $a \equiv b\ (m)$. In this case, assuming again that $\mathfrak{R}$ has a unity element, the congruence is also equivalent to $m \mid (a - b)$, or in other words to the existence of a $\lambda \in \mathfrak{R}$ with $a = b + \lambda m$. In particular, if $\mathfrak{R}$ is Euclidean (see §2.10), the division formula $a = qm + r$ shows that $a \equiv r(m)$, so that *numbers are congruent if and only if they leave the same remainder when divided by the modulus.* From the equality of the ideals $(m) = (\epsilon m)$ for every unit ϵ it follows that the congruences $a \equiv b(m_1)$ and $a \equiv b(m_2)$ are equivalent if $m_1 \sim m_2$. Finally, $a \equiv b\ (\text{mod}\ 1)$ is trivially true for arbitrary a, b. If $\mathfrak{a} = (0)$ is the zero ideal, then $a \equiv b(0)$ is equivalent to $a = b$, as can be seen at once.

Theorem 1 can be interpreted in another way. If we consider the residue classes mod $\mathfrak{a}$ as elements of a new set, to be denoted by $\mathfrak{R}/\mathfrak{a}$ (read $\mathfrak{R}$ with respect to $\mathfrak{a}$), then $\mathfrak{R}/\mathfrak{a}$ is a ring, for which we can define the sum of two residue classes $\hat{a}, \hat{b}$ as $\hat{a} + \hat{b} = \widehat{a + b}$, and the product by $\hat{a}\hat{b} = \widehat{ab}$, with $a \in \hat{a}$, $b \in \hat{b}$. Then theorem 1 states that these definitions are independent of the choice of the elements a and b from $\hat{a}$ and $\hat{b}$. The axioms for a ring can be verified at once (cf. IB5, §3.6). In the terminology of group theory, the additive group $\mathfrak{R}/\mathfrak{a}^+$ is precisely the factor group (or factor module) $\mathfrak{R}^+/\mathfrak{a}$ (see IB2, §6.3). The ring $\mathfrak{R}/\mathfrak{a}$ is called the *residue class ring modulo* $\mathfrak{a}$. The mapping $a \to \hat{a}$ is a ring homomorphism (see IB5, §3.6). If we choose one element from each residue class, the set of these representatives is called a *complete residue system* mod $\mathfrak{a}$. For example, $\{0, 1, ..., m - 1\}$ and $\{0, -1, -2, -3, ..., -(m - 1)\}$ are complete residue systems mod m in the ring $\mathfrak{C}$, the former being called the smallest positive residue system. In general, residue class rings have divisors of zero; for example, $\hat{2}, \hat{3}$ are divisors of zero in $\mathfrak{C}/(6)$, since $2 \cdot 3 \equiv 0(6)$, i.e., $\hat{2} \cdot \hat{3} = \hat{0}$. Below we shall also write a mod m instead of $\hat{a}$.

4.2. Let $\mathfrak{R}$ be a principal ideal ring (see IB5, §3.4). Then:
From $a \equiv b(m)$ *follows* $(a, m) = (b, m)$.

Proof: As was shown in 4.1, the congruence $a \equiv b(m)$ is equivalent to an equation of the form $a = b + \lambda m$. It follows that $(a, m) \mid b$ and $(a, m) \mid m$, so that $(a, m) \mid (b, m)$; similarly, $(b, m) \mid (a, m)$ and thus $(a, m) = (b, m)$.

The element $(a, m) = d$ *is called the greatest divisor of the residue class* $\hat{a} = a$ mod m. *If* $d = 1$, *the class* a mod m *is called a (relatively) prime residue class* mod m. *A system of representatives of the prime residue classes* mod m *is called a reduced residue system* mod m.

If m is a prime element, the nonzero residue classes coincide with the prime residue classes.

Theorem 2: *In a principal ideal ring* $\mathfrak{H}$ *the prime residue classes* mod m, *with arbitrary* m, *form a group* $\mathfrak{G}_m$, *the prime residue class group* mod m.

Proof: If $a \bmod m$ and $b \bmod m$ are prime residue classes, then $(a, m) = (b, m) = 1$, so that $(ab, m) = 1$, i.e., $ab \bmod m$ is also a prime residue class. Thus we need only show the existence of a residue class inverse to $a \bmod m$. From $(a, m) = 1$ follows the existence of elements x_0, y_0 in $\mathfrak{H}$ such that $ax_0 + my_0 = 1$, and thus $ax_0 \equiv 1(m)$, so that $x_0 \bmod m$ is the desired inverse residue class.

In $\mathfrak{C}$ the group $\mathfrak{G}_m$ obviously has finitely many elements; *the order* $|\mathfrak{G}_m|$ *of this group is denoted by* $\varphi(m)$ *and is called the Euler function.*

Theorem 2 states only that the congruence $ax \equiv b(m)$ is solvable for a and b prime to m. But if $(a, m) = 1$, the congruence $ax \equiv b(m)$ is *uniquely* solvable for arbitrary b; for it is obvious that $bx_0 \bmod m$ is a solution if $x_0 \bmod m$ is inverse to $a \bmod m$, and from $ax_1 \equiv ax_2 \equiv b(m)$ follows $a(x_1 - x_2) \equiv 0(m)$, so that $m \mid (x_1 - x_2)$, i.e., $x_1 \bmod m = x_2 \bmod m$ (see also §7.2).

If the order $|\mathfrak{G}_m|$ of the prime residue class group $\mathfrak{G}_m$ is finite, the group-theoretic theorem that the order of a subgroup is a factor of the order of the group (see IB2, §3.5) provides us with the (generalized) lesser *Fermat theorem*:

From $(a, m) = 1$ *follows* $a^{|\mathfrak{G}_m|} \equiv 1(m)$.

Proof: The element $a \bmod m = \hat{a} \in \mathfrak{G}_m$ generates the cyclic subgroup $(\hat{a}) = \{\hat{a}, \hat{a}^2, \dots, \hat{a}^{k-1}, \hat{a}^k = \hat{1}\}$; for its order k we have $k \mid |\mathfrak{G}_m|$, so that $\hat{a}^{|\mathfrak{G}_m|} = \hat{1}$ or, written as a congruence, $a^{|\mathfrak{G}_m|} \equiv 1(m)$.

The number k is also called the *order of the element* $\hat{a}$, or *of* a mod m.[22]

We have just now, and also earlier, made use of the obvious but essential fact that *congruences between numbers of the original ring* $\mathfrak{H}$ *are equivalent to equations between residue classes.* When we are passing from equations to congruences, we may choose arbitrary representatives of any given residue class, in view of the fact that by theorem 1 the sum and product of residue classes are independent of the choice of representatives; in other words: in a congruence any element of the ring may be replaced by any element congruent to it (when speaking of a power a^n we must of course consider n not as an element of the ring but as an operator).

In the ring $\mathfrak{C}$ the Fermat theorem obviously has the form

$$a^{\varphi(m)} \equiv 1(m), \qquad \text{if} \quad (a, m) = 1.$$

[22] Instead of "a has the order k mod m," it was customary in the older literature on the ring $\mathfrak{C}$ of integers to say "a belongs mod m to the exponent k."

In particular, if $m = p \neq 0$ is a prime, we have $a^{p-1} \equiv 1\,(p)$ for $p \nmid a$; and thus for all a without restriction,

$$a^p \equiv a(p), \qquad a \text{ an arbitrary integer,}$$

which for $p \nmid a$ is equivalent to $a^{p-1} \equiv 1\,(p)$, since in a congruence we may always cancel any number prime to the modulus (i.e., the inverse residue class exists).

If $a \bmod m$ is a nonprime residue class, so that $(a, m) = d \not\sim 1$, and if we set $a = a_1 d$, $m = m_1 d$, we have

$$am_1 \equiv a_1 dm_1 \equiv a_1 m \equiv 0(m), \qquad m_1 \not\equiv 0(m),$$

i.e., every nonprime residue class is a divisor of zero in $\mathfrak{H}/(m)$ and thus has no inverse. Since the prime residue classes coincide with the elements of $\mathfrak{H}/(m)$ that are not divisors of zero, we have the result:

In $\mathfrak{H}/(m)$ an element has an inverse if and only if it is not a divisor of zero.

4.3. Now let the module be a prime element p in $\mathfrak{H}$. Then the zero class is obviously the only divisor of zero in $\mathfrak{H}/(p)$, so that $\mathfrak{H}/(p)$ is an integral domain, and in fact a field, since all nonzero elements have inverses.

Theorem 3: *If $p \neq 0$ is a prime element in the principal ideal ring $\mathfrak{H}$, the residue class ring $\mathfrak{H}/(p)$ is a field,*[23] *the so-called residue class field modulo p.*

Furthermore (see §2.10) we have:

Theorem 4: *The polynomial ring $\mathfrak{H}/(p)[x]$, $p \neq 0$ and prime, is Euclidean.*

In the following discussion (leading up to the Wilson theorem) it must be remembered that a polynomial over a field cannot have a number of zeros greater than its degree, and that in the canonical factorization the factor $(x - \alpha)^m$ necessarily appears if α is a zero of mth order (see IB4, §2.2). The fact, emphasized in §4.2, that congruences are interchangeable with equations in the residue class ring means for polynomials $f(x) = \sum_{i=0}^{n} \hat{a}_i x^i$ and $g(x) = \sum_{i=0}^{s} \hat{b}_i x^i$ in $\mathfrak{H}/(m)[x]$[24] that the identity $f(x) = g(x)$, i.e., the equations $\hat{a}_i = \hat{b}_i$ $(i = 0, 1, ..., n)$, for the coefficients with $n = s$, has exactly the same significance as $f(x) \equiv g(x)(m)$, which means in turn that $n = s$ and $a_i \equiv b_i(m)$, $i = 0, 1, ..., n$[25] (comparison of coefficients mod m).

[23] If p is a unit, $\mathfrak{H}/(p)(= \mathfrak{H}/(1))$ consists of the zero class alone. Depending on whether we wish to consider the ring consisting of zero alone as a (trivial) field (the zero field) we will include or omit values for p that are units. The polynomial ring $\mathfrak{H}/(1)[x]$ also consists solely of the zero element.

[24] This result also holds in general for $\mathfrak{R}/\mathfrak{a}$, where $\mathfrak{a}$ is an arbitrary ideal in a ring $\mathfrak{R}$.

[25] See IB4, §2.1 for the difference in meaning between the statement "$f(x) = g(x)$" expressing the fact that $f(x)$ and $g(x)$ are the same elements (polynomials) in a

Now let $p > 1$ be a prime number. We consider the polynomial $x^{p-1} - \hat{1} \in \mathfrak{C}/(p)[x]$. By the Fermat theorem all the prime residue classes, which in the present case means all the nonzero residue classes, are zeros of this polynomial; if we ask how many of them there are, the answer is given by $\varphi(p) = p - 1 = \text{degree}\,(x^{p-1} - 1)$. Thus $x^{p-1} - 1 \bmod p$ splits into linear factors:

$$x^{p-1} - 1 \equiv \prod_{\nu=1}^{p-1} (x - \nu)(\text{mod } p). \tag{27}$$

Comparison of coefficients mod p in the absolute term shows that

$$-1 \equiv \prod_{\nu=1}^{p-1} (-\nu) \equiv (-1)^{p-1} \cdot \prod_{\nu=1}^{p-1} \nu \equiv (-1)^{p-1}(p - 1)!\,(\text{mod } p). \tag{28}$$

For $p > 2$, we have $p - 1 \equiv 0(2)$, so that $(-1)^{p-1} \equiv +1$; if $p = 2$, we still have $(-1)^{p-1} \equiv +1(2)$, since $+1 \equiv -1(2)$; thus (28) gives us the following theorem:

Wilson's theorem: *If p is a prime, then $(p - 1)! \equiv -1(p)$, and conversely.*

For if n is reducible, then $n \mid (n - 1)! + 1$ is obviously impossible. But if in (27) we compare the other coefficients (the coefficients on the right are the elementary symmetric polynomials of the zeros $1, 2, ..., p - 1$) (cf. IB4, §2.4), we see that they are all $\equiv 0(p)$, which gives the desired result. It is easy to generalize the Wilson theorem to principal ideal rings and groups $\mathfrak{G}_p$ of finite order ($p \neq 0$, $\nsim 1$ and prime).

4.4. *Prime Fields, the Characteristics of a Field*

Now let us suppose that there exists a nontrivial subfield $\mathfrak{K}$ of $\mathfrak{C}/(p)$ ($p > 1$ a prime); then certainly $\{\hat{0}, \hat{1}\} \subseteq \mathfrak{K}$, and therefore

$$\hat{2} = \hat{1} + \hat{1}, \hat{3} = \hat{1} + \hat{1} + \hat{1}, ..., \widehat{p - 1} = \sum_{\nu=1}^{p-1} \hat{1}$$

are also in $\mathfrak{K}$, so that $\mathfrak{K} = \mathfrak{C}/(p)$, in contradiction to the assumption. *Thus $\mathfrak{C}/(p)$ has no nontrivial subfield.*

A field $\mathfrak{P}$ is called a prime field if it has no nontrivial subfield. If for a given field $\mathfrak{K}$ there exists a positive integer n such that

$$\sum_{\nu=1}^{n} 1 = n \cdot 1 = 0 \qquad (1 \in \mathfrak{R}, \quad n \in \mathfrak{C}),$$

polynomial ring and the statement that $f(x) = g(x)$ for all x in a given set. For example, $x^2 + 1$ and $x + 1$, regarded as polynomials in $\mathfrak{C}/(2)\,[x]$, are distinct, although $x^2 + 1$ and $x + 1$ have the same values for all $x \in \mathfrak{C}/(2)$. The identity theorem which is valid in the real (or in the rational or complex) field is not valid in general.

then the smallest such number $n = \chi(\mathfrak{R})$ *is called the characteristic of* $\mathfrak{R}$. *If there is no such positive n, then the characteristic* $\chi(\mathfrak{R}) = 0$. *The characteristic of an integral domain with unity element is defined in the same way.*

The following theorem is now obvious:

Theorem 4: $\mathfrak{C}/(p)$ *is a prime field of characteristic p. The field* P *of rational numbers is a prime field and* $\chi(\mathsf{P}) = 0$.

As was pointed out in IB5, §1.11, the fields in theorem 4 are, up to isomorphism (cf. IB5, §1.13), the only possible prime fields, since a composite characteristic would imply the existence of divisors of zero. If $\chi(\mathfrak{R}) = p \neq 0$, then for arbitrary $\alpha \in \mathfrak{R}$ we obviously have $p \cdot \alpha(=\sum_{\nu=1}^{p} \alpha) = 0$. It is also immediately obvious that the *number of elements of a prime field* $\mathfrak{P}$ *is equal to* $\chi(\mathfrak{P})$, *if* $\chi(\mathfrak{P}) \neq 0$.

In an arbitrary field of prime characteristic p it is clear, since $p \left| \binom{p}{\nu} \right.$, $\nu = 1, 2, ..., p - 1$, that the binomial theorem takes the following simple form

$$(a + b)^p = a^p + b^p. \tag{29}$$

4.5. *Tests for Divisibility*

As can be shown by the division algorithm (§2.10), for every integer $g > 1$ a natural n has the uniquely determined digital representation

$$n = \sum_{i=0}^{k} a_i g^i = a_k a_{k-1} \cdots a_1 a_0\,, \qquad k = \left[\frac{\log n}{\log g}\right], \qquad a_i \textit{ integral},$$

$$0 \leqslant a_i \leqslant g - 1 \qquad (i = 0, 1, ..., k).$$

Here $q = q(n) = \sum_{i=0}^{k} a_i$ is called the *digital sum* of n, and $a = a(n) = a_0 - a_1 + a_2 - + \cdots + (-1)^k a_k$ is called the *alternating digital sum*. If we choose the decimal representation, i.e., $g = 10$, we have the following *criteria for divisibility:*

1) $3 \mid n$ *implies* $3 \mid q$ *and conversely,*
2) $9 \mid n$ " $9 \mid q$ " "
3) $11 \mid n$ " $11 \mid a$ " "
4) $2^\lambda \mid n$ " $2^\lambda \mid a_{\lambda-1} a_{\lambda-2} \cdots a_0$ *and conversely.*

Proof: Since $10 \equiv 1(3)$, we have $10^\nu \equiv 1(3)$ for all integers $\nu \geqslant 0$. Thus $\sum_{i=0}^{k} a_i 10^i \equiv \sum_{i=0}^{k} a_i (3)$, which implies 1). In the same way $10^\nu = 1(9)$ for all integers $\nu \geqslant 0$. As for 3), we have $10 \equiv -1(11)$, $10^2 \equiv +1(11)$ so that $10^{2\nu+1} \equiv -1(11)$, $10^{2\nu} \equiv +1(11)$ for $\nu \geqslant 0$, from which it follows at once that $a(n) = n(11)$. Finally, $2 \mid 10$, i.e., $10 \equiv 0(2)$, and since a congruence remains correct if both sides and the modulus are multiplied by the same number, we obtain $5 \cdot 2^\lambda \equiv 0(2^\lambda)$, and thus also $5^\lambda \cdot 2^\lambda \equiv 10^\lambda \equiv 0(2^\lambda)$ for $\lambda \geqslant 0$, so that $a_k a_{k-1} \cdots a \cdot 10^\lambda \equiv 0(2^\lambda)$, which implies the criterion 4.

Supplementary remarks: By combining one of the first three with the fourth of the criteria just given, we can obtain other tests, e.g.: from $6 \mid n$ follow $3 \mid n$ and $2 \mid n$, and conversely, etc.

In general, the first three criteria are based on the following lemma, which can easily be proved.

Criterion: Let $(d, 10) = 1$, and let h be the order of 10 mod d. Also let $10^i \equiv g_i(d)$ for $i = 0, 1, 2, ..., h - 1$. If

$$q_h(n) = a_0 g_0 + a_1 g_1 + \cdots + a_{h-1} g_{h-1} + a_h g_0 + a_{h+1} g_1 + \cdots$$

is the *generalized digital sum*, then $d \mid n$ implies $d \mid q_h(n)$ and conversely. For $d = 11$ we have $g_0 = 1$, $g_1 = -1$, where the g_i is chosen to have the least possible absolute value.

4.6. *Periodic Decimal Fractions*

The following remarks for the base $g = 10$ are equally valid for an arbitrary integer $g > 1$. By IB1 §4.1, every real number can be expanded in a unique way as a decimal fraction

$$a_{-l} a_{-l+1} \cdots a_{-1} a_0 \, , \, a_1 a_2 \cdots$$
$$(= a_{-l} \cdot 10^l + a_{-l+1} \cdot 10^{l-1} + \cdots + a_0 + a_1 \cdot 10^{-1} + \cdots),$$
$$0 \leqslant a_\nu \leqslant 9 \quad \text{for all} \quad \nu \geqslant -l,$$

if we agree on the *normalization* that $a = 9$ for all large ν is not permitted (i.e., there does not exist an N such that $a_\nu = 9$ for all $\nu \geqslant N$). If we start from a rational number $r = a/b \geqslant 0$ $(a \geqslant 0, b > 0, \{a, b\} \subset \mathfrak{C})$ and employ the division algorithm to obtain the successive digits a_ν, the fact that only the numbers $0, 1, ..., b - 1$ can occur as nonnegative remainders means that after at most $b + 1$ divisions two or more of the remainders must be equal to each other;[26] thus the sequence of digits must be repeated from a certain position on, so that we have a periodic decimal fraction: $r = a_{-l} \cdots a_0 \, , \, a_1 a_2 \cdots a_s \overline{a_{s+1} \cdots a_{s+P}}$, where as usual the periodic part is denoted by overlining. The number P is called a period, and every multiple of P is also a period. The smallest $P \geqslant 1$ is called the *primitive period*, and in the present section the word "period" always means the primitive period. If we choose $s \geqslant 0$ as small as possible, then $0 \, . \, a_1 \cdots a_s$ is called the *nonperiodic fractional part* and s is its *length*. The digits

[26] The division 5 : 2 shows that the number of steps $b + 1$ $(=3)$ can actually assume its maximum value. In the present discussion we regard terminating decimals as having the periodic part "$\overline{0}$."

$a_{-l} \cdots a_0 . a_1 \cdots a_s$ are called the *preperiod*, and $s + l + 1$ is its *length*. Also, $0. a_1 \cdots a_s \overline{a_{s+1} \cdots a_{s+P}}$ is called the *fractional part*, which is said to be *pure periodic* if $s = 0$, but otherwise *mixed periodic*.[27] In view of the normalization, the fractional part is smaller than 1, so that $[r] = a_{-l} \cdots a_0$; here $[r]$ is called the *integral part*, $k = l + 1$ is the number of digits before the decimal point, and r is an $(l + 1)$-place number.

From $10^l \leqslant r < 10^{l+1}$ it follows that $1 \leqslant \log r/\log 10 < l + 1$, so that $k = [(\log r)/(\log 10)] + 1$. Our main result is now as follows:

A necessary and sufficient condition for the pure periodicity of (a/b) *is* $(b, 10) = 1$. *The period P is then equal to the order of* 10 mod b.

Proof: If $a/b = \overline{0 . a_1 a_2 \cdots a_p}$, we obtain

$$\frac{a}{b} = \sum_{\lambda=0}^{\infty} \left(\frac{a_1}{10^{\lambda P+1}} + \frac{a_2}{10^{\lambda P+2}} + \cdots + \frac{a_P}{10^{\lambda P+P}}\right)$$

$$= \sum_{\lambda=0}^{\infty} \frac{1}{10^{\lambda P}} \left(\frac{a_1 \cdot 10^{P-1} + a^2 \cdot 10^{P-2} + \cdots + a_P}{10^P}\right)$$

$$= \frac{A}{10^P} \sum_{\lambda=0}^{\infty} \frac{1}{10^{\lambda P}} = \frac{A}{10^P} \cdot \frac{10^P}{10^P - 1} = \frac{A}{10^P - 1} \quad (A = a_1 \cdot 10^{P-1} + \cdots + a_P).$$

Since a/b was assumed to be in its lowest terms, we have $b \mid 10^P - 1$, so that $(b, 10) = 1$, and from $10^P \equiv 1(b)$ it follows that P is a multiple of the order P' of 10 mod b. But if $P' < P$, then $10^{P'} \equiv 1(b)$, i.e., $b \mid 10^{P'} - 1$, would imply that a/b can be brought into the form $a/b = A'/(10^{P'} - 1)$, and, since $A' < 10^{P'} - 1$, the numerator A' would then have the form $A' = a_0' 10^{P'-1} + \cdots + a_{P'}'$, so that we would have $a/b = \overline{0 , a_1 a_2 \cdots a_{P'}}$, in contradiction to the assumption that P is a primitive period. Conversely, if we assume $(b, 10) = 1$, then the order of 10 mod b, call it P, is such that $b \mid 10^P - 1$. But then, as has just been shown, a/b is periodic with period P. For an arbitrary fraction a/b ($a \geqslant 0$; $b > 0$, a and b integers) the length s of the pure periodic fractional part can be determined at once from the above theorem: $s(\geqslant 0)$ is *the smallest integer such that the denominator of the fraction* $a \cdot 10^s/b$ *in its lowest terms is coprime to* 10. For then the fractional part of $a \cdot 10^s/b$ is pure periodic.

The greatest possible value of P is $b - 1$, and this value is actually assumed, for example, for $\frac{1}{7} = 0.\overline{142857}$ ($P = 6$), whereas for $\frac{1}{11} = 0.\overline{09}$ we have $P = 2 < 10\,(=b - 1)$.

[27] These terms are sometimes applied to r itself.

5. Some Number-Theoretic Functions; The Möbius Inversion Formula

A function $f(x)$ is called a number-theoretic function if it is defined on a subset of $\mathfrak{C}$, e.g., for all natural numbers. The functions $\tau(n)$, $\sigma(n)$, $\sigma_k(n)$, defined in §2.12, and also the Euler function $\varphi(n)$, are number-theoretic functions; the summatory function of $f(n)$ is a number-theoretic function if $f(n)$ is such a function. A frequently occurring number-theoretic function is the *Möbius function* $\mu(n)$, defined by

$$\mu(n) = \begin{cases} 1 \textit{ for } n = 1, \\ (-1)^\lambda, \textit{ if } n = p_1 p_2 \cdots p_\lambda \textit{ is square-free } (p_1, ..., p_\lambda \textit{ primes}), \\ 0 \textit{ otherwise.} \end{cases}$$

In general, an integer g is said to be k-free ($k \geqslant 2$ integral) if $p^k \nmid g$ for every prime number $p > 1$; for $k = 2$ the number g is also said to be *square-free*. Let us also mention the *unity function* $\epsilon(n)$ defined by

$$\epsilon(n) = \begin{cases} 1 & \textit{for} \quad n = 1 \\ 0 & \textit{for} \quad n > 1. \end{cases}$$

For $\mu(n)$ we have the following theorem.

Theorem 5: *The function* $\mu(n)$ *is multiplicative* (see §2.12), *and for the summatory function of* $\mu(n)$ *we have*

$$\sum_{d|n} \mu(d) = \epsilon(n). \tag{31}$$

Proof: If one of the numbers n_1, n_2 is not square-free, then neither is $n_1 n_2$, so that $\mu(n_1 n_2) = 0 = \mu(n_1)\,\mu(n_2)$. If both n_1, n_2 are square-free and coprime, then the multiplicativity is evident. As for $n = 1$, the formula (33) is obvious; for $n > 1$ let $n = p_1^{\alpha_1} \cdots p_s^{\alpha_s}$ be the canonical factorization; a number $d \mid n$ then has the form $d = p_1 \cdots p_s$, $0 \leqslant \kappa_i \leqslant \alpha_i$ $(i = 1, 2, ..., s)$. Thus

$$\mu(d) = \mu(p_1^{\kappa_1} \cdots p_s^{\kappa_s}) = \prod_{i=1}^{s} \mu(p_i^{\kappa_i}),$$

and therefore

$$\sum_{d|n} \mu(d) = \sum_{\kappa_1=0}^{\alpha_1} \sum_{\kappa_2=0}^{\alpha_2} \cdots \sum_{\kappa_s=0}^{\alpha_s} \prod_{i=1}^{s} \mu(p_i^{\kappa_i}) = \sum_{\kappa_1=0}^{\alpha_1} \mu(p_1^{\kappa_1}) \sum_{\kappa_2=0}^{\alpha_2} \mu(p_2^{\kappa_2}) \cdots \sum_{\kappa_s=0}^{\alpha_s} \mu(p_s^{\kappa_s})$$

$$= \prod_{i=1}^{s} \sum_{\kappa_i=0}^{\alpha_i} \mu(p_i^{\kappa_i}) = \prod_{i=1}^{s} ((\mu(1) + \mu(p_i) + \mu(p_i^2) + \cdots + \mu(p_i^{\alpha_i}))$$

$$= \prod_{i=1}^{s} (\mu(1) + \mu(p_i)) = \prod_{i=1}^{s} (1 - 1) = 0.$$

The Möbius function enables us to prove an important relation between a function and its summatory function.

Theorem 6: *Every function $F(x)$ defined on the set $\mathfrak{Z}$ of all natural numbers $n > 0$ is the summatory function of a uniquely determined function $f(x)$ defined on $\mathfrak{Z}$, if the values of the function are the elements of an Abelian group $\mathfrak{G}$ (e.g., $\mathfrak{C}^+$ or $\mathsf{P}^\times$).*

Proof: Let $\mathfrak{G}$ be a module. For every $g \in \mathfrak{G}$ we then have $\mu(n)\,g = 0$, $+g$ or $-g$, according as $\mu(n) = 0, +1$, or -1.[28] For all integers $k > 0$ we now set

$$(32) \qquad f(k) = \sum_{d|k} \mu\left(\frac{k}{d}\right) F(d) \left(= \sum_{d|k} \mu(d)\, F\left(\frac{k}{d}\right)\right);$$

thus $f(k)$ is an element of $\mathfrak{G}$. For the summatory function of $f(k)$ we have

$$\begin{aligned}\sum_{d|n} f(d) &= \sum_{d|n} f\left(\frac{n}{d}\right) = \sum_{d|n} \sum_{d'|\frac{n}{d}} \mu\left(\frac{n}{dd'}\right) F(d') = \sum_{\substack{d,d'\\ dd'|n}} \mu\left(\frac{n}{dd'}\right) F(d')\\ &= \sum_{d'|n} \sum_{d|\frac{n}{d'}} \mu\left(\frac{n}{dd'}\right) F(d') = \sum_{d'|n} F(d') \sum_{d|\frac{n}{d'}} \mu\left(\frac{n}{dd'}\right)\\ &= \sum_{d'|n} F(d')\, \epsilon\left(\frac{n}{d'}\right) = F(n).\end{aligned}$$

It remains to prove the uniqueness. Let $g(x)$ be an arbitrary function such that $\sum_{d|n} g(d) = F(n)$. It follows from (32) that

$$\begin{aligned}f(n) &= \sum_{d|n} \mu(d)\, F\left(\frac{n}{d}\right) = \sum_{d|n} \mu(d) \sum_{d'|\frac{n}{d}} g(d') = \sum_{dd'|n} \mu(d)\, g(d')\\ &= \sum_{d'|n} g(d') \sum_{d|\frac{n}{d'}} \mu(d) = \sum_{d'|n} g(d')\, \epsilon\left(\frac{n}{d'}\right) = g(n).\end{aligned}$$

Theorem 6, together with (32), is called the *Möbius inversion formula:*

From $F(n) = \sum_{d|n} f(d)$ it follows that $f(n) = \sum_{d|n} \mu(n/d)\, F(d)$ and conversely.

[28] For every module $\mathfrak{Q}$ the ring $\mathfrak{C}$ may be regarded as an operator domain; we then define

$$n\gamma = \sum_{\nu=1}^{n} \gamma, \qquad \text{if} \qquad n \geqslant 0, \quad n \in \mathfrak{C}, \quad \gamma \in \mathfrak{Q}$$

and $(-n)\gamma = -(n\gamma)(n > 0)$, which is obviously in agreement with the above text. Compare also footnote 4, §2.5.

If the operation in $\mathfrak{G}$ is written multiplicatively, the formulas are as follows:

$$F(n) = \prod_{d|n} f(d), \qquad \text{and} \qquad f(n) = \prod_{d|n} F(d)^{\mu\left(\frac{n}{d}\right)}.$$

Let us now use these formulas to prove the following theorem for the Euler φ-function:

Theorem 7: *The function $\varphi(x)$ is multiplicative, and $\sum_{d|n} \varphi(d) = n$; moreover,*

$$\varphi(n) = n \prod_{p|n} \left(1 - \frac{1}{p}\right) = \prod_{i=1}^{s} p_i^{\alpha_i - 1}(p_i - 1) \qquad \left(n = \prod_{i=1}^{s} p_i^{\alpha_i} \textit{ canonical}\right).$$

Proof: Let $\mathfrak{M}_d$ be the set of all integers x with $1 \leqslant x \leqslant n$ and $(x, n) = d$. Then obviously

$$(33) \quad \bigcup_{d=1}^{n} \mathfrak{M}_d = \bigcup_{d|n} \mathfrak{M}_d = \{1, 2, \dots, n\} \text{ and } \mathfrak{M}_{d_1} \cap \mathfrak{M}_{d_2} = 0 \text{ for } d_1 \neq d_2,$$

and $\mathfrak{M}_d$ consists of exactly those multiples λd of d for which $(\lambda, n) = 1$ and $1 \leqslant \lambda \leqslant n/d$. The $\mathfrak{M}_d$ with $d \mid n$ thus contains $\varphi(n/d)$ numbers. From (33) it follows at once that:

$$n = \sum_{d|n} \varphi\left(\frac{n}{d}\right) = \sum_{d|n} \varphi(d).$$

Thus from the Möbius inversion formula for $n = ab$, $(a, b) = 1$, we have

$$\begin{aligned}
\varphi(ab) = \varphi(n) = \sum_{d|n} \mu(d) \frac{n}{d} &= \sum_{\substack{d_1 d_2 | ab \\ (d_1, d_2) = 1}} \mu(d_1 d_2) \frac{ab}{d_1 d_2} \\
&= \sum_{d_1|a} \sum_{d_2|b} \mu(d_1)\, \mu(d_2) \frac{ab}{d_1 d_2} \\
&= \sum_{d_1|a} \mu(d_1) \frac{a}{d_1} \sum_{d_2|b} \mu(d_2) \frac{b}{d_2} = \varphi(a)\, \varphi(b).
\end{aligned}$$

In view of this multiplicativity, for $n = \prod_{i=1}^{s} p_i^{\alpha_i}$ (canonical) we have at once

$$\varphi(n) = \prod_{i=1}^{s} \varphi(p_i^{\alpha_i}).$$

The numbers that are not prime to $p_i^{\alpha_i}$ are $2p_i, 3p_i, \ldots, p_i^{\alpha_i-1}p_i$, so that

$$\varphi(p_i^{\alpha_i}) = p_i^{\alpha_i} - p_i^{\alpha_i-1} = p_i^{\alpha_i}\left(1 - \frac{1}{p_i}\right).$$

6. The Chinese Remainder Theorem; Direct Decomposition of $\mathfrak{C}/(m)$

We now seek to find *all solutions x of the simultaneous system of congruences* $x \equiv a_i(m_i)$ $(i = 1, 2, \ldots, s)$, where the m_i are pairwise coprime; in other words, every solution must satisfy s congruences simultaneously.

The Chinese remainder theorem (fundamental theorem on simultaneous congruences) reads as follows: *the set of solutions of the simultaneous system of congruences $x \equiv a_i(m_i)$ $(i = 1, 2, \ldots, s)$ with $(m_\kappa, m_\lambda) = 1$ for $\kappa \neq \lambda$ consists of all the numbers in* a *uniquely determined residue class* mod $m_1m_2 \cdots m_s$.

Proof: If x_0 is a fixed solution and x_1 an arbitrary solution, it follows that $x_0 \equiv x_1 \equiv a_i(m_i)$ for $i = 1, 2, \ldots, s$, so that $m_i \mid (x_1 - x_0)$, and in view of the pairwise coprimality we also have $m = \prod_{i=1}^{s} m_i \mid (x_1 - x_0)$, i.e., $x_1 \equiv x_0(m)$; conversely, since any congruence remains correct when the module is replaced by any of its factors, $x' \equiv x_0(m)$ implies $x' \equiv x_0(m_i)$ for all i. Thus there can be at most one residue class mod m in which all the solutions are contained, and every number belonging to such a residue class is necessarily a solution. In order to prove the existence of solutions, we set $M_i = m/m_i$ $(i = 1, 2, \ldots, s)$; then obviously $(M_1, M_2, \ldots, M_s) = 1$; thus there exist numbers $y_1, y_2, \ldots, y_s$ such that $M_1y_1 + \cdots + M_sy_s = 1$. Let us suppose that also $e_i = M_iy_i$ $(i = 1, \ldots, s)$; then for all i we have the congruences:

(34) $$e_i \equiv 0(M_i), \qquad e_i \equiv 0(m_j) \quad \text{for} \quad j \neq i,$$

and thus

(35) $$1 \equiv e_1 + \cdots + e_s \equiv e_i(m_i), \qquad \text{i.e.,} \quad e_i \equiv 1(m_i) \qquad \text{and}$$

$$\sum_{i=1}^{s} a_ie_i \equiv a_ie_i \equiv a_i(m_i),$$

so that $a_1e_1 + \cdots + a_se_s$ is a solution of the system of simultaneous congruences.

The e_i have the important property that

(36) $$e_ie_j \equiv \begin{cases} 0(m), & i \neq j, \\ e_i(m), & i = j. \end{cases}$$

Proof: From (34) it follows that $e_i e_j \equiv 0(m_k)$ for $i \neq j$ and $k = 1, ..., s$, so that we also have $m \mid e_i e_j$ $(i \neq j)$. But this is exactly the first relation (36), and thus from $1 = e_1 + \cdots + e_s$ it follows by multiplication with e_i that

$$e_i = \sum_{j=1}^{s} e_j e_i \equiv e_i^2(m).$$

An element a of an arbitrary ring $\mathfrak{R}$ such that $a^2 = a$ is called *idempotent*. Then (36) states that in $\mathfrak{C}/(m)$ *the residue classes* $\hat{e}_1, ..., \hat{e}_s$ *form a system of orthogonal idempotents (i.e., idempotents such that the product of any two of them is equal to zero).*

For an arbitrary commutative ring we define the concept of direct sum as follows.

A ring $\mathfrak{R}$ *is the direct sum of the ideals* $\mathfrak{R}_i$ *(in* $\mathfrak{R}$*)* $(i = 1, ..., s)$, *or in symbols*:

$$\mathfrak{R} = \mathfrak{R}_1 \oplus \mathfrak{R}_2 \oplus \cdots \oplus \mathfrak{R}_s = \sum_{i=1}^{s} \oplus \mathfrak{R}_i ,$$

if it has the following property:

Every $r \in \mathfrak{R}$ *can be represented in exactly one way in the form*

$$r = r_1 + r_2 + \cdots + r_s , \qquad r_i \in \mathfrak{R}_i \; (i = 1, ..., s). \tag{37}$$

For an $r \in \mathfrak{R}_i \cap \mathfrak{R}_j$ $(i \neq j)$ it follows, since $r = r + 0 = 0 + r$ cannot be two different representations, that $r = 0$, i.e., $\mathfrak{R}_i \cap \mathfrak{R}_j = \{0\}$ $(i \neq j)$. For arbitrary $r_i \in \mathfrak{R}_i$, $r_j \in \mathfrak{R}_j$ $(i \neq j)$ the fact that all the $\mathfrak{R}_k$ are ideals implies at once that $r_i r_j \in \mathfrak{R}_i \cap \mathfrak{R}_j$, so that $r_i r_j = 0$. If for $\{a, b\} \subseteq \mathfrak{R}$ we have the representations $a = r_1' + \cdots + r_s'$, $b = r_1'' + \cdots + r_s''$ in the form (37), it follows that

$$ab = \sum_{i=1}^{s} r_i' r_i'' ,$$

so that the two elements may be multiplied componentwise. Since $r_i r_j = 0$ for $r_i \in \mathfrak{R}_i$, $r_j \in \mathfrak{R}_j$, $i \neq j$, the binomial theorem is valid in the simple form:

$$(r_i + r_j)^k = r_i^k + r_j^k \qquad (k \geqslant 0 \text{ integral}). \tag{38}$$

This decomposition of a ring is analogous to the direct product decomposition of a group or the direct sum decomposition of a module (see IB2, §8). If we do not insist on uniqueness in (37), we have simply the *sum*:

$$\mathfrak{R} = \mathfrak{R}_1 + \cdots + \mathfrak{R}_s .$$

If the ring $\mathfrak{R}$ has a unity element e and if $e = e_1 + \cdots + e_s$ is the representation (37), then by squaring we obtain

$$e = e^2 = e_1^2 + \cdots + e_s^2 = e_1 + \cdots + e_s ,$$

and thus, in view of $e_i^2 \in \mathfrak{R}_i$ and the uniqueness of (37), we have the analog of (36)

$$e_i e_j = \begin{cases} 0, & i \neq j, \\ e_i , & i = j. \end{cases}$$

Conversely, if we begin with a decomposition of the unity element $e = \sum_{i=1}^{s} e_i$ into a sum of orthogonal idempotents, then for an arbitrary $r \in \mathfrak{R}$, if we denote the principal ideals (e_i) by $\mathfrak{R}_i$, we have

$$r = re = \sum_{i=1}^{s} re_i = \sum_{i=1}^{s} r_i \qquad (r_i = re_i \in \mathfrak{R}_i , \qquad i = 1, 2, ..., s)$$

and for an arbitrary representation $r = \sum_{i=1}^{s} r_i' , r_i' = r_i'' e_i \in \mathfrak{R}_i$,

$$r_j = e_j r = e_j \sum_{i=1}^{s} r_i' = r_j'' e_j = r_j' ,$$

so that $r_i = r_i'$ for all i; that is, the representation is unique, so that $\mathfrak{R} = \sum_{i=1}^{s} \oplus \mathfrak{R}_i$. Thus (36) provides a direct decomposition of $\mathfrak{C}/(m)$. Furthermore, we have the following theorem.

Theorem 8: *The residue class ring* $\mathfrak{C}/(m)$ *admits the direct decomposition* $\mathfrak{C}/(m) = (\hat{e}_1) \oplus (\hat{e}_2) \oplus \cdots \oplus (\hat{e}_s)$ $(m = m_1 m_2 \cdots m_s)$, $(m_i , m_j) = 1$ *for* $i = j$), *and for the principal ideals* (e_i) *we have the ring isomorphism* $(\hat{e}_i) \cong \mathfrak{C}/(m_i)$.

Proof: It remains only to prove the isomorphism. For an $\hat{r}_i \in (\hat{e}_i)$ the congruence $r_i \equiv a_i e_i \equiv b_i e_i (m)$ obviously implies mod m that $r_i \equiv a_i e_i \equiv a_i \equiv b_i e_i \equiv b_i (m_i)$; thus the correspondence $r_i \bmod m \to a_i \bmod m_i$ is a one-valued mapping. In this correspondence every $\hat{a}_i \in \mathfrak{C}/(m_i)$ occurs as an image, since $\widehat{a_i e_i} \in (e_i \bmod (m))$ $(= (\hat{e}_i))$ has precisely this image. Thus the number of images is equal to m_i. We now show that the principal ideal ($e_i \bmod m$) contains at most m_i residue classes, and thus, since the number of images is precisely equal to m_i, this principal ideal contains exactly m_i residue classes, so that the mapping is one-to-one. From $x \equiv z(m_i)$ it follows by the definition of e_i that $xe_i \equiv xM_i y_i \equiv ze_i \equiv zM_i y_i (m_i)$, and in view of the fact that $m/m_i = M_i$, this congruence also holds mod m, so that the principal ideal $(\hat{e}_i)$ consists of the residue classes (possibly not distinct, so far as we have shown up to now) $0, e_i , 2e_i , ..., (m_i - 1) e_i \bmod m$, which already proves our assertion.

It is easy to show that this mapping preserves addition and multiplication, i.e., the images of the sum and the product of two residue classes are equal to the sum and the product, respectively, of the images. But these properties of one-to-oneness and the preservation of addition and multiplication are exactly the properties of a ring isomorphism.

By componentwise multiplication it follows from

$$(39) \qquad r \equiv r_i (\text{mod } m_i), \qquad (r_i \text{ mod } m) \subseteq (e_i \text{ mod } m),$$

that the mapping $(r \text{ mod } m) \to (r_i \text{ mod } m)$ is a (ring) endomorphism (i.e., a homomorphism of a ring into itself), and thus the mapping $(r \text{ mod } m) \to (r_i \text{ mod } m_i)$ $(= a_i \text{ mod } m_i)$ is a homomorphism, in view of the fact that $(\hat{e}_i) \cong \mathfrak{C}/(m_i)$.

If for m we choose the canonical decomposition, so that the m_i are now prime powers, then the structure of the residue class ring $\mathfrak{C}/(m)$ is already known if we know the structure of $\mathfrak{C}/(p^\alpha)$ (p a prime), as follows immediately from the fact that not only addition (trivially), but also multiplication, as we have shown, can be carried out componentwise. A similar special case occurs, of course, for the direct product of a group (see IB2, §8), a fact which is of interest in the present context for the prime residue class group $\mathfrak{G}_m$. For we see, first of all, that if $(r, m) = 1$ and $r \equiv r_1 + r_2 \cdots + r_s(m)$ is a decomposition mod m in accordance with (37), then $r \equiv r_i(m_i)$ and $(r, m_i) = 1$, so that we also have $(r_i, m_i) = 1$. Conversely, if $(r_i, m_i) = 1$ for all $i = 1, \ldots, s$, then it follows from $r \equiv r_i(m_i)$ that $(r, m_i) = 1$ for all i; thus also $(r, m) = 1$. If we now denote by $\mathfrak{U}_m$ the set of residue classes $\hat{r} \in \mathfrak{C}/(m)$ for which

$$r \equiv e_1 + \cdots + e_{i-1} + \lambda_i e_i + e_{i+1} + \cdots + e_s (\text{mod } m), \quad (\lambda_i, m_i) = 1,$$

it follows at once by componentwise multiplication that $\mathfrak{U}_m^{(i)}$ is a subgroup[29] in $\mathfrak{G}_m$. Thus we have the following theorem:

Theorem 9: *$\mathfrak{G}_m$ is the direct product of $\mathfrak{U}_m^{(i)}$:*

$$(40) \qquad \mathfrak{G}_m = \mathfrak{U}_m^{(1)} \times \mathfrak{U}_m^{(2)} \times \cdots \times \mathfrak{U}_m^{(s)} = \prod_{i=1}^{s} \mathfrak{U}_m^{(i)};$$

and we have the group isomorphism $\mathfrak{U}_m^{(i)} - \mathfrak{G}_{m_i}$.

Proof: Since

$$(e_1 + \cdots + e_{i-1} + \lambda_i e_i + e_{i+1} + \cdots + e_s)(e_1 + \cdots + e_{i-1} + \mu_i e_i + e_{i+1} + \cdots + e_s) \equiv e_1 + \cdots + e_{i-1} + \lambda_i \mu_i e_i + e_{i+1} + \cdots + e_s \; (\text{mod } m) \equiv \lambda_i \mu_i e_i \equiv \lambda_i \mu_i \; (\text{mod } m_i),$$

[29] In order to prove that a subset $\mathfrak{T}$ of a finite group $\mathfrak{G}$ is a subgroup it is sufficient to show that the product of every two elements in $\mathfrak{T}$ belongs to $\mathfrak{T}$; and the $\mathfrak{G}_m$ in the text is certainly finite.

the mapping

$$(e_1 + \cdots + e_{i-1} + \lambda_i e_i + e_{i+1} + \cdots + e_s) \quad \bmod m \to \lambda_i \bmod m_i$$

preserves addition and multiplication and is therefore a homomorphism of $\mathfrak{U}_m^{(i)}$ onto[30] $\mathfrak{G}_{m_i}$; thus $\mathfrak{U}_m^{(i)}$ has at least $\varphi(m_i)$ $(= |\mathfrak{G}_{m_i}|)$ elements. The one-to-oneness follows from the above result that, as in the proof of theorem 8, $(r, m) = 1$ is equivalent to $(r_j, m_j) = 1$ for all $j = 1, 2, \ldots, s$ (in the present case $r_j = 1$ for $j \neq i$, $r_i = \lambda_i$), and thus the homomorphism in question here is actually an isomorphism. By componentwise multiplication we further see that

$$\sum_{i=1}^{s} \lambda_i e_i \equiv \prod_{i=1}^{s} (e_1 + \cdots + e_{i-1} + \lambda_i e_i + e_{i+1} + \cdots + e_s) (\bmod m),$$

where the factors on the right-hand side belong successively to $\mathfrak{U}_m^{(1)}, \ldots, \mathfrak{U}_m^{(s)}$, so that $\mathfrak{G}_m = \mathfrak{U}_m^{(1)} \cdot \mathfrak{U}_m^{(2)} \cdots \mathfrak{U}_m^{(s)}$. Since $\varphi(m) = \prod_{i=1}^{s} \varphi(m_i) = |\mathfrak{G}_m| = \prod_{i=1}^{s} |\mathfrak{U}_m^{(i)}|$, the representation of the elements of $\mathfrak{G}_m$ as products of elements in $\mathfrak{U}_m^{(1)}, \mathfrak{U}_m^{(2)}, \ldots, \mathfrak{U}_m^{(s)}$ must be unique, so that the product decomposition of $\mathfrak{G}_m$ is direct.

If we again choose for m the canonical decomposition, we see that the structure of the groups $\mathfrak{G}_m$ is known if we know the structure of the prime residue classes for moduli which are powers of primes. Regarding the structure of these groups (see, e.g., Hasse [3], [4], Scholz-Schöneberg [1]) we have the following theorem:

Theorem 10: *For all primes* $p > 2$ *and all* $\lambda \geqslant 1$ *the groups* $\mathfrak{G}_{p^\lambda}$ *are cyclic of order* $\varphi(p^\lambda) = p^{\lambda-1}(p-1)$, *and the same statement is true (though trivial) for* $\mathfrak{G}_2$ *and* $\mathfrak{G}_{2^2}$. *For* $\lambda \geqslant 3$ *the group* $\mathfrak{G}_{2^\lambda}$ *is the direct product of two cyclic groups, one of which is always of order* 2; *e.g.,* $\mathfrak{G}_{2^\lambda} = (-\hat{1}) \times (\hat{5})$.

It is also easy to show that *every finite cyclic group can be represented as the direct product of cyclic groups of prime power order*. This fact, taken together with theorem 10, shows that in theorem 9 we have a decomposition of the group $\mathfrak{G}_m$ in the sense of the fundamental theorem for finite Abelian groups (see IB2, §9.2).

7. Diophantine Equations; Algebraic Congruences

7.1. If $y = f(x_1, x_2, \ldots, x_n)$ is an arbitrary function which is defined at least for all integers x_i, then $f(x_1, x_2, \ldots, x_n) = 0$ is called a *Diophantine*

[30] A mapping "onto" means that every element of $\mathfrak{G}_{m_i}$ appears as an image; in the present case this property, together with the uniqueness of the correspondence, is proved in the same way as in theorem 8.

equation, provided we are seeking *all integral solutions* $(x_1, ..., x_n)$. Closely related is the concept of an *algebraic congruence*, by which we mean the problem of determining all (integral) solutions of the congruences $g(x_1, x_2, ..., x_n) \equiv 0(m)$, where $g(x_1, x_2, ..., x_n)$ is a polynomial in the indeterminates $x_1, x_2, ..., x_n$ with integral coefficients. This congruence is obviously equivalent to the Diophantine equation $g(x_1, x_2, ..., x_n) + ym = 0$ in the unknowns $x_1, x_2, ..., x_n, y$. Since congruences can be added, subtracted, and multiplied, or in other words since $\mathfrak{C}/(m)$ is a ring, it follows from $x_i \equiv y_i(m)$ $(i = 1, ..., n)$ that $g(x_1, x_2, ..., x_n) \equiv g(y_1, ..., y_n)(m)$, so that every solution of an algebraic congruence (also called a *root* of the congruence) is an n-tuple of residue classes mod m. The definition of *systems of Diophantine equations* or *algebraic congruences* is similar. If $\mathfrak{L}(f_1, ..., f_k; m) = l(m)$ is the number of solutions of the system of congruences $f_i(x_1, ..., x_n) \equiv 0(m)$ $(i = 1, ..., k)$, it is obvious, from the trivial estimate $0 \leqslant l(m) \leqslant m^n$, that $l(m)$ exists. Furthermore, for fixed $f_i(x_1, ..., x_n)$ $(i = 1, ..., k)$ we have the following theorem:

Theorem 11: *$l(m)$ is a multiplicative function.*

Proof: Let $m = m_1 m_2$, $(m_1, m_2) = 1$, and let $\mathfrak{C}/(m) = \mathfrak{R}_1 \oplus \mathfrak{R}_2$ be the corresponding representation as a direct sum, so that $\mathfrak{R}_i \cong \mathfrak{C}/(m_i)$ $(i = 1, 2)$. If we decompose a solution $x_{10}, ..., x_{n0} (\text{mod } m)$ in the sense of (37): $x_{i0} \equiv x'_{i0} + x''_{i0}(m)$, $(\widehat{x'_{i0}}) \in \mathfrak{R}_1$, $(\widehat{x''_{i0}}) \in \mathfrak{R}_2)$ $(i = 1, ..., n)$, we obtain

$$f_i(x_{10}, ..., x_{n0}) \equiv f_i(x'_{10} + x''_{10}, \cdots) \ (\text{mod } m) \equiv \begin{cases} f_i(x'_{10}, ...) \equiv 0(\text{mod } m_1), \\ f_i(x''_{10}, ...) \equiv 0(\text{mod } m_2). \end{cases}$$

Conversely, if we start from solutions $x'_{10}, ...$ of $f_i \equiv 0(m_1)$ and $x''_{10}, ...$ of $f_i \equiv 0(m_2)$, $i = 1, 2, ..., m$, we see that $x'_{10}e_1 + x''_{10}e_2, ...$ is a solution of the system $f_i \equiv 0(m)$, since it follows from (34) and (35) that

$$f_i(x'_{10}e_1 + x''_{10}e_2, ...) \equiv \begin{cases} f_i(x'_{10}, ...) \equiv 0(m_1) \\ f_i(x''_{10}, ...) \equiv 0(m_2) \end{cases} \qquad (i = 1, 2, ..., n),$$

so that from $(m_1, m_2) = 1$ we have

$$m \mid f_i(x'_{10}e_1 + x''_{10}e_2, ...).$$

Summing up, we see that the solutions mod m are in one-to-one correspondence with the pairs of solutions mod m_1, mod m_2, so that $l(m) = l(m_1)\, l(m_2)$.

Thus it is sufficient to determine $l(p^\lambda)$ for primes p. For a polynomial $f(x)$ in one indeterminate, it is well known, since $\mathfrak{C}/(p)$ is a field, that $1(p) \leqslant \text{degree } f(x)$ (see IB4, §1.2). For *normalized polynomials* $f(x) = \sum_{\nu=0}^{n} a_\nu x^{n-\nu}$, i.e., with $a_0 = 1$, we have the general theorem: *if*

$f(x) = x^n + a_1x^{n-1} + \cdots + a_n \equiv 0(p)$ has no multiple zeros, then neither has $f(x) \equiv 0(p^\lambda)$, $(\lambda \geqslant 1)$, and $l(p) = l(p^\lambda)$; for $f(x) \equiv 0(m)$ $(m = \prod_{i=1}^{s} p_i^{\lambda_i}$ canonical) we thus have $l(m) = \prod_{i=1}^{s} l(p_i^{\lambda_i}) \leqslant n^s$ (see Scholz-Schöneberg [1]). The equality is possible.

7.2. The fact that the *linear congruence* $ax \equiv b(m)$ with $(a, m) = 1$ is uniquely solvable was already proved in §4.2. If $(a, m) = d$, then $d \mid b$ is obviously necessary for solvability and if we divide the result by d, producing the congruence $a/d \equiv b/d(m/d)$, we can easily show that the entire set of solutions consists of exactly d distinct residue classes. Thus it remains only to find a particular solution x_0 of $ax \equiv b(m)$ with $(a, m) = 1$. But $x_0 = ba^{\varphi(m)-1}$ is a solution, by the Fermat theorem. Another method of finding such a solution is as follows. By the Euclidean algorithm we calculate x_1, y_1 such that $(a, m) = 1 = ax_1 + my_1$. Then it is obvious that x_1b is also a solution. By §3.1 the Euclidean algorithm is closely connected with the expansion into a continued fraction; let $a/m = [b_0, b_1, ..., b_n] = A_n/B_n$; then certainly $a = A_n$, $m = B_n$ in view of the fact that $(a, m) = (A_n, B_n) = 1$, and thus §3.2 (21) gives for $\nu = n$ the result that $a = A_n$, $m = B_n$

$$A_nB_{n-1} - B_nA_{n-1} = (-1)^{n-1} = aB_{n-1} - mA_{n-1},$$

so that $x_1 = (-1)^{n-1} B_{n-1}$, $y_1 = (-1)^n A_{n-1}$ is a solution of $ax_1 + my_1 = 1$.

Since the congruence $ax \equiv b(m)$ is equivalent to the Diophantine equation $ax + my = b$, we see at once that for $(a, b) = 1$ the entire set of solutions can be obtained in the form $x = x_0 + \lambda m$, $y = y_0 - \lambda a$, as soon as we have a *particular solution* x_0, y_0, and that x_0 can be determined in either of the two ways described above.

7.3. A quadratic Diophantine equation that occurs in many mathematical contexts is the *Pell equation* $x^2 - dy^2 = h$ (see Weber [1]), especially with $h = 1$ and $h = 4$. If $d = a^2$, where a is an integer, and we set $ay = z$, we arrive at the easily solved equation $x^2 - z^2 = h = (x - z)(x + z)$. For if $h = h_1h_2$ is an arbitrary factorization of h, then from $x - z = h_1$, $x + z = h_2$ we obtain at once all possible solutions; thus if h_1, h_2 are either both even or both odd we already have all the integral solutions, and if $2 \mid h$ but $4 \nmid h$, then the problem is insoluble.

In general, it is easy to see that we may confine our attention to square-free d and coprime solutions x, y; if h is square-free, it is obvious that only coprime solutions exist. Then we can state an interesting connection with the theory of (regular) continued fractions (see Perron [2]):

For every solution $x > 0$, $y > 0$, $(x, y) = 1$ of $x^2 - dy^2 = h$ with $0 < |h| < |\sqrt{d}|$, $d > 1$ square-free, the fraction x/y is a partial quotient of the (periodic) continued fraction expansion of $\sqrt{d}$, and there exist infinitely *many solutions. If* $h = 1$ *and* $\sqrt{d} = [b_0, \overline{b_1, \ldots, b_s = 2b_0}]$,[31] *then from* $x/y = A_{\rho s-1}/B_{\rho s-1}$ *we obtain the entire set of coprime solutions by letting* ρ *run through all the integers* 0, 1, 2, ..., *for even s and through all the even integers* 0, 2, 4, ... *for odd s; moreover, the partial quotients satisfy the* (*Lagrange*) *relation*

$$A_{\rho s-1} + B_{\rho s-1}\sqrt{d} = (A_{s-1} + B_{s-1}\sqrt{d})^{\rho},$$

from which by expanding the right-hand side we obtain formulas for $A_{\rho s-1}$, $B_{\rho s-1}$ in terms of the *smallest positive* solutions A_{s-1}, B_{s-1}.

7.4. The question of the solvability of the Diophantine equation $x^n + y^n = z^n$, $n > 2$ has remained unanswered up to the present day, although the famous *Fermat conjecture* states that it is unsolvable. Clearly, we may consider only $(x, y, z) = 1$ and prime exponents $n = p > 2$, and it has become customary to divide the problem into the two cases; first case: $p \nmid xyz$; second case: $p \mid xyz$. In the first case the unsolvability is known for all $p < 253747889$, and with the help of electronic computers the second case has been settled up to $p < 4003$ (for some further remarks see §8.3).

For $n = 2$ it is obvious that

$$x = \lambda(u^2 - v^2), \quad y = 2\lambda uv, \quad z = \lambda(u^2 + v^2); \quad u, v, \lambda \text{ integers}, \tag{41}$$

namely, the so-called *Pythagorean triples*, are solutions of $x^2 + y^2 = z^2$ and it can be shown in various ways that they are the only solutions, e.g., as follows. If we set $\xi = x/z$, $\eta = y/z$, we obtain the equation of the unit circle $\xi^2 + \eta^2 = 1$, for which the parameter representation

$$\xi = \cos\varphi = \frac{1 - t^2}{1 + t^2}, \qquad \eta = \sin\varphi = \frac{2t}{1 + t^2}, \qquad \left(t = tg\,\frac{\varphi}{2}\right)^{32}$$

sets up a one-to-one correspondence, since $t = \eta/1 + \xi$, between rational t and the rational points of the circumference (i.e., points with rational coordinates). If we now introduce the homogeneous parameters: $t = v/u$ (u, v integers), we have

$$\frac{x}{z} = \xi = \frac{u^2 - v^2}{u^2 + v^2}; \qquad \frac{y}{z} = \eta = \frac{2uv}{u^2 + v^2}. \tag{42}$$

[31] It can be shown that for a square-free d the continued fraction expansion of $\sqrt{d}$ necessarily has the above form.

[32] This representation of the trigonometric functions as rational functions of the tangent of the half angle is useful in a well-known way for the integration of $f(\sin x, \cos x, \text{tg}\, x, \text{ctg}\, x)$ with rational f (half angle method).

From now on we may confine our attention to coprime x, y, z. Then, in view of the fact that $(2\gamma)^2 = (2\alpha + 1)^2 + (2\beta + 1)^2 \equiv 2(4)$, exactly one of the two integers x, y must be even, and without loss of generality we may say that y is even; furthermore, we may assume that $(u, v) = 1$, since otherwise we could cancel $(u, v)^2$. For a common prime divisor p of $u^2 - v^2$ and $u^2 + v^2$ we have $p \mid u^2 - v^2 \pm (u^2 + v^2)$, and thus $p \mid 2u$, $p \mid 2v$, so that from $(u, v) = 1$ it follows that $p = 2$; i.e., u, v are both odd, since otherwise $u^2 \pm v^2$ could not be even in view of the fact that $(u, v) = 1$. If we now cancel 2 from $2uv/(u^2 + v^2)$, the denominator of the reduced fraction must be odd, whereas we must have $y \equiv 0(2)$. Thus a prime divisor p of this sort cannot exist, so that in (42) we must have $x = u^2 - v^2$, $y = 2uv$, $z = u^2 + v^2$, as desired.

Parameter prepresentations other than (41) are obtained by unimodular transformations of u, v,

$$\begin{matrix} u = \alpha u' + \beta v' \\ v = \gamma u' + \delta v' \end{matrix}; \qquad \begin{vmatrix} \alpha & \beta \\ \gamma & \delta \end{vmatrix} = \pm 1; \qquad \alpha, \beta, \gamma, \delta \textit{ integral},^{33}$$

and only from such transformations, since only then is it true that the u', v' can also be expressed as integral linear forms in the u, v.

7.5. Of particular importance among the algebraic congruences are the *pure congruences* $x^n \equiv a(m)$, since the solution of such a congruence is equivalent to the determination of the nth roots in $\mathfrak{C}/(m)$. If $(a, m) = d = d_1^n d_2$ with n-free d_2, it is easy to show by setting $x = d_1 y$ that we may restrict our attention to the case $(a, m) = 1$. If $x^n \equiv a(m)$, $(a, m) = 1$ is solvable, a is called an nth *power residue*, and otherwise an nth *power nonresidue;* in particular, for $n = 2$ we speak of *quadratic residues*, denoted by QR, and *quadratic nonresidues*, denoted by NR.

For the pure congruences with $(a, m) = 1$ it is obvious that $(x_0, m) = 1$ for every solution x_0; thus by the Fermat theorem n can always be reduced modulo the order $\varphi(m)$ of the prime residue class group $\mathfrak{G}_m$, so that we may restrict attention to $1 \leqslant n \leqslant \varphi(m)$. Then $x^{\varphi(m)} - 1 \equiv 0(m)$ has exactly all the $\varphi(m)$ prime residue classes as its solutions. If $\mathfrak{G}_m$ is cyclic and g is a generating element (see §6, theorem 10), then every generating element of $\mathfrak{G}_m$ is a *primitive root* of the congruence. In g^ρ with $(\rho, \varphi(m)) = 1$ we obtain, as is easily shown, all the primitive roots of the congruence and the number of them is obviously equal to $\varphi(\varphi(m))$.

For cyclic $\mathfrak{G}_m$ it is obvious that the exponential congruence $b^x \equiv a(m)$, $(a, m) = (b, m) = 1$ is always solvable and the solution is unique mod $\varphi(m)$. In analogy with logarithms, we write $x \bmod (m) = \underset{b}{\text{ind}}\, a$ (to be read: *index of a*

[33] A matrix over an arbitrary ring $\mathfrak{R}$, $1 \in \mathfrak{R}$, is called *unimodular* if its determinant is a unit, so that in $\mathfrak{C}$ the determinant must be equal to ± 1.

to the base b). It is easy to show that the rules for calculation are the same as for logarithms.

For quadratic congruences $x^2 \equiv a(p)$, $p > 2$ a prime, $(a, p) = 1$, we define the Legendre symbol (a/p) (read: a by p) by setting

$$\left(\frac{a}{p}\right) = \begin{cases} 1, & \text{if } a \text{ is a } QR, \\ -1, & \text{if } a \text{ is an } NR. \end{cases}$$

If g is a primitive root of the congruence mod p, then $1 \equiv g^{p-1}(p)$, so that $(g^{(p-1)/2} - 1)(g^{(p-1)/2} + 1) \equiv 0(p)$, but $g^{(p-1)/2} \not\equiv 1(p)$, so that $g^{(p-1)/2} + 1 \equiv 0(p)$. Since $\mathfrak{G}_p$ is cyclic (see §6, theorem 10), there exists a λ such that $g^\lambda \equiv a(p)$. Thus we have

$$a^{\frac{p-1}{2}} \equiv g^{\lambda\frac{p-1}{2}} \equiv (-1)^\lambda(p).$$

For $\lambda = 2\mu$ it follows that $a \equiv g^\lambda \equiv (g^\mu)^2\,(p)$, i.e., $(a/p) = 1$ and $a^{(p-1)/2} \equiv 1(p)$. For odd λ we have $a^{(p-1)/2} \equiv -1(p)$, and $a \equiv g^{2\mu+1}(p)$ is an NR for the following reason. From $x_0{}^2 \equiv g^{2\mu+1}(p)$ it follows, if we set $x_0 \equiv g^\sigma(p)$, that $g^{2\sigma} \equiv g^{2\mu+1}(p)$ so that $2\sigma \equiv 2\mu + 1$ (mod $\varphi(p) = p - 1$), and thus it would follow from $2(p - 1)$ that $2 \mid (2 + 1)$, which is impossible. Summing up, we have the *Euler criterion:*

$$a^{\frac{p-1}{2}} \equiv \left(\frac{a}{p}\right) \ (\text{mod } p).$$

Since for even λ we obtain the QR, and for odd λ the NR, it follows that *there are exactly as many QR as NR.*

For $p > 2$ and $q > 2$ prime, we have the Gauss law of *quadratic reciprocity* (see e.g., Hasse [3], [4], Scholz-Schöneberg [1]):

$$\left(\frac{p}{q}\right) = (-1)^{\frac{p-1}{2}\cdot\frac{\lambda-1}{2}}\left(\frac{q}{p}\right).$$

Since (a/p) is *distributive* (*over* $\mathfrak{G}_m$), we see from the Euler criterion that as soon as we know the values of the symbols $(-1/p)$ and $(2/p)$ we can use the reciprocity law to reduce either of the following two questions to the other: "which p are QR mod q?" and "for which moduli q is p a QR?".

It is also convenient to introduce the *Jacobi symbol*

$$\left(\frac{a}{m}\right) = \prod_{i=1}^{s}\left(\frac{a}{p_i}\right)^{\lambda_i}, \qquad m = \prod_{i=1}^{s} p^{\lambda_i} \textit{ canonical}, \qquad (2, m) = 1;$$

then we have the *generalized law of reciprocity:*

(43) $$\left(\frac{a}{b}\right) = (-1)^{\frac{a-1}{2}\cdot\frac{b-1}{2}}\left(\frac{b}{a}\right), \qquad \textit{if } (ab, 2) = 1, \ a > 0, \ b > 0.$$

The definition (43) enables us to calculate the Legendre symbols in a practical way, since it is easily seen that the numerators in the symbols may be altered at will modulo the denominators (for the further theory see, e.g., Hasse [3], [4]).

8. Algebraic Numbers

8.1. Let $f(x)$ be an irreducible polynomial (see §2.6) in $\mathsf{P}[x]$ (where P is the field of rational numbers). Then in the field $\mathfrak{K}$ of complex numbers, $f(x)$ has exactly n zeros ϑ_i ($i = 1, ..., n$), where $n =$ degree $f(x)$, provided that every zero is counted according to its multiplicity ($=$ exponent of $x - \vartheta_i$ in the canonical factorization with respect to $\mathfrak{K}[x]$). Since the GCD$(f(x), f'(x)) \in \mathsf{P}[x]$, it follows from the irreducibility of f that $(f, f') = 1$, so that multiple zeros are impossible. If $f(x)$ is reducible and $f(\vartheta_i) = 0$ ($\vartheta_i \in \mathfrak{K}$), then ϑ_i is also a zero of an irreducible factor of $f(x)$. Thus it is sufficient to consider only the zeros of irreducible polynomials; they are called *algebraic numbers;* the nonalgebraic numbers in $\mathfrak{K}$ are called *transcendental.* If for $f(x) = \sum_{i=0}^{n} a_i x^i$ we set $h = n + \sum_{i=0}^{n} | a_i |$, then for a given h there exist only finitely many $f(x)$ with integral a_i (which is obviously no restriction of generality) and thus only finitely many algebraic numbers. Thus for $h = 1, 2, ...$ we obtain *all algebraic numbers in a countable sequence.* Since $\mathfrak{K}$ itself is not countable (see IB1, §4.8), it follows that there exist uncountably many transcendental numbers (for further details see III 13, §2).

If ϑ is a zero of the irreducible polynomial $f(x)$ with degree $f(x) = n$, then $f(x)$ is called the *defining polynomial* of ϑ, and ϑ is an *algebraic number of nth degree.*[34] It is easy to see that the defining polynomial is uniquely determined up to associates. If $f(x)$ is a normalized integral polynomial (i.e., all coefficients a_i are rational integers with $a_n = 1$), then the zeros are called *algebraic integers*, or simply *integers*.

Let $\mathsf{P}(\vartheta)$, where ϑ is algebraic, denote the intersection of all those extension fields of P in $\mathfrak{K}$ that contain ϑ.[35] Then $\mathsf{P}(\vartheta)$ is said to be an algebraic number field of degree n, if degree $\vartheta = n$. For example, $\mathsf{P}(\sqrt{2})$ consists of all numbers $\alpha + \beta\sqrt{2}$, $\{\alpha, \beta\} \subset \mathsf{P}$; and the numbers $a + b\sqrt{2}$, $\{a, b\} \subset \mathfrak{C}$, are the algebraic integers in $\mathsf{P}(\sqrt{2})$. In §2.10 this set of integers was given as an example of a (Euclidean) ring. In IB7, §2 we will prove in general that every number in $\mathsf{P}(\vartheta)$ is algebraic of not more than the nth degree and that $\mathsf{P}(\vartheta)$ is identical with the totality of all numbers of the form $\sum_{i=0}^{n-1} \rho_i \vartheta^i$ (all $\rho_i \in \mathsf{P}$). It is even true that this

[34] For the interesting continued fraction expansion of the algebraic numbers of second degree see §3.3.

[35] Thus $\mathsf{P}(\vartheta)$ is the smallest subfield of $\mathfrak{K}$ that contains P and ϑ.

representation is unique, since from $\sum_{i=0}^{n-1} \rho_i \vartheta^i = \sum_{i=0}^{n-1} \mu_i \vartheta^i$ it would otherwise follow that ϑ is a zero of $\sum_{i=0}^{n-1} (\rho_i - \mu_i) x^i$ and consequently of degree smaller than n. Thus in the sense of IB3, §1.3 the ring $\mathrm{P}(\vartheta)$ is an n-dimensional vector space with $\{1, \vartheta, \vartheta^2, ..., \vartheta^{n-1}\}$ as basis, and we can pass to other bases by linear transformations with rational coefficients and nonzero determinant. If $f(x) = \sum_{i=0}^{n} a_i x^i = 0$ is the defining equation of ϑ, then $\eta = a_n \vartheta$ is an algebraic integer, since it is a zero of the normalized polynomial $a_n^{n-1} f(x) = y^n + a_n a_{n-1} y^{n-1} + \cdots$ $(y = a_n x)$, so that, besides the rational integers, $\mathrm{P}(\vartheta)$ even contains algebraic integers of nth degree, and it also follows from $\vartheta = \eta / a_n$ that *every algebraic number can be represented as the quotient of two algebraic integers, where the denominator is an arbitrary rational integer*. Then it can be shown (see Hasse [3], van der Waerden [3]), that *the integers in* $\mathrm{P}(\vartheta)$ *form a ring*. For the theory of numbers in algebraic number fields the following theorem is of fundamental importance. *In* $\mathrm{P}(\vartheta)$ *there exist bases which consist entirely of integers and have the property that every integer can be represented as a linear combination with rational integers as coefficients; and conversely, every such linear combination is an algebraic integer.* Such a basis is called an *integer basis*. If $\omega_1, \omega_2, ..., \omega_n$ is an integer basis and if $\omega_i, \omega_i', \omega_i'', ..., \omega_i^{(n-1)}$ $(i = 1, ..., n)$ are the systems of numbers[36] conjugate to the ω_i, the determinant

$$\Delta = \begin{vmatrix} \omega_1 \omega_1', & ..., & \omega_1^{(n-1)} \\ \omega_2 \omega_2', & ..., & \omega_2^{(n-1)} \\ \vdots & & \vdots \\ \omega_n \omega_n', & ..., & \omega_n^{(n-1)} \end{vmatrix}^2$$

is called the *(field) discriminant of* $\mathrm{P}(\vartheta)$. Since integer bases can be obtained from one another only by unimodular transformations, the value of d, as follows from the rule for multiplication of determinants, has the same value for all integer bases and is thus a *field invariant:* the value of Δ is a rational integer, as can be shown from the developments of IB7, §7.

The existence of an integer basis for the ring of integers in $\mathrm{P}(\vartheta)$ shows that we are dealing here with a vector space, in symbols $\mathfrak{C}[\omega_1, \omega_2, ..., \omega_n]$. Since we have also defined multiplication for "vectors," the $\mathrm{P}(\vartheta)$ and $\mathfrak{C}[\omega_1, \omega_2, ..., \omega_n]$ are *algebras*,[37] and $\mathrm{P}(\vartheta)$ is also a division algebra.

[36] By a *system of conjugates* we mean the zeros of the defining polynomial, if its degree is equal to the degree of the field. If $\alpha \in \mathrm{P}(\vartheta)$ is of smaller degree, say k, it can be proved (see IB7, §6) that $k \mid n$, and then the system of conjugates is obtained by writing each zero n/k times.

[37] By an *algebra* we mean a vector space for which there has also been defined an associative and distributive multiplication; thus an algebra is a ring. It is clear that the multiplication is fully determined if we can write the products of the basis vectors as

Concerning the zeros of a polynomial with algebraic numbers as coefficients, it can be shown (by the fundamental theorem on symmetric polynomials) that there exist polynomials with rational integers as coefficients which have at least the same zeros. Furthermore, the zeros of a normalized polynomial with (algebraic) integers for its coefficients are again integers.

The product $\alpha\alpha' \cdots \alpha^{(n-1)}$ or, what is obviously the same, the absolute term multiplied by $(-1)^n$, of the normalized defining polynomial of α, is called the *norm of* α and is denoted by $N(\alpha)$. If α is an algebraic integer, then obviously $N(\alpha)$ is a rational integer (see also IB8, §1.2).

8.2. As is shown by the example $\mathsf{P}(\sqrt{-5})$ in §2.10, the integers in $\mathsf{P}(\sqrt{-5})$ do not from a u.f. ring. It was one of the greatest advances in the theory of numbers that Kummer found a substitute for the u.f. theorem by introducing the so-called ideal numbers. As an equivalent concept, Dedekind introduced the ideals defined in IB5, §3.1 (see also §2.5). But before we can formulate the theorem which takes the place of the u.f. theorem, we must make some preliminary remarks.

By the *product of the ideals* $\mathfrak{a}$ and $\mathfrak{b}$ of a ring $\mathfrak{R}$ we mean the ideal generated by all the products ab ($a \in \mathfrak{a}$, $b \in \mathfrak{b}$) or, in other words, the set of all finite sums $\sum_i a_i b_i$ ($a_i \in \mathfrak{a}$, $b_i \in \mathfrak{b}$). As in IB5, §3.6, an ideal $\mathfrak{p}$ is called a *prime* ideal if $\mathfrak{R}/\mathfrak{p}$ has no divisors of zero; if the prime ideal is a principal ideal, $\mathfrak{p} = (p)$, then p is called a *prime element*. Prime elements are irreducible as is easily shown (cf. IB5, §2.3), but the converse is not necessarily true; for example, in the ring $\mathfrak{G}$ of all even numbers, 30 is irreducible but not prime, since $6 \cdot 10 \equiv 0 \pmod{30}$. On the other hand, in a u.f. ring the irreducible elements are also prime, which is in fact the fundamental lemma of the theory of divisibility (see §2.6). Thus our terminology is in agreement with that of §2.6 and IB5, §2.6. Furthermore, calling an ideal $\mathfrak{a}$ *maximal* if no ideal $\mathfrak{b}$ exists with $\mathfrak{a} \subset \mathfrak{b} \subset \mathfrak{R}$ $(=(1))$, we have the theorem that *every maximal ideal is prime* (see van der Waerden [2]); but then again the converse is not necessarily true. However, *in the number rings* $\mathfrak{C}[\omega_1, ..., \omega_n]$ *this converse does hold for prime ideals* $\mathfrak{p}$ *that are distinct from the zero ideal and the unit ideal.*

If we now consider all numbers of the form $a + b\sqrt{-3}$, $\{a, b\} \subset \mathfrak{C}$, i.e., the ring $\mathfrak{C}(\sqrt{-3})$, we do not obtain all the integers in $\mathsf{P}(\sqrt{-3})$; for example, the zero $\frac{1}{2} + \frac{1}{2}\sqrt{-3}$ of $x^2 + x + 1$ is an integer but is not contained in $\mathfrak{C}[\sqrt{-3}]$. A ring $\mathfrak{G}$ of integers that is contained in $\mathsf{P}(\vartheta)$ is said to be *integrally closed* in its quotient field $\mathsf{P}(\vartheta)$ if all the zeros of a normalized polynomial (with coefficients in $\mathfrak{G}$) that are contained in $\mathsf{P}(\vartheta)$

---

linear combinations of these basis vectors: $w_i w_j = \sum_{\kappa=1}^{n} c_{ij\kappa} w_\kappa$; that is, if we know the *structure constants* $c_{ij\kappa}$ (cf. IB5, §3.9). If an algebra is a field, it is called a *division algebra* (cf. IB8, §3.4).

are already contained in $\mathfrak{G}$ itself.[38] Then by definition $\mathfrak{C}[\omega_1, \omega_2, ..., \omega_n]$ is integrally closed in $\mathsf{P}(\vartheta)$. Finally, it is of especial importance that in $\mathfrak{C}[\omega_1, ..., \omega_n]$ the maximal condition for ideals (see §2.9) is satisfied.

In a ring $\mathfrak{R}$ we say that the theorem of unique factorization into prime ideals, the u.f.p.i. theorem, is valid if every ideal can be represented in a unique way (apart from the order of the factors) as the product of powers of prime ideals; an integral domain is called a u.f.p.i. ring if the u.f.p.i. theorem holds and if $\mathfrak{a} \subseteq \mathfrak{b}$ ($\mathfrak{a}, \mathfrak{b}$ ideals) implies the existence of an ideal $\mathfrak{c}$ with $\mathfrak{a} = \mathfrak{b}\mathfrak{c}$. We then have the following fundamental theorem (see van der Waerden [3]).

An integral domain is a u.f.p.i. ring if and only if the following conditions are satisfied.

I. $\mathfrak{R}$ *is integrally closed in its quotient field.*[39]

II. *In* $\mathfrak{R}$ *the maximal condition* (see §2.9) *is satisfied.*

III. *Every prime ideal* $\mathfrak{p} \neq (0), \neq \mathfrak{R}$ *is maximal.*

Consequently, all rings $\mathfrak{C}[\omega_1, ..., \omega_n]$, (where $\omega_1, ..., \omega_n$ is an integral basis) are u.f.p.i. rings. It is not true that every u.f.p.i. ring is a u.f. ring, as was shown by the example of the ring $\mathfrak{C}[\sqrt{-5}]$ at the beginning of §8.2; and conversely, not every u.f. ring is a u.f.p.i. ring, as is shown by the example of the polynomial ring $\mathfrak{K}[x, y]$, where $\mathfrak{K}$ is a field, since in this field the prime ideal (x) is properly contained in the prime ideal (x, y) and therefore cannot be maximal.

8.3. If $\mathfrak{a}$ is an ideal in $\mathfrak{C}[\omega_1, ..., \omega_n]$ and $\alpha \in \mathfrak{a}$, then obviously $\{\alpha\omega_1, \alpha\omega_2, ..., \alpha\omega_n\} \subset \mathfrak{a}$, and the $\alpha\omega_i$ $(i = 1, ..., n)$ are linearly dependent over P, and therefore also over $\mathfrak{C}$, since the ω_i are linearly dependent over $\mathfrak{C}$. It can be shown (see van der Waerden [3]) that $\mathfrak{a}$ is a vector space over $\mathfrak{C}$, which we have just seen to be n-dimensional.

Two ideals $\mathfrak{a} = \mathfrak{C}[\alpha_1, ..., \alpha_n]$ and $\mathfrak{b} = \mathfrak{C}[\beta_1, ..., \beta_n]$ are *equivalent*, $\mathfrak{a} \sim \mathfrak{b}$, if there exists an algebraic ρ such that $\alpha_i = \rho\beta_i$ for all $i = 1, ..., n$. This relation is seen at once to be reflexive, symmetric, and transitive. Thus the set of all ideals falls into classes of equivalent ideals, the so-called *ideal classes*. It can be shown that the number h of ideal classes, the so-called *class number*, is finite (see Hasse [4], Landau [1], Vol. 3).

If ζ is a primitive nth root of unity, or in other words if ζ is of the form

[38] If $\mathfrak{G}$ has no unity element, then in place of ordinary polynomials we have expressions of the form $a_1x^{k-1} + a_2x^{k-2} + \cdots + a_{k-1}x + x^k + n_1x^{k-1} + \cdots + n_{k-1}x + n_k$, $a_h \in \mathfrak{G}$, $n_h \in \mathfrak{C}$.

[39] The property of being "integrally closed" is defined for rings of integers in $\mathsf{P}(\vartheta)$ in the same way as before. In the text algebraic numbers and the number field $\mathsf{P}(\vartheta)$ were defined as certain subsets of the field of all complex numbers, but now they are to be defined as elements of certain algebraic extension fields in the sense of IB7, §1.

$e^{2\pi ik/n}$ with $(k, n) = 1$, then $\mathsf{P}(\zeta)$ is called the nth *cyclotomic field.* We shall examine the special case that $n = p > 2$ is a prime.

A prime number $p > 2$ is said to be *regular* if $p \nmid h$, where h is the class number of the pth cyclotomic field $\mathsf{P}(\zeta)$. Kummer was led to the study of these fields by his investigations into the Fermat conjecture; he proved that $x^p + y^p = z^p$ *cannot be solved in integral quantities* $\mathsf{P}(\zeta)$ *if* p *is regular* (see Landau [1], Vol. 3).

Remark on §8. The theory of numbers in algebraic number fields can be developed in a way quite different from the ideal-theoretic discussion given here, namely, on the basis of the *theory of valuations* (see Hasse [3], [4].

9. Additive Number Theory

In our discussion up to now the foreground has been occupied by questions of divisibility, like the u.f. theorem and the u.f.p.i. theorem, which are sometimes called questions of the multiplicative theory of numbers. By additive number theory we mean questions that can be reduced to the following fundamental problems. Let there be given n sets $\mathfrak{A}_1, \mathfrak{A}_2, ..., \mathfrak{A}_n$ whose elements are non-negative integers. By the sum $\mathfrak{C} = \mathfrak{A}_1 + \mathfrak{A}_2 + \cdots + \mathfrak{A}_n = \sum_{i=1}^{n} \mathfrak{A}_i$ we mean the set of numbers $c = \sum_{i=1}^{n} a_i \ (a_i \in \mathfrak{A}_i)$ (Schnirelmann). If $\mathfrak{A}_1 = \mathfrak{A}_2 = \cdots = \mathfrak{A}_n = \mathfrak{A}$, we also write $\sum_{i=1}^{n} \mathfrak{A}_i = n\mathfrak{A}$. Let $\mathfrak{A}_i(x)$ $(i = 1, 2, ..., n)$ and $C(x)$ denote, respectively, the number of positive numbers $a_i \leqslant x$ $(a_i \in \mathfrak{A}_i)$ (or $1 \leqslant c \leqslant x$ $(c \in \mathfrak{C})$). The investigation of $C(x)$ is our first fundamental problem; the second consists of deciding in how many ways a $c \in \mathfrak{C}$ can be represented in the above form. It is easy to see that the definition of a sum can be extended to the case $n = \infty$; then if $0 \notin \mathfrak{A}_i$ for every i, the sum $\sum_{i=1}^{\infty} \mathfrak{A}_i$ is empty.

Let $\mathfrak{P}^{(2)} = \{0, 1, 3, 5, ..., p, ...\}$ be the set of all primes $p \geqslant 0$, $p \neq 2$; then the (unsolved) Goldbach conjecture states that $3\mathfrak{P}^{(2)} = \mathfrak{Z}$, where $\mathfrak{Z} = \{0, 1, 2, ...\}$ is the set of all non-negative integers. This conjecture is obviously equivalent to the conjecture that every positive even number can be expressed as the sum of two primes. If we agree to say that two sets $\mathfrak{A}$ and $\mathfrak{B}$ are asymptotically equal, $\mathfrak{A} \sim \mathfrak{B}$, if they coincide from some point on, i.e., if for a sufficiently large N they are identical in the interval (N, ∞), then it is known at the present time only that $4\mathfrak{P}^{(2)} \sim \mathfrak{Z}$ (Vinogradov).

A set $\mathfrak{B}$ is called a basis of kth order if $k\mathfrak{B} = Z$, $(k - 1)\,\mathfrak{B} \neq \mathfrak{Z}$, and $\mathfrak{B}$ is called an asymptotic basis of kth order if $k\mathfrak{B} \sim \mathfrak{Z}$, $(k - 1)\,\mathfrak{B} \not\sim \mathfrak{Z}$. Thus $\mathfrak{P}^{(2)}$ is an asymptotic basis of not more than fourth order and the Goldbach conjecture states that it is a basis of third order.

If we set $\mathfrak{Z}^{(n)} = \{0, 1^n, 2^n, ..., x^n, ...\}$, the *Waring problem* (solved by Hilbert) states that *$\mathfrak{Z}^{(n)}$ is a basis.* A special case is the theorem of Lagrange: *$\mathfrak{Z}^{(2)}$ is a basis of fourth order.*

The number $k(m; \mathfrak{A}_1, ..., \mathfrak{A}_n)$ of representations of m in the form $m = a_{1i_1} + a_{2i_2} + \cdots + a_{ni_n}$ $(a_{\lambda i_\lambda} \in \mathfrak{A}_\lambda, \ \lambda = 1, 2, ..., n)$ is called the *composition number* of m, and every such representation is called a *composition.* It is to be noted that under certain circumstances representations with the same summands but in different orders are to be counted a corresponding number of times in $k(m; \mathfrak{A}_1, ...)$; for example, if $\mathfrak{A}_1 = \{0, 3, 5\}$, $\mathfrak{A}_2 = \{3, 5\}$, $\mathfrak{A}_3 = \{2\}$, we have

$$\begin{aligned} 10 &= 3 + 5 + 2; \qquad 3 \in \mathfrak{A}_1, \qquad 5 \in \mathfrak{A}_2, \qquad 2 \in \mathfrak{A}_3, \\ &= 5 + 3 + 2; \qquad 5 \in \mathfrak{A}_1, \qquad 3 \in \mathfrak{A}_2, \qquad 2 \in \mathfrak{A}_3, \end{aligned}$$

so that $k(10; \mathfrak{A}_1, \mathfrak{A}_2, \mathfrak{A}_3) = 2$. If the order of the summands is explicitly disregarded, we use the term *partition* or *partition function* $p(n; \mathfrak{A}_1, ..., \mathfrak{A}_n)$. The difference is particularly noticeable in the case $\mathfrak{A}_1 = \mathfrak{A}_2 = \cdots = \mathfrak{A}_n = \mathfrak{A}$. If $n = \infty$ and $\mathfrak{A} = \mathfrak{Z}$, then $p(m; \mathfrak{Z}, \mathfrak{Z}, ...) = p(m)$ is simply called the number of partitions. The composition function $k(m; \mathfrak{Z}, \mathfrak{Z}, ...)$ is in finite for all m and therefore meaningless; thus in general $k(m; \mathfrak{A}, \mathfrak{A}, ...) = k(m, \mathfrak{A})$ is the number of representations of m with *positive* summands from $\mathfrak{A}$, account being taken of the order of the summands.

It is only in the rarest cases that the partition function or composition function can be explicitly calculated. For the most part we simply try to discover the asymptotic behavior of these functions for $m \to \infty$. Similarly, the question of the structure of the set of numbers represented by a sum can be answered in general only to the extent of finding the asymptotic behavior of $C(x)$ for $x \to \infty$, especially since for the most part we know nothing more about the summands than the asymptotic behavior of the functions $\mathfrak{A}_i(x)$ defined above. For example, the *prime number theorem* states that $\prod(x) \sim x/\log x$, where $\prod(x)$ is the number of prime numbers $\leqslant x$.

For the great majority of special problems it is necessary to employ the methods of analysis, in which case we speak of *analytic number theory* both for multiplicative and for additive problems.

A generalization of the prime number theorem should be mentioned here. *Let $\mathfrak{P}_{m,k}$ be the set of prime numbers $p \equiv k(m)$, $(k, m) = 1$; then*

$$\prod_{k,m}(x) \sim \frac{x}{\varphi(m) \log x} \qquad (\varphi(m) \textit{ Euler function}).$$

It follows that *every residue class* mod m *whose elements are coprime to m contains infinitely many prime numbers* (Dirichlet).

For a further discussion see also III13, §1.

The function $A(x)$ defined as the number of numbers $\leqslant x$ in a set $\mathfrak{A}$ is often described by means of comparison functions $\psi(x)$, a practice which has led to the introduction of the following concepts:

$$\delta(\mathfrak{A}, \psi(x)) = \underset{x=1,2,\ldots}{\underline{\text{fin}}} \frac{A(x)}{\psi(x)} \qquad (\psi(x)\textit{-density of } \mathfrak{A})^{40}$$

$$\delta_v(\mathfrak{A}, \psi(x)) = \underset{x=0,1,\ldots}{\underline{\text{fin}}} \frac{A(x)+1}{\psi(x+1)} \qquad (\textit{varied } \psi(x)\textit{-density},^{41} \quad \textit{if} \quad 0 \in \mathfrak{A}),$$

$$\delta^*(\mathfrak{A}, \psi(x)) = \underset{x\to\infty}{\underline{\lim}} \frac{A(x)}{\psi(x)} \qquad (\textit{asymptotic } \psi(x)\textit{-density}),$$

$$\bar{\delta}^*(\mathfrak{A}, \psi(x)) = \overline{\lim_{x\to\infty}} \frac{A(x)}{\psi(x)} \qquad (\textit{upper asymptotic } \psi(x)\textit{-density}).$$

If $\delta^* = \bar{\delta}^*$, we speak of the *natural* $\psi(x)$ *density* $\delta_*(\mathfrak{A}, \psi(x))$. Of particular importance is the case $\psi(x) = x$, for which we simply write $\delta(\mathfrak{A})$, $\delta_v(\mathfrak{A})$, and so forth. Then we have the following theorem, which is relatively simple and easy to prove.

Schnirelmann basis theorem: *If* $\delta(\mathfrak{A}) > 0$ *and* $0 \in \mathfrak{A}$ *(or equivalently* $\delta(\mathfrak{A}) > 0$ *and* $1 \in \mathfrak{A}$*), then* $\mathfrak{A}$ *is a basis.*

Proof: From $\delta(\mathfrak{A}) > 0$ it follows that $1 \in \mathfrak{A}$, and so we set $\mathfrak{A} = \{a_0 = 0, a_1 = 1, a_2, \ldots\}$, $a_1 < a_2 < \cdots$; if $a_\nu \in \mathfrak{A}$, then all $a_\nu + a_\rho$ with $a_\nu < a_\nu + a_\rho \leqslant a_{\nu+1} - 1$ belong to $2\mathfrak{A}$; from $a_\rho \leqslant a_{\nu+1} - a_\nu - 1$ it follows that the number of these $a_\nu + a_\rho$ is equal to $A(a_{\nu+1} - a_\nu - 1)$. If we define n by $a_n \leqslant x < a_{n+1}$, so that $A(x) = n$, and let $(2A)(x)$ denote the number of elements of $2\mathfrak{A} \leqslant x$, we obviously have

$$\begin{aligned}
(2A)(x) &\geqslant A(x) + \sum_{\nu=0}^{n-1} A(a_{\nu+1} - a_\nu - 1) + A(x - a_n) \\
&\geqslant A(x) + \sum_{\nu=0}^{n-1} \delta(\mathfrak{A})(a_{\nu+1} - a_\nu - 1) + \delta(\mathfrak{A})(x - a_n) \\
&= A(x) - n\delta(\mathfrak{A}) + \delta(\mathfrak{A}) \sum_{\nu=0}^{n-1} (a_{\nu+1} - a_\nu) + \delta(\mathfrak{A})(x - a_n) \\
&= A(x) - A(x)\,\delta(\mathfrak{A}) + \delta(\mathfrak{A})(a_n - a_0) + \delta(\mathfrak{A})(x - a_n)
\end{aligned}$$

[40] $\underline{\text{fin}}$ denotes the *limes inferior*. It is generally assumed that $\psi(x) > 0$ for $x \geqslant 0$, $\lim_{x\to\infty} \psi(x) = \infty$ and $\psi(x) = O(x)$; the last condition is obviously a natural one, since every function $A(x)$ must be such that $A(x) \leqslant x$, so that $A(x) = O(x)$.

[41] The function $A(x) + 1 = A(-1, x)$ enumerates all the $a \in \mathfrak{A}$ with $0 \leqslant a \leqslant x$, provided $0 \in \mathfrak{A}$, whereas $A(x)$ enumerates only the positive $a \leqslant x$.

$$= A(x)(1 - \delta(\mathfrak{A})) + x\delta(\mathfrak{A})$$
$$\geqslant x\delta(\mathfrak{A})(1 - \delta(\mathfrak{A})) + x\delta(\mathfrak{A})$$
$$= 2\delta(\mathfrak{A})\, x - \delta^2(\mathfrak{A})\, x,$$

and thus

$$\delta(2\mathfrak{A}) \geqslant 2\delta(\mathfrak{A}) - \delta^2(\mathfrak{A}) = 1 - (1 - \delta(\mathfrak{A}))^2.$$

Applying this formula to $2\mathfrak{A}$ instead of $\mathfrak{A}$, we obtain

$$\delta(4\mathfrak{A}) \geqslant 1 - (1 - \delta(2\mathfrak{A}))^2 \geqslant 1 - (1 - \delta(\mathfrak{A}))^4,$$

and thus in general by iteration

$$\delta(2^k\mathfrak{A}) \geqslant 1 - (1 - \delta(\mathfrak{A}))^{2^k}.$$

The case $\delta(\mathfrak{A}) = 1$ may be disregarded, since in this case $\mathfrak{A} = \mathfrak{Z}$; thus, since $0 < \delta(\mathfrak{A}) < 1$, there exists a k such that $(1 - \delta(\mathfrak{A}))^{2^k} \leqslant \frac{1}{2}$, and therefore $\delta(2^k\mathfrak{A}) \geqslant \frac{1}{2}$. From the following theorem we then have at once $\delta(2^{k+1}\mathfrak{A}) = 1$; that is, $2^{k+1}\mathfrak{A} = \mathfrak{Z}$, as desired.

From $A(x) + B(x) \geqslant x$ *for all* $x \geqslant 0$ (*as follows, for example, from* $\delta(\mathfrak{A}) + \delta(\mathfrak{B}) \geqslant 1$) and $0 \in \mathfrak{A} \cap \mathfrak{B}$ *it follows that* $\mathfrak{A} + \mathfrak{B} = \mathfrak{Z}$.

Proof: Assuming that there exists an x such that $x \notin \mathfrak{A} + \mathfrak{B}$, let us set $\mathfrak{B} = \{b_0 = 0, b_1, b_2, ...\}$ and determine n from $b_n < x < b_{n+1}$; such an n certainly exists since it follows from $x \notin \mathfrak{A} + \mathfrak{B}$ and $\mathfrak{A} \cup \mathfrak{B} \subseteq \mathfrak{A} + \mathfrak{B}$ that $x \notin \mathfrak{B}$. Since $x \notin \mathfrak{A} + \mathfrak{B}$, none of the n numbers $x - b_i$ $(i = 1, 2, ...)$ can belong to $\mathfrak{A}$, so that $A(x - 1) + n \leqslant x - 1$, and thus, since $n = B(x - 1) = B(x)$, we have

$$x - 1 \geqslant A(x - 1) + B(x - 1) = A(x) + B(x),$$

in contradiction to the assumption.

Now in order to prove the basis property for a set $\mathfrak{A}$ with *vanishing density*, it is sufficient, in view of the basis theorem, to prove *the existence of a number s such that* $\delta(s\mathfrak{A}) > 0$. It was in this way that Schnirelmann succeeded in proving for the first time the basis property of the set of prime numbers $\mathfrak{P}$, and in solving the Waring problem in a new way. By means of this method of Schnirelmann the additive theory of numbers has made great progress in recent times (see Ostmann [1]).

CHAPTER 7

Algebraic Extensions of a Field

Summary

The original problem of algebra was to find the "*solution*" of an algebraic equation by means of "*roots*" or, in modern terms, to represent the zeros of a given polynomial by "rational expressions" in the zeros of (irreducible) binomials $x^n - a$. After it had been recognized (Abel) that for the general polynomial (of higher than the fourth degree) such a representation is impossible, the problem took the form of representing the zeros of a given polynomial in such terms as are natural or unavoidable for the given case. For this new and more profound problem the Galois theory provides a completely satisfactory solution; it shows that the natural instruments for the representation of the zeros of a polynomial are determined by the structure, in terms of its subfields, of the smallest field (a splitting field) that contains all the coefficients and zeros of the polynomial. This structure (of the splitting field in terms of its subfields) is completely revealed by the structure of a finite group (the Galois group) uniquely determined by the polynomial; the Galois group consists of all those automorphisms of the splitting field (the isomorphisms of the field onto itself) that leave fixed all the elements of the coefficient field of the polynomial. From the properties of this group we can deduce the natural means of expression for the "solution" of the equation. For example, by examining the group alone we can decide whether a splitting field which belongs to this group can be constructed by means of "radicals" or, in other words, whether a polynomial that corresponds to such a splitting field is solvable "by radicals" (binomials). Thus in the present chapter we first prove for every polynomial the existence of a splitting field and its uniqueness up to isomorphism. Then the properties of such splitting fields are described and, more generally, the properties of a finite extension K' of an arbitrary field K, i.e. a field K' which arises from K by adjunction of (some or all) of the zeros of finitely many polynomials in $K[x]$. Every finite extension is seen to be a vector space of finite dimension over K. On the basis of these results we can construct the Galois group of a polynomial and set up the one-to-one correspondence of the intermediate fields (intermediate between K and a smallest splitting field) with the subgroups of the Galois group, and in particular the correspondence of the so-called conjugate subfields with the conjugate subgroups. By means of this correspondence we can then set up the above-mentioned group-theoretic criterion for "solvability by radicals" (cf. § 10.1).

In classical algebra the coefficients are usually assumed to be complex numbers, a restriction which does not correspond to the algebraic nature of the problem; thus we shall drop this restriction here and shall almost always consider polynomials with coefficients from an arbitrary field. Since we do not restrict ourselves to number fields it is clear that in the present chapter we are nowhere discussing the numerical "solution," or the numerical calculation of the zeros; our discussion is purely algebraic.

Exercises

1. Determine the zeros of the polynomial $x^2 - 5x + 3$ and express them rationally in terms of a zero of the binomial $x^2 - 13$.
2. If the polynomial is $x^2 - 7x - 3$, what binomial can then be chosen?
3. Let the polynomial be $x^2 + 0.8x + 5$, and the binomial $x^2 + 1$.
4. Let the polynomial be $x^2 + 2x + 8$, and the binomial $x^2 + 7$.
5. Let the polynomial $x^2 + x + 1$, and then what binomial?
6. If α is a zero of the binomial $x^3 - z$ and β of $x^2 + 3$, show that the polynomial $x^3 - 3x^2 + 3x + 17$ has the zeros $2\alpha + 1$, $2\alpha \cdot (-\frac{1}{2} + \frac{1}{2}\beta) + 1$, $2\alpha \cdot (-\frac{1}{2} - \frac{1}{2}\beta) + 1$.
(Hint. If x is replaced by $y + 1$ in the given polynomial, a simpler polynomial is obtained.)
7. Letting α and β have the same meaning as in ex. 6, show that the polynomial $x^3 + 6x + 2$ has the zeros
$$\alpha - \alpha^2, \quad -\tfrac{1}{2} \cdot (\alpha - \alpha^2) + \tfrac{1}{2}\beta \cdot (\alpha + \alpha^2), \quad -\tfrac{1}{2}(\alpha - \alpha^2) - \tfrac{1}{2}\beta(\alpha + \alpha^2).$$
8. Let β be a zero of $x^2 + 3$ and γ a zero of $x^3 - \frac{1}{2} \cdot (\beta - 1)$. Show that $\gamma + 1/\gamma$ is a zero of $x^3 - 3x + 1$.

1. The Splitting Field of a Polynomial

1.1 *Adjunction*

As before (cf. IB5, § 1.9), we let $J[x]$ denote the integral domain of the polynomials in the indeterminate x over the integral domain J, i.e. of the polynomials in x with coefficients in J, where J may, in particular, be a field K; and by $K(x)$ we denote the quotient field of $K[x]$. Also, in IB5, §3.10, we have proved the existence of (at least) one zero a of a polynomial $g(x) \in K[x]$ irreducible in $K[x]$; for we extended the field K to an extension field $K(a)$ by the adjunction of a zero a of $g(x)$. In general, we speak of the *adjunction* of a system S of (arbitrarily many new) *elements* $a, b, c,...$ to a given field K if we have constructed an extension field of K which includes S, i.e. all the elements of S; and in fact we have in mind the "*smallest*" such extension field, and we denote it by $K(S) = K(a, b, c, ...)$. As a *fundamental domain* in which the adjunction is carried out we assume the existence of a (fundamental) field K^* which is an extension field of K and contains S; thus the existence of such a K^* must be guaranteed in one way or another. Then the "smallest" extension field of K that contains S is to be defined as the (set-theoretic) intersection of all extension fields of K in K^* that contain S; since the intersection of arbitrarily many extension fields of K always exists and is again an extension field of K, the desired field $K(S)$ must exist. An element $w \in K^*$ belongs to $K(S)$ if and only if w can be represented as the value of a rational function (determined by w, though not uniquely) over K with arguments in S; for we see that every such element w belongs to every extension field K' of K that contains S; and the totality of such w, since it forms an extension field of K that contains S and is therefore contained in every K', must be the smallest field that contains both K and S. In particular: if the system S contains infinitely many elements, then for arbitrary $z \in K(S)$ there exists a *finite* subsystem S' of S such that $z \in K(S')$. Similarly, $J[S]$ denotes the intersection of all integral domains (in K^*) that contain J and S.

Example. The field $K' = \Pi^{(0)}(i, \sqrt{2}, \sqrt{3}, ..., \sqrt{p}, ...)$ is the smallest extension field of the field $\Pi^{(0)}$ of rational numbers that contains i and the square root of every prime p. As a fundamental field K^* here we may take the field of complex numbers.

Exercises

9. Let $K = \Pi^{(0)}$ and let K^* be either the field of all real numbers or the field of all real algebraic numbers. Show that the zeros of $x^2 - 5x + 3$ lie in a field which may be denoted by $K(\sqrt{13})$. (Cf. ex. 1.)
10. Let $K = \Pi^{(0)}$ and let K^* be either the field of all complex numbers or the field of all complex algebraic numbers. Show that the zeros of $x^2 + 0.8x + 5$ lie in $K(i)$. (Cf. ex. 3.)
11. Let K and K^* be as in ex. 10. Show that the zeros of $x^2 + x + 1$ lie in $K(i \cdot \sqrt{3})$. (Cf. ex. 5.)

12. Let K and K^* be as in the two preceding exercises. The binomial $x^3 - 2$ has a real zero, denoted by $\sqrt[3]{2}$. Show that all its zeros lie in $K(\sqrt[3]{2}, i \cdot \sqrt{3})$. Show that this field also contains all the zeros of $x^3 - 3x^2 + 3x - 17$ (cf. ex. 6) and of $x^3 + 6x + 2$ (cf. ex. 7).
13. Let $g(x) = x^3 - 3x + 1$ and let K and K^* be as in ex. 10.
    (a) Compute $g(-2)$, $g(0)$, $g(1)$, $g(2)$ and show that the polynomial g has three distinct zeros ξ_1, ξ_2, ξ_3.
    (b) Let β and γ be as in ex. 8. Show that ξ_1, ξ_2, ξ_3 are in $K^+ = K(\beta, \gamma)$.
    (c) Set $\overline{K} = K(\xi_1, \xi_2, \xi_3)$. Show that $\overline{K} = K(\xi_1) = K(\xi_2) = K(\xi_3)$ and that $\overline{K}$ is a *proper* subfield of K^+. (Hint. $\overline{K}$ contains only real numbers).
14. Let n be a positive integer with $n > 1$. Set $K = \Pi^{(0)}$ and let K^* be the field of real algebraic numbers. Set $p_n(x) = x^n - 2$ and let the positive zero of $p_n(x)$ be denoted by ξ_n (thus $\xi_n = \sqrt[n]{2}$). Let S be the set of all ξ_n with $n \geqslant 2$ and $\bar{S}$ the set of all ξ_n with $n \geqslant 10$. Show that $K(S) = k(\bar{S})$.

1.2. *Isomorphisms*

Two fields are said to be isomorphic to each other if they differ only in the notation (and meaning) for their elements and their "operations" (addition, multiplication); in other words, with a suitable change of the notation (and meaning) for the elements and the operations, a change which amounts to a one-to-one mapping (cf. IB1, §2.4 and IB5, §1.13), isomorphic fields may be *identified*. Thus "isomorphism" is an equivalence relation. From these remarks it is also clear what is meant by isomorphism of groups, integral domains, vector spaces, and so forth. The term *isomorphism* is synonymous with isomorphic mapping.

An isomorphism of a field K onto itself is called an *automorphism* of K. If U is a subfield common to K' and K'', and if f is an isomorphism of K' onto K'' in which each element of U remains fixed ($u = f(u)$ for every $u \in U$), then f is called an *isomorphism* of K' onto K'' *over* U or *with respect to* (*or relative to*) U.

Example. In $\Pi^{(0)}(i)$ (see the example in §1.1) let us map i onto $-i$ and every $r \in \Pi^{(0)}$ onto itself, so that $r_1 + ir_2$ is mapped onto $r_1 - ir_2$; then we obtain an automorphism of $\Pi^{(0)}(i)$ over $\Pi^{(0)}$.

Exercises

15. Let $K = \Pi^{(0)}$ and let K^* be the field of complex numbers, or else the field of all complex algebraic numbers. Denote the zeros of $x^3 - 2$ by ξ_1, ξ_2, ξ_3 and let $K_i = K(\xi_i)$ for $1 \leqslant i \leqslant 3$, $\overline{K} = K(\xi_1, \xi_2) = K(\xi_1, \xi_2, \xi_3)$. Prove

a) there exists an isomorphism from K_1 onto K_2 taking every rational number into itself
b) there exists an automorphism of $\overline{K}$ taking ξ_1 into ξ_2 and every rational number into itself.

16. Let ξ_1, ξ_2, ξ_3 and $\overline{K}$ be as in ex. 13. Prove that there exists an automorphism of $\overline{K}$ taking ξ_1 into ξ_2 and every rational number into itself.

1.3. *Irreducible Polynomials. Existence of a Zero*

Let K be a field and let $g(x) = b_n x^n + \cdots + b_0 \in K[x]$, where $K[x]$ is the integral domain of the polynomials in the indeterminate x with coefficients b_ν in K. The problem of finding the zeros of $g(x)$ then becomes the following *extension problem*: it is required to adjoin to K (one or more) zeros $a_1, a_2, \ldots$ of $g(x)$; that is, to adjoin elements a_i such that $g(a_i) = 0$. Essentially the same problem is solved if we construct an extension K' of K such that $g(x)$ has a linear factor (or several linear factors) in $K'[x]$. It turns out that such fields K' can be constructed in a purely algebraic way, i.e., essentially by computation with polynomials over K.

With respect to adjunction of a *single* zero of $g(x)$, where $g(x)$ is irreducible in $K[x]$, we have already (see IB5, §3.10) proved the following theorem.

Theorem 1. *Hypothesis. Let K be a field and let $p(x) \in K[x]$ be irreducible in $K[x]$ and be of degree $n \geqslant 2$.*

Conclusions. (1) *There exists (at least) one extension field L of K in which $p(x)$ has (at least) one zero a. The extension field $K(a)$ determined by the adjunction to K of a zero a of $p(x)$ is uniquely determined up to isomorphisms over K and is isomorphic to the field $K[x]/p(x)$ of the residue classes of $K[x]$ with respect to $p(x)$.*

(2) *Moreover, $K(a) = K[a]$, i.e., every $b \in K(a)$ is uniquely representable in the form $b = c_0 + c_1 a + \cdots + c_{n-1} a^{n-1}$, $c_\nu \in K$. The field $K(a)$ is a vector space over K of dimension n with the basis $a^0, a^1, \ldots, a^{n-1}$.*

Examples. Every rational-complex number g can be represented in exactly one way in the form $g = c_0 + c_1 i$, where c_0, c_1 are rational numbers. If $\Pi^{(0)}$ is the field of rational numbers, then every $g \in \Pi^{(0)}(\sqrt{2})$ can be represented in the form $g = c_0 + c_1 \cdot \sqrt{2}$, where $c_0, c_1 \in \Pi^{(0)}$ are uniquely determined by g.

1.4. *Arbitrary Polynomials. Existence of a Splitting Field*

By a *splitting field in the wider sense* (abbreviated: i.w.s.) of $g(x) \in K[x]$ we mean an extension Z of K such that in $Z[x]$ the polynomial $g(x)$ splits completely into linear factors. By a *smallest* splitting field Z' of $g(x)$ we then mean a splitting field i.w.s. such that none of its proper subfields are splitting fields i.w.s. of $g(x)$. In §1.3 we constructed, for a polynomial $p(x)$ irreducible in $K[x]$, an extension field $K' = K(a)$ such that $p(x) = (x - a)\,p_1(x)$ with $p_1(x) \in K'[x]$. Now if $g(x) \in K[x]$ is an arbitrary polynomial, and thus perhaps reducible in $K[x]$, any factor $p(x)$ of $g(x)$ that is irreducible in $K[x]$ has a zero a such that $g(x) = (x - a)\,g_1(x)$ with $g_1(x) \in K'[x]$, where $K' = K(a)$.[1] Then $g_1(x)$ is of degree $n - 1$, if $g(x)$ was of degree $n \geqslant 2$. For $n - 1 = 1$ the field K' is already a splitting field i.w.s. of $g(x)$. For $n - 1 \geqslant 2$ we can apply the same procedure to $K_1 = K'$ and to $g_1(x) \in K_1[x]$ as was just now applied to K and $g(x)$. After at most $n - 1$ repetitions of this procedure we obtain a splitting field i.w.s., call it Z, of $g(x)$. But the intersection of all splitting fields i.w.s. of $g(x)$ that are subfields of Z is a smallest splitting field of $g(x)$, which gives us the following theorem.

Theorem 2: Existence theorem. *For arbitrary $g(x) \in K[x]$ with arbitrary K there exists at least one smallest splitting field Z. If $g(x)$ can already be factored completely into linear factors in $K[x]$, then $K = Z$.*

We must now ask whether the factorization of $g(x)$ into linear factors is the "same" for all smallest splitting fields of $g(x)$; here the word "same" is meant in the sense that the number of different linear factors, i.e., the number of distinct zeros, is always the same, and the multiplicities are the same in every case. That this is so is the content of the *uniqueness theorem* proved below; the proof depends essentially on the following isomorphism theorem, which is also useful in other ways:

Theorem 3: Isomorphism theorem. *Hypotheses.* (1) *Let f be an isomorphic mapping of the field K' onto the field K'', and let $\bar{f}$ be the (uniquely determined) extension*[2] *of f to an isomorphic mapping of $K'[x']$ onto $K''[x'']$, where $x'' = \bar{f}(x')$.*

(2) *Let $p'(x') \in K'[x']$ be irreducible in $K'[x']$, so that*

$$p''(x'') = \bar{f}(p'(x')) \in K''[x'']$$

is irreducible in $K''[x'']$.

Conclusion. If a' and a'' are zeros of $p'(x')$ and $p''(x'')$, respectively, then there exists a uniquely determined isomorphism f^* of $K'(a')$ onto $K''(a'')$ which is an extension of f such that $a'' = f^*(a')$.

Remark. If $g'(x') = a_0' + a_1'x' + a_2'x'^2 + \cdots + a_n'x'^n \in K'[x']$, then $\bar{f}(g'(x')) = f(a_0') + f(a_1')\,x'' + f(a_2')\,x''^2 + \cdots + f(a_n')\,x''^n \in K''[x'']$.

[1] The case $a \in K$ is included.

[2] A mapping $\bar{f}$ of $\bar{A}$ into (onto) $\bar{B}$ is called an extension of the mapping f of A into (onto) B, where $A \subseteq \bar{A}$, $B \subseteq \bar{B}$, if $\bar{f}(a) = f(a)$ for every $a \in A$.

Proof. Let $p'(x')$ be of degree n, so that $p''(x'') = f(p'(x'))$ is also of degree n. Every element $b' \in K'(a')$ can be represented in the form $b' = g'(a')$, where the polynomial

$$g'(x') = a'_0 + a'_1 x' + \cdots + a'_{n-1}(x')^{n-1} \in K'[x']$$

is in one-to-one correspondence with the b'; in particular, $b' = a'$ is equivalent to $g'(x') = a'$, and the same remarks hold for $b'' \in K''(a'')$. Thus in order that f^* be an isomorphism of $K'(a')$ onto $K''(a'')$ and also be an extension of f, or, in other words, in order that $f^*(b') = f(b')$ for every $b' \in K'$ and $f^*(a') = a''$, it is necessary that $b'' = f^*(b')$, where $b' = g'(a')$ and $b'' = g''(a'')$ if and only if $\bar{f}(g'(x')) = g''(x'')$. For we have

$$\begin{aligned} b'' &= f^*(b') = f^*(a'_0 + a'_1 a' + \cdots + a'_{n-1}(a')^{n-1}) \\ &= f(a'_0) + f(a'_1)\, a'' + \cdots + f(a'_{n-1})(a'')^{n-1}; \end{aligned}$$

and since $g''(x'')$ is uniquely determined by $b'' = g''(a'')$, we get $g''(x'') = f(a'_0) + f(a'_1)\, x'' + \cdots + f(a'_{n-1})(x'')^{n-1} = \bar{f}(g'(x'))$. But this necessary condition for an isomorphism f^* is also sufficient; i.e., if in $\bar{f}(g'(x')) = g''(x'')$ we replace x' and x'' by a' and a'', respectively, we obtain an isomorphism f_0 such that $f_0(g'(a')) = g''(a'')$ and such that f_0 is an f^*. For if $\bar{f}$ is a one-to-one mapping, then so is f_0; also, $\bar{f}(g'(x')) + \bar{f}(h'(x')) = \bar{f}(g'(x') + h'(x'))$, so that $f_0(g'(a')) + f_0(h'(a')) = f_0(g'(a') + h'(a'))$, where $h'(x')$ is also of degree not greater than $n - 1$, and the corresponding remarks hold for multiplication; finally, $f_0(b') = f(b')$ for every $b' \in K'$ with $f_0(a') = a''$.

From this isomorphism theorem we now deduce the following uniqueness theorem.

Theorem 4: uniqueness theorem for smallest splitting fields. *All smallest splitting fields for an arbitrarily preassigned polynomial* $g(x)$ *over* K *are isomorphic relative to* K. More precisely:

(1) If K^* is a splitting field i.w.s. of polynomial $g(x)$ of degree $n \geqslant 2$ and if $g(x)$ has the zeros $a_1, \ldots, a_n$ in K^*, then $Z = K(a_1, \ldots, a_n)$ is the *unique* smallest splitting field of $g(x)$ contained in K^*.

(2) If $Z' = K(a'_1, \ldots, a'_n)$ and $Z'' = K(a''_1, \ldots, a''_n)$ are two smallest splitting fields of $g(x)$, with $g(a'_\nu) = g(a''_\nu) = 0$, $\nu = 1, \ldots, n$, there exists an isomorphism f of Z' onto Z'' over K such that $a''_\nu = f(a'_\nu)$, $\nu = 1, \ldots, n$, under suitable indexing of the a'_ν, a''_ν; thus in particular, every element of K remains fixed under f.

Corollary. The factorization of $g(x)$ *into linear factors is essentially the same in all splitting fields* (*i.w.s.*) *of* $g(x)$; in other words: if in a splitting field T' of $g(x)$ we have the factorization $g(x) = \prod_{\rho=1}^{r} (x - b'_\rho)^{k'_\rho}$, where all the b'_ρ are distinct and the k'_ρ are natural (positive) numbers, and if in the splitting field T'' of $g(x)$ we have the factorization $g(x) = \prod_{\tau=1}^{t} (x - b''_\tau)^{k''_\tau}$, where again all the b''_τ are distinct, then necessarily $r = t$ and, with a suitable indexing of the b''_τ, also $k'_\rho = k''_\rho$, $\rho = 1, \ldots, r$. (Here $n = k'_1 + \cdots + k'_r = k''_1 + \cdots + k''_r$.) If T', T'' are *smallest* splitting fields, there exists an isomorphism f of T' onto T'' over K with $f(b'_\rho) = b''_\rho$.

Proof: Proof of conclusion (1). The field Z is contained in every splitting field (i.w.s.) of $g(x)$ that is a subfield of K^*. As for conclusion (2), let $g(x) = g_{11}(x) \cdots g_{1t_1}(x)$ be the factorization of $g(x)$ in $K[x]$ into irreducible factors. Among these factors let $g_{11}(x)$ be of the highest degree $n_1 \leqslant n$. We index the a'_ν and the a''_ν in such a way that $g_{11}(a'_1) = g_{11}(a''_1) = 0$. On account of the isomorphism between $K'_1 = K(a'_1)$ and $K''_1 = K(a''_1)$ over K (cf. §1.4, Theorem 3), with a''_1 corresponding to a'_1, the factorizations of $g(x)$ in $K'_1[x]$ and $K''_1[x]$ into irreducible factors differ only in the notation; thus $g(x) = g_{21}(x; a'_1) \cdots g_{2t_2}(x; a'_1)$ in $K'_1[x]$ and $g(x) = g_{21}(x; a''_1) \cdots g_{2t_2}(x; a''_1)$ in $K''_1[x]$ (see §1.4, remark on Theorem 3). Again let $g_{21}(x; a'_1)$, and therefore also $g_{21}(x; a''_1)$, be a factor of the highest degree n_2, so that $n_2 \leqslant n - 1$. Furthermore, let a'_2 and a''_2 be zeros of $g_{21}(x; a'_1)$ and $g_{21}(x; a''_1)$, respectively. From the isomorphism of K'_1 and K''_1 over K it follows that $K'_2 = K'_1(a'_2)$ and $K''_2 = K''_1(a''_2)$ are isomorphic over K, where a_1 and a_2 correspond, respectively, to a_1 and a_2. Thus we can repeat the above procedure for the factorization of $g(x)$ in $K'_2[x]$ and $K''_2[x]$, and after at most n steps we arrive at $K(a'_1, \ldots, a'_n)$ and $K(a''_1, \ldots, a''_n)$, which are therefore isomorphic over K.

Exercises

17. Let $K = \Pi^{(0)}$ and let ξ be a zero of the binomial $x^2 - 13$. Then $K(\xi)$ is a smallest splitting field of this binomial, and also of the polynomial $x^2 - 5x + 3$. (cf. ex. 1).
18. On the analogy of ex. 17, construct other exericses from exs. 2 to 5.
19. Let $K = \Pi^{(0)}$ and let α, β as in exs. 6 and 7. Prove that $K(\alpha, \beta)$ is a smallest splitting field both for $x^3 - 3x^2 + 3x - 17$ and also for $x^3 + 6x + 2$ (cf. exs. 6, 7).
20. Let $K = \Pi^{(0)}$ and let ξ be a zero of $x^3 - 3x + 1$. Then $K(\xi)$ is already a splitting field. Prove that the other two zeros are $\xi^2 - 2$ and $-\xi^2 - \xi + 2$.

1.5. *Application to Equations of the Third Degree*

We now assume that K is of characteristic zero or $p > 2$, and that $g(x) = x^3 + b_2x^2 + b_1x + b_0 \in K[x]$, where K is not a splitting field of $g(x)$.

Theorem 5. (1) *If $g(x)$ is irreducible over K, then $g(x)$ is also irreducible over every extension K' of K which arises from K by adjunction of finitely many square roots.*

(2) *If $g(x)$ is reducible over K without being completely reducible to linear factors over K, then every smallest splitting field of $g(x)$ can be obtained by the adjunction of a suitable square root.* Here "square root" means a zero of a binomial of second degree that is irreducible over K, say $x^2 + c$ with $c \in K$.

Proof. If $g(x)$ is reducible over K without being completely reducible, then $g(x)$ contains exactly one prime factor of second degree $q(x) = ax^2 + bx + c$ with $a \neq 0$, which by completing the square we can write in the form of a binomial: $q(x) = a(x + (2a)^{-1}b)^2 + (c - (2a)^{-2}b^2 a)$; then $(2a)^{-1}$ exists ($2a \neq 0$, since $p \neq 2$). But if $g(x)$ is irreducible over K, though already reducible over $K_1 = K(a_1)$, where $a_1{}^2 + c_1 = 0$ with $c_1 \in K$, then $g(x) = (x - (d_1 + a_1 d_2))\, g_2(x; a_1)$ with $d_1, d_2 \in K$; and thus the polynomial $r(x) = (x - (d_1 + a_1 d_2))(x - (d_1 - a_1 d_2)) = (x - d_1)^2 + c_1 d_2{}^2$ belongs to $K[x]$, is of second degree, and has the linear factor $x - (d_1 + a_1 d_2)$ in common with $g(x)$, a fact which is inconsistent with the irreducibility of $g(x)$ over K, since the GCD is already determined in $K[x]$ by the Euclidean algorithm (IB5, §2.9 and IB6, §2.10). Now let K_n be obtained from K by successive adjunction of finitely many "square roots" $a_1, \ldots, a_n$, where a_ν is a square root over $K_{\nu-1} = K(a_1, \ldots, a_{\nu-1})$ and $K_0 = K$, $\nu = 1, \ldots, n$; in other words, $a_\nu{}^2 + c_\nu = 0$ with $c_\nu \in K_{\nu-1}$ and $x^2 + c_\nu$ is irreducible over $K_{\nu-1}$. Then if $g(x)$ is reducible over K_n (but irreducible over K_0), there exists a k with $0 \leqslant k \leqslant n - 1$ such that $g(x)$ is irreducible over K_k but reducible over K_{k+1}. Thus in the preceding argument we can replace K by K_k and a_1 by a_{k+1}, and finally K_1 by $K_{k+1} = K_k(a_{k+1})$, and in this way again arrive at a contradiction.

Examples. Problem of the *duplication of the cube*. The problem of constructing the edge of a cube with twice the volume of a given cube is not solvable with ruler and compass (such a construction corresponds algebraically to the solution of equations of the first and second degree). For this problem leads to the equation $a^3 - 2 = 0$; and $x^3 - 2$ is irreducible in $K[x]$, where K is the field of rational numbers (in general, if the polynomial $x^3 + b_2 x^2 + b_1 x + b_0$ has integral coefficients and is reducible over K, then it has an integral zero (cf. IB5, §4.4) which is a factor of b_0; but none of the numbers ± 1 and ± 2 is a zero of $x^3 - 2$).

Problem of the *trisection of an angle*. The problem of dividing a given angle into three equal parts cannot be solved for every angle with ruler and compass. For the problem leads to an equation $a^3 - 3a + s = 0$, where s is the length of the chord subtending the given angle, so that $0 < s < 2$. But for rational s the polynomial $x^3 - 3x + s$ is in general irreducible over the field K for rational numbers, e.g., for $s = 1$.

Exercises

21. Prove
    (*a*) the polynomial $x^3 + 6x + 2$ is irreducible over $\Pi^{(0)}$
    (*b*) if ξ_1 and ξ_2 are two distinct zeros, then $\Pi^{(0)}(\xi_1, \xi_2)$ is a smallest splitting field (cf. ex. 7).
22. Solve ex. 21 for the following polynomials:
    (*a*) $x^3 - 3x^2 + 3x - 17$,
    (*b*) $x^3 - 2$,
    (*c*) $x^3 - 3x + 1$.
    Prove in case (*c*) that $\Pi^{(0)}(\xi_1, \xi_2) = \Pi^{(0)}(\xi_1)$. (cf. ex. 13).
23. Determine whether each of the following polynomials is irreducible over $\Pi^{(0)}$, and for the reducible ones determine the smallest splitting field:
$$x^3 - 5x^2 + 3x + 1, \qquad x^3 - 2x + 1,$$
$$x^3 - x^2 - 8x + 12, \qquad x^3 - 5x^2 - 2x + 7.$$

2. Finite Extensions

For the Galois theory, to be developed in the following sections, certain other properties of the smallest extension fields of polynomials are of essential importance. In describing these properties we begin with the fact (cf. §1.3, theorem 1) that the extension $K(a)$ of K generated by the adjunction of a single zero a of a polynomial over K is a vector space over K with dimension n equal to the degree of the irreducible polynomial in $K[x]$ that has a for a zero. However, not only a but every $b \in K(a)$ is a zero of an (irreducible) polynomial in $K[x]$; in other words, every $b \in K(a)$ is *algebraic* over (or with respect to, or relative to) the field K. For the fact that $K(a)$ is a ring (and even a field) means that $b^1 = b \in K(a)$ implies $b^\nu \in K(a)$ for $\nu = 1, 2, ...,$ and also $b^0 = 1 \in K \subseteq K(a)$. Since $K(a)$ is of dimension n over K, the $n + 1$ elements b^μ, $\mu = 0, 1, ..., n$, are linearly independent over K (cf. §1.3, Theorem 1); thus there exist elements $c_\mu \in K$, not all equal to zero, for which $c_0 b^0 + c_1 b^1 + \cdots + c_n b^n = 0$. Thus b is a zero of a polynomial in $K[x]$ of degree not greater than n (the case $b = 0$ is included). For brevity we shall say that an *extension* K' of K is *algebraic* over K if every element $b \in K'$ is algebraic over K. If K' is algebraic over K, then every intermediate field T between K and K', i.e., every field T with $K \subseteq T \subseteq K'$, is also algebraic over K; and K' is itself algebraic over T. Finally, we shall say that an extension E of K is *finite* over K if E is a vector space of finite dimension over K. We now observe that our proof of the fact that $b \in K(a)$ is algebraic over K if a is algebraic over K made no use of any of the properties of $K(a)$ except that it is a vector space of finite dimension over K. We thus have the following theorem.

Theorem 1. *Every extension E of K that is finite over K is algebraic over K. In particular, $K(a)$ is algebraic over K if a is algebraic over K.*

Remark. An extension of K that is algebraic over K is not necessarily of finite dimension. For example, the field A of all algebraic numbers (over the field $\Pi^{(0)}$ of rational numbers) is not of finite dimension, since $\Pi^{(0)}[x]$ contains irreducible polynomials of arbitrarily high degree, e.g., $\Phi_p(x) = x^{p-1} + \cdots + x + 1$, for prime p (cf. §9.2), so that A cannot have a basis of finitely many elements over $\Pi^{(0)}$.

Examples of finite extensions are provided by the following theorem:

Theorem 2. (1) *Every extension of an arbitrary field K that can be generated by the adjunction of finitely many elements algebraic over K is finite (and therefore algebraic) over K. Conversely, every finite extension can be generated by the adjunction of finitely many elements algebraic over K.*

(2) *If the arguments of a rational function over K are algebraic over K, then the values of the function are also algebraic over K.*

Proof. (1) Let $K(a_1, ..., a_k)$ be the given extension, where the a_κ are algebraic over K, $\kappa = 1, ..., k$. For $k = 1$ the assertion is true by the preceding theorem. We now argue by complete induction on k. Assume that the assertion is already proved for all $k = 1, ..., m$ $(m \geqslant 1)$. Then it is also true for $k = m + 1$, since by the induction hypothesis $K' = K(a_1, ..., a_m)$ has a basis $\eta_1, ..., \eta_r$ and, if we set $a_{m+1} = a$, then for the case $K'(a) = K'$ there is nothing to prove; and otherwise the element a, since it is algebraic over K, is a zero of the polynomial $z(x) \in K[x] \subseteq K'[x]$ and is thus algebraic over K'. By the preceding theorem $K'' = K'(a)$ has a basis over K', say $\theta_1, ..., \theta_t$. For every $b \in K''$ we thus have $b = c_1\theta_1 + \cdots + c_t\theta_t$ with $c_\tau \in K'$. Moreover, $c_\tau = d_{\tau 1}\eta_1 + \cdots + d_{\tau r}\eta_r$ with $d_{\tau\rho} \in K$, so that $b = \sum_{\tau,\rho} d_{\tau\rho}\eta_\rho\theta_\tau$. Here the $r \cdot t$ elements $\eta_\rho\theta_\tau$ are linearly independent over K, since $b = 0$ implies $c_\tau = 0$ and thus $d_{\tau\rho} = 0$. Since the $b \in K''$ were arbitrary, the elements $\eta_\rho\theta_\tau$ form a basis for K'' over K.

The converse assertion follows from the fact that every finite extension, since it is algebraic, can be generated by the adjunction of the algebraic basis elements.

(2) Since $a_1, ..., a_n$ are algebraic over K, it follows that $K'' = K(a_1, ..., a_n)$ is algebraic over K (assertion (1)) and therefore contains all values of rational functions as described in assertion (2).

From the proof of assertion (1) of the preceding theorem it also follows that if r and t are the dimensions of K' over K and K'' over K', respectively, then K'' has the dimension $r \cdot t$ over K. The dimension m over K of a finite extension E of K is also called the *degree* of E over K, or in symbols: $m =$ degree $(E : K)$. Thus we have the following theorem:

Theorem 3. (I) *If E' or E'' is a finite extension of K or of E', respectively, then E'' is also a finite extension of K.*

(II) *Furthermore*, $\deg (E'' : K) = \deg (E'' : E') \cdot \deg (E' : K)$.

1. Remark. *If K' is algebraic over K, and K'' is algebraic over K', then K'' is also algebraic over K.* Proof: Every element of K'' is a zero of a polynomial with coefficients in K', and each of these finitely many coefficients is algebraic over K and thus belongs to a finite extension of K, so that the assertion follows from theorem 3 of §2.

2. Remark. If K is of characteristic $p = 0$ or $p > 2$ and if degree $(E : K) = 2$, then $E = K(a)$, where a is a zero of a polynomial of second degree $x^2 + c$ $(c \in K)$ that is irreducible over K.

Exercises

24. Let $K = \Pi^{(0)}$ and let ξ_1 and ξ_2 be two distinct zeros of $x^3 + 6x + 2$. Set $K(\xi_1) = K_1$ and $K_1(\xi_2) = \overline{K}$. Determine $\deg(K_1 : K)$, $\deg(\overline{K} : K_1)$, $\deg(\overline{K} : K)$. Now let η be a zero of $x^2 + 3$ and show that η can be chosen in $\overline{K}$, so that $K^+ = K(\eta)$ is thus a subfield of $\overline{K}$. Determine $\deg(\overline{K} : K^+)$ and $\deg(K^+ : K)$. (cf. exs. 7, 21.)
25. Again let $K = \Pi^{(0)}$. The polynomial $x^6 + 10x^3 + 125$ has six distinct zeros (in a suitable extension). Let $\overline{K} = K(\xi_1, \xi_2, \xi_3, \xi_4, \xi_5, \xi_6)$ and determine $\deg(\overline{K} : K)$. Prove that the zeros ϵ_1 and $\epsilon_2 = \epsilon_1^2$ of the polynomial $x^2 + x + 1$ can be chosen in $\overline{K}$ and that we may set $\xi_2 = \epsilon_1\xi_1$, $\xi_3 = \epsilon_1^2\xi_1$, $\xi_4 = 5/\xi_1$, $\xi_5 = \epsilon_1\xi_4$, $\xi_6 = \epsilon_1^2\xi_4$. Determine $\deg(\overline{K} : K(\xi_1))$, $\deg(K(\xi_1) : K)$, $\deg(\overline{K} : K(\epsilon_1))$, $\deg(K(\epsilon_1) : K)$ and show that $\overline{K} = K(\xi_1\epsilon_1)$.

3. Normal Extensions

A smallest splitting field Z of an irreducible polynomial in $K[x]$ has the further property that it is *normal* over K, or is a *normal* (or *Galois*) *extension* of K. Here an extension N of K is said to be normal over K if (a) N is algebraic over K and (b) every polynomial $g(x)$ that is irreducible in $K[x]$ can be factored completely into linear factors in $N[x]$ provided that N contains (at least) one zero of $g(x)$. The above assertion concerning Z is then a special case of the following theorem.

Theorem 1. *Criterion for normal (not necessarily finite) extensions.*

The extension P of K is normal over K if and only if P can be generated by the adjunction of all the zeros of (arbitrarily many) polynomials in $K[x]$.

1. Corollary. *In particular, every smallest splitting field of a polynomial in $K[x]$ is normal over K.*

Proof. (A) *Sufficiency.* (a) The field P is algebraic over K; for every $a \in P$ is contained (cf. §1.1) in an extension $K(b_1, ..., b_k)$ of K, where the b_κ are zeros of polynomials in $K[x]$, and therefore (§2, Theorem 2) the element a, and with it the whole field P, is algebraic over K.

(b) Let $g(x)$ be irreducible in $K[x]$ with $g(a) = 0$ and $a \in P$. Then we must show that $g(x)$ falls into linear factors in $P[x]$. But by (a) the field P contains elements $b_1', ..., b_m'$ algebraic over K, such that $a \in K(b_1', ..., b_m')$ and such that the irreducible polynomial $h_\mu(x) \in K[x]$ determined by $h_\mu(b_\mu') = 0$ falls into linear factors $(x - b_{\mu\tau})$ in $P[x]$, $\tau = 1, ..., t_\mu$; $\mu = 1, ..., m$. (For by the definition of P, if the b_μ are in P, then so are all the zeros of those irreducible polynomials in $K[x]$ for which the b_μ' are zeros.) Now let $h(x) \in K[x]$ be the product of all the $h_\mu(x)$, and let $c_1, ..., c_n$ be all the zeros of $h(x)$. Since

$$a \in K(b_1', ..., b_m') \subseteq K(c_1, ..., c_n) \subseteq P$$

we have $a = r(c_1, ..., c_n)$, where r denotes a rational entire function of n arguments with coefficients in K (§1.3, Theorem 1). If we permute the $c_1, ..., c_n$ in all $n!$ ways, we obtain $n!$ elements $a_\rho \in P$ from $r(c_1, ..., c_n)$. The coefficients of the polynomial $f(x) = (x - a_1) \cdots (x - a_{n!})$ are symmetric in the c_ν and thus belong to K (since $h(x) \in K[x]$) (see IB4, §2.4), where we agree to set $a_1 = a$. Since $g(a) = f(a) = 0$ and since $g(x)$ is irreducible in $K[x]$, it follows that $g(x)$ is a divisor of $f(x)$, so that the zeros of $g(x)$ are included among the a_ρ and consequently belong to P; but this means that $g(x)$ falls into linear factors in $P[x]$.

(B) *Necessity.* Let K' be a normal extension of K, so that K' is algebraic over K by definition. Then every $a \in K'$ is a zero of an irreducible polynomial $q(x) \in K[x]$ and all the zeros of $q(x)$ belong to K' (since K' is normal). Thus every $a \in K'$, and therefore the entire field K' itself, is obtained by adjunction of all the zeros of all the $q(x)$.

2. Corollary. *If K' is a finite normal extension of K, then K' can be obtained from K by the adjunction of all the zeros of a single polynomial.*

Consequence. Every finite extension E of K is contained in a normal extension N^* of K. For if $b_1, ..., b_r$ is a basis of E over K and if $h_\rho(b_\rho) = 0$ with $h_\rho(x) \in K[x]$ irreducible in $K[x]$, $\rho = 1, ..., r$, let us adjoin all the zeros of all the polynomials $h_\rho(x)$. By theorem 1, §3 the extension N^* of K obtained in this way is normal over K (and also over E), and furthermore $E \subseteq N^*$.

Consequently we make the following convention:

Convention. Whenever in any of the following sections we are dealing with a finite extension, we shall agree, even without explicit mention of this fact, that we are operating in a normal extension field N^* of the given finite extension.

Remark. In the sufficiency part §3(A) of the proof for theorem 1 the use of the theory of symmetric functions in (b) can be avoided if we have recourse instead to the theory of isomorphic mappings in §6, as follows. Let $g(x)$ be irreducible in $K[x]$ and let $g(a) = 0$ with $a \in K' = K(c_1, ..., c_n)$ (see the proof (A), (b)). Also let there exist a b not belonging to K', for which $g(b) = 0$. But $K(a)$ and $K(b)$ are isomorphic over K (see §6) and thus $K'(a) = K(a, c_1, ..., c_n)$ and $K'(b) = K(b, c_1, ..., c_n)$ are also isomorphic over K. In the isomorphic mapping of $K'(a)$ onto $K'(b)$ over K the elements $c_1, ..., c_n$ are only permuted among themselves. Also, since $a = r(c_1, ..., c_n)$, where r is a rational function in n arguments with coefficients in K, it follows that $b = \bar{r}(c_1, ..., c_n)$ with $\bar{r}$ rational over K, so that $b \in K'$, which contradicts the assumption.

Exercises

26. In exs. 19 to 25 determine which of the fields are normal extensions of $\Pi^{(0)}$.

4. Separable Extensions

For what follows it is often important to know under what conditions we may conclude that if a polynomial is irreducible in $K[x]$ all its zeros are distinct. If the zeros of an *irreducible* polynomial in $K[x]$ are all *distinct* (so that the number of distinct zeros of the polynomial is equal to its degree), the polynomial is said to be *separable over* K. (Examples of nonseparable polynomials will be given below.) If $g(x)$ is separable over K and if K' is an arbitrary extension of K, then the factors of $g(x)$ which are irreducible over K' are likewise separable (over K'). Every *zero of a polynomial that is separable over* K is also said to be *separable over* K, and similarly every *extension of K whose elements are all separable over K is said to be separable over* K; and every $a \in K$ is said to be separable over K. Separable elements and extensions are *algebraic* (over K) by definition. Moreover, *every finite separable extension* K' of K is a *simple extension*, i.e., can be represented in the form $K' = K(a)$, as will be proved below (cf. §6, theorem 3). We also make the following remark. The extension $K(a_1, ..., a_n)$ *is separable over* K *if* (*and only if*) *each of the elements* $a_1, ..., a_n$ *is separable over* K. (The proof, which will not be given here, depends on the fact that only for separable extensions is the number of isomorphisms of $L = K(a_1, ..., a_n)$ over K (cf. §6) equal to degree $(L : K)$.)

4.1. *Preliminary Remark. Computation in Integral Domains J with Characteristic $p > 0$*

(1) *For every $a \in J$ the sum $a + \cdots + a = p \cdot a = 0$ if the left-hand side has p summands a. In particular, it follows for $p = 2$ that $a = -a$.*

(2) *For arbitrary $a_1, \ldots, a_k \in J$ and an arbitrary integer $f > 0$ we have $(a_1 + \cdots + a_k)^r = a_1{}^r + \cdots + a_k{}^r$, if $r = p^f$.*

Proof. In view of (1) the desired assertion follows for $k = 2$ and $f = 1$ from the binomial theorem (IB4, §1.3) and the fact that the binomial coefficients are divisible by p (cf. IB6, §4.4 (29)). But if the assertion holds for $k = 2$ and for f, then it also holds for $k = 2$ and $f + 1$, since for $r' = p^{f+1}$ we have $(a_1 + a_2)^{r'} = ((a_1 + a_2)^r)^p = (a_1{}^r + a_2{}^r)^p = a_1^{r'} + a_2^{r'}$. Finally, if the assertion holds for a given k and (arbitrary) f, then it holds for $k + 1$ and f, since

$$(a_1 + \cdots + a_{k+1})^r = ((a_1 + \cdots + a_k) + a_{k+1})^r$$
$$= (a_1 + \cdots + a_k)^r + a_{k+1}^r = a_1{}^r + \cdots + a_k{}^r + a_{k+1}^r .$$

Furthermore,

$$(a_1 - a_2)^r = a_1{}^r - a_2{}^r . \tag{2a}$$

For it follows from (2) that $a_1{}^r = ((a_1 - a_2) + a_2)^r = (a_1 - a_2)^r + a_2{}^r$.
(3) *For $a \in K$ we have $a^p = a$ if (and only if) a belongs to the prime field $\Pi^{(p)}$ of K.* (Since every $a \in \Pi^{(p)}$ can be written as the sum of unit elements, it follows from (2) that $a^p = (1 + \cdots + 1)^p = 1 + \cdots + 1 = a$. Conversely, if $a^p = a$, then a is a zero of $x^p - x$; but this polynomial can have at most p distinct zeros, and by what has just been proved the $a \in \Pi^{(p)}$ already account for exactly p distinct zeros.

(4) *The derivative $h'(x)$ of a polynomial $h(x) \in J[x]$ of at least the first degree is the identically vanishing function if and only if J is of characteristic $p > 0$ and $h(x)$ is a polynomial in x^p with coefficients in J; in other words, $h(x) = g(x^p)$ with $g(x) \in J[x]$.*

Proof. Let $h(x) = a_0 + a_1 x + \cdots + a_n x^n, n \geqslant 1$, with $a_\nu \in J$ and $a_n \neq 0$. Then $h'(x) = a_1 + 2a_2 x + \cdots + n a_n x^{n-1}$. If $p = 0$, then $n a_n \neq 0$, so that $h'(x) \neq 0$. For $p > 0$, on the other hand, $h'(x) = 0$ if and only if $\nu a_\nu = 0$, $\nu = 1, \ldots, n$. But this condition is automatically fulfilled for $\nu \equiv 0 \pmod{p}$, so that $a_\nu = 0$ for $\nu \not\equiv 0 \pmod{p}$ is a sufficient condition for $h'(x) = 0$. But then $h(x) = \sum_{\rho=0}^{r} a_{\rho p} x_\rho{}^p$, from which the assertion follows.

4.2. If K is of characteristic zero, then every irreducible polynomial is separable, as follows from the next theorem.

Theorem 1. *Criterion for separability.* (I) *An irreducible polynomial* $g(x)$ *in* $K[x]$ *is separable over* K *if and only if either (a) the field* K *is of characteristic* 0 *or else (b) the field* K *is of characteristic* $p > 0$ *and* $g(x)$ *cannot be written as a polynomial in* x^p *over* K.

(II) *All the zeros of an irreducible polynomial* $n(x)$ *in* $K[x]$ *have the same multiplicity. For* $p > 0$ *and nonseparable* $n(x)$ *this multiplicity is a power* $q = p^e$, $e \geqslant 1$, *of the characteristic* p *of* K. *Moreover,* $n(x) = h(x^q)$, *where* $h(y) \in K[y]$ *is irreducible and separable over* K.

Proof. By IB4, §2.2 the polynomial $g(x)$ has at least one nonsimple zero a if and only if $g(x)$ and $g'(x)$ have a common zero and therefore are not coprime* (IB5, §2.9, especially (27)). Since $g'(x) \in K[x]$ and $g'(x)$ is of lower degree than $g(x)$ (in case $g'(x) \neq 0$), and since $g(x)$ is assumed to be irreducible in $K[x]$, we must have $g'(x) = 0$ if $g(x)$ has multiple zeros. But then the assertion (I) follows from §4.1, (4).

We now assume that $p > 0$ and that, if $g(x) = \sum_{\nu=0}^{n} a_\nu x^\nu$ has multiple roots, then $q = p^e$, with $e \geqslant 1$, is the highest power of p that divides all those ρp for which $a_{\rho p} \neq 0$. Then we have $g(x) = (x^q - b_1) \cdots (x^q - b_t) = h(x^q)$. Here the zeros $b_1, ..., b_t$ of $h(y)$ are distinct, since otherwise $h(y)$ would have multiple zeros and it would follow from (I) that $h(y) = k(y^p)$ and $g(x) = m(x^{pq})$, in contradiction to the definition of q. But now if d_τ is a zero of $x^q - b_\tau$, we have $g(x) = [(x - d_1) \cdots (x - d_t)]^q$, where all the d_τ are distinct, which proves the first part of (II). But $h(y)$ is irreducible over K, since $h(y) = h_1(y) \cdot h_2(y)$ implies $g(x) = h_1(x^q) \cdot h_2(x^q)$, which means that if $h(y)$ is irreducible, then so is $g(x)$. But then, since the zeros $b_1, ..., b_t$ of $h(y)$ are distinct, $h(y)$ is separable.

Example of a nonseparable polynomial. Let K^p be of characteristic $p > 0$, and let z be an indeterminate over K^p. Then the polynomial $x^p - z$ is irreducible over $K^p(z)$ (for the proof see, for example, Haupt [1], 13.3, theorem 3); on the other hand, it follows from $a^p - z = 0$ that $x^p - z = x^p - a^p = (x - a)^p$, so that $x^p - z$ has a zero of multiplicity p with $p \geqslant 2$.

Exercises

27. Let z be an indeterminate over $\Pi^{(5)}$, $K = \Pi^{(5)}(z)$. Prove that the polynomials $x^2 + x + z$ and $x^{10} + x^5 + z$ are both irreducible over K, but only the first one is separable.

* Two polynomials are said to be coprime if they have no common factor of positive degree.

5. Roots of Unity

5.1. *Definition of the* hth *Roots of Unity*

Introductory Remarks. As the *coefficient domain* we take an arbitrary field K. Then by an hth *root of unity* we mean a zero[3] of the polynomial

$$f_h(x) = x^h - 1, \tag{1}$$

where 1 is the unit element of K (h is a natural number).

The coefficients of $f_h(x)$ belong to the prime field Π of K, for which in the case of characteristic 0 we take the field $\Pi^{(0)}$ of rational numbers and in the case of characteristic p[4] the field $\Pi^{(p)}$ of residue classes mod p (in the ring of integers see IB5, §3.7 and IB6, §4.1).

From the algebraic point of view it is then natural to ask the following important *questions*:

1. *To what extent can $f_h(x)$ be factored over Π (and over K)?*

2. *Starting from Π (or from K), how do we obtain a smallest splitting field for $f_h(x)$? What can be said about the structure of this field?*

In sections 5, 8, and 9 we shall obtain far-reaching answers to these questions.

The roots of unity are of great importance for many problems in arithmetic and algebra. It is obvious that they are closely connected with the theory of "*pure equations,*" i.e., with the problem of determining the zeros of the polynomial

$$g_{h,a}(x) = x^h - a. \tag{2}$$

If $a \neq 0$,[5] we have the following theorem.

Theorem 1. *If α is a zero*[6] *of* (2), *and if ζ is an* hth *root of unity, then $\alpha \cdot \zeta$ is also a zero of* (2). *Moreover, if α_1 and α_2 are two (not necessarily distinct) zeros of* (2), *then α_2/α_1 is an* hth *root of unity. Thus we can obtain all the distinct zeros of* (2) *by multiplying any one of them with all the distinct* hth *roots of unity.*

In §5.2 we shall discuss the conditions under which the polynomial (1) has multiple zeros; the reader will have no difficulty in deriving the corresponding results for the polynomial (2).

[3] Belonging to K or to a suitable extension of K.
[4] In sections 5, 8, and 9 the number p is always a positive prime.
[5] The case $a = 0$ is of no interest.
[6] Belonging to K or to a suitable extension of K.

Note. It is well known that in the field of complex numbers the hth roots of unity can be represented in the following form (cf. IB8, (5)):

$$(3)\quad \zeta_{h,k} = e^{i\cdot k\cdot \frac{2\pi}{h}} = \cos k\cdot\frac{2\pi}{h} + i\cdot\sin k\cdot\frac{2\pi}{h}\qquad (k = 0, 1, \dots, h-1).$$

In the Gaussian plane they are represented by the vertices of a regular h-gon with its center at the origin and one of its vertices at the intersection of the postive real axis with the unit circle. Although this representation is advantageous for many purposes, we shall not make use of it here (in sections 5, 8 and 9), even for the case of characteristic 0, but shall develop a purely algebraic theory of the hth roots of unity.

5.2. *Multiplicity of the Zeros of* $f_h(x)$

As indicated above, we shall now undertake to find out when the polynomial (1) has multiple zeros. As a criterion for this purpose we make use of the following well-known theorem (cf. §4.2, beginning of the proof).

The polynomial $f(x)$ *in* $K[x]$ *has multiple zeros*[7] *if and only if* $f(x)$ *has a common factor of positive degree with* $f'(x)$; *in other words,* $f(x)$ *has no multiple zero if and only if it is coprime with* $f'(x)$.

We form the derivative

$$(4)\qquad f_h'(x) = h\cdot x^{h-1}.$$

It is obvious that in general this derivative has no common zero with (1) and is therefore coprime with (1). An exception occurs only if K is of prime characteristic p and p is a factor of h. Before dealing with this exceptional case we here present the main result.

Theorem 2. *If K is of characteristic* 0, *then* $f_h(x)$ *has only simple zeros. The same situation holds if K is of prime characteristic p but p is not a factor of h.*

If either one of the hypotheses of theorem 2 is satisfied, i.e., if the characteristic is equal to 0 or if the positive characteristic p is not a factor of h, we speak of the *principal case.*

Now let us turn to the *exceptional case*: the characteristic is p and $p \mid h$. Let the highest power of p that is a factor of h be p^f, so that $h = p^f\cdot\bar{h}$ and $p \nmid \bar{h}$ (with $f \geqslant 1$). Then, as we shall show in §8,[8] we have

$$(5)\qquad f_h(x) = [f_{\bar{h}}(x)]^{p^f}.$$

Since by theorem 2 the polynomial $f_{\bar{h}}(x)$ has only simple zeros, it follows that $f_h(x)$ has the same zeros as $f_{\bar{h}}(x)$, but each of them is of multiplicity p^f.

[7] In K or in a suitable extension.

[8] Independently, of course, of the *present* section.

Exercises

28. Let $K = \Pi^{(3)}$ and show that in $K[x]$ we have

$$f_4(x) = (x - 1) \cdot (x + 1) \cdot (x^2 + 1) \qquad \text{and} \qquad f_{36}(x) = [f_4(x)]^9.$$

5.3. *The Group* $\mathfrak{G}_h$ *of h*th *Roots of Unity*

In this subsection we confine ourselves to the *principal case*, so that by theorem 2 $f_h(x)$ has only simple zeros. Thus if both ζ_1 and ζ_2 are hth roots of unity (possibly with $\zeta_1 = \zeta_2$), then ζ_1/ζ_2 is also an hth root of unity, from which it readily follows that under multiplication the set of hth roots of unity forms an Abelian group of order h. We shall denote this group by $\mathfrak{G}_h$. Now let d be a (positive) factor of h. We first note that if (as we are now assuming) $f_h(x)$ comes under the principal case, then the same is true of $f_d(x)$. Thus we can speak of the group $\mathfrak{G}_d$. Also, since $\zeta^d = 1$ implies $\zeta^h = 1$, the group $\mathfrak{G}_d$ is a subgroup of $\mathfrak{G}_h$. The order of each element of $\mathfrak{G}_h$ is a factor of h. An hth root of unity ζ is called a *primitive h*th *root of unity* if it is of order h or, in other words, if no positive integer $g < h$ exists such that $\zeta^g = 1$. We shall denote the number of primitive hth roots of unity by $\psi(h)$.[9] Then, since the order of each element $\mathfrak{G}_h$ is a factor of h and every primitive dth root of unity (with $d \mid h$) is an hth root of unity, we have

$$\sum_{d \mid h} \psi(d) = h. \tag{6}$$

We now show that this result implies

$$\psi(h) = \varphi(h), \tag{7}$$

where $\varphi(h)$ is the Euler function (see IB6, §4.2). The proof of (7) is by complete induction, which we first carry through for characteristic zero.

I. It is obvious that there is exactly one primitive first root of unity, namely 1. Thus $\psi(1) = \varphi(1) = 1$.

II. Now assume that $h > 1$ and (7) is true for all smaller numbers. Then it follows from (6) that

$$\psi(h) = h - \sum_{d \mid h, d \neq h} \psi(d) = h - \sum_{d \mid h, d \neq h} \varphi(d). \tag{8}$$

But the Euler function (see IB6, §5, theorem 7) satisfies the well-known equality, corresponding to (6),

$$\sum_{d \mid h} \varphi(d) = h, \tag{9}$$

[9] It will turn out that this number (for the principal case) does not depend on K and is thus independent, in particular, of the characteristic of K.

from which we obtain

$$\varphi(h) = h - \sum_{d|h, d \neq h} \varphi(d), \tag{10}$$

so that the desired assertion follows directly from a comparison of (8) and (10).

It is easy to see that this method of proof is also valid for a prime characteristic p, provided (principal case) that h is not divisible[10] by p.

Thus (7) is completely proved for the principal case. From (7) it follows in particular that $\psi(h) > 0$, so that $\mathfrak{G}_h$ is a *cyclic* group (see IB2, §5) of order h. Thus the structure of the group $\mathfrak{G}_h$ is completely determined, and we have the following theorem.

Theorem 3. *The multiplicative group $\mathfrak{G}_h$ of the* hth *roots of unity is a cyclic group of order h. Thus it is isomorphic to the additive group (the module) of residue classes of the ring of integers* mod h.

Exercises

29. Let K be of characteristic $\neq 2$. Prove that there exist two primitive fourth roots of unity and that they are the zeros of the polynomial $x^2 + 1$. Construct $\mathfrak{G}_4$.
30. Let K be of characteristic $\neq 2$, $\neq 3$. Prove that there exist four primitive twelfth roots of unity and that they are the zeros of the polynomial $x^4 - x^2 + 1$. Construct $\mathfrak{G}_{12}$. Prove that if ξ is a primitive twelfth root of unity, then ξ^2 is a primitive sixth roots and ξ^3 is a primitive fourth root.

5.4. *The Cyclotomic Polynomial $\Phi_h(x)$.*

In this subsection we again restrict ourselves to the principal case. We now introduce the polynomial, called a *cyclotomic polynomial*, whose zeros are the primitive hth roots of unity (each with multiplicity 1) and whose leading coefficient is 1. We shall denote this polynomial by $\Phi_h(x)$ or also, in order to emphasize its dependence on the characteristic of the field K, by $\Phi_h^{(0)}(x)$ or $\Phi_h^{(p)}(x)$. We now prove the following theorem.

Theorem 4. *The coefficients of $\Phi_h(x)$ belong to the prime field Π of K.*

We first note that the argument leading to (6) indicates the following connection between the polynomials $f_h(x)$ and the cyclotomic polynomials:

$$\prod_{d|h} \Phi_d(x) = f_h(x). \tag{11}$$

[10] If h is divisible by p, we can set $\psi(h) = 0$.

It is easy to see that each of the factors in the product on the left-hand side of (11) is a factor of $f_h(x)$, that any two of these factors are coprime, and finally that every zero of $f_h(x)$ is a zero of one of these factors.

The proof of theorem 4 now follows by complete induction, where again in the case of characteristic p we must restrict ourselves to integers h that are not divisible by p.[11]

I. We have $\Phi_1(x) = f_1(x) = x - 1$, so that the coefficients of $\Phi_1(x)$ belong to Π.

II. Now assume that $h > 1$ and that the assertion is true for all smaller numbers. For characteristic p we again restrict ourselves to the case $p \nmid h$. From (11) we have

$$\Phi_h(x) \cdot \prod_{\substack{d \mid h \\ d \neq h}} \Phi_d(x) = f_h(x). \tag{12}$$

For abbreviation, we set the second factor on the left-hand side of (12) equal $P_h(x)$, so that (12) becomes

$$\Phi_h(x) \cdot P_h(x) = f_h(x). \tag{13}$$

From the induction hypothesis it is easy to see that all the coefficients of $P_h(x)$ belong to Π and that the same is true for $f_h(x)$, and consequently (for example, by the division algorithm) the desired statement is also true for $\Phi_h(x)$. Thus the proof of theorem 4 is complete.

For characteristic 0 this theorem can be sharpened as follows.

Theorem 5. *If K is a field of characteristic* 0, *then $\Phi_h(x)$ has integral coefficients.*

It is only necessary to repeat the steps of the above proof and to note that in the application of the division algorithm the coefficient of the highest power of x in $P_h(x)$ is 1.

We now investigate the relationship between $\Phi_h^{(0)}(x)$ and $\Phi_h^{(p)}(x)$. Of fundamental importance here is the fact that there exists exactly one homomorphism $H^{(p)}$ of the ring of integers G onto $\Pi^{(p)}$. The homomorphism $H^{(p)}$ is obtained by setting the integer g in correspondence with the residue class mod p determined by g. It is easy to show that (with respect to addition and multiplication) this correspondence is a homomorphism (cf. IB5, §3.7, and IB6, §4.1) and that no other homomorphism can exist (the image of the number 1, which must be the unit element of $\Pi^{(p)}$, already determines the image of every integer). We now have the following theorem.

[11] For h divisible by p we can set $\Phi_h(x) = 1$.

Theorem 6. *If the homomorphism $H^{(p)}$ is applied to the coefficients of $\Phi_h^{(0)}(x)$, the polynomial $\Phi_h^{(0)}(x)$ becomes $\Phi_h^{(p)}(x)$.*

For the proof it is only necessary to repeat the steps of the proof of theorem 4.

5.5. *Computation of the Cyclotomic Polynomials.* We note that this proof, in particular the steps (12) and (13), actually enables us to calculate the cyclotomic polynomials. As an example we calculate the case $h = 12$, where we must avoid the characteristics 2 and 3. Since 12 has the factors 1, 2, 3, 4, 6, 12, the individual steps are as follows (we leave the multiplication and division to the reader).

I. $\Phi_1(x) = h_1(x) = x - 1$

II. $P_2(x) = \Phi_1(x) = x - 1; \quad \Phi_2(x) = x + 1$

III. $P_3(x) = \Phi_1(x) = x - 1; \quad \Phi_3(x) = x^2 + x + 1$

IV. $P_4(x) = \Phi_1(x) \cdot \Phi_2(x) = x^2 - 1; \quad \Phi_4(x) = x^2 + 1$

V. $P_6(x) = \Phi_1(x) \cdot \Phi_2(x) \cdot \Phi_3(x) = x^4 + x^3 - x - 1; \quad \Phi_6(x) = x^2 - x + 1$

VI. $P_{12}(x) = \Phi_1(x) \cdot \Phi_2(x) \cdot \Phi_3(x) \cdot \Phi_4(x) \cdot \Phi_6(x) = x^8 + x^6 - x^2 - 1;$
$\Phi_{12}(x) = x^4 - x^2 + 1$

These relations hold for all characteristics other than 2 and 3. The reader may illustrate theorem 6 for the case $h = 12, p = 7$.

For characteristic 2 every twelfth root of unity is also a third root, and for characteristic 3 every twelfth root is also a fourth root, so that in the first case we may use formulas I and III, and in the second, I, II, and IV.

The following representation, in terms of the Möbius function $\mu(n)$ (cf. IB6, §5) also holds for the principal case:

$$\Phi_h(x) = \prod_{d \mid h} [f_d(x)]^{\mu\left(\frac{h}{d}\right)}, \tag{14}$$

but we shall make no further use of it. For comparison we calculate $\Phi_{12}(x)$ from (14). Since $[\mu(1) = \mu(6) = 1, \mu(2) = \mu(3) = -1$, and $\mu(4) = \mu(12) = 0$, we have

$$\begin{aligned}\Phi_{12}(x) &= [(x^{12} - 1) \cdot (x^2 - 1)] : [(x^6 - 1) \cdot (x^4 - 1)] \\ &= (x^{14} - x^{12} - x^2 + 1) : (x^{10} - x^6 - x^4 + 1) \\ &= x^4 - x^2 + 1.\end{aligned}$$

Exercises

31. Let K be of characteristic $\neq 2, \neq 3$. Determine $\Phi_{24}(x)$. How is $\Phi_{24}(x)$ related to $\Phi_{12}(x)$ and $\Phi_6(x)$? (cf. ex. 30.) Prove that

$$\Phi_{24}(x) = \frac{(x^{24} - 1) \cdot (x^4 - 1)}{(x^{12} - 1) \cdot (x^8 - 1)}.$$

5.6. *Concluding Remark*

We return to the two questions raised in §5.1. Let ζ be a primitive hth root of unity (in a suitable extension of K); then $K(\zeta)$ is obviously the smallest splitting field not only of the polynomial $f_h(x)$ but also of $\Phi_h(x)$. Thus if ζ_1 and ζ_2 are two primitive hth roots of unity, then $K(\zeta_1) = K(\zeta_2)$. As a result, $\Phi_h(x)$ is either irreducible over K or splits into irreducible polynomials, all of which are of the same degree. In §8 we deal with the case that K is a finite field. In §9 it will be shown that $\Phi_h(x)$ is irreducible over the field $\Pi^{(0)}$ of the rational numbers.

6. Isomorphic Mappings of Separable Finite Extensions

The Galois theory of separable polynomials is based on a study of the isomorphic mappings J *relative to* K (or *over* K) of a separable finite extension E of K *in* a separable normal extension field N^* of K (cf. the end of §3, convention). Thus J leaves every element of K fixed, while every element of E is mapped onto an element of N^*; so we may write J more explicitly in the form $J(E : K; N^*)$ or $J(E : K)$. The images a' of an element $a \in E$ under the mappings $J(E : K)$ are called the *conjugates of a with respect to E over K*; or in symbols, $a' = \text{conj}(a; E : K)$. By the *number* of $\text{conj}(a; E : K)$ we mean the number of mappings $J(E : K)$, even if the $\text{conj}(a; E : K)$ are not all distinct (compare the following examples).

Examples. (1) If $E = K(a)$ is separable over K, it follows from §1.4, theorem 3, that all the mappings $J(E : K)$ are obtained by replacing a with each of the zeros (in N^*) of the irreducible (in $K[x]$) polynomial $g(x)$ for which $g(a) = 0$. The number n of the $J(E : K)$ is thus equal to degree $(E : K)$, or to the maximal number n for which the $a^0, \ldots, a^{n-1}$ are linearly independent over K, or finally to the degree n of $g(x)$. For since E is separable over K, the number n is equal to the number of zeros of $g(x)$, in view of the fact that these zeros, and thus also the $\text{conj}(a; E : K)$, are all distinct. But if, for example, $b \in K$, then the $\text{conj}(b; E : K)$ are all equal, since they are all equal to b.

(2) Now let $K' = K(a)$, or $E = K'' = K'(b)$, be separable over K or over K' respectively, where $n' = \text{degree}(K' : K) \geqslant 2$ and $n'' = \text{degree}(K'' : K) \geqslant 2$. The number of the $J(E : K)$ is again equal to $\text{degree}(E : K) = n' \cdot n'' = n$ (cf. §2, theorem 3). But among the $\text{conj}(a; E : K)$ there are only n' distinct elements; for under each of the mappings $J(E : K)$ the polynomial $g(x)$ with $g(a) = 0$ (cf. example (1)) is mapped onto itself, so that a is mapped onto one of the n' zeros of $g(x)$, or in other words onto one of the $\text{conj}(a; E : K)$. Thus the $\text{conj}(a; E : K)$, the number of which is equal to $n = n' \cdot n''$, fall into n' classes of n'' equal elements each.

From the isomorphism theorem (§1.4, theorem 3), we obtain the following generalization of the results in example (2).

Theorem 1. *Hypothesis. The element* $a_{\nu+1}$ *is algebraic over* $K_\nu = K(a_0, \ldots, a_\nu)$, $\nu = 0, \ldots, n-1$, $a_0 \in K$, $K_0 = K$. *Moreover,* a_ν *has exactly* k_ν *distinct* $\text{conj}(a_\nu; K_\nu : K_{\nu-1})$, $\nu = 1, \ldots, n$.

Conclusion. (1) *There exist exactly* $k = k_1 \cdots k_n$ *isomorphic mappings* $J(K_n : K; N^*)$.[12] *Thus for every* $b \in K_n$ *the number of* (*not always distinct*) $\text{conj}(b; K_n : K)$ *is equal to* k.

Conclusion. (2) *If* K_n *is separable over* K, *then* $k = \text{degree}(K_\nu : K_{\nu-1})$, *so that* $k = \text{degree}(K_n : K)$.

If E is an extension of K with $E = K(a)$, then a is said to be a *primitive element* of E (over K). We then have the following theorem.

Theorem 2. *If E is a finite separable extension of K, then $b \in E$ is a primitive element of E over K if and only if all the* $\text{conj}(b; E : K)$ *are distinct.*

Proof. *Necessity.* If b is primitive, the assertion follows from the preceding theorem or from example (1).

Sufficiency. If the $\text{conj}(b; E : K)$ are all distinct, then the number n of them is equal to $\text{degree}(E : K)$ (compare the preceding theorem, conclusion (2)). On the other hand, the number of distinct $\text{conj}(b; E : K)$ is not greater than the degree k of the irreducible (in $K[x]$) polynomial $g(x)$ for which $g(b) = 0$. Thus $n \leqslant k$. But $n = \text{degree}(E : K)$ is equal to the dimension of E over K. Thus $k \leqslant n = \text{degree}(E : K)$, from which $k = n$. Thus $b^0, ..., b^{n-1}$ is a basis of E, so that $E = K(b)$, as was to be proved.

Furthermore, we have the following important theorem.

Theorem 3. *Every separable finite extension E of K has primitive elements and can thus be represented as a simple extension $E = K(b)$ of K.*

Proof. (A) If K is a *finite* field, then E is also finite. Thus the assertion follows from §8.2.

(B) If K contains *infinitely many* elements, then by §2, theorem 2, we have $E = K(a_1, ..., a_n)$ for (finitely many) suitably chosen elements $a_1, ..., a_n \in E$ algebraic over K. For $n = 1$ the assertion is true with $b = a_1$. Arguing by complete induction on n, we now assume that the assertion is true for $n = 1, ..., k$, and that $E = K(a_1, ..., a_k, a_{k+1})$ or $E = E_k(c)$ for $E_k = K(a_1, ..., a_k)$ and $c = a_{k+1}$. By the induction assumption we may set $E_k = K(b)$, so that it only remains to prove the assertion for $K(b, c)$. If $b_1 = b, ..., b_r$, or $c_1 = c, ..., c_t$, form the complete set of $\text{conj}(b; K(b) : K)$ or $\text{conj}(c; K(c) : K)$ respectively, then (because of the separability of E over K) all the $b_i(i = 1, ..., r$ are distinct from one another, and similarly all the $c_j(j = 1, ..., t$ (compare theorem 1, conclusion (2)). If we set $a_{\rho\tau} = b_\rho + dc_\tau$ with $d \in K$, and $a = a_{11}$, then the $\text{conj}(a; E : K)$ are included among the $a_{\rho\tau}$, since every $J(E : K)$ maps b or c onto a b_ρ or c_τ, respectively, and maps d onto itself; furthermore, to every $J(E : K)$ there corresponds exactly one ρ and exactly one τ (since the $\text{conj}(b; K(b) : K)$ and also the $\text{conj}(c; K(c) : K)$ are all

[12] In this conclusion (1) it is not necessary for N^* to be separable.

distinct), so that the mappings $J(E : K)$ correspond in an one-to-one way with certain pairs (ρ, τ). Thus if there exists a $d \in K$ for which $a_{\rho\tau} \neq a_{\mu\nu}$, provided $\rho \neq \mu$ or $\tau \neq \nu$, then all the $\text{conj}(a; E : K)$ are distinct, and the assertion follows from theorem 2. The existence of such a $d \in K$ follows from the fact that on the one hand the finitely many equations $a_{\rho\tau} = a_{\mu\nu}$ in the unknown d for all ρ, τ, μ, ν with $\rho \neq \mu$ or $\tau \neq \nu$ are at most of first degree in d and thus each of them has at most one solution, and on the other hand K is infinite.

Corollary. If K'' is a finite separable extension of K, and if K' is an extension field of K contained in K'', or, in other words, if K is a so-called intermediate field, then there exist $a', a'' \in K''$ such that $K' = K(a')$ and $K'' = K'(a'')$; since K' and K'' are finite and separable over K and K' respectively.

As an extension of this result we have the following theorem.

Theorem 4. *Hypothesis. Let E be a finite separable extension of K, and for $a \in E$ assume that the set of* $\text{conj}(a; E : K)$ *contains r distinct elements.*

Conclusion (1). *It follows that* $r = \text{degree}(K(a) : K)$ *and* $r \cdot t = \text{degree}(E : K)$, *where* $t = \text{degree}(E : K(a))$.

Conclusion (2). *Moreover, $a \in K$ if and only if all the* $\text{conj}(a; (E : K)$ *are equal.*

Proof. For (1). This conclusion follows from theorem 1. For (2). *Necessity.* The necessity is clear. *Sufficiency.* By conclusion (1) we have $r = \text{degree}(K(a) : K) = 1$, from which it follows that $a \in K$.

Exercises

32. Let $K = \Pi^{(0)}$. Let ξ_1 be a zero of $x^6 + 10x^3 + 125$ and ϵ_1 a zero of $x^2 + x + 1$ (cf. ex. 25). Set $K^+ = K(\epsilon_1)$, $\overline{K} = K(\epsilon_1, \xi_1)$, $K_1 = K(\xi_1)$ and determine $\deg(K^+ : K)$, $\deg(\overline{K} : K)$, $\deg(\overline{K} : K^+)$, $\deg(K_1 : K)$, $\deg(\overline{K} : K_1)$. In each case, determine the relative isomorphisms, i.e. the $J(K^+ : K)$, $J(\overline{K} : K)$, $J(\overline{K} : K^+)$, $J(K_1 : K)$, $J(\overline{K} : K_1)$, and the number of these isomorphisms. What are the images a) of ξ_1, b) of ϵ_1 under the isomorphisms $J(\overline{K} : K)$. Prove that $\overline{K} = K(\epsilon_1 + \xi_1)$ so that $\epsilon_1 + \xi_1$ is a primitive element of $\overline{K}$ (over K).

7. Normal Fields and the Automorphism Group (Galois Group)

Summary of the argument in §7. In the introduction to the present chapter we have seen that the "solution of an algebraic equation" corresponds to the construction of a smallest splitting field N for the polynomial over K that defines the equation. The field N is obtained by adjunction of finitely many a_κ, $\kappa = 1, ..., k$, such that

$$K \subset K(a_1) \subset \cdots \subset K(a_1, ..., a_k) = N.$$

Now instead of asking for such elements a_κ it is obvious that we can also seek an increasing chain of (suitable) intermediate fields Z_κ, for example $Z_\kappa = K(a_0, ..., a_\kappa)$, with degree $(Z_\kappa : Z_{\kappa-1}) > 1$ $(Z_0 = K)$. The decisive feature of this change in the form of what we are seeking is the fact that (cf. theorem 2) the total number of all the intermediate fields between N and K is *finite* (whereas, in general, there are infinitely many possibilities for the choice of the elements $a_1, ..., a_k \in N - K$). The same theorem (theorem 2 below) also states that under the isomorphic mappings $J(N : K)$ the field N is mapped onto itself and the entire set of mappings $J(N : K)$ constitute a finite group $G(N : K)$ of order degree$(N : K)$ Moreover, the intermediate fields between N and K and the subgroups of $G(N : K)$ are in one-to-one correspondence with each other in such a way that a larger intermediate field corresponds to a smaller subgroup. Finally (cf. theorem 3), conjugate subgroups correspond to conjugate intermediate fields (i.e., to fields that are mapped onto each other by one of the $J(N : K)$); thus, in particular, the normal subgroups of the group correspond to intermediate fields that are normal over K. The increasing chains of intermediate fields Z_κ that provide the desired solution are then seen to be chains for which the corresponding subgroups U_κ are "maximal" *relative* normal subgroups, i.e., such that $U_\kappa \subset U_{\kappa-1}$ and U_κ is a "maximal" normal subgroup of $U_{\kappa-1}$, so that Z_κ is thus a "minimal" normal field over (relative to) $Z_{\kappa-1}$. The fact that U_κ is a maximal normal subgroup of $U_{\kappa-1}$ and thus that Z_κ is a minimal normal field over $Z_{\kappa-1}$ has the following significance: the chain of subgroups, and of corresponding intermediate fields, cannot be refined by the insertion of additional relative normal subgroups and relative normal intermediate fields. It may be said that such a chain provides all the successive adjunctions that are indispensable for the solution of the problem.

Details of the argument. We now proceed to the detailed proofs. Again we consider only finite separable normal extensions N of an (arbitrary) field K, so that we may set $N^* = N$ (cf. end of §3, convention). In view of the existence of a primitive element a of N over K, we set $N = K(a)$. We first prove the following theorem.

Theorem 1. (1) *The isomorphisms $J(N : K)$ of N relative to K map N onto itself and are thus automorphisms of N over K.*

(2) *These automorphisms form a finite group of order $n =$* degree$(N : K)$, *the Galois group $G(N : K)$ of N over K.*

Proof. (I) *Assertion* (1). Since N is normal over K, all the conj$(a; N:K)$ lie in N and (for primitive a) are distinct, and the number of them is $n = \text{degree}(N:K)$. We now set $J = J(N:K)$ and let $N' = J(N)$ and $a' = J(a)$ be the images of N and a respectively under J. Since J is an isomorphism, it follows that $N' = K(a')$ and N' is normal over K; thus, in view of the fact that $a' \in N$, it follows that $N' \subseteq N$. Since we also have $N \subseteq N'$, it follows that $N = N'$.

(II) *Assertion* (2). By (I) the number of mappings $J(N:K)$ is $n = \text{degree}(N:K)$. Furthermore, the converse mapping of a $J(N:K)$ is again a $J(N:K)$, and the identical mapping, and also the product of two $J(N:K)$ (i.e., the result of their successive application) is again a $J(N:K)$, as was to be proved.

We now turn to the correspondence between the subgroups of $G(N:K)$ and the intermediate fields Z between K and N. For abbreviation, the system of all those automorphisms $J(N:K)$ under which a Z remains elementwise fixed will be denoted by $U(Z)$, so that $U(Z)$ is a subgroup of $G(N:K)$, and to every Z there corresponds a unique $U(Z)$. Similarly, the system of all those elements of N that remain fixed under all the automorphisms of a subgroup V of $G(N:K)$ will be denoted by $T(V)$, so that $T(V)$ is an intermediate field and to every V there corresponds a unique $T(V)$. Then $U(Z)$ produces a unique correspondence between the intermediate fields Z and certain subgroups $U(Z)$, and similarly $T(V)$ produces a unique correspondence between the subgroups V and certain intermediate fields $T(V)$. We now show that this set of subgroups $U(Z)$ contains *all* the subgroups, and similarly the set of intermediate fields $T(V)$ contains *all* the intermediate fields. In fact, we have the following fundamental theorem.

Theorem 2. *Hypothesis. Let N be a finite separable normal extension of K.*

Conclusion. The correspondence $U = U(Z)$ assigns, in a one-to-one way, to every intermediate field Z a subgroup U, and similarly $T = T(V)$ assigns, in a one-to-one way, to every subgroup V (of $G(N:K)$) an intermediate field T. These two correspondences are inverse to each other: $Z = T(U(Z))$ and $V = U(T(V))$ for all Z and all V. Thus the set of intermediate fields is mapped in a one-to-one way onto the set of subgroups. To every subgroup U there corresponds the greatest intermediate field Z that remains elementwise fixed under the automorphisms of U $(U = U(Z))$ and V is the greatest subgroup whose automorphisms leave $T(V)$ elementwise fixed. Thus V is the Galois group of N over $T(V)$; that is, $V = G(N:T(V))$. The group $G(N:K)$ is finite, and thus N contains only finitely many fields between K and N.

Proof. *For* $Z = T(U(Z))$ *with arbitrary* Z. By definition, $U(Z)$ is the (greatest) subgroup of $G(N : K)$ that leaves Z elementwise fixed. Since N is also normal over Z, the group $U(Z)$ must be the Galois group $G(N : Z)$. But $G(N : Z)$ leaves exactly the elements of Z fixed (§6, theorem 4).

For $V = U(T(V))$ *with arbitrary* V. In any case we have $V \subseteq G(N : T(V)) = U(T(V))$. Thus if we denote by v the order of V and by g the order of $G(N : T(V))$, then $v \leqslant g$. But now we shall also prove that $g \leqslant v$. To this end we set degree$(N : K) = k$ and $N = K(a)$. The k elements conj$(a; N : K)$ are all distinct; let us denote them by $a_1 = a, \ldots, a_k$, so that $N = K(a_\kappa)$, $\kappa = 1, \ldots, k$. (Compare §6, theorem 2.) If $J_1, \ldots, J_v$ are the automorphisms of V, then (with a suitable enumeration of the a_κ) we have $a_r = J_r(a_1)$, $r = 1, \ldots, v \leqslant k$. But the a_r are only interchanged among themselves by the mappings J_r, since $J_t(a_r) = J_t(J_r(a_1)) = (J_tJ_r)(a_1)$ and $J_tJ_r \in V$, in view of the fact that V is a group. Thus the coefficients of $p(x) = (x - a_1) \cdots (x - a_v)$ are invariant under the mappings $J_r \in V$ and consequently belong to $T(V)$, so that $p(x) \in T(V)[x]$. Since v is the degree of $p(x)$, it follows that degree$(K(a) : T(V)) =$ degree$(N : T(V)) \leqslant v$. But by theorem 1 we have $g =$ degree$(N : T(V))$, so that $g \leqslant v$. Consequently $g = v$ and $V \subseteq U(T(V))$ implies $V = U(T(V))$. Thus the $U(Z)$ and $T(V)$ each give rise to a one-to-one mapping of the set of all the Z onto the set of all the V, as follows from the uniqueness of the $U(Z)$ and $T(V)$ and from the fact that $Z = T(U(Z))$ and $V = U(T(V))$.

In continuation of the outline at the beginning of the present section, we now proceed as follows: we show that the one-to-one correspondence just proved between the subgroups and the intermediate fields allows us to deduce the structure of the normal field N over K from the structure of the Galois group. For this purpose we first introduce some terminology: by the *intersection* of given subgroups or given intermediate fields we mean the greatest subgroup, or the greatest intermediate field, that is contained in all the given groups, or fields; by the *union* (compositum) of a given set of groups or of intermediate fields we mean the smallest subgroup, or the smallest intermediate field, in which all the given subgroups, or fields, are contained. (We note that the intersection is at the same time the set of all mappings, or field elements respectively, that belong to every one of the given subgroups, or to every one of the given intermediate fields. We now have the following theorem.

Theorem 3. (1) *In the correspondence* (*given by the preceding theorem*) $V = U(Z)$ *or* $Z = Z(V)$ *the intersection, or the union, of a set of subgroups corresponds to the union or the intersection, respectively, of the corresponding intermediate fields, and conversely.*

(2) *The intermediate field Z' is isomorphic over K to the intermediate field Z'' if and only if $U(Z')$ is conjugate in $G(N : K)$ to $U(Z'')$.*

(2*a*) *Thus in particular every normal subgroup of $G(N : K)$ is the image of an intermediate field that is normal over K, and conversely.*

In accordance with (2), we say that two intermediate fields Z', Z'' isomorphic over K are *conjugate* in N over (with respect to) K; or in symbols, $Z'' = \text{conj}(Z'; N : K)$.

Proof. For (1). We denote the compositum and the intersection of the subgroups V', V'' by $V' \vee V''$ and $V' \wedge V''$, respectively (in IB2 we wrote $\langle V' \cup V'' \rangle$ and $V' \cap V''$) and correspondingly for the intermediate fields. From the definition of $V' = U(Z')$, $V'' = U(Z'')$ we have the result: to $Z = Z' \vee Z''$ the correspondence $V = U(Z)$ assigns the greatest subgroup V under which both Z' and Z'' remain elementwise fixed, so that $V \subseteq V' \wedge V''$. On the other hand, both Z' and Z'' remain elementwise fixed under every subgroup contained in $V' \wedge V''$, so that $V = V' \wedge V''$. In the same way, $Z' \wedge Z'' = T(V' \vee V'')$.

For (2). The subgroups V', V'' are conjugate (in G) if and only if there exists a $J \in G$ with $V'' = J^{-1}V'J$, and thus $V' = JV''J^{-1}$, where the symbol $J^{-1}J'J$ with $J' \in V'$ means that J' is to be carried out after J and J^{-1} after J', or in other words that the operations are to be read from right to left.

We now let $V' = U(Z')$, $V'' = U(Z'')$ and $Z' = J(Z)$, $V = U(Z)$ so that, for example, $Z'' = T(V'')$, $Z = J^{-1}(Z')$. Then we must show that $Z = Z''$. The proof proceeds as follows. Let $J' \in V'$ be arbitrary. Under J the field Z is mapped onto Z'; under J' the field Z' remains fixed and under J^{-1} it is mapped back onto Z, in such a way that Z remains elementwise fixed under $J^{-1}J'J$. Thus $J^{-1}V'J \subseteq V$. If we now interchange J with J^{-1} and correspondingly Z' with Z and V' with V, we likewise have $JVJ^{-1} \subseteq V'$, so that $V \subseteq J^{-1}V'J$. Thus $V = J^{-1}V'J$ and thus $V'' = V$, and therefore $Z = T(V) = T(V'') = Z''$. Conversely, it now follows from the one-to-oneness of $V = U(Z)$ that if $V'' = U(Z'')$ and $V' = U(Z')$, and also $Z' = J(Z'')$, then $V'' = J^{-1}V'J$.

Remark. Under the operations of formation of the compositum and the intersection, the system v of all the subgroups, and similarly the system z of all the intermediate fields, becomes a lattice (not necessarily distributive). Then under the correspondence $V = U(Z)$ it follows from the first conclusion of the above theorem that v and z correspond dually to each other (cf. IB9, §1).

From the above results we now have the following two theorems.

Theorem 4. *Let $Z' \subseteq N$ be normal over Z'', and write $G' = G(N : Z')$ and $G'' = G(N : Z'')$. Then G' is a normal subgroup of G'' and the factor group G''/G' is isomorphic to $G(Z' : Z'')$.*

For we see that the residue classes of G'' with respect to G' consist of exactly those $J \in G''$ which generate the same automorphism A of Z' over Z'', and the product of two residue classes corresponds to the product of the corresponding A.

Theorem 5. *Every composition series of* $G(N:K)$ (cf. IB2, §12) *corresponds to a "composition series" of N over K, namely to an increasing sequence of intermediate fields* $Z_0 = K \subset Z_1 \subset \cdots \subset Z_k = N$ *in such a way that* Z_κ *is a smallest (in N) normal field over* $Z_{\kappa-1}$, $\kappa = 1, \ldots, k$, *and conversely. All these composition series of N have the same length, and the groups* $G(Z_\kappa : Z_{\kappa-1})$ *are simple (they have no proper normal subgroups).*

Exercises

33. Let $K, K_1, \overline{K}, K^+, \xi_1, \xi_2, \eta$ be as in ex. 24. Also let ξ_3 be the third zero of $x^3 + 6x + 2$ and set $K(\xi_2) = K_2$ and $K(\xi_3) = K_3$. Prove that K^+ and $\overline{K}$ are normal over K. Consider the Galois groups $G(\overline{K} : K)$, $G(\overline{K} : K^+)$ and $G(K^+ : K)$. Prove that $G(\overline{K} : K)$ is isomorphic to S_3 (the symmetric group on three elements). Determine the fields into which K_1 is taken by the automorphisms in $G(\overline{K} : K)$.
34. Let K, K^+ and $\overline{K}$ be as in ex. 32, and (for $1 \leqslant i \leqslant 6$) set $K(\xi_i) = K_i$ (the ξ_i as in ex. 25). Show that K^+ and $\overline{K}$ are normal over K. Investigate the structure of $G(\overline{K} : K)$, $G(\overline{K} : K^+)$ and $G(K^+ : K)$.
    Prove
    (*a*) $G(\overline{K} : K^+)$ is isomorphic to S_3,
    (*b*) the zeros η_1, η_2, η_3 of the polynomial $x^3 - 15x + 10$ can be chosen in $\overline{K}$,
    (*c*) the binomial $x^2 - 3$ is also reducible over $\overline{K}$.
35. In exs. 33, 34 determine
    (*a*) the subgroups of $G(\overline{K} : K)$,
    (*b*) the intermediate fields in $\overline{K}$ over K
    (*c*) the correspondence between the subgroups and the intermediate fields.
36. In exs. 33, 34 construct a composition series in each case (in ex. 34 there are several possibilities) and determine the corresponding sequence of intermediate fields.

8. Finite Fields

8.1. *Preliminary Remark. Simple Relations in Fields of Characteristic p*

Some remarks on the importance and the historical development of finite fields are to be found at the end of the present section. The discussion in the section itself is given in modern form, corresponding to the general contents of the present volume on algebra.

Every finite field has a positive characteristic, so that we are now in a position to assemble some relations that hold in fields in characteristic p.[13]

In the first place, it is well known that in integral domains (in particular, in fields) of characteristic p we have

$$(a + b)^p = a^p + b^p. \tag{1}$$

By repeatedly raising this equation to the pth power we readily obtain

$$(a + b)^{p^f} = a^{p^f} + b^{p^f} \qquad (f = 1, 2, 3, \ldots). \tag{2}$$

Setting $a = c - d$ and $b = d$ in (2) gives $(c - d)^{p^f} = c^{p^f} - d^{p^f}$, or, changing the letters,

$$(a - b)^{p^f} = a^{p^f} - b^{p^f}. \tag{3}$$

From (2) we have

$$(a + b + c)^{p^f} = (a + b)^{p^f} + c^{p^f} = a^{p^f} + b^{p^f} + c^{p^f}.$$

By complete induction on n it follows that

$$(a_1 + a_2 + \cdots + a_n)^{p^f} = a_1^{p^f} + a_2^{p^f} + \cdots + a_n^{p^f}. \tag{4}$$

By $\Pi^{(p)}$ we again denote the prime field of characteristic p. Then

$$a^p = a, \qquad \text{if} \quad a \in \Pi^{(p)}, \tag{5}$$

since for integral $n > 0$ we have

$$n^p \equiv (1 + 1 + \cdots + 1)^p \equiv 1^p + 1^p + \cdots + 1^p \equiv n \pmod{p}.$$

Now let $g(x)$ be a polynomial over $\Pi^{(p)}$, say

$$g(x) = b_m x^m + b_{m-1}x^{m-1} + \cdots + b_0,$$

where all the b_μ belong to $\Pi^{(p)}$. Then from (5) and from the application of (4) for $f = 1$ to the integral domain $\Pi^{(p)}[x]$ we have

$$\begin{aligned}[g(x)]^p &= b_m x^{pm} + b_{m-1}x^{p(m-1)} + \cdots + b_0 \\ &= b_m(x^p)^m + b_{m-1}(x^p)^{m-1} + \cdots + b_0 = g(x^p),\end{aligned}$$

and thus

$$[g(x)]^p = g(x^p), \text{ if } g(x) \text{ is a polynomial over } \Pi^{(p)}. \tag{6}$$

[13] Cf. also §4.1.

If we apply (3) to the integral domain $\Pi^{(p)}[x]$, we obtain

$$(x^{\bar{h}} - 1)^{p^f} = x^{\bar{h}p^f} - 1,$$

which is exactly the equality used in §5 and denoted there by (5).

Exercises

37. Give the decomposition over $\Pi^{(3)}$ of:
(*a*) $x^{36} - 1$, (*b*) $x^{72} - 1$.

38. Decompose the following polynomials into linear factors over $\Pi^{(p)}$: (*a*) $x^p - x$, (*b*) $x^{p-1} - 1$, (*c*) $x^{p^{f+1}} - x^{p^f}$, (*d*) $x^{(p-1)\cdot p^f} - 1$.

8.2. *Fundamental Theorems on Finite Fields*

By a *finite field* we mean a field that contains only finitely many elements. The number of elements will be denoted by q, where we assume $q > 1$. The finite field with q elements will be denoted by F_q[14] and the characteristic, which must of course be positive, will be denoted by p. Obviously F_q is a *finite*[15] and therefore *algebraic*[16] extension of $\Pi^{(p)}$. We now prove the following theorem.

Theorem 1. *The field F_q is separable*[16] *over $\Pi^{(p)}$, and therefore separable over every intermediate field Z.*

For if α is an element of F_q and $h(x)$ is the irreducible polynomial in $\Pi^{(p)}[x]$ for which α is a zero, then if $h(x)$ were inseparable it could be represented in the form $h(x) = g(x^p)$, where $g(x)$ is a polynomial in $\Pi^{(p)}[x]$, and it would follow from (6) that $h(x) = [g(x)]^p$, so that $h(x)$ would be reducible, which completes the proof of the theorem.

Theorem 2. *The field F_q can be represented as a simple extension of the prime field $\Pi^{(p)}$, and therefore of any intermediate field Z. The multiplicative group of the field F_q is cyclic.*

For the proof we consider the multiplicative group of the field F_q, consisting of all the nonzero elements of F_q. Since its order is $s = q - 1$. for every nonzero element ξ of F_q we have

$$\xi^s - 1 = 0. \qquad (7)$$

[14] The notation is reasonable, since we shall see that the structure of the field depends only on q. The letter F is used to suggest the word "field." In the literature a finite field is often called a "Galois field" and is denoted by $GF(q)$ instead of F_q.

[15] Cf. §2.

[16] Cf. §4.

The polynomial[17]

$$f_s(x) = x^s - 1 \tag{8}$$

thus has exactly s distinct zeros in F_q. But these zeros are precisely the s roots of unity, and since they are distinct from one another it follows, as was proved in §5, that s is coprime to p and that (at least) one primitive sth root of unity ζ exists. Consequently $F_q = \Pi^{(p)}(\zeta)$, since the element 0 already belongs to $\Pi^{(p)}$ and every nonzero element ξ can be represented in the form ζ^k. Then it follows easily that $F_q = Z(\zeta)$, which completes the proof not only of theorem 2 but also of the fact that every primitive sth root of unity ζ is a generating element of F_q.

Theorem 3. *There exists a natural number* ν *such that*

$$q = p^\nu. \tag{9}$$

In other words: *the number* q *of elements of a finite field is a power of the characteristic.*

For we again let ζ be a primitive sth root of unity and let the corresponding irreducible polynomial in $\Pi^{(p)}[x]$ have the degree ν; then every α in F_q can be represented uniquely in the form

$$\alpha = a_0\zeta^0 + a_1\zeta^1 + \alpha_2\zeta^2 + \cdots + a_{\nu-1}\zeta^{\nu-1}, \tag{10}$$

where the coefficients belong to $\Pi^{(p)}$. Conversely, every such expression with coefficients in $\Pi^{(p)}$ is obviously an element of F_q. Since there are p possible values for each of the coefficients, the result (9) follows at once.

Exercises

39. Investigate the polynomial $x^9 - x$ corresponding to $F_3[x]$. Show: a) it falls into linear factors in $F_9[x]$; b) its zeros (except for zero) are the eighth roots of unity; c) it falls into linear and quadratic factors in $F_3[x]$; d) two of its quadratic factors (let them be denoted by $g(x)$ and $h(x)$) have the primitive eighth roots of unity for their zeros; e) if ξ_1 is a zero of $g(x)$ (in F_9), then $\xi_1{}^3$ is the other zero; also, $\xi_1{}^5$ and $\xi_1{}^7$ are the zeros of $h(x)$.
40. (Cf. ex. 39). Every element of F_9 can be represented in the form $a\xi_1 + b$, where a and b are elements of F_3. Set up the multiplication table. E.g., what is $(\xi_1 + 2) \cdot (2\xi_1 + 1)$? (Hint. Use one of the polynomials $g(x)$ and $h(x)$ in ex. 39.)

[17] Cf. §5(1).

41. (*a*) Determine the polynomials of third degree in $F_3[x]$ which are irreducible over F_3 (the coefficient of x^3 is always to be taken equal to unity). (Hint. There are eight of them; every polynomial of third degree in $F_3[x]$ which is reducible over F_3 has a zero in F_3.
(*b*) Compute $\Phi_{13}^{(3)}$ and $\Phi_{26}^{(3)}$ and factor these polynomials into their irreducible (over F_3) factors. (Hint. The irreducible factors are the irreducible polynomials determined in *a*)).
(*c*) Factor the polynomial $x^{27} - x$ into its irreducible (over F_3) factors.
(*d*) Investigate the structure of F_{27} with the aid of one of the irreducible polynomials determined in *a*).

8.3. *Existence and Uniqueness of F_{p^ν} for Arbitrary p and Arbitrary ν*

Theorem 4. *Let there be given an arbitrary prime p and an arbitrary natural number ν. Then there exists (at least) one finite field with p^ν elements. All finite fields with the same number of elements are isomorphic to one another.*

We set $q = p^\nu$. The proof depends on the remark that if the field F_q exists, its nonzero elements are exactly the zeros of $x^{q-1} - 1$, from which it follows at once that the polynomial $x^q - x$ has all the elements of F_q for its zeros, and only these. Thus we have proved (for the time being under the hypothesis that the field F_q exists) that:

A. *The polynomial*

$$g_q(x) = x^q - x \qquad (q = p^\nu) \tag{11}$$

has only distinct zeros, namely all the elements of the field F_q.

B. *The field F_q is the smallest splitting field of $g_q(x)$.*

We now discard the hypothesis that the field F_q exists, since we wish to prove its existence. Since $g_q'(x) = -1$, the polynomial $g_q(x)$ has no zero in common with its derivative, and thus $g_q(x)$ has only distinct zeros. We now show that the zeros of $g_q(x)$ already form a field. For let α and β be two zeros of $g'_q(x)$, so that $\alpha^q = \alpha$ and $\beta^q = \beta$. Then from (2) and (3) it follows that $(\alpha + \beta)^q = \alpha + \beta$ and $(\alpha - \beta)^q = \alpha - \beta$, and then also that $(\alpha \cdot \beta)^q = \alpha \cdot \beta$ and, if $\beta \neq 0$, then $(\alpha/\beta)^q = \alpha/\beta$.

Thus the proof of theorem 4 is complete and at the same time we have shown the general validity of the theorems A and B. Since F_q is the smallest splitting field of $g_q(x)$, it is uniquely determined up to isomorphisms.[18] More precisely: the field F_q (with $q = p^\nu$) admits, as we shall show in §8.5, exactly ν automorphisms, i.e., isomorphic mappings onto itself. It follows that if F_q and $\bar{F}_q$ are two finite fields with the same number of elements,

[18] Cf. §1.

then an isomorphic mapping of one onto the other is possible in exactly ν different ways.

8.4. *The Subfields of the Field F_q*

Let F_{q_1} be a subfield of F_q. Since F_{q_1} is a finite field with the same characteristic p, it follows from theorem 3 that there exists a natural number ν_1 such that

(12) $$q_1 = p^{\nu_1}.$$

On the other hand, by theorem 2 the field F_q is a simple extension of F_{q_1}, so that, if k is the degree of F_q with respect to F_{q_1}, then the same proof as in theorem 3 gives

(13) $$q = q_1^{\ k}.$$

From (12) and (13) it follows, since $q = p^\nu$, that

(14) $$\nu = k \cdot \nu_1.$$

Conversely, let ν_1 be an arbitrary factor of ν, so that (14) holds. We now set

(15) $$s = q - 1 \text{ and } s_1 = q_1 - 1, \text{ with } q_1 = p^{\nu_1}.$$

Then s_1 is a factor of s, from which it follows that the polynomial $f_{s_1}(x) = x^{s_1} - 1$ is a factor of the polynomial $f_s(x) = x^s - 1$. Since $g_q(x) = x \cdot f_s(x)$, and correspondingly $g_{q_1}(x) = x \cdot f_{s_1}(x)$, it follows that $g_q(x)$ is divisible by $g_{q_1}(x)$. The set of q elements of the field F_q, which constitute all the zeros of $g_q(x)$, thus contains all the q_1 zeros of $g_{q_1}(x)$, which (cf. §8.3, theorem B) form a field of q_1 elements. It is also clear that F_q can contain only one subfield with the fixed number of elements q_1. Thus we have proved the following theorem.

Theorem 5. *Let F_q be a finite field with $q = p^\nu$ elements. Then if ν_1 is an arbitrary factor of ν, the field F_q contains exactly one subfield with $q_1 = p^{\nu_1}$ elements. If ν_1 runs through all the factors of ν, we obtain all the subfields of F_q.*

Example. Let us take $q = 3^6 = 729$. Then, apart from itself, the field F_q contains the following subfields: one with $3^1 = 3$ elements, one with $3^2 = 9$ elements, one with $3^3 = 27$ elements, and no other subfield.

8.5. *The Automorphism Group of the Field F_q*

By theorem 1 the field F_q is separable. Moreover, by theorem B (§8.3), F_q is the smallest splitting field of $g_q(x)$, from which it follows that F_q

is a normal extension of $\Pi^{(p)}$.[19] Since the finite fields have a particularly simple structure, their Galois groups are also particularly easy to describe.

We first note that the prime field $\Pi^{(p)}$ obviously admits only one automorphism, namely the identity,[20] and that $\Pi^{(p)}$ remains elementwise fixed under every automorphism of F_q. Thus the Galois group $G(F_q : \Pi^{(p)})$ includes all the possible automorphisms of F_q. Since F_q is of degree ν with respect to $\Pi^{(p)}$, there are exactly ν of these automorphisms, so that the order of the Galois group is ν. We may now state the following theorem.

Theorem 6. *The Galois group $G(F_q : \Pi^{(p)})$ is a cyclic group of order ν. As a generating automorphism we may take the mapping $\alpha \to \alpha^p$.*

We first show that this mapping is one-to-one. For by (3) it follows from $\alpha^p = \beta^p$ that $(\alpha - \beta)^p = \alpha^p - \beta^p = 0$, so that $\alpha = \beta$. Furthermore, it follows from (1) and from $(\alpha\beta)^p = \alpha^p\beta^p$ that this mapping is an isomorphism, and therefore an automorphism. Since the order of the group is obvious, it is only necessary to prove that this mapping actually is of order ν in the Galois group. But its ith power is obviously the mapping $\alpha \to \alpha^{p^i}$ and since the equation $x^{p^i} = x$ has at most p^i solutions, ν is the smallest positive integral value of i for which the mapping $\alpha \to \alpha^{p^i}$ becomes the identity.

The correspondence between the subfields of F_q and the subgroups of the Galois group is now obvious: the subgroup $G(F_q : F_q)$ corresponding to the subfield F_{q_1} (with $q_1 = p^{\nu_1}, \nu = k \cdot \nu_1$) contains the k automorphisms $\alpha \to \alpha^{q_1^i}$ $(i = 0, 1, ..., k - 1)$, which are exactly those automorphisms of $G(F_q : \Pi^{(p)})$ that leave all the elements of F_{q_1} individually fixed.

In the proof of theorem 6 we have incidentally shown that in every finite field of characteristic p the pth root of every element exists and is unique; and then the same remark readily follows for the p^nth root (n a natural number). The uniqueness of the pth root is obvious for an arbitrary field K of characteristic p, even when K is not finite. But in the case of an infinite field we cannot always conclude from $\alpha \in K$ that $\sqrt[p]{\alpha} \in K$.

Exercises

42. Prove
    (*a*) the polynomial $x^2 + x + 2$ falls into linear factors over F_9 but is irreducible over F_{27};
    (*b*) the polynomial $x^3 + 2x^2 + 1$ falls into linear factors over F_{27} but is irreducible over F_9;

[19] Cf. §3.
[20] Cf. the remarks at the end of §8.3.

(*c*) consequently, both polynomials fall into linear factors over F_{729}, so that their zeros may be chosen in F_{729};

(*d*) if the zeros of the polynomial in (*a*) are denoted by ξ_1, ξ_2, and those of the polynomial in (*b*) by η_1, η_2, η_3, then $F_9 = F_3(\xi_1)$, $F_{27} = F_3(\xi_1)$, $F_{729} = F_3(\xi_1, \eta_1)$.

43. With the notation of ex. 42 investigate the structure of the five Galois groups $G_1 = G(F_9 : F_3)$, $G_2 = G(F_{27} : F_3)$, $G_3 = G(F_{729} : F_3)$, $G_4 = G(F_{729} : F_9)$, $G_5 = G(F_{729} : F_{27})$. In particular, determine how the automorphisms in G_1, G_3, G_4, G_5 act on ξ_1, ξ_2 and how the automorphisms in G_2, G_3, G_4, G_5 act on η_1, η_2, η_3.

8.6. *Decomposition of the Cyclotomic Polynomial $\Phi_h(x)$ over Finite Fields*

The particularly simple structure of the finite fields depends partly on the fact that F_q is a normal extension of its prime field $\Pi^{(p)}$ and of every intermediate field (cf. beginning of §8.5). If F_{q_1} is a subfield of F_q, the degree k of F_q with respect to F_{q_1} is equal to the order of the Galois group $G(F_q : F_{q_1})$ (see the next-to-last paragraph of §8.5).

We now turn to the problem of factoring the cyclotomic polynomial $\Phi_h(x)$.[21] Obviously we must assume that h is coprime to p. We let η denote a primitive hth root of unity. Since the elements of F_q, apart from its zero element, consist of all the $(q - 1)$th roots of unity, the element η will belong to F_q if and only if h is a factor of $q - 1$. But (cf. IB6, §4.2) we also have

$$p^{\varphi(h)} \equiv 1 \mod h.$$

Now let e be the exact exponent to which p belongs mod h, namely the smallest positive integer satisfying the congruence

$$p^x \equiv 1 \mod h;$$

then η is obviously contained in F_{p^e} but not in any proper subfield of F_{p^e}. Thus η is of degree e with respect to $\Pi^{(p)}$. Since this statement is true for every primitive hth root of unity, we have the following theorem.

Theorem 7. *Let* $(h, p) = 1$ *and let p belong to the exponent e* modulo h. *Then $\Phi_h(x)$ splits into irreducible polynomials of degree e over $\Pi^{(p)}$.*

In order to answer the question how $\Phi_h(x)$ splits over an arbitrary F_q, we need only investigate how a polynomial that is irreducible over a given finite field splits in a finite extension field. Here we have the following theorem.

[21] Cf. §5.4.

Theorem 8. *Let F_q be an extension field of F_{q_1}, let $q = p^\nu$, and assume* (12) *and* (14). *Let $f(x)$ be a polynomial of degree n irreducible over F_{q_1}. Let the greatest common factor of n and k be d, and finally let $n = d \cdot \bar{n}$. Then $f(x)$ splits over F_q into d irreducible polynomials of degree $\bar{n}$.*

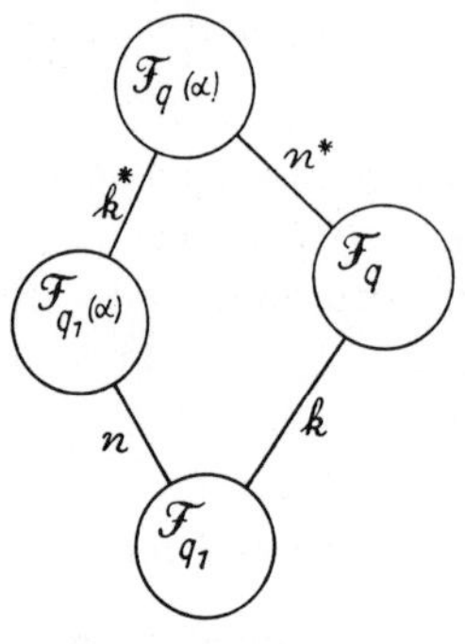

Proof. Let α be a zero of $f(x)$ in anextension field of F_q. Then we consider the four fields F_{q_1}, F_q, $F_{q_1}(\alpha)$, and $F_q(\alpha)$. In the diagram these fields are represented by circles. Where two fields are joined by a straight line in the diagram, the upper is an extension field of the lower, and the relative degree is written beside the line. In particular, n^* denotes the degree of $F_q(\alpha)$ with respect to F_q. Then we must show that $n^* = \bar{n}$. We set $k = d \cdot \bar{k}$. From $(n, k) = d$ it then follows that $(\bar{n}, \bar{k}) = 1$.

The number of elements in $F_{q_1}(\alpha)$ is $p^{\nu_1 n}$ and the number in $F_q(\alpha)$ is $p^{\nu n^*}$. Since $F_{q_1}(\alpha)$ is a subfield of $F_q(\alpha)$, it follows from §8.4 that $\nu_1 \cdot n \mid \nu \cdot n^*$. But $n = d \cdot \bar{n}$, $\nu = \nu_1 \cdot k = \nu_1 \cdot d \cdot \bar{k}$, so that $\nu_1 \cdot d \cdot \bar{n} \mid \nu_1 \cdot d \cdot \bar{k} \cdot n^*$ and therefore $\bar{n} \mid \bar{k} \cdot n^*$. But then from $(\bar{n}, \bar{k}) = 1$ we have $\bar{n} \mid n^*$, so that we can set $n^* = c \cdot \bar{n}$.

Now $F_q(\alpha)$ is obviously the *smallest* common extension field of $F_{q_1}(\alpha)$ and F_q. But $F_q(\alpha)$ has $p^{\nu c^*} = p^{\nu c \bar{n}}$ elements. Thus by §8.4 the field $F_q(\alpha)$ contains a subfield K with $p^{\nu \bar{n}}$ elements. Then, since $\nu \mid \nu\bar{n}$, it follows from §8.4 that the field K contains a subfield with p^ν elements, which must therefore be a subfield of $F_q(\alpha)$. But by §8.4 the field $F_q(\alpha)$ cannot contain any field other than F_q with p^ν elements. Consequently F_q is a subfield of K. In the same way we show that $F_{q_1}(\alpha)$ is a subfield of K. To this end we must first show that $\nu_1 \cdot n \mid \nu \cdot \bar{n}$. But this fact is immediately obvious, since $n = d \cdot \bar{n}$ and $\nu = \nu_1 \cdot d \cdot \bar{k}$. Since $F_q(\alpha)$ cannot contain any field other than $F_{q_1}(\alpha)$ with $p^{\nu_1 n}$, it follows that $F_{q_1}(\alpha)$ is also a subfield of K. On the other hand, $F_q(\alpha)$ is the *smallest* common extension field of F_q and $F_{q_1}(\alpha)$; thus we must have $c = 1$ and therefore $n^* = \bar{n}$. Since this result holds for every zero α of $f(x)$, the proof of theorem 8 is complete. The results of theorems 7 and 8 provide the solution of the problem dealt with in the present section. By Theorem 7 (with the same notation as in that theorem) the polynomial $\Phi_h(x)$ splits over the prime field $\Pi^{(p)} = F_p$ into irreducible polynomials of degree e. From Theorem 8 we see that we must set $\nu_1 = 1$, $n = e$. Thus $k = \nu$ and $d = (e, \nu)$. We thus have the following theorem.

Theorem 9. *Let $(h, p) = 1$, let p belong* mod *h to the exponent e, let $q = p^\nu$, and finally let $(e, \nu) = d$. Then $\Phi_h(x)$ splits over F_q into irreducible polynomials of degree e/d.*

Remark. It is easy to see that the number e/d in theorem 9 is the exact exponent to which q belongs mod h.

Examples. 1. We set $h = 12$ throughout, so that we must exclude the characteristics 2 and 3. By §5.5 of the present chapter $\Phi_{12}(x) = x^4 - x^2 + 1$. For every number a coprime to 12 we have $a^2 \equiv 1 \bmod 12$; thus the prime numbers different from 2 and 3 can be divided into two classes, those belonging to the exponent 1 mod 12 and those belonging to the exponent 2. If p is in the first class ($p = 13, p = 37, \cdots$), then $\Phi_{12}(x)$ splits into linear factors in $\Pi^{(p)}[x]$; if p belongs to the second class ($p = 5, 7, 11, \cdots$), then $\Phi_{12}(x)$ splits in $\Pi^{(p)}[x]$ into irreducible polynomials of the second degree, and if ν is even these latter polynomials split over F_{p^ν} into linear factors, whereas for odd ν they remain irreducible.

2. We now set $q = 3^6 = 729$ throughout. Then we have the factorization $q - 1 = 728 = 2^3 \cdot 7 \cdot 13$. If $h \mid 728$, then $3^6 \equiv 1 \bmod h$, so that $e \mid 6$ and $(e, 6) = e$. In the case $h \mid 728$ the polynomial $\Phi_h(x)$ splits into linear factors over F_{729}, a result which can also be obtained from the fact that F_{729} contains all 728th roots of unity and consequently all the hth roots of unity with $h \mid 728$. We now choose $h = 65$. Since 3 belongs to the exponent 4 mod 5 and to the exponent 3 mod 13, it therefore belongs to the exponent 12 mod 65. Thus $d = (12,6) = 6$, so that by theorem 9 the polynomial $\Phi_{65}(x)$ splits over F_{729} into irreducible polynomials of degree $\frac{12}{6} = 2$.

Theorem 10. *There exist irreducible polynomials of arbitrary degree n in $F_q[x]$. Every such polynomial splits completely in $F_{q^n}[x]$ and is thus a divisor of $x^{q^n} - x$ and, apart from the single irreducible polynomial x (for $n = 1$), every such polynomial is even a divisor of $f_S(x) = x^S - 1$, where for abbreviation we have set $S = q^n - 1$. Conversely, the polynomials $x^{q^n} - x$ and $f_S(x)$ split in $F_q[x]$ into irreducible polynomials whose degrees are divisors of n.*

We give the proof of the first assertion. By theorem 4 (§8.3) there exists a field K with $q^n = p^{\nu n}$ elements; by theorem 5 (§8.4) this field K contains a subfield L with $q = p^\nu$ elements. If $K = L(\alpha)$ (cf. §8.2, theorem 2), then the irreducible polynomial in $L[x]$ corresponding to α is of degree n. Since F_q is isomorphic to L by theorem 4 of §8.3, it follows that $F_q[x]$ also contains an irreducible polynomial of degree n. The other assertions of theorem 10 can then be derived without difficulty from the results of §8.

Exercises

44. What is the degree of the cyclotomic polynomial $\Phi_{175}^{(3)}$? Into what factors does it split over F_{3^ν} $(1 \leqslant \nu \leqslant 12)$?

8.7. *Closing Remarks*

Our knowledge of finite fields is due to the genius of Galois.[22] Galois

[22] Evariste Galois, born in 1811, met his death in a duel in 1832.

obtained the finite field with p^ν elements by starting from a polynomial $f(x)$, irreducible mod p, with integral coefficients and of degree ν mod p, and then adjoining an "*imaginary*" zero of $f(x)$ to the field of residue classes mod p.[23] The reader will realize from §8.5 how closely this procedure is associated with the revolutionary ideas that are now called the "Galois theory."

The original Galois procedure of symbolic adjunction of an "*imaginary*" naturally gave rise to the idea of altering his method by considering *congruences with respect to the double modulus p and* $f(x)$ in the domain of polynomials with integral coefficients. It is easy to see that this method, which was developed by R. Dedekind in 1857,[24] also leads to the Galois field $GF(p^\nu)$.

Dedekind also made use of the finite fields (or in other words, of the theory of higher congruences) for the investigation of algebraic number fields. In this connection we may point out that in the ring of integers of an algebraic number field (IB6, §8) the residue classes with respect to a prime ideal constitute a finite field.

The importance of the finite fields for the study of groups of linear substitutions was made clear by L. E. Dickson in a publication of fundamental importance.[25]

If an indeterminate t is adjoined to a finite field F_q, the resulting field $F_q(t)$ shows far-reaching analogies with the field of rational numbers but its arithmetical structure is simpler in many respects, and by algebraic extension of $F_q(t)$ we obtain fields that correspond to the algebraic number fields. The study of these fields has led to the development of a general theory[26] of fields and ideals, and they have proved very useful for illustrating the abstract theorems of such theories with simple but nontrivial examples.

9. Irreducibility of the Cyclotomic Polynomial and Structure of the Galois Group of the Cyclotomic Field over the Field of Rational Numbers

9.1. *Lemmas on the Connection between the Polynomial Rings* $G[x]$ *and* $\Pi^{(0)}[x]$

Again we let $\Pi^{(0)}$ be the field of rational numbers and G the integral domain of integers. Then $\Pi^{(0)}[x]$ contains the polynomials with rational

[23] Œuvres mathématiques, published by E. Picard, Paris, 1897.

[24] Abriss einer Theorie der höheren Kongruenzen in Bezug auf einen reellen Primzahl-Modulus. Journal f. d. reine u. ang. Math., Vol. 54, pp. 1–26.

[25] Linear groups with an exposition of the Galois field theory, Leipzig, 1901.

[26] Cf. the important work of E. Steinitz [1].

coefficients and $G[x]$ contains the polynomials with integral coefficients, so that $G[x]$ is contained in $\Pi^{(0)}[x]$. (The ideas and theorems developed in §9.1 apply only to *nonzero* polynomials, as will be tacitly assumed throughout.) A polynomial in $G[x]$ is said to be *primitive* if there is no factor (except 1 and -1) common to all its coefficients; moreover, if the coefficient of the highest power of x is positive, the polynomial is said to be *normed*.[27] Then we have the well-known theorem of Gauss:[28] *the product of two primitive polynomials is primitive, from which it readily follows that the product of two normed polynomials is normed. If $g(x)$ is a polynomial in $\Pi^{(0)}[x]$, there exists exactly one representation*

$$g(x) = \frac{a}{b} g^{+}(x), \tag{1}$$

where $g^{+}(x)$ is a normed polynomial and a and b are coprime integers, with b positive. When we speak of normed polynomials, we shall always mean polynomials from $G[x]$.

Theorem 1. *Let $f^{+}(x)$ be a normed polynomial which admits the factorization*

$$f^{+}(x) = g(x) \cdot h(x) \cdot k(x) \cdots, \tag{2}$$

where $g(x)$, $h(x)$, $k(x)$, are finitely many polynomials of positive degree in $\Pi^{(0)}[x]$. Then $f^{+}(x)$ also admits the factorization

$$f^{+}(x) = g^{+}(x) \cdot h^{+}(x) \cdot k^{+}(x) \cdots,$$

where $g^{+}(x)$, $h^{+}(x)$, $k^{+}(x)$, ... are the normed polynomials corresponding to $g(x)$, $h(x)$, $k(x)$, ..., respectively. If the coefficient of the highest degree of x in $f^{+}(x)$ is equal to 1, *then the same is true for $g^{+}(x)$, $h^{+}(x)$, $k^{+}(x)$ ... .*

We give the proof for the case of two factors $g(x)$ and $h(x)$. By (1) we have

$$g(x) = \frac{a}{b} g^{+}(x), \qquad h(x) = \frac{c}{d} h^{+}(x),$$

and thus

$$f^{+}(x) = \frac{ac}{bd} g^{+}(x) \cdot h^{+}(x).$$

But if $g^{+}(x)$ and $h^{+}(x)$ are normed, then by the theorem of Gauss the polynomial $g^{+}(x) \cdot h^{+}(x)$ is also normed, from which it easily follows that

$$f^{+}(x) = g^{+}(x) \cdot h^{+}(x).$$

[27] For brevity we say "normed" for "normed and primitive."

[28] Cf. IB5, §2.11 and 12.

If we now set

$$f^+(x) = a_m x^m + \cdots, \qquad g^+(x) = b_n x^n + \cdots, \qquad h^+(x) = c_r x^r + \cdots,$$

we have $a_m = b_n \cdot c_r$. If $a_m = 1$, it follows that $b_n = c_r = 1$.

The generalization to the case of more than two factors offers no difficulty.

9.2. *Proof of the Irreducibility of $\Phi_n(x)$ over $\Pi^{(0)}$*

In §5.4 of the present chapter we introduced $\Phi_n(x)$ as that polynomial (with leading coefficient 1) which has exactly the primitive nth roots of unity as its zeros. Since in §9 the characteristic is always taken to be 0, the polynomial $\Phi_n(x)$ has integral coefficients (§5, theorem 5), and since the coefficient of the highest power of x is equal to 1, the polynomial $\Phi_x(x)$ is normed. By theorem 1 of §9.1. it then follows: if $\Phi_n(x)$ is reducible at all, there must exist a factorization of the form

$$\Phi_n(x) = g_1(x) \cdot g_2(x) \cdots g_k(x), \tag{3}$$

where $g_\kappa(x)$ (for $\kappa = 1, 2, \ldots, k$) is a *normed*[29] polynomial, irreducible over $\Pi^{(0)}$, with leading coefficient 1. On the other hand, if $\Phi_n(x)$ is irreducible, then in the representation (3) we must obviously take $k = 1$. Thus the purpose of the present section is to prove that $k = 1$. This purpose is almost completely attained by the following lemma.

Lemma. *If ζ is a zero of $g_1(x)$ and p is a prime number that does not divide n, then ζ^p is also a zero of $g_1(x)$.*

Proof. Since p does not divide n, the element ζ^p is also a primitive nth root of unity; thus there exists an i such that ζ^p is a zero of $g_i(x)$. (Our assertion is equivalent to the statement that $i = 1$.) We set

$$g_i(x) = x^l + b_{l-1}x^{l-1} + \cdots + b_1 x + b_0 . \tag{4}$$

Since ζ^p is a zero of $g_i(x)$, it follows that

$$(\zeta^p)^l + b_{l-1}(\zeta^p)^{l-1} + \cdots + b_1\zeta^p + b_0 = 0. \tag{5}$$

But this equation means that the polynomial

$$g_i(x^p) = x^{pl} + b_{l-1}x^{p(l-1)} + \cdots + b_1 x^p + b_0 \tag{6}$$

has ζ as a zero. Since $g_1(x)$ is the irreducible polynomial corresponding to ζ, it follows that

$$g_1(x) \mid g_i(x^p), \tag{7}$$

[29] From now on the superscript $^+$ for normed polynomials will be omitted.

and thus

$$g_i(x^p) = q(x) \cdot g_1(x), \tag{8}$$

where $q(x)$ is also a normed polynomial in $G[x]$ with leading coefficient 1.

Again letting $H^{(p)}$ denote the homomorphism of G onto $\Pi^{(p)}$ (cf. §5.4), so that the number g in G is mapped onto the corresponding residue class $\bar{g}$ mod p, we now extend $H^{(p)}$ to the homomorphism $H^{(p)}$ of $G[x]$ onto $\Pi^{(p)}[x]$ by agreeing that $f(x) = a_r x^r + \cdots + a_0$ is to be mapped onto $\bar{f}(x) = \bar{a}_r x^r + \cdots + \bar{a}_0$. As a result of $H^{(p)}$ it follows from (8) that

$$\bar{g}_i(x^p) = \bar{q}(x) \cdot \bar{g}_1(x). \tag{9}$$

Moreover, from (3) and §5, theorem 6, we have

$$\Phi_n^{(p)}(x) = \bar{g}_1(x) \cdot \bar{g}_2(x) \cdots \bar{g}_k(x), \tag{10}$$

where the polynomials $\bar{g}_1(x), \ldots, \bar{g}_k(x)$ are not necessarily irreducible. Then by formula §8 (6) we have

$$\bar{g}_i(x^p) = [\bar{g}_i(x)]^p. \tag{11}$$

From (9) it thus follows that

$$[\bar{g}_i(x)]^p = \bar{q}(x) \cdot \bar{g}_1(x). \tag{12}$$

Now let $\bar{\zeta}$ be a zero of $\bar{g}_1(x)$ (in an extension field of $\Pi^{(p)}$). Then (12) shows that $\bar{\zeta}$ is also a zero of $[\bar{g}_i(x)]^p$ and consequently of $\bar{g}_i(x)$. Since $p \nmid n$, it follows from §5 that $\Phi_n^{(p)}(x)$ can have no multiple zeros, from which we see that $i = 1$, so that the proof of the lemma is complete.

Now it is easy to prove the following theorem.

Theorem 2. *$\Phi_n(x)$ is reducible over $\Pi^{(0)}$.*

Proof. Let η be a primitive nth root of unity and let ζ be a zero of $g_1(x)$. Then $\eta = \zeta^r$, with r coprime to n. Let

$$r = p_1 \cdot p_2 \cdots p_m \tag{13}$$

be the factorization of r into prime numbers. Every p_μ is coprime to n. We now set

$$\eta_1 = \zeta^{p1}; \qquad \eta_2 = \eta_1^{p2}; \ldots; \qquad \eta = \eta_m = \eta_{m-1}^{pm}. \tag{14}$$

Then by an m-fold application of the lemma we see that η_1 is a zero of $g_1(x)$, and that $\eta_2, \ldots, \eta = \eta_m$ are also zeros of $g_1(x)$. Since every primitive nth root of unity is thus a zero of $g_1(x)$, it is clear that $k = 1$ in (3), or in other words that $\Phi_n(x)$ is irreducible.

9.3. *Structure of the Galois Group of the Field $\Pi^{(0)}(\zeta)$*

Since $\Phi_n(x)$ is irreducible by §9.2, it follows that the field $\Pi^{(0)}(\zeta)$ is of degree $\varphi(n)$ over $\Pi^{(0)}$, where ζ again denotes a primitive nth root of unity, and since $\Pi^{(0)}(\zeta)$ contains all the zeros of $\Phi_n(x)$, it is necessarily a normal field. Let $\eta_1, \eta_2, \ldots, \eta_u$ (with $u = \varphi(n)$) be the entire set of primitive nth roots of unity, including ζ. Then the u automorphisms of the Galois group $\mathfrak{G}$ of the field $\Pi^{(0)}(\zeta)$ with respect to $\Pi^{(0)}$ can be represented in the following way:

$$\zeta \to \eta_1; \qquad \zeta \to \eta_2; \ldots; \qquad \zeta \to \eta_u. \tag{15}$$

Here $\eta_i = \zeta^{a_i}$, where the number a_i coprime to n is determined only mod n.

Theorem 3. *The Galois group $\mathfrak{G}$ of the field $\Pi^{(0)}(\zeta)$ with respect to $\Pi^{(0)}$ is isomorphic to the multiplicative group of the relatively prime residue classes* mod n (for this group see IB2, §1.2.10).

Proof. For $\zeta \to \eta_i$ we can write $\zeta \to \zeta^{a_i}$. If $\zeta \to \eta_j$ is also an automorphism of $\mathfrak{G}$, we can write it correspondingly in the form $\zeta \to \zeta^{a_j}$. If we apply the two automorphisms successively, we obviously obtain the automorphism $\zeta \to \zeta^{a_i a_j}$. Thus the theorem is proved.

It follows that $\mathfrak{G}$ is an Abelian group. If p is an odd prime, then in the cases $n = p^k$ and $n = 2p^k$ it even follows that the group $\mathfrak{G}$ is cyclic, and similarly for $n = 4$. On the other hand, for example, the group $\mathfrak{G}$ is no longer cyclic for $n = 8$ and $n = 15$.

As will be shown in the following section, the possibility of dividing a given circle into n equal parts by means of ruler and compass depends on the structure of the Galois group $\mathfrak{G}$ of the field $\Pi^{(0)}(\zeta)$. It turns out that *this construction is possible if and only if the order $\varphi(n)$ of $\mathfrak{G}$ is a power of* 2.

Exercises

45. Let ξ be a primitive nth root of unity. Investigate in detail the structure of the Galois group of the field $\Pi^{(0)}(\xi)$ with respect to $\Pi^{(0)}$ for the following values of n: 8, 9, 12, 18, 24, 36, 72, 35, 175.

10. Solvability by Radicals. Equations of the Third and Fourth Degree

10.1. By means of the Galois theory (§7) we can answer the question of the *solvability* of "*an equation by radicals*" or better of "*a polynomial by radicals.*" By a *radical of degree k over the field T* we mean a zero of a binomial $x^k + c$ of degree k which is irreducible over T. Also, we shall say that a polynomial $g(x)$ irreducible over K is *solvable by radicals* (*over* K) if there exists a splitting field in the wider sense $Z_R = K(b_1, ..., b_m)$ such that b_μ is a radical over $K_{\mu-1} = K(b_1, ..., b_{\mu-1})$; here $K_0 = K$ and $\mu = 1, ..., m$. We then have the following theorem (which will not be proved here; see e.g., Haupt [2], p. 515).

Theorem 1. *Let K be a field which (for simplicity) we take to be of characteristic zero. A polynomial $g(x) \in K[x]$ is solvable by radicals if and only if the (simple) factor groups of a composition series* (cf. IB2, §12) *of the Galois group $G(N : K)$ are all of prime order; here N denotes the smallest splitting field of $g(x)$ over K.*

Thus in the case of solvability by radicals the smallest splitting field N (which of course is contained in Z_R) can be generated by successive adjunction of zeros of normal[30] polynomials of prime degree with cyclic group; each of these zeros can then itself be represented by radicals. Since in general N is a proper subfield of Z_R, the solution by radicals is in general not a "natural" process of solution; in other words, it is not a process which corresponds to the structure of N.

10.2. By means of §10.1 we can decide, for example in the case of characteristic zero, for which degrees n a "*general*" polynomial $g(x)$ is solvable in radicals; here a polynomial $g(x) = x^n + a_{n-1}x^{n-1} + \cdots + a_0$ is said to be *a general polynomial* (*over* K) *if the coefficients a_ν are also indeterminates over K.*[31] *The zeros of a general polynomial $g(x)$ over K are indeterminates over K* (see Haupt [2], p. 177), *and thus are distinct*; *on the other hand, since $g(x)$ is irreducible over $K(a_0, ..., a_{n-1})$* (see Haupt [2], p. 256) *the polynomial $g(x)$ is separable. The group of the normal field of a general polynomial is the "largest possible"*; *in other words* (see Haupt [2], p. 556), *it is isomorphic to the symmetric group S_n of degree n* (see IB2, §1.2.5 and §15.2), *and is thus of order $n!$*. But for $n \geqslant 5$ the only possible composition series consists of S_n, A_n, E, where E is the subgroup consisting of the unit element alone and A_n is the so-called alternating group, with order $2^{-1}n!$ (for $n = 5$, cf. IB2, §15.4; in general, see e.g., Haupt [2], p. 560 ff.). For $n \geqslant 5$ the orders of the factor groups are therefore 2 and $2^{-1}n!$, the latter of which is not a prime number.

For $n = 4$ a composition series of the group $S_n = S_4$ consists of S_4, A_4, N_4, Z_1, E, where A_4 is again the alternating group, N_4 is the Abelian subgroup of 4th order, which is generated by the 3 products P_j of two transpositions without common element, and Z_1 is the cyclic group

[30] A polynomial $h(x) \in K[x]$ is said to be *normal over* K if there exists a zero α of $h(x)$ such that $h(x)$ splits completely into linear factors in $K(\alpha)[x]$.

[31] Thus in a general polynomial $g(x)$ the $a_0, ..., a_{n-1}, x$ are indeterminates over K. For this concept see IB4, §2.3.

of 2nd order generated, say, by P_1. Thus for $n = 4$ the orders of the factor groups are 2, 3, 2, 2. Finally, for $n = 3$ a composition series consists of S_3, A_3 and E with 2 and 3 as the orders of the factor groups. Thus we have the result: *the general polynomial of nth degree is solvable in radicals if and only if* $n = 2, 3,$ *or* 4.

10.3. From §10.1 we also obtain a *necessary and sufficient condition* that *a regular n-gon* P_n, *inscribed in a circle* (of radius 1), *is constructible with ruler and compass.* If we represent P_n in the complex plane, the problem is seen to be equivalent to expressing the zeros of the polynomial $x^n - 1$ over the field $\Pi^{(0)}$ of rational numbers by means of radicals of the 2nd degree; for the vertices of P_n are the images of the nth roots of unity, which are representable as powers of a primitive nth root of unity, say ζ; and therefore $N = \Pi^{(0)}(\zeta)$ must be normal over $\Pi^{(0)}$, since it is the smallest splitting field of the irreducible (over $\Pi^{(0)}$) cyclotomic polynomial $\Phi_n(x)$ of degree $\varphi(n)$. Since the Galois group $G_n = G(N : \Pi^{(0)})$ (§9, theorem 3) is Abelian, the number ζ can be represented by radicals, in view of the fact that the factor groups of a composition series of an Abelian group are all (cyclic) of prime order (cf. IB2, §12.1). The product of these prime orders is equal to the order of G_n, and is thus equal to $\varphi(n)$. *For* $\Phi_n(x)$ *to be solvable in radicals of the 2nd degree it is therefore necessary that* $\varphi(n) = 2^k$. *But this condition is also sufficient for solvability by radicals of the 2nd degree.* For if $\varphi(n) = 2^k$, the increasing sequence of relative normal fields K_r corresponding to a composition series of G_n has the following property: K_{r+1} is of 2nd degree over K_r, so that $K_{r+1} = K_r(a)$, where a is a radical of 2nd degree over K_r. We now let $n = 2^r p_1^{t_1} \cdots p_m^{t_m}$, with $r \geqslant 0$, $t_\mu \geqslant 1$, $p_\mu > 2$, where the p_μ (and also the number 2 in the case $r > 0$) are the only prime numbers dividing n; then, since $\varphi(n) = 2^q p_1^{t_1-1} \cdots p_m^{t_m-1}(p_1 - 1) \cdots (p_m - 1)$ (see IB6, §5), with $q = 0$ for $r \leqslant 1$, it follows from $\varphi(n) = 2^k$ that $t_\mu = 1$ and $p_\mu = 2^{s_\mu} + 1$, $\mu = 1, \ldots, m$. We thus have the result: *the n-gon* P_n *is constructible with ruler and compass if and only if* $n = 2^t$ *with* $t \geqslant 1$, *or else* $n = 2^k p_1 \cdots p_m$, *where* $k \geqslant 0$, $m \geqslant 1$ and p_μ *is a prime with the property that* $p_\mu = 2^{s_\mu} + 1$, $\mu = 1, \ldots, m$. (Not every number $2^s + 1$ is a prime number.) Examples: the 4 smallest primes p_μ with this property are 3, 5, 17, 257. Thus for an n that contains, for example, one of the factors, 7, 11, 13, 19, the corresponding P_n cannot be constructed with ruler and compass.

Exercises

46. Take $K = \Pi^{(0)}$, and
    (*a*) $g(x) = x^3 - 3x^2 + 3x + 17$ (cf. exs. *6*, *12*, *19*, *22*),
    (*b*) $g(x) = x^3 + 6x + 2$ (cf. exs. *7*, *12*, *19*, *24*, *33*),

(*c*) $g(x) = x^3 - 3x + 1$ (cf. exs. *8*, *13*, *16*, *20*, *22*),
(*d*) $g(x) = x^3 - 2$ (cf. exs. *15*, *22*),
(*e*) $g(x) = x^6 + 10x^3 + 125$ (cf. exs. *25*, *32*, *34*),
(*f*) $g(x) = x^4 - x^2 + 1$ (cf. ex. *30*),
(*g*) $g(x) = \Phi_{24}(x)$ (cf. exs. *31*, *45*).

CHAPTER 8

Complex Numbers and Quaternions

1. The Complex Numbers

1.1. *Geometric Representation*

Given a Cartesian coordinate system in the real Euclidean plane, the entire set of real numbers can be put in one-to-one correspondence with the dilatations of the plane, with the origin as center, in such a way that the real number a corresponds to the mapping

$$x' = ax, \qquad y' = ay. \tag{1}$$

Here (x, y), (x', y') are the coordinates of a point and its image, and (though often excluded in other contexts) the *zero dilatation* with $a = 0$ is here included. This mapping of the set of real numbers onto the set of dilatations is an isomorphism with respect to multiplication; for it is obvious that the product of two real numbers is mapped onto the product, i.e., the successive application, of the corresponding dilatations.

For $a = -1$ the mapping (1) represents a rotation through 180 degrees. If we now wish to extend the domain of real numbers in such a way as to include an element i with $i^2 = -1$, it is natural to define i as the rotation through 90 degrees, since its square (repeated application) is exactly a rotation through 180 degrees.[1] Thus it is appropriate for us to include the rotations (again with origin as center), or in other words the mappings

$$x' = ax - by, \qquad y' = bx + ay \tag{2}$$

with $a^2 + b^2 = 1$; here the angle of rotation φ is determined by $a = \cos\varphi$, $b = \sin\varphi$. Then in order to have unrestricted multiplication we must

[1] Of course, the same remark also holds for the rotation through 270 degrees, which could just as well be taken as the definition of i.

also include arbitrary products of a dilatation with a rotation, i.e., the *dilative rotations*, which are obtained by simply dropping the restriction $a^2 + b^2 = 1$ in (2).

By (2) the dilative rotations are set in one-to-one correspondence with the vectors $\mathfrak{e}_1 a + \mathfrak{e}_2 b$, where $\mathfrak{e}_1$, $\mathfrak{e}_2$ are the basis vectors of the coordinate system, i.e., the vectors with coordinates (1, 0), (0, 1). It is obvious that the relationship between a dilative rotation and the corresponding vector can also be described in the following way: the dilative rotation takes the vector $\mathfrak{e}_1$ into the vector corresponding to the rotation. Thus the complex numbers we are seeking may be defined either as the dilative rotations or as the vectors. A plane in which the complex numbers are so represented is called the *Gauss plane*. If we choose the first possibility, multiplication of the dilative rotations at once defines multiplication of complex numbers and shows that the nonzero complex numbers form a commutative group with respect to multiplication. Addition of complex numbers is simply defined as the addition of the corresponding vectors: if the complex numbers z, z' correspond to the vectors $\mathfrak{z}$, $\mathfrak{z}'$, then that dilative rotation which takes $\mathfrak{e}_1$ into $\mathfrak{z} + \mathfrak{z}'$ is denoted by $z + z'$. Thus the complex numbers also form a commutative group with respect to addition. In order to show that the set of complex numbers with these definitions of addition and multiplication constitutes a field, it only remains to prove the distributive law of multiplication. For the proof we first note that if

$$z' = \alpha z, \tag{3}$$

then the complex number (the dilative rotation) α takes the vector corresponding to z into the vector corresponding to z', as can be seen at once by applying αz (first z, and then α) to $\mathfrak{e}_1$. Thus (3) represents the dilative rotation α as a left-multiplication $z \to \alpha z$ in the domain of complex numbers. Now if $\mathfrak{z}_1$, $\mathfrak{z}_2$ are the vectors corresponding to the complex numbers z_1, z_2, then the vectors corresponding to αz_1, αz_2, $\alpha(z_1 + z_2)$ are given by the images of $\mathfrak{z}_1$, $\mathfrak{z}_2$, $\mathfrak{z}_1 + \mathfrak{z}_2$ under α. But the sum of the images of $\mathfrak{z}_1$, $\mathfrak{z}_2$ under the affine mapping α is the image of the sum $\mathfrak{z}_1 + \mathfrak{z}_2$ and by the definition of addition the vector corresponding to $\alpha z_1 + \alpha z_2$ is the sum of the vectors for αz_1, αz_2, and it follows that $\alpha(z_1 + z_2)$ and $\alpha z_1 + \alpha z_2$ have the same corresponding vector: in other words, $\alpha(z_1 + z_2) = \alpha z_1 + \alpha z_2$, so that multiplication is distributive.

The correspondence between the dilatation (1) and the real number a is obviously an isomorphism of the field of real numbers onto the set of dilatations. Thus we may equate the real number a with the dilatation (1), or in other words with that complex number to which the vector $\mathfrak{e}_1 a$ is in correspondence (cf. the procedure in IB1, §4.4 in (62)). The dilative rotation corresponding to the vector $\mathfrak{e}_2$, i.e., the rotation through

90 degrees has already been denoted by i. Then the dilative rotation ib is the image of $\mathfrak{e}_2$ under the dilatation with the factor b, so that the dilative rotation ib corresponds to $\mathfrak{e}_2 b$ and the dilative rotation $a + ib$ therefore corresponds to the vector $\mathfrak{e}_1 a + \mathfrak{e}_2 b$. Thus the complex number $a + ib$ is precisely the dilative rotation represented by (2). In this complex number $\alpha = a + ib$ the *real part* $a = Re\,\alpha$ and the *imaginary part* $b = Im\,\alpha$ of α are, of course, uniquely determined. In polar coordinates $a = r\cos\varphi$, $b = r\sin\varphi$ $(r \geqslant 0)$ where the angle φ is the oriented angle between $\mathfrak{e}_1$ and $\mathfrak{e}_1 a + \mathfrak{e}_2 b$ and is thus the angle of rotation of the dilative rotation $a + ib = r(\cos\varphi + i\sin\varphi)$. The fact that under successive rotations the angles of rotation are simply added is now expressed by the equation

$$(4) \quad (\cos\varphi + i\sin\varphi)(\cos\varphi' + i\sin\varphi') = \cos(\varphi + \varphi') + i\sin(\varphi + \varphi');$$

multiplying out on the left and comparing real and imaginary parts on both sides, we see that this equation is merely a combination of the theorems of addition for cosines and sines. Complete induction on n yields from (4) the *de Moivre formula*

$$(5) \qquad (\cos\varphi + i\sin\varphi)^n = \cos n\varphi + i\sin n\varphi$$

for all natural numbers n. In view of the fact that $(\cos n\varphi + i\sin n\varphi)^{-1} = \cos n\varphi - i\sin n\varphi = \cos(-n)\,\varphi + i\sin(-n)\,\varphi$ this formula also holds for negative integers n.

The number $r = \sqrt{a^2 + b^2}$ in polar coordinates is the length, or modulus, of the vector $\mathfrak{e}_1 a + \mathfrak{e}_2 b$ and is thus also called the *modulus* $|\alpha|$ (or the *absolute value*) of the complex number α. Since $\sqrt{a^2} = |a|$ (see IB1, (66)), this definition is in agreement for real numbers a with the definition of absolute value in IB1, §3.4. The triangle inequality in vector algebra shows that the modulus of complex numbers also satisfies the inequality IB1, (53), and thus we also have the following consequence (IB1, (54)):

$$||\alpha| - |\beta|| \leqslant |\alpha + \beta| \leqslant |\alpha| + |\beta|.$$

Since $|\alpha|$ is obviously the dilatation factor for the dilative rotation α, we also obtain (in agreement with IB1, (52)) the equation $|\alpha\beta| = |\alpha|\,|\beta|$.

The addition of complex numbers was defined above as the addition of the corresponding vectors. Thus the vector space of complex numbers is isomorphic to the two-dimensional (geometric) vector space (cf. IB3, §3.1). We will now derive the relations that hold between the multiplication of complex numbers and the two familiar vector multiplications.

From the above isomorphism

$$z = a + bi \rightarrow \mathfrak{z} = \mathfrak{e}_1 a + \mathfrak{e}_2 b$$

we see that for the vectors $\mathfrak{x} = \mathfrak{e}_1 x_1 + \mathfrak{e}_2 x_2$, $\mathfrak{y} = \mathfrak{e}_1 y_1 + \mathfrak{e}_2 y_2$ we have the following relation between the inner product $\mathfrak{x} \cdot \mathfrak{y} = x_1 y_1 + x_2 y_2$ (see IB3, §3.2) and the outer (or exterior) product (see IB3, §3.3) $[\mathfrak{x}, \mathfrak{y}] = x_1 y_2 - x_2 y_1$ (with $\bar{\mathfrak{x}}$ as the complex conjugate of $\mathfrak{x}$)

$$\mathfrak{x}\mathfrak{y} = \mathfrak{x} \cdot \mathfrak{y} + i[\mathfrak{x}, \mathfrak{y}].$$

Taking complex conjugates on both sides of this equation (§1.2), we have $\mathfrak{x}\bar{\mathfrak{y}} = \mathfrak{x} \cdot \mathfrak{y} - i[\mathfrak{x}, \mathfrak{y}]$. From these two equations we obtain

$$\mathfrak{x} \cdot \mathfrak{y} = \tfrac{1}{2}(\bar{\mathfrak{x}}\mathfrak{y} + \mathfrak{x}\bar{\mathfrak{y}}),$$

$$[\mathfrak{x}, \mathfrak{y}] = \frac{1}{2i}(\bar{\mathfrak{x}}\mathfrak{y} - \mathfrak{x}\bar{\mathfrak{y}}).$$

Thus the inner and the outer product of vectors in the plane have been reduced to the multiplication of complex numbers.

1.2. *Algebraic Methods of Introducing the Complex Numbers*

The geometric introduction of the complex numbers has the advantage that the operations of addition and multiplication for complex numbers are reduced to well-known geometric operations (addition of vectors, multiplication of mappings), but this procedure takes no account of the fact that the construction of the complex numbers is a purely algebraic question. To realize the truth of this statement, we have only to replace the dilative rotations by the corresponding matrices

$$\begin{pmatrix} a & -b \\ b & a \end{pmatrix}. \tag{6}$$

This set of matrices is closed with respect to addition, subtraction, and multiplication of matrices (see IB3, §2.2) and thus forms a ring under these operations. The commutativity of multiplication is easily shown. Since the determinant of (6) has the value $a^2 + b^2$, such a matrix (unless it is the zero matrix) has the inverse

$$\begin{pmatrix} ac^{-1} & bc^{-1} \\ -bc^{-1} & ac^{-1} \end{pmatrix}, \quad \text{with} \quad c = a^2 + b^2.$$

Thus the matrices (6) even form a field. Essential for the proof of this latter statement is the following property of the real numbers:

(7) *If $a \neq 0$ or $b \neq 0$, then $a^2 + b^2 \neq 0$,*

which follows immediately from the order properties of the real numbers (see IB1, §3.4).

In order to regard every real number as a complex number, it is only necessary to identify a with $\begin{pmatrix} a & 0 \\ 0 & a \end{pmatrix}$. If we then define $i = \begin{pmatrix} 0 & -1 \\ 1 & 0 \end{pmatrix}$, the matrix (6) can also be written in the form $a + ib$:

$$a + ib = \begin{pmatrix} a & 0 \\ 0 & a \end{pmatrix} + \begin{pmatrix} 0 & -1 \\ 1 & 0 \end{pmatrix}\begin{pmatrix} b & 0 \\ 0 & b \end{pmatrix} = \begin{pmatrix} a & -b \\ b & a \end{pmatrix}.$$

The determinant $c = a^2 + b^2 = | a + ib |^2$ of (6) is called the *norm* $N(a + ib)$ of $a + ib$.

This purely algebraic definition of complex numbers shows at the same time that the only property required of the real numbers, apart from the fact that they form a field, is the property (7). Now for a field K the property (7) is equivalent to

$$(8) \qquad c^2 \neq -1 \textit{ for all } c \in K;$$

for if $a = c$, $b = 1$, then (8) follows at once from (7), and conversely, given (8) and, let us say, $a \neq 0$, we may set $c = ba^{-1}$, from which $a^2 + b^2 \neq 0$ from $c^2 \neq -1$. Thus the complex numbers can be introduced for any field K with property (8): the result is an extension field of K whose elements can all be expressed rationally (and even linearly) in terms of i and the elements of K, so that this extension field (by IB7, §1.1) can be denoted by $K(i)$.

The field $K(i)$ is a vector space of dimension 2 over K with the basis $\{1, i\}$, since $a + ib$ uniquely determines the pair (a, b). Thus in forming $K(i)$ we can dispense with the matrices; we simply take a two-dimensional vector space over K; for example, the space of pairs (a, b) with $a, b \in K$ (cf. IB3, §1.2). If $\{\mathfrak{e}_1, \mathfrak{e}_2\}$ is a basis of this vector space with, let us say, $\mathfrak{e}_1 = (1, 0)$, $\mathfrak{e}_2 = (0, 1)$, we then seek to introduce a distributive multiplication in such a way that $\mathfrak{e}_1$ is the unit element and $\mathfrak{e}_2{}^2 = -\mathfrak{e}_1$. If we also require that $(\mathfrak{e}_\nu a)(\mathfrak{e}_\mu b) = (\mathfrak{e}_\nu \mathfrak{e}_\mu)(ab)$ for $\nu, \mu = 1, 2$ and for all $a, b \in K$, the multiplication must be as follows:

$$(9) \quad (\mathfrak{e}_1 a_1 + \mathfrak{e}_2 a_2)(\mathfrak{e}_1 b_1 + \mathfrak{e}_2 b_2) = \mathfrak{e}_1(a_1 b_1 - a_2 b_2) + \mathfrak{e}_2(a_1 b_2 + a_2 b_1).$$

In particular, if we have constructed the vector space of the pairs (a_1, a_2) and have taken $\mathfrak{e}_1 = (1, 0)$, $\mathfrak{e}_2 = (0, 1)$, then (9) can be written more simply as

$$(a_1, a_2)(b_1, b_2) = (a_1 b_1 - a_2 b_2, a_1 b_2 + a_2 b_1).$$

We must now show that under the multiplication defined by (9) the vector space is actually a field. The calculations necessary for this purpose become considerably more concise if we introduce matrices, so that it is not desirable to dispense with them entirely. Finally we must still show that $x \to \mathfrak{e}_1 x$ is an isomorphism of the field K onto the set of multiples of $\mathfrak{e}_1$, so that we may set $x = \mathfrak{e}_1 x = x$ without giving rise to difficulties as a result of this identification of the operations in K and in the new field. There is a further danger consisting in the fact that $(\mathfrak{e}_1 a_1 + \mathfrak{e}_2 a_2)\, b$ is already defined in the vector space, namely as $\mathfrak{e}_1(a_1 b) + \mathfrak{e}_2(a_2 b)$. But since exactly the same value is determined by (9) when b is replaced by $\mathfrak{e}_1 b$, everything is in order.

Finally the introduction of complex numbers can be subsumed under the procedure described in IB5, §3.10: in the polynomial ring $K[x]$ we we compute mod $x^2 + 1$; that is, we form the residue classes[2] $\overline{f(x)}$ of $f(x) \in K[x]$. Again, it is not altogether necessary to define $\overline{f(x)}$ as the set of polynomials $\equiv f(x) \bmod x^2 + 1$; alternatively, in accordance with the procedure described in another connection in IB1, §2.2, we may consider $\overline{f(x)}$ as a new symbol formed from a polynomial $f(x)$ with the conventions: $\overline{f(x)} = \overline{g(x)}$ if and only if $f(x) \equiv g(x) \bmod x^2 + 1$; $\overline{f(x)} + \overline{g(x)} = \overline{f(x) + g(x)}$, $\overline{f(x)}\,\overline{g(x)} = \overline{f(x)\, g(x)}$. Then the $\overline{f(x)}$ form a commutative ring, namely the ring of residue classes of $K[x]$ with respect to $x^2 + 1$. Since $x^2 + 1$, by (8), has no zero in K, this polynomial is irreducible and the ring of residue classes is thus a field (see IB5, §3.10). If $f(x) \equiv a \bmod x^2 + 1$ with $a \in K$, we set $a = \overline{f(x)} = a$. The division algorithm (IB6, §2.10) shows that to every $f(x) \in K[x]$ there correspond uniquely determined elements $a, b \in K$ with $f(x) \equiv a + bx \bmod x^2 + 1$, so that $\overline{f(x)} = a + b\bar{x}$. Thus we need only set $\bar{x} = i$ (then $i^2 = -1$, since $x^2 \equiv -1 \bmod x^2 + 1$) in order to write the field of residue classes in the form $K(i)$.

We now investigate the automorphisms of $K(i)$ with respect to K (cf. IB7, §1.2). If σ is such an automorphism, then $i^2 = -1$ naturally implies $\sigma(i)^2 = -1$. Thus, since $x^2 + 1 = (x - i)(x + i)$, we must have either $\sigma(i) = i$ or $\sigma(i) = -i$. Since $\sigma(a + ib) = a + \sigma(i)\, b$ we have in the first case the identical automorphism and in the second case the mapping $a + ib \to a - ib$. The fact that this mapping is also an automorphism of $K(i)$ with respect to K follows either from the general theory (IB7, §6) or from the fact that in calculations involving the sum and product of complex numbers only the special property $i^2 = -1$ is utilized (beyond the general rules for addition and multiplication) and this property holds for $-i$, since $(-i)^2 = -1$. The element $a - ib$ is called the *complex conjugate* $\overline{a + ib}$ of $a + ib$. Obviously $\alpha \in K$ is equivalent to $\alpha = \bar{\alpha}$, and in general $\alpha + \bar{\alpha} = 2 \operatorname{Re} \alpha$, $\alpha - \bar{\alpha} = 2i \operatorname{Im} \alpha$. Since the passage to complex conjugates is an automorphism and $N(a + ib) = (a + ib)(a - ib)$, it follows that

$$N(\alpha\beta) = N(\alpha)\, N(\beta); \tag{10}$$

[2] Of course, the overbar here has nothing to do with the notation for a complex conjugate.

for we have $N(\alpha\beta) = \alpha\beta\bar{\alpha}\bar{\beta} = \alpha\bar{\alpha}\beta\bar{\beta}$. Incidentally, the equation (10) enables us to write the product of sums of two squares again as the sum of two squares; for example, $(1^2 + 2^2)(3^2 + 4^2) = 5^2 + 10^2$. Finally, we remark that for $\alpha \neq 0$ the equation $\alpha\bar{\alpha} = N(\alpha)$ implies $\alpha^{-1} = N(\alpha)^{-1}\bar{\alpha}$.

2. Algebraic Closedness of the Field of Complex Numbers

2.1. *The Intermediate Value Theorem*

A real function (i.e., a mapping of a set of real numbers, namely of the domain of definition of f, into the set of real numbers) is said to be *continuous at the point* x if for every real number $\epsilon > 0$ there exists a real number $\delta > 0$ such that for all x' in the domain of definition of f with $|x - x'| < \delta$ we have $|f(x) - f(x')| < \epsilon$. It is shown in analysis (see II2, §1) that the rational entire functions (for their definition see IB4, §1.1) are everywhere (i.e., at every point) continuous in the field of real numbers.

The *intermediate value theorem* now states: *if the real function f is defined and continuous at every point x with $a \leqslant x \leqslant b$ $(a < b)$ and if*[3] $f(a) < C < f(b)$, *then there exists a value c with $a < c < b$ such that* $f(c) = C$.

Proof. Let M be the set of x with $a \leqslant x \leqslant b$ and $f(x) < C$. Since $f(a) < C$, the set M is not empty. Thus there exists a real number c, with $a \leqslant c \leqslant b$, which is the least upper bound of M. If $f(c) < C$, and thus in particular $c < b$, we set $C - f(c) = \epsilon$ and then, in view of the continuity of f, we can determine δ with $c < c + \delta \leqslant b$ such that $|f(x) - f(c)| < \epsilon$ for all x with $c \leqslant x < c + \delta$. For these values of x we would then have $f(x) = f(c) + f(x) - f(c) < C$, in contradiction to the fact that c is an upper bound of M. On the other hand, if $f(c) > C$, and thus in particular $a < c$, we set $f(c) - C = \epsilon$ and determine δ with $a \leqslant c - \delta < c$ such that $|f(x) - f(c)| < \epsilon$ for all x with $c - \delta < x \leqslant c$. For these values of x we would then have $f(x) = f(c) + f(x) - f(c) < C$, so that $c - \delta$ $(<c)$ would be an upper bound of M, in contradiction to the definition of c as the least upper bound of M.

In view of the above-mentioned continuity of the rational entire functions, we have thus proved the *intermediate value theorem for polynomials* over the field of real numbers, a theorem which for greater simplicity we now write in the form of a theorem on the zeros of a polynomial:[4]

[3] We could also assume $f(a) > C > f(b)$, which would amount to the above case if we replace f, C by $-f$, $-C$.

[4] By applying this theorem to the polynomial $f(x) - C$ we obtain the general intermediate value theorem (for polynomials).

If the polynomial $f(x)$ satisfies the inequality $f(a) < 0 < f(b)$ with $a < b$, then $f(x)$ has a zero c with $a < c < b$.

This theorem holds not only for the field of real numbers but also for certain other ordered fields. For example, we obtain such a field if we restrict ourselves to the real algebraic numbers. These numbers actually form a field A; for it was shown in IB7, §2 that for two elements that are algebraic over a field K (here the field R of rational numbers) the difference and the quotient of the two elements are also algebraic over K. If $\sum_{\nu=0}^{n} a_\nu c^\nu = 0$, $a_\nu \in A$, $a_n \neq 0$, the real number c is algebraic over $R(a_0, ..., a_n)$ and thus (by IB7, §2) also over R, and is therefore an algebraic number. Consequently, we have also proved the intermediate value theorem for polynomials over A. *If the intermediate value theorem holds for polynomials over a field K, we shall say for brevity*: *the intermediate value theorem holds in K.*

From the intermediate value theorem for polynomials we have the *Sturm theorem*, which allows us in general to state the number of distinct zeros c with $a < c < b$: for if we are given the polynomial $f(x) \in K[x]$, let us form the *Sturm chain* $(f(x), f'(x), f_1(x), ..., f_r(x))$, where the $f_k(x)$ are determined by $f_{k-1}(x) = q_k(x) f_k(x) - f_{k+1}(x)$ $(k = 0, ..., r-1)$, $f_{-1}(x) = f(x), f_0(x) = f'(x), f_{r-1}(x) = q_r(x) f_r(x)$, with polynomials $q_k(x)$ and degree $f_k(x) <$ degree $f_{k-1}(x)$ $(k = 1, ... r)$; and if now for $u \in K$ we denote the number of changes of sign in the sequence

$$(f(u), f'(u), f_1(u), ..., f_r(u))$$

by $w(u)$, where zeros are disregarded, it follows that if $f(a), f(b) \neq 0$, $a < b$, then $f(x)$ has exactly $w(a) - w(b)$ distinct zeros c with $a < c < b$. We shall not prove this theorem here[5] but will merely illustrate it for the polynomial $x^2 - 1$. Here the Sturm chain is $(x^2 - 1, 2x, 1)$, and we have $w(u) = 2$ for $u < -1$, $w(u) = 1$ for $-1 \leqslant u \leqslant 1$, $w(u) = 0$ for $u > 1$. Thus the assertion of the Sturm theorem actually follows for the polynomial $x^2 - 1$.

2.2. *Real-Closed Fields*

A field K is said to be *real-closed* if it can be ordered in such a way that the intermediate value theorem (for polynomials) is valid. We now wish to deduce for real-closed fields a certain characterization in which there is no mention of the order. By IB1, §3.4 we know that in every ordered field K, and thus in every real-closed field, we have

$$\sum_{k=1}^{n} a_k^2 \neq -1, \qquad \textit{if} \quad a_1, ..., a_n \in K. \tag{11}$$

[5] A proof is to be found, e.g., in van der Waerden [2], §69.

We also have, in every real-closed field:

(12) *If* $a \in K$ *and* $a \neq b^2$ *for all* $b \in K$, *there exists a* $c \in K$ *with* $-a = c^2$.

For the proof we assume that K is ordered in such a way that the intermediate value theorem holds in K. If $a \geqslant 0$, there exists an element $b \in K$ with $a = b^2$, as was already deduced in IB1, §4.7 from the intermediate value theorem. Thus, under the hypothesis in (12), we must have $a < 0$ and therefore $-a > 0$, so that, as we have just remarked, there must exist an element $c \in K$ with $-a = c^2$.

Finally, we have the following theorem for a real-closed field K:

(13) *Every polynomial of odd degree in* $K[x]$ *has a zero in* K.

For the proof we may restrict our attention to a polynomial $f(x) = x^n + \sum_{k=0}^{n-1} a_k x^k$ $(a_k \in K)$. If we let b denote the maximum of $1, 1 - a_0, ..., 1 - a_{n-1}$, we have $b^k > 0$, $a_k \geqslant -(b-1)$ and thus $f(b) \geqslant b^n - (b-1)\sum_{k=0}^{n-1} b^k = 1$. Applying the same procedure to the polynomial $-f(-x)$, whose leading term must also be x^n in view of the fact that n is odd, we obtain an $a \leqslant -1$ with $f(a) \leqslant -1$. By the intermediate value theorem there must exist a zero c of $f(x)$ with $a < c < b$.

But we have herewith obtained the desired characterizing properties: *a field* K *is real-closed if and only if it has the properties* (11), (12), (13).

It remains only to prove that (11), (12), (13) imply that K can in fact be ordered in such a way that the intermediate value theorem holds in K. We first show that on the basis of (11), (12) the field K can be ordered in exactly one way. By IB1, §3.4 we know that a domain of positivity in K will in any case contain all the squares a^2 with $a \neq 0$, $a \in K$. Thus the desired result will follow if for the set P of these squares we can deduce the characterizing properties IB1, (44) of a domain of positivity. The relations IB1, $(44_{1,4})$ are obvious and IB1, (44_2) follows at once from (12). But if $a^2 + b^2$ were not a square (so that in particular $a, b \neq 0$), then by (12) there would exist an element $c \in K$ with $-(a^2 + b^2) = c^2$, which would imply $-1 = (a/c)^2 + (b/a)^2$ in contradiction to (11). Since (11) also implies the impossibility of $a^2 + b^2 = 0$ for $a \neq 0$, we have thus completed the proof of IB1, (44_3). The proof that under the ordering defined by this domain of positivity (in which every positive element is a square) the intermediate value theorem follows from (13) will be postponed to §2.3, where the investigation of real-closed fields will be based not on the original definition of this concept but only on the properties (11), (12), (13).

As was proved in IB11, §3, the field K can be ordered if and only if (11) holds; in this case the field is said to be *formally real.* A field K is real-closed if and only if K itself is formally real but no proper algebraic extension of K has this property. We shall not give a proof of this fact, which is often used as a definition of real-closed fields. Likewise without proof we mention that every real-closed algebraic extension of the field of rational numbers is isomorphic to the field A of real algebraic numbers introduced in §2.1; in this way the latter field is characterized (up to isomorphism) in a purely algebraic way, and thus in particular without any use of the field of real numbers. For the proofs of these statements we refer the reader to van der Waerden [2], §71.

2.3. *Algebraic Closure of a Real-Closed Field*

Let the field K have the properties (11), (12), (13). Since (11) implies (8), we can form the extension field $K(i)$, as was done in §1.2. We now prove the following basic theorem.

A polynomial of positive degree in $K[x]$ has a zero in $K(i)$. Let us write the degree n of the polynomial $f(x)$ in the form $n = 2^l m$ (m odd) and prove the assertion by complete induction on l. For $l = 0$ the polynomial $f(x)$ has a zero in K itself, by (13). Thus we need only prove the assertion for $l > 0$ under the induction hypothesis, i.e., under the assumption that the assertion holds for polynomials whose degree is divisible by 2^{l-1} but not by 2^l. By IB7, §1.5 there exists an extension $L = K(i, \alpha_1, ..., \alpha_n)$ with $f(x) = c \prod_{\nu=1}^{n} (x - \alpha_\nu)$ $(c \in K)$. We now set $N = n(n-1)/2$ and form the polynomials of degree N in $L[x]$:

$$f_h(x) = \prod_{1 \leqslant \nu < \mu \leqslant n} (x - (\alpha_\nu \alpha_\mu + h(\alpha_\nu + \alpha_\mu))) \qquad \text{for} \quad h = 0, ..., N.$$

Since the coefficients of these polynomials are obviously the values of symmetric polynomials (in n indeterminates with rational integers as coefficients) for the arguments $\alpha_1, ..., \alpha_n$, it follows from the fundamental theorem on the elementary symmetric polynomials (see IB4, §2.4) and from the equations IB4, (25), (26) that these coefficients are already contained in K. Since $N = 2^{l-1}m(n-1)$ and $m(n-1)$ is odd, we can apply the induction hypothesis to $f_h(x)$: one of the zeros of $f_h(x)$ is already contained in $K(i)$. Thus for every value of $h (= 0, ..., N)$ there exists a pair of numbers (ν_h, μ_h) with $1 \leqslant \nu_h < \mu_h \leqslant N$ and a pair of elements $a_h, b_h \in K$ such that

$$\alpha_{\nu_h}\alpha_{\mu_h} + h(\alpha_{\nu_h} + \alpha_{\mu_h}) = a_h + ib_h .$$

But by the Dirichlet pigeonhole principle (see IB1, §1.5) the mapping $h \to (\nu_h, \mu_h)$ cannot be invertible, since the set of preimages contains $N + 1$ elements but the set of images contains at most N elements. Thus there exist two distinct values h, k with $\nu_h = \nu_k = \nu$, $\mu_h = \mu_k = \mu$. A simple calculation then shows[6]

$$\alpha_\nu \alpha_\mu = (h - k)^{-1}(h(a_k + ib_k) - k(a_h + ib_h)),$$

$$\alpha_\nu + \alpha_\mu = (h - k)^{-1}((a_h + ib_h) - (a_k + ib_k)).$$

Consequently, the coefficients of the polynomial $x^2 - (\alpha_\nu + \alpha_\mu) x + \alpha_\nu \alpha_\mu$, which has the zeros α_ν, α_μ, are contained in $K(i)$. Thus in order to show $\alpha_\nu, \alpha_\mu \in K(i)$ it only remains to prove that every element $a + ib$ of $K(i)$, and in particular the discriminant of the above polynomial, is a square in $K(i)$. But from the ordering of K (which by §2.2 is unique) we have $a^2 + b^2 \geqslant 0$, and thus there exists a $c \in K$ with $a^2 + b^2 = c^2$. Since c or $-c \geqslant 0$, we may also assume $c \geqslant 0$. Then $c + |a| \geqslant 0$, and from $(c + |a|)(c - |a|) = b^2 \geqslant 0$ it follows that $c - |a| \geqslant 0$, so that $c + a$, $c - a \geqslant 0$. Thus there exist elements $u, v \in K$ with $u^2 = (c + a)/2$, $v^2 = (c - a)/2$, $uvb \geqslant 0$, and then in view of $(2uv)^2 = b^2$ we have $2uv = b$ and therefore $(u + iv)^2 = (u^2 - v^2) + 2uvi = a + ib$.

On the basis of the theorem that has just been proved, we can show that $K(i)$ is algebraically closed, i.e., every polynomial $f(x)$ of positive degree in $K(i)[x]$ has a zero in $K(i)$. For if by $\bar{f}(x)$ we denote the polynomial whose coefficients are the complex conjugates of the corresponding coefficients of $f(x)$, a short calculation shows that the coefficients of $g(x) = f(x)\bar{f}(x)$ are identical with their complex conjugates and are thus contained in K. Consequently there exists an $\alpha \in K(i)$ with $g(\alpha) = 0$, so that $f(\alpha) = 0$ or $\bar{f}(\alpha) = 0$. In the second case the passage to complex conjugates shows at once that $f(\bar{\alpha}) = 0$, so that the proof of the theorem is complete.

From the fact that the field L is algebraically closed it follows in general that every polynomial of positive degree $L[x]$ splits into linear factors in this ring; that is, the polynomial is the product of linear polynomials. For if $\alpha_1, \ldots, \alpha_s$ are the distinct zeros of $f(x)$ in L with multiplicities $m_1, \ldots, m_s$, then by IB4, §2.2 there exists a polynomial $h(x) \in L[x]$ with $f(x) = h(x) \prod_{k=1}^{s} (x - \alpha_k)^{m_k}$. By the definition of the multiplicity of a zero (see IB4, §2.2) we have $h(\alpha_k) \neq 0$, so that $h(x)$ has no zero in K and must therefore, since L is algebraically closed, be of degree 0; in other words, $h(x)$ lies in L, which completes the desired factorization.

If the coefficients of $f(x)$ are already contained in K, then passage to complex conjugates shows that the equation $f(\alpha) = 0$ with $\alpha \in K(i)$ implies $f(\bar{\alpha}) = 0$: in other words, if the zero α of $f(x)$ is in $K(i)$ but not

[6] Since K can be ordered, it follows that if the integers h, k are distinct, then $h - k$ as *element of the field* (which for $h > k$ is a sum of $h - k$ summands, each equal to the unit element of K) is actually $\neq 0$.

in K, then $\bar{\alpha}$ is another zero of $f(x)$. If in the linear factorization of $f(x)$ we combine the factors $x - \alpha$, $x - \bar{\alpha}$, then in view of

$$(x - \alpha)(x - \bar{\alpha}) = x^2 - 2(\operatorname{Re} \alpha)\, x + N(\alpha),$$

we obtain a factorization of $f(x)$ into two factors in $K[x]$, one of which is quadratic. By complete induction on the degree of $f(x)$ it follows that every polynomial of positive degree in $K[x]$ can be factored into linear and quadratic factors. But now we can make good the omission in §2.2, by proving the intermediate value theorem: for if $f(a)$, $f(b)$ are of different sign, then at least one of the factors in the factorization of $f(x)$ must be of different sign for the values a and b; but for a quadratic factor this is impossible in view of

$$(x - \alpha)(x - \bar{\alpha}) = (x - \operatorname{Re} x)^2 + (\operatorname{Im} \alpha)^2,$$

and for the linear factor $x - c$ it means that $a - c < 0 < b - c$; in other words, $f(x)$ actually has a zero $c \in K$ with $a < c < b$.

If for K we now take the field of real numbers and then the field of complex numbers, we have the following result:

In the ring of polynomials in one indeterminate[7] *over the field of real numbers every polynomial of positive degree can be factored into linear and quadratic factors; and over the field of complex numbers every such polynomial can be factored into linear factors.*

This theorem is often called the *fundamental theorem of algebra*. The name was justified as long as algebra was confined to the study of the field of complex numbers and the fields and rings contained in it. But today the field of complex numbers has lost its central importance for algebra. It would be better to call this theorem the *fundamental algebraic theorem for complex numbers*.[8]

3. Quaternions

In §1 we have constructed the field of complex numbers as an extension of the field of real numbers. It is now natural to ask whether we can proceed in the same way beyond the complex numbers. Such an extension is possible, as we shall show below, if we abandon the commutativity of multiplication; in §3.4 we shall discover to what extent such a weakening of the axioms is in fact necessary.

[7] Even for only two indeterminates the theorem is no longer true; the quadratic polynomial $1 + xy$ in the indeterminates x, y cannot be factored into linear factors over any field.

[8] As the *fundamental topological theorem* for complex numbers we could consider the Cauchy convergence criterion, i.e., the fact that every fundamental sequence (see IB1, §4.4) of complex numbers has a limit.

3.1. *Quaternions as Hermitian Dilative Rotations*

In order to repeat for the complex numbers the step which in §1 led us to them from the real numbers we now introduce into the affine complex plane a *Hermitian metric*; in other words the length of a vector with the complex coordinates z_1, z_2 (in a given coordinate system) is now defined as the square root of the positive real number $z_1\bar{z}_1 + z_2\bar{z}_2$. By *Hermitian rotations* about the origin we now mean affine length-preserving mappings with determinant[9] 1, leaving the origin fixed. Then how are these mappings to be expressed? An affine mapping with the origin as fixed point is given by

$$\begin{aligned} z_1' &= \alpha z_1 + \gamma z_2 \\ z_2' &= \beta z_1 + \delta z_2, \end{aligned} \tag{14}$$

and the fact that lengths are preserved means that

$$\begin{aligned} z_1\bar{z}_1 + z_2\bar{z}_2 = (\alpha\bar{\alpha} + \beta\bar{\beta})\, z_1\bar{z}_1 + (\alpha\bar{\gamma} + \beta\bar{\delta})\, z_1\bar{z}_2 + (\bar{\alpha}\gamma + \bar{\beta}\delta)\, \bar{z}_1 z_2 \\ + (\gamma\bar{\gamma} + \delta\bar{\delta})\, z_2\bar{z}_2 . \end{aligned} \tag{15}$$

If in (15) we set $z_1 = 1$, $z_2 = 0$ and then $z_1 = 0$, $z_2 = 1$, we obtain the following equations

$$\alpha\bar{\alpha} + \beta\bar{\beta} = 1, \qquad \gamma\bar{\gamma} + \delta\bar{\delta} = 1. \tag{16}$$

For $z_1 = z_2 = 1$ and $z_1 = 1$, $z_2 = i$ the equations (15) and (16) imply

$$(\alpha\bar{\gamma} + \beta\bar{\delta}) + (\bar{\alpha}\gamma + \bar{\beta}\delta) = 0, \qquad (\alpha\bar{\gamma} + \beta\bar{\delta}) - (\bar{\alpha}\gamma + \bar{\beta}\delta) = 0,$$

from which it follows that

$$\bar{\alpha}\gamma + \bar{\beta}\delta = 0. \tag{17}$$

Conversely, the preservation of length, or in other words (15), follows from (16), (17). Taken together with the requirement on the determinant $\beta\gamma - \alpha\delta = -1$ and the first equation (16), the equation (17) means that $\gamma = -\bar{\beta}$, $\delta = -\bar{\alpha}$, so that by (14) the Hermitian rotations have the form

$$\begin{aligned} z_1' &= \alpha z_1 - \bar{\beta} z_2, \\ z_2' &= \beta z_1 + \bar{\alpha} z_2, \end{aligned} \tag{18}$$

with $\alpha\bar{\alpha} + \beta\bar{\beta} = 1$. If after a mapping of this sort we apply a dilatation with a real factor (which may $= 0$), we again obtain a mapping of the form (18), but now without the requirement $\alpha\bar{\alpha} + \beta\bar{\beta} = 1$. Conversely,

[9] It would actually be enough to require that the determinant be real and positive.

it is obvious that every mapping (18) can be produced by the successive application of a Hermitian rotation and a dilatation (with the real dilatation factor $\alpha\bar{\alpha} + \beta\bar{\beta}$). Thus the mappings (18) are called *Hermitian dilative rotations.*

But for the sake of simplicity we shall operate below not with these mappings but with the matrices

$$A = \begin{pmatrix} \alpha & -\bar{\beta} \\ \beta & \bar{\alpha} \end{pmatrix}, \tag{19}$$

which (if the coordinate system is fixed) are in one-to-one correspondence with such mappings. As can easily be seen, these matrices form a subring of the ring of all two-rowed square matrices; in other words, the sum, difference, and product of two matrices (19) are again of the same form. Then the mapping

$$a \to \begin{pmatrix} a & 0 \\ 0 & a \end{pmatrix}$$

is seen at once to be an isomorphism of the field of real numbers into the ring of matrices, so that without fear of misunderstanding we may set

$$a = \begin{pmatrix} a & 0 \\ 0 & a \end{pmatrix} \tag{20}$$

for all real numbers. In this manner our subring of the ring of matrices becomes an extension ring Q of the field of real numbers. For $A \neq 0$ we have $|A| = \alpha\bar{\alpha} + \beta\bar{\beta} \neq 0$, in view of the fact that $\alpha\bar{\alpha} + \beta\bar{\beta}$, as the sum of four squares, can be equal to zero only if all four summands, and thus also α, β, are equal to zero. Since $|A|$ is real, it is easy to show that A^{-1} again lies in Q. Thus Q is actually a skew field (see the definition in IB3, §1.1). If we set[10]

$$j = \begin{pmatrix} 0 & i \\ i & 0 \end{pmatrix}, \quad k = \begin{pmatrix} 0 & -1 \\ 1 & 0 \end{pmatrix}, \quad l = \begin{pmatrix} i & 0 \\ 0 & -i \end{pmatrix}, \tag{21}$$

$$\alpha = a_0 + ia_3, \quad \beta = a_2 + ia_1, \tag{22}$$

it follows from (19) and (20) that

$$A = a_0 + ja_1 + ka_2 + la_3. \tag{23}$$

[10] It is customary to write i, j, k instead of j, k, l but we have intentionally avoided this notation, since the square of each of the quaternions j, k, l is equal to -1; thus, if we wish, we may take any one of these quaternions to be the complex number i, if on the basis of an isomorphism of the field of complex numbers into the skew field of quaternions we set the complex numbers equal to certain quaternions.

Since $a_0 + ja_1 + ka_2 + la_3 = 0$ (a_0, a_1, a_2, a_3 real) implies $\alpha = \beta = 0$ and thus $a_0 = a_1 = a_2 = a_3 = 0$, the $1, j, k, l$ are linearly independent over the field of real numbers, so that Q is a vector space of dimension 4 over this field. It is for this reason that Q is called the *skew field of quaternions* and its elements are called *quaternions* (Latin *quaternio* = set of four).

By (20) the real number 1 is to be replaced in the multiplication of quaternions by the unit matrix, and thus it is the unit element of Q. From (19), (22) and $A^* = \begin{pmatrix} \bar{\alpha} & \bar{\beta} \\ -\beta & \alpha \end{pmatrix}$ we have

$$AA^* = A^*A = \alpha\bar{\alpha} + \beta\bar{\beta} = \sum_{\nu=0}^{3} a_\nu^2, \qquad A^* = a_0 - ja_1 - ka_2 - la_3 .$$

The number $\sum_{\nu=0}^{3} a_\nu^2$ is called the *norm* $N(A)$ of A, and A^* the *quaternion conjugate* of A. Since A^* arises from A by transposition (see IB3, §2.6) and passage to complex conjugates, the invertible mapping $A \to A^*$ of Q onto itself is an *antiautomorphism*; i.e., $(A + B)^* = A^* + B^*$ as in the case of an automorphism, but now $(AB)^* = B^*A^*$. In exact analogy with the equation (10) we have*

$$N(AB) = AB(AB)^* = ABB^*A^* = AA^*N(B) = N(A)\,N(B).$$

Thus we can express the product of sums of four squares as the sum of four squares; for example,

$$(1^2 + 2^2 + 3^2 + 4^2)(5^2 + 6^2 + 7^2 + 8^2) = 12^2 + 24^2 + 30^2 + 60^2.$$

For $A \neq 0$ we obtain, in the same way as for complex numbers: $A^{-1} = N(A)^{-1}A^*$.

From (21) it follows that

$$(24) \qquad \begin{array}{llll} j^2 = k^2 = l^2 = -1, & jk = l, & kl = j, & lj = k, \\ & kj = -l, & lk = -j, & jl = -k. \end{array}$$

Thus multiplication is not commutative, so that the quaternions do not form a field. However, by (20) we have $aA = Aa$ for every real number a and $A \in Q$. Thus, exactly as in §1.2, we can construct the skew field Q by choosing a four-dimensional vector space with a basis $\{\mathfrak{e}_0, \mathfrak{e}_1, \mathfrak{e}_2, \mathfrak{e}_3\}$ and defining for it a distributive multiplication in such a way that $(\mathfrak{e}_\nu a)(\mathfrak{e}_\mu b) = (\mathfrak{e}_\nu \mathfrak{e}_\mu)(ab)$ for $\nu, \mu = 0, 1, 2, 3$ and arbitrary a, b, where e_0 is the unit element and equations (24) hold for $\mathfrak{e}_0$, $\mathfrak{e}_1$, $\mathfrak{e}_2$, $\mathfrak{e}_3$ in place of $1, j, k, l$.

* The third equality follows from the fact that $aA = Aa$ for every quaternion A and every real number a, and $N(B)$ is real.

In the preceding discussion we could equally well replace the field of real numbers by any other field K in which $\sum_{\nu=0}^{3} a_\nu^2 = 0$ implies $a_\nu = 0$ ($\nu = 0, 1, 2, 3$); for in order to show that Q is a ring we do not need to impose any conditions on the field K, and Q is a skew field if and only if $N(A) = |A| = \sum_{\nu=0}^{3} a_\nu^2 = 0$ implies $A = 0$, so that $a_0 = a_1 = a_2 = a_3 = 0$. The above condition on K is equivalent to the condition that -1 is not a sum of two squares of elements from K. For on the one hand the equation $1 + a^2 + b^2 = 0$ $(a, b \in K)$ obviously violates the given condition, and on the other hand if -1 is not the sum of the squares of two elements from K and if a_ν ($\nu = 0, \ldots, 3$) in K are not all $= 0$, then in order to prove $\sum_{\nu=0}^{3} a_\nu^2 \neq 0$ we may without loss of generality take $a_2 \neq 0$. In view of the fact that $(a_1/a_2)^2 \neq -1$ we then have $a_1 + a_2 \neq 0$, so that $(a_0^2 + a_3^2)(a_1^2 + a_2^2)^{-1} = N(\alpha\beta^{-1})$ by (22). Since a norm, as the sum of the squares of two elements in K, must always be $\neq -1$, we thus have $\sum_{\nu=0}^{3} a_\nu^2 \neq 0$.

Let us also remark that the concept of the quaternion skew field can be generalized in the following way: in a field K let there exist elements c, d with $c \neq 0$ and $-d \neq x^2 + y^2 c$ for all $x, y \in K$; the construction of Q, starting with a four-dimensional vector space over K, is now altered by replacing[11] (24) with

$$j^2 = -c, \quad k^2 = -d, \quad l^2 = -cd, \quad jk = l, \quad kl = jd, \quad lj = kc,$$
$$kj = -l, \quad lk = -jd, \quad jl = -kc$$

These generalized quaternion skew fields arise if we ask for those skew extensions of K in which every element u satisfies the equation $ua = au$ for all $a \in K$ and for every element u there exist certain $b, b' \in K$ with $u^2 + bu + b' = 0$.[12] It is easy to see that the generalized quaternion skew fields just defined possess this property: for $u = a_0 + ja_1 + ka_2 + la_3$ we must set $b = -2a_0$, $b' = a_0^2 + a_1^2 c + a_2^2 d + a_3^2 cd$. If K satisfies the conditions (11), (12), then the general case can be reduced to the special case $c = d = 1$; in fact, there then exists (up to isomorphism) exactly one (generalized) quaternion skew field over K.

3.2. *Quaternions and Space Rotations*

From the complex affine plane in which in §3.1 we considered the Hermitian dilative rotations we now turn to the complex projective line; i.e., we consider z_1, z_2 as the homogeneous coordinates of a point, with $z_1/z_2 = z$ as the inhomogeneous coordinate in the case $z_2 \neq 0$, while $z_2 = 0$ (with $z_1 \neq 0$) gives the ideal point. In this way the complex

[11] If K has characteristic 2, then the definition is somewhat different; in this case the above definition provides only a field.

[12] For details see e.g. Pickert [3], Section 6.3.

numbers correspond in one-to-one fashion with the proper points of the complex projective line. We extend the set of complex numbers by a new element ∞, which we put in correspondence with the ideal point and use as its coordinate. For convenience we sometimes identify the points with their coordinates z_1/z_2 or ∞. The mapping (18) then becomes the projectivity on the projective line described as follows, with z, z' as the coordinates of a point and its image:

$$(25)\qquad \begin{aligned} z' &= \frac{\alpha z - \bar{\beta}}{\beta z + \bar{\alpha}}, && \textit{if}\ \ \beta z + \bar{\alpha} \neq 0, \quad z \neq \infty; \\ z' &= \infty, && \textit{if}\ \ \beta z + \bar{\alpha} = 0, \quad z \neq \infty; \\ z' &= \alpha/\beta, && \textit{if}\ \ z = \infty, \quad \beta \neq 0;^{13} \\ z' &= \infty, && \textit{if}\ \ z = \infty, \quad \beta = 0. \end{aligned}$$

Since a common real factor for α, β is here of no importance, we may restrict ourselves to the case $\alpha\bar{\alpha} + \beta\bar{\beta} = 1$.

We now map the complex projective line onto a sphere of radius 1/2 in the following way: to the proper point $z = x + iy$ we first assign the point in space with the coordinates $(x, y, -\frac{1}{2})$ (in a given Cartesian coordinate system) and then project this point (stereographically) from the point $(0, 0, \frac{1}{2})$ onto the sphere with the equation $\sum_{\nu=1}^{3} x_\nu^2 = \frac{1}{4}$ for the coordinates x_1, x_2, x_3 of its points. To the point ∞ we assign the point $(0,0,\frac{1}{2})$ on the sphere, which is not an image under the stereographic projection. A sphere used in this way for the representation of the complex numbers is called a *Riemann sphere*. The equations connecting $z(\neq \infty)$ and the corresponding point of the sphere (with the coordinates x_1, x_2, x_3) are easily calculated to be

$$(26)\qquad \begin{aligned} x_1 + ix_2 &= z(z\bar{z} + 1)^{-1}, \qquad x_1 - ix_2 = \bar{z}(z\bar{z} + 1)^{-1}, \\ 2x_3 &= (z\bar{z} - 1)(z\bar{z} + 1)^{-1}. \end{aligned}$$

We now assert that under this mapping the projectivities (25) (with $\alpha\bar{\alpha} + \beta\bar{\beta} = 1$) become the entire set of rotations of the sphere.[14]

We first consider the special case $\beta = 0$. If we set $\rho = \alpha/\bar{\alpha}$, the projectivity (25) becomes $z' = \rho z$ with the fixed point ∞. Since $\alpha\bar{\alpha} = 1$, we may set $\alpha = \cos \varphi/2 + i \sin \varphi/2$ and thus obtain $\rho = \cos \varphi + i \sin \varphi$. We then have $z'\bar{z}' = z\bar{z}$, so that (26) leads to

$$x_1' + ix_2' = (\cos \varphi + i \sin \varphi)(x_1 + ix_2), \qquad x_3' = x_3$$

[13] As follows from (14) if we set $Z_2 = 0$, and similarly for the second and fourth cases.

[14] In fact, it was the study of rotations in space that led Hamilton (1844) to the definition of the multiplication of quaternions.

for the coordinates x'_ν of the point on the sphere corresponding to z'. But this equation represents the rotation of the sphere through the angle φ around the axis oriented by the vector $\mathfrak{e}_3$, where we consider the whole space as oriented by the triple $(\mathfrak{e}_1, \mathfrak{e}_2, \mathfrak{e}_3)$ of basis vectors of the coordinate system.

We next consider the special case in which α, β are real and $\beta \neq 0$. Multiplication by the complex conjugate of the denominator in (25) leads, for $z \neq -\alpha/\beta, \infty$ with $z' = x' + iy'$ and the abbreviation

$$d = (\beta\bar{z} + \alpha)(\beta z + \alpha) = \beta^2 z\bar{z} + 2\alpha\beta x + \alpha^2,$$

to the equations

$$x'd = (\alpha^2 - \beta^2)\, x + \alpha\beta(z\bar{z} - 1),$$

$$y'd = y, \qquad z'\overline{z'}d = (\alpha z - \beta)(\alpha\bar{z} - \beta) = \alpha^2 z\bar{z} - 2\alpha\beta x + \beta^2.$$

From (26) we thus have

$$x'_3 = (\alpha^2 - \beta^2)\, x_3 - 2\alpha\beta x_1, \quad x'_1 = 2\alpha\beta x_3 + (\alpha^2 - \beta^2)\, x_1, \quad x'_2 = x_2,$$

and the same equations can easily be shown to hold for the two cases excluded above, namely, $z = -\alpha/\beta$ and $z = \infty$. Since we may set $\alpha = \cos \varphi/2$, $\beta = \sin \varphi/2$, we here obtain the rotations about the axis oriented by the vector $\mathfrak{e}_2$.

We now consider the general case, where we may assume $\beta \neq 0$. Then the projectivity σ defined by (25) does not leave the point ∞ fixed. Consequently, its fixed points ζ are obtained from the quadratic equation

$$\beta\zeta^2 + (\bar{\alpha} - \alpha)\,\zeta + \bar{\beta} = 0. \tag{27}$$

But this equation may be written (by passage to complex conjugates) in the form

$$\beta(-\bar{\zeta}^{-1})^2 + (\bar{\alpha} - \alpha)(-\bar{\zeta}^{-1}) + \bar{\beta} = 0$$

and thus, if ζ is a fixed point, so also is $-\bar{\zeta}^{-1}(\neq \zeta)$. Thus (25) has two distinct fixed points, and by (26) they give two diametrically opposite points of the sphere. The line joining these two points can now be brought into the plane $x_2 = 0$ by a rotation about the x_3 axis and then, by rotation about the x_2 axis, can be made to coincide with the x_3 axis. These two rotations taken together produce a rotation δ which, by the special cases dealt with above, corresponds to a projectivity τ taking the fixed points of σ into 0 and ∞. Thus the projectivity $\sigma_0 = \tau\sigma\tau^{-1}$ (first τ^{-1}, then σ, then τ) has the fixed points $0, \infty$; in other words, it belongs to the case $\beta = 0$. Thus it corresponds to a rotation δ_0 about the x_3 axis, and therefore the projectivity $\sigma = \tau^{-1}\sigma_0\tau$ corresponds to the rotation $\delta^1 = \delta^{-1}\delta_0\delta$.

Conversely, every rotation is obtained in this way. For let $\mathfrak{d} = \sum_{\nu=1}^{3} \mathfrak{e}_\nu d_\nu$ with $\sum_{\nu=1}^{3} d_\nu^2 = 1$ be the vector orienting the axis of rotation, where we may exclude the cases $\mathfrak{d} = \pm \mathfrak{e}_3$. Then the fixed points of the rotation on the sphere have the coordinates $(d_1/2, d_2/2, d_3/2)$ and $(-d_1/2, -d_2/2, -d_3/2)$, so that by the equation

$$z = (2x_1 + 2x_2 i)(1 - 2x_3)^{-1},$$

which follows from (26), the corresponding complex numbers are

$$(28) \qquad \zeta_1 = (d_1 + id_2)(1 - d_3)^{-1}, \qquad \zeta_2 = -(d_1 + id_2)(1 + d_3)^{-1}.$$

If in (25) we set $\alpha = (1 + \zeta_2\bar{\zeta}_2)^{-1}$, $\beta = \bar{\zeta}_2(1 + \zeta_2\bar{\zeta}_2)^{-1}$, we obviously obtain a projectivity τ taking ζ_2 into 0 and (since $\bar{\zeta}_2 = -\zeta_1^{-1}$) taking ζ_1 into ∞. Thus the corresponding rotation δ takes the vector $\mathfrak{d}$ into $\mathfrak{e}_3$, so that $\delta_0 = \delta\delta'\delta^{-1}$ is a rotation about the axis defined by $\mathfrak{e}_3$ with the same angle of rotation as δ' and therefore corresponds to a projectivity σ_0. Consequently, the rotation $\delta' = \delta^{-1}\delta_0\delta$ corresponds, as desired, to the projectivity $\sigma = \tau^{-1}\sigma_0\tau$.

On the basis of these results we can now express the α, β for σ in terms of the $\mathfrak{d}_\nu$ and the angle of rotation of δ'. Since the projectivity σ_0 has the fixed points $0, \infty$, it takes z into ρz with a fixed factor ρ of absolute value 1. If we now apply $\tau\sigma = \sigma_0\tau$ to the point ∞, we obtain, assuming $\beta \neq 0$ (and thus $\zeta_2 \neq 0$)

$$(\alpha - \beta\zeta_2)(\alpha\bar{\zeta}_2 + \beta)^{-1} = \rho\bar{\zeta}_2^{-1},$$

so that

$$(29) \qquad \rho = (\alpha - \beta\zeta_2)(\alpha - \beta\zeta_1)^{-1}.$$

From (27), (28) we obtain

$$2d_3(d_1 - id_2)^{-1} = \zeta_1 + \zeta_2 = (\alpha - \bar{\alpha})\,\beta^{-1},$$

or, in the notation of (22),

$$d_3(d_2 + id_1)^{-1} = a_3(a_2 + ia_1)^{-1}.$$

Thus there exists a real number c with $a_\nu = cd_\nu$ $(\nu = 1, 2, 3)$ and therefore (since $\alpha\bar{\alpha} + \beta\bar{\beta} = 1$, $\sum_{\nu=1}^{3} d_\nu^2 = 1$) with $a_0 + c^2 = 1$, so that we may set

$$(30) \qquad a_0 = \cos \varphi/2, \qquad a_\nu = d_\nu \sin \varphi/2 \qquad (\nu = 1, 2, 3).$$

From (29), (30) a short calculation shows that

$$\rho = \cos \varphi + i \sin \varphi.$$

In view of the significance of ρ for σ_0 and the earlier discussion of the special case $\beta = 0$, it follows that φ is the angle of rotation of δ_0 and thus also of δ'. Consequently, the equation (30), which obviously holds for the special case $\beta = 0$, represents the desired connection between quaternions and rotations in space. We note that to every rotation the equation (30) associates exactly two quaternions with norm 1; for if we replace φ by $\varphi + 2\pi$ or $\mathfrak{d}$ by $-\mathfrak{d}$ and φ by $-\varphi$ the numbers a_ν $(\nu = 0, 1, 2, 3)$ become $-a_\nu$.

3.3. *Quaternions and Vector Algebra*

For a fixed Cartesian coordinate system with the basis vectors $\mathfrak{e}_1, \mathfrak{e}_2, \mathfrak{e}_3$ the mapping

$$\sum_{\nu=1}^{3} \mathfrak{e}_\nu x_\nu \to jx_1 + kx_2 + lx_3$$

is obviously an isomorphism of the three-dimensional vector space into the four-dimensional vector space of quaternions. So we may make the identification

$$\sum_{\nu=1}^{3} \mathfrak{e}_\nu x_\nu = jx_1 + kx_2 + lx_3 .$$

With $\mathfrak{a} = \sum_{\nu=1}^{3} \mathfrak{e}_\nu a_\nu$ the quaternion

$$A = a_0 + ja_1 + ka_2 + la_3$$

can then be written in the simple form $a_0 + \mathfrak{a}$. Thus a_0 is called the *scalar part* and $\mathfrak{a}$ the *vector part* of A. A simple calculation on the basis of (24) then shows that

$$\text{(31)} \qquad \mathfrak{a}\mathfrak{b} = -\mathfrak{a} \cdot \mathfrak{b} + \mathfrak{a} \times \mathfrak{b};$$

here $\mathfrak{a}\mathfrak{b}$ denotes the product of the vectors regarded as quaternions, and $\mathfrak{a} \cdot \mathfrak{b}$ and $\mathfrak{a} \times \mathfrak{b}$ are the inner (scalar) and vector products, respectively. If we pass to the conjugate quaternions on both sides of (31) (whereby the scalar part is not changed and the vector part is multiplied by -1) and note that such a passage is an antiautomorphism, we see that $\mathfrak{b}\mathfrak{a} = -\mathfrak{a} \cdot \mathfrak{b} - \mathfrak{a} \times \mathfrak{b}$. Together with (31) this result shows that:

$$\text{(32)} \qquad \mathfrak{a} \cdot \mathfrak{b} = -\tfrac{1}{2}(\mathfrak{a}\mathfrak{b} + \mathfrak{b}\mathfrak{a}),$$

$$\text{(33)} \qquad \mathfrak{a} \times \mathfrak{b} = \tfrac{1}{2}(\mathfrak{a}\mathfrak{b} - \mathfrak{b}\mathfrak{a}).$$

Thus (in analogy to the equations at the end of §1.1 for the two-dimensional case) we may express the scalar and the vector products in terms of quaternion multiplication.[15]

By means of (32), (33) the rules for calculation with these two vector products can easily be proved. For example, for the rule

$$(34)\qquad (\mathfrak{a} \times \mathfrak{b}) \times \mathfrak{c} = \mathfrak{b}(\mathfrak{a} \cdot \mathfrak{c}) - \mathfrak{a}(\mathfrak{b} \cdot \mathfrak{c})$$

we have by (33)

$$4(\mathfrak{a} \times \mathfrak{b}) \times \mathfrak{c} = \mathfrak{abc} - \mathfrak{bac} - \mathfrak{cab} + \mathfrak{cba} = (\mathfrak{abc} - \mathfrak{cab}) - (\mathfrak{bac} - \mathfrak{cba}).$$

By (32) we also have

$$\mathfrak{abc} - \mathfrak{cab} = (\mathfrak{abc} + \mathfrak{acb}) - (\mathfrak{acb} + \mathfrak{cab}) = 2\mathfrak{b}(\mathfrak{a} \cdot \mathfrak{c}) - 2\mathfrak{a}(\mathfrak{b} \cdot \mathfrak{c}),$$

and by interchange of a with b we obtain

$$\mathfrak{bac} - \mathfrak{cba} = 2\mathfrak{a}(\mathfrak{b} \cdot \mathfrak{c}) - 2\mathfrak{b}(\mathfrak{a} \cdot \mathfrak{c}).$$

These three equations taken together lead, after division by 4, directly to (34).

If we regard the unit vector $\mathfrak{d}$ as a quaternion, the corresponding rotation is, by §3.2, the rotation δ_π around the axis determined by $\mathfrak{d}$ through the angle π. Now let δ be an arbitrary rotation (with the origin as fixed point) and let A be the corresponding quaternion. Then the quaternion $A\mathfrak{d}A^{-1}$ corresponds to the rotation $\delta'_\pi = \delta\delta_\pi\delta^{-1}$. But the latter is a rotation through the angle π and its axis is determined by the image $\mathfrak{d}'$ of $\mathfrak{d}$ under δ, since $\delta'_\pi\delta = \delta\delta_\pi$ implies that the metric relations valid for $\mathfrak{d}$, $\mathfrak{x}$, $\delta_\pi(\mathfrak{x})$ also hold for $\mathfrak{d}'$, $\delta(\mathfrak{x})$, $\delta'_\pi(\delta(\mathfrak{x}))$. Thus $A\mathfrak{d}A^{-1}$ is again a quaternion with scalar part 0 and vector part $\mathfrak{d}'$ or $-\mathfrak{d}'$. For an arbitrary vector $\mathfrak{x}$ (which may always be written as a scalar multiple of a unit vector $\mathfrak{d}$) and its image vector $\mathfrak{x}' = \delta(\mathfrak{x})$ we thus have

$$(35)\qquad \mathfrak{x}' = \pm A\mathfrak{x}A^{-1}.$$

But the sign here cannot depend on $\mathfrak{x}$, since from

$$\mathfrak{x}'_1 = A\mathfrak{x}_1A^{-1},\ \mathfrak{x}'_2 = -A\mathfrak{x}_2A^{-1}$$

and $\mathfrak{x}_1, \mathfrak{x}_2 \neq 0$ it follows that

$$\mathfrak{x}'_1 + \mathfrak{x}'_2 = A(\mathfrak{x}_1 - \mathfrak{x}_2)A^{-1} = \pm(\mathfrak{x}_1 - \mathfrak{x}_2)' = \pm(\mathfrak{x}'_1 - \mathfrak{x}'_2),$$

[15] Originally the scalar part and the vector part of the quaternion product were actually used in place of these vector products. Gibbs (1884) was the first to introduce vector products in modern notation, independently of quaternion multiplication. The concept of inner product had already occurred in the works of Grassmann (1844), whose exterior product (see IB3, §3.3) is also in close relationship with the above vector product.

which is impossible for any choice of sign. With a minus sign the equation (35) does not represent a rotation. For if we write $A = a + \mathfrak{a}$, the condition $\mathfrak{x}' = \mathfrak{x}$ becomes $\mathfrak{x}(a + \mathfrak{a}) = -(a + \mathfrak{a})\,\mathfrak{x}$, or in other words $a\mathfrak{x} = 0$, $\mathfrak{a} \cdot \mathfrak{x} = 0$, and for $a \neq 0$ these equations are correct only for $\mathfrak{x} = 0$, whereas for $a = 0$ they are correct for all $\mathfrak{x} \perp \mathfrak{a}(\neq 0)$, while for a rotation ($\neq 1$) the fixed vectors form a one-dimensional subspace.[16] So we have

$$(36) \qquad \mathfrak{x}' = A\mathfrak{x}A^{-1}.$$

Since $N(A) = 1$, we could of course replace A^{-1} by the quaternion A^* conjugate to A. But we let A^{-1} stand here, because (36) then represents a rotation for an arbitrary quaternion $A \neq 0$: we need only write $A = \sqrt{N(A)}\, A_0$, so that $\mathfrak{x}' = A_0\mathfrak{x}A_0^{-1}$, $N(A_0) = 1$.

If $\mathfrak{d} = \sum_{\nu=1}^{3} \mathfrak{e}_\nu d_\nu$ is the orienting vector of the axis of rotation and φ is the angle of rotation in (36), then by (30) we may write

$$A = \cos \varphi/2 + \mathfrak{d} \sin \varphi/2, \qquad A^* = \cos \varphi/2 - \mathfrak{d} \sin \varphi/2.$$

If we now transform (36) by means of (31) and note that $(\mathfrak{d} \times \mathfrak{x}) \cdot \mathfrak{d} = 0$ and also, as follows from (34), $(\mathfrak{d} \times \mathfrak{x}) \times \mathfrak{d} = \mathfrak{x} - \mathfrak{d}(\mathfrak{x} \cdot \mathfrak{d})$, we have the *Rodrigues* formula for rotations:

$$(37) \qquad \mathfrak{x}' = \mathfrak{x} \cos \varphi + \mathfrak{d}(\mathfrak{x} \cdot \mathfrak{d})(1 - \cos \varphi) + (\mathfrak{d} \times \mathfrak{x}) \sin \varphi.$$

This equation provides another proof of the fact that the $\mathfrak{d}$, φ appearing in (30) actually have the significance assigned to them above: for we need only prove that $\mathfrak{d}' = \mathfrak{d}$, $\mathfrak{x} \cdot \mathfrak{d} = \mathfrak{x}' \cdot \mathfrak{d}$ and also, with the abbreviations

$$\mathfrak{y} = \mathfrak{x} - \mathfrak{d}(\mathfrak{x} \cdot \mathfrak{d}),\ \mathfrak{y}' = \mathfrak{x}' - \mathfrak{d}(\mathfrak{x} \cdot \mathfrak{d}), \quad \text{that} \quad |\mathfrak{y}| = |\mathfrak{y}'|,$$
$$\mathfrak{y} \cdot \mathfrak{y}' = |\mathfrak{y}|^2 \cos \varphi, \quad (\mathfrak{y} \times \mathfrak{y}')\,\mathfrak{d} = |\mathfrak{y}|^2 \sin \varphi.$$

3.4. *The Theorem of Frobenius*

From §1.2 and §3.1 we now see that the two extensions of K, namely, the quaternion skew field and the field $K(i)$ (with $i^2 = -1$) have the following properties in common: each of them is a skew field with K as subfield; also $az = za$ for every element z and all $a \in K$; and finally, they are both vector spaces of finite dimension over K. We are thus led to the concept of a *division algebra of rank n over K*, namely, a skew extension field L of K with $a\alpha = \alpha a$ for all $a \in K$, $\alpha \in L$, which is of dimension n regarded as a vector space over K (see IB3, §1.2). The latter property requires the existence of elements $\rho_1, \ldots, \rho_n \in L$ such that for every $\alpha \in L$ there is exactly one n-tuple $(a_1, \ldots, a_n)$ with $a_\nu \in K (\nu = 1, \ldots, n)$ and $a = \sum_{\nu=1}^{n} \beta_\nu a_\nu$.

[16] Thus (35) with the minus sign gives the reflections ($a = 0$) and the rotatory reflections ($a \neq 0$).

The field K itself is, of course, a division algebra of rank 1 over K, where for the basis element β_1 we may take any element $\neq 0$ of K. When the rank of the division algebra is not stated, we speak of a *division algebra of finite rank*.[17]

The outstanding algebraic importance of the field of complex numbers and of the quaternion skew field now depends on the fact that they are the only division algebras of finite rank > 1 over the field of real numbers, a result that follows from the *theorem of Frobenius*:

A division algebra L of finite rank over a real-closed field K is either K itself or else (up to isomorphism) is the field $K(i)$ (with $i^2 = -1$) or the quaternion skew field over K.

For the proof we first deduce the following property of L:[18] *for every $\alpha \in L$ there exist $r, s \in K$ with*

$$(38) \qquad \alpha^2 = r\alpha + s.$$

Since L is of dimension n over K, the elements $1, \alpha, \ldots, \alpha^n$ are linearly dependent; in other words, there exist $a_\nu \in K$, not all zero, such that $\sum_{\nu=0}^{n} a_\nu \alpha^\nu = 0$. Thus the polynomial $f(x) = \sum_{\nu=0}^{n} a_\nu x^\nu$ is of positive degree and can therefore, by §2.3, be split into linear and quadratic factors. Since $f(\alpha) = 0$ and L has no divisors of zero, one of these factors* will become 0 when x is replaced by α. If this factor is quadratic, we already have the desired equation (38). But otherwise $\alpha \in K$ and we have (38) with $r = \alpha$, $s = 0$.

Now by (11) we see that $1 \neq -1$, so that $1 + 1 \neq 0$, and thus we may divide by 2 ($= 1 + 1$), so that (38) becomes

$$(39) \qquad (\alpha - r/2)^2 = r^2/4 + s.$$

Since we may assume $L \neq K$, there exists an element in L that is not in K. For this element the right-hand side of (39) cannot be the square of an element $t \in K$, since otherwise we would have $\alpha = r/2 + t$ or $\alpha = r/2 - t$. Thus by (12) there exists a $t \in K$ with $r^2/4 + s = -t^2 = 0$, so that from (39) we obtain $j^2 = -1$ for $j = (\alpha - r/2)\,t^{-1}$. If K is not already exhausted by the adjunction to K of a zero of $x^2 + 1$, there must exist a

[17] Such an algebra is seen at once to be an algebra in the sense of IB5, §3.9. By the words "division algebra" alone we mean a skew extension field L of K with $a\alpha = \alpha a$ for all $a \in K$, $\alpha \in L$. In some investigations the concept "division algebra" is defined more generally; i.e., so as to include the alternative fields of §3.5, in which case the division algebras defined here must be called "associative division algebras."

[18] It is only here that any use is made of the hypothesis of finite rank.

* What we need here is the fact that replacement of x by α gives a homomorphism of $K[x]$, but by IB4, §2.1, this follows from the easily proved fact that, for every given α, the $g(\alpha)$ with $g(x) \in K[x]$ are the elements of a *commutative* subring of L.

$\beta \in L$ that is not of the form $a + jb$ ($a, b, \in K$). We now make the following assertion, which will also be useful later on: for $\xi, \eta \in L$ there exist $a, b, c \in K$ with

$$\xi\eta + \eta\xi = a\xi + b\eta + c. \tag{40}$$

For the proof of this assertion we need only write the equations corresponding to (38) for $\xi, \eta, \xi + \eta$ and take account of the fact that $\xi\eta + \eta\xi = (\xi + \eta)^2 - \xi^2 - \eta^2$. In particular, we now use (40) with $\xi = j, \eta = \beta$:

$$j\beta + \beta j = aj + b\beta + c. \tag{41}$$

Then for arbitrary $u, v \in K$ we have

$$(uj + v\beta)^2 = -u^2 + uvc + uvaj + uvb\beta + v^2\beta^2.$$

From the equations corresponding to (38)

$$\begin{aligned} \beta^2 &= r'\beta + s' \qquad (r', s' \in K), \\ (uj + v\beta)^2 &= r''(uj + v\beta) + s'' \qquad (r'', s'' \in K), \end{aligned} \tag{42}$$

we further have

$$r''uj + r''v\beta + s'' = uvaj + (uvb + v^2r')\,\beta + (-u^2 + uvc + v^2s').$$

Since β is not a linear combination of $1, j$ with coefficients from K, the elements i, j, β are linearly independent over K, so that comparison of coefficients in the last equation gives

$$r''u = uva, \qquad r''v = uvb + v^2r'.$$

For $u, v \neq 0$ we thus have

$$ub = v(a - r').$$

If in this equation we first set $u = v = 1$ and then $u = 1$, $v = -1$, it follows, since a, b, r' do not depend on u, v, that

$$b = 0, \qquad a = r'. \tag{43}$$

With $x, y, z \in K$ we now write $k = x\beta - yj - z$, and from (41), (42), (43) we calculate

$$\begin{aligned} k^2 &= x^2s' - y^2 + z^2 - xyc + (xa - 2z)(x\beta - yj), \\ jk + kj &= (xc + 2y) + (xa - 2z)\,j. \end{aligned}$$

These equations suggest that we should require

$$xc + 2y = 0, \qquad xa - 2z = 0,$$

as may obviously be done for arbitrary $x \neq 0$, so that $k^2 \in K$ and $jk + kj = 0$. But since $x \neq 0$ and $1, j, \beta$ are linearly independent, the element k cannot lie in K, so that k^2 is not the square of an element of K, and thus by (12) the element $-k^2$ is such a square. Thus we can choose x in such a way that $k^2 = -1$. But then the elements j, k satisfy those equations (24) in which l does not occur. If we now set $l = jk$, we have $l^2 = jkjk = -jkkj = jj = -1$, $kl = -kkj = j$, $lk = jkk = -j$, $lj = jkj = -kjj = k, jl = jjk = -k$, so that (24) is completely satisfied.

From $l = a + jb + kc$ with $a, b, c \in K$ it would follow, by left-multiplication with j that $-k = ja - b + lc$, so that

$$-k = j(a + cb) + (ac - b) + kc^2$$

and thus, on account of the linear independence of $1, j, k$, we would have $c^2 = -1$, which is impossible by (11). Thus $1, j, k, l$ are linearly independent. In order to show that L is the quaternion skew field over K, we need only express every element $\xi \in K$ as a linear combination of $1, j, k, l$ with coefficients from K. For this purpose we note that, under the single condition $j^2 = -1$, we have from (41), (43) the formula $j\beta + \beta j = aj + c$ with $a, c \in K$; it is true that in the proof of (43) we made the further assumption that β is linearly independent of $1, j$, but if this is not the case our assertion follows at once from (41) (though with other values of a, c). We thus have the equations

$$\begin{aligned} j\xi + \xi j &= aj + c, \\ k\xi + \xi k &= a'k + c', \\ l\xi + \xi l &= a''l + c'', \end{aligned}$$

and consequently

$$l(j\xi + \xi j)\,k - k(k\xi + \xi k) - l(l\xi + \xi l) = (-a + a' + a'') - jc - kc' - lc''.$$

Since the left-hand side of this equation is equal to 2ζ, the desired representation for ξ is thereby obtained.

3.5. *Cayley Numbers*

In the theorem of Frobenius the only one of the "usual rules for calculation" that is given up is the commutativity of multiplication, and as a result we obtain, together with the field of complex numbers, the quaternion skew field. It is now natural to ask what new structures we could obtain by giving up further rules of calculation, or perhaps by only weakening them. A complete answer to this question can be given in the following special case: the associative law for multiplication, which is required in skew fields, is replaced by the rules

(44) $$a(ab) = (aa)\,b, \qquad (ab)\,b = a(bb).$$

Algebraic structures of this kind are called *alternative fields*, a name which is explained by the fact that if the other rules (for example the distributive laws) are retained, the rules (44) have the following consequence: the *associator* $(ab)\,c - a(bc)$ is changed to $-(ab)\,c + a(bc)$ when any two of the elements a, b, c are interchanged. If in the definition of "division algebra" we replace "skew field" by "alternative field," the theorem of Frobenius must now be extended, to the effect that (up to isomorphism) there exists exactly one further division algebra, which is of rank 8.[19] It is called a *Cayley algebra*, and its elements are called *Cayley numbers* or *octaves*.[20] We can obtain these numbers most conveniently if we start from the quaternion skew field Q over K and in the set of pairs (A, B) with $A, B \in Q$ we define addition and multiplication in the following way, where the conjugate quaternion is again denoted by an asterisk:[21]

(45) $$\begin{aligned}(A_1, B_1) + (A_2, B_2) &= (A_1 + A_2, B_1 + B_2),\\ (A_1, B_1)(A_2, B_2) &= (A_1A_2 - B_2B_1^*, A_2B_1 + A_1^*B_2).\end{aligned}$$

Since $A \to (A, 0)$ is then an isomorphism of Q, we may set $A = (A, 0)$, so that we actually have an extension of K, and even of Q. The fact that multiplication is not associative is seen from the following example, in which for abbreviation we have set $(0, 1) = \mathsf{E}$:

$$(\mathsf{E}j)\,k = (0, -l),\ \mathsf{E}(jk) = (0, l).$$

By (45) it is easy to see that the mapping $(A, B) \to (A^*, -B)$ is an antiautomorphism of the Cayley algebra. Then $\Gamma^* = (A^*, -B)$ is called the *conjugate Cayley number* of $\Gamma = (A, B)$. The real number

$$\Gamma\Gamma^* = \Gamma^*\Gamma = N(A) + N(B)$$

is called the *norm* $N(\Gamma)$ of Γ. In the same way as for complex numbers and quaternions, we have $N(\Gamma_1\Gamma_2) = N(\Gamma_1)\,N(\Gamma_2)$, so that we may write

[19] The fact that only the ranks 1, 2, 4, 8 are possible can be proved without (44). For the case that K is the field of real numbers, this proof was already given by J. Milnor (Ann. of Math. 68, 444–449, 1958). From this fact, by using a result of A. Tarski (A Decision Method for Elementary Algebra and Geometry, 2nd ed., Univ. of California Press, 1951) we can then prove, for every natural number $n \neq 1, 2, 4, 8$, that the rank n is impossible for an arbitrary real-closed K.

[20] In analogy with the quaternions (of rank 4, whereas here the rank is 8).

[21] For details see, e.g., Pickert [3], Section 6.3.

the product of sums of 8 squares again in form of 8 squares.[22] Since multiplication is no longer associative, we cannot prove this assertion in the same way as for quaternions, namely, from the fact that passage to conjugates is an antiautomorphism. Nevertheless, it is easy to show from (45) that

$$N(\Gamma_1\Gamma_2) = N(\Gamma_1)\,N(\Gamma_2) - (CB_2^* + (CB_2^*)^*) + (B_2^*C + (B_2^*C)^*)$$

with $C = A_1A_2B_1$, and since the scalar part of the product of two quaternions is independent of the order of the factors, we thus have the desired assertion.

Let us note that the passage from quaternions to Cayley numbers is quite analogous to the procedure described in §3.1 for passing from the complex numbers to the quaternions; in order to display this analogy we have only to replace the quaternion A in (19) by the pair (α, β), whereupon the matrix multiplication becomes

$$(\alpha_1, \beta_1)(\alpha_2, \beta_2) = (\alpha_1\alpha_2 - \bar{\beta}_1\beta_2, \beta_1\alpha_2 + \bar{\alpha}_1\beta_2).$$

The concept of a Cayley algebra can be generalized in the following way. We take for Q a generalized quaternion skew field, choose an element $c \in K$ that is not the norm of an element of Q, and in the second equation (45) replace the term $-B_2B_1^*$ by $cB_2B_1^*$. These (generalized) Cayley algebras are again alternative fields. Their great importance lies in the fact, not discovered until 1950/51, that every alternative field which is not already a skew field must be such a Cayley algebra.[23] The alternative fields play an important role in the study of projective planes: just as the incidence axioms for the projective plane and the theorem of Desargues allow us to construct a plane coordinate geometry over a skew field, so the little Desargues theorem, i.e., the special case with incidence of center and axis, and the incidence axioms for plane projective geometry allow us to construct a coordinate geometry over an alternative field (R. Moufang, 1933). One can also say (in a sense that can be made quite precise): in exactly the same way as the Desargues theorem corresponds to the associative law for multiplication, the little Desargues theorem corresponds to the alternative law (44).[24]

[22] Numerical example:

$$(1^2 + 2^2 + 3^2 + 4^2 + 5^2 + 6^2 + 7^2 + 8^2)(9^2 + 10^2 + 11^2 + 12^2 + 13^2 + 14^2 + 15^2 + 16^2)$$
$$= 36^2 + 38^2 + 54^2 + 62^2 + 72^2 + 108^2 + 112^2 + 474^2.$$

[23] For details see, e.g., Pickert [3], Section 6.3.

[24] See, e.g., Pickert [3], p. 134 (theorem 27), and p. 187 (footnote 1).

CHAPTER 9

Lattices

Introduction

For the connectives $\wedge$ (and) and $\vee$ (or) between statements and the connectives $\cap$ (intersection) and $\cup$ (union) between sets we have the following rules (see IA, §§2, 7, 9):

Commutative laws: $a \sqcap b = b \sqcap a$; $a \sqcup b = b \sqcup a$;

Associative laws: $a \sqcap (b \sqcap c) = (a \sqcap b) \sqcap c$; $a \sqcup (b \sqcup c) = (a \sqcup b) \sqcup c$;

Absorption laws: $a \sqcap (a \sqcup b) = a$; $a \sqcup (a \sqcap b) = a$.

Here the symbols $\sqcap$, $\sqcup$ are intended to suggest $\wedge$, $\vee$ and $\cap$, $\cup$ but to be distinct from them.

If a, b, c are subgroups of a group and if $a \sqcap b$ is the greatest subgroup common to a and b, and $a \sqcup b$ is the smallest one of the subgroups containing both a and b, then the rules listed above still hold, but the further rules for $\wedge$, $\vee$ and $\cap$, $\cup$ (see IA, loc. cit.) are in general no longer valid. The same remark holds for the normal subgroups of a group, the subrings of a ring, the ideals of a ring, the subfields of a field, the sublattices of a lattice, and in general (cf. IB10) for the subconfigurations of an (algebraic) configuration with respect to a given structure. The concepts and theories of the present chapter are thus of importance for a general theory of structure, the basic features of which will be discussed in the next chapter.

A set with two connectives satisfying the above laws was called a *dual group* by Dedekind (1897); today it is customary to use the name *lattice* introduced by G. Birkhoff in 1933; (the corresponding French name *treillis* is due to Châtelet (1945) and the German *Verband* to Fritz Klein-Barmen (1932); the reason for the choice of such a name is given on page 487.

Quite apart from its importance for the general theory of structure, there is a certain attractiveness in studying an algebraic structure in which the convenient rules of commutativity and associativity are satisfied and there is complete duality between the two connectives; moreover, this structure is of quite different character from other well-known structures; e.g., neither of the two connectives is uniquely invertible. The study of Boolean lattices is even of practical importance for the construction of electric circuits (circuit algebra).

In §1 we make use of the lattice *Pe* of all subsets of a given set *e* to introduce the fundamental concepts of the theory of lattices, in §2 we give examples, the large number of which indicates the importance of the theory, and then in the following sections we describe some of the characteristic properties of a few particular kinds of lattices.

In the present chapter we can give only a first introduction to the theory. For further study we recommend the outstanding textbook:

G. Birkhoff, *Lattice Theory*, Amer. Math. Soc. Coll. Publ., 2nd ed., 1948, supplemented by the later work of the same author:

Proceedings in Pure Mathematics, Vol. II, *Lattice Theory*, Amer. Math. Soc., Providence, R.I., 1961.

Other important textbooks are:

M.L. Dubreil-Jacotin, L. Lesieur, R. Croisot, *Leçons sur la théorie des treillis, des structures algébriques ordonnées et des treillis géometriques*, Paris 1953.

H. Hermes, *Einführung in die Verbandstheorie*, Verl. Springer, Berlin, Göttingen and Heidelberg, 1955.

G. Szasz, *Einführung in die Verbandstheorie*, Budapest, 1962.

In many respects our discussion is based directly on the book of Hermes, to which we shall often refer below.

The first comprehensive account was given in the encyclopedia article.

H. Hermes and G. Köthe, *Theorie der Verbände*, Enz. d. Math. Wiss., 2nd ed., I, 13, 1939.

This article also includes important historical information.

Since the theory of lattices is closely related to structure theory and consequently to mathematical logic, we use the logical symbols introduced in IA:

$\neg$ not	$\wedge$ and	$\vee$ or
$\bigwedge_s$ for all s	$\bigvee_s$ there exists an s such that	
$\rightarrow$ if ... then	$\leftrightarrow$ if and only if	

1. Properties of the Power Set

Let e be a set with elements denoted by lower-case Greek letters and subsets by lower-case italic letters. Let Pe be the set of subsets of e, the so-called *power set*. It is with this power set that the discussion in the present section deals almost exclusively; the reader should keep in mind that the elements of Pe are the sub*sets* of e.

Pe has the following properties (cf. IA, §9):

I. In Pe there is defined a two-place relation, namely inclusion $\subseteq$. This relation is

(1r)	reflexive:	$a \subseteq a$,
(1t)	transitive:	$a \subseteq b \wedge b \subseteq c \rightarrow a \subseteq c$,
(1i)	identitive:	$a \subseteq b \wedge b \subseteq a \rightarrow a = b$,

and is thus an *order* (in the sense of $\leqslant$). We use the word *order* here instead of "semiorder" or "partial order." If it is also true, which generally will not be the case in the present section, that

(1l) $$\bigwedge_a \bigwedge_b (a \subseteq b \vee b \subseteq a),$$

then we will speak of a *linear order* (cf. IA, §8.3).

Problem: For which sets e is Pe linearly ordered by inclusion?

Definition: A set in which an order is defined is called an ordered set, and a linearly ordered set is also called a *chain*.

In general, we shall denote an order relation by $\leqslant$ (read "smaller than or equal to"), reserving $\subseteq$ for inclusion of sets. The relation "smaller than or equal to" for real numbers will be denoted by $\leqslant_z$. By $a < b$ we mean $a \leqslant b \wedge a \neq b$, and by $a \geqslant b$ we mean $b \leqslant a$.

To describe an ordered set, or (later) a lattice, it is necessary to state the ordering relation, so that such a set is completely described by symbol like $(M, \leqslant)$. But where no misunderstanding can arise, we shall speak simply of the "ordered set M" or of the "lattice M."

The order of a finite set M can be described by an *order diagram*, also called a *Hasse diagram*. For such a diagram we first make the following definition.

Definition: a is said to be a *lower neighbor* of b if $a < b$ and there is no element of M between a and b; in other words, if $\neg \bigvee_c (a < c \wedge c < b)$; i.e.,

$$a \leqslant c \leqslant b \rightarrow c = a \vee c = b.$$

The elements of M are then denoted by points in a plane in such a way that if a is a lower neighbor of b, the point a is lower on the page than the point b and is joined to b by a straight line. For example, Figure 1 indicates an ordered set with the following relations and no others:

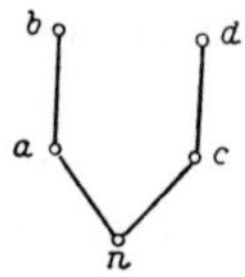

Fig. 1

$$n \leqslant n, \quad n \leqslant a, \quad n \leqslant b, \quad n \leqslant c, \quad n \leqslant d,$$
$$a \leqslant a, \quad a \leqslant b, \quad b \leqslant b, \quad c \leqslant c, \quad c \leqslant d, \quad d \leqslant d.$$

II. In *Pe* we have defined two connectives, which to a pair (a, b) of subsets assign the *intersection* $a \cap b$ and the *union* $a \cup b$, respectively. In terms of the elements of e these connectives can be defined as follows (cf. IA, §7):

$$\xi \in a \cap b \leftrightarrow \xi \in a \wedge \xi \in b$$
$$\xi \in a \cup b \leftrightarrow \xi \in a \vee \xi \in b,$$

but they can also be defined by the order alone without reference to the elements of e; namely, $a \cap b$ is the *greatest common lower element*, or the *greatest lower bound*, and $a \cup b$ is the *least common upper element*, or *least upper bound*. In place of the pair of elements *Pe*, namely the two subsets a, b of e, we now consider an arbitrary subset N of an ordered set M.

Definition: s is called an *upper bound* of N if $\bigwedge_{x \in N} x \leqslant s$; and v is called the *least upper bound* of N if

(2v, 1) v is an upper bound of N, so that $\bigwedge\limits_{x \in N} x \leqslant v$, and

(2v, 2) v is smaller than or equal to every upper bound of N:

$$\bigwedge_s (\bigwedge_{x \in N} x \leqslant s \rightarrow v \leqslant s).$$

Not every subset of an ordered set has a least upper bound or even an upper bound; for example, in the set with the order diagram of Figure 1 the subset consisting of the elements a, n, c has no upper bound, and in the set of positive rational numbers ordered by $\leqslant_z$ the subset of numbers with $x^2 \leqslant_z 2$ has upper bounds, but no least upper bound.

From (2v, 2) it follows that a set N cannot have more than *one* least upper bound. (If v and w are least upper bounds, then $v \leqslant w$ and $w \leqslant v$.) It is this fact that justifies our speaking of *the* least upper bound of N. It is also called the smallest common upper element or the *join* and is denoted by $\sqcup_{x \in N} x$ or $\sqcup_N x$. In case N contains only two elements, we write $v = a \sqcup b$. For this case let us write out the definition once again:

$$\text{(3v)} \qquad v = a \sqcup b \leftrightarrow a \leqslant v \wedge b \leqslant v \wedge \bigwedge_s [(a \leqslant s \wedge b \leqslant s) \rightarrow v \leqslant s],$$

in other words: $a \sqcup b$ is characterized by the following relations:

$$\text{(3v')} \quad a \leqslant a \sqcup b \wedge b \leqslant a \sqcup b \wedge \bigwedge_s [(a \leqslant s \wedge b \leqslant s) \rightarrow a \sqcup b \leqslant s].$$

Correspondingly, the *least upper bound* (*greatest common lower element, intersection*, or *meet*) is defined as follows:

(2d) $$d = \bigsqcap_{x \in N} x \leftrightarrow \bigwedge_{x \in N} d \leqslant x \wedge \bigwedge_t \Big(\bigwedge_{x \in N} t \leqslant x \rightarrow t \leqslant d\Big),$$

or for two elements:

(3d) $$d = a \sqcap b \leftrightarrow d \leqslant a \wedge d \leqslant b \wedge \bigwedge_t [(t \leqslant a \wedge t \leqslant b) \rightarrow t \leqslant d],$$

or

(3d′) $$a \sqcap b \leqslant a \wedge a \sqcap b \leqslant b \wedge \bigwedge_t [(t \leqslant a \wedge t \leqslant b) \rightarrow t \leqslant a \sqcap b].$$

Definition: An ordered set in which a meet and a join exist for every pair of elements is called a *lattice*; if for every subset of the ordered set there exists a meet and a join (in this case it is more usual to say "greatest lower bound" and "least upper bound"), the lattice is said to be *complete*. If only one of the two elements "meet" and "join" is known to exist, we speak of a *semilattice*.

The name *lattice* is explained by the order diagrams (e.g., Figures 2 to 4, p. 492), in which each of the elements is joined by a straight line segment after the manner of a latticework for vines.

The lattice *Pe* is complete, since in this case the join and meet are simply the union and intersection respectively of the subsets. The set-theoretic connectives will be denoted by round signs to distinguish them from the lattice-theoretic connectives with rectangular signs, and the logical connectives with acute-angled signs.

From the definition of join and meet we have the following rules: The commutative laws:

(4, 1, d) $a \sqcap b = b \sqcap a$, (4, 1, v) $a \sqcup b = b \sqcup a$,

the associative laws:

(4, 2, d) $a \sqcap (b \sqcap c) = (a \sqcap b) \sqcap c$, (4, 2, v) $a \sqcup (b \sqcup c) = (a \sqcup b) \sqcup c$,

the absorption laws:

(4, 3, d) $a \sqcap (a \sqcup b) = a$, (4, 3, v) $a \sqcup (a \sqcap b) = a$.

Remarks: 1. On the basis of the associative laws we may omit the brackets when the connectives are applied to finitely many elements (cf. IB1, §1.3).

2. The definitions and these rules for calculation are *dual* in the following sense: if in a valid theorem involving $\leqslant$, $\sqcap$. $\sqcup$ we interchange $\leqslant$ with $\geqslant$

and $\sqcap$ with $\sqcup$, the resulting theorem is also valid. The principle of duality in projective geometry is a special case (cf. §2.2).

Instead of using an order relation, we may also define a lattice by means of two connectives with the properties (4, 1–3). In order to make this statement completely clear, we introduce a new symbol: a set M in which two connectives $\top,\bot$ are defined,[1] satisfying the rules $a \top b = b \top a$ and so forth, as in (4), is called (after Dedekind) a *dual group*. The above remarks (if the proofs for the rules (4) are carried out) show that *every lattice is a dual group*. We now assert conversely: *every dual group is a lattice*.

For the proof we must show that in every dual group we can introduce an order with the property that for every two elements a, b there exists a least common upper element $a \sqcup b$ and a greatest common lower element $a \sqcap b$. In fact, we can arrange matters so that $a \sqcup b = a \bot b$ and $a \sqcap b = a \top b$.

From (4) we first deduce two further relations:

(5d) $a \top a = a,$ (5v) $a \bot a = a;$

(6) $$a \bot b = b \leftrightarrow a \top b = a.$$

Here we shall give only the proof for (5d):

In (4, 3, v) we set $b = a$:

(i) $$a \bot (a \top a) = a.$$

In (4, 3, d) we set $b = a \top a$:

(ii) $$a \top (a \bot (a \top a)) = a.$$

Then (i) and (ii) imply (5d).

We now define

(7) $$a \leqslant b \leftrightarrow a \top b = a.$$

In order to preserve the duality we also define

(7′) $$a \geqslant b \leftrightarrow a \bot b = a.$$

From (6) we then have: $a \leqslant b \leftrightarrow b \geqslant a$.

We must now prove:

a) $\leqslant$ is an order relation; i.e., we have (1r), (1t), (1i). As an example, we give the proof of (1t). In view of (7) the assertion reads:

$$(a \top b = a \wedge b \top c = b) \rightarrow a \top c = a.$$

[1] We here allow ourselves a temporary misuse (in the present subsection only) of the symbols $\top$, $\bot$, introduced by Bourbaki in a different sense (in which they are used below in IB10).

Proof: $a \top c = (a \top b) \top c = a \top (b \top c) = a \top b = a.$

b) With $\top$ instead of $\sqcap$ we have (3d′) and thus $a \top b = a \sqcap b$.

c) With $\perp$ in place of $\sqcup$ we have (3v′) and thus $a \perp b = a \sqcup b$.

(For b) it is convenient to use (7), and for c) to use (7′).)

It is interesting to note that a given algebraic configuration can be defined either on the basis of an order relation or on the basis of connectives (cf. IB10, §1).

III. In *Pe* we have the distributive laws

$$\text{(8d)} \qquad a \sqcap (b \sqcup c) = (a \sqcap b) \sqcup (a \sqcap c),$$

$$\text{(8v)} \qquad a \sqcup (b \sqcap c) = (a \sqcup b) \sqcap (a \sqcup c).$$

These laws do not follow from the above laws. For example, they do not hold in the lattice ($\bar{\text{d}}$) with the order diagram in Figure 2 (page 492). Here we have

$$a \sqcap (b \sqcup c) = a\,,$$

$$(a \sqcap b) \sqcup (a \sqcap c) = n \sqcup n = n.$$

However, either of the two laws (8) follows from the other; for example, (8v) from 8d):

$$\begin{aligned}(a \sqcup b) \sqcap (a \sqcup c) &= [(a \sqcup b) \sqcap a] \sqcup [(a \sqcup b) \sqcap c] && \text{by (8d)}\\ &= a \sqcup [(a \sqcup b) \sqcap c] && \text{by (4, 3, d)}\\ &= a \sqcup (a \sqcap c) \sqcup (b \sqcap c) && \text{by (8d) and (4, 2, v)}\\ &= a \sqcup (b \sqcap c).\end{aligned}$$

Furthermore, in every lattice we have the distributive inequality

$$\text{(8d′)} \qquad (a \sqcap b) \sqcup (a \sqcap c) \leqslant a \sqcap (b \sqcup c)$$

since

$$a \sqcap b \leqslant a \qquad \text{and} \qquad a \sqcap b \leqslant b \leqslant b \leqslant b \sqcup c$$

and

$$a \sqcap c \leqslant a \qquad \text{and} \qquad a \sqcap c \leqslant c \leqslant b \sqcup c.$$

In order to prove (8d) we thus need only prove that

$$\text{(8d″)} \qquad a \sqcap (b \sqcup c) \leqslant (a \sqcap b) \sqcup (a \sqcap c).$$

Definition: a lattice in which the distributive laws hold is called a *distributive lattice*.

We cannot enter here upon the subject of infinite distributive laws in complete lattices, although they are of fundamental importance for the representability of a lattice as the lattice of subsets of a set (cf. Hermes [1], and in particular §24.)

IV. The set Pe has a least and a greatest element, namely the empty set and e.

Definition: an element of an ordered set is called a

$$\textit{least element} \text{ or } \textit{zero element } (n), \qquad \text{if } \bigwedge_{x \in M} n \leqslant x,$$

$$\textit{greatest element} \text{ or } \textit{unit element } (e), \qquad \text{if } \bigwedge_{x \in M} x \leqslant e.$$

V. Definition: If a is an element of the lattice M with zero element n and unit element e, then a' is the *complement* of a if

$$a \sqcap a' = n \quad \text{and} \quad a \sqcup a' = e. \tag{9}$$

A lattice in which every element has a complement is said to be *complemented.*

A distributive complemented lattice is called a *Boolean lattice*. For example, Pe is a Boolean lattice. Boolean lattices are discussed in IA, §9, so that we shall not deal with them here.

VI. The set Pe has a class of distinguished elements, namely the subsets of e that consist of exactly one element ξ. In the lattice, they are the upper neighbors of the zero element. Every element of Pe is a union of such elements.

In general, in an ordered set with zero element n the upper neighbors of the zero element are called *atoms* (occasionally also *points*). A lattice is called *atomic* if every element other than n is an upper bound for at least one atom.

Thus Pe is a complete atomic Boolean lattice, and these properties characterize Pe in the following sense:

Theorem. *Every complete atomic Boolean lattice is isomorphic to the lattice of subsets of a set* (*namely the set of its atoms*).

For the proof we must refer again to Hermes [1].

2. Examples

2.1. *Lattices of Subgroups*

In the set e let there be defined a connective (to be denoted by simple juxtaposition) with respect to which e forms a group. Instead of Pe we now consider the set Ue of all subgroups of the group e. In Ue also there is an order defined by (set-theoretic) inclusion, and it is well known that in this case the symbols $a \subseteq b$ or $a \leqslant b$ can be read as "a is a subgroup of b." Also it is well known that the intersection $a \cap b$ is likewise a subgroup of e and is thus in Ue, and in fact $a \cap b$ is the greatest subgroup of e contained in both a and b; thus

$$a \cap b = a \sqcap b.$$

But the set-theoretical union $a \cup b$ is not always a subgroup of e, and thus is not always contained in Ue. Nevertheless, for two subgroups, or even for arbitrarily many subgroups, there always exists a smallest subgroup of e that contains them all, $a \sqcup b$ or $\sqcup_N a$. Thus *Ue forms a complete lattice.*[2]

It is also true that the set of subrings or the set of ideals of a ring e or the set of subfields of a field e form a complete lattice. A general statement in this direction is given in IB10, §2.3.

The existence of the join in Ue is a consequence of the following general theorem.

Theorem on the least upper bound. *If an ordered set M has the following two properties*:

1. *every non-empty subset of M has a greatest lower bound, and*
2. *the subset $N \subseteq M$ has an upper bound,*

then N has a least upper bound v, and in fact v is the greatest lower bound of all the upper bounds of N.

Proof. Let S be the set of upper bounds of N. By property 2, the set S is not empty, and thus by property 1 the greatest lower bound v of S exists. For v we have by definition

$$\text{(V1)} \quad \bigwedge_{s \in S} v \leqslant s \qquad \text{and} \qquad \text{(V2)} \quad \bigwedge_{s \in S} w \leqslant s \rightarrow w \leqslant v.$$

The assertion is that v is the least upper bound of N; or in other words

$$\text{(B1)} \quad \bigwedge_{X \in N} x \leqslant v \qquad \text{and} \qquad \bigwedge_{X \in N} x \leqslant t \rightarrow v \leqslant t.$$

Proof for (B1). For every $x \in N$ we have $\bigwedge_{s \in S} x \leqslant s$, so that $x \leqslant v$ follows from (V2).

Proof for (B2). If $\bigwedge_{x \in N} x \leqslant t$, then $t \in S$, so that $v \leqslant t$ by (V1).

The lattice of subgroups of a group is not necessarily distributive. For example, if e is the Klein four-group (see IB2, §15.3.3), then Ue has the order-diagram ($\bar{d}$), Figure 2.

[2] In IB2, §3.4, we wrote $\langle a \cup b \rangle$ for $a \sqcup b$ and $\langle \mathfrak{R} \rangle$ for $\sqcup_N a$, with $\mathfrak{R} = \bigcup_{a \in N} a$.

If e is the commutative group with generators α, β and relations $\alpha^4 = \epsilon$, $\beta^2 = \epsilon$ (where ϵ is the neutral element of e), and is thus the direct product of two cyclic groups of orders 2 and 4, then the subgroups are

$$n = \{\epsilon\},\ \ a = \{\epsilon, \alpha^2\},\ \ b = \{\epsilon, \beta\},\ \ p = \{\epsilon, \alpha, \alpha^2, \alpha^3\},\ \ q = \{\epsilon, \alpha^2, \beta, \alpha^2\beta\}, e.$$

The order-diagram is given in Figure 3. The juxtaposed numbers will be

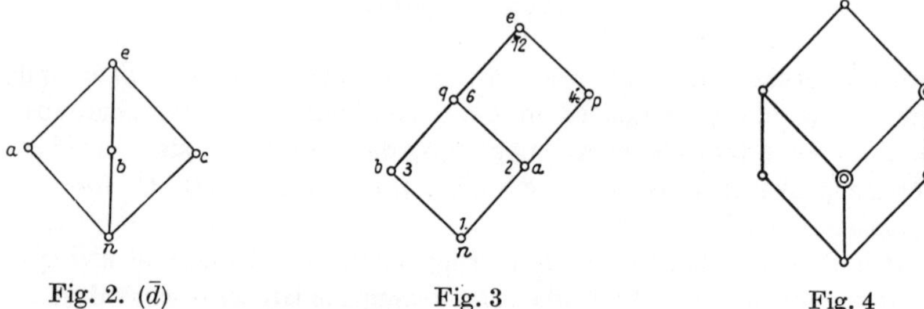

Fig. 2. $(\bar{d})$ Fig. 3 Fig. 4

explained in §2.4. This lattice is distributive but not complemented, since a and q have no complements.

Thus we see that various types of lattices can occur as lattices of subgroups. At the present time it is not known what properties a lattice V must have in order that there may exist a group e for which Ue is isomorphic to V, or in other words, under what conditions is V representable as a lattice of subgroups (cf. M. Suzuki, "Structure of a Group and the Structure of its Lattice of Subgroups." Ergebnisse d. Math. Neue Folge, Heft 10. Springer Verlag, Berlin, Göttingen and Heidelberg, 1956.)

For a group e we have $Ue \subseteq Pe$. The order relation in Ue is same as in Pe. But, although Ue thus forms a lattice under the same ordering as Pe, we do not call Ue a *sublattice* but only as *ubband* (from the German name *Teilbund*, introduced by Schwan).

For the concept of a sublattice we do not use the definition of a lattice in terms of order but in terms of the connectives (in other words, we make use of the definition of a dual group); if M is a lattice (dual group) with the connectives $\sqcap$, $\sqcup$ and N is a subset of M, then N is called a *sublattice* of M if N forms a lattice with respect to the same connectives.

A further example is given in Figure 4. If we omit the two points that are twice circled, we obtain a sublattice, but if we omit only the central point, we obtain only a subband.

The reader is invited to prove:

1. N is a sublattice of M if and only if N is closed with respect to $\sqcap$ and $\sqcup$; in other words, if and only if N contains $a \sqcap b$ and $a \sqcup b$ for every $a, b \in N$.

2. Every sublattice is a subband.

2.2. *Vector Spaces. Projective Geometry*

The vector subspaces of a vector space can be interpreted geometrically as the linear subspaces of a projective space.

Let the set e consist of the points of the (for convenience, three-dimensional) projective space, and let Le be the set of its linear subspaces, so that Le consists of the empty set n, the points, lines, planes, and e itself. Let the order be defined by inclusion. Then $a \sqcap b$ is the set of elements common to a and b, or in other words the intersection of these subspaces, and $a \sqcup b$ is the smallest linear space which includes a and b, or in other words it is the subspace spanned by a and b.

This lattice is complemented: for every subspace a of e there exists a disjoint subspace a' ($a \sqcap a' = n$) which together with a spans the whole space e ($a \sqcup a' = e$). Every element has infinitely many complements.

The set Le is not distributive. For example, if a is a plane and b, c are two points not on a, then $a \sqcap (b \sqcup c)$ is the point of intersection of the plane a with the line determined by b and c, but, on the other hand,

$$(a \sqcap b) \sqcup (a \sqcap c) = n \sqcup n = n.$$

2.3. *Every Linearly Ordered Set is a Lattice*, with $a \sqcap b$ as the smaller, and $a \sqcup b$ as the greater of the two elements a and b. We write $a \sqcap b = \min (a, b)$, $a \sqcup b = \max(a, b)$.

Every such lattice is distributive, since

$$a \sqcap (b \sqcup c) = \min (a, \max (b, c)),$$
$$(a \sqcap b) \sqcup (a \sqcap c) = \max (\min (a, b), \min (a, c)).$$

In verifying this statement the reader may assume $b \leqslant c$ (since b and c occur symmetrically) and may therefore confine his attention to the three cases $a \leqslant b \leqslant c$, $b \leqslant a \leqslant c$, $b \leqslant c \leqslant a$.

A special case is the set Z of natural numbers (with or without 0) with $\underset{z}{\leqslant}$ as the order: $(Z, \underset{z}{\leqslant})$.

In preparation for the next example we introduce the general concept of a *direct product*: let $(M_1, \leqslant)$, $(M_2, \leqslant)$ be two lattices. (Whether the ordering is "the same" in both, and in fact just what such a question would mean, is of no importance to us here.) We form the set $M = M_1 \times M_2$; its elements are the pairs (a_1, a_2), $a_1 \in M_1$, $a_2 \in M_2$. In M we define an order by

$$(a_1, a_2) \leqslant (b_1, b_2) \leftrightarrow a_1 \leqslant b_1 \wedge a_2 \leqslant b_2 .$$

Then M is a lattice with

$$(a_1, a_2) \sqcap (b_1, b_2) = (a_1 \sqcap b_1, a_2 \sqcap b_2),$$
$$(a_1, a_2) \sqcup (b_1, b_2) = (a_1 \sqcup b_1, a_2 \sqcup b_2).$$

It is easy to see that the direct product of two distributive lattices is a distributive lattice.

We may allow the case that M_1 and M_2 are the same lattice. Furthermore, we can form the direct product of arbitrarily many factors.

2.4. *Divisibility*

Let $M = Z$ be the set of natural numbers, not including 0. The relation $a \mid b$ (a divides b) is an order, $a \sqcap b$ is the greatest common divisor, and $a \sqcup b$ is the smallest common multiple of a and b. Then $(Z, \mid)$ is a lattice, which is distributive. The distributivity is closely connected with the unique factorization of a number into prime factors. Let $p_1, p_2, \ldots$ be the prime numbers in their natural sequence, and let

$$a = \prod_\nu p_\nu^{\alpha_\nu}, \qquad b = \prod_\nu p_\nu^{\beta_\nu}, \qquad c = \prod_\nu p_\nu^{\gamma_\nu};$$

for $\alpha_\nu, \beta_\nu, \gamma_\nu$ the value 0 is also allowed; the product is to be extended only up to the highest prime dividing a, b, c. Then $b \sqcap c = \prod_\nu p_\nu^{\min(\beta_\nu, \gamma_\nu)}$, $a \sqcup b = \prod_\nu p_\nu^{\min(\alpha_\nu, \beta_\nu)}$ and the assertion reads: for every ν

$$\min(\alpha_\nu, \max(\beta_\nu, \gamma_\nu)) = \max(\min(\alpha_\nu, \beta_\nu), \min(\alpha_\nu, \gamma_\nu)).$$

The proof is the same as in §2.3. By assigning a to the system $(\alpha_1, \alpha_2, \ldots)$ we map the lattice $(Z, \mid)$ isomorphically onto the direct product of infinitely many factors $(Z, \underset{z}{\leqslant})$.

A finite sublattice of $(Z, \mid)$ is formed, for example, by the factors of the number 12. Its order diagram is presented in Figure 3. It has a zero element $n = 1$ and a unit element $e = 12$. It is not complemented.

2.5. *Circuit Algebra*

An electric circuit consists of conductors to which, by neglecting resistance, we may assign the conductivity 1 or 0 according to whether or not a current can flow between their endpoints. The conductivity of a circuit in which conductors with conductivity a, b are joined in series or in parallel is denoted by $a \sqcap b$ and by $a \sqcup b$, respectively. The values of $a \sqcap b$ and $a \sqcup b$ are determined from those of a and b according to the following tables:

$a \sqcap b$:	$a =$	$b = 1$	$b = 0$
	1	1	0
	0	0	0

$a \sqcup b$:	$a =$	$b = 1$	$b = 0$
	1	1	1
	0	1	0

which are the same as the truth tables for logical conjunction and alternative (IA, §2.4). We are dealing here with a Boolean lattice.

These examples show that various types of lattices play an important role in various parts of mathematics. In the following sections we give some important theorems for a few types of lattices.

3. Lattices of Finite Length

In algebra an important role is played by a certain finiteness condition, namely, the *factor chain condition*.[3] In the simplest case this condition refers to the natural numbers as follows: if, beginning with a natural number z_0, we form a chain of factors, i.e., a sequence $z_0, z_1, z_2, \ldots$ with $z_{\nu+1} \mid z_\nu$, the chain contains only finitely many distinct elements, a fact which is the basis, for example, for the factorization of every natural number into prime factors (though not of the uniqueness of this factorization).

The factor chain condition can be formulated as a lattice property in the following way:

Definition: a lattice $(M, \leqslant)$ is said to be *of finite descending length* if every descending chain $a = x_0 \geqslant x_1 \geqslant x_2 \geqslant \ldots$ contains at most finitely many distinct elements. The concept of *finite ascending length* is defined analogously. If a lattice is of finite length both ascending and descending, it is said to be *of finite length*.[4]

The lattice $(Z, |)$ is of finite descending length but not of finite asceuding length. The lattice of linear subspaces of a projective space is of finite length if the dimension of the space is finite. But one of the advantages of a lattice-theoretic treatment of projective geometry is that it includes spaces of infinite dimension.

In $x_0 \leqslant x_1 \leqslant \cdots \leqslant x_l$, the number l is called the *length* of the chain. Statements about the length of chains are generally of interest only for *proper* chains, i.e., for chains containing only distinct elements.

If a lattice is of finite length, it does not necessarily follow that the lengths of the proper chains have an upper bound. For example, in the lattice with the order diagram of Figure 5 the kth chain is of length k.

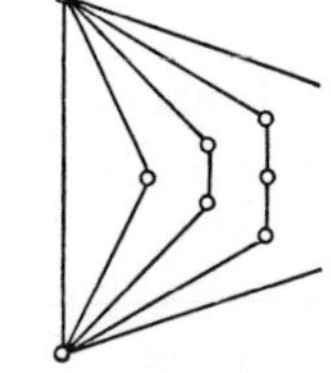

Fig. 5

[3] Cf. IB6, §2.9. Maximality condition for ideals.

[4] For our present purposes we have given a stronger (or more restrictive) meaning to "of finite length" than the one given in Hermes ([1], page 73). A lattice is there said to be of finite length if every chain joining any two elements is of finite length. Every lattice of finite length in our sense is of finite length in the sense of Hermes, but not conversely.

Theorem. *Every non-empty lattice M that is of finite descending length contains a zero element.*

Proof. Let x_0 be an arbitrary element of M. Then we have

either $\bigwedge_x x_0 \leqslant x$; here x_0 is the zero element;

or $\bigvee_x \neg (x_0 \leqslant x)$. Here it is not necessarily true that $x < x_0$, but then we have $x_1 = x \sqcap x_0 < x_0$. We then proceed with x_1 in exactly the same way as with x_0. In this way we obtain a descending proper chain, which by hypothesis has only finitely many elements, breaking off say with the element x_n. But this means that for x_n we have $\bigwedge_x x_n \leqslant x$, so that x_n is the desired zero element.

Thus it is easy to see that every lattice of finite descending length is atomic (definition in §1, VI). But an atomic lattice is not necessarily of finite descending length; for example, if e is infinite, then Pe is atomic but is not of finite descending length. Thus atomicity represents a weakening of finiteness of length, just as finiteness of length is a weakening of finiteness. The fact that important consequences can be drawn from the property of atomicity is clear from §1, VI.

For lattices of finite descending length there is an analogue to the factorization of a number into prime factors. We first replace the concept of product, which is foreign to the lattice $(Z, |)$, by the concept of least common multiple: every number can be represented as the least common multiple of finitely many prime powers. The prime powers can be characterized (in a lattice-theoretic way) by the property that they cannot be represented as the least common multiples of other elements. Then the factorization theorem can be expressed as follows for lattices:

Definition. An element q of a lattice M is said to be $\sqcup$-*irreducible*, or *primary* (the latter term being due to a certain analogy with the concept of "primary ideal" in rings (IB5, §3.6)), if q cannot be represented as the join of two other elements, or in other words if

$$q = x \sqcup y \rightarrow q = x \vee q = y.$$

It may happen that a lattice M consists entirely of primary elements, as will be the case, for example, if M is linearly ordered. It can also happen that a lattice has no primary elements at all; for example, the lattice of the pairs of rational numbers (a_1, a_2) with $0 \underset{z}{<} a_i \underset{z}{<} 1$ under the ordering

$$(a_1, a_2) \leqslant (b_1, b_2) \leftrightarrow a_1 \underset{z}{\leqslant} b_1 \wedge a_2 \underset{z}{\leqslant} b_2,$$

in which for every given pair (a_1, a_2) there exist pairs (b_1, b_2) with $b_1 \underset{z}{<} a_1$, $b_2 \underset{z}{<} a_2$, and $(a_1, a_2) = (b_1, a_2) \sqcup (a_1, b_2)$.

Under what conditions is it true that every element in a lattice can be represented as the join of primary elements (or as we shall also say, can be factored into primary elements)? Here we shall discuss only representation as the join of *finitely many* primary elements and not the question of representation as the least upper bound of arbitrarily many elements.

If a itself is a primary element, we shall regard it as a factorization.

Lemma. If a is an element that cannot be represented as the join of finitely many primary elements, then there exists an $a_1 < a$ with the same property.

For since a is not a primary element, there exists a factorization $a = a_1 \sqcup b_1$ with $a_1 < a$ and $b_1 < a$. But then at least one of the two elements a_1, b_1 cannot be factored into finitely many primary elements, since otherwise there would exist such a factorization for a itself.

Thus if there is an element a in M that cannot be factored into finitely many primary elements, there is a descending chain $a > a_1 > \ldots$ of infinite length. So we have the following theorem.

In a lattice of finite descending length every element can be represented as the join of finitely many primary elements ($\sqcup$*-irreducible elements*).

The fact that the condition "of finite descending length" is sufficient but not necessary is shown by the example of an infinite linearly ordered set.

Such a factorization is not necessarily unique, as can be seen from the nondistributive lattice ($\bar{d}$), Figure 2. In the next section we shall see that in a distributive lattice the factorization is unique.

4. Distributive Lattices

In the preceding section we described the effect of a finiteness condition; in the present section and the next one we discuss the effect of certain special rules of calculation.

We have already mentioned the theorem: *in a distributive lattice every element can be represented in at most one way as the join of finitely many primary elements without superfluous elements.*

The phrase "without superfluous elements" is necessary, since, e.g., in the lattice of the divisors of 12 we have

$$12 = 4 \sqcup 3 = 2 \sqcup 4 \sqcup 3.$$

We define: in a representation $a = q_1 \sqcup \cdots \sqcup q_k$ the factor q_i is said to be *superfluous* if the representation contains a factor q_j with $q_i \leqslant q_j$, $i \neq j$.

The proof is exactly analogous to the proof for the uniqueness of the factorization of natural numbers into prime factors. In the latter proof an important role was played by the lemma:[5] if p is a prime number, then $p \mid ab$ implies $p \mid a$ or $p \mid b$. We here prove the following lemma.

Lemma 1. *If M is a distributive lattice and p is primary, then*

$$p \leqslant a \sqcup b \rightarrow p \leqslant a \vee p \leqslant b.$$

Proof. The relation $p \leqslant a \sqcup b$ means that $p = p \sqcap (a \sqcup b) = (p \sqcap a) \sqcup (p \sqcap b)$ (by the distributive law). Since p is primary, it follows that $p = p \sqcap a$ or $p = p \sqcap b$, or in other words $p \leqslant a$ or $p \leqslant b$.

Successive application of the lemma gives: if p is primary and $p \leqslant x_1 \sqcup x_2 \sqcup \cdots \sqcup x_k$, then there exists an x_j with $p \leqslant x_j$.

Now if

$$a = p_1 \sqcup \cdots \sqcup p_r = q_1 \sqcup \cdots \sqcup q_s$$

are two decompositions of a into primary elements without superfluous elements, then for every p_i there exists at least one q_j with $p_i \leqslant q_j$ and for this q_j again a p_k with $q_k \leqslant p_k$. But from $p_i \leqslant q_j \leqslant p_k$ it follows that $i = k$, since otherwise p_i would be superfluous. Thus $p_i = q_j$. For every p_i there exists an equal q_j and conversely.

We now have the following converse. *If every element in M can be factored into finitely many primary elements without superfluous elements, then M is distributive.*

For the proof we make use of an analogue to lemma 1.

Lemma 2. *If every element in M is uniquely decomposable into primary elements (up to superfluous elements) and if p is a primary element, then*

$$p \leqslant a \sqcup b \rightarrow p \leqslant a \vee p \leqslant b.$$

Proof. If $a = q_1 \sqcup \cdots \sqcup q_r$, $b = q_1' \sqcup \cdots \sqcup q_s'$ are the unique decompositions of a and b, then

$$a \sqcup b = q_1 \sqcup \cdots \sqcup q_r \sqcup q_1' \sqcup \cdots \sqcup q_s'$$

is the unique decomposition, after cancellation of superfluous elements, of $a \sqcup b$. From $p \leqslant a \sqcup b$ it follows that

$$p \sqcup a \sqcup b = a \sqcup b = p \sqcup q_1 \sqcup \cdots \sqcup q_r \sqcup q_1' \sqcup \cdots \sqcup q_s',$$

which fails to be a second decomposition of $a \sqcup b$ only if p is superfluous, or in other words if $p \leqslant q_i$ or $p \leqslant q_j'$.

In order to show the distributivity we must prove

$$a \sqcap (b \sqcup c) \leqslant (a \sqcap b) \sqcup (a \sqcap c).$$

[5] Fundamental lemma of the theory of divisibility. IB6, §2.5.

If $a \sqcap (b \sqcup c) = q_1 \sqcup \cdots \sqcup q_t$, then for each of these elements q we have

$$q \leqslant a \wedge q \leqslant b \sqcup c,$$

i.e.

$$q \leqslant a \wedge (q \leqslant b \vee q \leqslant c),$$

$$(q \leqslant a \wedge q \leqslant b) \vee (q \leqslant a \wedge q \leqslant c),$$

$$q \leqslant a \sqcap b \vee q \leqslant a \sqcap c,$$

and therefore

$$q \leqslant (a \sqcap b) \sqcup (a \sqcap c).$$

Thus distributivity (in lattices of finite descending length) is characteristic for the uniqueness of decomposition into primary elements. In complemented lattices it is also characteristic for the uniqueness of the complement.

Theorem. *In a distributive lattice every element has at most one complement.*

We prove somewhat more, namely:
From $a \sqcup u = a \sqcup v$ and $a \sqcap u = a \sqcap v$ it follows that $u = v$.

Proof. $u = (a \sqcup u) \sqcap u = (a \sqcup v) \sqcap u = (a \sqcap u) \sqcup (v \sqcap u)$
$= (a \sqcap v) \sqcup (u \sqcap v) = (a \sqcup u) \sqcap v = (a \sqcup v) \sqcap v = v.$

What we have proved here can be expressed somewhat differently if we introduce a new concept, as follows.

It is easy to prove that if $a \leqslant b$, then the elements x with $a \leqslant x \leqslant b$ form a sublattice, which is called the *closed interval* b/a (to be read: b over a). An element x' with $x \sqcup x' = b, x \sqcap x' = a$ is called the *relative complement* of x in b/a. A lattice is said to be *relatively complemented* if every interval is complemented.

We have proved: *in a distributive lattice every interval contains at most one relative complement for a given element.*

Conversely: *if in a relatively complemented lattice M every relative complement is uniquely determined, then M is distributive.*

The last result follows from the fact that every nondistributive lattice contains a sublattice of type $(\bar{d})$ or $(\bar{m})$ (see §5). We shall not give the proof here.

We have now made the first approach to a question of fundamental importance, namely the *representation of a lattice as a set-lattice*. In many applications the elements of a given lattice V are subsets of a set e: $V \subseteq Pe$. Often the order in V coincides with inclusion in Pe, so that V is a subband of Pe. It is natural to ask: when is V a sublattice of Pe? A sublattice of Pe is called a *set-lattice* because in this case the meet is the intersection-set and the join is the union-set. For an arbitrary lattice V, whose elements are not assumed in advance to be subsets of a set e, the question at issue can be stated precisely as follows: what conditions must V satisfy in order that there may exist a set e such that V is isomorphic to a sublattice of Pe?

"Isomorphic" here means: there exists an invertible (i.e., one-to-one) mapping φ of V onto a subset φV of Pe which satisfies the homomorphism conditions:

(H1) $\varphi(a \sqcap b) = \varphi a \cap \varphi b,$ (H2) $\varphi(a \sqcup b) = \varphi a \cup \varphi b.$

A necessary condition can be stated at once: every sublattice of Pe is distributive, since the distributive law holds for all elements of Pe and thus also for the elements that form part of a subset of Pe (cf. IB10, §2.3). Consequently, V must be distributive. This necessary condition is also sufficient, although the proof requires methods that we do not develop here.[6] However, our present methods enable us to prove the theorem in the special case that V is of finite descending length, so that every element is uniquely decomposable into primary elements.

As the desired set e it is natural to consider the set of primary elements of V, which we denote by Q. It is also natural to assign to every element a of V the set φa of primary elements $p \leqslant a$:

$$p \in \varphi a \leftrightarrow p \in Q \wedge p \leqslant a.$$

The set φa is not empty, since a is decomposable into primary elements.

The proof of (H 1) is very simple:

$$p \in \varphi(a \sqcap b) \leftrightarrow p \leqslant a \sqcap b \leftrightarrow p \leqslant a \wedge p \leqslant b, \quad \text{i.e.,} \quad p \in \varphi a \wedge p \in \varphi b.$$

For the proof of (H2) we require lemma 1:

$$p \in \varphi(a \sqcup b) \leftrightarrow p \leqslant a \sqcup b \rightarrow p \leqslant a \vee p \leqslant b, \quad \text{i.e.,} \quad p \in \varphi a \vee p \in \varphi b.$$

(by lemma 1)

In the opposite direction we have at once the following result:

$$p \in \varphi a \cup \varphi b \leftrightarrow p \leqslant a \vee p \leqslant b \rightarrow p \leqslant a \sqcup b, \qquad \text{i.e.,} \quad p \in \varphi(a \sqcup b).$$

Thus the set of images φV forms a sublattice of PQ. We assert that the mapping of V onto φV is one-to-one and is thus an isomorphism, and in fact we obtain the inverse mapping by assigning to φa the element $\bigsqcup_{p \in \varphi a} p$. We must then prove:

$$\bigsqcup_{\varphi a} p = a.$$

a) For all $p \in \varphi a$ we have $p \leqslant a$, so that $\bigsqcup_{\varphi a} p \leqslant a$.

[6] Hermes [1], page 106 ff.

b) The element a can be represented as the join of primary elements:

$$a = p_1 \sqcup \cdots \sqcup p_k = \bigsqcup_{\kappa=1}^{k} p^\kappa .$$

Here every $p_\kappa \leqslant a$, so that $p_\kappa \in \varphi a$, and therefore $a = \bigsqcup_{\kappa=1}^{k} p_\kappa \leqslant \bigsqcup_{\varphi a} p$. The assertion then follows from a) and b).

Thus V has been mapped isomorphically onto a sublattice of PQ. Consequently we have the theorem: *every distributive lattice of finite descending length is isomorphic to a set-lattice*: or as we may also say, *can be represented as a set-lattice.*

In a certain sense we have thus made a survey of all possible distributive lattices of finite descending length.

From the above theorem we have: *every distributive lattice of finite descending length is isomorphic to a sublattice of a Boolean lattice*, or in other words: *can be imbedded in a Boolean lattice.*

5. Modular Lattices

5.1. In the lattice *Le* of linear subspaces of the (three-dimensional) projective space the distributive law does not hold; for example, $a \sqcap (b \sqcup c) > (a \sqcap b) \sqcup (a \sqcap c)$, if a is a plane and b, c are two points not on the plane. But if c lies in the plane a, then $a \sqcap (b \sqcup c) = c$, $(a \sqcap b) \sqcup (a \sqcap c) = n \sqcup c = c$. Thus in this case the distributive law is satisfied under the additional assumption $c \leqslant a$. The law in this weakened form

(10) $c \leqslant a \rightarrow a \sqcap (b \sqcup c) = (a \sqcap b) \sqcup c$ (note that $a \sqcap c = c$)

is called the *modular identity*, and a lattice whose elements satisfy the modular identity is said to be a *modular lattice.*

The modular identity is *self-dual*, which means that if $\sqcap$ is exchanged with $\sqcup$ and $\leqslant$ with $\geqslant$ the result is the same as before (with interchange of the letters a and c, which is of no importance).

The modular identity does not hold in every lattice, as is seen from the example $(\bar{m})$, Figure 6. However, in every lattice we do have the inequality

(10′) $c \leqslant a \rightarrow a \sqcap (b \sqcup c) \geqslant (a \sqcap b) \sqcup c$,

so that for the proof of (10) we need only show that

(10″) $c \leqslant a \rightarrow a \sqcap (b \sqcup c) \leqslant (a \sqcap b) \sqcup c$.

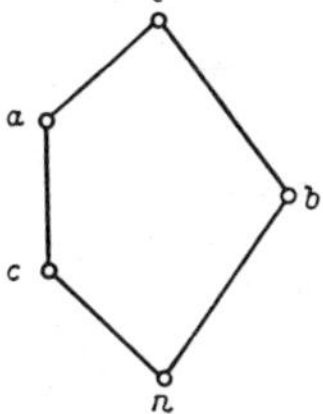

Fig. 6. $(\bar{m})$

5.2. An important class of modular lattices is characterized by the following theorem.

Theorem. *The normal subgroups of a group form a modular lattice with the ordering* $\leqslant$: "*subgroup of.*"

For the proof we must determine the meaning of $a \sqcap b$ and $a \sqcup b$ in this case.

If a, b are normal subgroups of e, then $a \cap b$ is also a normal subgroup of e and is thus the largest subgroup of e contained in a and b; or in other words:

$$a \sqcap b = a \cap b.$$

As for $a \sqcup b$, which is the smallest normal subgroup of e that contains a and b, it consists exactly of those elements ξ of e that can be represented in the form

$$\xi = \alpha\beta, \qquad \alpha \in a, \qquad \beta \in b$$

(see Part B, Chapter 2, §6.4).

For the proof of the theorem we must now prove (10″), or in other words

$$c \leqslant a \rightarrow [\xi \in a \sqcap (b \sqcup c) \rightarrow \xi \in (a \sqcap b) \sqcup c].$$

But $\xi \in a \sqcap (b \sqcup c)$ means that there exist $\alpha \in a$, $\beta \in b$, $\gamma \in c$ such that

$$\xi = \alpha = \beta\gamma.$$

Since $c \leqslant a$, we have $\gamma \in a$, so that $\beta = \alpha\gamma^{-1} \in a$, or in other words $\beta \in a \sqcap b$. Thus ξ is represented as the product of an element $\beta \in a \sqcap b$ and an element $\gamma \in c$, so that $\xi \in (a \sqcap b) \sqcup c$.

A commutative group is also called a module. By the above theorem the lattice of submodules of a module is a modular lattice, which explains the choice of the word "modular".

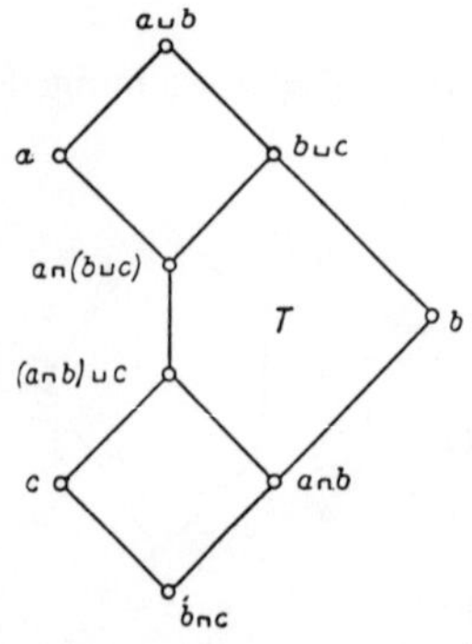

Fig. 7

5.3. The example of the nonmodular lattice $(\bar{m})$ is characteristic in the following sense.

Theorem. *Every nonmodular lattice contains a sublattice* $(\bar{m})$.

Proof. If M is nonmodular, it follows from (10′) that M contains at least three elements a, b, c with $c \leqslant a$ and $(a \sqcap b) \sqcup c < a \sqcap (b \sqcup c)$. Thus M also contains the following necessarily distinct elements: $a \sqcap b$, $a \sqcup b$, $c \sqcap b$, $c \sqcup b$, $a \sqcap (b \sqcup c)$, $(a \sqcap b) \sqcup c$. Their order relations are represented in Figure 7. This order diagram provides us with a sublattice T of type $(\bar{m})$, so that it only remains to prove that if two elements of T were equal, we would have $a \sqcap (b \sqcup c) \leqslant (a \sqcap b) \sqcup c$. From symmetry we need consider only the following cases:

a) $a \sqcap (b \sqcup c) = b \sqcup c$. Then it follows that

$$a \sqcap (b \sqcup c) \sqcap b = (b \sqcup c) \sqcap b,$$

and thus by the absorption law

$$a \sqcap b = b, \qquad (a \sqcap b) \sqcup c = b \sqcup c = a \sqcap (b \sqcup c).$$

b) $b \sqcup c = b$. Then it follows that 1. $a \sqcap (b \sqcup c) = a \sqcap b$; 2. $c \leqslant b$; and from $c \leqslant a$, we have $c \leqslant a \sqcap b$ and therefore $(a \sqcap b) \sqcup c = a \sqcap b$.

c) $a \sqcap (b \sqcup c) = b$. Then it follows that

$$(a \sqcap b) \sqcup c = [a \sqcap (a \sqcap (b \sqcup c))] \sqcup c = [a \sqcap (b \sqcup c)] \sqcup c \geqslant a \sqcap (b \sqcup c).$$

5.4. By §5.3 a modular lattice is characterized by the fact that it contains no sublattice of type $(\bar{m})$. This fact can be interpreted in the following way. *If in a modular lattice two elements a, b have an upper neighbor in common (which is then of course $a \sqcup b$), they also have a lower neighbor in common ($a \sqcap b$).*

Proof. If c is an element between $a \sqcup b$ and a: $a \sqcap b \leqslant c \leqslant a$, then we assert that $c = a \sqcap b$ or $c = a$.

We form $c \sqcup b$. From the definition of $\sqcup$ and $c \leqslant a$ it follows that $b \leqslant c \sqcup b \leqslant a \sqcup b$, so that, since $a \sqcup b$ is an upper neighbor of b, we have only the two possibilities: *first* $c \sqcup b = b$; then $c \leqslant b$, and since also $c \leqslant a$, therefore $c \leqslant a \sqcap b$; and on the other hand, we had $a \sqcap b \leqslant c$, so that $c = a \sqcap b$; *second* $c \sqcup b = a \sqcup b$; then it follows from (10), under the additional assumption $a \sqcap b \leqslant c$, that $a \sqcap (b \sqcup c) = (a \sqcap b) \sqcup c = c$, so that $c = a \sqcap (a \sqcup b) = a$.

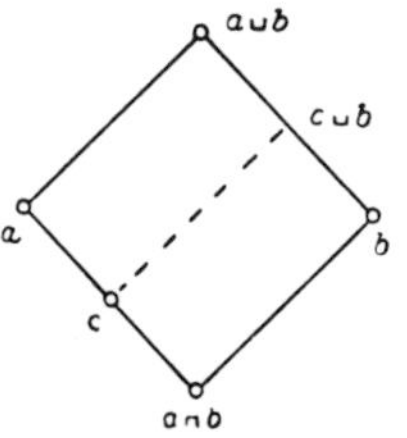

Fig. 8

Definition. A lattice in which every two elements that have a common upper neighbor also have a common lower neighbor is said to be *semimodular below*. The concept of *semimodular above* is defined dually.

Every modular lattice is semimodular below and above. Figure 4 (page 492) shows a lattice that is semimodular below but not above, so that it is not modular. If we omit the twice-circled points we obtain a sublattice $(\bar{m})$. If a lattice is semimodular below and above, we cannot at once conclude that it is also modular. For example, it may happen that no element has a neighbor, in which case the lattice is to be considered as semimodular, although trivially so. The additional condition "of finite length" is sufficient, but we do not give the proof here.

5.5. The theorem on semimodularity can be extended to larger complexes of elements by the *chain theorem* of Dedekind, as follows.

Definition. A proper chain between a and b

$$a = x_0 < x_1 < \cdots < x_l = b$$

is called a *maximal chain* if it cannot be properly refined, i.e., if $x_i \leqslant y \leqslant x_{i+1}$ implies either $y = x_i$ or $y = x_{i+1}$.

The *chain theorem* states: *if in a lattice that is semimodular either above or below there exists a maximal chain of length l joining a and b, then every maximal chain between a and b is of length l.*

We give here an outline of the proof (see Figure 9). Let

$$a = x_0 < x_1 < \cdots < x_l = b$$

and

$$a = y_0 < y_1 < \cdots < y_m = b$$

be two maximal chains between a and b. The assertion is that $m = l$.

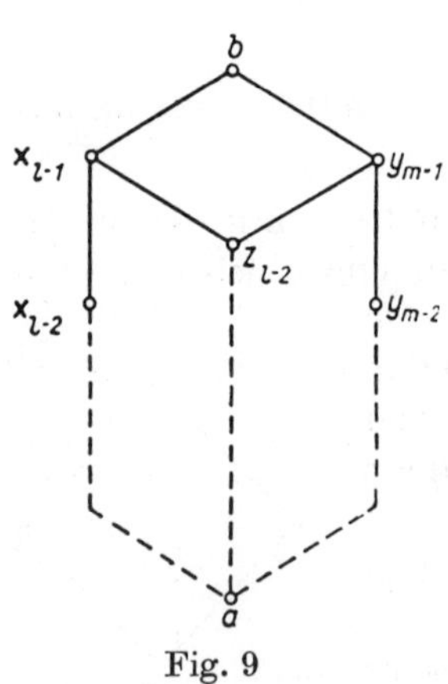

Fig. 9

For $l = 1$ this assertion is correct, and also for $l = 2$ on account of the semimodularity. We now argue by complete induction. If $x_{l-1} = y_{m-1}$, the proof is clear. Otherwise the two elements have the common upper neighbor b, and thus, if we assume semimodularity below, they have the common lower neighbor z_{l-2} (which may coincide with x_{l-2} or with y_{l-2}, or with both of them). In any case the maximal chain $a < \cdots < x_{l-2} < x_{l-1}$ is of length $l - 1$, so that by the induction hypothesis the maximal chain $a < \cdots < z_{l-2} < x_{l-1}$ is of the same length, and thus $a < \cdots < z_{l-2}$ is of length $l - 2$. But then

$$a < \cdots < z_{l-2} < y_{m-1}, \quad \text{and therefore} \quad a < \cdots < y_{m-2} < y_{m-1},$$

is also of length $l - 1$.

The chain theorem stands in close analogy with the Jordan-Hölder theorem (IB2, §12), but for the latter theorem we have the peculiar difficulty that it is based on the relation "normal subgroup of," which is not transitive and thus is not an ordering. For details we refer again to Hermes [1].

5.6. For lattices of finite length the chain theorem allows us to introduce the concept of *dimension*: to every element a we may assign as its dimension δa the length of a maximal chain from n to a. (The reader will see that there must exist at least one such maximal chain.) In the lattice Le the geometric dimension is given by $\delta a - 1$. In arguments involving this concept an important role is played by the following *dimensional equation*

$$\delta(a \sqcup b) + \delta(a \sqcap b) = \delta a + \delta b,$$

which is a consequence of the isomorphism theorem: *in a modular lattice the interval* $a \sqcup b/a$ *is isomorphic to the interval* $b/a \sqcap b$.

Let us give the main part of the proof of this theorem.

To every element x in $a \sqcup b/a$ the mapping φ with $\varphi x = x \sqcap b$ assigns an element in $b/a \sqcap b$, since $a \leqslant x \leqslant a \sqcup b$ implies

$$a \sqcap b \leqslant x \sqcap b \leqslant (a \sqcup b) \sqcap b = b.$$

This mapping is one-to-one. Every element y in $b/a \sqcap b$ is the image of exactly one element in $a \sqcup b/a$, namely $y \sqcup a$. So we now prove the following three statements:

1) $a \leqslant y \sqcup a \leqslant a \sqcup b$;
the first part is clear, and the second follows from $y \leqslant b$.

2) $\varphi(y \sqcup a) = y$;
by the modular identity, we have $\varphi(y \sqcup a) = (y \sqcup a) \sqcap b = y \sqcup (a \sqcap b)$, which is equal to y, since $a \sqcap b \leqslant y$.

3) If x, z are elements of $a \sqcup b/a$ and if $\varphi x = \varphi z$, then $x = z$;
but from $x \sqcap b = z \sqcap b$ it follows that $(x \sqcap b) \sqcup a = (z \sqcap b) \sqcup a$, so that by the modular identity we have $x \sqcap (b \sqcup a) = z \sqcap (b \sqcup a)$, and thus $x = z$, since both are $\leqslant b \sqcup a$.

The remainder of the proof of the isomorphism theorem is left to the reader, as well as the proof of the dimensional equation, which is now easy.

By means of the isomorphism theorem it can also be shown that two finite chains between the same elements have isomorphic refinements.

It must be emphasized that a rigorous proof of the theorems in §§5.5 and 5.6, for which we have only given outlines, depends upon a considerable number of details left unmentioned here.

6. Projective Geometry

In a k-dimensional projective space e the lattice Le of the linear subspaces is modular, complemented, and of finite length; in fact, the length of the chains is even bounded, namely, $\underset{z}{\leqslant} k + 1$. Thus *Le* is also atomic. The atoms are the points.

Conversely, let M be a modular complemented lattice of finite length which is not empty and does not consist of the zero element n alone. We assert that it can be interpreted as the lattice of the linear subspaces of a projective space.

Since M is of finite length, there exist upper neighbors of zero, which we shall call *points*. Then it is possible that M consists of n and one point p. But if M contains other elements (at least one), we assert that there exist at least two points. For p must have at least one complement p'. From p' there is a maximal chain leading to n; this chain contains a point q, and if we had $q = p$, it would follow that $p \sqcap p' = p \neq n$.

Two distinct points p, q have n as common lower neighbor, and thus they have a common upper neighbor $g = p \sqcup q$. Consequently there exist elements (at least one) of dimension 2; we call them *lines*. Through two points there passes at least one line.

Let h be a second line through p, q ($p \leqslant h$, $q \leqslant h$). Then from the definition of $\sqcup$ it follows that $g \leqslant h$. But $\delta g = \delta h$, so that $g = h$. Thus we have

(P1) *Through two points there passes exactly one line.*

We further assert:

(P2) *If a, b, c, d are distinct points and if the lines $a \sqcup b$ and $c \sqcup d$ have exactly one point* (s) *in common, then the lines $a \sqcup c$ and $b \sqcup d$ have a point* (t) *in common* (Figure 10).

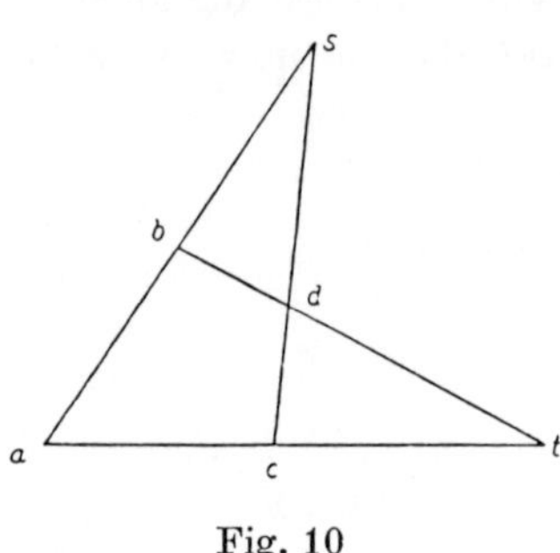

Fig. 10

For the proof we use the dimensional equation:

$$\begin{aligned}\delta((a \sqcup c) \sqcap (b \sqcup d)) &= \delta(a \sqcup c) + \delta(b \sqcup d) - \delta(a \sqcup c \sqcup b \sqcup d)\\ &= \qquad 4 \qquad\qquad - \delta(a \sqcup c \sqcup b \sqcup d).\end{aligned}$$

On the other hand,

$$\begin{aligned}\delta(a \sqcup b \sqcup c \sqcup d) &= \delta(a \sqcup b) + \delta(c \sqcup d) - \delta((a \sqcup b) \sqcap (c \sqcup d))\\ &= \qquad 4 \qquad\qquad - 1 = 3,\end{aligned}$$

since we have assumed the existence of a point of intersection of $a \sqcup b$ and $c \sqcup d$. Thus $\delta((a \sqcup c) \sqcap (b \sqcup d)) = 1$, as was to be proved.

If the four points are not all distinct, the theorem must be formulated somewhat differently, as follows: if the five points a, b, c, d, s have the property that a, b, s are collinear (i.e., lie on one line) and c, d, s are also collinear, then there exists a point t for which a, c, t are collinear and also b, d, t. In the proof it is necessary to discuss the various cases.

Finally we show:

(P3) *On a line g there lie at least two distinct points.*

In the first place there certainly exists at least one point p on g. A second point is provided by a complement p' of p. Naturally we will expect that $p' \sqcap g = q$ is a second point on g. We have $q \leqslant g$. Furthermore,

$$p \sqcup q = p \sqcup (p' \sqcap g)$$

is equal, by the modular identity, to

$$g \sqcap (p \sqcup p') = g.$$

Thus $q \neq n$ and $q \neq p$, since $g \neq p$; and also $q \neq g$, since otherwise we would have $p \leqslant q \leqslant p'$. Consequently, q is a point on g distinct from p.

In geometry we usually require a sharpening of (P3) which is not a consequence of (P1)—(P3), namely: on every line there lie at least three distinct points. In lattice-theoretic language this axiom corresponds to the property that the lattice M cannot be represented as a direct product.

In this way statements about modular complemented lattices of finite length are translated into incidence axioms of projective geometry. Further details of the lattice-theoretic interpretation of geometry, including geometries of infinite dimension, cannot be given here.

Let us close with a remark about the significance of lattice theory. It is the task of the theory to formulate and prove general statements in the generality appropriate to them, without any unnecessary special assumptions. Thus the chain theorem does not depend on whether we are dealing with normal subgroups of a group, with the factors of a number or with other objects; it depends only on the existence of an ordering and of $\sqcap$ and $\sqcup$, and on the modular identity and the concept of finite length.

CHAPTER 10

Some Basic Concepts for a Theory of Structure

Introduction

1. The theory of lattices, as described in the preceding chapter, is one of the technical means at our disposal for giving the full appropriate generality to statements of general import in mathematics. But if we wish to approach this problem systematically, we must first construct and investigate the necessary general concepts. We must not base our study on any special branch of mathematics, even though well-known mathematical facts will serve as guidelines for the construction of concepts; however, it will be necessary to make use of logic, and in fact the theory of structure is exactly the place where the importance of mathematical logic for mathematics as a whole is most clearly seen.

2. Let us begin with the example of a group, i.e., of a *set* in which there is defined an operation (or composition, as we shall call it in the more general setting below) satisfying certain rules, known as the axioms of the group. Here the particular concrete group is defined by its actual elements, together with the operation. In this sense we speak of a *configuration* (or *mathematical system*; in German, *Gebilde*). But abstractly considered, it is characterized as a *group* by the existence of an operation and by the axioms, a characterization which has nothing whatever to do with the special set of elements defining the concrete group. In this sense we speak of *structure*. We now wish to describe these concepts in the greatest possible generality. Our discussion will be based on the work of P. Lorenzen [1].

1. Configurations

1.1. *Definition*

A set M for whose elements (denoted by lower-case italic letters) there is defined a finite sequence **P** of relations $R_1, ..., R_n$ (upper-case italic letters, with or without indices) is called a *configuration* (M, **P**).

Note. This terminology has nothing to do with the concept of an "analytic configuration" in the theory of functions of a complex variable.

Examples. A set of points with the three-place relation: "the points x, y, z are collinear." The set of natural numbers with the two-place relation: "x is smaller than y." The same set with the three-place relation: "z is the sum of x and y."

One may ask how a relation can be "given" in a concrete case. Here are two examples. 1) In a finite set a multiplication, for instance, can be defined by setting down the product of any two elements in a table; and similarly, for any two-place relation in a finite set we can write down the pairs of elements for which the relation holds and the pairs for which it does not hold. 2) For the natural numbers addition and multiplication, for instance, are defined recursively (see IB1, §1, and IA, §10).

1.2. The following kinds of relation are of particular importance:

1) Correspondences between elements of two sets N_1, N_2:

$$R(x_1, x_2) \text{ cannot hold unless } x_1 \in N_1, x_2 \in N_2.$$

In the sense of our definition we must take $M \supseteq N_1 \cup N_2$ if we are to interpret the whole system as a configuration, which is not always desirable. The correspondence is called a *mapping* (also a *function*; see IA, §8.4) if it is unique with respect to the second element, i.e., if

$$\text{(U)} \qquad R(x_1, x_2) \wedge R(x_1, x_2') \rightarrow x_2 = x_2'.$$

In general we speak of a mapping *from* N_1 *into* N_2. If every element of N_1 has an image, we speak of a mapping *of* N_1, and if every element of N_2 is the image of an element of N_1, we speak of a mapping *onto* N_2. In the present chapter we shall usually denote mappings by lower-case Greek letters, and the existence of a correspondence φ between x_1 and x_2 will then be written in the form $\varphi x_1 = x_2$ (cf. also IA, §8.4).

We note that by definition a mapping is always one-valued, so that there will be no need to mention this fact from now on. Thus for a mapping φ we always have

$$\bigwedge_x \bigwedge_y (x = y \rightarrow \varphi x = \varphi y).$$

Of course, it can happen that $\varphi x = \varphi y$ and $x \neq y$. If this situation does not occur, i.e., if

$$\wedge_x \wedge_y (\varphi x = \varphi y \rightarrow x = y),$$

we see that the passage from φx to x is also a mapping. As the *inverse mapping* of φ we denote it by φ^{-1} and say that in this case φ is *invertible*, or *one-to-one*.

2) An $(n + 1)$-place relation is called an n-place *inner composition* (German *Verknüpfung*, French *composition*) if for every n-tuple $x_1, \ldots, x_n$ of elements in M there exists exactly one element z in M for which $R(x_1, \ldots, x_n, z)$ holds, i.e., if

$$\text{(E)} \qquad \wedge_{x_1} \cdots \wedge_{x_n} \vee_z R(x_1, \ldots, x_n, z) \qquad \text{and}$$

$$\text{(U)} \quad \wedge_{x_1} \cdots \wedge_{x_n} \wedge_z \wedge_{z'} (R(x_1, \ldots, x_n, z) \wedge R(x_1, \ldots, x_n, z') \rightarrow z = z').$$

In other words: a composition is a mapping of $M \times M \times \cdots \times M = M^n$ *into* M. Thus we may write $z = \rho(x_1, \ldots, x_n)$ instead of $R(x_1, \ldots, x_n, z)$.

For example, addition and multiplication are two-place compositions. As a general symbol for a two-place inner composition it is customary to adopt the Bourbaki symbol $\top$, and to write this symbol between the arguments to which it applies: $x \top y = z$.

3) According to this definition the formation of the least upper bound, or of the greatest lower bound, in a complete lattice is not a composition; for instead of the n-tuple $x_1, \ldots, x_n$ we are dealing here with an arbitrary subset of M, and the mapping in question is from the power set PM into M. But for the purposes of the present chapter we shall include mapping of PM into M among the compositions, calling it a *nonelementary* composition (since formation of the power set is not part of elementary logic).

4) Another extension is necessary if we wish to include, for example, the S-multiplication in vector spaces (multiplication of a vector with a scalar) (IB3, §1.2). In addition to the set M (the vector space) we now have a domain of scalars S (in general we may speak of it as a domain of operators Ω) and to each pair (ω, x), $\omega \in \Omega$, $x \in M$ there is assigned an element z of M. Under the relevant assumptions (E) and (U) a correspondence of this sort is called an *outer operation*, and for it we use the symbol $\perp$: $\omega \perp x = z$, again following Bourbaki. In place of x we could also have an n-tuple or a subset of M. The domain Ω may also coincide with M.

5) Finally, for all these compositions we could allow the set of images to be not M but some other set, as is the case, for example, with the tensor

product and the outer product of vectors (IB3, §3.3). But in the present chapter we shall exclude this generalization of the concept of a composition.

Definition. If all the defining relations of a configuration are compositions, the configuration is called a *composition-configuration* or an *abstract algebra* (IA, §8.5).

Configurations of other kinds are, for example, the ordered sets (defined by an ordering), and the topological spaces, defined by a relation "U is a neighborhood of x" between an element x of X and an element U of a subset $\mathfrak{U}$ of PX ($M = X \cup \mathfrak{U}$, cf. 1.2, §1); in this case X may, for example, be the set of points of a plane, where $\mathfrak{U}$ is the set of open circular disks. It is remarkable that a dual group (IB9, §1) is a composition-configuration, whereas a lattice is not.

1.3. *Homomorphism and Isomorphism*

Some of the most usual (algebraic) concepts have to do with configurations, and others with the notion of structure defined below. In particular the concepts of homomorphism, isomorphism, and congruence have to do with configurations. The first two of these deal with mappings of a configuration $(M, \mathbf{P})$ into another configuration $(M', \mathbf{P}')$. Here it is not only the elements of M that are mapped but also the defining relations. For the latter it is assumed that every relation R_i' in $\mathbf{P}$ corresponds to exactly one relation R_i' in $\mathbf{P}'$ and conversely. We then say that the configurations are *homologous*. For the most part we shall be interested only in the subset of M that actually undergoes a mapping; in other words, we usually deal with a mapping of $(M, \mathbf{P})$ into $(M', \mathbf{P}')$.

A mapping φ of $(M, \mathbf{P})$ into $(M', \mathbf{P}')$ is called a *homomorphism* if for every relation

$$R(x_1, ..., x_n) \to R'(\varphi x_1, ..., \varphi x_n).$$

For a two-place inner composition $\top$ this condition means that

$$x \top y = z \to \varphi x \top' \varphi y = \varphi z$$

or

$$\varphi x \top' \varphi y = \varphi(x \top y).$$

For an outer composition we must take account of the domain of operators, which is done in the same way as for the set of relations: we assume that there exists a one-to-one correspondence between the elements of Ω and those of Ω', or we may at once assume that the two configurations have the same operator domains. The condition for φ to be a homomorphism is, in the simplest case:

$$\varphi(\omega \perp x) = \omega \perp' \varphi x.$$

Definition. If φ is a one-to-one mapping of $(M, \mathbf{P})$ *onto* $(M', \mathbf{P}')$ such that its inverse is also a homomorphism, then φ is called an *isomorphism*.

This last condition (namely that the inverse mapping must also be a homomorphism) does not follow, in general, from the other conditions, as may be shown by the example of the mapping of an ordered set with the order diagram I (Figure 1) onto the set with the order diagram II. But for compositions the latter condition can be deduced from the others, since if we are given elements x', y', there must exist x, y with $x' = \varphi x$, $y' = \varphi y$, and thus we have

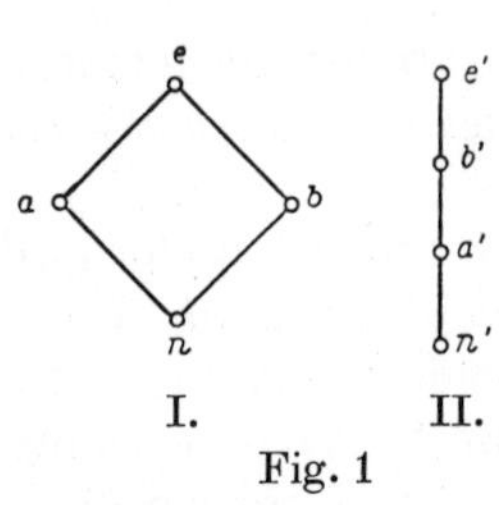

Fig. 1

$$\begin{aligned}\varphi^{-1}(x' \top y') &= \varphi^{-1}(\varphi x \top \varphi y)\\ &= \varphi^{-1}(\varphi(x \top y))\\ &= x \top y = \varphi^{-1}x' \top \varphi^{-1}y'.\end{aligned}$$

Definition. A homomorphism of M into M itself is called an *endomorphism*, and an isomorphism of M onto M itself is called an *automorphism*.

1.4. *The Automorphism Group*

Successive application of two automorphisms produces a uniquely determined automorphism and is thus a composition. It is easy to see that the automorphisms of a configuration form a group under this composition, the *automorphism group of the configuration*.

Conversely, we have the theorem: For every group G there exists a configuration $(M, \mathbf{P})$, and in fact a composition-configuration, whose automorphism group is isomorphic to G.

Birkhoff's proof.[1] Let the elements of G be denoted by lower-case Latin letters. The set M will consist of the elements of G and the pairs of elements of G:

$$M = G \cup (G \times G).$$

Let the defining relations be: an outer composition with G as operator domain, defined by

$$c \perp a = c, \qquad c \perp (a, b) = a,$$

[1] On the structure of abstract algebras. Proc. Cambridge Phil. Soc. 31, 1935, pp. 434–454.

and an inner composition $\top$, defined by

$$(a, b) \top (a', b') = b \cdot b'^{-1},$$
$$a \top a' = a \top (a', b') = (a', b') \top b = 1,$$

or in words: when $\top$ is applied to two pairs of elements, it produces the quotient of the two second elements, and otherwise $\top$ always produces 1 (the neutral element of G).
Assertion. The automorphism group of the configuration defined in this way is isomorphic to G. For the proof we must put every automorphism φ into one-to-one correspondence with an element of G in such a way that the correspondence is an isomorphism. To this end we examine the effect of an automorphism φ on the elements of M.

1) If $c \in G$, then for every automorphism φ we have: $\varphi c = c$. (G is elementwise fixed under every automorphism.)

Proof. From $c \perp a = a$ it follows that

$$\begin{aligned}\varphi c = \varphi(c \perp a) = c \perp \varphi a &= c, \qquad \text{if} \quad \varphi a = a' \in G;\\ &= a', \qquad \text{if} \quad \varphi a = (a', b') \in G \times G.\end{aligned}$$

In each case we have $\varphi c \in G$; since this statement holds for every element of G, the second case does not arise.

2) $\varphi(a, b) = (a', b')$, or in words: the image of a pair is again a pair. For if we had $\varphi(a, b) = c$, then for the inverse mapping we would have $\varphi^{-1}c = (a, b)$, in contradiction to 1).
More precisely we have: $\varphi(a, b) = (a, b')$; i.e., an automorphism leaves the first element of a pair unchanged.

Proof. Let $\varphi(a, b) = (a', b')$.
From $c \perp (a, b) = a$ and $\varphi a = a$ it follows that

$$a = \varphi a = \varphi(c \perp (a, b)) = c \perp \varphi(a, b) = c \perp (a', b') = a'.$$

3) By $\varphi(1, 1) = (1, c)$ an element c in G is assigned to each automorphism. We must show that conversely the automorphism is uniquely defined by c. For this purpose we need only express the b' in $\varphi(a, b) = (a, b')$ in terms of a, b, and c.
From $(a, b) \top (1, 1) = b$ it follows that

$$b = \varphi b = \varphi(a, b) \top \varphi(1, 1) = (a, b') \top (1, c) = b' \cdot c^{-1},$$

so that $b' = bc$.

Up to now we do not know whether any automorphism of our configuration actually exists. We have only shown: if φ is an automorphism, then φ determines an element $c \in G$ by the equation $\varphi(1, 1) = (1, c)$, and for arbitrary a, b

$$(*) \qquad \varphi a = a, \qquad \varphi(a, b) = (a, bc).$$

But it is easy to verify that:

1) For every $c \in G$ the mapping of M into itself defined by (*) is an automorphism.

2) The correspondence between φ and c is one-to-one.

3) If the group elements c_1, c_2 correspond to two given automorphisms, then the group element c_1, c_2 corresponds to the automorphism determined by successive application of the given automorphisms.

1.5. *Congruence Relations*

A congruence relation $\equiv$ is an equivalence relation that is *consistent* with the defining relations R of the configuration; that is, for a congruence relation we have

$$R(x_1, ..., x_{n+1}) \wedge x_1 \equiv x_1' \wedge \cdots \wedge x_{n+1} \equiv x_{n+1}' \rightarrow R(x_1', ..., x_{n+1}').$$

If R is an inner composition, so that $R(x_1, ..., x_n, z)$ can be written in the form

$$z = \rho(x_1, ..., x_n),$$

then the condition for consistency can be written:

$$x_1 \equiv x_1' \wedge \cdots \wedge x_n \equiv x_n' \rightarrow \rho(x_1, ..., x_n) \equiv \rho(x_1', ..., x_n'),$$

and in the case of an outer two-place composition:

$$x \equiv y \rightarrow \omega \perp x \equiv \omega \perp y.$$

A congruence relation gives rise, as an equivalence relation, to a partition into classes. Let the class defined by x be denoted by κx. From the set κM of classes we construct a configuration homologous to $(M, \mathbf{P})$ by defining

$$R(\kappa x_1, ..., \kappa x_n) \leftrightarrow R(x_1, ..., x_n)$$

or for compositions $\rho(\kappa x_1, ..., \kappa x_n) = \kappa(\rho(x_1, ..., x_n))$.

It is possible to make these definitions only because the right-hand sides depend, in view of the condition for consistency, only on the classes and not on our choice of representatives; that is, if we choose other representatives from the same classes, we obtain the same results. It is easy to see that the mapping x into κx is a homomorphism. Thus *every congruence relation corresponds to a homomorphism.*

For a composition-configuration we can prove the converse. *To every homomorphism* φ *of* $(M, \mathbf{P})$ *into the configuration* $(M', \mathbf{P})$ *there corresponds a congruence relation*, namely, the relation $\equiv$ defined by $a \equiv b \leftrightarrow \varphi a = \varphi b$. For we need only prove that such a relation is reflexive, transitive, and symmetric and is consistent with the compositions. As an example, we will demonstrate the consistency for the case of a two-place inner composition:

Hypothesis: $x \equiv x'$, i.e., $\varphi x = \varphi x'$; $y \equiv y'$, i.e., $\varphi y = \varphi y'$.

Assertion: $x \top y \equiv x' \top y'$, i.e., $\varphi(x \top y) = \varphi(x' \top y')$.

Proof: $\varphi(x \top y) = \varphi x \top \varphi y = \varphi x' \top \varphi y' = \varphi(x' \top y')$.

We have seen that to every congruence relation for groups there corresponds a normal subgroup, and for rings an ideal; the corresponding question for configurations, namely, whether to every congruence relation there corresponds a subconfiguration, is a question of an entirely different sort, to be answered affirmatively only for certain very special configurations.

2. Structure

2.1. *Definition*

A set is made into a configuration by means of defining relations; and we then say that the set carries a *structure*. Our present purpose is to give a definition of structure. Certainly this concept must be independent of the particular set under consideration, much in the same way as the concept of an architectural "style" is independent of any particular edifice erected in the given style. The concept of "structure" will refer to the properties of the defining relations.

For example, let us consider a defining relation R which is three-place and has the following four properties:

(1) $$\bigwedge_x \bigwedge_y \bigvee_z R(x, y, z),$$

(2) $$\bigwedge_x \bigwedge_y \bigwedge_z \bigwedge_{z'} (R(x, y, z) \wedge R(x, y, z') \rightarrow z = z');$$

so that R is a composition. We write $xy = z$ for $R(x, y, z)$.

(3) $$\bigwedge_x \bigwedge_y \bigwedge_z ((xy)\, z = x(yz)),$$

(4) $$\bigvee_e \bigwedge_x (ex = x \wedge \bigvee_{\bar{x}} \bar{x}x = e).$$

(In other words, R defines the concept of a group, cf. IB2, §1.1.)

These formulas are constructed from individual variables and the relation symbols (of which there is here only one) with the logical particles $\wedge$, $\vee$, $\rightarrow$ and the quantifiers $\bigwedge$, $\bigvee$, where the quantifiers are applied only to the individual variables, not to the relation symbols, and *all* the individual variables are bound by the quantifiers. These formulas become statements when the individual variables are replaced by the elements of a set M and the relation symbols by the defining relations of a configuration. Such a system of formulas is called an *axiom system*. (In this choice of terminology there is no reference to the philosophical meaning of the word "axiom" or to the question of self-evidence.)

If the formulas of an axiom system $\mathbf{\Sigma}$ become *valid* statements when a configuration $(M, \mathbf{P})$ is inserted in the way just described, the configuration $(M, \mathbf{P})$ is said to be a *model* of the axiom system $\mathbf{\Sigma}$. We say that $(M, \mathbf{P})$ has a *structure*, which is described by $\mathbf{\Sigma}$.

If an axiom system $\mathbf{\Sigma}'$ is logically equivalent to an axiom system $\mathbf{\Sigma}$ (that is, if $\mathbf{\Sigma}'$ follows from $\mathbf{\Sigma}$ and conversely) we say that $\mathbf{\Sigma}'$ describes the *same* structure as $\mathbf{\Sigma}$, and thus we define a *structure* as a class of logically equivalent axiom systems.

In general, a configuration will carry various structures; for example, a configuration of the rational numbers with the compositions of addition and multiplication carries the structures of a ring, an integral domain, and a field. One might be tempted to try to find a "comprehensive" structure, from which would follow all valid statements about this configuration. But it is a consequence of the incompleteness theorem of Gödel (IA, §10.5) that for the configuration of the rational numbers there cannot exist an axiom system of this sort that is finite or recursively enumerable.

Another kind of similarity between structures is illustrated by the concepts of lattice and dual group (IB9, §1). Here we have two types of defining relations, on the one hand the order $\leqslant$, and on the other the compositions $\sqcap$, $\sqcup$. These relations can be defined in terms of each other, as follows.

On the one hand: $a \leqslant b \leftrightarrow a \sqcap b = a$,

on the other: $d = a \sqcap b \leftrightarrow d \leqslant a \wedge d \leqslant b \wedge \bigwedge_{d'}((d' \leqslant a \wedge d' \leqslant b) \rightarrow d' \leqslant d)$.

With these definitions the two systems of axioms are seen to be equivalent. Every model of either of them is also a model of the other. In the present chapter we cannot enter upon a precise general description of this situation. Two structures that are similar in this way are said to belong to the same *structure type*.

2.2. *Subconfigurations*

In our discussion of lattice theory (IB9, §1, II) we at first used the two names "dual group" and "lattice" to distinguish the two structures defined in different ways, namely, the dual group by the two (dual) compositions and the lattice by the order relation, and it is desirable to have these two distinct names. Then, as is customary, we have used the name "lattice" for the structure type to which both the structures belong.

But now the subsets which under the given compositions form a dual group (they are usually called sublattices) are not always the same as the subsets (called subbands) that form a lattice (in the sense of the structure) with respect to the given order. Thus we must define the concept "subconfiguration" in terms of the structure and not in terms of the structure type.

Definition. Let $(M, \mathbf{P})$ be a configuration carrying the structure $\mathbf{\Sigma}$. Assume that under the same relations as for M and under outer compositions with the same operator domains a given subset N satisfies the axiom system $\mathbf{\Sigma}$; then $(N, \mathbf{P})$ is said to be a *subconfiguration of* $(M, \mathbf{P})$ *with respect to* $\mathbf{\Sigma}$.

Remarks. 1) The phrase "with respect to $\mathbf{\Sigma}$" is omitted if no misunderstanding can arise.

2) The conditions (E), (U) that characterize a relation as a composition, belong to $\mathbf{\Sigma}$. If the condition (U) is satisfied in M for a given relation R, it is also satisfied in any subset N of M. If the condition (E) is satisfied in N with respect to R, then N is said to be *closed* with respect to the composition R; e.g., for a two-place inner composition $\top$ the subset N is closed if $a, b \in N$ implies $a \top b \in N$. For a composition-configuration the closedness of a subset N is a necessary but not sufficient condition for N to be a subconfiguration; for example, the integers with addition as the composition form a group for which the set of positive integers is closed with respect to addition but is not a subgroup.

3) Every two-place inner composition in a set M can be regarded as an outer composition with M as the operator domain, and this may be done in two ways, depending on whether we regard the left or the right factor as an operator. Corresponding possibilities exist for many-place compositions, which we need not discuss here. If we interpret multiplication in a ring as an outer composition in this sense, then the subconfigurations of this structure are not the subrings but the left and right ideals, respectively (cf. IB6, §2.5).

4) In a group $(G, .)$ the forming of inner automorphisms can be regarded as an outer composition with G as operator domain: $g \perp x = g^{-1}xg$.

In this case, and more generally for any set of operators that give rise to endomorphisms, we speak of a *group with operators*. The subconfigurations are called *admissible subgroups*. In the present case these are exactly the normal subgroups (IB2, §6.1).

Vector spaces are also groups with operators. Here the group composition is addition; the S-multiplication gives rise to endomorphisms (cf. IB3, §1.2); the admissible subgroups are the vector subspaces. Since the theorem in IB9, §5.2 also holds for admissible subgroups of a group with operators, and since in a commutative group every subgroup is a normal subgroup, it follows that the lattice of vector subspaces of a vector space is modular.

The importance of the remarks 3) and 4) lies in the fact that they illustrate the great generality of the theorem of §2.3 (that the subconfigurations... form a complete lattice).

In each case where we have used the names "group," "ring," "field" the reader should consider whether we have been referring to a structure or to a structure type. The answer may be different from case to case, and it appears that up to the present no one has given an interpretation of the situation that will command universal assent.

2.3. *The Lattice of Subconfigurations*

In the set $\mathfrak{U}(G)$ of subconfigurations of a configuration $(G, \mathbf{P})$ an order is defined by inclusion. If the intersection of arbitrarily many subconfigurations is again a subconfiguration, it follows from the theorem on the least upper bound (IB9, §2.1) that $\mathfrak{U}$ is a complete lattice. A structure which is "bequeathed" by any set of subconfigurations to their meet is said to be meet-hereditary, and we then have the following theorem.

Theorem. *The subconfigurations of a configuration with meet-hereditary structure form a complete lattice.*

We now ask: which structures, or in other words which sets of axioms, are meet-hereditary? Here the answer depends on the logical form of the axioms. For example, the following are meet-hereditary:

1) Axioms in which only $\wedge$-quantifiers occur; for if such an axiom holds for G, then it holds for *every* subset of G.

Here it must be assumed that the axiom has already been brought into the so-called prenex normal form, in which all the quantifiers appear in non-negative form at the beginning of the formula and apply throughout up to the end of the formula, a situation that can always be attained by logical transformations. (Otherwise we could always arrange to have only $\wedge$-quantifiers by replacing $\vee_x \mathsf{A}(x)$ with $\neg \wedge_x \neg \mathsf{A}(x)$.

2) Axioms in whose prenex normal form the quantifiers appear in the successive order

(E′) $\bigwedge_x \bigvee_y \bigwedge_z A(x, y, z)$, $A(x, y, z)$ free of quantifiers,

only if the uniqueness statement

(U′) $\bigwedge_x \bigwedge_{y_1} \bigwedge_{y_2} \bigwedge_z [A(x, y_1, z) \wedge A(x, y_2, z) \rightarrow y_1 = y_2]$

is valid, i.e., is either an axiom or a consequence of the axioms.

Thus we are asserting that if N_1, N_2 are subsets of M with $N_1 \cap N_2 = D$ and if (E') holds when the domains of the variables are restricted to N_1 or to N_2, then (E') also holds when the domains of variables are restricted to D, i.e.,

$$\bigwedge_{x\in D} \bigvee_{y\in D} \bigwedge_{z\in D} A(x, y, z).$$

Proof. If we assume $x \in D$, we have

$$\bigvee_{y_1\in N_1} \bigwedge_{z\in N_1} A(x, y_1, z) \quad \text{and} \quad \bigvee_{y_2\in N_2} \bigwedge_{z\in N_2} A(x, y_2, z).$$

If for z we choose an element in D, it follows that

$$A(x, y_1, z) \wedge A(x, y_2, z),$$

so that (U′) implies $y_1 = y_2 \in N_1 \cap N_2$.

The proof is also valid if the x, y, z are replaced by systems $x_1, ..., x_p$; $y_1, ..., y_q$; $t_1, ..., t_r$.

The proof becomes simpler if z and the corresponding quantifier do not occur. For then in place of (E′), (U′) we have exactly (E), (U) as on page 510, and thus we obtain: *closedness with respect to operations is meet-hereditary.*

If x and the corresponding quantifier do not occur, we must take account of the possibility that D is empty, and then the proof runs somewhat differently. From (E′) and (U′) we have: there exists exactly one y in M with the property $\bigwedge_{z\in M} A(y, z)$. The restriction of (E′) to N_i, $(i = 1, 2)$, means that $\bigvee_{y_i\in N_i} \bigwedge_{z\in N_i} A(y_i, z)$. Thus N_i is not empty; consequently there exists a z_i in N_i with $A(y, z_i)$ and $A(y_i, z_i)$. So by (U′) we have $y = y_i$ for $y = 1, 2$, and thus $y \in D$, which completes the proof of (E′) for D. This result shows, for example, that the existence of the neutral element of a group is meet-hereditary.

Groups, rings, fields, and lattices are examples of meet-hereditary structures. So it is natural to ask what properties of a configuration correspond to given properties of the lattice of its subconfigurations. For the particular case of groups, this question has been the subject of many profound investigations, of which M. Suzuki has recently given a connected account.[2]

The above theorem admits the following converse: *if V is a complete lattice, there exists a configuration with meet-hereditary structure whose subconfigurations form a lattice isomorphic to V.*

[2] Structure of a group and the structure of its lattice of subgroups. Ergebnisse d. Math., Neue Folge, Heft 10, 1956.

Birkhoff's proof.[3] We take $M = V$ and define an outer composition, with V as operator domain, which to an element a of $V(= \Omega)$ and a subset $B \subseteq V$ assigns an element $x \in V$:

$$a \perp B = a \sqcap \bigsqcup_B b.$$

Since V is a complete lattice, this correspondence is actually a composition, i.e., the conditions (E) and (U) are satisfied for the corresponding three-place relation. We now take $\mathbf{\Sigma}$ to consist of these axioms alone. Then the configuration $(M, \perp)$ has a meet-hereditary structure, so that its subconfigurations form a complete lattice. We assert that this lattice is isomorphic to V. The proof runs as follows.

The subset $N \subseteq M$ is a subconfiguration if

$$a \in V \wedge B \subseteq N \rightarrow a \sqcap \bigsqcup_B b \in N.$$

We consider $\bigsqcup_N y = c$. Since V is complete, there exists an element c in V. In fact, c even belongs to N, since $c = c \sqcap c = c \sqcap \bigsqcup_N y$, and $N \subseteq N$. By the definition of the least upper bound, $x \in N \rightarrow x \leqslant c$; and conversely, $x \leqslant c \rightarrow x \in N$ since

$$x = x \sqcap c = x \sqcap \bigsqcup_N y.$$

Consequently: for every subconfiguration N there exists an element c with the property that N consists exactly of the elements $x \leqslant c$. This set A_c is called the *segment* of c.

Conversely, every segment A_c is also a subconfiguration. For if $B \subseteq A_c$, we have: $b \in B \rightarrow b \leqslant c$, so that for every $a \in V$:

$$a \sqcap \bigsqcup_B b \leqslant c, \qquad \text{i.e.,} \in A_c .$$

Obviously there exists a one-to-one correspondence between the segments A_c and the elements c of V such that

$$A_c \subseteq A_d \leftrightarrow c \leqslant d.$$

It follows that the segments, and consequently also the subconfigurations, form a lattice isomorphic to V, so that the proof is complete.

In the present chapter we have not been able to give more than the first steps in a theory of structure. In the theorems on the automorphism group and the lattice of subconfigurations we have tried to prove some of the first results. They illustrate the importance of the concepts of group and lattice.

[3] "On the Combinations of Subalgebras," *Proc. Cambridge Phil. Soc.* 29, 1933, pp. 441–464.

Let us mention some other questions without attempting to answer them here.

How can other configurations be constructed from a given configuration? For example, how should subconfigurations be constructed; or direct products? (On the same question for lattices see IB9, §2.3.) Which of the properties of a configuration are preserved in passing to a subconfiguration or a direct product; or to a homomorphic configuration?

How can we classify systems of axioms, i.e., structures, on the basis of the parts of logic that are employed? For example, Lorenzen distinguishes *pure-elementary* structures (essentially those that we have used here, but not including the nonelementary compositions), *elementary-arithmetical*, in which arithmetic is used, e.g., in the Archimedean axiom for the calculus of line segments $\bigwedge_{y>0} \bigwedge_{x>0} \bigvee_n n \cdot x > y$, ($n$ a natural number), and further: *elementary-logical* and *non-elementary*, which are characterized by the fact that the relation symbols occur as variables (e.g., in the induction axiom for the Peano system). The last two types are distinguished by the linguistic-logical means employed, in a way which cannot be described here for lack of space.

Can it happen that a system of axioms uniquely characterizes a configuration up to isomorphism, such a structure being called *monomorphic*? The answer here is affirmative, as is shown by the example of a vector space of given dimension over the field of rational numbers (which is an elementary-arithmetical structure); but the most important algebraic structures, such as group, ring, field, and lattice, are not monomorphic.

Here we have tried to give some indication of possible questions in a theory of structure. What we have said is perhaps enough to show that we are dealing here with new points of view, from which an attempt is made to survey the whole of mathematics.

CHAPTER 11

Zorn's Lemma and the High Chain Principle

The present chapter deals with two maximal principles in the theory of sets: *Zorn's lemma*, which has been used very frequently in recent times, since it simplifies many former proofs; and the *high chain principle* (cf. §4), which, although trivially equivalent to Zorn's lemma, has the advantage of being intuitively plausible. The key position of the high chain principle in this type of argument appears to have remained unnoticed up to now.[1]

1. Ordered Sets

We first give some definitions and a few of their immediate consequences (cf. also IB9, §1; IA, §8.3).

By an *ordered set*, or an *order*, we mean a set $M = \{a, b, c, ...\}$ together with a two-place relation $\leqslant$ defined on it, with the following properties:

Reflexivity: $a \leqslant a$ *for all* a.

Identivity: $a \leqslant b$ *and* $b \leqslant a$ *imply* $a = b$.

Transitivity: $a \leqslant b$ *and* $b \leqslant c$ *imply* $a \leqslant c$.

An ordered set is said to be *totally* (*or linearly*) *ordered*, or to be a *chain*, if it has the following additional property:

Comparability, or connexity: for any two elements a, b *we have*

$$a \leqslant b \qquad \textit{or} \qquad b \leqslant a.$$

The terminology varies in the literature. It is also common to refer to our ordered set as a "partially ordered set," and to restrict the term "ordered set"

[1] For other set-theoretic maximal principles and their equivalence to the axiom of choice see §4, exercise 12, and, for example [12], [1], [6], in the bibliography at the end of the chapter.

to chains. Moreover, a distinction is often made between an ordered set and an order, the latter term being used to refer only to the relation defined on an ordered set.

Examples of orders are given by the "set-orders": if S is an arbitrary system of subsets of a set A, then S becomes an ordered set under the relation of inclusion ($\subseteq$). This class of examples already includes, up to isomorphism, all possible orders: every order M is isomorphic to a set-order. For if $a \in M$ and we let $\underset{\sim}{a}$ be the set of elements $x \in M$ with $x \leqslant a$, then the mapping $a \to \underset{\sim}{a}$ is an isomorphism of M onto a set-order.

In an ordered set we write $a < b$ to mean "$a \leqslant b$ with $a \neq b$," and $a \geqslant b$ or $a > b$ to mean $b \leqslant a$ or $b < a$, respectively.

Let T be a subset of an ordered set M. Then T itself is an ordered set under the same relation $\leqslant$. In particular, a subset of a chain is also a chain. The statement "$x \leqslant a$ (or $x < a$) for all $x \in T$" is abbreviated to $T \leqslant a$ (or $T < a$). We now make the following definitions:

To say that s is an upper bound of T means that $T \leqslant s$.
If here $s \notin T$ (i.e., $T < s$), *then s is a proper upper bound of T.*
To say that g is a greatest element of T means that $g \in T$ and $T \leqslant g$ (i.e., *g is an upper bound of T contained in T*). *Thus an upper bound of T is either a proper upper bound of T or a greatest element of T.*
To say that m is a maximal element of T means that $m \in T$ and that there exists no $x \in T$ with $x > m$.

A subset T need not necessarily have an upper bound or a greatest element or a maximal element. But obviously T can have at most one greatest element, though it may have several maximal elements, and also, of course, several upper bounds. A greatest element is always a maximal element, but the converse is in general false, although the two concepts coincide if T is a chain.

The concepts *lower bound, least element, minimal element* of T are defined dually (i.e., with $\geqslant$ in place of $\leqslant$).

Exercises

1. Every finite ordered set has at least one maximal element.
2. (*a*) If M is an ordered set in which every two elements have an upper bound, then every maximal element of M is also a greatest element of M.
   (*b*) If M is finite, the converse of (*a*) also holds (cf. ex. 1).
3. If $a < b$ and there exists no χ with $a < \chi < b$, then a is called a *lower neighbor* of b, and b is an *upper neighbor* of a. Does there exist a chain in which every element has an upper neighbor but infinitely many elements have no lower neighbor?

4. Prove that in every infinite chain in which every nonempty subset has a smallest element there exists an ascending sequence, i.e. a sequence $a_1, a_2, a_3 \cdots$ with $a_1 < a_2 < a_3 \cdots$.

5. An ordered set in which every nonempty subset has a least and a greatest element must be a finite chain (and conversely).

6. In every ordered set M the following statements are equivalent:
   I. Every nonempty subset of M has at least one maximal element.
   II. Every nonempty chain has a greatest element.
   III. ("Ascending chain condition.") There exists no ascending sequence, i.e. for every $a_1 \leqslant a_2 \leqslant a_3 \leqslant \cdots$ there exists an n with $A_n = A_{n+1} = A_{n+2} = \cdots$.
   III′. Every "finite-below" chain is also "finite-above;" i.e., if for every element a in a given chain there are only finitely many elements below a, then for every element b in the chain there are only finitely many elements above b.
   III″. Every finite-below chain is finite.

7. Construct (e.g. by drawing their "order diagrams" as in the chapter on lattices) all the ordered sets with fewer than five elements. (There are exactly 25 such sets, apart from isomorphism; five with 3 elements, and sixteen with 4.)

8. Let a set $M = \{a, b, c, ...\}$ be said to be *ordered* if there is given on M a two-place relation $<$ with the two properties:

   Irreflexivity: $a < a$ for all a.

   Transitivity: $a < b$ and $b < c$ imply $a < c$.

   Prove that this definition of order is equivalent to the one given above, in the following sense: if a given relation $<$ is reflexive, identive and transitive, then the relation $<$, defined by $a < b$ if $a < b$ and $a = b$, is irreflexive and transitive, and conversely, if a given relation $<$ is irreflexive and transitive, then the relation $\leqslant$, defined by $a \leqslant b$ if $a < b$ or $a = b$, is reflexive, identive and transitive.

2. Zorn's Lemma

After these preliminary remarks we now formulate *Zorn's lemma.*[2]

Z. *An ordered set in which every chain has an upper bound contains a maximal element.*

The role of Zorn's lemma may be described as follows: in arguments involving infinite sets, the older proofs often made use of the well-ordering

[2] The name Kuratowski's lemma would be more correct (cf. [8] 1922, statement (42), [21] 1935.

theorem and transfinite induction (for these concepts see IA, §7.4, appendix to IB1, §§2, 3, 5 and for example, [6], [12], [16]). In general, the well-ordering used in the proof has nothing to do with the underlying structure of the set or with the theorem to be proved; the well-ordering theorem merely provides a proof that the set in question admits at least one well-ordering, the particular nature of which is unknown and irrelevant, and this well-ordering is made the basis of a transfinite induction. But in spite of its correctness such a procedure is usually felt to be unsatisfactory. In many cases Zorn's lemma allows us to avoid these unsatisfactory arguments and to replace them by a more natural method of proof; for the most part, the proofs become much clearer and shorter.

Some examples of proofs by Zorn's lemma will be given in the next section. The proof of the lemma itself is given in §4.

In most applications Zorn's lemma is used in the following special form, which refers to set-orders and makes a sharper assumption on the upper bound:

Z′. *Let S be a nonempty system*[3] *of subsets of a set A which with every nonempty chain contains its union. Then S contains a maximal element* (i.e., a subset of A that is maximal in S).

It is to be noted that in Z (and thus also in Z′) the assertion can be sharpened:

Sharpened form of Z or of Z′: *under the same assumptions as in* Z *or* Z′, *for every element there exists a maximal element over it.*

For let M be an ordered set satisfying the assumptions of Z, let $a \in M$ and let N be the subset of $x \in M$ with $x \geqslant a$. Then it is obvious that N is also an ordered set satisfying the assumptions of Z and the assertion follows by the application of Z to N.

3. Examples of the Application of Zorn's Lemma

We shall now prove three algebraic theorems by means of Zorn's lemma. Further examples of proofs based on Zorn's lemma are easy to find in the recent literature on topics in algebra or topology.

Theorem 1. *In a commutative ring R with unit element every ideal distinct from R is contained in a maximal ideal.*[4]

An ideal M in R is said to be *maximal* if $M \neq R$ and there is no other ideal between M and R (in other words, if M is a maximal element in the set-order of the ideals $\neq R$).

[3] The word "system" will be used as a synonym of "set."

[4] For the definitions of "ring" and "ideal of a ring" see IB5, §1.2, §3.1.

Proof of theorem 1. Let J_0 be an ideal in the given ring R with $J_0 \neq R$. Let S be the set of ideals J with $J \supseteq J_0$, $J \neq R$.

We show that S satisfies the assumptions of Z′ (with $A = R$). The assertion then follows by application of Z′ to S.

Since $J_0 \in S$ we see that S is nonempty. Let K be a nonempty chain in S and let V be the union of K (i.e., the set-theoretic union of all the ideals in K). Then we must show that $V \in S$; that is to say,

a) V is an ideal in R,

b) $V \supseteq J_0$,

c) $V \neq R$.

As for *a*): arguments similar to the proof about to be given for *a*) occur everywhere in the applications of Zorn's lemma; we give such an argument in detail here once for all: if $a, b \in V$ and if $r \in R$, then there exist $J, J' \in K$ with $a \in J$, $b \in J'$, and since K is a chain, we have $J \subseteq J'$ or $J' \subseteq J$. Without loss of generality we may assume $J' \subseteq J$. Then $a, b \in J$, and therefore $a - b$, $ra \in J$ (since J is an ideal) and thus also $\in V$, so that V is an ideal.

As for *b*): since K is nonempty, there exists a $J \in K$, and for this J we have $J_0 \subseteq J \subseteq V$, from which it follows that $V \supseteq J_0$.

As for *c*): for the ideals J of a ring R with unit element it is clear that $J = R$ if and only if $1 \in J$.

From $V = R$ it would follow that $1 \in V$, so that there would exist a $J \in K$ with $1 \in J$ and then for this J we would have $J \in S$ and $J = R$, in contradiction to the definition of S.

Remark on theorem 1. For a not necessarily commutative ring R with unit element it is obvious that the corresponding statements for left ideals, right ideals, and two-sided ideals can be proved in exactly the same way.

Theorem 2. *Every vector space has a basis.*

More precisely, we show that every (not necessarily finite-dimensional) vector space V over a skew field K has a basis.[5]

A (not necessarily finite) subset T of V (more precisely, an indexed subset) is said to be *linearly independent* if each of its finite subsets is linearly independent (in the usual sense). The set T is called a *generating system* for V if T is not contained in any proper subspace of V. By a *basis* of V we mean a linearly independent generating system of V.

Proof of Theorem 2. We may assume that V does not consist of the zero vector alone (otherwise the empty set is a basis of V). Let S be the

[5] For the definitions of "skew field," "vector space," "subspace (vector subspace)" and "linearly independent (for finite sets of vectors)" see IB3, §1.1–1.4.

aggregate of all linearly independent subsets of V. Then S obviously satisfies the assumptions of Z', with $A = V$ (cf. the remarks under a) in the proof of theorem 1), so that by Z' there exists a maximal linearly independent subset of V[6]. Thus it only remains to show:

Every maximal linearly independent subset B of V is a generating system of V (and thus also a basis of V).

If we assume that there exists a subspace T of V with $T \neq V$ and $B \subseteq T$, then $T \neq V$ would mean that there exists a vector $\mathfrak{n} \in V$ with $\mathfrak{n} \in T$. Let $B' = B \cup \{\mathfrak{n}\}$. Then it is easy to see that B' would also be linearly independent, so that B would not be a maximal linearly independent subset.

Theorem 3 (Theorem of Artin-Schreier). *Every formally real field can be ordered* (*is orderable*).[7]

By a *domain of positivity* of a field K we mean a subset P of K with the following properties (here $-P$ denotes the set of all $-x$ with $x \in P$):

1) $a, b \in P$ imply $a + b$, $ab \in P$,

2) $0 \notin P$,

3) $-P \cup \{0\} \cup P = K$.

Not every field has a domain of positivity; for example, a field of characteristic $\neq 0$ cannot have one; on the other hand, there exist fields that have several.

A field K with at least one domain of positivity is said to be *orderable*. If one of the domains of positivity in an orderable field is distinguished, we speak of an ordered field. More precisely: an *ordered field* is a pair K, P consisting of a field K and a domain of positivity P in K.

In an ordered field K, P a relation $a < b$ is defined by $b - a \in P$. For $a = 0$ it follows from this definition of $<$ that P is the set of elements > 0, a fact which explains the name "domain of positivity" for P.

If in an ordered field K, P we set $R = P \cup \{0\}$, then $a, b \in R$ imply $a + b$, $ab \in R$, and we have $-R \cap R = \{0\}$ and $-R \cup R = K$; and if the relation $a \leqslant b$ (i.e., $a < b$ or $a = b$) is defined by $b - a \in R$, then $\leqslant$ is a total order on K which is compatible with addition and multiplication in K (i.e., $a \leqslant b$ implies $a + c \leqslant b + c$ for all c and implies $ac \leqslant bc$ for all c with $0 \leqslant c$. These facts enable us to provide equivalent definitions of an ordered field, in the following way:

A field is said to be *formally real* if -1 cannot be represented as the

[6] From the sharpened form of Z' we see that every linearly independent subset of V can be extended to a maximal linearly independent subset of V.

[7] For theorem 3 see also IB1, §2.5, §3.4 and IB8, §2.2.

sum of squares; or equivalently, if $a_1^2 + a_2^2 + \cdots + a_n^2 = 0$ implies $a_1 = a_2 = \cdots = a_n = 0$.

We note that the converse of theorem 3 is trivial; for in an ordered field every square, and consequently every sum of squares, is $\geqslant 0$, but -1 is < 0. Theorem 3 thus gives an "algebraic" characterization (i.e., a characterization in terms of the operations $+$ and $\cdot$ alone) of the orderable fields: a field is orderable if and only if it is formally real.

Proof of theorem 3. Let K be a formally real field. Let Q be the set of all nonzero sums of squares of elements in K. Let S be the set of all those subsets T of K that contain Q and have the properties 1), 2).

It is obviously enough to show:

a) S satisfies the assumptions of Z′, with $A = K$.

b) Every maximal element of S has the property 3).

As for *a*): since K is formally real, Q is exactly the set of all $\sum_{i=1}^n a_i^2$ with $a_i \in K$, $a_i \neq 0$, $n \geqslant 1$. Thus $Q \in S$, so that S is not empty. The fact that with every nonempty chain the set S also contains its union is proved in the same way as under *a*) in the proof of theorem 1.

As for *b*): for $T \in S$ let $T_0 = \{0\} \cup T$. Then *b*) is a consequence of the following lemma.

Lemma. If $T \in S$ and $r \notin -T \cup \{0\} \cup T$, then $T' = T + rT_0$ (the set of all $a + rb_0$ with $a \in T$, $b_0 \in T_0$) is an element of S properly containing T. (Thus T is not maximal).

Proof of the lemma. Obviously $T \subseteq T'$ (we may choose $b_0 = 0$), and thus $Q \subseteq T$ implies $Q \subseteq T'$.

Since $1 \in Q \subseteq T$ and $-r \in T$ (for otherwise $r \in -T$) we have $-r \neq 1$, so that $r + 1 \neq 0$. Thus

$$r = \left(\frac{2r}{r+1}\right)^2 + r\left(\frac{r-1}{r+1}\right)^2,$$

so that $Q \subseteq T$ implies $r \in T'$. Since $r \notin T$, it follows that $T' \neq T$, so that $T \subset T'$. Thus we need only verify the properties 1), 2) for T'.

1) For any two elements $a + rb_0$, $c + rd_0 \in T'$ it follows from $Q \subseteq T_0$ and from the additive and multiplicative closedness of T and T_0 that

$$(a + rb_0) + (c + rd_0) = (a + c) + r(b_0 + d_0) \in T'$$

and

$$(a + rb_0)(c + rd_0) = (ac + r^2 b_0 d_0) + r(ad_0 + b_0 c) \in T'.$$

2) $0 \in T'$ would imply $0 = a + rb_0$ with $a \in T$, $b_0 \in T_0$; $a = -rb_0$; $b_0 \neq 0$ (since $a \neq 0$), so that $b_0 = b$ with $b \in T$; and thus $-r = a/b = (1/b)^2$

ab with $a, b \in T$; $-r \in T$ (since $Q \subseteq T$ and T is multiplicatively closed); consequently $r \in -T$, in contradiction to the assumption of the lemma.

4. Proof of Zorn's Lemma from the Axiom of Choice

In this section we introduce the high chain principle mentioned in the introduction. Zorn's lemma at once turns out to be nothing but a more complicated form of the high chain principle. In the rest of this section, a simple proof of the high chain principle (and thus of Zorn's lemma) from the axiom of choice is given in full detail.

Let us first give the definition and some trivial properties of the operation "roof" (denoted by ^), on which this section will depend.

Let M be an ordered set. Here and below the word "chain" will always refer to a subchain of M. The elements of M will be denoted by $a, b, c, \ldots, x, y$, the subsets of M by A, B, the chains by C, K, L and the empty set by $\emptyset$.

For every subset A of M let $\hat{A}$ be the set of all elements x with $A < x$. Thus $\hat{A}$ is the set of all proper upper bounds of A.

Obviously $A \cap \hat{A} = \emptyset$, and $A \subseteq B$ implies $\hat{B} \subseteq \hat{A}$.

(1) *If A, B are arbitrary subsets of M, at least one of the two sets $A \cap \hat{B}$, $\hat{A} \cap B$ is empty.*

For otherwise there would exist elements a, b, with $a \in A$, $B < a$ and $b \in B$, $A < b$, which would imply $b < a$ and $a < b$, in contradiction to the identivity.

(2) *If A, B are subsets of M with $A \subseteq B \cup \hat{B}$ and $B \subseteq A \cup \hat{A}$, then $A \subseteq B$ or $B \subseteq A$.*

For by (1) we have $A \cap \hat{B} = \emptyset$, so that $A \subseteq B$, or $\hat{A} \cap B = \emptyset$, so that $B \subseteq A$.

(3) *If K is a chain and $C \subseteq K$, then: $\hat{C} = \hat{K}$ is equivalent to $\hat{C} \cap K = \emptyset$.*

Proof. $\hat{C} = \hat{K}$ implies $\hat{C} \cap K = \hat{K} \cap K = \emptyset$. Conversely, from $\hat{C} \cap K = \emptyset$ we have, in succession: for every $x \in K$ it is untrue that $C < x$; for every $x \in K$ there exists a $c \in C$ with $c < x$ untrue, i.e., with $x \leqslant c$ (since c, x are comparable, being elements of the same chain K); $\hat{C} \subseteq \hat{K}$; $\hat{C} = \hat{K}$ (for $C \subseteq K$ always implies $\hat{C} \supseteq \hat{K}$).

We say that a chain K is *high*, and we call it a *high chain* (with respect to M) if $\hat{K}$ is empty. Thus a high chain is a chain that has no proper upper bound, i.e., a chain with no element properly over it, a chain that cannot be continued upward.

Note that a high chain need not be a maximal chain, i.e. maximal in the set-order consisting of the chains of M (although, of course, every maximal chain is a high chain). For example, if m is a maximal element of M, the chain constisting of m alone is a high chain, but in general it will not, of course, be a maximal chain.

This difference between the concept of high chain and maximal chain marks the difference between the high chain principle and the Hausdorff-Birkhoff maximal chain principle.

Let us now formulate the

High Chain Principle. Every Ordered Set Contains a High Chain.[8]

This maximal principle makes no hypothesis about the given ordered set, and it has an intuitive acceptability which is independent of any proof — both in contrast to Zorn's lemma. Nevertheless, it is in fact identical with Zorn's lemma, as we shall now see.

There are two kinds of high chains: high chains without upper bound, and high chains with upper bound. *The high chains without upper bound are precisely the chains without upper bound.* The upper bounds of high chains are precisely the greatest elements of high chains, and so precisely the maximal elements of M. Thus *the high chains with upper bound are precisely the chains that contain a maximal element* (*of* M). These remarks show at once that:

Zorn's Lemma and the High Chain Principle are Equivalent.

For in an arbitrary ordered set M the following statements are equivalent (the first one being Zorn's lemma and the last one the high chain principle):

If every chain has an upper bound, there exists a maximal element.
There exists a chain without upper bound or there exists a maximal element.
There exists a high chain without upper bound, or there exists a high chain with upper bound.
There exists a high chain.

The proof of the high chain principle from the axiom of choice, which we now give, is the last step in a gradual development beginning with Zermelo's first proof of the well-ordering theorem ([19], 1904). For example, Kneser's proof of Zorn's lemma ([7], 1950), and Weston's outline of a proof ([17], 1957), which forms the basis of the proof to be given here, are steps in this development toward simplicity.

The proof makes use of the so-called *Axiom of Choice* (cf. IA, §7.6, supplement to IB1, §5):

[8] Of course, the high chain principle can be sharpened to the statement that in an ordered set every chain K is an initial segment of a high chain (it is only necessary to apply the high chain principle to the ordered set $\hat{K}$).

Axiom of Choice. *For every system S of nonempty sets there exists a choice function, i.e., a function f which to every set $N \in S$ assigns an element of N: thus $f(N) \in N$.*

Proof of the high chain principle. Let M be an ordered set. By the axiom of choice there exists a choice function defined on the system of all nonempty sets $\hat{C}$ (where C is a chain). Let f be such a function. Then $\hat{C} \neq \emptyset$ implies $f(\hat{C}) \in \hat{C}$.

The proof depends on the concept of an f-chain. A chain K is called an *f-chain* if it has the following property:

(*) *$C \subseteq K$ and $\hat{C} \cap K \neq \emptyset$ imply that $f(\hat{C})$ is the least element of $\hat{C} \cap K$, i.e., that $f(\hat{C}) \in \hat{C} \cap K$ and $f(\hat{C}) \leqslant \hat{C} \cap K$.*

In other words: If C is a subchain of K with proper upper bound in K, then $f(\check{C})$ is the least of these proper upper bounds.[9]

In view of (3) and $f(\hat{C}) \in \hat{C}$ the property (*) is equivalent to

(**) *$C \subseteq K$ and $\hat{C} \neq \hat{K}$ imply that $f(\hat{C}) \in K$ and $f(\hat{C}) \leqslant \hat{C} \cap K$.*

The proof consists in deriving two rules for the creation of f-chains ((i), (ii)) and applying them to the set-theoretic union of all f-chains.

(i) *Continuation of f-chains: if K is an f-chain with $\hat{K} \neq \emptyset$, then $K^* = K \cup f(\hat{K})$ is an f-chain* (and, of course, $K^* \nsubseteq K$).

Proof. $K < f(\hat{K})$ implies that K^* is a chain with greatest element $f(\hat{K})$, and $K^* \nsubseteq K$.

Assume that $C \subseteq K^*$ and that $\hat{C} \cap K^*$ is nonempty. Let $s \in \hat{C} \cap K^*$. Then it follows successively that $C < s \leqslant f(\hat{K})$; $f(\hat{K}) \notin C$; $C \subseteq K$.

If now $\hat{C} = \hat{K}$, then $f(\hat{C}) = f(\hat{K})$ and $\hat{C} \cap K^* = \hat{K} \cap K^* = f(\hat{K})$.

On the other hand, if $\hat{C} \neq \hat{K}$, it follows from (**) that $f(\hat{C}) \in K$ (so that $f(\hat{C}) < \hat{K}$ and therefore $f(\hat{C}) < f(\hat{K})$) and $f(\hat{C}) \leqslant \hat{C} \cap K$.

Thus in every case $f(\hat{C}) \in K^*$ and $f(\hat{C}) \leqslant \hat{C} \cap K^*$.

The crucial point in the proof of (ii) is the following lemma.

Lemma. If K, L are f-chains, then $L \subseteq K \cup \hat{K}$ (and, of course, also $K \subseteq L \cup \hat{L}$).

Proof of lemma. For $L \subseteq K$ there is nothing to prove. Consequently, assume $L \nsubseteq K$ and let y be an arbitrary element with $y \in L$, $y \notin K$. The assertion is that $y \in \hat{K}$.

[9] The function f provides us with a "rule for climbing" that not only allows us to climb from a given element to a greater element but also to surmount, with one jump, a whole infinite chain; and the f-chains are the "upward paths" created in this way. Taking $C \neq \emptyset$ we see, in particular, from (*) that every nonempty f-chain begins with $f(\hat{\emptyset})$.

Let C be the set of all x with $x \in L \cap K$ and $x \leqslant y$.[10] Then $C \leqslant y$ and (in view of $y \notin K$) $y \notin C$, so that $C < y$, i.e., $y \in \hat{C}$.

Since $C \subseteq L$, $y \in \hat{C} \cap L$ we have from (*): $f(\hat{C}) \in L$ and $f(\hat{C}) \leqslant y$.

Since $C \subseteq K$, the hypothesis $\hat{C} \neq \hat{K}$ would (by (**)) imply $f(\hat{C}) \in K$ and therefore (by the definition of C) $f(\hat{C}) \in C$. So $\hat{C} = \hat{K}$ and thus $y \in \hat{K}$ (since $y \in \hat{C}$).

From this lemma and (2) we get the *comparability of f-chains*: *if K, L are f-chains, then* $K \subseteq L$ *or* $L \subseteq K$.

(ii) *Union of f-chains*: *the union F of an* arbitrary set of f-chains is also an f-chain.

Proof. The comparability of f-chains shows that F *is a chain*. From the lemma it follows further that $F \subseteq K \cup \hat{K}$ *for every f-chain K*.

Now let $C \subseteq F$ and $\hat{C} \cap F$ be nonempty. Let x be an arbitrary element of $\hat{C} \cap F$. Then, since $x \in F$, there exists an f-chain K with $x \in K \subseteq F$, and it follows that $x \in \hat{C} \cap K$, so that $\hat{C} \cap K \neq \emptyset$.

Since $C \subseteq F$, and $F \subseteq K \cup \hat{K}$, we have $C \subseteq K \cup \hat{K}$. Since $\hat{C} \cap K \neq \emptyset$, it follows from (1) that $c \cap \hat{K} = \emptyset$, so that $C \subseteq K$.

Since K is an f-chain, it follows from $C \subseteq K$, $\hat{C} \cap K \neq \emptyset$ that $f(\hat{C}) \in K$ and $f(\hat{C}) \leqslant \hat{C} \cap K$, so that $f(\hat{C}) \in F$ and $f(\hat{C}) \leqslant x$.

Now let V be the *union of all f-chains*. By (ii) we see that V is an f-chain and consequently by (i) that $\hat{V} = \emptyset$, i.e., V is a high chain. For if we had $\hat{V} = \emptyset$, then by (i) there would exist an f-chain V^* with $V^* \nsubseteq V$, in contradiction to the definition of V.

Remark. Let us denote the axiom of choice by A and the high chain principle by H. A trivial application of Z′ gives A, so that we have proved the implications A → H → Z → Z′ → A. Thus, the set-theoretic maximal principles A, H, Z, Z′ are equivalent.

Exercises

We first give 3 definitions.

(i) A subset A of an ordered set M is called an *initial segment* of M if for every $x \leqslant a$ that xA.

(ii) For every subset S of an ordered set M the corresponding set S of lower elements is defined as the set of elements xM for which there exists an element sS with $x \leqslant s$.

(iii) An ordered set M is said to be *well-ordered* if every subset of M has a least element.

1. The relation of being an initial segment is transitive; i.e. every initial segment of an initial segment of an initial segment of an ordered set M is an initial segment of M.

[10] In fact, $C = K$.

2. The intersection, and also the union, of an arbitrary number of initial segments of an ordered set M is an initial segment of M.
3. Prove that for every subset S of an ordered set M:
   (*a*) $\underline{S}$ is the initial segment of M generated by S; namely, $\underline{S}$ is an initial segment of M that contains S, and $\underline{S}$ is the intersection of all initial segments of M that contain S.
   (*b*) From (*a*) it follows that S is an initial segment of M if and only if $S = \underline{S}$.
   (*c*) $\underline{S}$ is an initial segment A of M with $S \subseteq A$ and $\hat{S} = \hat{A}$. (Thus, in the definition of an f-chain we could take C to be an *initial segment*.)
   (*d*) $\underline{S} \cap \hat{S} = \phi$.
4. In every subset S of a *totally* ordered set K we have $\underline{S} \cup \hat{S} = K$, and therefore the following three statements are pairwise equivalent (cf. 3(*b*), (*d*)): S is an initial segment of K; $S = \underline{S}$; $S \cup \hat{S} = K$.
5. A subset L of a chain K of an ordered set M is an initial segment of K if and only if $K \subseteq L \cup \hat{L}$.
6. (*a*) Every well-ordered set is totally ordered.
   (*b*) For finite sets the converse also holds.

In exs. 7 to 11 below, the assumptions are the same as in the proof of the high chain principle; i.e., M is an ordered set and f is a choice function on the system of all nonempty sets $\hat{C}$ (where C is a subchain of M). Then M has the following properties (7–11):

7. The intersection of arbitrarily many f-chains is an f-chain.
8. The set of all f-chains is well-ordered with respect to inclusion.
9. Every f-chain is well-ordered.
10. A subset L of an f-chain K is an f-chain if and only if it is an initial segment of K (use ex. 5 and the lemma of §4).
11. The f-chains are precisely the initial segments of the union V of all f-chains.
12. Consider the following statements:
    (*a*) axiom of choice A
    (*b*) high chain principle H (cf. §2)
    (*c*) Zorn lemma Z
    (*d*) special case Z′ of the Zorn lemma (cf. §2)
    (*e*) well-ordering theorem W: "every set can be well-ordered"
    (*f*) Hausdorff-Birkhoff maximal chain principle M: "in every ordered set there exists a maximal (with respect to inclusion) chain" i.e. a chain which ceases to be a chain if any further element of the ordered set is adjoined to it.

Prove the following implications:

A → H → Z → W → A and H → Z → Z′ → M → H.

In other words, the statements A, H, Z, Z′, W, M are pairwise equivalent.
Hints for the proofs.
For A → H and H → Z cf. §4.
The implications Z → Z′ and M → H are specializations W → A: the union V of the given system S of nonempty sets can, by W, be well-ordered; for a fixed well-ordering of V choose the smallest element from each set of S. Z′ → M: the entire aggregate of chains of an ordered set M forms a system S satisfying the assumptions of Z′ (with $A = M$). Z → W: for an arbitrary set M let Ω be the set of all well-ordering relations defined on subsets of M. For $\omega_1, \omega_2 \in \Omega$ let $\omega_1 \leqslant \omega_2$ be defined as follows: the domain of definition T_1 of ω_1 is contained in the domain of definition T_2 of ω_2, on T_1 the two relations ω_1 and ω_2 coincide, and T_1 is an initial segment of T_2 with respect to ω_2. With this relation $\leqslant$ the set Ω is an ordered set in which every subchain has an upper bound. Then Z states that Ω has a maximal element. But every maximal element ω of Ω must be defined on the whole of M, since an element of M not contained in the domain of definition of ω could be "adjoined to ω from above," thereby producing an $\omega' > \omega$.

13. In the proofs of §3 it is possible, of course, to use the high chain principle instead of the Zorn lemma. For the proof of Theorem 2, for example, one may first prove (without the Zorn lemma and thus independently of the axiom of choice): the union of a high chain in the ordering of all linearly independent subsets of vector space V is a basis of V. Then Theorem 2 follows immediately from this theorem and the high chain principle. What is the corresponding "quintessence" (i.e., formulation independent of the axiom of choice) of Theorem 1 (of Theorem 3)?

5. Questions Concerning the Foundations of Mathematics

In the present chapter we have up to now taken the so-called "naive point of view" concerning sets. (cf. IA, §1.4, §7.1, §7.2). But everything we have said here, and in particular the proof we have given for the high chain principle, could also be formulated in the usual axiomatic set theories (cf. IA, §7.1 and §7.6). In this sense the proof we have given in §4 for Z is correct and can be verified even by an intuitionist or a constructivist.

But a constructivist would regard an axiomatic (formalistic) interpretation of the concept of "set" as meaningless and therefore without interest;[13] he would admit only constructive interpretations, and from this point of view (cf. IA, §1.4, §1.5) he would find two mistakes, or at least gaps, in the proof given above in §4:

1) In one place in the proof we made use of the axiom of choice without actually constructing a choice function (on this question see, for example, [4], Chapter II, §4).

2) The set V was defined as the union of all f-chains, but it turned out later that V is itself an f-chain. Thus we have defined an object (the set V) by means of a concept (f-chain) under which the object itself is included.

More precisely, from the constructive point of view the situation is somewhat as follows: the f-chains are (in general, infinite) subsets. The only possibility of constructing an infinite subset is to construct a "representing property" for it (namely, a propositional form in a suitable language). Every construction of representing properties for sets must be carried out by means of certain linguistic tools, which must either be given or constructed in advance. With more linguistic tools at our disposal we can construct more properties and thus represent more sets. But the totality of all linguistic tools can never have been constructed (for if we were to assume that this is the case, we could proceed to use these linguistic tools in order to create further ones), and thus, in a constructive interpretation, the expression "all f-chains" can never have an *absolute* meaning but only a relative one; it can only be understood in the sense of all f-chains "*representable in a given language S*." If we now form the union V of this relative totality of f-chains, we do not know whether a representing property of V can be found in the language S, i.e., whether V itself belongs to this totality. But precisely this fact was used in the above proof, namely when we said: "if $\hat{V} \neq \emptyset$, then $V^* = V \cup \{f(\hat{V})\}$ is an f-chain, and thus $V^* \subseteq V$." For in order to draw the conclusion that $V^* \subseteq V$, we must know that V^* is an f-chain *representable in S*. Since it is obvious that V^* is representable in S if and only if V is representable in S (the two sets differ only by a single element), we see that a constructive interpretation of our proof in a language S is correct only if the union V of all f-chains that are representable in S is itself representable in S.

"Impredicative definitions," like this definition of V, occur in many places in mathematics in its usual form, e.g., in the introduction of the real numbers (cf. IB1, §4.3 and [18]). An objection of the type 2) above was already raised by Poincaré against Zermelo's first proof of the well-ordering theorem (cf. [11], [19], and Russell's "vicious circle principle" in

[13] Even if it were proved that the underlying formalized set theory is free of contradictions.

the introduction to [13]; see also [4] and the literature given there). A more precise examination of the whole question in the framework of P. Lorenzen's operational mathematics is given in [10] (cf. IA, §10.6 and [9]).

In operational mathematics every set is countable in a suitable language level. So let us note here that, for a countable ordered set, a constructive proof of the high chain principle can easily be given by complete induction.

Bibliography

[1] Becker, O.: Grundlagen der Mathematik in geschichtlicher Entwicklung. Freiburg, München 1954.

[2] Bourbaki, N.: Théorie des ensembles, chap. III. Paris 1956.

[3] Bourbaki, N.: Sur le théorème de Zorn. Arch. Math. 2 (1951), pp. 434–437.

[4] Fraenkel, A. A. and Bar-Hillel, Y.: Foundations of set theory. Amsterdam 1958.

[5] Hermes, H.: Einführung in die Verbandstheorie. Berlin-Göttingen-Heidelberg 1955.

[6] Kamke, E.: Mengenlehre. Third ed., Berlin 1955.

[7] Kneser, H.: Eine direkte Ableitung des Zornschen Lemmas aus dem Auswahlaxiom. Math. Z. 53 (1950), pp. 110–113.

[8] Kuratowski, C.: Une méthode d'élimination des nombres transfinis des raisonnements mathématiques. Fund. Math. 3 (1922), pp. 76–108.

[9] Lorenzen, P.: Einführung in die operative Logik und Mathematik. Berlin-Göttingen-Heidelberg 1955.

[10] Lorenzen, P.: Über den Kettensatz der Mengenlehre. Arch. Math. 9 (1958), pp. 1–6.

[11] Poincaré, H.: Les mathématiques et la logique. Revue de Métaphysique et de Morale 14 (1906), pp. 307–317.

[12] Redei, L.: Algebra. Part I, Leipzig 1959.

[13] Russell, B. and Whitehead, A. N.: Principia mathematica, Vol. 1. Cambridge 1910.

[14] Teichmüller, O.: Braucht der Algebraiker das Auswahlaxiom? Deutsche Math. 4 (1939), pp. 567–577.

[15] Tukey, J. W.: Convergence and uniformity in topology. Ann. Math. Studies, Princeton 1940.

[16] van der Waerden, B. L.: Algebra. Part I, Fifth ed., Berlin-Göttingen-Heidelberg 1960.

[17] Weston, J. D.: A short proof of Zorn's Lemma. Arch. Math. 8 (1957), p. 279.

[18] Weyl, H.: Das Kontinuum. Leipzig 1918.

[19] Zermelo, E.: Neuer Beweis für die Möglichkeit einer Wohlordnung. Math. Ann. 65 (1908), pp. 107–128.

[20] Zorn, M.: A remark on method in transfinite algebra. Bull. Amer. Math. Soc. 41 (1935), pp. 667–670.

Index

A number, an, 77
Abelian group, 111n33, 168
Absolute values, 128, 310n27, 458
Absorption laws, 53
Abstract algebra, 65, 91, 92, 511
Abstract science, 27
Abstraction, 5, 65
Abundant number, 371
Addition, 98; algorithm for, 34; of integers, 111; of matrices, 251; monotonicity of, 100; recursive definition of, 98n12; of transformations, 247
Additive group of a ring, 318, 357
Adjoint, 315
Adjunction, 411
Admissible subgroups, 518
Affine complex plane, 468
Aleph-zero, 55
Algebra(s), 345, 402–403n37; abstract, 65, 91, 92, 511; associative, 402–403n37; Boolean, 9; Cayley, 481; circuit, 484, 494; division, 402–403n37, 478n17; entire rational function in the sense of, 301; of finite rank, 478; fundamental theorem of, 467; of sets, 53
Algebra of Logic (Boole's), 42
Algebraic: complement of a subdeterminant, 283; congruence, 396; elements, conjugate, 431; extension, 418, 420; functions, 306; integers, 330, 401; manifold, 354; number, 401
Algorithm, 7, 32, 79; for addition, 34; consistent, 39; Euclidean, 32, 332, 365; for multiplication, 34; rule of an, 33
Algorithmic: derivation, 33; proof, 33
Alphabet, 39
Alternating, 314n30; group, 227; product, 235, 275
Alternative, 13
Amicable numbers, 372
Analysis: entire rational function in the sense of, 301; ramified, 40
Analytic number theory, 406
"And," 12, 67, 483
Angle, trisection of, 417
Antiautomorphism, 470
Antinomy(ies), 9, 51, 80; of Burali-Forte, 57; Grelling's, 85; of the Liar, 76, 81; Russell's, 59, 81; semantic, 81; syntactic, 81; of the universal set, 57
Antireflexivity, 100
Application, successive, 115
Approximation, best, 377
ARCHIMEDES, axiom of, 127n57
Archmidean ordering, 127
Argument(s): entire rational function of n, 305; from n to $n + 1$, 94
ARISTOTLE, 10, 20, 83
Arithmetic, 5; incompleteness of, 40
Arithmetization, 36
Ars indicandi, 49
Ars inveniendi, 42, 49
ARTIN-SCHREIER, theorem of, 527
Ascending chain condition, 524
Assertion, 43
Associate(s), 328, 356
Associative, 98; division algebras, 478n17; laws, 53
Associator, 481
Assumption(s), 42, 43; -elimination, 45; -introduction, 45
At most countable, 55
Atom(s), 69, 79, 80, 490
Atomic, 490

Attribute, 21
Aut, 13
Autologic, 85
Automorphism, 148, 191, 412, 512; group, 191; group, of the configuration, 512
Autonomous: notation, 10n; system of axioms, 27, 29, 67
Axiom(s), 4–5; of Archimedes, 127n57; autonomous system of, 27, 29, 67; of cancellation, 181; categorical system of, 30; of choice, 60, 103n20, 164, 530, 534; complete system of, 29; of comprehension, 58; dual, 68; for the empty set, 60; heteronomous system of, 27; of induction, 94; of infinity, 60; monomorphic system of, 30; Peano system of, 93; power set, 60; replacement, 60; schema, 26, 34; self-contradictory system of, 31; for sets with one element, 60; system, 516; for unions (first and second), 60; well-ordering, 164; Zermelo's (of choice), 164

Babylonians, 4, 27
Base, 129
Basis, 242, 526; condition, 339; of an ideal, 339; integer, 402; of *k*th order, 405; theorem (Schnirelmann's), 407; vectors, 234
Bernays, R., 58
Bernstein, equivalence theorem of, 54
Best approximation, 377
Betti number, 303
Bézout, theorem of, 354
Bilinear: form, 264, 268; form, Hermitian, 271; mapping, 234
Binary operation, 167
Binomial: coefficients, 295; theorem, 295
Birkhoff, G., 483, 520; Hausdorff-, maximal chain principle, 530, 533
Bolzano, Bernard, 20
Boole, George, 174n; *Algebra of Logic*, 42
Boolean: algebra, 9; lattice(s), 67, 484, 490, 495; ring, 69
Bound: greatest lower, 68, 132, 486; least upper, 68, 486; lower, 523; proper upper, 523; upper, 486, 523; variable, 17, 23, 60
Bourbaki, N., 510
Brouwer, L. E. J., 6
Burali-Forte, antinomy of, 57
Burnside, W., 217

Calculus(i), 9, 32
Cancellation: axioms of, 181; first rule, 319; second rule, 323
Canonical: decomposition, 225; factorization, 332
Cantor: diagonal procedure, 152; diagonal procedure (first and second), 55; fundamental sequences, 133; fundamental theorem, 163
Cardinal number(s), 54, 94
Cardinality (of a set), 54
Categorical system of axioms, 30
Cauchy: convergence criterion of, 467n81; sequences, 139
Cayley, 223; algebra, 481; conjugate number, 481; numbers, 481; octaves, 481
Center, 218, 226
Centralizer, 218
Chain(s), 485, 522, 529; divisor, 330; f-, 531; above and below, 524; length of a, 495; maximal, 504; principle, Hausdorff-Birkhoff maximal, 530, 533; proper, 495; Sturm, 463; theorem (Dedekind), 504
Chain condition: ascending, 524; divisor, 360; factor, 495; high, 529; principle, 522, 530; principle, sharpened, 530n
Characteristic: equation of a matrix, 286; of a field, 324
Châtelet, 483
Chinese remainder theorem, 391
Choice: axiom of, 60, 103n20, 530, 535; axiom of Zermelo, 164; function, 535
Church, A., 49
Circle principle (Russell), vicious, 535
Circuit algebra, 484, 494
Circularity, restriction against, 43
Class(es): equality of, 58; equation, 218; equivalence, 65; ideal, 404; number, 56, 404; in set theory, 58; universal, 59
Class residue, 109, 341, 461; left and right, 186; ring, 342, 381
Classical: ideal theory, 343; logic, 4
Closed, 517; interval, 499
Coefficients, 299, 304; binomial, 295; comparison of, 296; leading, 299
Cogredient, 264
Column rank of a matrix, 253
Combination, linear, 239
Common divisor, 358; greatest, 332, 358
Communication, 9
Commutative, 98; group, 111n33, 168; laws, 53; ring, 117, 317
Commutator group(s), 197, 217
Comparability, 522
Comparable, 158
Comparativity, 65

Comparison of coefficients, 296
Complement, 52, 62, 490; relative, 499
Complementary divisor, 356
Complementation, laws for, 53
Complemented, 490; lattice, 68; relatively, 499
Complete, 487; induction, 57, 94, 117; induction starting from k, 101; lattice, 138n76, 491; ordered module, 138; residue system, 381; system of axioms, 29; system of rules of inference, 41
Completeness: of predicate logic, 42; theorem (Gödel), 42
Complex: conjugate, 461; plane, affine, 468; -product, 182, 266; projective line, 471
Complex numbers: fundamental algebraic theorem for, 467; fundamental topological theorem for, 467n8; left-multiplication in the domain of, 457
Complexes of a group, 182
Componentwise multiplication, 392
Composition, 406; configuration, 511; factors, 216; of fields, 436; of groups, 436; inner, 510; nonelementary, 510
Composition series, 216, 354; of fields, 438; of groups, 438
Comprehension, axiom of, 58
Computable function, 36
Concepts, fundamental, 22, 26
Condition(s): ascending chain, 524; basis, 339; divisor-chain, 360; factor chain, 495; maximal, 360; normality, 265; orthogonality, 265
Configuration(s), 251, 508, 509; automorphism group of the, 512; composition-, 511
Congruence(s); algebraic, 396; pure, 399; relation, 65, 514; modulo an ideal, 341; a subgroup, 380
Congruent, 270
Conjecture: Fermat, 11–12, 37, 398; Goldbach, 10, 76, 405
Conjugate(s), 183; algebraic elements, 431; Cayley number, 481; complex, 461; fields, 437; quaternion, 475; system of, 402n36; transposed matrix, 270
Conjunction, 12
Connectives: lattice-theoretic, 487; logical, 487; set-theoretic, 487
Connex, 63
Connexity, 522
Consequence, 8, 20, 24, 46
Consistency, 6, 109; relative, 31; semantic, 31; syntactic, 31
Consistent algorithm, 39
Constant function, 291
Constants: propositional, 12; structure, 402–403n37
Constructibility of a regular polygon, 454
Constructive (point of view), 7
Constructivist, 534–535; school, 7–8, 40
Continuous function, 462
Continuum, 55; hypothesis, 60; hypothesis, special, 60
Contradiction, 23, 80
Contragredient, 264
Contraposition, 47
Contravariant vector, 264
Convention, Einstein summation, 239
Conventionalism, 6
Convergence criterion (Cauchy), 467n8
Convergent of a continued fraction, 373
Converse relation, 62
Coordinates, 240
Coprime, 332
Coset: left, 186; right, 186
Countable, 55, 151; at most, 55
Countably infinite, 151n90
Covariant: tensor, 264; vectors, 264
Criterion(a): convergence (Cauchy), 467n8; for divisibility, 385; irreducibility (Eisenstein), 347; for multiplicity of zeros, 426; for a quadratic residue function (Euler), 400; for separability, 424; for subgroups, 184
Crystallography, 204
Cube, duplication doubling of, 417
CURRY, 40
Cut, Dedekind, 50, 133, 135
Cycle, 224
Cyclic, 191; groups, 191; groups, fundamental theorem for, 192
Cyclotomic: field, 405; polynomial, 428, 430

DE MOIVRE, formula of, 458
Decidable, 36
Decimal, infinite, 130
Decomposition: canonical, 225; into partial fractions, 368
DEDEKIND, RICHARD, 72, 403, 448, 483, 488; chain theorem of, 504; cut, 50, 133, 135; definition of infinity, 54
Deduction, 41
Deficient number, 371
Definite, positive, 273
Definition, 20–21
Degree, 299; of an algebraic extension, 420; of a representation, 220
Denominator, 122; lowest common, 360
Denotation (*bedeutung*), 10

Density, 407
Dependent, linearly, 240
Derivation, 41; algorithmic, 53
Derivative, 303; of a polynomial, 423
DESARGUES, little theorem of, 482
Description operator, 12, 18
Determinant(s), 235, 279; expansion of a, 282; multiplication of, 280; Sylvester, 349
Diagonal: form of a matrix, 260; procedure (Cantor's), 152; procedure (Cantor's), first and second, 55; sequence, 152
Diagram(s), 186; of Hesse, 485; order, 485
DICKSON, L. E., 448
Difference: left, 161; right, 161
Division algebra(s), 402–403n37, 477; associative, 478n17; of finite rank, 478
Digital sum, 385; alternating, 385; generalized, 385
Dilatation(s): of the plane, 456; zero, 456
Dilative rotations, 457; Hermitian, 469
Dimension, 240, 505
Dimensional equation, 505
DIRAC, δ-function, 5
Direct: product, 198, 493; product of groups, 394; sum of (the ideals), 392
Directed, 64; set, 64
DIRICHLET, 406; pigeon-hole principle, 102, 463
Discriminant, 352, 402
Disjoint, 52
Disjunction, 13
Distributions, 5
Distributive: lattice, 68, 489; laws, 53, 99, 115; laws, infinite, 490
Divisibility: criteria for, 385; fundamental lemma of the theory of, 359
Division: two-sided, 180
Divisor(s): common, 358; complementary, 356; greatest common, 332, 358; prime, 358; proper, 357; trivial, 357; of zero, 119, 293, 322; of zero, nilpotent, 342
Divisor-chain, 330; condition, 360; proper, 330
Domain(s), 62; first, 62; of a function, 50; fundamental, 411; image, 247; of individuals, 21; integral, 119n47, 323; of integrity, 323; operator, 511; of positivity, 120, 464, 527; of scalars, 235; second, 62; of transitivity, 221
Dual, 487; axiom, 68; group, 483, 488, 517; self-, 501; space, 263; vector space, 234
Duality, principle of, 68
Duplication doubling of the cube, 417
Dyadic fractions, 130n60
Echelon matrix, 259
Eigenvalues, 286
Eigenvectors, 286
EINSTEIN, summation convention of, 239
EISENSTEIN, irreducibility criterion of, 347
"Either-or," 13
Element(s), 50; axiom for sets with one, 60; conjugate algebraic, 431; exponent of a group, 193; G, order of the, 193; greatest, 490, 523; greatest common lower, 486; identity, 167; "imaginary," 448; inverse, 111; least, 490, 523; least common upper, 486; maximal, 164, 523; minimal, 523; neutral, 111, 167, 237; order of a group, 193; permutable, 168; prime, 403; of a set, 51; superfluous, 497; unit, 66, 116, 167, 179, 321, 490; unity, 321; zero, 179, 490
Elementary: -arithmetical structure, 521; -logical structure, 521; ornament, 204; predicate logic, 73; symmetric functions, 302n23; symmetric polynomials, 307
Elimination, 353; assumption-, 45; ideal, 354
Empty: relation, 62; set, 52, 103; set, axiom for, 60; word, 231
Endomorphism(s), 114, 128, 512; monotone, 146; multiplication of, 115; ring of, 116; sum of, 114
Entire rational function, 292; in the sense of algebra, 301; in the sense of analysis, 301; of n arguments, 305
Enumerable, recursively, 33, 35
Enumerability, 35
Equality, 108; of classes, 58; of value, 122
Equation(s): characteristic, of a matrix, 286; class, 218; dimensional, 505; Pell, 397
Equivalence, 14; class, 65; of matrices, 254; relations, 29, 65, 108; of sets, 103; theorem (Bernstein), 54
Equivalent, 54
EUCLID, 28
Euclidean: algorithm, 32, 332, 365; rings, 332, 361
EULER: criterion for a quadratic residue, 400; function, 382
Even transpositions, 227
Excluded middle, law of, 8
Existence-introduction, 48n23
Existential quantifier, 17, 18
Expansion: of a determinant, 282; Laplace, 283

Exponent, 445; of a group element, 193; of a root of unity, 445
Exponential function, 150
Expression, relevant, 76n
Extended: matrix, 259; predicate logic, 23n, 42, 72
Extension(s), 51; algebraic, 418; algebraic, degree of an, 420; Galois, 420; field, 297n11, 413; finite, 418; normal, 420; problem, 413; ring, 297; of a set, 51; separable, 422
Extensionality, principle of, 51, 58

F-chain, 531
Factor, 75; chain condition, 495; composition, 216; group, 196; proper, 328
Factorization: canonical, 332; rings, theorem for, 343; rings, unique, 331
False, 10, 23
FERMAT: conjecture, 11–12, 37, 398; number, 372; theorem, 382
Field(s), 124, 323; alternative, 481; characteristic of a, 324; composition of, 436; composition series of, 438; conjugate, 437; cyclotomic, 405; extension (or subfield), 297n11, 413; finite, 440; formally real, 464, 527; intersection of, 436; invariant, 402; skew, 235, 324, 526n; skew, of quaternions, 470; multiplicative group of a, 324, 357; ordered, 527; power series, 311; prime, 324; quotient, 125, 325; radical over a, 452; of rational numbers, 124; real-closed, 464; of real numbers, 141; of relations, 63; of sets, 53; union of, 436
Field splitting, 413; smallest, 413; uniqueness theorem for smallest, 415; in the wider sense, 413
Figure, 172; group of a, 173
Fin, symbol, 134
Finis superior (or *supremum*), 134n
Finished proof, 46
Finitary, 40
Finite, 55, 102; ascending length, 495; above and below chain, 524; descending length, 495; extension, 418; field, 440; group, 168; length, 495; rank, division algebra of, 478; set, Dedekind definition of, 103n21; system of generators, 185
Finitely generated, 185
First: axiom for unions, 60; Cantor diagonal procedure, 55; domain, 62
Fix-group, 221
Flagged variables, 43
Fonction polynome, 301
"for all," 12, 96
Form(s): bilinear, 264, 268, 271; diagonal, of a matrix, 260; fundamental, 270; Hermitean, 271; Hesse normal, 30; linear, 234, 262, 263, 345n48; multilinear, 264, 268; prenex normal, 518; propositional, 11, 22, 94n2, 535; quadratic, 268; signature of a, 272
Formalists, 6
Formalization, 9
Formally real (field), 464, 527
Formula: of de Moivre, 458; inversion, of Möbius, 389; for rotations, Rodrigues', 477
Fraction(s), 122, 325; dyadic, 130n60; partial, 368; partial, decomposition into, 368; proper, 368
continued, 333n29, 373; convergent of a, 373; Hurwitz, 379; regular, 333n29, 373
Free: group, 231; square-, 388; torsion-, 200; variables, 17; renaming of variables, 44n18
FREGE, G., 10, 51, 72
FROBENIUS, theorem of, 478
Function, 64, 509; algebraic, 306; choice, 535; computable, 36; constant, 291; continuous, 462; Dirac δ–, 5; of a domain, 50; elementary symmetric, 307n23; Euler, 382; exponential, 150; identical, 291; inverse, 64; Möbius, 288; partition, 406; product, 38; range of a, 50; recursive, 35, 38; signs, 12, 15; sum, 38; summatory, 371, 388; unity, 388; unity, 388; zero of a, 293. *See also* Entire rational function
Fundamental: concepts, 22, 26; domain, 411; form, 270; lemma of the theory of divisibility, 359; sequences, 139; sequences of Cantor, 133; system of solutions, 257; tensor, 270; theorem for cyclic groups, 192

G, element, order of a group, 193
GALOIS, 447; extension, 420; group, 409, 434; theory, 409
GAUSS: number, 372; plane, 457
Gaussian integers, 316
Gebilde, 508
General polynomial, 453
Generalization of the prime number theorem, 406
Generalized: digital sum, 386; predicate variable, 72

Generated, 185; finitely, 185
Generating system, 526
Generators, 185; finite system of, 185
GENTZEN, 43, 76; and QUINE, rules of inference, 43
GIBBS, JOSIAH WILLARD, 476n
Glide reflections, 205
GÖDEL, 31, 38; completeness theorem, 42; incompleteness theorem, 72; index, 36; numbers, 76
Gödelization, 36
GOLDBACH, conjecture of, 10, 76, 405
Graphs, 186
GRASSMAN, 275
Greatest: common divisor, 332, 358; common lower element, 486; element, 490, 523; lower bound, 68, 132, 486
GRELLING, antinomy of, 85
Ground set, 62
Group(s), 167; Abelian, 111n33, 168; additive, of a ring, 318, 357; alternating, 227; automorphism, 191; automorphism, of the configuration, 512; commutative, 111n33, 168; commutator, 197, 217; complexes of a, 182; composition of, 436; composition series of, 438; cyclic, 191; cyclic, fundamental theorem for, 192; direct product of, 394; dual, 483, 488, 517; factor, 196; of a figure, 173; fix-, 221; free, 231; Galois, 409, 434; Hamiltonian, 194; Klein four-, 491; length of a, 216; of motions, 172; multiplication, 28; multiplication table for a, 188; multiplicative, of a field, 324, 357; nilpotent, 220; with operators, 518; order of a, 168; P-, 219, 220; planar rotation, 214; power of a, 192; quaternion, 194; simple, 196; structure problem for, 191; symmetric, 171; theory, 28; topological, 232; torsion, 200; type problems for, 191; union of, 436
Group element: exponent of a, 193; order of a, 193

Half-turns, 205
HAMEL, G., 146n88
Hamiltonian groups, 194
HANKEL, permanence principle of, 105n26
HAUSDORFF-BIRKHOFF, maximal chain principle, 530, 533
Hemihedrism, 209
HERBRAND, 38
Hereditary, 93; meet-, 518
HERMES, H., 484
Hermitian (HERMITE): bilinear form, 271; dilative rotations, 469; form, 271; matrix, 271; metric, 468; rotations, 468
HERTZ, H., 6
HESSE: diagram, 485; normal form, 30
Heterologic, 85
Heteronomous system of axioms, 27
High, 529. *See also* Chain, high
HILBERT, D., 6, 21, 406
HÖLDER (and JORDAN), theorem of, 216, 354, 504
Holohedrism, 209
Homogeneous, 305; system, 235
Homologous, 511
Homomorphism, 107, 212, 511; theorem, 212; theorem, for rings, 342
HORNER, rule of, 294, 295
HURWITZ, continued fraction of, 379
Hypercomplex system, 345
Hypothesis: continuum, 60; continuum, special, 60; induction, 95

Ideal(s), 338; basis of an, 339; classes, 404; congruence modulo an, 341; direct sum of, 392; elimination, 354; left and right, 357; manifold of zeros of an, 354; maximal, 403; primary, 342; prime, 342, 403; principal, 339; principal, ring, 339; theory, classical, 343; two-sided, 357; unit, 338; zero, 338
Idealism, 3
Idempotent, 69n, 392; ring, 69
Identical: function, 291; permutation, 170
Identification, 5
Identitive, 63; law, 53
Identity, 62, 321; element, 167; modular, 501
"If and only if," 14
"If-then," 14
Image, 64, 212; domain, 247; pre-, 64; space, 247
"Imaginary" element, 448
Imaginary part, 458
Implication, 14
Impredicative, 85; definitions, 535
Improper: real number, 136; subgroup, 184
Inclusion, 62
Incompleteness: of arithmetic, 40; of extended predicate logic, 42; theorem of Gödel, 72
Indecomposable, 199
Independent: indeterminates, 304; linearly, 240, 526; transcendents, 304
Indeterminates, 297; independent, 304

Index: Gödel, 36; kernel-, notation, 246; of a subgroup, 186
Indirect proof, 42–43
Individual(s), 21; domain of, 21
Induction: Λ-, 43; axiom of, 94; complete, 57, 94, 117; complete starting from *k*, 101; hypothesis, 95; mathematical, 94; modified principle of, 101; schema, 75; step, 95; transfinite, 57
Inductive set, 164
Inertia, Sylvester's law of, 272
Inference, 41; system of natural, 42
Inference, rules of, 41; complete system of, 41; of Gentzen and Quine, 43
Infimum (or greatest lower bound), 132
Infinite: countably, 151n90; decimal, 130; distributive laws, 490
Infinitely distant points, 5
Infinitesimal, 139
Infinity: actual, 7; axiom of, 60; Dedekind definition of, 54; potential, 7
Initial: case, 95; intervals, 156; segments, 156, 532
Inner: composition, 510; product, 234, 266, 269
Integer(s), 109, 368; addition of, 111; algebraic, 401; algebraic, integral domain of, 330; basis, 402; Gaussian, 316; module of, 112, 120; negative, 113; positive, 113; ring of, 120
Integral domain(s), 119n47, 323; of algebraic integers, 330
Integrally closed ring, 403
Intermediate value theorem, 462
Interpretation(s), 20, 22; isomorphic, 30
Intersection, 52, 59, 62, 483, 486, 487; of fields, 436; of subgroups, 436
Interval(s): closed, 499; initial, 156; nested, 133
Into, 64, 509, 510
Intramathematical, 5
Introduction: assumption-, 43, 45; existence-, 48n23
Intuitionist(s), 6, 534
Intuitive theory of sets, 51
Invariant, 183; field, 402; subgroups, 194
Inverse, 111n34, 123, 167; element, 111; function, 64; left, 167n2; mapping, 510; right, 167n2
Inversion formula of Möbius, 389
Invertible, 510; mapping, 64; mapping, one-to-one, 64; transformation, 249
Irrational numbers, 152n
Irrationality of $\sqrt{2}$, 47
Irreducible, 328, 357
Irreducibility criterion of Eisenstein, 347
Isobaric, 350
Isolated (ordinal number), 161
Isomorphic: interpretations, 30; relations, 64
Isomorphism, 117, 156, 190, 412, 512; order-preserving, 144; theorem, 214

JACOBI, symbol, 400
Join, the, 486, 491
JORDAN-HÖLDER, theorem of, 216, 354, 504

k-place predicate, 16
KANT, IMMANUEL, 4
Kernel, 212; -index notation, 246
KLEIN-BARMEN, FRITZ, 483; (Klein) four-group, 491
KNESER, H., 530
KRONECKER, symbols of, 245
KUMMER, 403
KURATOWSKI, lemma of, 524

LAGRANGE, JOSEPH LOUIS, 186, 406; (Lagrange) relation, 398
Λ-introduction, 80
Language(s): layer, second, 136n74; natural, 4, 9, 85
LAPLACE, expansion, 283
Lattice(s), 68, 360, 483, 487, 517; Boolean, 67, 484, 490, 495; complemented, 68; complete, 138n76, 491; distributive, 68, 489; modular, 501; points, 204; semi-, 487; set-, 499; sub-, 492, 517; -theoretic connectives, 487; theory, 28
Law: of the excluded middle, 8; of inertia (Sylvester), 272; reflexive, 53
Leading coefficients, 299
Least: common multiple, 359; common upper element, 486; element, 490, 523; upper bound, 68, 486
Left: cosets, 186; difference, 161; ideals, 357; inverse, 167n2; -multiplication in the domain of complex numbers, 457; residue classes, 186
LEIBNIZ, GOTTFRIED WILHELM, 42
Lemma: fundamental, of the theory of divisibility, 359; Kuratowski's, 524; Zorn's, 164, 522, 525
Length: of a chain, 216; finite, 495; finite ascending, 495; finite descending, 495; of a group, 216
Levels of real numbers, 8
Lexicographic ordering, 130
Lexicographically ordered, 159
Liar: Antinomy of the, 76, 81; Paradox of the, 77

Limit: number(s), 56, 161; of a sequence, 132
Line(s), 506; complex projective, 471
Linear: combination, 239; mappings, 234; order, 485; transformation, 246; transformation, ring of, 249
Linear form(s), 234, 262, 345n48; module of, 263; multi-, 264, 268
Linearly: dependent, 240; independent, 240, 526; ordered, 522
Little Desargues theorem, 482
Logarithm, 151
Logic: *Algebra of Logic* (Boole), 42; classical, 4; of the first order, 73; history of, 9; operator in, 16; of the second order, 23n72. *See also* Predicate, logic
Logical: connectives, 487; matrix (truth table), 12; symbols, 484
Logicism, 51
Logics, many valued, 10
Longitudinal reflections, 205
LORENZEN, P., 40, 72, 79, 94n3, 536
LÖWENHEIM, and SKOLEM, theorem of, 71
Lower: bound, 523; bound, greatest, 68; element, greatest common, 486; neighbor, 485, 523
Lowest: common denominator, 360; terms, 360

Manifold(s): algebraic, 354; of zeros of an ideal, 354
Many-place properties, 21
Mapping(s), 64, 509; bilinear, 234; inverse, 510; invertible, 64; invertible, one-to-one, 64; linear, 234; normalization of a, 280; onto, 395n; rigid, 172n
Mathematical: induction, 94; system, 508
Matrix(ces), 234; addition of, 251; characteristic equation of a, 286; coefficient, 259; conjugate transposed, 270; diagonal form of a, 260; echelon, 259; equivalence of, 254; extended, 259; Hermitian, 271; logical, 12; multiplication of, 177, 252; rank of a, 255, 260, 284; column rank of a, 253; skew-symmetric, 314; square, 177; of a transformation, 250; unit, 273
Maximal, 525; condition, 360; element, 164, 523; ideal, 403; segment, 157; subgroup, 186
Maximal chain, 504; principle, Hausdorff-Birkhoff, 530, 533
Mechanics, quantum, 10
Meet, 487; -hereditary, 518
MERSENNE numbers, 371
Meta-metalanguage, 85
Metalanguage, 84
Metamathematics, 3–4
Metric: Hermitian, 468; space, 270; structure, 270
Minimal: element, 523; subgroup, 186
MÖBIUS: function, 388; inversion formula, 389
Model, 23, 516
Modified principle of induction, 101
Modular: identity, 501; lattice, 501; semi-, 503; semi-, above, 503; semi-, below, 503
Module, 111, 318n4; of integers, 112, 120; of linear forms, 263; ordered, 120; complete ordered, 138; property, 338
Modulo, congruence, an ideal, 341; a subgroup, 380
Modulo *n*, reduced, 174
Modulus, 458
Modus ponens, 41
Monomorphic, 73, 521; system of axioms, 30
Monotone endomorphism, 146
Monotonic law for multiplication, 120
Monotonicity of addition, 100
Motions: group of, 172; proper, 214; spiral, 205
MOUFANG, R., 482
Multilinear forms, 264, 268
Multiple: least common, 359; zero, 303
Multiplication, 99; algorithm for, 34; componentwise, 392; of determinants, 280; of endomorphisms, 115; group, 28; left-, in the domain of complex numbers, 457; of matrices, 177, 252; monotonic law for, 120; of ordinal numbers, 160; of real numbers, 146; table, 345; table, for a group, 188; of transformations, 248; of vectors, 267
Multiplicative: group of a field, 324, 357; semigroup of a ring, 357
Multiplicity of a zero, 302; criterion for, 426

Naive set theory, 51, 534
Natural: inference, system of, 42; languages, 4, 9, 85; number, 72; numbers, totality of, 72; science, 27
Negation, 13
Negative integers, 113
Neighbor: lower, 485, 523; upper, 523
Nested intervals, 133
Neutral element, 111, 167, 237

Nilpotent: divisor of zero, 342; groups, 220
Noetherian rings, 339
Nominalism, 3
Nonelementary: composition, 510; structure, 521
Nonseparable polynomial, 424
Norm, 403, 460, 470, 481
Normal, 183; extensions, 420; form, prenex, 518; polynomial, 453; subgroups, 188, 194, 518
Normality conditions, 265
Normalization of a mapping, 280
Normalized, 370
Normalizer, 218
Normed, 449
"not," 12, 13
Notation: autonomous, 10n; kernel-index, 246
*n*th root, 148
Number(s): A, 77; abundant, 371; algebraic, 401; amicable, 372; Betti, 203; cardinal, 54, 94; Cayley, 481; Cayley conjugate, 481; class, 56, 404; deficient, 371; Fermat, 372; Gauss, 372; Gödel, 76; irrational, 152n; limit, 56, 161; Mersenne, 371; natural, 72; natural, totality of, 72; and numerals, 39n14; perfect, 7, 371; sequence of, 104; sum of, 95; theory, analytic, 406; torsion, 203; transcendental, 401
Number, rational, 122; field of, 124; positive, 125
See also Complex, numbers; Ordinal number(s); Prime, number; Real, number(s)
Numerals, and numbers, 39n14
Numerator, 122

O-place predicates, 16
Octaves, Cayley, 481
One-to-one, 510; (invertible) mapping, 64
Onto, 64, 509; mapping, 395n
Operation: binary, 167; outer, 510
Operator(s): description, 12, 18; domains, 511; group with, 518; in logic, 16; of set formation, 12
"Or," 12, 13, 67, 483
Order(s), 32, 485, 522; basis of *k*th, 405; diagram, 485; of element G, 193; of a group, 168; of a group element, 193; linear, 485; -preserving isomorphism, 144; set-, 523; type, 56
Orderable, 527
Ordered, 524; field, 527; linearly, 522; module, 120; module, complete, 138; pairs, 52; ring, 120. *See also* Sets, ordered
Ordering(s), 63; Archmidean, 127; lexicographic, 130; partial, 63; quasi-, 356; semi-, 63. *See also* Well-ordering(s)
Ordinal number(s), 56, 153, 158; isolated, 161; multiplication of, 160; of the first kind, 161; of the second kind, 161; sum of, 159
Ornament(s), 204; elementary, 204
Orthogonal, 273, 285; transformations, 273
Orthogonality conditions, 265
Outer: operation, 510; product, 235, 275

P-group(s), 219; Sylow, 220
Pairs, ordered, 52
Paradox, 80; of the Liar, 77
Part: imaginary, 458; preperiodic, 378; real, 458; scalar, 475; vector, 475
Partial: fractions, 368; orderings, 63; well-orderings, 64
Partition function, 406
PASCAL, triangle, 296
PEANO, GIUSEPPE, 72, 94; system, 9, 30; system of axioms, 93
PELL, equation, 397
Perfect, 198; number, 7, 371
Period, primitive, 378
Permanence principle of Hankel, 105n26
Permutable element, 168
Permutation(s), 169; identical, 170
PHILON, 14
Pigeon-hole principle, 102, 465
Planar rotation group, 214
Plane: affine complex, 468; dilatations of the, 456; Gauss, 457; rotation of the, 456
PLATO, 4
POINCARÉ, H., 535
Point(s), 169, 490, 506; infinitely distant, 5; lattice, 204
Polygon, constructibility of a regular, 454
Polynomial(s), 299, 304; cyclotomic, 428, 430; derivative of a, 423; nonseparable, 424; elementary symmetric, 307; general, 453; normal, 453; primitive, 335, 449; ring, 299, 304; separable, 422; value of a, 300; zero of a, 302
Positive: definite, 273; integers, 113; rational numbers, 125
Positivity, domain of, 120, 464, 527
Potential infinity, 7
Power, 125; of a group, 192; rule, 200; set, 52, 54, 485; -series, 308; -series field, 311; -series ring, 311; set axiom, 60

Pre-image, 64
Predicate(s), 12, 15–16, 21; k-place, 16; O-place, 16; two-place, 16; variable, 11, 22; variable, generalized, 72
Predicate logic: completeness of, 42; elementary, 73; extended, 23n, 72; incompleteness of extended, 42
Prenex normal form, 518
Preperiod, 387
Preperiodic part, 378
Primary, 496; ideal, 342
Prime, 329; divisor, 358; element, 403; field, 324; ideal, 342, 403; number, 75; number theorem, generalization of, 406; regular, 405
Primitive: period, 378; polynomial, 335, 449; root, 399; root of unity, 427
Principal ideal(s), 339; ring, 339
Principia Mathematica (Whitehead and Russell), 42
Principle: of duality, 68; of extensionality, 51, 58; Hausdorff-Birkhoff maximal chain, 530, 533; high chain, 522, 530; high chain, sharpened, 530n; permanence (Hankel), 105; pigeonhole, 102, 465; of recursion, 96; Russell's vicious circle, 535; of two-valuedness, 10, 21
Problem(s): type and structure, for groups, 191; Waring, 406; word, 35, 232
Product(s), 167; alternating, 235, 275; complex-, 182, 266; direct, 198, 493; direct, of groups, 394; function, 38; inner, 234, 266, 269; outer, 235, 275; quaternion, 475; relative, 62; scalar, 234, 266, 269; tensor, 235, 273; vector, 266
Projection, 255
Projective line, complex, 471
Proof, 41, 43; algorithmic, 33; finished, 46; indirect, 42–43
Proper: chains, 495; divisor, 357; divisor chain, 330; factor, 328; fraction, 368; motions, 214; segments, 156; subset, 52, 95n5; upper bound, 523; variable, 34
Properly simple, 196
Property(ies), 21; many-place, 21; module, 338; representing, 535; two-place, 21
Proposition, relevant, 29
Propositional: constants, 12; form(s), 11, 22, 94n2, 535; variables, 15
Propositions, 10, 11
"Protologic," 79
Pure: congruences, 399; -elementary structures, 521
Pythagorean: theorem, 4, 27; triples, 398

Quadratic: form, 268; reciprocity, 400; residues, 399; residue, Euler criterion for a, 400
Quantifier(s), 12, 16; existential, 17, 18; universal, 17, 18
Quantum mechanics, 10
Quasi-ordering, 356
Quaternion(s), 317n1, 470; conjugate, 470; group, 194; product, 475; skew field of, 470
QUINE, W. V., and GENTZEN, 43
Quotient(s), 125; field, 125, 325; ring, 368

Radical(s): over a field, 452; solvability by, 452
Ramified analysis, 40
Range (of a function), 50
Rank: of a matrix, 255, 260, 284; column, of a matrix, 253; of a transformation, 247
Rational numbers, 122; field of, 124; positive, 125
Real: -closed, 463; part, 458
Real number(s), 134, 135; field of, 141; improper, 136; levels of, 8; multiplication of, 146
Realism, 3
Reciprocal, 123
Reciprocity, quadratic, 400; generalized law of, 400
"Rectangular array," 250
Recursion, principle of, 96
Recursive: definition, 96; definition of addition, 98n12; function, 35, 38
Recursively enumerable, 33, 35
Reduced: modulo n, 174; remainder, 174
Reducible, 328
Reflection(s), 205; glide, 205; longitudinal, 205; rotatory, 206; transverse, 205
Reflexive, 63; law, 53
Reflexivity, 29
Regressions, 157
Regular, 253; continued fraction, 333n29, 373; prime, 405; representation, 223
Relation(s), 21, 509; congruence, 65, 514; connex, 63; converse, 62; empty, 62; equivalence, 29, 65, 108; field of, 63; identitive, 63; isomorphic, 64; Lagrange, 398; reflexive, 63; successor, 32; symmetric, 63; theory of, 9, 61; transitive, 63; universal, 62; void, 62

Relative: complement, 499; consistency, 31; product, 62
Relatively complemented, 499
Relevant: expression, 76n; proposition, 29
Remainder: reduced, 174; theorem (Chinese), 391
Replacement axiom, 60
Representation, 220; degree of a, 220; regular, 223; transitive, 222
Representing property, 535
Residue(s): class ring, 342, 381; classes, 109, 186, 341, 461; quadratic, 399; quadratic, Euler criterion for a, 400; system, complete, 381
Restriction against circularity, 43
Resultant, 349
RIEMANN, sphere, 472
Right: cosets, 186; difference, 161; ideals, 357; inverse, 167n2; residue classes, 186
Rigid mapping, 172n
Ring(s), 116; additive group of a, 318, 357; Boolean, 69; commutative, 117, 317; of endomorphisms, 116; Euclidean, 332, 361; extension, 297; factorization theorem for, 343; homomorphism theorem for, 342; of integers, 120; integrally closed, 403; of linear transformations, 249; multiplicative semigroup of a, 357; Noetherian, 339; ordered, 120; polynomial, 299, 304; power-series, 311; principal ideal, 339; quotient, 368; residue class, 342, 381; theory, 28; unique factorization, 331
RODRIGUES, formula for rotations, 477
Roof, 529
Root, 302n15; *n*th, 148; primitive, 399; square, 148; of unity, 425; exponent of a, 445; primitive, 427
Rotation(s), 205; dilative, 457; dilative, Hermitian, 469; Hermitian, 468; of the plane, 456; Rodrigues formula for, 477; of the sphere, 472
Rotatory reflections, 206
Rule(s): of an algorithm, 33; Horner's, 294, 295; of inference, 41, 43; power, 200; of separation, 41
RUSSELL, BERTRAND, 22, 51; antimony, 59, 81; vicious circle principle, 535; and A. N. Whitehead, 42

Scalar(s): domain of, 235; part, 475; product, 234, 266, 269
Schema: axiom, 26, 34; induction, 75
SCHNIRELMANN, 405; basis theorem of, 407
SCHOUTEN, 246, 262
SCHREIER, ARTIN-, theorem of, 527
Science: abstract, 27; natural, 27
Second: axiom for unions, 60; Cantor diagonal procedure, 55, domain, 62; language layer, 136n74
Segment(s), 102, 156, 520; initial, 156, 532; maximal, 157; proper, 156
Self-contradictory system of axioms, 31
Self-dual, 501
Semantic(s), 20; antinomy, 81; consistency, 31
Semi-orderings, 63
Semilattice, 487
Semimodular: above, 503; below, 503
Sense (*Sinn*), 10
Separability, criterion for, 424
Separable: extensions, 422; polynomial, 422
Separation, rule of, 41
Sequence(s), 64; fundamental, of Cantor, 133; Cauchy or fundamental, 139; diagonal, 152; limit of a, 132; of numbers, 104; zero, 140
Series: composition, 216, 354; of fields and groups, 438; power-, 308; power-, field and ring, 311
Set(s), 50; algebra of, 53; axiom for, with one element, 60; cardinality of a, 54; directed, 64; element of a, 51; empty, 52, 103; empty, axiom for the, 60; equivalence of, 103; extension of a, 51; field of, 53; finite (Dedekind definition), 103n21; formation, operator of, 12; ground, 62; -lattice, 499; -orders, 523; power, 52, 54, 485; power, axiom, 60; -theoretic connectives, 487; universal, 52, 57; universal, antinomy of the, 57
Sets, ordered, 50, 56, 522; linearly, 522; totally, 522, 533; well-, 153, 155, 532
Sets, theory of, 3–4, 9; class in, 58; naive or intuitive, 51, 534
Sharpened high chain principle, 530n
Signature of a form, 272
Signs, function, 12, 15
Similar, 56, 221, 226, 262
Similarity, 156
Simple, 228; group, 196; properly, 196
Skew field, 235, 324, 526n; of quaternions, 470
Skew-symmetric: matrix, 314; tensor, 276
SKOLEM, and LÖWENHEIM, theorem of, 71
Smallest splitting field, 413; uniqueness theorem for, 415
Solutions, fundamental system of, 257
Solvability, by radicals, 452

Solvable, 216
Space, 169; dual, 263; image, 247; metric, 270; vector, 233, 238, 526n; vector, dual, 234
Spanned subspace, 243
Special continuum hypothesis, 60
Sphere: Riemann, 472; rotations of the, 472
Spiral motions, 205
Splitting field, 413; smallest, 413; smallest, uniqueness theorem for, 415; in the wider sense, 413
Square: -free, 388; matrix, 177; root, 148
Statements, valid, 516
Stoics, 14
Structure(s), 5, 508, 515, 516; constants, 402–403n37; elementary-arithmetical, 521; elementary-logical, 521; metric, 270; nonelementary, 521; problem for groups, 191; pure-elementary, 521; type, 516
STURM: chain, 463; theorem, 463
Subband(s), 492, 517
Subconfiguration, 517
Subdeterminant, algebraic complement of a, 283
Subfield, 297n11
Subgroup(s), 184; admissible, 518; criterion for, 184; congruence modulo a, 380; improper, 184; index of a, 186; intersection of, 436; invariant, 194; maximal, 186; minimal, 186; normal, 188, 194, 518; proper, 184; trivial, 184
Subideal, 339n
Subject(s), 12, 15, 21; variables, 11
Sublattice, 492, 517
Subring, 322, 338
Subset, 52; proper, 52, 95n5
Subspace, 526n; spanned, 243; vector, 243, 526n
Substitution, 168, 300
Successive application, 115
Successor, 38, 56, 93, 94; relation, 32
Sum, 133; digital, 385, 386; direct, of the ideals, 392; of endomorphisms, 114; function, 38; of numbers, 95; of ordinal numbers, 159
Summation convention, Einstein, 239
Summatory function, 371, 388
Superfluous elements, 497
SUZUKI, M., 519
SYLOW: p-groups, 220; theorem, 219
SYLVESTER: determinant, 349; law of inertia, 272
Symbol(s): fin, 134; Jacobi, 400; Kronecker, 245; logical, 485
Symmetric, 63, 108, 270, 306; group, 171; skew-, matrix, 314; skew-, tensor, 267
elementary: functions, 307n23; polynomials, 307
Symmetry, 29
Syntactic: antinomy, 81; consistency, 31
Syntax, 20

TARSKI, A., 20
Tautology, 23, 24–25
Tensor: covariant, 264; fundamental, 270; product, 235, 273; skew-symmetric, 276
Terms, lowest, 360
Tertium non datur, 43, 47, 49
Tetartohedrism, 209
THALES, 27
Theorem: fundamental, of algebra, 467; Artin-Schreier, 527; Bernstein equivalence, 54; Bézout, 354; binomial, 295; Cantor fundamental, 163; chain, of Dedekind, 504; Chinese remainder, 391; completeness, of Gödel, 42; for complex numbers, fundamental algebraic, 467; for complex numbers, fundamental topological, 467n8; fundamental, for cyclic groups, 192; little, of Desargues, 482; factorization, for rings, 343; Fermat, 382; of Frobenius, 478; incompleteness (Gödel), 72; intermediate value, 462; isomorphism, 214; Jordan-Hölder, 216, 354, 504; of Löwenheim and Skolem, 71; prime number, 406; Pythagorean, 4, 27; Schnirelmann basis, 407; Sturm, 463; Sylow, 219; Uniqueness, for smallest splitting fields, 415; well-ordering, 56, 525; Wilson's, 384
Topological: group, 232; theorem for complex numbers, fundamental, 467n8
Torsion: -free, 200; group, 200; numbers, 203
Totality of natural numbers, 72
Totally ordered, 522, 533
Trace, 286
Transcendent(s), 296; independent, 304
Transcendental number, 401
Transfinite, 55; induction, 57; inductive definition, 57
Transformation(s): addition of, 247; invertible, 249; linear, 246; linear, ring of, 249; matrix of a, 250; multiplication of, 248; orthogonal, 273; rank of a, 247

Transforms, 183
Transitive, 63, 108; law, 53; representation, 222
Transitivity, 29, 100; domain of, 221
Translations, 205
Transposition(s), 226; even, 227
Transverse reflections, 205
Treillis, 483
Triangle, 92; Pascal, 296
Triples, Pythagorean, 398
Trisection of an angle, 417
Trivial: divisors, 357; subgroup, 184
True, 10, 23
Truth: table (logical matrix), 12; -value, 107, 109
TURING, 36
Two-place: predicate, 16; property, 21
Two-sided: division, 180; ideals, 357
Two-valuedness, principle of, 10, 21
Type: order, 56; problems for groups, 191; structure, 516

Uncountable, 55, 151
Unimodular, 399n
Union(s), 52, 62, 483, 486; first and second axioms for, 60; of fields, 436; of groups, 436
Unique factorization rings, 331
Uniqueness theorem for smallest splitting fields, 415
Unit(s), 327, 356; element, 66, 116, 167, 179, 321, 490; ideal, 338; matrix, 273; vector, 285
Unitary, 273
Unity: element, 321; function, 388 root of, 425; exponent of a, 445; primitive, 437
Universal: class, 59; quantifier, 17, 18; relation, 62; set, 52, 57; set, antinomy of, 57
Upper: bound, 486, 523; bound, least and proper, 68, 486, 523; neighbor, 523

Valid statements, 516
Valuation, 310n, 405
Value: absolute, 128, 310n, 458; equality of, 122; of a polynomial, 300; theorem, intermediate, 462; truth-, 107, 109
Variable(s), 11; bound, 17, 23, 60; flagged, 43; free, 17; free renaming of a, 44n18; predicate, 11, 22; predicate, generalized, 72; proper, 34; propositional, 15; subject, 11
Vector(s), 233; basis, 234; contravariant, 264; covariant, 264; multiplication of, 267; part, 475; product, 266; space, 233, 238, 526n; space, dual, 234; subspace, 243, 526n; unit, 285
Vel, 13
Verband, 483
Verknüpfung, 510
Vicious circle principle (Russell's), 535
VINOGRADOV, 405
Void relation, 62
VON NEUMANN, 51, 58

WARING, problem, 406
Well-ordered sets, 153, 155, 532
Well-ordering(s), 64, 101; axiom, 164; partial, 64; theorem, 56, 525
WESTON, J. D., 530
WHITEHEAD, A. N., and RUSSELL, 42
WILSON, theorem of, 384
Word(s), 6, 230; empty, 231; problem, 35, 232

ZERMELO, E., 51, 535; axiom of choice, 164
Zero(s), 111; aleph-, 55; divisors of, 119, 293, 322; nilpotent divisor of, 342; element, 179, 490; of a function, 293; ideal, 338; manifold of, of an ideal, 354; multiple, 303; multiplicity of a, 302, 426; of a polynomial, 302; sequence, 140
ZORN, lemma of, 164, 522; sharpened form of, 525